World Geography

WORLD

McGraw-Hill Book Company

New York
St. Louis
San Francisco
Düsseldorf
Johannesburg
Kuala Lumpur
London
Mexico
Montreal
New Delhi
Panama
Rio de Janeiro
Singapore
Sydney
Toronto

GEOGRAPHY *Third Edition*

Edited by JOHN W. MORRIS

World Geography

Copyright © 1958, 1965, 1972, by McGraw-Hill, Inc. All rights reserved. Printed in the United States of America. No part of this publication may be reproduced, stored in a retrieval system, or transmitted, in any form or by any means, electronic, mechanical, photocopying, recording, or otherwise, without the prior written permission of the publisher.

Library of Congress Catalog Card Number 72-172658
007-043138-8
1234567890FLBP 798765432

This book was set in Baskerville by Black Dot, Inc., printed by Federated Lithographers-Printers, Inc., and bound by The Book Press, Inc. The designer was Merrill Haber; the drawings were done by Harry Scott. The editors were Janis Yates and Helen Greenberg. Matt Martino supervised production.

Cover Credit

The world map used on the cover was found in a manuscript containing a commentary on the Apocalypse, which was composed by Beatus of Liebena, edited in the eighth century, and dedicated to Eutherus. The manuscript was composed about 1150 in the monastery of Santa Domingo de Silos, in the diocese of Burgos in ancient Castille. CNA/Editorial Photocolor Archives, Inc., 663 Fifth Ave., New York.

Contents

Preface

1 **An Introduction to World Geography** 2
   John W. Morris, *University of Oklahoma*

2 **The United States East of the Rocky Mountains** 18
   John W. Morris, *University of Oklahoma*

3 **The United States West of the Great Plains** 64
   Hallock F. Raup, *Kent State University*

4 **Canada and Greenland** 102
   J. Lewis Robinson, *University of British Columbia*

5 **Middle America** 138
   Don R. Hoy, *University of Georgia*

6 **South America** 174
   Herbert L. Rau, *Chicago State University*

7 **Northwestern and Central Europe** 218
   Lawrence M. Sommers, *Michigan State University*

8 **Southern Peninsular Europe** 268
   Lucile Carlson, *Case Western Reserve University*

Contents

9 Eastern Europe   308
   Charles Bajza, *Texas A. and I. University*

10 Union of Soviet Socialist Republics   344
   W. A. Douglas Jackson, *University of Washington*
   Warren E. Hultquist, *Sacramento State College*

11 The Middle East: North Africa and Southwest Asia   386
   Gerry A. Hale, *University of California, Los Angeles*

12 Africa South of the Sahara   426
   James W. King, *University of Utah*

13 Central Eastern Asia   470
   The Late Herold J. Wiens, *University of Illinois*

14 South Asia   512
   Joseph E. Schwartzberg, *University of Minnesota*

15 Southeast Asia   554
   Thomas F. Barton, *Indiana University*

16 Australasia, Oceania, and Antarctica   594
   Tom McKnight, *University of California, Los Angeles*

Conclusion: This Changing World   620

Glossary   629

Index   635

Preface

As every intelligent individual realizes, ours is a complex world. The complexities, however, are largely of man's own making because of his misunderstanding of his natural environment and general distrust of his fellow men. As its prime purpose, *World Geography* aims to present essential facts and ideas about the natural and cultural environments in such a way that the student may use these to develop concepts, generalizations, and ideas about his total world environment.

The text is organized to give information about regions, groups of nations, or individual nations in order to help students understand various geographic situations. The study of familiar geographic areas has been chosen as the primary approach, because it is believed that this organization, often used by the news and television media, is best known. Such terms as "Southeast Asia" and "the Soviet Union" are seen or heard almost daily.

World Geography is divided into sixteen chapters. Chapter 1 presents some ideas of what geography is, plus some examples of the kinds of concepts that may be developed by a study of geography. This chapter also gives a brief summary of the earth's principal natural features, followed by an introduction to the problems created by the unequal distribution of world population, and concludes with a statement about the lifelayer, or biosphere. Chapters 2 and 3 both deal with the United States, since we believe that the student needs a greater knowledge of and more information about his home country. The remaining chapters deal with other countries and regions. The last chapter is followed by a summary that points up certain international problems and competitive factors that influence the activities of most countries today.

Each chapter has been written by an experienced teacher who has traveled widely in and

Preface

gained a comprehensive knowledge of the region, area, or country discussed. Each author has presented the information he believes will give the truest picture of the area; accordingly, the chapters vary in approach. For example, urbanization is stressed as a factor in the Northeastern United States, manufacturing in West-Central Europe, underdevelopment in Africa, and overpopulation and poverty in South Asia. To understand these and other factors, however, related historical, economic, and anthropological factors are also presented as necessary to the explanation of present conditions.

Throughout the text, human activities have been related to the earth's relief features, climatic regions, and natural resources. A discussion of the natural environment is followed by an analysis of the cultural environment. Man succeeds most easily when and where he acts in harmony with nature; but modern man is no longer dominated by his physical environment, for his advancing technology has aided him in overcoming many natural handicaps. A study of world geography gives the student information about the broad patterns, population distribution, important areas of commodity production, and other items of interest. It contributes greatly to knowledge that will help in understanding, evaluating, and reaching decisions about current world problems. The text, then, provides both concepts about the field of geography and facts about world regions.

The teacher may use this book in a variety of ways. It can, of course, be used as the basis for a standard lecture-discussion course. The authors hope however, that it will not be used for the memorization of statistics and statements showing current conditions and trends, since such figures are always subject to change. Because human, as well as natural, resources are so important in geography today, the writers hope that the student will use this text to help him understand the context in which current world conditions and events are evolving. Concepts can thus be developed and generalizations about specific regions made. Should the teacher desire to use the "case method" approach, the facts and figures are available to serve as a base for such study. For example, a case study about world agriculture, lands available for settlement, population densities, or some other subject on a worldwide basis could be developed by using the relevant material available from the various chapters.

A Study Guide has been prepared for use with *World Geography*. Each unit in this manual is correlated with a text chapter. Several behavioral objectives are listed and a glossary of important terms given. The activity section consists of four parts: (1) map exercise, (2) review of concepts and definitions, (3) discussion questions, and (4) self-text of multiple-choice and matching exercises.

This third edition of *World Geography* is the product of many people. The basic idea for the book was developed by and the first edition was produced under the editorship of the late Professor Otis W. Freeman. Numerous individuals have read portions of the current edition and have offered helpful suggestions. Many persons, public agencies, and representatives of foreign countries have supplied photographs and information used in the book. For this assistance, the authors and the editor extend their sincere thanks.

John W. Morris

World Geography

1

Introduction to World Geography

Since its formation the earth has been undergoing changes. There have been upheavals of mountains, vast outpourings of molten rock, downwarps to form inland seas, and reduction by erosion of entire mountain chains to plains. Large areas of some landmasses have been covered by glaciers; numerous areas have been inundated by oceans. These and other changes, such as the development of minerals and the variations of climate that are constantly in progress, are but a few of the factors affecting the occupancy of the earth by man.

For centuries man had little influence upon that part of the earth where he lived. His numbers were so small and his technology so insignificant that he was largely controlled by his environment. As his numbers increased, as his

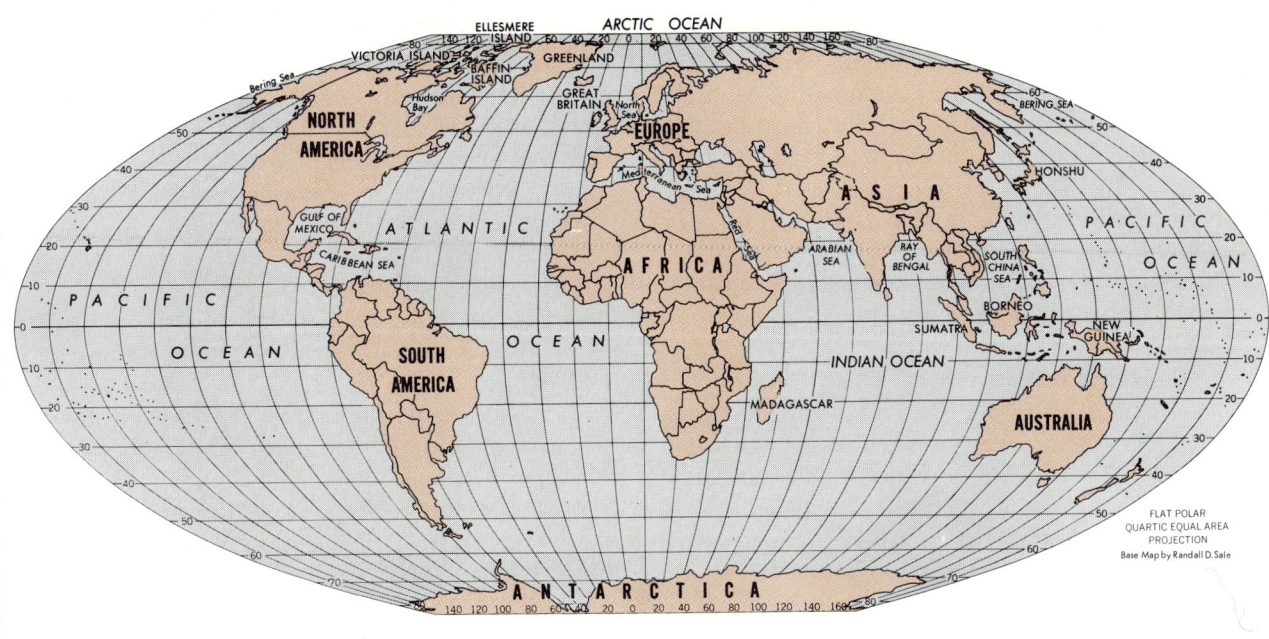

knowledge and understanding of the environment grew, he began to take advantage of certain aspects of that environment to make a better livelihood for himself. In time, he caused changes or adaptations in certain animals and plants so that they would work for him. He also started to use certain minerals and learned that, for specific purposes, some soils were better than others. Although man was no longer largely controlled by his environment, he was still greatly influenced by it, especially by climate, atmospheric conditions, soils, and water supply. However, as the influence of the natural environment became less important, the influence of cultural factors increased not only by means of technology but also by increasingly complex societal organization.

During the nineteenth century, man's technology increased to such an extent that he made a series of inventions which resulted in tremendous changes in his way of life. The steam turbine, the telegraph and telephone, the rotary printing press, the electric trolley car and the locomotive, steel, the gasoline engine, numerous types of textile-making machines, and various kinds of agricultural implements were but a few of the many inventions developed during the Industrial Revolution. The need for natural resources, especially minerals, brought about the exploration and development of most previously unexplored areas, and thereby the spread of European ideas, political institutions, and economic activities to various other parts of the world. As a result, most of Africa and much of Asia, Australia, and many of the Pacific islands came under the control of the more aggressive European nations. For a period of time during the early nineteenth century, most of Latin America was also controlled by European countries, as well as that area of Anglo-America not forming a part of the United States. Along with the European nations, the United States profited greatly by the Industrial Revolution but did not expand into other continents, primarily because of its own small population and large area.

The first half of the twentieth century was an era of even greater activity. During these fifty years there were two World Wars and many minor wars, revolutions, and "police actions." As a result, several pre–World War I empires disintegrated; some small independent nations were formed and existed for only a few years before being annexed by larger neighbors; in some areas, as in Indochina, formerly large political units were fragmented; also, many countries changed their form of government. World War II resulted in the downfall of some strong nations and the changing of boundaries of several others. This half century also saw the beginning of the end for most colonial empires and the development of atomic energy—two items that have had a great impact upon the interrelationships of all nations. From this period the two leaders of the current politically different worlds emerged: the United States and the Union of Soviet Socialist Republics.

Since 1950 the political situation of the world has been an unsettled one. Although no major wars have thus far developed, there have been many areas of stress and tension, especially in Korea, the Middle East, Vietnam, Czechoslovakia,. Nigeria, and along the border between China and the Soviet Union. All, or most, of the former colonies of Belgium, France, Italy, the Netherlands, Spain, and the United Kingdom have emerged as independent nations. During the 1950–1970 period, fifty-five new nations were formed, with thirty-eight of these in Africa. Most of the new nations are economically and educationally underdeveloped, some are overpopulated, and many are politically unstable. China, with its vast population, has become a third world power, challenging both the Soviet Union and the United States. In spite of the political conflicts, scientific technology has continued to increase. In 1957 the Soviet Union launched the first man-made satellite to circle the earth; in 1969 the United States landed the first men on the moon to begin explorations in outer space.

It is apparent, then, that a knowledge of geography is fundamental to understanding human activities in the modern world. When one scans a daily paper, reads a newsmagazine, or views a newscast on television, he indicates an interest in national and international affairs. All parts of the world are within a few hours travel time of one another, and most are only minutes apart or less in communication time. The activities of any one group of people may greatly influence the work and life of numerous other groups. A definite interrelationship exists among all the peoples or nations of the world. To understand why certain groups act as they do, why people prosper in some areas and are poor or stagnant in others, a knowledge of the various environments, both physical and cultural, in which these differences and/or likenesses develop is needed.

Regional Geography

The word "geography" is derived from the Greek words *gē* (earth) and *grapho* (I write or describe); hence geography means literally a "description of the earth." With the development of the subject has come a series of more sophisticated definitions for the term. Among

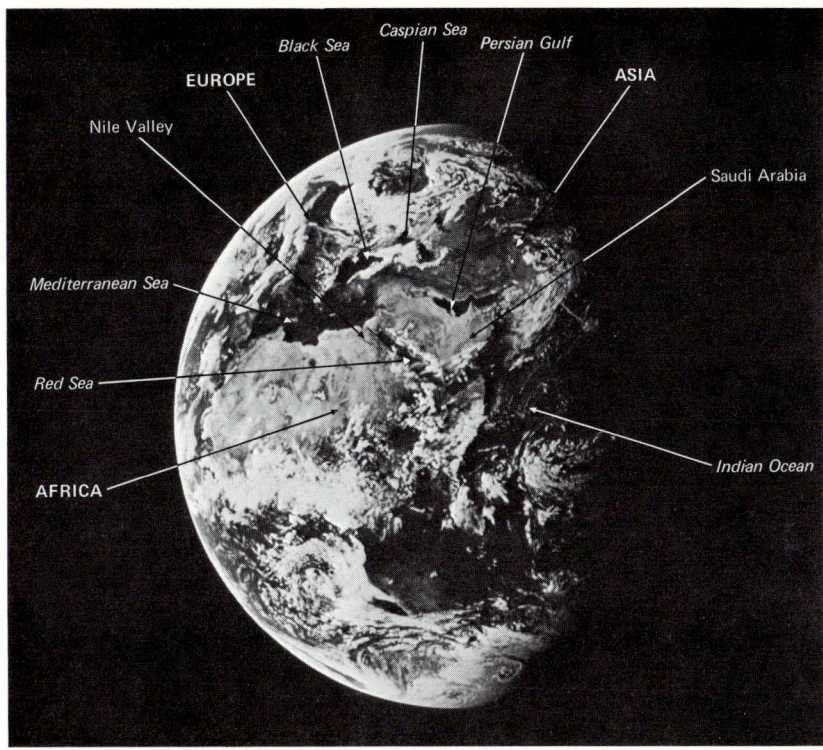

Fig. 1-1 The world as seen by the astronauts from about 98,000 nautical miles in space. (Courtesy of NASA)

these are such statements as the following: "Geography is human ecology." Geography is "the science of areal differentiation of the earth's surface." Geography is the "significance of differences from place to place." Geography "has for its special charge to study the changing expressions which, according to the locality, the appearance of the earth assumes." Geography is the "study of the relationship of man to his environment." The emphasis in "modern geography is on the study of spatial organization, expressed as both pattern and process." From these and other definitions, two common factors may be noted—the earth and man. The geographer, then, studies interrelationships between various areas or regions in order to understand something of the nature and pattern of human activities; thus, it is still one of the prime functions of geography to describe various areas or regions of the earth, or the earth as a whole.

In order to comprehend man's activities in any given area of the complicated world of today, it is necessary to have some specific knowledge about the physical landscape (landforms, vegetation, climate) as well as the cultural development (history, economics, ethnography) of that area. For one to understand the area being studied, it is necessary that specific place locations be visualized and that essential facts about that area be understood. By using these facts, one may then form a series of concepts or generalizations. A concept is defined as a general notion or idea of something formed by mentally combining all its characteristics or particulars—that is, a construct. Among the major geographic concepts, briefly stated, are the following:[1]

1. *Regionalism* The earth may be divided in many ways, depending upon the elements selected. Regardless of the way the divisions are made, however, there must be a central theme to unite the whole. In some instances, more than one theme may be of equal importance. The geographer selects the quantitative and qualitative elements needed for the type of region he chooses to form. In this book the outer limits of the area designated are based on the boundaries of nations, since most people are more familiar with political divisions than with other distinguishing geographic phenomena.

2. *Areal distinctions* After an area or region is delimited, the variety of likenesses and differences therein soon become apparent. The people and environments within a region may differ from place to place and from time to time, even though there is a central theme common to the whole. The geographer needs to consider where and why such variations exist and the influences of one variation on another.

3. *Interrelationships* Just as there are distinctions within a region, there are also differences and likenesses between regions. No one region or subregion of the earth, no one person or group of people is completely independent of the other regions and peoples of the earth. Men, nations, and nature are interdependent, for each influences and is influenced by the actions and reactions of the others. The geographer, then, is interested in the causes and effects of this spatial interaction or interrelationship.

4. *Constant change* The concept that time, place, and man are constantly changing must be fully appreciated. The people living in an area decide how, why, and where the resources of a region will or will not be used. Man, by his stage of development, selects which natural resources will become a part of his cultural environment. By so doing, he is constantly changing the environment in which he lives. Thus, the development of the earth has been one of constant change, geographically as well as historically, because man is the chooser of the ways he attempts to control his environment.

5. *Life layer* That part of the earth on which man lives is limited and encompasses a relatively small bit of land, air, and water. It is in this area that man is paramount, but his control of an area varies greatly from place to place. Sometimes, when man and nature are not in agreement, the environmental change may be harmful (pollution of water and air, waste of minerals); but, on the other hand, the change may be beneficial (improvement of plant and animal life, the conquest of certain diseases). Geographers, then, are also concerned with man's use of this life layer and the ways he has changed it.

6. *Resource limitation* The advancement of any region depends to a great extent upon the natural resources available and the ability of those living there to use them. Only within the last half century, however, has mankind begun to appreciate fully the fact that natural resources such as minerals and soil are limited in both quantity and quality, and not until the present decade has a large proportion of mankind begun to realize the importance of conserving fresh air, pure water, and natural energy in making and maintaining a livable environment. In fact, the three resources air, water, and energy largely determine the quality of life on earth, for without any one of these three life as we know it today is impossible.

The field of geography is frequently divided into two distinct categories known as "systematic geography" and "regional geography." Systematic geography is a study of some specific aspect of the subject, such as the physical, historical, human, or economic, largely from a single

---

[1] For a discussion of geographic concepts, see Henry J. Warman, "Major Concepts in Geography," in Wilhelmina Hill (ed.), *Curriculum Guide for Geographic Education*, National Council for Geographic Education, Normal, Illinois, 1964, pp. 9–27.

point of view. Regional geography, however, approaches the study of the subject by selecting an area of space that has some degree of common identity. The idea of regionalization, then, implies homogeneity within a region and heterogeneity among regions. In general, regional geography is complementary to systematic geography in that it attempts to bring the various systematic factors together to form a whole. There are, therefore, almost infinite possibilities of dividing the earth into regions, depending upon the type of region to be formed. Systematic geography is usually the product of an analytical study; regional geography, a product of synthesis.

An understanding of the *regional* geography of an area, then, requires the student to bring together the various factors of both the physical and the cultural environment of the region under consideration. The physical environment evolves, of course, from such influences as the landforms, waterforms, climate, soils, natural vegetation, minerals, and other natural factors. The cultural environment necessitates an understanding of such factors as the sequent occupance of the region and the diffusion of ideas and technology therein, as well as a perception of the resources and handicaps of the area under consideration.

The Earth's Spheres

The surface of the earth is approximately 196,950,000 square miles in area; of this, 139,440,000 square miles is water and 57,510,000 square miles is land. Although the land area above water covers only about 29 percent of the earth's total surface, to man (who lives on the land) the continents and islands are, at present, of greater significance than are the oceans that occupy the remaining 71 percent of the planet's surface. Actually, since they include the continental shelves as well as the land above water, the huge masses of land called "continents" cover about one-third of the earth, and the ocean basins account for about two-thirds. However, since the volume of water exceeds the capacity of the basins, ocean water has so encroached upon the margins of the continents that it has reduced the exposed land area. In reality, the continents and oceans are only comparatively slight elevations or depressions in the earth's crust. The range from the bottom of the Mariana Deep (35,958 feet below sea level) to the summit of Mt. Everest (29,028 feet above sea level) amounts only to about 12.3 miles—which is a small difference in relation to the earth's diameter.[2] Encircling both the land- and water masses is the vast layer of atmosphere, which not only extends out into space but also penetrates both the land and the water of the planet.

THE LITHOSPHERE

The land areas, or solid portions, of the earth's surface are called the "lithosphere." Within the lithosphere, seven major landmasses are recognized as continents. Asia, Africa, and Europe —sometimes referred to as the Eastern Hemisphere continents—form a continuous landmass. Australia, which lies to the southeast of this mass, is also considered a part of the Eastern Hemisphere. North and South America, known as the Western Hemisphere continents, are connected by a narrow isthmus. Antarctica, which surrounds the South Pole, is isolated from the other continents.

There are many hypotheses and several theories as to how the continents were formed. The Continental Drift Theory, first suggested by Francis Bacon in the 1500s but later refined and published by the German scientist Alfred

---

[2]Because of rotation the earth is slightly flattened at each pole. As a result, the equatorial diameter is 7,927 miles; the polar diameter, 7,900 miles.

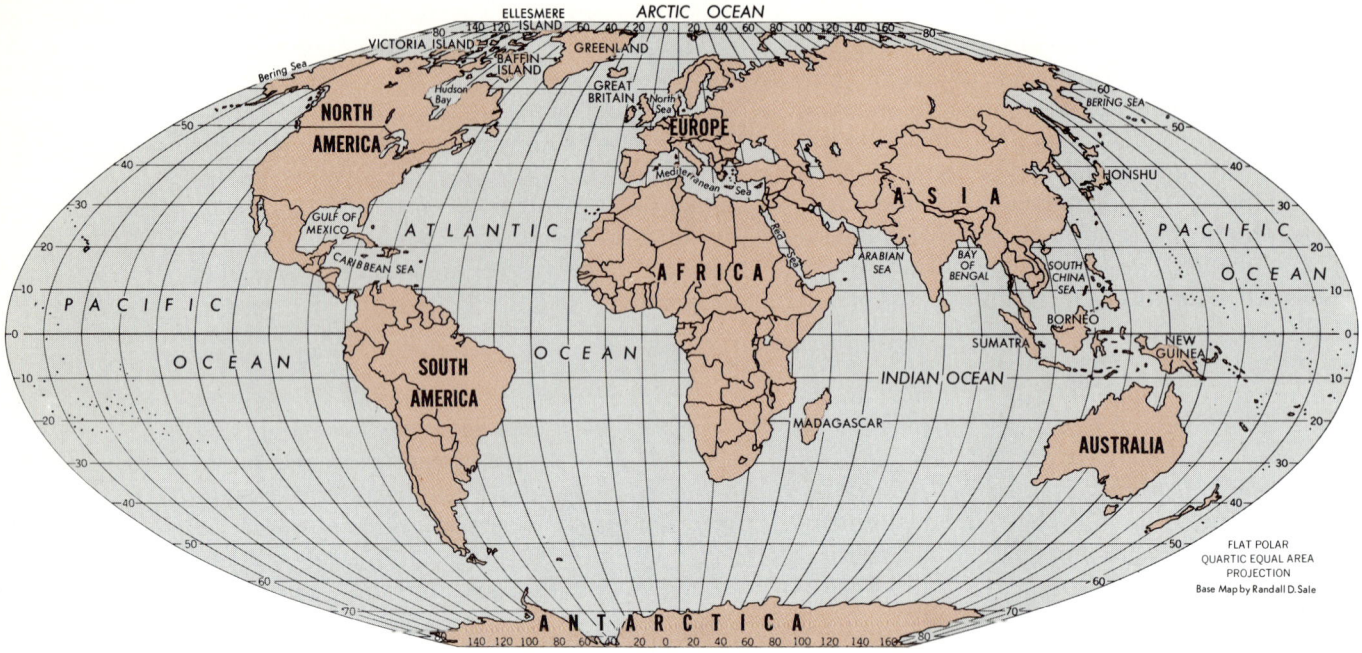

Fig. 1-2 Of the 197 million square miles of the earth's surface, about 57 million is land. The population of the earth, however, is concentrated largely on about 22.5 million square miles, mostly in the Northern Hemisphere.

Wegener in 1915, is now being given considerable study by numerous well-known geologists. This theory holds that at one time all the present-day continents may have formed a single landmass. Through time, owing to various geologic processes, the single original mass has broken into segments that drifted to their present locations. Some factors which tend to support this theory are: (1) the continental shapes, such as the similar contour of the east coast of South America when compared with that of the west coast of Africa; (2) the discovery of coal and certain fossils in Antarctica, which indicates that that continent was previously attached to Africa; (3) the finding of possible glacier markings that extend from Morocco to Libya and Chad in the windswept parts of the Sahara; and (4) a recently completed study by marine geologists of a ridge of mountains that extends for thousands of miles across the ocean floor and has an outline comparable to the edges of the continents.

The division of the landmasses into continents as recognized today is largely a matter of history. There is little justification for separating North and South America at the international boundary of Panama and Colombia except that the Isthmus of Panama ends at that location. Frequently the Americas are divided on the basis of dominant language differences: the land north of the United States–Mexico boundary is thus called Anglo-America, and that to the south of the boundary, including the West Indies, is known as Latin America. Nor is there firm justification, other than historical background, for the boundary between Europe and Asia being drawn to follow along the Ural Mountains, Ural River, Caspian Sea, and Caucasus Mountains. It is not uncommon for maps to show both continents with the single name of Eurasia. Asia is separated from Africa at the Sinai Peninsula, where a common landscape extends in both directions from this man-made boundary. Australia, the smallest of the continents (2.9 million square miles), is only about 3.5 times larger than Greenland, the largest of the world's islands.

Continental Topography

Each continent has its areas of mountains, hills, plains, and plateaus. Frequently the continental shape is determined by the location of the younger mountains or the older, more resistant rock masses. Both North and South America have high mountain ranges along their western sides. Each also has older, worn-down mountains near its eastern borders. Wide plains fill the spaces between the mountain masses. Only narrow coastal plains are adjacent to the western mountains of the American continents; in many places, however, fertile coastal plains are located along the eastern borders of the landmasses.

A long mountain axis extends across south-central Eurasia. South from this axis extend such plateaus as Iberia, Anatolia, Arabia, Iran, and the Deccan. To the north and northwest of this mountain core lies the great lowland plain of Europe and Asia; to the east are the densely populated coastal plains of China and Vietnam. Within the heart of this landmass are the high and dry Tibet Plateau and the Gobi, as well as the semiarid plains of the central part of the Soviet Union. Because of the numerous peninsulas extending from the European part of the landmass, the coastline of Europe is longer than that of any other continent.

Africa is frequently referred to as a plateau continent, since much of its interior is a great central plateau. With the exception of the Atlas Mountains in the northwest, most of the African highlands are scattered along the eastern side of the continent from the Red Sea to the Cape of Good Hope. Large plains are found only in the borderlands of the Sahara; for the most part, the coastal plains are narrow. The abrupt changes in elevation between the coastal plains and the central plateau, along with tropical climates and diseases, deterred the exploration of the African interior for many decades.

Australia has three dominant landforms: the Eastern Highlands, the plains area in the east-central part of the continent, and the western plateaus. The continent has a lower average elevation than any of the other continental landmasses. Narrow coastal plains, south and east of the highlands, are the principal populated regions of Australia.

Little is definitely known about the detailed topography of Antarctica, since most of the continent's surface is continually covered with snow and glaciers. Among the topographical features known to exist there are several rugged mountain ranges and peaks.

Islands

The exact number of islands in the world is not known. Estimates place more than 25,000 in the Pacific alone. Islands range in size from Greenland, some 840,000 square miles in area, to those having areas of less than an acre. Some islands, such as Newfoundland and Great Britain, are referred to as "continental islands" because they are parts of the continental shelf that remain above water. Numerous islands are the result of volcanic activity—Hawaii, Tahiti, Martinique, and Iceland, among others—being the tops of volcanoes that have been built up from the ocean floor. In many tropical areas, coral islands and/or atolls have formed around many of the dormant volcanoes. The greatest concentrations of islands are in the southwestern Pacific and the west-central Atlantic oceans.

The West Indies form an arc extending from the United States to Venezuela. The East Indies, or Indonesian Islands, also form a large arc, which extends from Southeast Asia to Australia. Several of the world's largest islands are located near the continental shorelines in the Arctic Ocean. Small isolated islands or island groups are found in all the oceans.

THE HYDROSPHERE

The water parts of the earth—oceans, seas, bays, gulfs, lakes, rivers, etc.—are called collectively the "hydrosphere." The most important of these water bodies, because of their great size, are the oceans. There are three major oceans, the Pacific, Atlantic, and Indian, and one minor ocean, the Arctic. Some maps may show an Antarctic Ocean, or Southern Ocean; but this so-called ocean is essentially the southern zones of the world's three major water bodies. Adjacent to the oceans are several mediterranean ("between the land") seas. Some 74 percent of the earth's surface in the Southern Hemisphere and 62 percent in the Northern Hemisphere is covered by water.

Oceans and Man

The world's oceans, like the land, influence man and his activities in many ways. From the oceans man secures food, furs, minerals, and numerous other useful items. Also, the oceans are the primary source of one of man's chief needs—water. It is from the oceans that the atmosphere secures most of the moisture which eventually falls as rain or some other form of precipitation.

Because of differences in temperature, variations in salinity, the rotation of the earth, and winds, the ocean waters have definite movements, called "drifts" and "currents." Warm currents are streams of water that are warmer than the body of water through which they are flowing. For the most part, warm currents and drifts move generally toward the polar regions. In the Atlantic the warm Gulf Stream, flowing northward along the east coast of the United States, crosses the ocean under the influence of the westerly winds and moves northward along the northwest coast of Europe as the North Atlantic Drift. There the warmed waters help produce a more moderate climate much farther north than is found along the eastern sides of the continental masses. The Japanese Current flowing in the North Pacific warms the western coast of North America. Cold currents are, in many ways, the opposite of the warm currents: they are colder than the water through which they are flowing, usually flow toward the equator, and often adversely affect the climate of continental areas past which they move, by causing dryness or lower temperatures. The Labrador Current flowing south along the eastern Canadian coast, the Humboldt (Peruvian) Current moving north along the west coast of South America, and the California Current, which flows south along the California coast from San Francisco, are typical examples of cold currents.

Lakes and Rivers

Lakes and rivers, like oceans and seas, form an important part of the hydrosphere. Numerous lakes of varying size are found on all continents except Antarctica. Formed in depressions caused by various geologic activities, large lakes are located at elevations ranging from 1,296 feet below sea level (Dead Sea) to 12,507 feet above sea level (Lake Titicaca). The world's largest lake, the Caspian Sea, is a landlocked body of salt water that has an elevation of 86 feet below sea level. Its chief tributary is the Volga, Europe's longest river. Africa and North America have the greatest number of very large lakes. The Great Lakes system of central North America, along with the connecting rivers and the St. Lawrence River outlet, has been aided by the building of canals and locks that make it the greatest inland-waterway transportation system in the world. The lakes of eastern Africa are of prime importance for transportation, water supply, and food to the nations bordering them. Many northern Eurasian lakes, as well as those in Canada and the Great Lakes system, are closed to transportation during the winter. Several of the world's larger lakes are big enough to influence weather and climatic conditions in their surrounding areas.

Rivers were long man's chief means of transport. The longest of the world's rivers is the Nile, which flows northward from the lakes and swamps of Central Africa to the Mediterranean Sea. The Amazon, flowing eastward from its source in the Andes Mountains of Peru across Brazil and emptying into the Atlantic near the equator, carries a greater volume of water than any other river. Many of the large rivers of the Soviet Union and Canada flow northward to the Arctic and are thereby of little use for transport because of the long winter season. In west-central Europe the chief rivers (Rhone, Rhine, Danube, Seine) have been connected by a series of canals; thus, they form a part of an international transportation system. The Missouri-Mississippi-Ohio river system serves the heartland of the United States in a variety of ways. The Hwang, Yangtze, and Mekong, Salween, Ganges, and Indus rivers, located in east-central and southern Asia, all benefit densely populated areas. Many rivers in the underdeveloped countries of Africa and South America are now being surveyed and studied for possible ways of modern utilization.

THE ATMOSPHERE

The layer of the atmosphere that rests on and penetrates the land and water surfaces of the earth is called the "troposphere." This layer, which contains more than 80 percent of the mass of the atmosphere, extends outward from the earth for a distance of about 4 miles at the poles to approximately 15 miles over the equator. The two principal permanent gases in the atmosphere are nitrogen and oxygen. In a "dry-air" situation nitrogen, an inert gas, makes up 78 percent of the air by volume; oxygen, the gas that supports life and combustion, makes up 21 percent. The remaining 1 percent is composed of several variable constituents—chiefly water vapor, carbon dioxide, and argon. Of the variables, water vapor is the most important. In very humid situations, the water vapor may expand to make up as much as 2 or 3 percent of the total air mass.

It is in the troposphere that weather originates and the zones of climate are formed. Weather is the result of the constant change in atmospheric conditions: namely, temperature, atmospheric pressure, wind direction and velocity, rainfall, relative humidity, cloud cover, and so on. When weather conditions are observed over a period of years and the data are plotted on maps, fairly well-defined general climatic patterns can be identified.

Climate and Man

The climate of an area affects, at least to some extent, what man may do to make a living. Most of the grasslands or steppes of the world are used for grazing. Humid continental interior areas are usually good farmlands; the marine west coasts are, in many respects, ideal for forests. Where minerals are found, where natural good harbors exist, or where manufacturing has been developed, climate has less influence in determining the activities of man. In the not-too-distant future, technological advances may enable man to desalt economically and bring water for irrigation from the oceans, to seed clouds for rainfall when and where needed, to track and dissipate tornadoes, and in numerous other ways to adjust the climate to better fit human needs.

The Earth's People

The activities and attitudes of each nation are related either directly or indirectly to the activities and attitudes of every other nation. No nation stands completely alone or is entirely independent. All nations, or groups of peoples, are confronted by two outstanding problems: (1) developing the correct interrelationship between man and his physical environment, that is, wise management of resources (minerals, soil, air, water, natural vegetation, etc.) on which that nation depends; and (2) developing and maintaining peaceful and just interrelationships between their own nation and the other nations or peoples of the earth. Should the nations of the world fail in the solution of these two major problems, it matters little what they do about most of the other problems confronting them.

Past generations have thought largely in terms of material matters; human resources are the focus of attention today. In most parts of the modern world, people are seeking something better, something new. They are trying to find what will make for greater happiness, what they believe will make life more worth living. Most people seek more recognition, more freedom, a greater sense of personal worth. Today the entire world is involved in a greater revolution; not only are patterns of the past being broken, but there is also a never-ending struggle to find new patterns for the future. Instead of being the unknown world it was a century ago, the world today has become a vast neighborhood created by broad revolutionary advances. Currently the nations of the world are on the threshold of the most startling revolution of all, one allied with atomic energy, jet propulsion, and interplanetary exploration and travel.

However, both in spite of and because of the technological advances and the accumulation of knowledge, the peoples and nations of today's world may be more divided than they have been during any previous time in history. Although the world's people have always been somewhat divided by such physical factors and forces as rivers, mountains, climates, and soils, they have been able to overcome many of these natural handicaps. It is, therefore, such man-made or man-induced conditions or forces as poverty, hunger, ignorance, greed, tradition, and hate that now present the greater problems needing solution. To understand why certain groups act as they do, why people or nations prosper in some areas and are poor or stagnant in others, a knowledge of the various environments, both physical and cultural, in which these differences develop is needed.

The location of man's original home on earth is not definitely known. Some anthropologists believe it was somewhere in western Asia; others think it may have been in central Africa. From his place of origin, however, man has spread over the entire habitable part of the world. Representatives of each of the major races of mankind (Caucasoid, Mongoloid, Negroid) now reside permanently on all of the continents except Antarctica.

The Caucasian peoples are classified into several groups, among whom are the Indians of Hindustan and Pakistan, the Semites of North Africa and southwestern Asia, the Nordics of northwestern Europe, the Alpine type in Central Europe, and the Slavs in the eastern and southeastern parts of Europe. The primitive Ainus of Japan are among those included in the Caucasian group. The Mongoloid peoples are found in China, Japan, and Central Asia; also included in this group are the Indians and Eskimos of the Americas as well as the Malays, who are natives of Burma, the Malay Peninsula,

Indonesia, the Philippines, and other Pacific island groups. Negroes and Negroid peoples live largely in Africa south of the Sahara, in New Guinea, in isolated portions of Southeast Asia, and in the Pacific islands as far east as Fiji, as well as in most islands of the West Indies. Many Negroes also live in the more densely populated parts of the United States and Europe, especially in urban centers.

Through intermarriage, many racial characteristics have been fused. The Polynesians are a mixed race living on Pacific islands from Hawaii to New Zealand. The Malagasy of Madagascar are related to the Polynesians. The Hamites of the Sudan and Ethiopia are another example of mixed ancestry. Thus, it is difficult to describe a "typical" German, Korean, Greek, Brazilian, American, or a great number of other peoples.

Population Increase

One of the most important problems facing the nations of the world today is the "population explosion." It is believed that man has inhabited

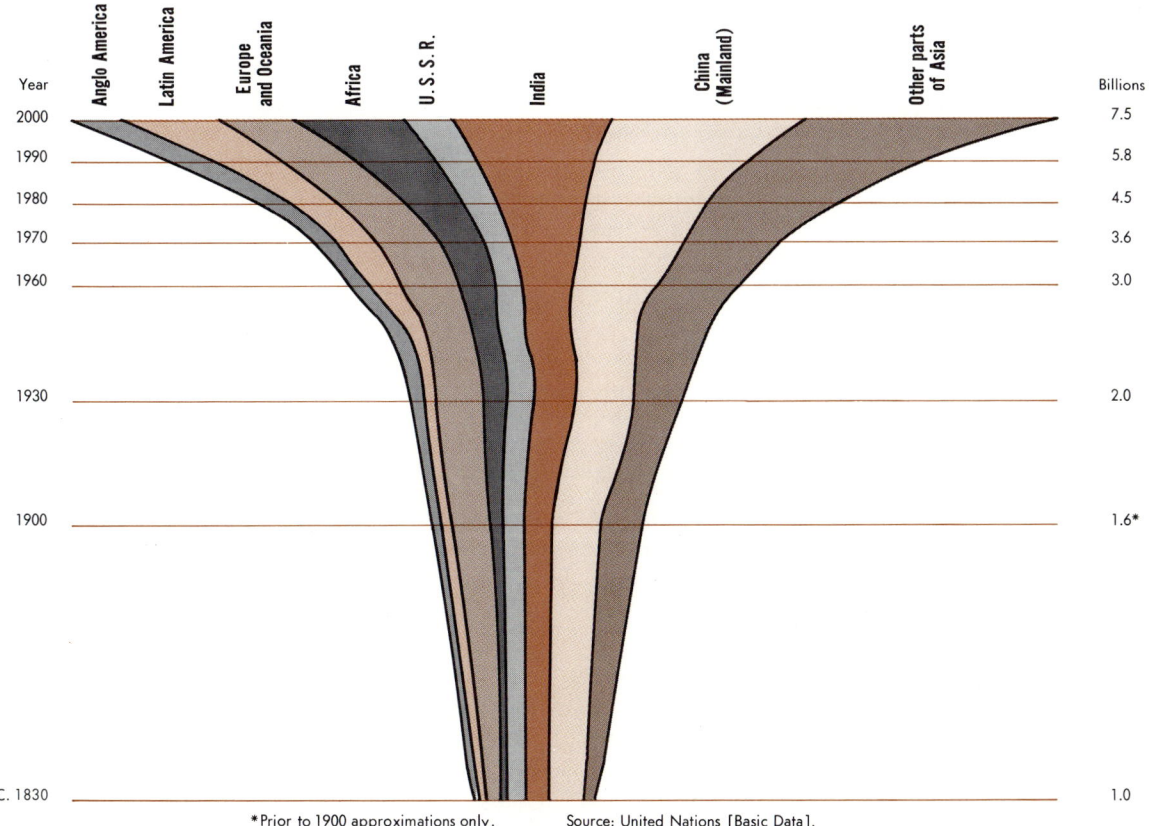

Fig. 1-3 It is estimated that the world's population will more than double between 1970 and 2000. More than half will live in Eastern, Southern, and Southeastern Asia.

POPULATION INCREASE 1830 TO 2000

the earth for more than 2.5 million years. During most of that period the world's population was small, probably being less than 5 million persons some 8,000 years ago. At the time of Christ it is estimated that the total population had increased to some 300 million persons. It was not until 1830 A.D., however, that the population totaled 1 billion persons. During the next 100 years (1830–1930), the population doubled to 2 billion; then, in less than half that time, between 1930 and 1960, it increased to 3 billion.

In spite of the fact that numerous world leaders have pointed out the disaster that may result from overpopulation, most people and nations continue to pay little attention to the warnings being given. Estimates made by the United Nations have indicated that, with the continued present rate of growth, the total population of the world will reach the 4-billion mark no later than 1980 and may exceed 7 billion by the end of this century. The present rate of world growth is estimated at 1.9 percent per year. The population of the United States is increasing at the rate of 1 percent per year. If such a growth rate continues, the 1970 population of the United States will be doubled by the year 2040.

The increase in population is the result of several factors: increase in food production, improvement in the standard of living, and advances in medical technology that have resulted in lowering the mortality rate and increasing the life-span. As the population has increased, however, the man-land ratio has decreased. If all the land in the world could be used for food production, there would be slightly more than 12 acres per person. Some land, of course, is too dry, too cold, too rough, or too bare of soil to support plant growth. This reduces the number of acres per person to about 4; but of this amount, only about 1 acre per person is now used for food production. Should the world population growth continue at 1.9 percent per year, it has been estimated by some that by the year 2000 the earth's arable land may no longer be able to feed all the people of the world.

Distribution and Density

In spite of the fact that the earth is facing a "population explosion," there remain substantial areas of empty space in the world. The most sparsely populated parts are located in deserts, rugged mountainous and high plateau areas, or regions too cold to support much plant or animal life. A few isolated areas in the Sahara and Arabian deserts, a part of the Takla Makan Desert and the Tibet Plateau in China, and the continent of Antarctica are without permanent population. A large part of North America and much of Asia north of 50° north latitude has fewer than 2 persons per square mile. Even so, some of this area is overpopulated. Large portions of South America and all of western and central Australia, as well as South-West Africa, are low-density areas. Four high-density population areas are the east-central part of Anglo-America, Western Europe, Eastern Asia, and South Asia. In these few areas live about 75 percent of the world's population.

The distribution of population depends upon numerous factors, both cultural and physical. Availability of water, temperature extremes, rough topography, and length of the growing season are the chief natural, or *physical*, limiting factors. The absence of communication and transportation systems, lack of medical and educational resources, instability of government, and, in some places, the desire of the people in an area to keep others out are examples of *cultural* limiting factors.

The countries with the largest total populations are China and India. The population, however, is not evenly distributed throughout their territory. Because of various physical factors, the people of China are concentrated on the narrow southeastern coastal plains, along the valleys of the Hsi, Yangtze, Hwang, and numerous smaller rivers and streams, and in some instances on the sides of terraced highlands. Several sectors of the interior of China are too dry and far too elevated to support a large population. In India the situation is similar; the Ganges Valley and the narrow coastal plains are densely populated, but the rougher parts of the Deccan and drier interior areas are less densely settled.

When applied to the country as a whole, population density means little; when considered for the areas of arable land, it is a significant statistic. For example, the population density for all of China is about 200 persons per square mile. Since the different parts of the country vary greatly in their potential agrarian production, however, the actual density ranges from zero in the deserts to more than 2,500 persons per square mile in certain arable areas adjacent to the great rivers. Thus, in some arable areas, there is a density of 4 persons per acre, or probably 2.5 times the number who can be fed by agriculture alone. In the United States, at least 2.5 acres of arable land per person are needed to maintain the present standard of living. In many ways, India and China are typical or classic examples, since they emphasize the problem of too many people being dependent on too small an amount of arable land.

Parts of the northeastern quarter of the United States and west-central Europe have large areas in which the density of rural population exceeds 250 persons per square mile. In each of these areas, not only is there a considerable amount of productive land, but other factors such as mineral resources, suitable climate for particular crops, and sufficient water supply are also favorable. People in some areas do not take as much advantage of these natural factors as they might; nevertheless, the potentiality for development is theirs.

The industrial districts of Japan, Western and Central Europe, and the Northeast United States are among the notable centers of high population density. In certain parts of some industrial centers such as Tokyo, New York, London, Brussels, Chicago, Berlin, and other large cities, it is not unusual for the population density to exceed 50,000 persons per square mile during certain hours of the working day. Where large industries exist in a series of cities and towns, as around the Inland Sea of Japan, in the megalopolis of the Northeast United States, and in the Benelux countries (Belgium, Netherlands, Luxemberg) and West Germany, the urban agglomerations are frequently adjacent to each other, so that it is extremely difficult to tell where one city begins and another ends. In such industrialized regions, even the less dense urban areas often exceed 5,000 persons per square mile.

Monaco, the most densely populated country in the world, has an area of about 0.6 square mile and a total population of 23,000, thus making a population density of 38,300 persons per square mile. In general, the larger the country, the smaller the average population density (U.S.S.R., 29; United States, 55; Canada, 5; Australia, 5; Brazil, 26); and the smaller the country, the greater the density (Barbados, 1,508; Belgium, 816; Singapore, 8,752; Burundi, 312). As noted above, the important point to be considered, however, is the density per square mile of arable land rather than the density per square mile for the country as a whole.

An understanding of where the population of the United States fits into the world population picture may be stated as follows:[3]

Perhaps a more appropriate feel for population and related data can be obtained if one imagines the world's population compressed into a single city of 1,000 people. In this imaginary place fewer than 60 of the 1,000 would be American citizens, and the remaining group would represent the combined population of

---

[3]Arthur H. Doerr, *An Introduction to Economic Geography*, Wm. C. Brown Company Publishers, Dubuque, Iowa, 1969, p. 24.

all other nations. In this imaginary city the United States citizens would receive about half of the town's income.

The Americans, representing somewhat less than 6 percent of the population, would consume about 15 percent of the town's food supply; on a per capita basis would use twelve times as much coal as the others; twenty-one times as much oil; fifty times as much steel and fifty times as much general equipment.

Of the 1,000 people in the town, about 300 would be Christian, and 700 would have some other faith or none at all. Of the population about 300 would be white and 700 nonwhite. The Americans would have a life expectancy of about seventy years, whereas the remaining groups would have a life expectancy of only about forty years.

The Life Layer

That part of the earth's crust, waters, and atmosphere where living organisms can subsist, the life layer, is known as the "biosphere." In spite of the fact that there are 57.5 million square miles of land on the earth, only about half of that amount is suitable for intensive human occupancy. As stated previously, some areas are too cold, and others too dry or mountainous to support more than a few thousand people. For the most part, for permanent habitation man is also confined to elevations of less than 10,000 feet. Most animal life exists within the same elevations. In the oceans and seas, most marine life lives within 600 feet of the surface. The entire environment in which man now lives exists in a precarious balance.

Man has learned much about his environment, but in so doing he has been extremely wasteful. In some of the developed countries, it now appears that frequently man has not considered himself subject to the laws of nature.

In several highly industrialized countries such as the United States, Germany, and Japan, technology has been developed to such an extent that many natural resources have been wasted and/or destroyed. Numerous streams and lakes have been partly filled with deadly waste products that have killed much of the plant and animal life therein. Waste products of numerous factories, automobiles, and other man-made machines have been emptied into the troposphere. As a result, the mixture of gases necessary to sustain life at its best has been polluted.

With uncanny precision, the mixture has been maintained by plants, animals, and bacteria, which use and return the gases at equal rates. The result is a closed system, a balanced cycle in which nothing is wasted and everything counts. For example, about 70 percent of the earth's oxygen is produced by ocean phytoplankton—passively floating plants and animals. This entire living system modified temperatures, and curbed floods and nurtured man about 5 million years ago. Only if the biosphere survives can man survive.[4]

The balances existing between the various ecosystems of the biosphere have been and are being disturbed to such an extent that man is beginning to realize adjustments and changes must be made in his way of life and technology. People in general are becoming aware of the fact that natural resources are not unlimited, and that man must find ways to conserve essential nonreplaceable minerals, to deter further destruction of soils and wildlife, and to prevent greater pollution of the earth's atmosphere and waters.

The discussions in the chapters that follow focus on the cultural and physical similarities and diversities of various areas or regions, and on the spatial organization and processes being carried on within the regions. These chapters

---
[4]*Time*, Feb. 2, 1970, p. 57.

examine the choices man has made in his use of land and other resources as a reflection of his culture. The choices are, to a great extent, a blending of man and land, space and time. Since there are approximately 150 nations in the world, ranging in size from the 8.6 million square miles of the Soviet Union to the 100.8 acres (a little less than 1/6 square mile) of Vatican City, it is not feasible to study each nation by itself. Accordingly, in the textual material, the earth has been divided into fourteen areas or regions. Each area, regardless of the number of nations included therein, has some distinct cultural and physical features that apply to the area as a whole. For ease of delimiting each study area, political boundaries of the nations included are used as regional boundaries. It must be realized, however, that political boundaries very seldom constitute limiting factors for either cultural tradition or physical features; thus, the patterns of an area will probably grade through a "zone of transition" from one region to another.

In this day and age, since the space barrier has been practically removed by time, "the need for men to carry on the task of supplementing ignorance with knowledge, or displacing prejudice with sympathy born of understanding, is truly great."[5] Geography alone cannot accomplish such a goal, but a development of geographic concepts and principles, based upon the facts given for each of the regions or areas, may well contribute more toward an understanding of this objective than will any other subject.

REFERENCES

Broek, Jan O. M.: *Geography: Its Scope and Spirit,* Charles E. Merrill Books, Inc., Columbus, Ohio, 1965.
Clarke, John I.: *Population Geography,* Pergamon Press, New York, 1965.
Demko, George J., Harold M. Rose, and George A. Schnell (eds.): *Population Geography: A Reader,* McGraw-Hill Book Company, New York, 1970.
Erhlich, Paul R., and Anne H. Erhlich: *Population/Resources/Environment,* W. H. Freeman and Company, San Francisco, 1970.
———, John P. Holdren, and Richard H. Holm: *Man and the Ecosphere,* W. H. Freeman and Company, San Francisco, 1971.
Gullion, Edmund A.: *Uses of the Seas,* Prentice-Hall, Inc., Englewood Cliffs, N.J., 1968.
Hartshorne, Richard: *Perspectives on the Nature of Geography,* Rand McNally & Company, Chicago, 1959.
Murphy, Rhoads: *The Scope of Geography,* Rand McNally & Company, Chicago, 1966.
Spar, J.: *Earth, Sea, and Air,* Addison-Wesley Publishing Company, Inc., Reading, Mass., 1962.
Strahler, Arthur N.: "The Life Layer," *Journal of Geography,* vol. 69, pp. 70–76, February, 1970.
Taaffe, Edward J. (ed.): *Geography,* Prentice-Hall, Inc., Englewood Cliffs, N.J., 1970.

---

[5] J. Russell Whitaker, *Geography in School and College,* Peabody College Bureau of Publications, Nashville, Tenn., 1948, p. 27.

2

The United States East of the Rocky Mountains

The United States ranks fourth among the nations of the world in both area and total population; the Soviet Union, Canada, and China each cover more territory, and China, India, and the Soviet Union each have a greater number of people. The U.S.S.R. has an area almost 2.5 times that of the United States, but Canada and China are both comparable in size to this nation. China has over 3.5 times and India 2.5 times the population of the United States. India has its vast numbers crowded into an area only a little more than one-third that of the United States, while Canada has only about one-tenth the population of this nation. Although each of these four nations are of prime importance in the world today, none approaches the United States as a producer of raw materials, foods, or

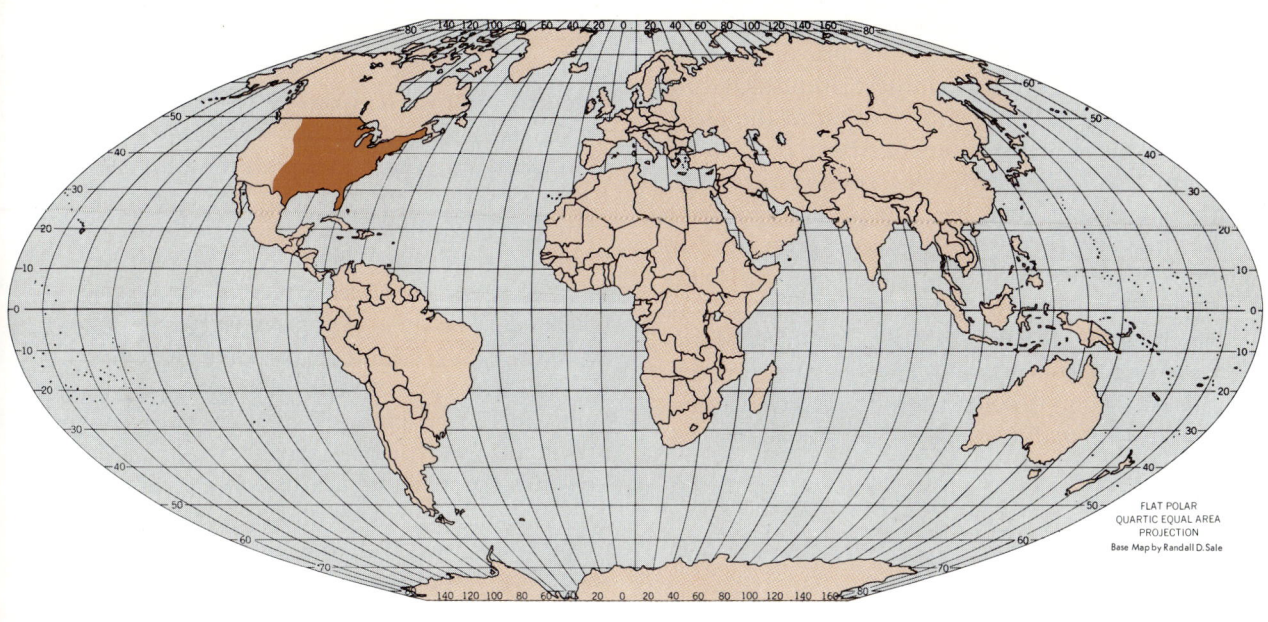

FLAT POLAR
QUARTIC EQUAL AREA
PROJECTION
Base Map by Randall D. Sale

manufactured goods, and each lags behind the United States in medical research and technological development.

It takes many things besides size and numbers of people to make a nation great. The United States, fortunately, is located largely in the mid-latitudes between 25° and 50° north. A large variety of landforms, variations of favorable climates, productive soils located where the growing season is long enough to grow most essential crops, an abundant supply of numerous minerals, and comparative isolation gave the early European settlers numerous advantages not enjoyed by those migrating to several other parts of the world. The ability of the early settlers to use expediently the resources at hand, plus their willingness to try new methods and accept new ideas, established the foundation upon which this nation was and is built.

The study of the United States is divided into two chapters—the first dealing with the United States east of the Rocky Mountains and the second with the United States west of the Great Plains—so that a more complete and intensive study may be made of this nation. The boundary between the Great Plains and the Rocky Mountains is one of the most definite physical boundaries within the United States, since the transition zone between the two, in most instances, is limited. The area east of the Rocky Mountains is dominated largely by plains and hill lands, with a few low plateaus and low mountains. In this eastern area, which comprises about 55 percent of the total area of the United States, lives more than 75 percent of the nation's total population. Here, also, are located most of the great agricultural, industrial, and urban areas of the country. West of the Great Plains, including the states of Alaska and Hawaii as well as the western conterminous area, the dominant landforms are large plateaus and high mountains. In this western area, crop production is highly specialized and generally very intensive, confined largely to those sections where water for irrigation is available. Manufacturing is concentrated in a few areas around the larger metropolitan centers there. In general, its population is widely dispersed.

The Northeast

The Northeast, as designated in this study, extends southward from northern Maine to the Potomac River and westward from the Atlantic Coast to western New York, Pennsylvania, and West Virginia. This Northeastern division differs from the other parts of the United States by reason of its vast urban development. Of its total population of almost 60 million persons, nearly 80 percent live in approximately 1,500 incorporated places. The only state in the area where more than half of the population does not live in urbanized areas is Vermont; no more than 40 percent of its people are so classified. Of the ten largest cities in the United States, having populations of 750,000 or more, four (New York, Philadelphia, Baltimore, Washington) are in the Northeastern part of the nation. In this part of the United States the incorporated places dominate the landscape and have formed the "American megalopolis," the largest continuous urban area in the world.

PHYSICAL SETTING

The principal physical region in the Northeastern area is the Appalachian System, which extends southward from northern Maine to northern Alabama. In New England the highland areas are continuous with those of eastern Canada. After being formed, the Appalachian System was eroded to a rolling surface. Since then the system has been uplifted and extensively dissected by erosion, and its northern part glaciated. The upland areas of rough, irregular topography are inland, at no point extending to the sea. The Coastal Plain, beginning with Cape Cod and Long Island, extends only a few miles inland in the Northeastern area but increases in width toward the south. The Adirondack Mountains, lying between Lakes Champlain and Ontario and north of the Mohawk Valley, are separated from the Canadian section of the Canadian Shield by the St. Lawrence Lowland. A northeastward extension of the Central Lowlands occupies a small area south of Lake Ontario.

Relief Features

New England New England has diverse relief, with hills predominating; there are large areas of highlands and smaller areas of lowlands. In general, these upland surfaces exceed 2,000 feet in elevation, with a few peaks attaining heights in excess of 5,000 feet. Many of the lowlands are located where those rocks offering the least resistance to glaciation occur. Most of the agriculture in New England is carried on in these valleys and basins; and in them are located also a great majority of the important New England cities. The entire surface of New England has been glaciated. The tops of many hills and mountains were scraped off; weak rocks were gouged away. In some places rivers were dammed; elsewhere glacial till, or moraines, was spread over large areas. Glacial

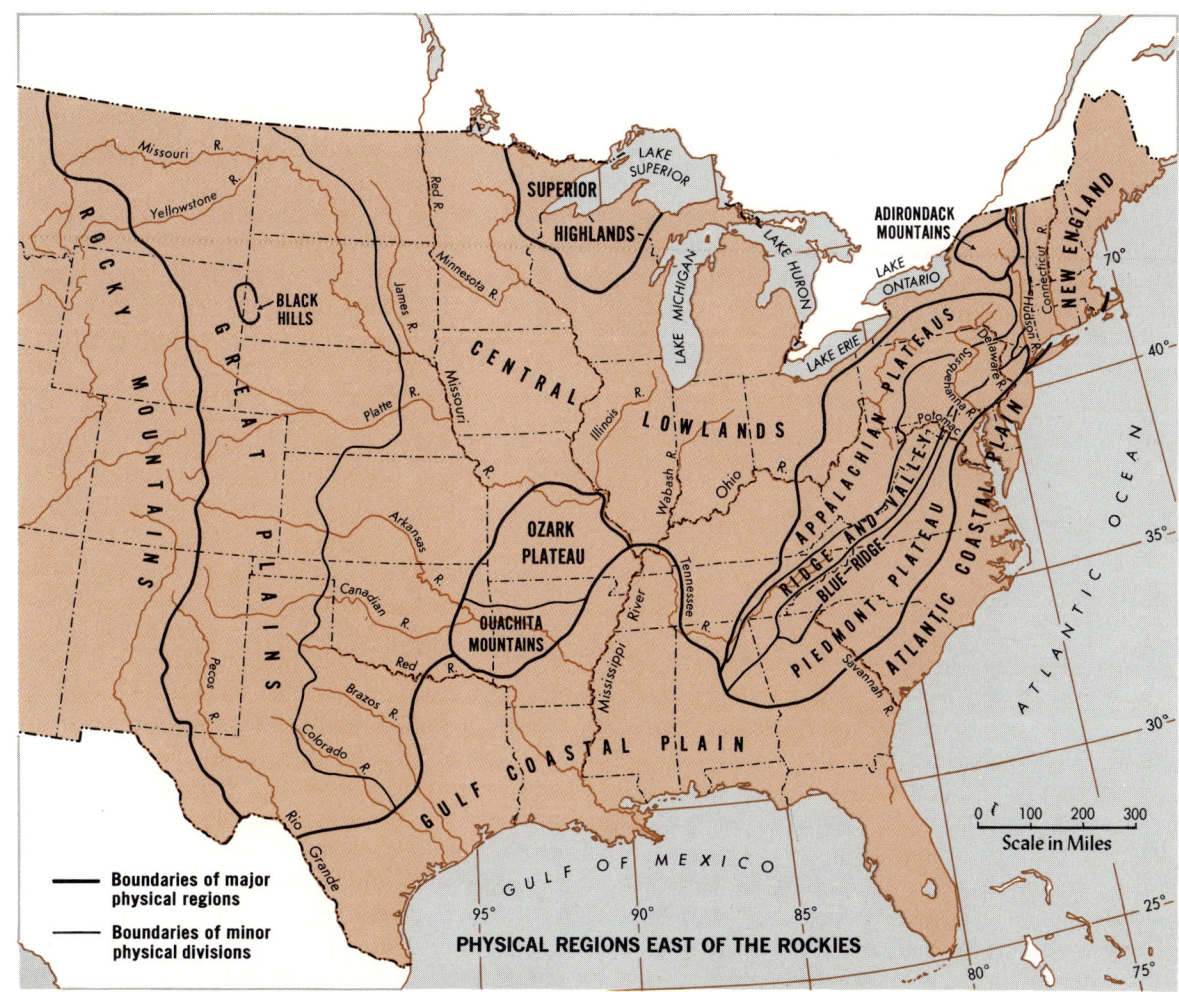

Fig. 2-1 The physical landscape of the United States east of the Rocky Mountains is characterized by plains and low mountains.

lakes are numerous, but only a few are large enough to serve as sources of municipal and industrial water supply. On the other hand, these lakes do regulate stream flow, serve as recreation sites, and add beauty to the landscape. Glaciers and glacial deposits often changed the course of rivers and caused waterfalls or rapids to form at outcrops of resistant rocks. Many factory towns were founded at such natural power sites.

New York Lowlands and the Adirondacks
New York state is crossed by two lowland areas. Along its eastern side the state is separated from New England by the Hudson River–Lake Champlain Lowland. Extending east-west

across the state and separating the Adirondacks from the Appalachian Plateau is the Mohawk Lowland. The two lowlands, which are at right angles to each other, form the much-used land and water transportation route from New York to Buffalo and the interior of the United States. The Hudson River is deep enough for ocean-going ships to navigate inland as far as Albany. Smaller ships can navigate beyond to Troy, or through a canal to Lake Champlain and then by way of the Richelieu River to the St. Lawrence.

The chief tributary of the Hudson River is the Mohawk, which flows east to join the major stream at Cohoes. Through the Mohawk Valley and the lowlands to its west was built the Erie Canal. Between 1905 and 1918, the Erie Canal and the canalized Mohawk River were deepened and widened to form the New York State Barge Canal, which extends from Troy on the Hudson River to Tonawanda on the Niagara River, from which point it connects with Lake Erie. Branch canals provide access to Lake Ontario ports. Along this same route has also developed an extensive system of railways and highways.

During the glacial period the ice sheets that covered the Adirondacks removed the surface materials, smoothed and rounded the bedrock, and deepened the valleys. As the ice retreated, the area was left covered with till (glacial debris), which deranged many of the streams and caused the forming of waterfalls, rapids, and lakes. Although some of the lakes have disappeared as a result of filling or draining, the Adirondacks are still a region noted for its lakes.

Atlantic Coastal Plain The Atlantic Coastal Plain has a nearly level surface and rises toward the interior in a series of low terraces. The great irregularity of the coastline is due to the drowning of river mouths by the invasion of ocean water. The resulting estuaries, or tidewater bays, include the Delaware and Chesapeake bays and the Potomac River. Inland a scarp marks the contact of the Coastal Plain with the Piedmont. Where streams cross this scarp, there occur rapids or falls, according to differences in the hardness of the rocks. The resulting feature, known as the Fall Line, determines the head of navigation on most major rivers of the area, as well as the site of numerous water power developments. Many Eastern cities, including Trenton, Baltimore, and Washington, are located along the Fall Line, and their elevations vary from near sea level to over 300 feet.

Northern Appalachian Highlands The Northern Appalachian Highlands, like their counterpart in New England and eastern Pennsylvania, greatly influenced the early activities in the region. The Ridge and Valley area, with its curving depressions, in addition to the width of the rugged Allegheny Plateau, presented many problems to the development of transportation systems and lines of communications.

The northern Allegheny Plateau includes the Catskill and Pocono Mountains as well as the Finger Lakes district of New York. Along with portions of northwest Pennsylvania and northeast Ohio, this area has been glaciated and is more extensively eroded than the land farther south. Many lakes are scattered among the forest-covered hills. In the Poconos, Catskills, and Finger Lakes areas a thriving tourist industry has developed.

Climate

The climate of the Northeastern area is of a continental type. Near the coast it is somewhat modified by the winds that blow from the sea part of the time. Away from the coast, toward the interior, the length of the growing season decreases markedly and rapidly; for instance, it is only 159 days in Burlington, Vermont, as compared with 199 days in Boston. The length of the growing season also increases as one pro-

Fig. 2-2 The population and, therefore, the number and size of incorporated communities decrease westward from the Atlantic Coastal Plain to the Rocky Mountains.

ceeds southward; it is as much as 200 days in some parts of Lancaster County, Pennsylvania. Rainfall occurs during every month of the year, with the heaviest rainfall usually coming during the summer. Temperatures range from below freezing in the winter months to well above 80°F during the summer.

ECONOMIC DEVELOPMENT

The people of the Northeastern area make their living in numerous ways. Agriculture, forestry, fishing, and the various recreational activities are all important sources of income. Manufacturing and the commercial activities associated directly with it, however, bring more wealth into the area and employ more people than do all other kinds of economic development combined.

Agriculture

Farming in New England, although handicapped by cold winters, short cool summers, thin stony soils, poor transportation, and a limited market, survived as a major activity until the opening of the Erie Canal, the building of railroads, and the development of manufacturing. Importation of cheap wheat and other farm products from regions farther inland brought about the abandonment of many Northeastern farms, especially those situated at high elevations; fields reverted to scrub and forest, and farmhouses settled into ruin. Some of this erstwhile farmland is now a part of state and national forests. Some abandoned farms were later reoccupied by European immigrants and by cityfolk desiring to own land or a country home.

In many sections of New England, specialized farming prevails. Dairy farming is highly developed, and dairy products normally account for about 30 percent of the region's cash income from agriculture. During the cool wet summers, hay and pasture grasses grow well even on steep slopes with thin leached soils. From 80 to 90 percent of the cropland harvested in the Vermont and New Hampshire uplands is devoted to hay. Corn is grown for grain and silage, and near the large industrial cities the dairy industry uses large quantities of imported feed concentrates. Improved handling of fresh milk and the large urban demand for milk products have enabled dairy farmers who are remote from markets to operate successfully. The Lake Champlain Lowland of western Vermont ships milk and cream to Boston, other southern New England cities, and New York.

Specialty crops are locally important. Sumatra leaf tobacco, which is used for cigar wrappers, is grown on the terraces along the lower Connecticut River. This crop is shielded from the sun by cloth netting supported with poles and wire; whereas tobacco for cigar fillers is grown in the open. In the Aroostook Valley of Maine the sandy, porous soils and the cool rainy summers delimit an area where potatoes are grown; because of their superior quality and the care used in production and grading, many of these Maine potatoes are sold for seed. Onions and a variety of market vegetables come from suitable land near the cities. On Cape Cod, cranberries improved from the wild variety are cultivated in swampy mucklands. Maple syrup is another New England specialty product, especially in Vermont, where the long cool spring season, with freezing nights and thawing days, results in a flow of sap sufficiently prolonged to justify the labor and expense of tapping the sugarbush.

Agriculture also has long been an important activity on the Coastal Plain. Much of the area is covered with light sandy soils. The ground is easily tilled, and numerous truck farmers produce early- and late-season vegetables and fruits for the urban markets. On drained swampland and loam soils in northern New Jersey, the truck farms are meticulously tended and yield abundantly. Other sections of New Jersey and peninsular Maryland are also intensively farmed, and much fertilizer is used in their cultivation in order to increase production. In the Delmarva Peninsula poultry raising is of prime importance, for that area is one of the nation's leaders in the broiler industry. Besides fresh fruit, vegetables, and poultry sped by truck to the cities, great quantities are canned or frozen. More than 400 canneries and other food-processing plants operate in and around Baltimore; other hundreds operate in the Camden-Philadelphia district.

Forests

A view of the New England landscape from an airplane or mountain peak gives the impression of extensive forests broken by small cultivated areas and settlements. Often small communities are half-hidden by the tall shade trees that give added charm to the fine old dwellings. Over 70 percent of New England is under some sort of forest cover. These forests provide the recreation playgrounds and sports areas for the large urban communities of Northeastern United States. Besides the woodlot found on nearly every farm, commercial forests cover large areas.

For over 200 years the region supplied lumber for buildings, wood-product and paper manufacturing, shipbuilding, and fuel. After the virgin forests had been cut, the new growth was useful chiefly for pulpwood and cordwood. Today the pulp and paper industry is the largest consumer of timber, although the forests still produce other materials in reduced amounts. Improved forest management is a definite regional need, although some of the paper companies manage their large forest holdings so efficiently that their pulpwood requirements can continue to be supplied permanently. National, state, and community forests, along with some privately owned land, are being developed for growing timber. Usually pulp and paper mills are located along rivers, which can be used for power, log storage, and other manufacturing needs, just as the sawmills were built on comparable sites before the timber was largely exhausted.

Fisheries

In the shallow waters adjacent to New England is one of the major fishing grounds of the world. Here the banks, usually less than 100 fathoms (600 feet) in depth, are important feeding grounds for cod, halibut, haddock, mackerel, and herring. Fisheries have been an important source of income for the region since colonial days. Although the catch varies, it is approximately a billion pounds annually, and two-thirds of it enters through two Massachusetts ports, Gloucester and Boston. Fish are sold fresh in local markets and are also quick-frozen, canned, salted, dried, or smoked. Shore fisheries supply clams, oysters, crabs, and lobsters. Some by-products of the industry are fish meal, fertilizer, and vitamin oil.

Fish have always been an important source of food along the Coastal Plain. The Chesapeake and Delaware bays lead all other areas of the United States in the production of shellfish. Oysters grow there in the warm shallow water, which is relatively free from silt and mud. Overexploitation of the beds and the pollution of streams flowing into the bays, as well as the bay water itself, has led to the abandonment of fishing in many formerly profitable areas. As a result, oyster "farms" have been developed where oysters can be propagated scientifically. Herring, shad, flounder, mackerel, and bluefish are taken locally by offshore fishermen, and quantities of menhaden are netted; though inedible, the last-named is used in the fertilizer industry. Owners of small craft often take parties of vacationers "deep-sea fishing" in the shallow waters of the Continental Shelf.

Minerals

Mining is not a major activity in New England, since the area contains no coal or petroleum, and deposits of metals are of small importance. Building stone, however, is quarried extensively. Barre, Vermont, is famed for granite, and marble is produced in several localities. Stone is now used mostly for decorative purposes and monuments. It has been largely replaced by concrete, steel, and glass in large structures because these materials cost less, require less labor, and so make building more economical. Iron ore is mined in small quantities in the Adirondacks and in Pennsylvania. Near Syracuse, salt deposits are exploited by the chemical industries.

Pennsylvania and West Virginia are important producers of mineral fuels; the principal anthracite coal mines of the United States are located near Scranton and Wilkes-Barre. Anthracite is mined from those parts of the Appalachian ridges which were subject to intense pressure. Because of competition from petroleum and natural gas as fuels, mining of anthracite has declined in the last three decades.

In western Pennsylvania and West Virginia lies the most productive part of the Appalachian coalfield, an area of high-grade bituminous coal. The availability of this source of power is the most important factor in the location of the manufacturing belt in the United States, the area known especially for manufacture of iron and steel and other heavy goods. Here the towns are generally of two types: farm-market centers, with fine old houses and shady streets, and mining towns, in many of which the mine-owners have built all the houses alike. These latter towns are frequently ugly, devoid of trees, and situated in a landscape dominated by huge piles of waste rock from the mining operation. Bituminous coal production is highly mechanized in both strip and underground mines. Cutting, loading, sorting, cleaning, chemical treatment, and blending of the coal is all done by machines.

Most of the coal-mining areas of West Virginia, as well as those in the Southern Appalachians, have in recent years been classed as chronically depressed areas. Many mines have been closed because of the increasing cost of operation, greater use of petroleum products as fuel for transportation, development of waterpower for the generation of electricity by the building of large dams, and the decreasing grade of the coal available for mining. The increasing use of machinery has replaced many laborers. As a result, hundreds of miners are out of work. Since most are middle-aged or older, they do not desire to move from their home areas, and most fear they are too old to find employment elsewhere. Most of the younger people from these regions who have migrated to industrial centers find themselves in the lower income brackets and often living in conditions as bad or worse than those they left. Many of the older miners have lung diseases of various kinds, caused by the conditions under which they worked. For numerous families now living in the coal-mining areas of the Northeast, the chief source of income is a monthly subsistence or relief check paid by the mine workers' union or some branch of government.

Federal and state governments as well as the union are now showing greater concern for the people of this region, as well as for the region itself. Better educational facilities for both young and old are being developed; in some places medical and health centers have been established. Conservation programs for the land as well as mining activities are underway.

Petroleum and natural gas are two other sources of available power. The first oil well in the United States was drilled in 1859 in northwestern Pennsylvania. The names of Oil City and Titusville, Pennsylvania, were famous in the early days of the oil industry. Much high-grade petroleum is still pumped from the oil fields of Pennsylvania and West Virginia.

RECREATION

Recreation in the Northeastern part of the United States is favored by the beauty and variety of the landscapes and by the nearness of the scenic outdoor playgrounds to New York, Washington, Boston, and other large cities. The area offers mountains, hills, forests, lakes and rivers, and irregular coasts varying from sandy beaches to rock-grit harbors; there are hunting, fishing, and boating facilities, deep long-lying snows for winter sports, interesting historical sites, and national, state, and community parks, all made accessible by a closely woven net of highways and railroads. There are varied accommodations for people of different means. Thousands of people are supported by the outdoor recreation industry, which brings to New England alone an estimated income of more

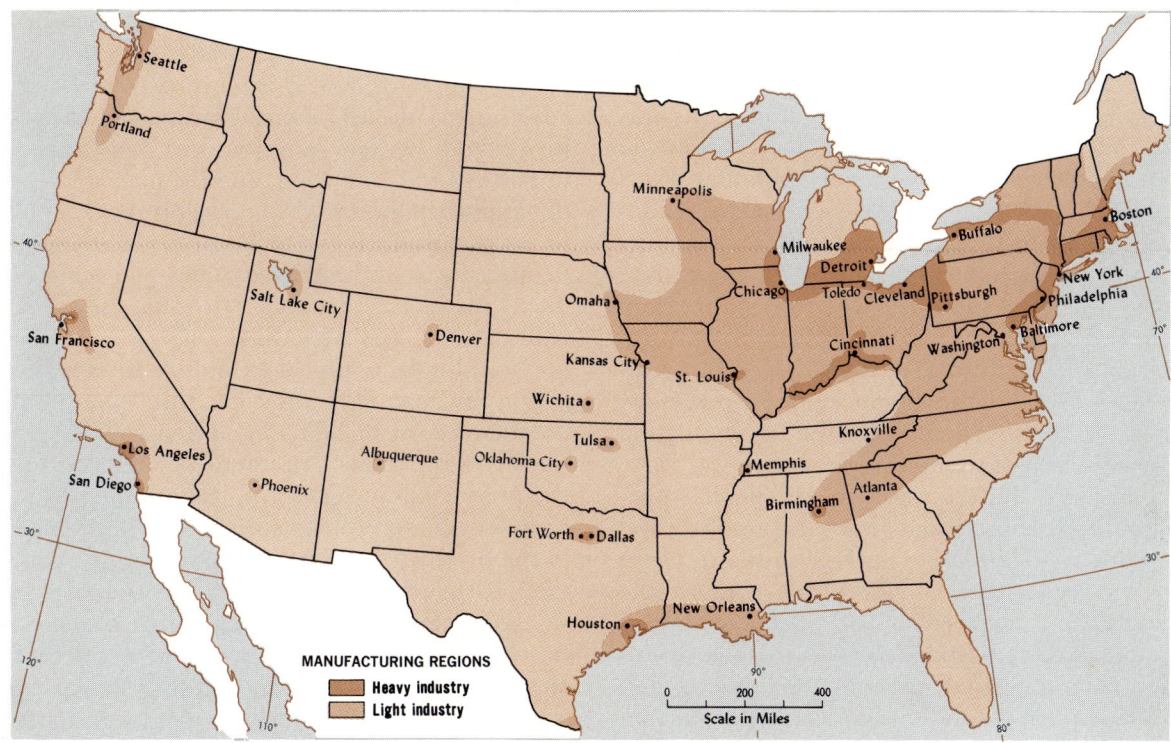

The United States East of the Rocky Mountains 27

Fig. 2-3 The principal manufacturing region of the United States is located in the northeastern quarter of the nation. An ample supply of fuel, reserves of iron ore nearby, and natural deepwater transportation furnish an adequate base to make this one of the world's most productive industrial areas.

than three-quarters of a billion dollars annually.

Upper New York state, north of the Mohawk Valley, is a rugged, mountainous country. A region of peaks, forests, lakes, and great scenic beauty, the Adirondacks serve as an important recreation area, and many hunters and fishermen are attracted to the region. The Adirondack Forest Reserve covers much of the section. Resort hotels have been built along the shores of the region's numerous bodies of water; during the winter season, the resorts remain open for such activities as skiing, sledding, and ice skating.

Along the Coastal Plain the tourist and recreation industry is also big business, with revenues amounting to several hundred million dollars annually. Summer resorts have developed along the Mid-Atlantic seaboard, where sandy beaches are within easy reach of the larger Eastern cities. The best-known among these resorts are Atlantic City, with its famed seven-mile-long boardwalk, Asbury Park, Jones Beach, and Coney Island.

CITIES AND INDUSTRIES

From Portland, Maine, to northern Virginia stretches an almost continuous series of cities. Functioning as seaports and railroad terminals, or as manufacturing, financial, and political centers, they form the American megalopolis. The focal point of the cities in this strip is New

York. Between the major cities of this urbanized zone are situated mills, factories, and refineries at sites convenient to transportation, great numbers of suburban communities, and thousands of country homes. Transportation by private automobile, trucks, buses, and commuter trains makes living in the country or suburbs as convenient as living in the big cities. So many people prefer homes outside the city limits that, since 1920, the suburbs have grown faster than the cities themselves. Within the Northeastern area, in 1970, were 29 of the 100 largest metropolitan areas of the nation. Many cities, however, have recorded decreases in population during the past twenty-five years.

Because the Appalachian Highlands separate this urbanized zone from the populous and productive Ohio Valley and the southern Great Lakes region, the corridors that connect the coastal and interior cities are of major importance. Railroads and highways cross the mountain barrier from Washington, Baltimore, and Philadelphia; but the lowest-level route, one that is without steep grades, crosses the Hudson-Mohawk lowlands between New York and Buffalo. This route extends north from New York to Albany and then turns west to Lake Erie. Indians and fur traders first used the route, and wagons followed. In 1825 the Erie Canal, now part of the New York State Barge Canal, was opened, and New York became the seaboard terminus of the only low-level water route from the rich and rapidly developing interior lowlands. In 1851, a through railroad was completed between Buffalo and New York. The combination of canal and rail services so stimulated commerce and settlement that New York quickly outgrew its former rivals, Boston, Philadelphia, and Baltimore, in trade and population. Along the Hudson-Mohawk route, an urbanized belt containing scores of cities has developed. The excellent transportation has led to establishment of many manufacturing plants.

A large majority of the people of New England live in cities, where they work in factories or engage in commerce, trade, and service occupations or follow some craft or profession. Also important are financial businesses such as investment brokers and banking, insurance, importing and exporting, and the management of large corporations. The region is a hive of industry, producing nearly 10 percent of all manufactures of the United States—an achievement all the more noteworthy since few raw materials are secured locally.

New England produces little wool and no cotton raw materials; yet it leads the country in the fabrication of woolens and manufactures much cotton cloth. Although not notable for the production of hides and leather, the region leads in making women's shoes and ranks high in the production of men's shoes. Large quantities of electronic and electrical materials, as well as numerous specialty items, are also manufactured in the villages, towns, and cities of New England. It has been said that "few parts of the world make so much, so well, with so few local resources." Among the factors accounting for New England's industrial importance are an early start and a reputation for producing excellent goods, skilled labor, efficient management, proximity to local markets, and capital for investment.

Although coal, oil, and natural gas are imported, New England has much waterpower. Before the development of electric power transmission, mills and factories were located at waterfalls and dams, where machinery could be run directly by waterwheels; as a result, many New England towns were crowded into narrow valleys where waterpower was available. Manufactures are heavily concentrated, for as an industry succeeds, competitors follow, since their enterprises are favored by the skilled labor that has been trained for certain methods of production. Industrial concentration, however, also has disadvantages. Too many factories along a stream may pollute the water to such an extent that it is no longer usable. If a factory that is the main support of a community ceases operations, moves to some other site, or works only on a part-time basis, the entire community suffers. Competition from Southern textile mills, which have expanded until they now surpass New England mills in output of cottons, has closed many New England mills. Some cities affected have been able to secure new industries, but others have suffered from unemployment and loss of population.

Many industrial cities have developed in the Mohawk Valley. Started because of some particular advantage of site, these cities have continued to grow because of improved transportation and their concentration upon producing industrial specialties. Some developed by utilizing the special skills of early settlers (Gloversville) or by building a significant industry upon a local mineral deposit. (Syracuse built up a chemical industry based upon local salt deposits.)

Buffalo, the second largest city in New York, is located at the eastern terminus of waterborne traffic on the four western Great Lakes and at the western end of the Mohawk Lowland. This situation is excellent for receiving iron ore, limestone, and grains from upper lake ports. Coal comes from the Allegheny Plateau. Buffalo's lakeshore harbor, which has connections with the Atlantic Ocean via the St. Lawrence Seaway, has been extended and improved to accommodate large lake vessels, and docks and wharves are mechanized to speed loading and unloading. The western terminus of the New York State Barge Canal is an integral part of the waterfront. Originally primarily a transshipment point, Buffalo is today a manufacturing city, utilizing power from coal and from Niagara Falls for flour and cereals manufacture, steel fabrication, and chemical, electrical, and other industries.

The Pittsburgh district leads the world in the manufacture of iron and steel. The abundance of high-grade coking coal, the ease with which iron ores are imported from the Lake Superior region and Canada via the St. Lawrence Seaway, the early start of the industry, and the huge demand from other nearby industries help to account for the regional importance of iron and steel manufacturing. Pittsburgh is located at the junction of three navigable rivers, the Ohio, the Monongahela, and the Allegheny, on a usable water route to the interior and ultimately to the Gulf of Mexico, and it also has excellent railroad and highway connections. Although Pittsburgh is located in the Northeastern area, its activities and those of Buffalo are closely related to other manufacturing centers in the American Midlands.

American Megalopolis

In spite of the importance of truck farming, fishing, and the tourist industry, most of the people living on the Coastal Plain in New Jersey, Pennsylvania, Delaware, and Maryland, as well as those residing in the cities and towns of Connecticut, Rhode Island, and eastern Massachusetts, are employed in manufacturing, commerce, retail trade, and various service occupations. The American megalopolis covers less than 2 percent of the area of the United States, but in this area is done nearly a third of the nation's business and here live approximately 10 percent of its people. Situated on the eastern edge of the manufacturing belt, the area makes at least some of almost every product manufactured in the United States. Most large cities of the Coastal Plain developed along the Fall Line, where waterpower, protection from storms and the open sea, and the necessity for bulk breaking and transshipment led to early settlement and continued growth. Several of these cities have become major seaports, and nearly all are manufacturing centers. The large population of the region provides a huge market for all kinds of goods.

The low relief of the plain further aided manufacturing by permitting easy construction of railroads and highways, but the broad tidewater bays in many places prevented construc-

Fig. 2-4 A section of Kennedy Airport, New York. Rapid transportation for large numbers of people and certain types of freight is a requirement in the rapidly changing world of today. The Boeing 747 is the largest passenger plane of the superjet era; small jets are shown at the left. (Courtesy of Trans World Airlines)

tion of overland routes along the coast. This problem has led in modern times to the building of great highway bridges over several of the bays—bridges rising so high above the water that oceangoing ships pass freely underneath their spans. Some examples are the Delaware, Chesapeake Bay, Potomac, and York bridges, in addition to those across the Hudson River. Combined with the turnpikes, such bridges have greatly reduced travel time from New York to other Northeastern cities, as well as to those to the south and west. In addition, numerous railroads, highways, and airways connect the East Coast industrial and port cities with the uplands and interior plains.

This large-scale development of metropolitan regions and urban areas has been called the "main street and crossroads of the nation." The almost continuous urban buildup accounts for the greatest concentration of population in the United States. New York State, ranking thirtieth in area, has the second largest population total among the states. New Jersey has the greatest population density, 942 persons per square mile; Rhode Island is second, with 872 per square mile. The density of population in the District of Columbia is approximately 13,500 persons per square mile.

Boston, New York, Philadelphia, Baltimore, and Washington are the centers around which the American megalopolis functions. These cities, like all the larger and many of the smaller cities of the United States, are continually faced with a series of problems—physical, political, economic, and humanitarian. When these five places were founded, no one could envision the immense size they were to attain, either in area or in population, nor that each would be classed among the important cities of the world as well as the nation. No early plan for the development of Boston, New York, or Baltimore was conceived; each just grew. An early city plan for Philadelphia was made by William Penn; at the beginning of the nineteenth century, Pierre L'Enfant was employed to make the master plans for Washington.

The chief physical problems facing the modern cities are site, air pollution, water supply, and water pollution. With the exception of Philadelphia, each city was somewhat handicapped by its original site. Boston, Baltimore, and Washington had swamps adjacent to their original locations. New York was isolated by being situated on an island. In all except Washington the streets were narrow; and in Boston, New York, and Baltimore the topography, swamps, and streams caused the thoroughfares to be crooked. The lack of suitable space for outward expansion forced the cities eventually to build upward; thus, the narrow, crooked streets became lined with tall buildings, many of these being skyscrapers, which now prohibit the widening of streets and thereby slow the movement of traffic. None of these cities now has a sufficient supply of water in its immediate vicinity. Each day over 1,200 million gallons of water are used by the industries and people of New York. The cost of constructing aqueducts and tunnels to bring the city water from sources east of the Hudson and from the Catskills has exceeded the cost of the Panama Canal or the Grand Coulee Dam project. The modern factories that use fossil fuels, plus the exhaust of motor vehicles, cause air pollution that keeps the cities in fog, haze, or smog much of the time. In addition, the great amount of refuse to be destroyed or moved each day adds to the pollution of the air by burning, or of the sea by dumping.

Perhaps the greatest need in each city is that of humanitarianism. The great agglomerations have long been noted as the gathering places for immigrants. New York has its Chinatown, Boston is famous as the home of the Irish, and Philadelphia is the home of many who are descended from German or Dutch ancestry. Since 1940, however, these cities have become the collection centers for people from minority groups and/or depressed areas of this country. Many Negroes, Puerto Ricans, Spanish-Americans, and even Indians have migrated to Washington, New York, and other large urban centers not only in the Northeast but throughout the nation. In numerous instances these migrants have settled in ghettos where housing is poor, playground space is inadequate, educational facilities are understaffed, and jobs are scarce. As a result, many ghettos have become noted as areas of violence and centers for riots. Urban renewal is helping to correct some of the problem by destroying pest-infected dwellings and replacing them with modern buildings that furnish more room, better sanitary facilities, and satisfactory living conditions in general. Many of the smaller cities in the American megalopolis, as well as the larger centers, have established vocational schools, improved law-enforcement systems, and started programs to aid in the assimilation of the newly arrived groups. The movement of many city residents to the suburbs, in order to secure more favorable living conditions, has lessened the number of taxpayers in and increased the financial problems of these Eastern cities.

Boston has an excellent harbor, with deep entrance channels and an extensive frontage for berthing ships. Its seaborne imports are largely coal and petroleum products, together with a variety of raw materials and commodities that are either consumed in New England or used in manufacturing there. Imports passing through Boston far exceed exports, both in tonnage and in value.

Massachusetts Institute of Technology and

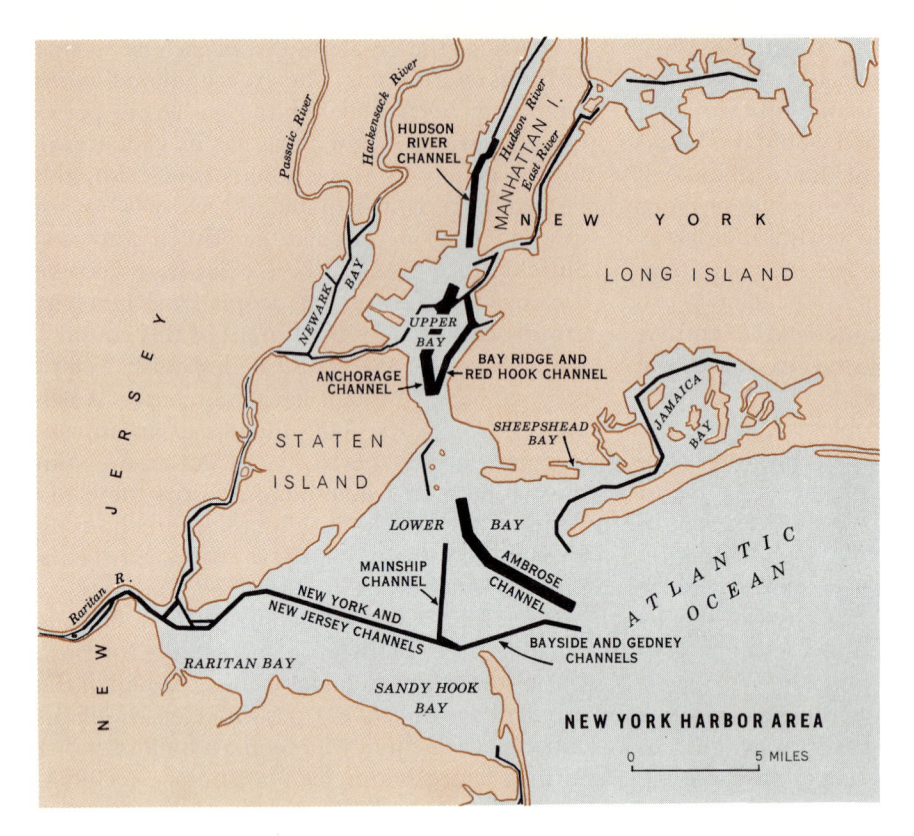

Fig. 2-5 Map of the New York harbor area showing major rivers, bays, and ship canals.

Harvard University, both located in nearby Cambridge, as well as Tufts University, Boston University, and Boston College are noted local institutions of higher learning. (Many other noted universities and colleges, among them Yale, Princeton, Columbia, Syracuse, Clark, University of Pennsylvania, Pennsylvania State, Johns Hopkins, and New York University, are located in other Northeastern urban centers.) On the faculties of these colleges and universities are some of the outstanding research workers and teachers of the nation. Research is carried on in medicine, space exploration, weather, food, and various other technological fields. Numerous foundations such as Ford, Carnegie, the National Science Foundation, and the Office of Naval Research contribute generously to the support of research activities in which people are paid to think. The discoveries resulting from such work have aided not only the people of the United States but the people of the entire world.

New York, the largest city in the Western Hemisphere, has developed from the settlement founded as New Amsterdam in 1626 at the southern end of Manhattan Island. Now covering approximately 300 square miles and having a population of about 7.8 million, the city is the core of the Greater New York metropolitan area. Within this area of over 2,600 square miles live some 14.1 million persons, who reside in approximately 400 different municipalities. The metropolitan area includes such cities as Newark, Hoboken, and Paterson in New Jersey,

Stanford in Connecticut, and Yonkers and New Rochelle in New York. New York City now covers Manhattan and Staten Islands, part of Long Island, and some territory on the mainland. Administratively, the city is divided into the five boroughs of the Bronx, Brooklyn, Manhattan, Queens, and Richmond.

Manhattan Island is the heart of the New York business district. Upon this island has been built the greatest group of skyscrapers in the world, many buildings exceeding fifty stories in height. Within these great buildings are housed the headquarters of the leading commercial, industrial, and financial enterprises of the United States. The density of population in the entire area of New York City exceeds 25,000 per square mile. During the daytime working period, in many of the Manhattan business districts the density is greater than 100,000 persons per square mile.

The intermingling of land and water hampers the movement of goods and people. Vehicular and railroad tunnels plunge under the water; enormous and costly bridges stretch high above, while tugs, barges, lighters, and ferryboats navigating the surface facilitate the movement of both people and goods. Ten major railroads serve the port of New York, although only three of them have terminals on Manhattan Island. Since most of the streets are narrow and sometimes winding, traffic congestion makes delivery by truck slow and costly. Most New York residents do not try to drive cars to work; but even so, vehicular tunnels, subways, and throughways for automobiles have relieved the traffic only in part. Hundreds of thousands of officeworkers, professional persons, and laborers commute daily to Manhattan Island from Long Island, New Jersey, Connecticut, and Hudson River Valley points, and at rush hours the subways, suburban trains, ferries, and buses are filled to capacity in transporting suburbanites into or out of the metropolis.

Most of the world's great business firms maintain offices in New York, and numerous large corporations are managed from the city. As the nation's financial center, New York dominates the banking, investment, and insurance fields. Industries such as garment and jewelry manufacturing, which produce high-value, high-style items in relatively small space, remain in midtown Manhattan, where wholesale and retail buyers congregate. The city leads in the styling and production of clothes. It makes or controls the manufacture of most toys, novelties, pharmaceuticals, books, and periodicals. New York managers present the country's major concerts, plays, and musical attractions and book most of those sent on tour.

As a result of congestion and high land values, industries requiring sizable space for their operations have sought sites in neighboring areas. Many of these outlying cities have other complementary and supplementary functions in the metropolitan area, such as providing residential sections for people who work in New York, terminals and freightyards for railroads that do not go into Manhattan, dockage for coastal vessels and tramp steamers, and storage facilities. Several cities that share the environmental advantages of New York have become centers of commerce and industry in their own right.

One of the principal reasons for the great development in the New York vicinity is the excellent harbor. A marvelous system of protected and interconnected waterways, well sheltered from but easily accessible to the sea, is the focus of routes to and from the ocean and the continental interior. The Hudson River has a channel deep enough to accommodate the largest ships, and tides and currents are strong enough to help keep it scoured. There are no natural obstacles to shipping, since the tidal range is small and the waters are ice-free during the entire year. Ten thousand or more vessels utilize the port facilities annually. The New York area normally handles 40 to 50 percent of the nation's imports and 30 to 40 percent of its exports. Its total of well over 150 million short tons yearly is more than twice that of any other United States port.

Philadelphia, the fourth largest city in the nation in population, is located at the junction of the Delaware and Schuylkill rivers. This city has continued to develop according to the plan approved by its founder, William Penn. City Hall Square forms the heart of the business district, and a gridiron pattern of streets extends from there to all parts of the city. Philadelphia contains many historical shrines: Independence Hall, where the Declaration of Independence was signed, still stands; nearby are such buildings as Betsy Ross House, Carpenter's Hall, where the first Congress met, and Christ Church, where many of the nation's founders worshiped. Modern Philadelphia is among the top-ranking seaports of the nation. The Delaware River, navigable for oceangoing ships as far upstream as Trenton, ranks second to New York City harbor facilities in commercial tonnage. Proximity to coal and other resources has furthered Philadelphia's growth. The city has a great diversity of industrial enterprises—chemicals, printing, tobacco, hats, rugs, foodstuffs, and radios, among others—which have benefited from excellent transportation, access to markets both at home and abroad, and the early start of the city. Just across the Delaware River is Camden, New Jersey, which is an important producer of food and electronic items. Nearby, in Delaware, is Wilmington, the center for chemical industries.

Baltimore, on the west coast of Chesapeake Bay, is an old, historic city as well as a modern commercial and manufacturing center. Along its waterfront are numerous shipbuilding facilities and drydocks, grain elevators, oil-storage facilities, and ore piers. Sparrows Point, a suburb, has one of the largest steel mills in the world and an enormous shipyard. Transportation in the Baltimore area is excellent. Chesapeake Bay is connected to Delaware Bay by a canal.

Washington is unique among the cities of the nation. Although its urban population exceeds 750,000 persons, manufacturing is of minor importance. The city has just one important industry—government. Many varied service enterprises care for the thousands of government workers and the numerous tourists and visitors. For the most part, the city is a place of beauty. Wide avenues radiate from the Capitol somewhat like the spokes of a wheel, and in addition the city is overlaid by a gridiron pattern. Trees shade many of the main avenues as well as streets in residential areas. Large parks, the Mall, various monuments and shrines, museums, and national archives are all places of interest. The present city has outgrown the District of Columbia proper and expanded into Maryland and across the Potomac River into Virginia.

In the Northeastern states, as in the other parts of the nation, the problems of urbanization are critical. Good housing, slum clearance, an adequate road and highway net for rapid ingress and egress, sufficient and properly spaced parks and recreational areas, necessary supplies of pure and fresh water, and good educational and health programs are basic requirements for modern city development. So important is a satisfactory solution to these community problems that the federal government now has nearly fifty major programs affecting urban development and redevelopment. Practically all cities have planning boards, and many work in conjunction with county or metropolitan boards.

The Midlands

American society evolved during the expansion of settlement from the Atlantic fringe westward across the continent to the Pacific shore. An important factor in the development of this changing society was the frontier. Before the new nation could be fully developed, the land had to be transformed. In America the frontier was a zone between a comparatively stable and well-settled region and one hardly touched by civilization. It was thinly occupied by Indians and offered abundant opportunities for development, although at the cost of risk and hardship. As the zone of settlement was extended, forests were cleared, the sod broken, crops and orchards planted, homes and villages built, railroads and highways constructed, and a savage wilderness and a seemingly endless grassland were transformed into farms and cities.

For many years the American frontier existed in the old Northwest Territory. An important route of migration from the seaboard colonies followed the Hudson-Mohawk Lowlands to Buffalo and along the southern shore of Lake Erie. Other routes, such as the Braddock and Forbes roads, crossed the Appalachians by following rivers and low divides to terminate in Pittsburgh at the confluence of the Allegheny and Monongahela rivers, and from there continued down the Ohio River southwestward. The National, or Cumberland, Road also crossed the plateau and headed west over the plains toward St. Louis. During later decades the Mormon Trail, Oregon Trail, and Sante Fe Trail originated in the Midwest, and along these overland routes the frontier continued to move westward across the plains to the mountains. The advancing frontier brought immigrants from many lands, people who were used to hardships, work, and want. As each adapted his methods to the varying natural environment, progress was made, and the area grew into one of the most productive and wealthy parts of the world. Such descriptive superlatives as "Center of American Industry" and "Breadbasket of the Americas" have been applied to Midland United States.

The "Heart of the Nation" is located on the extensive Interior Plains region, the largest physical region in North America. The Plains extend westward from the Appalachians to the Rockies and southward from the Arctic Ocean to the Rio Grande. In the United States the area between the Appalachians and the Mississippi River slopes gently westward; that between the Rockies and the Mississippi River slopes eastward. The two chief tributaries of the Mississippi, the Ohio and Missouri rivers, mark the approximate southern boundary of the area once covered by continental glaciation. The region is usually divided into two large subregions, the Central Lowlands and the Great Plains, both of which have a moderately rolling surface. The Bluegrass area of central Kentucky, the Nashville Basin of middle Tennessee, and the Osage Plains of Oklahoma and Texas are southward extensions of the Central Lowlands that, because of historical background and cultural development, are considered a part of the American South. The southern extension of the Great Plains is also historically connected with the South.

CENTRAL LOWLANDS

The Central Lowlands extend from the Eastern Highlands to the Great Plains, and from the Canadian border to the Ozarks and the Gulf Coastal Plain. Their surface is level to moderately rolling; the soil is generally fertile; and the Great Lakes offer advantageous inland transportation. The region has a continental climate, with adequate rainfall ranging from 25 to over 45 inches annually, and a growing season sufficiently long (about 200 days along the southern border to 140 along the northern) for midlatitude crops such as corn, oats, and soybeans. Minerals found in the region include large deposits of bituminous coal, petroleum and natural gas, iron ore, copper, lead, zinc, and building materials. There is some available waterpower and water supply for industries. The original vegetation cover was varied. Large forests of conifers predominated around the upper Great Lakes; a mixed forest of broadleaved deciduous species was found in the middle United States; and tall-grass prairies extended from Illinois westward to the Great Plains.

In utilizing these natural resources, man has developed numerous and varied activities. Within the Central Lowlands is the world's largest corn-growing area, known familiarly as the Corn Belt. Few other parts of the world equal this region in the production of pork, beef, and dairy and poultry products. Large crops of hay, soybeans, wheat, oats, and other grains are harvested; manufactures are also notable, especially the output of steel and machines. Rail transportation is highly developed, and the largest freshwater freighters in the world ply the Great Lakes. Within this region is the nation's center of population as well as the center of its agricultural activity.

Agriculture

Although the Central Lowlands rank high in manufacturing and mining, agriculture was the foundation of the region's development. Since the area is so large, diversities in the length of the growing season, soils, topography, and distance from markets have brought about differences in agricultural practices, crops, and livestock. The northern half of the region was repeatedly covered by continental glaciers, which strongly modified the surface features; glacial erosion and the damming of drainage by debris dropped by the melting ice resulted in the formation of tens of thousands of lakes, ponds, and swamps. When drained, the beds of shallow lakes and swamps usually make superior farmland, especially good for growing corn, sugar beets, celery, and other varied crops. Glacial soils may be stony, but they are fertile, for they have been formed of mixtures of rock waste and humus derived from many sources; moreover, the limestone and other soluble materials have not been leached as much as in unglaciated soil. The best soils south of the limits of glaciation are those of alluvial origin, usually located on the valley floors, and the prairie soils, where the topsoil is deep and an abundance of humus has been derived from plant stems and roots. Soils which developed under a cover of deciduous trees contain more humus and are more fertile than those formed in areas of pine forests, where the soils are apt to be leached and low in usable plant foods.

During the past half century there has been a tremendous change in American agriculture. In 1920 there were approximately 6.5 million farms in the conterminous United States, which averaged 147 acres each in size. By 1970 the total number of farms in the United States, including Alaska and Hawaii, had decreased to about 2.9 million, and the average farm size had increased to 377 acres. It is not unusual today for the larger wheat farms to cover 640 to 1,280 acres (1 to 2 square miles) and for ranches to exceed 6,400 acres (10 square miles) in area. Approximately 350 million of the nearly 2,000 million acres in the conterminous states are now tilled (in production); other hundreds of millions of acres are used for grazing lands. It has been estimated that, in all, 455 million acres are arable (capable of being cultivated).

Along with the increase in farm size has come a decrease in farm population. The 1940 census listed more than 30 million persons as farm population; in 1970 fewer than 10 million were so listed. Although the proportion of agriculture workers varies widely in different parts of the nation, it is recognized that the total farm population, not only in the United States but in all parts of the world, is decreasing. Even so, the importance of agriculture is not lessening.

Important to the growth of American agriculture has been the invention of tractors, trucks, combines for harvesting cereals, cotton-picking machines, and other machinery that replaced animal power and reduced the need for manpower. Land once used to grow feed for draft animals now is available to grow food for people. No country in the world grows as

The United States East of the Rocky Mountains   37

AGRICULTURAL REGIONS OF THE UNITED STATES

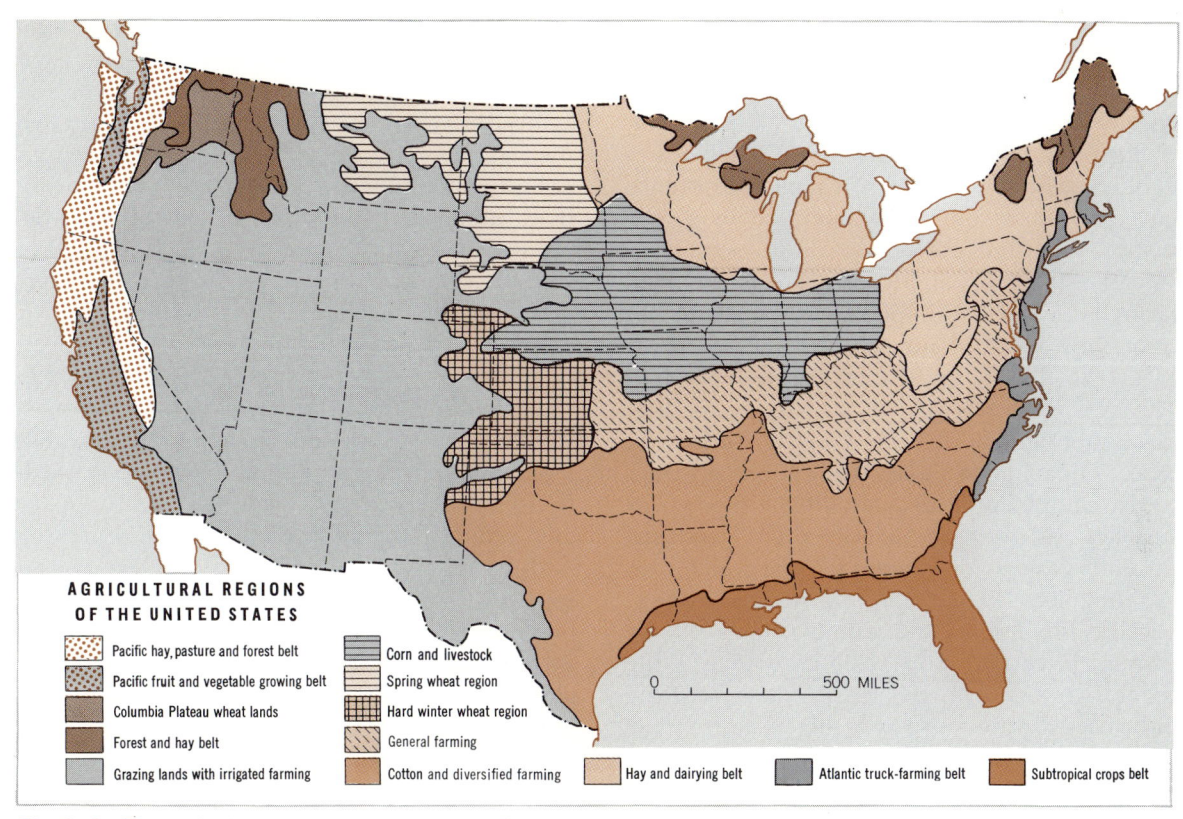

- Pacific hay, pasture and forest belt
- Pacific fruit and vegetable growing belt
- Columbia Plateau wheat lands
- Forest and hay belt
- Grazing lands with irrigated farming
- Corn and livestock
- Spring wheat region
- Hard winter wheat region
- General farming
- Cotton and diversified farming
- Hay and dairying belt
- Atlantic truck-farming belt
- Subtropical crops belt

Fig. 2-6   The agricultural regions of the United States are the result of both physical and cultural activities of these areas. Although one or two principal crops may predominate, each region grows many crops and each has a large variety of land uses.

much farm produce per manpower-hour as does the United States. Today less than one-seventh of the population are farmers, yet they feed the 205 million inhabitants of this nation and have surplus commodities for export. Part of this productivity results from the use of fertilizers, such improvements in plant strains as hybrid corn, improved breeds of livestock, and more efficient handling of crops that tends to reduce waste and loss. Dairy cows are milked by machines, and pigs and poultry are raised under efficient and sanitary conditions. Transportation is adequate; nearly every farm is on a good-to-passable road; and few farms in the Midwest are farther than 5 or 10 miles from a railroad. Refrigeration is available via both truck and railway to preserve perishable produce during shipment.

In spite of all the technical advances that have been made, the farming industry is still much dependent upon nature. The crops that can be grown successfully in paying quantities in any given area are limited by the length of the growing season, the amount and distribution of rainfall unless water for irrigation is available, the range of temperature during the growing season, and also the texture and fertility of the soil. For the profitable use of large farm machinery, the land needs to be level to gently rolling. Farmers must prepare their land so as to prevent as much soil erosion and depletion as possible. Should a prolonged drought develop in a specific area, there is little that the farm operator can do.

Economically, farming in the United States is big business. Except for small subsistence farms or those sometimes operated by sharecroppers and tenants, it takes an investment of at least $20,000 for an individual to start a small farm; large, well-equipped farms and ranches frequently involve investments in excess of $100,000. Farming is not the independent activity it was a few years ago. Numerous large farms operate successfully from a solid credit base. Today, most successful farmers buy their seeds, equipment, and many of their everyday supplies, as well as market their products, through granges or cooperatives of which they are members. Also, many farm activities—especially, for instance, the number of acres to be planted in certain crops—are regulated by the federal government. Like large manufacturing industries, farmers are also subject to the laws of supply and demand.

Owing to the diversification of crops and the increased consumption of meats and fruits, the delimiting of the agricultural regions of the United States is much more complex today than it was three or more decades ago. Although certain crops, such as wheat, corn, vegetables, and hay, may be more widely grown in the specific region, farmers no longer place all "their eggs in one basket." Various combinations of crops, or crops and animals, give greater assurance of a satisfactory income. With the variations in crop patterns has also come a change in rural cultural patterns. Today the patterns change gradually from place to place; thus, the boundaries between the various agricultural regions are in reality transition zones rather than definite lines of demarcation.

The Corn Belt, actually an area of diversified agriculture, extends from western Ohio to eastern Nebraska and from central Minnesota to central Missouri. In this area, glaciers generally leveled the topography and improved the soils. Summers are rainy and hot—just what corn needs for best growth—and the region has assured markets since it is close to the center of the nation's population. Corn is grown extensively outside this central area, but nowhere else does it so dominate the cultural landscape. Nevertheless, although more land is planted in corn than in any other crop, little corn is marketed as grain; much of it is fed to hogs and beef animals and marketed in the form of meat. Also, many food products are made from corn. In addition to corn, other crops such as oats, wheat, barley, soybeans, and hay are planted, making the area important for its diversification. Frequently, such legume crops as clover and alfalfa, because of the nitrogen-carrying nodules that develop on their roots, are plowed under to enrich the soil. This favorable combination of climate, soil, and topography, along with man's judicious use of these natural resources, has resulted in (1) intensive land use for agriculture, (2) a large percentage of the rural area under cultivation in crops, (3) conversion of much of the crop output into livestock production, and (4) high land values. Good farm practices, such as crop rotation and soil conservation, and income from a variety of crops and livestock are basic to Corn Belt farming.

Excellent and extensive groups of farm buildings are characteristic of most Midwest farms. Large, two-story homes fully equipped with electricity and running water are common. Butane gas is frequently used for fuel. Big barns, silos, equipment sheds, chickenhouses, hogpens, and feedlots occupy the barnyard; near this complex may be the garden plot and possibly a small orchard. Since the eastern part of the Corn Belt was originally covered with a forest of mixed hardwoods, an adjunct of most farms is the woodlot. Woodlots add beauty and variety to the scenery, supply fuel and posts, provide a refuge for birds and small game, and furnish some saw logs for sale. Where the prairies once were, the country is still open, but

Fig. 2-7 Red River Valley, North Dakota. On the rich soils of this area are grown abundant crops of cereals and sugar beets. The land is carefully cultivated, and windbreaks help prevent erosion. (Courtesy of North Dakota Business and Industrial Development Department)

cultivated crops have replaced the native grasses. Trees were frequently planted to shelter the houses, and groves around the farmsteads stand out on the level land. On the Great Plains, however, even the farmsteads may be treeless.

Winter wheat, planted in the fall, is the chief crop southwest of the Corn Belt; to the northwest spring wheat, sown in the spring, is most widely grown. These two crops predominate in the zone of transition between the Central Lowlands and the Great Plains, where the rainfall is not sufficient to produce good corn or where the growing season is too short for its maturing.

Northward from the Corn Belt the winters increase in length and severity. The summers are rainy and long enough for excellent hay and forage crops, and although the growing season is too short for corn to mature, much corn is grown for use as silage. Pastures are generally good, and most areas of swampy, hilly, and cutover land can be used for grazing. In this region of rather short summers, farmers get better results by using land for forage crops and hay rather than for grains. Small amounts of rye and barley are usually grown as cash crops since these are hardier than wheat or corn.

Dairying is a primary type of farming north of the Corn Belt. Jerseys, Holsteins, Brown Swiss, and other breeds of dairy cattle are numerous. Not only are climatic conditions very favorable for forage, but the region is close to the markets of Minneapolis, Chicago, Detroit, and other large cities. Cooperative creameries and processed cheese factories, the latter being a specialty of Wisconsin, provide an additional market for cream, milk, and milk products that can be stored and distributed throughout the year. Milk-condensing factories are also numerous. Throughout the dairy region are big, well-built barns that protect the animals during the long, cold winters and also provide storage space for hay and other feeds. This is the area where a young couple beginning as dairy farmers are told, "Build your barn first, and your barn will build your house." Near cities, much milk and cream are marketed fresh. Increasing distance from market is reflected in more specialization.

Of special importance is fruit. Numerous orchards have been planted along the shores of the larger lakes. The eastern shore of Lake Michigan is one of the principal fruit-growing areas of the nation; large quantities of noncitrus fruits are shipped from this area each year. Practically all kinds of mid-latitude berries (gooseberries, loganberries, dewberries, raspberries) are grown. The lakes have a strong effect upon the climate and influence the kind of crops grown. In the spring, cool westerly winds blowing off the lakes retard plant growth and thus prevent damage by late frost. During the fall the lakes warm the winds and delay the killing frosts, so that the farmer has time to gather his ripened crop and prepare his orchard for winter. The effects of these winds extend from 20 to 25 miles inland. The Door Peninsula of Wisconsin and the southern shore of Lake Erie are also famous for fruits. The fruit business is helped by the nearby large urban markets. Although some fruit is sold fresh, much of the crop is processed by freezing, canning, or preserving for marketing throughout the year.

Minerals

The Central Lowlands are well endowed with fuels, iron, and other minerals. The Eastern Interior coalfields of Illinois, southwestern Indiana, and western Kentucky rank next in production after the Appalachian fields, and some coal is also mined in Kansas, Missouri, and Iowa. Although most of this bituminous coal is not so suitable for coking as that from the Appalachian fields, it is nevertheless an excellent power resource. The iron and steel manufacturing centers are so located that imports from mines in Pennsylvania, West Virginia, and eastern Kentucky are easily accessible.

Petroleum and natural gas, produced locally, have contributed much to the industrial progress of the Central Lowlands. The principal producing fields entirely within the area are the southern Illinois and Kentucky fields and various pools in Indiana, Ohio, and Michigan. The Mid-Continent field, largest in the nation, extends northward from Oklahoma into Kansas. Some natural gas is produced in connection with the petroleum, especially in the Mid-Continent field. Large oil refineries that receive petroleum from these and other fields are located in many places; some of the more important are in Whiting and East Chicago, Indiana, and in Wood River, Illinois.

The iron ore deposits in Minnesota, Wisconsin, and Michigan are actually in the Superior Highland section of the Canadian Shield rather than in the Central Lowlands. The Superior Highlands, the most important source of iron ore deposits in the United States, supply about 85 percent of the nation's output of approximately 90 million tons per year. The Mesabi Range in Minnesota, furnishing over 50 percent of the total output, is the most productive single source in the United States. Most of the mining is from deep open pits. The Cuyuna and Vermilion Ranges, also in Minnesota, and the deep mines of the Marquette, Iron Mountain, and Menominee Ranges of Michigan and Wisconsin are other sources. Enormous deposits of taconite, a low-grade ore, have been developed in Minnesota. By concentrating this material, it will be possible to maintain production long after the high-grade hematite ore has been exhausted.

Rock, clay, sand, and gravel deposits are widely distributed. Limestone for building is quarried in central Indiana, for flux in steelmaking in northern Michigan, and for cement in practically all Midwestern States. Clay is used in making brick, tile, and other ceramic products. Brick and tile plants are common because of the amounts of good clay available and the large markets. Glacial deposits of sand and gravel are used in roadmaking, building construction, and railroad maintenance.

Forests

Originally forests covered much of the eastern part of the Central Lowlands; in the western part of the region, vegetation graded from forests to prairies to short grasses, with trees being found only along streams in the latter areas. The mixed hardwood forest extended westward from Pennsylvania to Illinois; there were also minor stands of pine. Much of this forest has now been cleared, and the land is now used for agriculture.

The northern forests of Michigan, Wisconsin, and Minnesota consisted of white pine, spruce, fir, and other conifers, with some hardwoods, especially maple, oak, birch, and aspen. Most of the white pine has been cut, and great inroads have been made on other species. Many cutover areas have been replanted, and numerous tree farms have developed. Some states have zoning regulations that prevent certain types of areas from being used for purposes other than forests. Although the quality of the second-growth timber is usually poorer than the original, it is suitable for use as pulpwood. Many large paper mills, such as those in International Falls and Grand Rapids, Minnesota, make paper of varying grades. Where suitable lumber can be cut, furniture factories continue to be of local importance. Grand Rapids, Michigan, is nationally known for its manufacture of quality furniture.

Transportation

The five Great Lakes and their connecting waterways form the busiest inland water route in the world. The long lake freighters equal many ocean carriers in size and carrying capacity. The large lakeport cities, long shorelines, and rich hinterlands make the Central Lowlands a populous and busy commercial and industrial region.

Traffic on the Great Lakes totals over 200 million tons of freight annually. The four commodities of iron ore, coal, limestone, and grains make up 90 percent of the tonnage. Lying between producing and consuming regions, the waterway provides the world's cheapest freight rates per ton-mile for transporting materials in bulk. The major movement is down the lakes, from western Lake Superior to the southern shores of Lake Michigan and Lake Erie, with cargoes of iron ore and grains. Limestone is moved from quarries near the water's edge in the vicinity of Alpena, Michigan, to dockside steel mills. Coal from Lake Erie ports is transported up the lakes.

As ice usually closes the lakes to navigation for three to five months of the year, ships make as many trips as possible during the open season. To facilitate quick turnaround, the docks, loading and unloading devices, and the ships themselves have been especially designed. Channels have been dredged and four big locks built to form the Soo Canal, so that ships may be lifted and lowered between Lakes Superior and Huron. To facilitate the automobile and truck traffic between Canada and the United States, a large bridge has been built over and a tunnel dug under the Detroit River. A gigantic bridge has been erected over the Straits of Mackinac. A few car ferries, built to break through the ice floes, operate across Lake Michigan all winter.

More than 60 million tons of iron ore are moved over the Great Lakes each year. At the upper lake ports of Duluth, Superior, Two Harbors, Marquette, and Escanaba, ore trains run along docks, on high trestles that are 2,000 and more feet long. The ore is dumped into pockets, from which chutes carry it into the open holds of vessels. Ships of 12,000 to 16,000 tonnage can be loaded, if necessary, in an hour or two. At lower lake ports such as Conneaut, Cleveland, Erie, and Buffalo, huge clamshell unloaders bite the ore out of the holds with almost equal speed. Cargoes of coal, limestone,

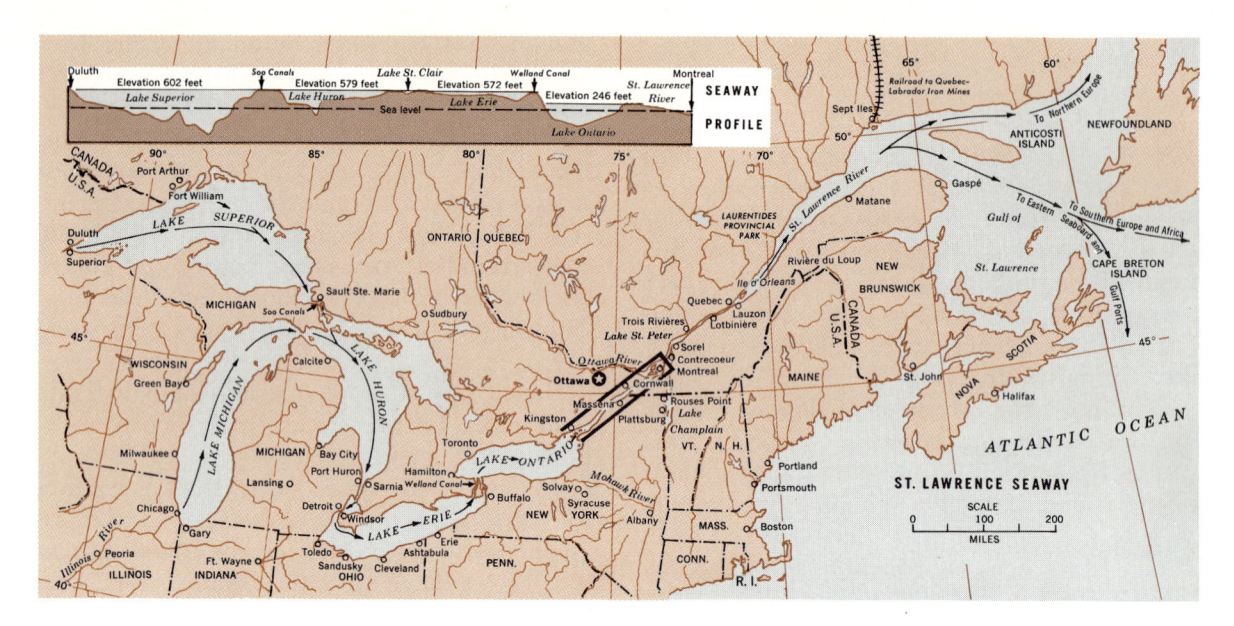

Fig. 2-8 The St. Lawrence Seaway was developed by both Canada and the United States. A series of six locks and three dams was constructed between Montreal and Lake Ontario. Ships are lifted from 22 feet above sea level at Montreal to 246 feet in Lake Ontario. The Welland and Soo (Sault Sainte Marie) Canals make it possible for ships to move as far inland as Duluth.

and grain are likewise loaded and unloaded in remarkably short times.

The St. Lawrence–Great Lakes Seaway offers an opportunity for oceangoing vessels to dock at inland ports. Large modern docks have been built in Chicago, Cleveland, and other cities. These cities, although far inland, are important ocean ports. Small craft can reach the Mississippi River from Chicago, via a canal and the Illinois River from Lake Michigan.

Inland transportation facilities in the Central Lowlands are better than those in any other region of the United States. The area is crisscrossed by a railway network that extends to practically every large town and city as well as to numerous villages and hamlets. Chicago is recognized as the leading railway center of the nation. Its location near the southern tip of Lake Michigan and the level topography of the area make the city the focus of rail lines from west, south, and east. In addition to the Great Lakes waterway, which serves most of the larger cities of the region, the Ohio and Mississippi Rivers and some of their tributaries are important arteries of transportation. Coal, sand and gravel, and other bulk goods, in addition to barge loads of automobiles and oil, are frequently moved by water. Wide paved highways connect most of the cities, and the towns are connected by all-weather roads. Toll roads, freeways, and bypasses now enable the traveler to drive long distances without passing through congested areas. All the larger cities are served by national and international airlines.

Industries and Cities

A large part of the manufacturing belt of the United States is located in the northeastern part of the Central Lowlands. The industries range in size from the one-man or one-family activity, such as broom making, to the large auto assembly plant at River Rouge, near Detroit. So varied are the manufactures that they are frequently grouped under three headings: (1) heavy industries dependent upon adequate mineral supplies and cheap transportation, such as the steel mills in Gary, Detroit, Cleveland, and Duluth; (2) manufacturing that is directly related to agricultural activities, such as flour milling in Minneapolis and Milwaukee, making of farm equipment in Moline and Indianapolis, and meatpacking in Chicago, Omaha, and Cincinnati; and (3) medium and light industries such as the manufacture of glass, scales, rubber goods, furniture, paper products, and plastics. In some parts of the manufacturing belt, industries of certain types tend to be grouped together, for example, heavy industry in the Youngstown area; but in other sections, as in central Illinois, a variety of industries may be found.

In the Midlands, as in the Northeast, there are numerous metropolitan areas. The Midlands, however, do not have a megalopolis to compare with that extending from Boston to Washington. In the not-too-distant future, though, it is conceivable that urbanization will expand across the rich agricultural lands between such large cities as Buffalo, Cleveland, Toledo, and Detroit, so that farms, hamlets, villages, and towns will form an almost continuous built-up area along the shore of Lake Erie and the Detroit River. Another Midlands megalopolis, now developing about the southern part of Lake Michigan, includes the territory from Milwaukee through Chicago to Gary.

Along the southern shore of Lake Erie, from Buffalo to Toledo, lies one of the chief manufacturing regions of the nation. Although each city makes a variety of products, many are also noted for some specialty. The Detroit area, to the west of the Detroit River and Lake St. Clair, is the heart of the world's automobile industries. Huge plants in this vicinity and in Flint, Pontiac, South Bend, and Toledo have turned out more than 6 million cars within a year. A host of Middle Western cities manufacture parts for automobiles, trucks, and aircraft.

Around southern Lake Michigan, from the Door Peninsula on the west to Grand Rapids on the east, is a manufacturing area that processes vast quantities of manufactured agricultural products made from items grown in the Central Lowlands. Large meatpacking plants are located in the Chicago area, great quantities of grains are used by the Milwaukee brewing industry, and most towns and cities have creameries and canneries.

Smaller manufacturing areas have developed around the larger river cities. Minneapolis and St. Paul, the so-called Twin Cities, together form the greatest market, manufacturing, and distribution center in the upper Mississippi Valley. St. Louis began as a fur-trading post, in 1764, near the junction of the Missouri and Mississippi rivers. Founded by the French, it later attracted so many German people that the pronunciation of its name was changed. St. Louis is a river port as well as an important rail and air center. Its population exceeds 600,000, and the metropolitan district has a population of over 2 million people. Louisville and Cincinnati are both important Ohio River ports. Kansas City and Omaha are the largest industrial centers on the Missouri River; each has large stockyards, meatpacking plants, and flour mills.

Chicago, with a population of over 6 million people in its metropolitan area, is the largest city in the Central Lowlands and the second largest in the United States. The city is frequently referred to as the transportation center of the world; it is the focus of numerous major highway routes, a railway center where trunk-line railroads and many terminal lines meet, site of an international airport, and also an important waterway terminus, since it has canal and river connections with the Mississippi River and the Gulf of Mexico, as well as with the Atlantic Ocean via the St. Lawrence route. With such adequate transportation, the city has developed

Fig. 2-9 Many older sections of the larger American cities such as Chicago, Illinois, have been redeveloped under various urban renewal programs. Apartments as well as retail outlets, offices, and financial institutions have been constructed in and near the central business district (CBD). (Courtesy of United Airlines)

as the natural processing place for the produce of the Corn Belt and the hay and dairy region. Chicago is a leading meatpacking center, a manufacturer of flour and other food items, an important maker of agricultural machinery, and the retailing and wholesale center for the Middle West. The city is also noted for its cultural advantages (Natural History Museum, Shedd Aquarium, Chicago Art Institute) as well as for several outstanding universities.

Detroit, the fifth largest city in the nation, has approximately 4 million people in its metropolitan area, or one-half of the total population of Michigan. Located on the west bank of the Detroit River, the city developed because of its strategic position between Lake Huron and Lake Erie. Among the first manufactures were carriages. Partly because of the early start in this type of industry and inventions by its citizens, Detroit became the leader in the automobile industry. Other numerous and varied industries such as saltworks and the production of chemicals, aircraft, drugs, and furniture have also developed.

Cleveland, the largest city on the shores of Lake Erie, has a metropolitan population of almost 2 million. The city is near the center of the Lake Erie industrial region, which extends, almost without break, between Ashtabula and Toledo. Cleveland serves as a transshipment point for Mesabi and Canadian iron ore going to the Pittsburgh and Youngstown areas, as well as for coal being shipped westward from the Appalachian field. Large steel mills, oil refineries, machine shops, and food-processing plants are among the city's various industries. Like Chicago and Detroit, the principal retail district is near the lakeshore.

The large cities of the Central Lowlands, like those located in the Northeast, are faced with many urban problems. Each has its crowded slums inhabited largely by minority groups or foreign "colonies" speaking the language of their native country. Many have law-enforcement problems. One of the most serious problems faced even by cities in which only small manufacturers are located is that of water pollution. So much manufacturing waste has been emptied into rivers and lakes that many have become more or less sewers and cesspools. Lake Erie is said to be one of the most polluted lakes in the world. In numerous smaller streams, most aquatic life has been killed. The search for pure water is indeed critical. All the larger manufacturing centers and most of the smaller ones have air pollution problems. New regulations for industrial activities, improved housing, educational and recreational

Fig. 2–10 River Rouge Plant, Detroit. A single industrial complex has various kinds of plants or factories working toward a single end product. Centered on the boatslip are coke ovens, blast furnaces, and assembly areas. (Courtesy of Ford Motor Company)

facilities for the depressed areas of the cities, and long-range planning are being formulated or are in progress.

Recreation

The vacation and tourist industry brings hundreds of millions of dollars in income to residents of the Great Lakes states and lesser amounts to the people in other parts of the Central Lowlands. Excellent beaches, forested surroundings, and fishing are among the area's attractions. Along the shoreline of the Great Lakes region are hundreds of resorts. Inland Michigan, Minnesota, and most of Wisconsin have been strongly glaciated, with the resulting formation of thousands of lakes—17,000 in Minnesota alone. Hundreds of private vacation cabins and many resorts have been built along the lakeshores and riverbanks. In the fall the forests of northern Michigan, Wisconsin, and eastern Minnesota are visited by hunters after deer and other game. The amounts spent on recreation support many communities that, after the decline of the lumber industry, would have been depopulated without this outside income. Many state parks, such as Itasca State Park at the source of the Mississippi River in Minnesota, have been established. Isle Royale National Park, the only national park in the region, is an island in Lake Superior. Winter recreation areas with ski jumps and ski trails have been developed in several places where the topography is suitable.

GREAT PLAINS

The Indians who lived on the broad and grassy Great Plains depended in large part upon wild animals, principally the bison, or buffalo, for their food, shelter, and clothing. The meat from these creatures was a staple food that could be dried for future use. Hides formed the Indian tepee, and hairy robes made warm beds and clothing. Dried droppings, called "buffalo chips," were the common fuel used by pioneers for cooking in this nearly treeless country. Other game and furbearing animals supplemented the buffalo. The Plains tribes were nomadic, and after the introduction of the horse about 1550 the tribesmen were able to hunt bison from horseback. Horses were also used to pull burdens placed on a set of poles called a "travois"; this was a pair of shafts stretched from the back of the horse and dragged over the ground.

Physical Setting

The Great Plains slope eastward from the Rocky Mountains, dropping from an elevation of 4,000 to 6,000 feet to about 500 to 1,600 feet where they merge with the Central Lowlands. In this subregion, frequently deficient in moisture, the plant life is dominated by shortgrass, bunchgrass, tumbleweeds, and other subhumid vegetation. Numerous sectional names are applied to parts of this area: the northern section, in Montana, North Dakota, and South Dakota, is frequently referred to as the Missouri Plateau; the Nebraska Sandhills cover a large area in the north and west-central part of that state; and the part of the southern section of the Great Plains generally west of the 101st meridian is known as the High Plains, where alluvial materials washed from the Rockies were deposited on top of an older surface, causing a greater elevation. In general, the surface of all the High Plains appears almost flat, being broken only by the streams that cross it. The Great Plains extend southward into Texas and as such are a part of the South.

The High Plains are broad, relatively level, treeless uplands that extend for miles. Over this level surface, winds having an average velocity between 10 and 15 miles per hour blow much of the time. At varying intervals the High Plains are cut by narrow, steep-sided valleys that trend in general from west to east. In these valleys are to be found the narrow rows of trees —cottonwood, locust, cedar, and a few others —that are typical of the region. Some of the rivers, such as the Platte, Republican, and Arkansas, have developed wide valleys containing rich alluvial soil.

Two areas in the Great Plains, the Bad Lands and the Black Hills of western South Dakota, contrast vividly with their surroundings. The Bad Lands have been formed in an area of easily eroded shale bedrock. The core of the Black Hills is an uplift of igneous rock, largely granite, around which the sedimentary rocks have been eroded. The Black Hills are high enough above the surrounding plain to receive sufficient rain to support forests of spruce and pine. The sandstones that outcrop around the central core serve as the source of water and as aquifers for much of the artesian water used in eastern South Dakota.

The weather of the Great Plains is of the semiarid steppe or continental type, with severely low temperatures in winter during cold waves, though there is a high frequency of sunlit days. Well-defined warming winds called "chinooks" often sweep southeastward down these plains in winter and spring. The chinooks are welcomed by the ranchers, for they rapidly melt snow and ice by evaporation and thus make it possible for the animals to graze. When the strong winter windstorms called "blizzards" occur, with their drifting snow, transportation is impeded and grazing becomes difficult. Summer days may be hot, with temperatures above 100°F, but nights generally are cool.

Rainfall of about 10 to 20 inches a year necessitates irrigation or special dry-farming methods of growing crops. If precipitation on the plains were dependable, farming would be more profitable. When farmers must obtain water from mountain streams or reservoirs for irrigation, they must make a great expenditure of time and money. Nevertheless, many reservoirs have been built in the shallow natural basins and the low-banked river channels, and the water is used to irrigate sugar beets, alfalfa, and other crops. The more elevated surfaces, locally called "benches" in Montana, are devoted to dry farming. In the more favored areas, the farms are smaller and the population correspondingly larger than in the dry-farming regions.

The native grasses, before destruction by overgrazing or cultivation, provided a moderate income since they were suitable for large-scale pasturage; but so much grassland has been destroyed that each year sees less of it available for commercial purposes. Once destroyed, the grass is difficult and expensive to replace. For this reason, conservation of vegetation and soil in the semiarid Great Plains is of prime importance to a grazing economy—a matter all too often ignored or misunderstood by those who live in the humid realm of the United States.

Agriculture

In the mid-nineteenth century, the white men began to displace the Indians and to replace the herds of buffalo with herds of cattle. Sheepherding became a secondary activity where native fodder was too sparse for grazing cattle. This extensive grazing economy prevailed until the last quarter of the nineteenth century, when, after a series of humid seasons and high prices for wheat, much of the native grass was plowed under and the land was planted in grain. For a time the change to extensive wheat farming paid, but it took only a few years of subnormal precipitation to make this type of agriculture unprofitable for the region. Eventually the plowed land was abandoned; then, after a period of humid years, the cycle was repeated.

Kansas and Oklahoma are the principal winter-wheat states, and the Dakotas lead in spring-wheat production. The Wheat Belts continue eastward into the Central Lowlands. Wheat farms are large and become greater in size toward the west as the amount of yearly rainfall decreases. Wheat is grown as an annual crop or in rotation with grain sorghums or other suitable crops.

The Wheat Belts form a contrast to the Corn Belt; here agriculture is extensive and less varied. Farmhouses are far apart, and the villages are smaller and have fewer functions. Grazing is an important source of income in the Winter-Wheat Belt. The wheat, after being planted in the fall, comes up and may be pastured by feeder stock during the winter. In the spring, these animals are sold through various markets. The wheat then develops and is harvested. Practically all work is done by machines. Groups of combines, with their crews, start working the wheat harvest as the crop begins to ripen first along the southern margin. The crews then move north with the season, many of them starting the harvest in Oklahoma or Texas and ending it in the Prairie Provinces of Canada. Numerous grain elevators are located along railroad sidings far from hamlets or villages; and during the harvest season, activity around these elevators is great.

Since grain farming brings uncertain profits, and since neither wheat nor beef cattle can support many people, throughout most of the rural Great Plains the population density is less than 10 persons per square mile, and vast areas cannot support more than 2 per square mile. In some areas the wheat farmers no longer live on their farms. Many have moved into nearby villages or towns, where they engage in other business activities; others employ some individual to look after routine farm activities and return to the farm only two or three times a year. Such farmers are frequently referred to as "suitcase" farmers. Once the crop is planted there is little to do until the following harvest season. The area in which the feeder stock is to be grazed can be easily controlled by movable electric fences.

In the Great Plains, grazing competes with wheat growing as the chief user of the land. Beef cattle have always been the dominant domestic animal of the area. Even prior to and during the days of the first permanent settlement, the large cattle companies of Texas, Montana, and other places used the land for grazing. Many famous early cattle trails crossed the Great Plains to such well-known railhead cattle towns as Abilene and Dodge City. Since about 1925, the number of cattle in the area has varied with (1) climatic conditions, (2) the market demand for beef, and (3) the market demand for wheat. Ensuring a sufficient supply of water is always a problem. Most ranches have strategically placed windmills or pumps in addition to the water in rivers, creeks, or ponds. Animals are seldom kept more than a mile from water. Even though sheep were brought into the Great Plains at an early date, they never became an important part of the economy, partly because of the attitude of the cattlemen, but

Fig. 2-11 Custer Trail Ranch, North Dakota. Large ranches require a considerable expenditure for homes, animal shelters, fences, and various types of machinery and other equipment. A successful modern rancher must be a businessman who understands finances as well as cattle and employs professionals such as veterinarians and accountants as well as cowboys. (Courtesy of North Dakota Business and Industrial Development Department)

largely because cattle brought greater financial returns.

Although grazing has always been a leading industry in the northern Great Plains, the settlers soon recognized that the large variability in the amount of annual rainfall would present serious water problems. By 1900, dams and reservoirs were in operation or planned, whereby floodwaters might be impounded and later distributed to the fertile areas east of the Rocky Mountains. An early project of this type centered in the vicinity of Greeley, Colorado, and there have been many since. Subsequently, under the auspices of the Bureau of Reclamation and other federal agencies, the building of irrigation projects has allowed a more intensive use of the land. One of the largest constructions is the Fort Peck Dam on the Missouri River in Montana; built to control floods, it also impounds water for irrigation. Other projects are located on the Yellowstone River near Billings; the Milk River Plain in Montana, near Williston, North Dakota; and north of the Black Hills, on the Belle Fourche River. Alfalfa, sugar beets, and grains are important irrigated crops. The rural population has increased in density on the reclaimed lands.

Water supplies are precarious in the Great Plains south of the Nebraska Sand Hills, where evaporation frequently exceeds precipitation. Irrigation projects in Nebraska, Kansas, and Colorado have reclaimed some land. Water is taken from the Platte, Arkansas, and other rivers, as well as from lakes and deep wells, to irrigate such crops as grain sorghums, sugar beets, alfalfa, melons, and vegetables, especially onions and beans. Although these crops do not occupy large acreages, they are of special economic importance to the area in which they are grown.

Minerals

The most important mineral resources of the Great Plains are (1) the petroleum and natural gas deposits in both the northern and southern parts of the region, (2) the rock-salt deposits of Kansas, and (3) the lignite deposits of the Dakotas and Kansas. Although the mineral industries employ hundreds of persons and add materially to the economy of the area in which these are mined, their economic contribution to the Great Plains is much less than that of the agricultural and ranching activities.

The Williston Basin, located in North Dakota and Montana, is the chief northern petroleum-producing area. The great Mid-Continent petroleum field of the South extends northward from Texas and Oklahoma into Kansas. The largest natural gas field in the United States, the Hugoton, extends southward from near Garden City, Kansas, into the Texas Panhandle. Much gas is piped from this field to urban areas in other parts of the nation for use as fuel.

Salt mines have been developed near Hutchinson, Kansas. The salt deposits came from a great inland sea that covered much of central Kansas about 275 million years ago. Salt deposition began when an isolated arm of the sea began to dry up through evaporation. The salt deposits formed into a thickening mass about 45 miles wide and 100 miles long. During later geologic periods, sediments buried the salt and compressed it into veins of dense rock salt about 300 feet thick. The deposit was discovered in 1889 when oilmen drilled into the salt. Mining was begun in the early 1920s, when the first shaft was dug. It has been estimated that this deposit holds enough salt to meet the needs of the United States for 250,000 years.

Settlement

Villages, towns, and cities in the Great Plains are usually small, widely spaced, and bear strong resemblance to one another. Communities sprang up, at first, in connection with railroad lines crossing the plains, because of the need for farm and ranch service centers. The region is not noted for its large cities or manufactures. Towns are frequently located near stream channels, where water was obtainable, or along railway lines, via which the inhabitants could maintain contacts with distant places. Each town or village usually has a small commercial center with a few stores that can supply the local necessities. In many places, one or two large general stores will sell everything, from food and clothing to drugs and machine parts. In many, the two dominant places of activity will be the stockyards and the grain elevators, especially during the marketing season. A large consolidated school, a few churches, and some modest homes are usually found adjacent to the business district.

By far the largest and most important city of the Great Plains is Denver, which not only commands the plains but also lies so near the Southern Rockies that it serves both these extensive regions. It is particularly important for wholesale and retail trade. Several main highways and transcontinental railways focus on Denver. It is also district headquarters for numerous agencies of the federal government. Tourism is a major industry of this mile-high city.

Recreation

Living in the Great Plains has been picturesque if not always easy. The landscape is wide-open and without the polluted air of large manufacturing cities. Though ranches are very large, they can be operated with relatively few laborers. In the early history of the plains, laborsaving devices such as barbed wire, the high windmill, and mechanical pumps aided the rancher in changing the landscape from cattle kingdom to wheat field. More recently, the radio, automobile, and electric pump have been important to the farmers. Distinctive features of the Great Plains cultural environment have spread throughout the United States. Perhaps the greatest contribution to recreation from the area has been cowboy music, certain words and characteristic phrases, and Western-type clothing, all derived from the pattern of the Great Plains cattle ranch of the late nineteenth century.

Set in the midst of the northern Great Plains, the dome of the Black Hills offers sharp contrast to the surrounding lands; here are forests used for their timber supply. The rugged mountains —including Mt. Rushmore, famous for its giant carved heads of four presidents—and the relatively pleasant summer weather and the forests combine to make this part of South Dakota attractive to tourists.

Several other neighboring areas are becoming nationally recognized as places of interest for tourists. A short distance west of the Black Hills is a national monument known as Devils Tower. This oddity, visible for many miles, rises for almost a thousand feet above the surrounding plain. It is the central core of an ancient volcano. An excellent highway has been built through the Dakota Bad Lands, making it possible for tourists to view the striking erosional features from several points. The area is also frequented by fossil hunters, and many well-preserved fossils of prehistoric animals have been discovered. Various Indian reservations located near the Bad Lands also attract thousands of tourists each year.

The South

That part of the United States commonly referred to as the South is probably the least-understood section of the nation. Far too many people think only of the South's historical past rather than of the modern, progressive developments in the area. At present the South is in a state of transition, changing from cotton growing to a great variety of agricultural activities, from rural to urban life, from manpower to mechanical power, from a static to a mobile population, and from conservatism to liberalism. True, other parts of the nation have gone through similar changes, but not within so short a period of time. More changes have occurred in the South during the past quarter century than took place in the Northeast in over a century.

The South is a sprawling area extending from the Atlantic Ocean westward into the Great Plains and northward from the Gulf of Mexico to the Ohio and Potomac rivers and the Ozark Plateau. Prior to the discovery of America, the area was inhabited by Indians in various stages of civilization, ranging from the Plains tribes in the west, who followed the buffalo for sustenance, to those living in the southeast, who had developed a primitive type of agriculture and planted such crops as corn, beans, and squash.

For many decades the South has been a region of economic maladjustment. The history of its people and their activities has been a record of turbulence. Before the Civil War the economy was based on cotton plantations and slavery; after the war, on cotton and tenant farmers or sharecroppers. Various political conditions, along with declining soil fertility and poor farming methods, made recovery slow; but by 1900 a semblance of agricultural adjustment had occurred. Because of the high cotton prices during and immediately after World War I, however, much eroded and infertile soil was planted in cotton. During the Depression years of the 1930s the South, along with the rest of the nation, found its economy again shattered and many of its people in want. In several respects the South, with its dominant one-crop system and tenant farmers, became the economic problem of the nation.

Since 1935 a new South has been in the process of formation. Recognizing the need for diversification of crops, better soil and forest conservation, and the development of industries, the people of the South, with the aid of the federal government, again began an economic adjustment. No longer is the region just a "land of cotton" and small farms. Although cotton is still widely grown and there are numerous small farms, the tendency now is toward consolidation of holdings; increased attention to livestock, with greater acreages planted in soybeans, peanuts, and hay crops; terracing, contouring, and fertilizing of the soil; and the development of industries to use the natural resources of the region.

There are now three large minority groups —the Negroes, Spanish-Americans, and Indians— living in the United States, and the South is the home area for large numbers of each. Each of these minority groups presents a series of sociological and economic problems (low educational achievement, poor health standards, inadequate housing facilities, insufficient income, etc.), but most of these problems are neither unique to these minority groups nor confined to the South.

The Negro, now the most migrant of the three groups, was for many decades virtually confined to the rural South. Since 1940 large numbers of rural blacks have sought to improve themselves economically and educationally by moving to the larger industrial centers of the nation. It appears that those living in the Carolinas, Georgia, and Virginia migrated generally into the larger metropolitan centers from Washington to Boston; those from Alabama, Mississippi, Tennessee, and Kentucky moved generally toward the Milwaukee-Chicago-Detroit area; a majority of those leaving Louisiana, Texas, and Arkansas migrated to the West Coast, with a smaller segment moving to the Midwest. Few Negroes have left the rural South for other rural areas. New York now has a larger Negro population than any other state, but Mississippi has the greatest percentage of Negro citizens.

The Spanish-Americans are concentrated along the United States-Mexican boundary. A greater number now reside in southern and western Texas than in other states, but this group also forms a large percentage of the populations of New Mexico, Arizona, and southern California. Although large numbers of Spanish-American farm workers travel about the nation each summer following agricultural harvests, most consider the border regions their home and return there each winter.

Oklahoma today has the largest Indian population among the states; pockets of Indian settlements are still to be found in some of the Mississippi River. The number of Indians is small when compared with the number of Negroes or Spanish-Americans. Most Indians continue to live near their traditional homes. Although a small number have migrated to the cities, very few plan to make the city a permanent place of abode.

PLAINS OF THE SOUTHLAND

Physical Setting

Most of the Atlantic and all of the Gulf Coastal plains are in the South. Varying in width from less than 50 miles to approximately 500 miles where it follows the Mississippi River northward to southern Illinois, the plain is a region of relatively recent sedimentary deposits. Much of the Coastal Plain is a nearly flat lowland that slopes gently toward the ocean. The Virginia and North Carolina coastline is very irregular because of the drowning of river mouths by the sinking of the coast and the consequent invasion of ocean water. In places, low hills marking the outcrop of more resistant rocks rise above the lowlands. Florida is a peninsula built mostly of limestone, in which many lakes have been formed in natural depressions or by solution. From the southern end of the peninsula a festoon of keys extends into the Gulf of Mexico. Along and near the shore of the Coastal Plain are several great swamps, such as the Dismal Swamp in Virginia and North Carolina, Okefenokee in Georgia, the Everglades in Florida, and much of the lower Mississippi Delta in Louisiana. Just offshore, in various locations, long narrow sand islands have been formed.

The Mississippi Embayment, which separates the East Gulf Coastal Plain from the West Gulf Coastal Plain, has been filled by river sediments. In these lowlands the Mississippi meanders over a wide alluvial floodplain, and along each side of the river are low bluffs. The vast amount of sediment that the river carries constantly extends the large delta, making it hard to tell where the land ends and the sea begins. The Mississippi, with its tributaries, broad floodplain, and delta, is the major river and the dominant feature that divides the Gulf Coastal Plain into two nearly equal parts.

Florida and the Gulf Coast enjoy a humid, subtropical climate with little or no frost. Inland, during two or three winter months, freezing temperatures may occur, and many plants become dormant. Rainfall is usually fairly evenly distributed and generally totals 40 to 50 inches annually. At some Gulf stations the annual rainfall may exceed 80 inches; but inland, to the west and northwest, the total decreases to 25 or 30 inches. The principal weather handicaps are the hurricanes that occasionally affect the coasts, winter cold waves that bring freezing temperatures injurious to fruits and vegetables, and occasional droughts. In general, the northern boundary of the Coastal Plains roughly coincides with the 200-day growing season.

Parts of the Interior Plains extend from the Midwest into the South. The section of the Central Lowlands in middle Tennessee and central Kentucky contains the famous Bluegrass region and the Nashville Basin, the latter being surrounded by the Highland Rim. In general, the area is undulating and its soils are rich. West of the Interior Highlands, the Osage Plains section of the Central Lowlands extends across Oklahoma into central Texas. For the most part, the area is a rolling plain, with the native vegetation changing from forests and prairies to short grass toward the west. In southern Oklahoma the remains of two eroded mountain groups, the Arbuckles and the Wichitas, contrast in roughness of topography and elevation with the surrounding plain.

The Great Plains extend southward across the Panhandle of Oklahoma and western Texas to the Rio Grande. The southern extension of the High Plains part of the Great Plains is comparable to the area in Kansas and Nebraska. South of the Canadian River, the High Plains are called the Llano Estacado (or Staked Plain); the southern part is known as the Edwards Plateau. The eastern edge of the High Plains is marked by an escarpment, or a narrow zone of rough topography, known as the "Break of the Plains."

The climate of the Bluegrass region – Nashville Basin, as well as of the eastern part of the Osage Plains, is of the humid, continental, long-summer type. Summers range from warm to hot, with short periods of cold weather in winter. In the eastern areas, rainfall is adequate, but in the Osage Plains rainfall decreases to about 25 inches yearly toward the west. Usually the maximum rainfall occurs in early summer.

Most of the Southern High Plains are handicapped by the lack of rainfall, with the yearly average being between 15 and 20 inches. Evaporation is so rapid that in some years the meager rainfall is inadequate to support good pastures or dry croplands. Usually winters are moderate to mild, although freezing blizzards are not unknown. Summers are warm to hot.

Agriculture

Agriculture is the basic occupation in much of the Coastal Plain. The long growing season and abundant rainfall favor corn, cotton, and many other crops, such as peanuts, soybeans, tobacco, melons, and sunflowers. Only the lower

Rio Grande Valley requires regular irrigation. The soils are highly variable. Those in the river bottomlands and deltas and in the zones of marls and silts inland from the coasts are very fertile; in contrast are the poor soils of the pine-flats and sandy hills. When drained, muck soils of the Everglades and along parts of the Gulf Coast are well suited for vegetables and sugarcane. All through the South, marked changes are taking place in agriculture, particularly in the use of machinery, in marketing methods, and in the variety of products grown.

For over 150 years, cotton was so constantly the money crop that the region is often called the Cotton Belt. The best soils planted in cotton are those of alluvial origin along the Mississippi, Yazoo, Arkansas, and other rivers. The very fertile, black waxy prairie soils of Texas are also excellent producers of cotton. The southern Piedmont and a zone of sandy loams on the inner coastal plain extending from Georgia into North Carolina are traditionally of great importance for cotton growing. Since 1900, however, production in these eastern sections has declined because of soil erosion or depletion, infestation by the boll weevil, replacement by other farm products (especially corn, fodder, and pasture crops), and competition for labor from industry. Cotton lands are increasingly dependent on the use of fertilizers.

Since 1920 a westward shift in cotton production has come about largely owing to the growing of cotton in irrigated areas in Texas, New Mexico, Arizona, and California. So discontinuous is cotton acreage in the old Cotton Belt today that some publications are referring to the area as the "Shattered Cotton Belt," while others have called it the "Historical Cotton Belt." During the past quarter century the cotton farmer, like the wheat, corn, and peanut farmers, has been limited in the number of acres he may plant in cotton by government regulations that he voted to follow.

The High Plains in western Texas and southwestern Oklahoma have largely replaced the southeastern areas in cotton production. Near Lubbock, Texas, is the nation's leading cotton-growing region. Water is pumped from deep wells for irrigation. Mechanization in caring for and picking the crop, together with the development of varieties suited to a drier climate, has made possible the expansion of the industry into these western lands. Since cotton must be picked in dry weather, only small amounts are grown along the southern coast, where autumn rains may delay or prevent the harvest. In the nonmechanized areas, much labor is needed to grow a good cotton crop. After planting, the farmer must cultivate and/or chop (hoe) the growing crop for thinning and weeding. Until the 1930s the picking of cotton by hand required much cheap labor, but since then cotton-picking machines have been increasingly used, especially in the newer growing areas in the west, where the farms are larger than in the older districts. Cottonseed, a by-product used for edible oils, soap, and other items, often has greater cash value than the fiber. In addition to the large quantities of fiber consumed by mills in the United States, cotton is exported to several textile-manufacturing countries, for the South (Piedmont, Coastal Plain, and High Plains of west Texas) grows approximately 25 percent of the world's cotton. The location of principal seaports, such as Houston and New Orleans, which are close to the cotton fields, is advantageous for export.

Although citrus fruits are also grown along the outer margins of the Coastal Plain, central Florida and the lower Rio Grande Valley are so seldom reached by frosts that they produce most of the South's commercial crop. Oranges, grapefruit, tangerines, and limes are most important in Florida; grapefruit is a specialty in Texas. Normally, citrus fruits come on the market in quantity only at certain seasons; to prolong the marketing season and preserve the products, therefore, nearly one-half of the crop is processed into canned fruit and juice, frozen concentrates, and marmalade. Over 60 million gallons of orange juice are frozen and processed yearly in Florida.

Sugar, rice, corn, peanuts, tobacco, and winter vegetables are among other Southern crops. Sugarcane is grown on the lower Mississippi Delta, especially in southern Louisiana, and to some degree in Florida. Although there is danger of crop injury from cold, the development of cold-resistant hybrid varieties has lessened this danger and has also doubled acreage yields. Planting and caring for the sugar crop, harvesting and transporting the cane, and extracting the sugar at the mills and refineries require large capital investment. Southern rice, like wheat, is planted and harvested by machine methods. Level land underlaid by impervious clay is surrounded by dikes to retain the water in which rice must stand during much of its growing season. Louisiana, Texas, and Arkansas produce over 60 percent of the domestic rice crop. In the South more acres are planted in corn than in any other crop, for corn is the common grain crop of the South. As cornbread, hominy, and grits, corn is widely used for food as well as feed for livestock. Peanuts, which grow well on sandy soils, are used for food, for processing into other food products, and as raw material by chemical industries.

Livestock is of increasing importance, notably on the depleted soils of the older settled sections of the South. Fodder crops such as lespedeza, kudzu, and cowpeas are good stock feeds and grow well on wornout land, which they enrich while they help hold eroding soils in place. The recent growth in production of cattle, pigs, and poultry east of the Mississippi is noteworthy. The cattle industry benefits from the mild temperatures; the even rainfall that facilitates rapid growth of hay and forage crops makes pasturage available most of the year. Because pork products are popular in the South, corn and hog raising has prospered. The poultry business has expanded, especially in northeastern Georgia; quantities of fowls are processed for shipment to Northern and Eastern markets. Mild temperatures and available feed are favorable factors; but the "know-how" of hatching, feeding, processing, and marketing poultry is even more important. Although natural conditions favor dairying, its development has lagged in the South, partly because many farmers lack experience in this area. An effort is being made to develop breeds of dairy cows that will be unaffected by the long hot summers. Beef-cattle production also is increasing, and crossbreeding of Hereford and Angus stock with the Brahmas of India is providing good beef cattle that can withstand the heat of the long summers. Many former cotton plantations are now prosperous livestock farms.

Fig. 2-12 Grain sorghum harvest, High Plains. Many wheat farmers, because of surplus wheat crops, have changed to growing grain sorghums and other feed crops to supply the increasing number of feedlots where cattle are fattened for market. (Courtesy of Santa Fe Railway)

The High Plains of Texas have long been noted for raising of beef animals. Ranches are numerous, and ranch life here does not differ greatly from that of the northern Great Plains. The milder winters minimize the danger of animal loss from freezing. Ranches often include more than twenty-five sections of land. The King Ranch, the largest in the United States, located in southeast Texas, covers a discontinuous area as large as the state of Rhode Island. On the brush-covered Edwards Plateau, the grazing of sheep and Angora goats is of prime importance. The goats produce mohair.

An agricultural transition zone is found between the Corn and Cotton Belts from the Cumberland Plateau to the Mississippi River bottoms. This hilly region includes the Ohio Valley and central and western Kentucky and Tennessee. Some fertile valleys and wide basins occur. Much of the hill land is wooded, and the landscape is varied. Frequently steep slopes have been cleared and planted in corn and tobacco, but heavy rains on the bare cultivated slopes have caused much erosion and contributed to the poverty of the hill farmers. In the fertile river valleys and the Bluegrass and Nashville Basin areas, farming is generally very successful. There is a striking difference between the big mansions, blooded livestock, and well-tilled, productive fields of the fertile areas and the unpainted shacks and the eroded, weed-infested plots characteristic of some of the hill country. Tobacco is the major cash crop. Though total production is large, heavy labor demands, exacting soil requirements, and government allotments restrict the amount of land planted in tobacco. The Kentucky Bluegrass region and the Nashville Basin, being areas of fertile limestone soils, are regions of superior pastures, where purebred cattle and blooded horses are raised.

Forests

The South is a natural woodland area. Because of its long growing season and ample rainfall, trees attain double the annual growth of those in northern climates. Pines predominate on the flat plains; hardwoods are found in the river bottoms; and in areas of hilly land the two are frequently mixed. Except on natural prairies in Texas and Oklahoma, abandoned farmland is quickly seeded from the nearby woods and is in a short time covered with a second-growth forest. In 30 to 40 years, the pines become large enough for saw logs. A chief obstacle to reforestation in the coastal sections is the large number of forest fires. Few national forests have been established in this area, but large acreages under private ownership are scientifically handled and form the basis for permanent lumber and pulp or paper enterprises. Logging is possible throughout the year, and the logs are delivered to the mills by truck and rail. Many mills are located at seaports or so near them that lumber can be easily exported. Mississippi and Louisiana are the leaders, but lumber is marketed from every Southern state. In total output of lumber, the South is second only to the Pacific Northwest. Lumber used for construction, naval stores, and wood pulp is the principal product of Southern forests. The sawing of hardwood lumber is especially important in Tennessee and Arkansas.

Fisheries

Shrimp, sponges, oysters, and crabs provide the principal source of income from the Southern coastal fisheries. Inshore waters, especially along the lagoons of the Mississippi Delta, supply small shrimp and crabs. Large shrimp are secured by trawlers, fishing at night, in the deeper Gulf waters. Corpus Christi and Galveston are important shrimp ports. From these and several other ports, shrimp are shipped fresh or as frozen or canned products. Oysters are taken mainly from the lagoons on the coasts of Louisiana and Florida. The center of the sponge industry is Tarpon Springs, Florida. Sponges are taken largely by divers.

Minerals

In total value the South produces over 30 percent of the nation's minerals. Petroleum, coal, iron ore, sulfur, phosphate, and salt are produced in large quantities; the South exceeds all the rest of the nation in the production of petroleum. Building stones and numerous minerals of lesser value such as asphalt, glass sand, and kaolin are of considerable local importance. One of the most important factors in the industrial development of the South has been this large and varied mineral reserve.

The Mid-Continent oil field, the most important producer of petroleum in the United States, is largely in the South. Included in this field are producing areas in Arkansas, northern Louisiana, Oklahoma, northern and western Texas, and southeastern New Mexico. The Gulf Coast field, the second most important petroleum field in the nation, is located along the Gulf Coast of Texas and Louisiana. Because of the large production of this field, along with that of the Mid-Continent, Texas ranks first among all states in petroleum production, California second, Louisiana third, and Oklahoma fourth.

Seemingly insuperable difficulties have been overcome in developing oil fields along and in the Gulf of Mexico. Much oil lies beneath coastal swamps and even under the ocean floor. Drilling operations proceed from massive platforms in quagmires or in open waters. Transporting men and materials over swamps and water requires different types of vehicles and amphibious boats. Some producing wells are 10 miles offshore in the shallow Gulf waters, and the seaward limit of the oil fields has not yet been determined.

More than 40 oil refineries are located along the Gulf Coast of Texas and Louisiana; among these are 4 of the 5 largest in the United States. Two refineries in Port Arthur and one in Beaumont, Texas, have a daily crude capacity of over 200,000 barrels each. The largest American refinery, in Baton Rouge, has a daily capacity greater than 350,000 barrels.

Natural gas, usually associated with petroleum, is found in large quantities in the South. Formerly, much of this excellent fuel was wasted because there were no markets nearby. The invention of spiral steel pipe and machines to lay and weld the sections quickly has made possible delivery of gas from the South to Northeastern and Midwestern industrial and seaport cities. Natural gas is considered the ideal fuel for heating and cooking in homes. Two large helium plants have been built in the Hugoton Gas Field, one each in Texas and Oklahoma. Natural gas is also used as a raw material in certain chemical industries and is considered the best fuel for making glass.

Five Southern states—Alabama, Arkansas, Georgia, Oklahoma, and Texas—have reserves of bituminous coal. Arkansas and Texas also have deposits of lignite. Excellent bituminous coal, which can be used to make coke, is mined in Alabama near the southern end of the Appalachian Plateau and is used in the iron and steel industries of that state. Coking coal is mined in eastern Oklahoma and shipped in large quantities directly to Japan.

The largest production of sulfur in the world is on the Gulf Coast of Louisiana and Texas. Water heated under pressure to temperatures higher than 300° F is forced into the sulfur deposits. At such temperatures the sulfur becomes soluble. The liquid is then forced to the surface by compressed air, run into large storage bins, and allowed to cool and solidify. Gulf Coast sulfur is shipped to chemical plants throughout the world. The chief product made from sulfur is sulfuric acid, but the mineral is also used in making drugs, paper, fertilizer, and a variety of other products.

Phosphate, iron ore, and salt are mined in the South. Phosphate rock, used for fertilizer, is mined in the Tampa area. This part of Florida supplies 80 percent of the phosphate used in the United States and also exports a large amount. The chief iron ore mining region is near Birmingham and is close to good coking coal and limestone. These resources provide the basis for the largest iron and steel production in the South.

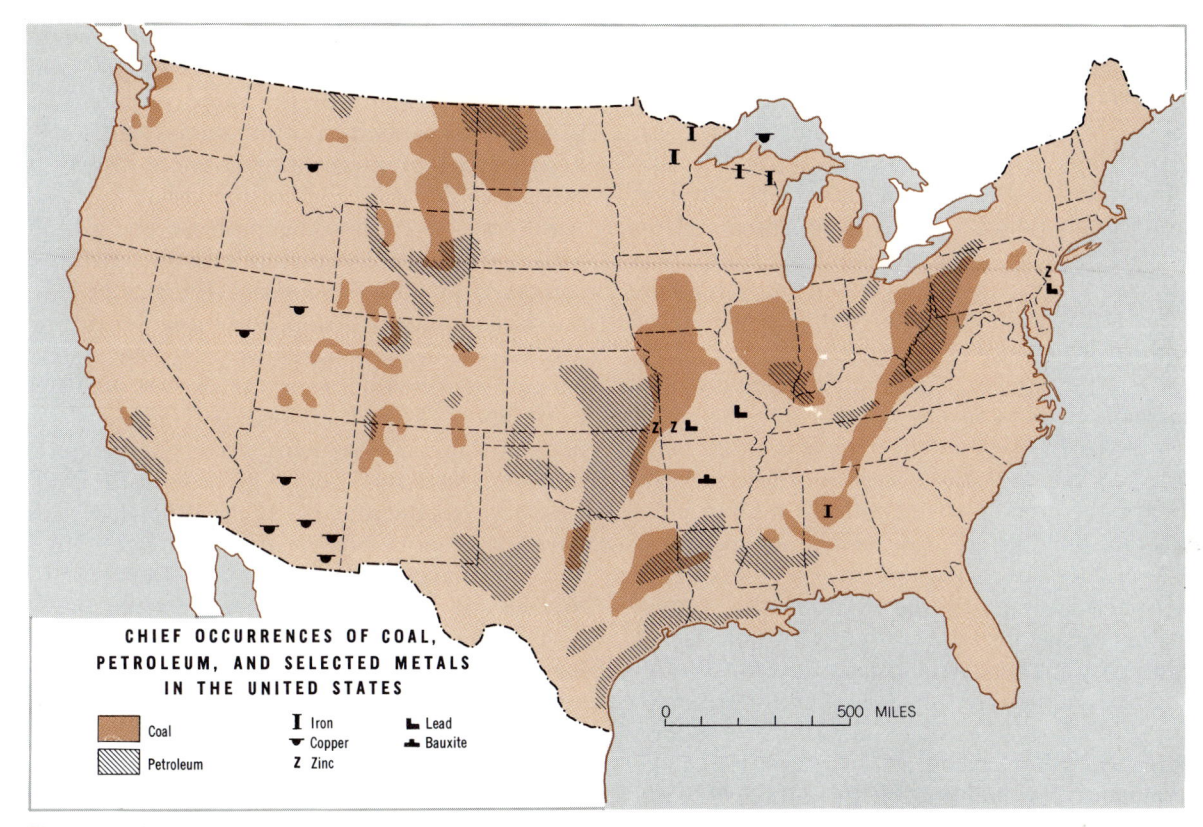

CHIEF OCCURRENCES OF COAL, PETROLEUM, AND SELECTED METALS IN THE UNITED STATES

- Coal
- Petroleum
- I Iron
- ▼ Copper
- Z Zinc
- L Lead
- ▲ Bauxite

Fig. 2-13 With the exception of copper, most of the essential minerals for American industry are mined east of the Rocky Mountains.

Manufacturing and Transportation

During the past twenty-five years the number of manufacturing establishments in the South has greatly increased. Factors that favor the industrial development of this region are (1) vast reserves of petroleum, natural gas, and coal that can be used both for power and for chemical raw materials; (2) resources of timber, minerals, and agricultural products available for processing; (3) a large supply of available labor; and (4) good transportation by sea and land. Among the handicaps are (1) a shortage of local capital and lack of industrial experience; (2) the competition of Northern mills that have established reputations; (3) the lack of skilled labor; and (4) the greater distance to the large urban markets.

Although the South is less densely populated than the Northeast, it has an adequate supply of potential industrial workers. Some come from the farms, where mechanization, soil depletion, and changes in crop systems have reduced the number of laborers needed. In general, living costs less in the South, but wages have usually been below those of the North. Many branch plants of Northern-owned corporations have been located in the South to supply the demands of the region for nationally distributed articles. Examples include assembly plants for automobiles, farm machinery, and heavy goods, on which important savings in freight charges from the home plants to the South can be made.

The pattern of manufacturing in the Coastal Plain is characterized by wide distribution and great diversity. The Texas Gulf Coast area, particularly in and around Houston, is especially important for chemicals, machinery, oil refining, and shipbuilding. Elsewhere, Birmingham leads the South in steelmaking, and Memphis processes agricultural products and lumber. The Dallas-Fort Worth area has a variety of industries, including meatpacking, the making of wearing apparel, automobile assembly, and airplane manufacture.

Transportation in the region is easily managed. Except for the necessity of bridging the rivers, the broad lowlands offer few hindrances to the building of railroads and highways. There are many navigable rivers, of which the Mississippi is the major artery for bulky freight. A few canals are noteworthy. Houston, an inland city, has become a great seaport by dredging and improving a shallow river and developing the Houston Ship Canal. A dredged waterway connecting the Mississippi and Lake Pontchartrain has increased waterfront land available to industrial development in the vicinity of New Orleans.

Cities and Towns

The uniform flat surface of the rural Coastal Plain favors a fairly even distribution of population. Only the swamps and areas with the poorest of sandy soils have few inhabitants. Zones of superior soil, such as the black, waxy prairie in east-central Texas and the Yazoo bottomlands of northwestern Mississippi, support many prosperous farms. Rural hamlets and villages are numerous throughout the region. In most of them a consolidated school and a post office serve the immediate vicinity; a small general store or two, filling stations, and possibly a drugstore, garage, and cotton gin comprise the local business establishments. Two or three small churches are found in almost every village. Many of the people drive to the larger towns or cities to buy most of their needs.

The large urban areas of the South are confronted with many of the same problems as urban areas in the rest of the nation. In some, a large part of the central business district is being rebuilt with the aid of the Urban Renewal Authority. New housing projects, being built with both public and private capital, are under construction. Each Southern city has at least one large minority group, and those west of the Mississippi usually have two. In the Texas cities, Spanish-Americans far exceed Negroes in number. As in the Northeast, each city has its ghetto areas. Problems of the ghettos here, however, have not been as great as in the Midwest and Northeast, since the Southern ghettos have many more private or semiprivate homes and fewer large apartment houses. In the South there is still space for urban expansion.

Houston, with a population of over a million, has developed rapidly during the past decade into one of the principal cargo ports of the nation. The chief exports are petroleum, sulfur, cotton, grain, and chemical products. Numerous imports from the Latin American countries are received. The city is noted for its petrochemical industries and other manufactures based upon the natural resources of the region. In annual tonnage of trade, Houston ranks third among the ports of the nation. One of the most noted of all federal projects is the Manned Spacecraft Center located in the Houston metropolitan area. From this center all manned space flights, including the moon flights, are controlled.

New Orleans, located 90 miles upstream from the mouth of the Mississippi River, is one of the principal southern seaports and also ranks among the top ten ports of the United States in annual tonnage handled. Water transportation is important; along the banks of the Mississippi are many docks, wharves, grain elevators, cotton warehouses, and numerous other commercial and industrial activities. Much Latin American trade moves through New Orleans.

Texas has more large cities than any other Southern state. Eleven places, including Houston, have populations in excess of 100,000. Near the Balcones Escarpment, forming the boundary between the Coastal Plain and the Interior Plains, such major cities as San Antonio, Austin, and Dallas, as well as several smaller cities, have developed. This physical feature, in some respects like the Fall Line, formed a breakpoint in transportation routes; also, fresh water was available. Dallas is a major cotton market and a leader in women's fashions. The city is the principal banking and insurance center in the southwestern part of the nation. San Antonio, established in 1718, is a tourist and ranch center. Its pleasant climate, its wealth of historical associations—the Alamo, the Spanish Governor's Palace, the San Antonio River that winds picturesquely through the city—and its large Spanish-American population have all played a part in the city's development.

Recreation

Tourists and vacationists spend hundreds of millions of dollars on travel through the South and in sojourning at coast resorts from the Carolinas to Texas. Both the Atlantic and Gulf Coasts of Florida are major resort areas. More visitors come in the winter season, but many are encouraged to vacation in the summer by the cheaper rates of the off-season. Miami Beach stretches for 9 miles on an island just offshore from the mainland city of Miami. It boasts of approximately 400 hotels, with a capacity to care for visitors far in excess of most larger cities. The east coast of Florida, between Jackson and Key West, situated at the end of a noted overseas highway that follows a row of coral islets, has numerous large resort areas. The central Florida lake district is becoming increasingly popular with vacationers and retired Northerners. The warm sunshine, freedom from frost and snow, sandy beaches for sun and saltwater bathing, and sport fishing are the leading attractions. Many retired people have moved to Florida.

SOUTHERN APPALACHIAN HIGHLANDS

The southern part of the Appalachian Highlands is almost a direct contrast to the adjacent Coastal Plain. Because of variations in soils, relief, and mineral resources, the distribution of population is very irregular. Compared with the adjacent lowlands, the highlands have few large cities, since factors favoring urban growth are seldom present. Forests and woodlands occupy one-half of the area, and many of the higher ridges or more rugged slopes are uninhabited. In some areas deer, bear, and other wild game are common.

The Appalachians have had marked effects on transportation, since railroads and highways had to be built in the valleys and over the passes that afford the lowest or easiest routes. Until roads were constructed, travel was so slow and difficult that settlement was largely restricted to the coastal lowlands and the Piedmont. The eighteenth century was more than half gone before frontiersmen searched out the gaps in the mountain barrier and began settling in Tennessee, the Bluegrass region, and the Ohio Valley. These early settlers were generally farmers. Today, although there are some highly productive farming regions, many of the farms are small, and income from them is often considerably below the national average, thus resulting in large rural depressed areas.

Piedmont

Extending eastward from the Blue Ridge or other front ranges of the Appalachian Mountains to the Fall Line is a hilly belt or low plateau, known as the Piedmont, whose rolling surface is underlaid by hard, crystalline bedrock. The region is approximately 50 miles wide in northern Virginia and increases to 120 miles in South Carolina. On the east, at the Fall Line, elevation of the Piedmont is generally 200 to 400 feet; it rises to 1,400 feet near the Blue Ridge. Many streams cross the region, and the valleys that have been eroded into the bedrock offer numerous sites for generating electricity.

The Piedmont is productive. Its wooded hills and broad cultivated slopes are attractive to both residents and visitors. Agriculture was originally favored by the generally fertile, reddish loam soils, an abundant rainfall of 40 to 50 inches annually, and a long growing season ranging from 180 to 240 days. After the forests had been cleared and row crops of corn, cotton, and tobacco were grown, erosion of the hilly land became a problem. This is especially true in the Carolinas. In some areas, because of the long frost-free and rainy season, three-fourths of the topsoil has been removed and great gullies have been formed, thus hindering farming operations on the sloping land. Many upland farms have been abandoned. Damage by floods has increased, and reservoirs built for the storage of water for hydroelectric plants are being filled with silt. Some owners are now rehabilitating gullied land by seeding permanent pastures. This has resulted in a shift from tilled crops to the raising of beef and dairy cattle. The Piedmont in Virginia has long been important for the raising of livestock, especially cattle.

Much of the Piedmont is included in the manufacturing belt of the United States. Coal can be secured from the Appalachian Plateaus, and much waterpower has been developed, especially in North Carolina. Numerous manufacturing centers are located here. Atlanta, near the southern end of a high mountain barrier, serves both the Coastal Plain and the Piedmont; it is the inland railroad and distributing center of the South, a busy commercial hub that has also a large variety of manufactures. Cheap waterpower, an abundant labor supply, and raw materials available nearby are among the factors favoring the expansion of manufacturing. The southern Piedmont is the leading cotton-textile manufacturing region in the world, both in the number of active spindles and in fabric production. Much rayon is also manufactured.

Appalachian Ridges and Valleys

The Appalachian Mountains are composed of closely folded sedimentary rocks that were once deposited on the floor of an ancient sea. The rocks—mainly sandstone, limestone, and shale—differ considerably in hardness. After millions of years of prolonged erosion, the more resistant rocks now form a series of ridges; the exposures of weaker rocks have been removed to such an extent that they form valleys. As a result, the region is characterized by alternate rows of ridges and valleys that extend northeast to southwest from New York state to Georgia and Alabama. The ridges are discontinuous and may end with the dying-out of the fold that caused them. In some instances, a river may cut through a ridge and form a water gap. Such gaps, or passes, were important in determining the routes followed by the travelers on early trails and by the pioneer settlers, and later by the canals, highways, and railroads. In the generally fertile valleys, farming is frequently very successful. The ridges are used for forestland and have few permanent inhabitants. Forestry, mining, and tourism are the leading activities.

The Blue Ridge, the name applied to the eastern-facing front of the Appalachians, is the longest uplift. At about the Virginia–North Carolina boundary a series of ridges and ranges forms a rough mountain area. Just west of the Blue Ridge are such mountain groups as the Unakas and the Great Smokies. In North Carolina and Tennessee, the Great Smokies and the Blue Ridge attain a width of 70 miles. In this section many peaks exceed 6,000 feet in elevation; Mt. Mitchell (6,694 feet) is the highest peak in the eastern part of the United States. These mountain masses formed important barriers to the east-west movement of goods and people. Areas of hardwoods, as well as forests of pine and fir, extend the length of these mountains and support a lumber industry. More important, however, is the resort and tourist industry, which is favored by the scenic beauty of the region and its coolness in summer. The Great Smoky Mountains National Park, which is the most-used of all national parks, Blue Ridge Parkway, Cumberland Gap, and other national monuments have been established in this region.

The Great Valley, which lies west of the Blue Ridge, is the most important depression in the Appalachian Ridge and Valley area. It includes parts of eight states, is about 1,000 miles long, and averages about 20 miles wide. Locally the valley is known by various names, among them the Shenandoah, the Valley of East Tennessee, and the Cumberland. The state of Virginia calls it the Valley of Virginia. Although the Great Valley reaches 2,000 feet in elevation in southern Virginia, the floor is generally flat; the divides between rivers are so gentle that there are no obstructions of importance. The bedrock is a limestone in which many caves and the Natural Bridge of Virginia have been formed. This valley has served as an important artery for travel and commerce for the last 2 centuries. Soils in the region are fertile, and diversified farming with dairying and livestock production is dominant. Since the valley slopes are relatively free from harmful frosts, apples and other orchard fruits are grown on a large scale, especially in the Shenandoah Valley.

Between the Great Valley and the Appalachian Plateaus are the many ridges and interwoven valleys that form the remainder of the region. In certain sections, forest-covered ridges dominate over valleys and may be without permanent population. In other places the valleys, although short and narrow, are floored with fertile soil and support productive farms similar to those in the Great Valley. Some lumbering is done.

The Tennessee Valley Authority essentially covers the watershed of the Tennessee River. Federal government agencies, with the approval of Congress, organized and developed the project. The TVA served as a pilot plan for regional development and as a guide for future planning in other regions such as the Arkansas, Missouri, and Columbia river valleys. With the construction of nine dams across the main channel, the river has become almost a series of connected lakes. Chemical, aluminum, and various manufacturing plants have located in or near Knoxville and Chattanooga because of the available low-cost hydroelectric power. The first atomic energy plant was developed at Oak Ridge.

Cumberland Plateau

West of the Appalachian Ridge and Valley area is a plateau extending southward from central New York to northern Alabama, usually referred to as the Appalachian Plateaus. The southern part of this area, southward from the Kanawha River in West Virginia, is known as the Cumberland Plateau, and its steep eastern edge is called the Cumberland Escarpment. In spite of the hilly aspect of the area, if one looks over the country from a high viewpoint, it is apparent that most summits are of nearly the same height and that they were once joined to form a relatively flat surface. The bedrock is similar to that in the Ridge and Valley area; however, it was only gently warped during its uplift instead of being closely folded. The hills, then, are the result of erosion of the uplifted plateau by thousands of streams. In general, the plateau is so deeply and widely dissected that only remnants of the former surface remain. Southwestward the plateau gradually merges with the Interior Plains and the Coastal Plains. Small areas of flatland occur along the river bottoms and, with the gentler slopes of the usually steep-sided valleys, are the principal sites of farmland. More than half the plateau is wooded, and although the best timber has been cut, some logs and mine props are still produced.

Coal is the principal mineral resource of the Cumberland Plateau. The mines of eastern Kentucky, western Virginia, and southern West Virginia rank second only to those in Pennsylvania and northern West Virginia. Most of the coal mined in this region is shipped either to the Great Lakes ports or the Hampton Roads area. Petroleum is also an important resource. Although tillable areas are small, subsistence farming is the chief support of many people. There are no large cities, and except for the mining communities, the towns are few and small. The population is predominantly rural.

INTERIOR HIGHLANDS

The Interior Highlands, composed of the Ouachita (pronounced "Washita") Mountains, part of the Arkansas River Valley, and the Ozark Plateau, are located largely in Arkansas and Missouri, with a smaller area in Oklahoma. This region of folded mountains and dome-shaped plateaus is completely surrounded by the plains of the Central Lowlands and the Coastal Plain. In many respects, the region is much like the Appalachian Ridge and Valley area and the Cumberland Plateau.

Settlers began moving to the Ouachitas and Ozarks soon after the United States acquired the land through the Louisiana Purchase, usually migrating from the older Southern states. These early settlers established subsistence farms and grazed livestock, which was driven to market. The Oklahoma part of the Ozarks, however, was set aside as Indian land for the Cherokees, and the Ouachita area in Oklahoma was given to the Choctaws. In many instances, the Indians made more progress in their development than did the whites. The boundary line between Arkansas and Oklahoma has been referred to as a "cultural fault line," since it so definitely marks the division between people of two races and two cultures.

Ouachita Mountains and Arkansas River Valley

The Ouachita Mountains are formed by a series of steeply folded sandstone ridges that have elevations up to 3,000 feet and extend in a general east-west direction. Because of these ridges, the rainfall of the region is considerably greater than that of the surrounding plains, with several stations recording over 60 inches annually. Forestry and grazing are the principal activities here. Coniferous trees, especially pines, grow well on the poorer, sandy soils. A variety of deciduous trees, dominantly oak and hickory, is found in the valleys. The Ouachita National Forest covers about one-fifth of the total area, and there are several state-supervised forests in the rest of the region. Both hardwoods and softwoods are cut, sawed, dried, and marked for flooring, siding, and general construction. Much wood is used in the making of kraft paper. Fenceposts are also a common product. In many parts of the mountains, cattle graze over the open range. The only important source of bauxite, the ore of aluminum, within the United States is in the Ouachitas; the region mines about 98 percent of the domestic supply. The large open pits are near Benton, Arkansas.

The Arkansas River, flowing in a general easterly direction, separates the Ouachita Mountains from the Ozark Plateau. The valley floor is flat and the soil is rich; much of the land is in pasture or is used for specialty crops. There are coal mines within the area. The most important function of the valley is as a transportation route through the adjacent higher, rougher areas. Since 1960 many dams, with locks, have been constructed, and the channel of the river has been dredged. Water transportation is now available upstream as far as Tulsa.

Ozark Plateau

The Ozark Plateau is a dome-shaped uplift with streams radiating from the broadly rolling central portion, deepening and widening their valleys as they approach the perimeter. The southern edge of the Ozarks, known as the Boston Mountains, is higher and has a rougher topography than the rest of the area.

Although much of the more rugged area is forest-covered, and lumbering is an important industry, agriculture and grazing are of prime importance. In northwestern Arkansas the broiler industry has been highly developed; many acres are planted in vegetables, and vineyards are common. Wineries and canneries in the nearby communities buy and process the crops. The rocky hills in both Arkansas and Oklahoma produce large quantities of strawberries. The Springfield Plain in southwestern Missouri is among the important dairy regions of the nation.

Lead, zinc, and limestone are mined or quarried in the Ozarks. The Joplin district of Oklahoma, Kansas, and Missouri mines much lead and was for many years one of the principal zinc-producing areas of the world.

Recreation is rapidly becoming the most important industry in many parts of this area. With the construction of large dams for flood control, power, and recreation, artificial lakes such as Lake-O'-The Cherokees, Lake of the Ozarks, and others have been formed. Each of these lakes is larger than many of the large natural lakes in the eastern part of the United States. Several limestone caves and big springs also add to the tourist attractions.

REFERENCES

Aiken, Wallace E.: *The North Central United States,* D. Van Nostrand Company, Inc. Princeton, N.J., 1968.
Alexander, Lewis M.: *The Northeastern United States,* D. Van Nostrand Company, Inc., Princeton, N.J., 1967.
Beardwood, Roger: "The Southern Roots of the Urban Crisis," *Fortune,* vol. 78, pp. 80–87, August, 1968.
Borchert, John R.: "American Metropolitan Evolution," *Geographical Review,* vol. 57, pp. 301–332, July, 1967.
Chapman, John D., and John C. Sherman: *The United States and Canada: Oxford Regional Economic Atlas,* Oxford University Press, London, 1967.
Fenneman, Nevin M.: *Physiography of Eastern United States,* McGraw-Hill Book Company, New York, 1938.
Gottmann, Jean: *Megalopolis: The Urbanized Northeastern Seaboard of the United States,* The Twentieth Century Fund, New York, 1961.
Gregor, Howard F.: "The Large Industrialized American Crop Farm: A Mid-Latitude Plantation Variant," *Geographical Review,* vol. 60, pp. 151–175, April, 1970.
Hart, John Fraser: "Loss and Abandonment of Cleared Farm Land in the Eastern United States," *Annals of the Association of American Geographers,* vol. 58, pp. 417–440, September, 1968.
———: *The Southeastern United States,* D. Van Nostrand Company, Inc., Princeton, N.J., 1967.
Hunt, Charles B.: *Physiography of the United States,* W. H. Freeman and Company, San Francisco, 1967.
Kenyon, James B.: "Elements in the Inter-Port Competition in the United States," *Economic Geography,* vol. 46, pp. 1–24, January, 1970.
Lewis, G. M.: "The Distribution of the Negro in the Conterminous United States," *Geography,* vol. 54, pp. 410–418, November, 1969.
Starkey, Otis P., and J. Lewis Robinson: *The Anglo-American Realm,* McGraw-Hill Book Company, New York, 1969.
Webb, Walter Prescott: *The Great Plains,* Blaisdell Publishing Company, Waltham, Mass., 1959.

3

The United States West of the Great Plains

The western American states are lands of contrast, from the summit of Mt. McKinley at 20,300 feet, to the depths of Death Valley at −282 feet. Rainfall in Olympic National Forest in the state of Washington exceeds 100 inches annually, but Yuma in southern Arizona receives a meager 3.2 inches. The January thermometer at Rogers Pass, Montana, may go as low as −70°F, whereas the temperature in Death Valley has reached 134°F in July. The coast north of San Francisco holds the record for protracted high winds, averaging 34.5 miles per hour for May. On that coast are the tallest redwood trees (368 feet) and those of greatest girth (50 feet, 6 inches); but farther to the southeast, vegetation is almost absent from the Death Valley landscape. Complex urban and industrial environments prevail at Portland, San Francisco, and Albuquerque; in contrast, a Wyoming or Colorado cattle ranch is a comparatively simple form of settlement. Densely peopled stretches of land in southern California, along with intensively farmed irrigated lands, often adjoin uninhabited and untilled deserts, wilderness, or the broad reaches of land devoted to grazing range cattle.

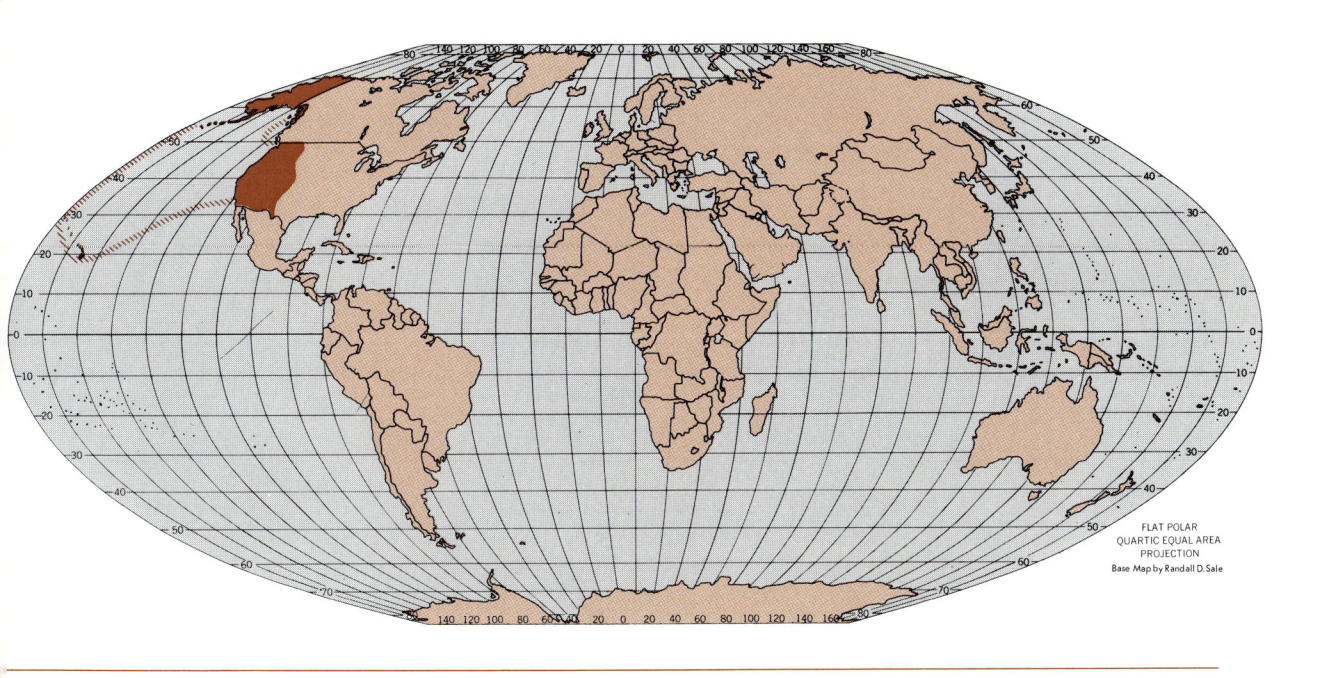

California is the most populous state; Alaska, the least though the largest in area. Alaska, Arizona, Nevada, New Mexico, Utah, and Wyoming, with almost a third of the land area of the nation, have only about 2 percent of the farmland. Some Westerners live in near-tropical locations; others face an Arctic sea. Some Western residents, those living in the Hawaiian archipelago, are islanders, but others in Utah or Idaho may never glimpse the ocean. Certainly, contrast would seem to be a key word in describing the geography of the United States west of the Great Plains for the traveler.

Rocky Mountains

North America's backbone—the Rocky Mountains—is a natural feature so impressive to the traveler that it becomes a milestone on any journey across the United States by train, automobile, or air. Here, in imposing grandeur,

65

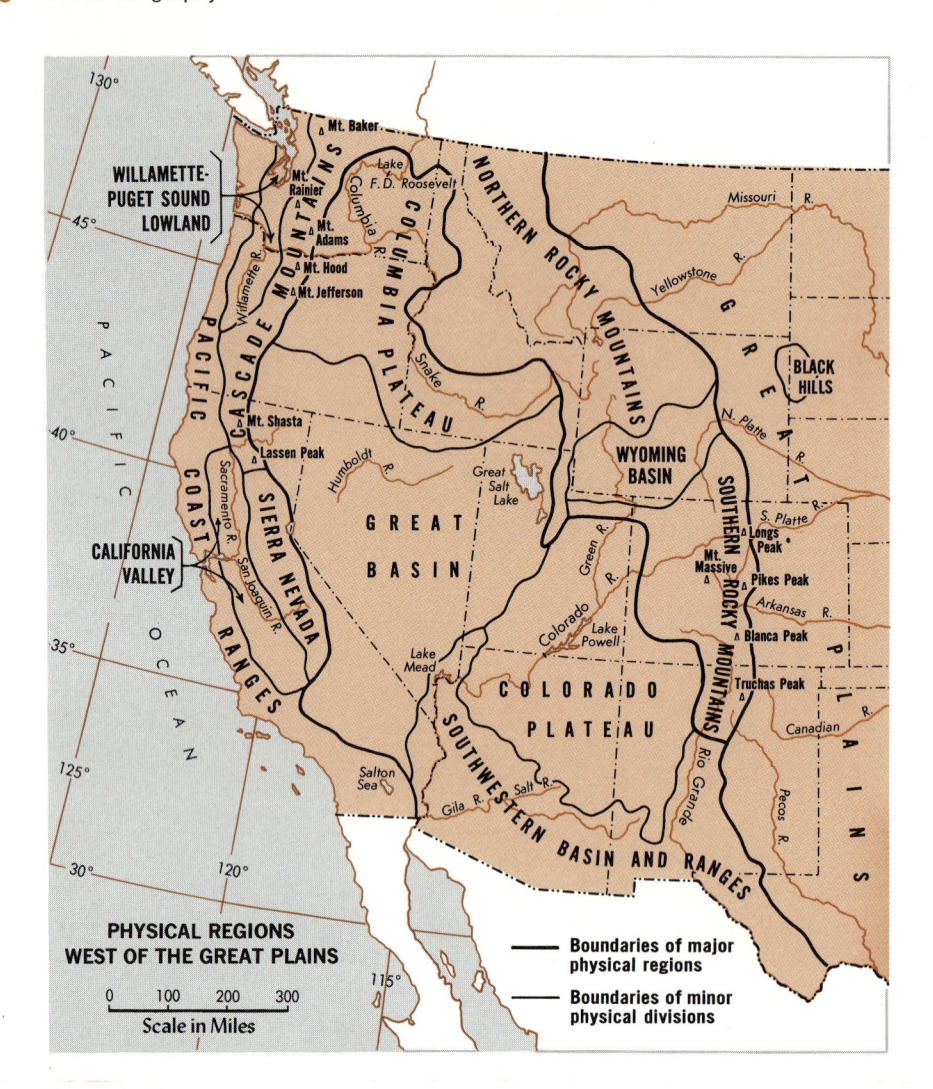

Fig. 3-1 The physical landscape of the United States west of the Great Plains is characterized by mountains and plateaus.

from Canada to Mexico stretches a wilderness of peaks and valleys, separating the streams that flow west to the Pacific from those which flow east to the Atlantic. Rugged mountains, deep snows, narrow canyons, and wide vistas make crossing these mountains a memorable experience.

Northern Rocky Mountains

The Rocky Mountains are divided conveniently into northern and southern sections, with the centrally located Wyoming Basin as the point of partition. The Rockies reach their greatest width in Montana, Idaho, and Colorado, where their trend is generally northwest to southeast; farther south their alignment is almost north-south. Some ranges—for example, the Big Horn and Wind River Mountains of Wyoming—are detached from the main system. Individual ranges often are separated from each other by long narrow valleys called "trenches," the longest of which is the Rocky Mountain Trench, extending for over 1,200 miles from the Bitterroot Valley of Montana northwestward into Canada. In some sections the mountains resemble deeply eroded plateau surfaces rather than ridges, especially in parts of Idaho and northeastern Utah.

The great width, about 500 miles, and ruggedness of the range have made this part of the United States inaccessible in many places, so parts of the Northern Rockies remain underdeveloped in spite of rich timber and mineral resources. Though most of these mountains appear to be well wooded, the best and most accessible commercial timber has been cut, and the remaining forests often lie within the protection of national parks or forests. Widespread glacial action has eroded the higher parts of the Northern Rockies, where even today snow remains throughout the summer season, providing a constant flow of water for the streams in the canyons of the Columbia, Snake, Missouri, Yellowstone, and other rivers. Because canyons make good reservoir sites, large amounts of hydropower have been developed in this region, and the reservoirs supply water for irrigation and impound floodwaters to prevent serious flooding.

Mining has been a great source of wealth in the Northern Rockies and is still one of the leading occupations. Western Montana and parts of Idaho and Utah have been noted for production of copper ores, gold, lead, silver, and other metallic minerals. The Colorado Rockies have been an excellent source of gold, silver, lead, and molybdenum. Thus, the present-day settlements have many residents whose principal incomes have been derived from mining: Butte, Anaconda, and Virginia City in Montana; the Coeur d'Alene district of the Idaho panhandle; and Leadville, Aspen, and Cripple Creek in Colorado. Not all of these noted mining centers are still productive.

Fig. 3-2 Northern Rocky Mountains, Idaho. In the fall, cattle are moved from mountain pastures to feedlots for fattening prior to marketing. (Courtesy of Idaho Department of Commerce and Industry)

Grazing of sheep and beef cattle occupies many people throughout the Rockies but is particularly important in the more southerly sections—especially in Colorado, Wyoming, and northern New Mexico—where the climate is generally too dry to provide extensive forest cover. Although pastures on the upper mountain slopes are used in summer, the animals must then be brought to lower elevations where winter winds and heavy snow interfere less with grazing. Much pasturage of this type occurs on government reservations. The carrying capacity of the mountaintops tends to be low, however, and in places the forage is so poor that excessive grazing induces serious soil erosion.

Lumbering may be more profitable in the Northern Rockies, where lower altitudes and greater water supply encourage forest growth. Most trees are needle-leaved evergreens of good quality, but processing and marketing of the timber is sometimes handicapped by distances to the consuming markets and by the expense of building highways through steep-walled canyons to tap the cutting operations.

Settlements in the Rocky Mountains generally are small; this is not a region of advancing urbanization. Some communities serve as ranch-supply centers; others have developed at foci of land-transport routes; some serve the rail or highway employees; and others have originated as lumber towns or resort centers. But limited types or quantities of natural resources, difficulty of access, severe winter weather, or other unfavorable environmental conditions have prevented the growth of large cities. Cities of moderate size, such as Missoula, Butte, Helena, or Coeur d'Alene in Montana, usually service large tributary areas.

Impressive mountain scenery justifies three national parks in the Northern Rocky Mountains, and the Mountain states have more land used for public parks than any other part of the nation. Active volcanism, with the accompanying hot springs, geysers, and lava surfaces, is present in Yellowstone National Park; this part of northwestern Wyoming is one of the most popular destinations for American tourists. A little to the south, the Grand Tetons, Jackson Lake, and the Jackson Hole country provide a different type of mountain scene. In northwestern Montana, on the Canadian border, visitors seek out Glacier National Park with its ice-scoured peaks, mountain lakes and glaciers, and waterfalls.

Tourism is a major source of income in this part of America; here, rugged landscapes appeal to the traveler who prefers to spend his vacation in hunting, fishing, camping, or sightseeing. Some of the more favored spots for camping are adjacent to lakes. Among these are Priest, Pend Oreille, and Coeur d'Alene lakes in Idaho; Flathead Lake in Montana; and Yellowstone and Jackson lakes in Wyoming.

Wyoming Basin

In central and southwestern Wyoming, the Wyoming Basin separates the northern from the southern section of the Rockies. The geography of this basin resembles that of the Great Plains, from which it is nearly isolated by the Big Horn and Laramie ranges. Rainfall is deficient (7.74 inches annually, on the average, at Rawlins); the Basin therefore is mainly a section of open, grass-covered slopes suited principally for grazing beef cattle and sheep on large ranches, though locally there are some irrigated tracts. Small supply towns are strung along the transcontinental rail lines and highways, but the period of population growth has slowed. Coal is mined in the southwest corner of the state at Rock Springs, and petroleum is of growing importance in most parts of the Wyoming Basin and accounts in part for settlement. Casper and Rawlins represent the region's larger settlements.

This open basin, because it nearly severs the Northern and Southern Rockies, has been an important transport route for rail lines and highways for more than a century. Travelers on the Oregon Trail made full use of its springs

Fig. 3-3 In general, the incorporated communities in the Western United States are smaller than those in the East. Only one Western city exceeds a million in population.

The United States West of the Great Plains

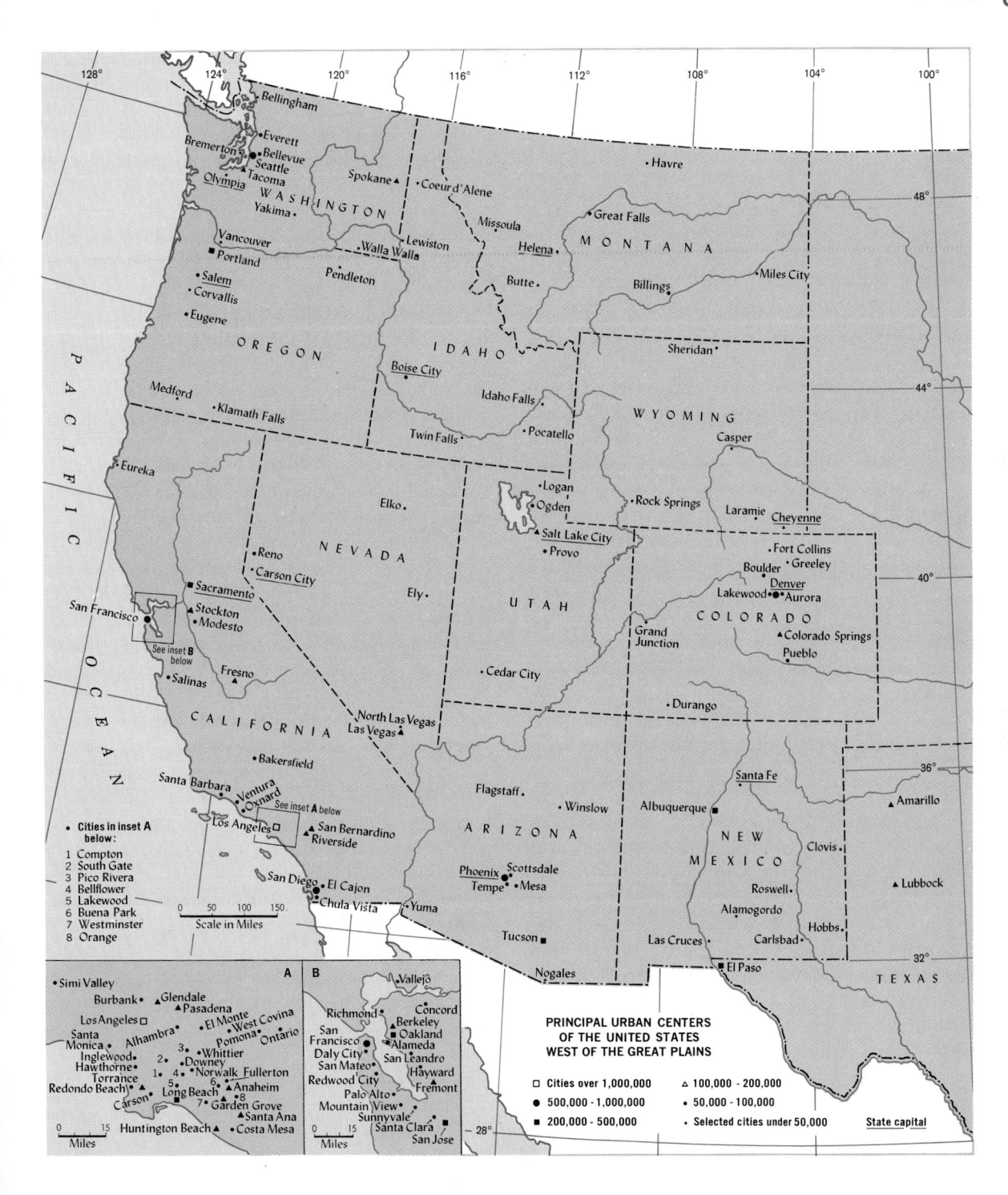

PRINCIPAL URBAN CENTERS OF THE UNITED STATES WEST OF THE GREAT PLAINS

- □ Cities over 1,000,000
- ● 500,000 – 1,000,000
- ■ 200,000 – 500,000
- △ 100,000 – 200,000
- ▲ 50,000 – 100,000
- • Selected cities under 50,000
- State capital (underlined)

Cities in inset A below:
1. Compton
2. South Gate
3. Pico Rivera
4. Bellflower
5. Lakewood
6. Buena Park
7. Westminster
8. Orange

and its meager-flowing streams, and by 1869 it proved to be the easiest transmontane route to the west by way of the Union Pacific Railroad.

Southern Rocky Mountains

South of the Wyoming Basin, the long, narrow trenches of the Northern Rockies are replaced by the high-level, basinlike North, Middle, South, and San Luis parks (elevated grassy tracts surrounded by peaks) of Colorado, and the Estancia, Roswell, and Tularosa basins of New Mexico. In Colorado, the Front Range rises sharply to elevations of more than 14,000 feet at Grays Peak, Longs Peak, Pikes Peak, and other summits. Farther west is the Sawatch Range, and then a confused maze of mountains in southwestern Colorado, the most prominent of which are the San Juan Mountains. The Wasatch Range, with its eastern extension in the Uinta Mountains, dominates northeastern Utah. East-west land travel through these ranges and basins is difficult.

Where water can be obtained from nearby mountain sources, as it is near Grand Junction, Colorado, intensive irrigated farming may be practiced, with correspondingly greater population densities; but in the grazing areas, population density is low and ranch settlements are widely dispersed.

Agriculture in these mountains is almost impossible because of the rugged terrain and lack of water, but streams coming from the summit snows and rains and flowing eastward do provide excellent water supplies for the western parts of the Great Plains. Water from the western mountain slopes normally drains into tributaries of the Colorado or the Rio Grande; from these and other streams some water for irrigation is obtained. In some places, as along the Fryingpan and Arkansas rivers, water is diverted from western to eastern slopes for irrigation, as well as to provide urban water supplies for Pueblo, Colorado Springs, and several Great Plains cities.

Minerals, including molybdenum and uranium ores, have been exploited extensively, and many towns depend upon mining as a chief source of income. The paucity of other natural resources, however, prevents great economic expansion. Most of the larger urban centers related to the Southern Rockies are located outside the mountains proper; these include Cheyenne, Boulder, Denver, Colorado Springs, Pueblo, Santa Fe, and Salt Lake City.

The altitudes of the Southern Rockies inhibit east-west transport; mountain passes at heights greater than 10,000 feet are common in parts of Colorado. Raton Pass (7,834 feet) crosses the Southern Rockies between Trinidad, Colorado, and Raton, New Mexico, and is traversed by main-line rail and highway routes. Lack of a convenient pass west of Denver was partly overcome by construction of the Moffat Tunnel through the mountains 50 miles west of that city. The Arkansas River crosses the southern end of the Front Range through Royal Gorge, a deep, narrow canyon that is a very impressive tourist attraction. West of Denver lies Rocky Mountain National Park; with its great glaciated heights of Longs Peak (14,255 feet), this is another major scenic attraction during summer, when its highways are free of snow and ice.

In New Mexico, the Southern Rockies are reduced to lower ranges; their wide intervening basins become of importance for grazing and generally are too arid for other activity. Despite the completion of many reclamation projects, including Elephant Butte Reservoir, farming in southern New Mexico basins is mainly limited by the amount of water obtainable from the upper Rio Grande or its tributaries. Prospects for expanding the extent of cultivated land are not bright, though population continues to increase—especially in Albuquerque, New Mexico's largest city, with about a quarter-million people.

In the western elbow of Texas, the Rocky Mountains are even more broken than in New Mexico. In the Davis Mountains they reach heights greater than 8,000 feet in a section so dry that agriculture is precarious and extensive cattle ranching is the main economic activity. For this part of the country, El Paso is the urban and transportation center as well as the gateway southward to Mexico. The city is important as a trading center and for smelting local ores. Downstream from El Paso, the Rio Grande enters mountainous country so impressive that the Big Bend district has been set aside as a national park, regardless of the fact that it is less accessible than most.

Intermontane Plateaus, Basins, and Ranges

Between the Rocky Mountains and the combined Sierra Nevada–Cascades is a large region generally lacking in rainfall. It includes all or parts of nine states and is sparsely populated. Elevations vary from below sea level to over 12,000 feet; land relief is diversified and includes broad plateaus, high mountain ranges, and wide basins. Most of Nevada, half of Utah, and large expanses in other states have no rivers that reach the sea. Climates vary from near-tropical desert to continental extremes of temperatures. Vegetation is mainly of the desert and steppe types, though higher plateaus and mountain summits may support coniferous forests because of the greater rain and snowfall. Much of the region is grazed over by cattle and sheep, but some sections toward the north are moist enough for growing wheat; in others, there is irrigation. This intermountain realm with its wide variations in landscapes is divided into four parts: the Colorado Plateau, the southwestern Basin and Range province, the Great Basin, and the Columbia Intermontane province. Much of the region is a barrier to be crossed by rail lines and highways leading to the cities and populated valleys of the Pacific Coast, but locally there are important centers of settlement in the interior.

Colorado Plateau

The Colorado Plateau is centered around the junction of the four states of Arizona, New Mexico, Colorado, and Utah. It was formed from uplifted rock strata reaching 5,000 to 10,000 feet in elevation; it has a surface undergoing erosion by streams under arid conditions, with the result that deep canyons, arroyos (gullies), gorges, badlands, and other evidences of extreme erosion distinguish the plateau landscape.

The most striking effect of stream trenching occurs in the northern part of Arizona, where the Colorado River has eroded the Grand Canyon by cutting its way through many strata of sedimentary and some igneous rocks, thereby forming a vast gorge 12 miles or more in width and a mile in depth. This spectacular trenching attracts many tourists each year, and the canyon floor and its walls have been set apart as Grand Canyon National Park. Other nearby canyons are only slightly less impressive in magnitude of scenery. Beautiful as the river and its canyons may be, the streams contribute little real wealth to the plateau itself, since the trenching is too deep to permit water from the canyon bottoms to be used for irrigation, though it may be impounded for hydropower; as an example, Glen Canyon Dam, situated upstream from Grand Canyon National Park near the Arizona-Utah boundary, was completed in 1963 and was designed to control the river floods, provide electricity, and furnish some water for irrigation. Its reservoir, 186-mile-long Lake Powell, provides a recreation area of striking proportions.

Downstream from the Grand Canyon, the river itself is important because it provides water and power for urban centers of southern California and irrigation water for the Imperial Valley and Arizona—an accomplishment made

Fig. 3-4 Grand Canyon of the Colorado River, Arizona. A prime tourist attraction, the Grand Canyon is 217 miles long, up to 18 miles wide, and over a mile deep. (Courtesy of the Santa Fe Railway)

possible by several dams and reservoirs such as Hoover Dam and Lake Mead. Other national parks have been established in southern Utah at Zion and Bryce canyons, both spectacular examples of erosion in multicolored sedimentary rocks. In addition, elsewhere on the Colorado Plateau are several national monuments, such as Cedar Breaks in Utah, and numerous sites of historical and archaeological interest in Arizona, for this is prehistoric cliff-dwelling country. In southwestern Colorado, Mesa Verde National Park contains outstanding examples of these ancient structures.

Since the altitude and latitude of the area combine to prevent large amounts of precipitation except on the very highest plateaus and mountains, a steppe climatic condition generally prevails; that is, summers are warm to hot, winters are cold and snowy. Moisture is insufficient for agriculture except where occasional springs provide a small flow of water, or along canyon bottoms in alluvial material where occasional "flood farming" is practiced by those who make their homes in this harsh environment.

Though the surroundings seem repressive to human development, the Colorado Plateau was the setting for the early Indian settlements called "pueblos." The high cultural level prevailing among their inhabitants was apparent from the excellence of their stonework, basketry, pottery, and weaving. In a land deficient in rainfall, the Indians carried on irrigation by flooding, depending upon maize as their staple crop. After the introduction of sheep, cattle, and horses by the Spaniards, their food supplies were enriched. Today the picturesque life of the pueblo villages attracts tourists and is a source of income for this region. Nonetheless, on those lands of northeastern Arizona which have been set aside for the exclusive use of the Indians there is only a meager water supply, and living conditions are difficult. Though there may be some income from flocks or herds of livestock, poverty generally is widespread here.

Except for one transcontinental rail line across northern Arizona and some good highways, few overland routes traverse the Colorado Plateau. Some towns have been built to service the rail facilities; from the former mining center of Kingman on the west through Williams, Flagstaff (a popular winter-sports center), Winslow, Holbrook, and Gallup eastward to Albuquerque, the communities tend to be tied to railroad activity, or in more recent years to highways. Distances between settlements are great, and outlying centers may take on the aspect of Indian trading posts, with a general store, service station, and a few necessary services such as the agencies of the Bureau of Indian Affairs, which administers relations between the U.S. government and members of the Indian nations. Perhaps the most sparsely settled part of the United States—except for Alaska's north coast—is the rough terrain surrounding the "Four Corners," where Arizona, Utah, Colorado, and New Mexico meet.

For permanent dwellers, the resources of the Colorado Plateau are limited. Petroleum and natural gas are produced in some localities, and New Mexico leads in production of uranium ore and potash. Commercial timber is available only on the higher lands. Grazing is the most important occupation throughout the area and is widespread among both white and Indian residents. Water, or its absence, governs the distribution of settlements. Small farms are found in deep canyons, where temperatures are high all year, or on the plateau surface where steadily flowing springs provide sufficient water. However, these enterprises are few in number; more common are the large ranches with substantial herds of cattle or flocks of sheep.

Southwestern Basin and Range Province

The southern rim of the Colorado Plateau descends to lower altitudes in a series of cliffs, in a line roughly from west to east halfway across Arizona and western New Mexico. Southwest of these steep escarpments lies an arid and semiarid land marked by tilted-block ranges caused by geological faults. These fault-block mountains alternate with basins or "valleys" covered with debris washed down from the mountains.

For decades southern Arizona was regarded as unfit for habitation by white men and was left to its primitive inhabitants; but wherever water is available, its fertile soils can be made productive. Thus, in the last half century, large irrigation projects have been completed in the valley of the Salt River, a tributary of the Gila River. Here, near Tempe and Mesa, where prehistoric people are known to have irrigated their crops, modern irrigation works have been built to make the valley a prosperous producer of winter-grown vegetables and citrus fruits. Phoenix, the state capital, is also the regional capital, and the city and its suburban satellites form one of the most rapidly growing urban centers in the entire nation. The area has gained greatly from winter tourist trade, numerous retirement centers, and the establishment of many light industries, with emphasis on the field of electronics.

Tucson, located in southeastern Arizona and settled since ancient times, also has made rapid

Fig. 3-5 Hoover Dam. Built across the Colorado River, on the boundary between Arizona and Nevada, Hoover Dam is 726 feet high and has a crest length of 1,244 feet. It resulted in the formation of Lake Mead, which covers 247 square miles and has a shoreline of 550 miles. (Courtesy of Trans World Airlines)

economic advances in the last decade; its clear, sunny, and almost frostless winter weather attracts tourists, but the summers are excessively hot. Here, too, diversified light industry has been established successfully, and Tucson is a university center as well. Farther south are the copper-mining towns of Bisbee, Ajo, Globe, and Morenci. Nogales, a border town, is dependent upon income from transportation. In the southwest part of the state, Yuma is served by one transcontinental rail line and highways that serve the needs of southern Arizona and New Mexico. Most residents are bilingual, because numbers of Mexican-Americans make up a large part of the local population and have left an indelible stamp on the culture of this part of the Southwest.

In spite of desert landscapes, relics of former civilizations, and pleasant winters, this is a region of limited economic potential, resembling the Mexican Plateau across the international boundary. Although summer temperatures rise well above 100°F, low relative humidity keeps the weather endurable, and for those able to afford air-conditioning devices there is little difficulty in living in such an environment. Rainfall is only 3 to 5 inches a year, but erratic cloudbursts sometimes occur. No large population has developed except in irrigated valleys, and small settlements are more characteristic than urban areas. Without irrigation, these drought-ridden lands have a climate too unfavorable for the support of many people. The importance of water supply cannot be overemphasized; when it is available, high summer temperatures and fertile soils provide excellent growing conditions for specialized crops, which can be marketed in distant places if rapid transport is at hand.

The most extreme subtropical desert in the United States is located in the Imperial Valley, a region of interior drainage in southeastern California. This lowland represents the ancient sea floor of the Gulf of California, long ago separated from the main body of water by the encroachment of the Colorado River delta. The valley is so hot and dry in summer (57 days a year may have temperatures higher than 100° F) that before 1900 it was regarded as worthless land. Though desert conditions are still apparent, water for irrigating a half-million acres is now obtained from the Colorado River and has transformed an otherwise barren landscape into a "hothouse" for whose produce the rest of the nation pays high out-of-season prices.

The early-day agriculture of cotton fields in the Imperial Valley and Gila Valley has been replaced by high-value intensive farming of carrots, melons, citrus fruit, dates (in Coachella Valley), alfalfa, and tomatoes. Land values have increased correspondingly, and this formerly barren desert now includes some of the highest-priced farmland in the entire nation. This transformation has been accomplished with considerable expense and effort by means of installations such as the All-American Canal, the principal distributing system for water from the Colorado River. The prevailing agricultural economy is based on air and rail transport and on refrigeration facilities that preserve perishable commodities shipped to markets outside California. The northern part of the basin (Coachella Valley) specializes in grapefruit and dates. Also located in the region, the city of Palm Springs has become a popular winter resort because of its sunshine and mild weather during that season, and other similar resorts now flourish in Coachella Valley. Between Coachella and Imperial valleys is a salt lake, the Salton Sea; about 30 miles long, it was formed by an accidental diversion of the Colorado River, which took place in 1905.

Great Basin

The Great Basin is a region of interior drainage enclosed by the Wasatch Mountains on the east and the Sierra Nevada on the west. The term "basin" is misleading because the area is broken by many fault-block ranges aligned north-south and separated by intervening basins filled with debris washed in from higher altitudes. In structure, both the Sierra Nevada and Wasatch ranges are only larger fault-blocks. The Mojave Desert of California is a southern extension of the Great Basin, whose northern section, often called the Basin and Range province, merges on the north with the Columbia Intermontane province.

Since most of the Great Basin is in a rain-shadow position, its mountains and basins receive insufficient precipitation for agriculture without irrigation, and the lower parts of the basins are so completely sheltered from sources of rain and snow that their character is usually that of a mid-latitude desert. Their deep deposits of alluvial material often become storage basins for quantities of water accumulating from mountain snows; thus some water can be obtained by sinking artesian wells or by pumping from underground sources. The floors of many basins can be used for year-round pasturage of beef cattle or may even have sufficient water to support small amounts of irrigated hay crops; but agriculture generally is not a profitable enterprise in most of the Great Basin. Ranches devoted to grazing operations account for much of the region's dispersed settlement and for settlement nuclei. Many settlements are highly specialized, such as ranch centers, mining towns, tourist accommodations, and recreation centers.

On the western side of the Great Basin the largest community is Reno, a university, trading, and transport center that also depends on gambling for part of its income, since gambling is legal in Nevada. To the south, Las Vegas, famed for gambling and lavish resorts, is also important for chemical manufactures. In relation to land area, population is still small in the state of Nevada, though the total is almost a half million and population increase during the 1960s approached 70 percent. The land will support few people, but with care its residents can carry on grazing with profit. Once-famous mining centers such as Virginia City, Goldfield, Bullfrog, Rhyolite, and Tonopah no longer extract their gold and silver ores, but have become such picturesque reminders of the past that they attract visitors.

Salt Lake City, dominating the eastern edge of the Great Basin, is located on a broad piedmont between Great Salt Lake and the western face of the Wasatch Range. The city provides commercial, educational, religious, touring, and transport services for a large area and is the state capital of Utah. Its prosperity is based partly on irrigated grain and fruit lands extending north and south along the piedmont. It has a plentiful and excellent water supply from the melting snows accumulated in great depth during the winter on the Wasatch summits to the east.

To the north is Ogden, a rail and supply center, and west are the salt flats of the Salt Lake Desert and the saline deposits of ancient Lake Bonneville, ancestor of the present Great Salt Lake. This whole region lacks agricultural activity, but some grazing is possible. Its mineral wealth is important, especially the copper ores at Bingham, Utah, and Ely and Yerington, Nevada. Provo, Utah, is near sources of coal and iron ore and operates an iron and steel plant. Other mines in this sector produce silver, lead, and zinc.

The Humboldt River, although not large when compared with most major streams of the nation, is the principal river in the Great Basin. The stream flows westward from the higher lands of northeastern Nevada across the northern part of the state to disappear in the Humboldt Sink. In the mid-1800s the California Trail closely followed the river, and today the larger communities of northern Nevada are located near it. In general, in view of presently known resources, the economic future of this part of the United States seems limited.

To the southwest, along the eastern face of the Sierra Nevada in California, at an altitude of more than 3,500 feet lies the long troughlike depression known as Owens Valley. Here, in a mid-latitude steppe, there was a profitable grazing economy until 1909, when the demand for water by residents of the city of Los Angeles prompted the purchase of water rights from the valley. This forced the eventual abandonment of Owens Valley grazing and what little farming had developed there. The water was taken southward to the city by aqueduct, a distance of about 200 miles, bypassing the interior basin of Searles Lake (actually a marsh), which is a good source of commercial borax, potash, and other industrial materials. In the Mojave Desert, farther southwest, conditions are more favorable for agriculture, particularly in its western angle, Antelope Valley, which obtains water from pumped wells.

Southeast of Owens Valley is a similar but much deeper trough, Death Valley, reaching a depth of −282 feet. Of all places in the United States it seems least suited for human occupation, but pleasant winter weather (almost rainless) attracts some travelers, and the desert scenery is so impressive that the valley has been set aside as a national monument. Exceedingly high summer temperatures discourage visitors in that season.

Columbia Intermontane Province

The Columbia Intermontane province, between the Rocky Mountains on the east and north and the Cascade Range on the west, grades into the Great Basin to the south. Sometimes called the Columbia Plateau, this is a region of diverse relief features—plains, plateaus, ridges, and mountains rising to more than 10,000 feet in places. In southern Idaho it is called the Snake River Plains; in eastern

Washington and Oregon and western Idaho, it is the Columbia Basin. The two sections are separated by the Blue Mountains of northeastern Oregon. The Snake River Plains occupy a broad crescent-shaped basin extending westward from Yellowstone Park, drained by the Snake River, which has eroded a deep canyon into the basaltic flows that form the principal rock material. The Columbia Basin, farther northwest, resembles the Snake River Plains in many ways, but it is drained by the Columbia River, into which the Snake River flows. In the northeastern part of the Columbia Basin are the "channeled scablands," formed during the glacial period by floods of meltwater that eroded shallow channels across the land surface. As the glacial ice disappeared, these were abandoned, leaving deep, rock-walled gorges in the "scabrock," as the exposed basalt is called locally. The largest of these dry channels is the Grand Coulee. In eastern Washington near Idaho are the Palouse Hills, covered with fine fertile loess; this area receives sufficient precipitation for raising winter or spring wheat and peas.

The Snake River Plains in southern Idaho are the most highly developed and the most populated part of that state. Their climate is continental, with low winter and high summer temperatures. Rainfall is often less than 10 inches a year, and irrigation is necessary for most crops. Originally the plains were covered with grass and sagebrush, used for grazing cattle and sheep. This economic activity continues, but wherever it has been possible to in-

Fig. 3-6 About one-third of the potential hydropower available from the Columbia River and its tributaries has been generated by means of the dams and power plants already operating on the rivers. They have also prevented serious flooding and have provided some water for irrigation.

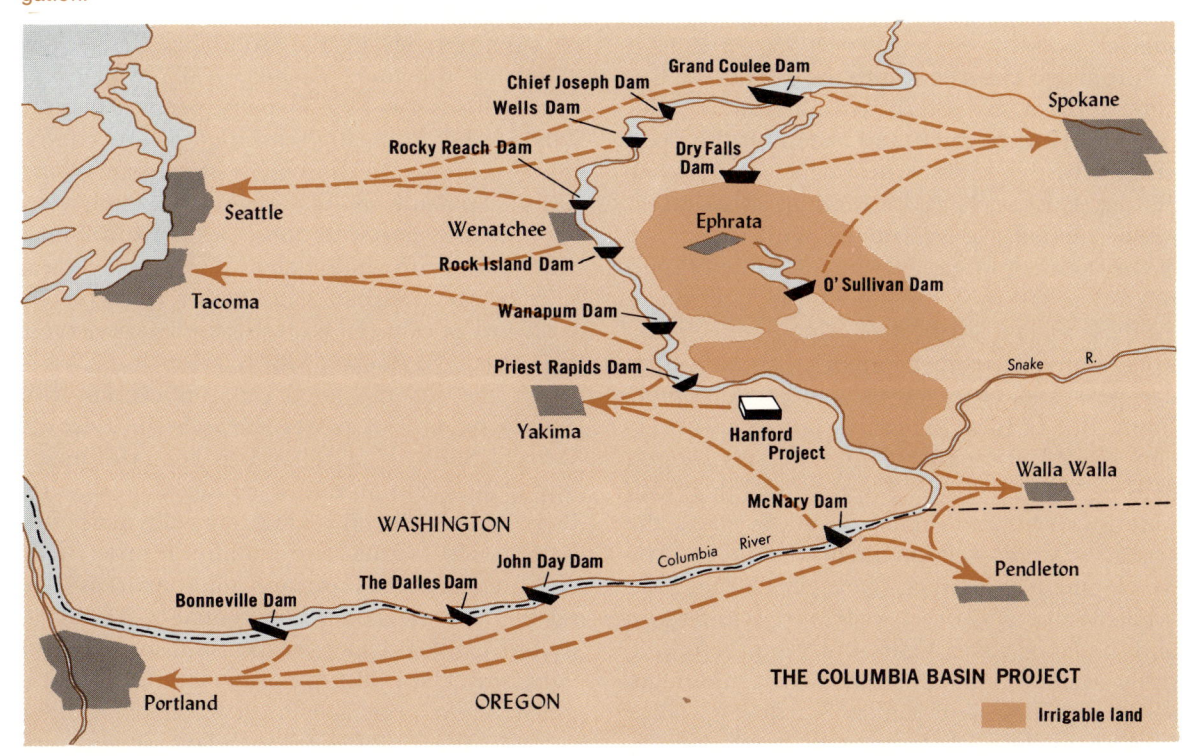

THE COLUMBIA BASIN PROJECT
Irrigable land

stall dams and reservoirs providing water for irrigation, the land grows alfalfa, wheat, beans, sugar beets, hardy fruits, and the famous Idaho potatoes, which comprise a large part of the United States' yearly potato production. Prosperous modern cities line the Snake Valley, from Idaho Falls through Pocatello and American Falls as far west as Twin Falls. Elsewhere irrigation projects are less promising, and the surface is poor for farming except in lowlands near the Oregon-Idaho boundary, where small cities such as Caldwell dominate a local deciduous fruit district. The Idaho state capital, Boise, was once a supply center for gold and silver mines in the nearby mountains.

In the nearby Columbia Basin, precipitation is greatest in the subhumid higher rim and decreases to an arid condition on the plains eastward from the Cascade Range. Maximum precipitation occurs in winter. The temperature range is intermediate between that of the marine West Coast and the continental interior.

From the late 1860s into the 1880s, the Columbia Basin in eastern Washington was devoted to grazing cattle and sheep on the open range. In the 1880s, wheat replaced livestock on the Palouse Hills soil and in other areas where rainfall was sufficient. With the arrival of the railways, wheat growing expanded rapidly, and today this "Inland Empire" produces over 70 million bushels of winter and spring wheat annually from 2 million acres, as well as dry peas and other crops. In the Columbia Basin, dry farming of grains produced uncertain crops. Irrigation, however, was begun before 1900; thereafter it was known that, with water, the soils and climate in valleys east of the Cascades were well adapted to fruits. The development of Yakima, Wenatchee, Okanogan, Walla Walla, and other valleys in Washington and the Hood River Valley in Oregon followed, using water obtained from nearby mountain sources. These and similar Northwest valleys furnish one-fifth of the nation's apple crop, as well as quantities of pears, apricots, cherries, and peaches. Yakima Valley farmers also grow much alfalfa, sugar beets, and potatoes.

Since the plains in the central part of the Columbia Basin lie too high above the entrenched Columbia River to benefit from its abundant flow, this region long remained a thinly populated and poor grazing country. In the 1930s the federal government built the multiple-purpose Grand Coulee Dam, which supplied hydroelectric power for the development of atomic energy materials, processing of aluminum, and other metallurgical plants during the 1940s. By the 1950s, irrigation aspects of this mammoth reclamation project were completed, with some 5,000 farms of 10 to 40 acres each, growing alfalfa, sugar beets, beans, potatoes, melons, and other fruits. In contrast, the wheat farms of 1,000 acres or more in the Palouse Hills receive only enough precipitation for the dry farming of wheat, with the aid of high-powered farm machinery.

Communities of the Columbia Basin vary widely in function: Yakima, Walla Walla, and Wenatchee serve local farmers' needs; Moscow, Idaho, and Pullman, Washington, are educational centers. In the northeastern part of the basin, Spokane is the regional capital and supplies the needs of those who live in small valleys in nearby mountains. This city is a creation of the railroad age; passes immediately eastward permit rail lines and highways to cross the Northern Rockies with relative ease. Westward, the city has good rail connections with Seattle by way of passes through the Cascade Range. Spokane, a rail and air center of major importance, has an industrial mix that includes flour milling, meatpacking, sawmills, pulp and paper manufacture, and aluminum processing.

The Columbia Basin and adjacent areas have enormous waterpower potential. Dams at Grand Coulee, Chief Joseph, McNary, Bonneville, and The Dalles are projects of the Federal government. Dams at Priest Rapids and other sites on the Columbia and Snake rivers were financed from other sources. The water-storage facilities and power plants of the Columbia River system provide for more facilities than along any other river in the nation. New plants are constantly planned or under construction.

This part of the nation faces a difficult geographical problem. The relatively great distances from population centers and the larger urban districts necessarily imposes a degree of isolation upon the Columbia Basin, and poses a problem that cannot be entirely overcome. The first great route into this area by way of the Oregon Trail in the 1840s failed to bring many permanent settlers, for they were headed for the more desirable farmlands west of the mountains. The opening of the basin to settlement by people of European stock was delayed until the completion of the great railroads to the Pacific Northwest. Population growth began to be noticeable during the 1880s, but in those years the length and difficulty of the rail trip still inhibited settlement. Other settlers arrived with the coming of the automobile and good roads in the 1920s, but today much of the traffic in and out of this part of the country depends on air transport; thus, the isolation arising from geographic location has been partly overcome.

Cascade Mountains and Sierra Nevada

These mountain ranges are sufficiently high throughout their north-south extent to serve as a barrier to eastward-flowing surface winds, which might otherwise penetrate deeply into the heart of the continent. Instead, the moving air is forced to such altitudes in crossing the mountains that it loses the greater part of its moisture on the western-facing slopes. Thus, while heavy rains support a dense forest, mainly of softwoods, on the western windward side, the eastern leeward slopes are nearly barren, except where water is available from streams such as the Yakima, Wenatchee, Hood, Truckee, and Owens rivers.

Together, the Cascades and Sierra Nevada extend from the Canadian border southward across Washington and Oregon as far as southern California. Despite differences in landscape, structure, and resources, they function as a geographic divide, since they separate two very different regions. West of the mountains lie fertile, well-watered and well-populated lowlands and valleys. East of the dividing crest is the semiarid and desert intermountain region previously described. The railway lines and highways that cross the Rockies, the intermountain basins, and the Cascade-Sierra region were built primarily to serve Pacific Coast cities and favored lowlands like the Willamette Valley. Since few people stopped along the route to settle in mountains or deserts, population density is low, and only small amounts of freight and few passengers originate there today. The Cascades and Sierras are impressively scenic and have resources of timber, minerals, and waterpower, but they do not support a large population. Their resources are exploited mainly by those who live near the Pacific Coast, and their natural wealth has been used to help the growth of West Coast cities and industries.

Cascade Range

The Cascades are formed mainly from volcanic material—a broad platform of rock above which snowcapped peaks of volcanic origin tower to great heights. Only the southernmost of these, Lassen Peak, has erupted in this century, though others of the range may not be entirely quiescent. The mountains vary from 50 to 100 miles wide and extend from Lassen Peak more than 500 miles northward into British Columbia. Summits exceed 10,000 feet, with Mt. Rainier topping the whole chain at 14,408 feet. River and glacial erosion has been active on all the peaks, and the mountains are scarred by steep canyons, rugged ridges, and many glacial lakes. A few small glaciers remain as relics of the Ice Age, most of them being on Mt. Rainier. Much of the Cascade Range is included in national forests, ensuring conservation of its timber reserves and the maintenance of the tree cover needed to control the flow of streams used for power, irrigation, and urban water supply.

The Cascade Range is of such height that it forms a major barrier to east-west communication except by air. The mountains function negatively in the geographic pattern, since they pose land transport problems that are overcome only at great cost. Principal traffic movement is east-west, at right angles to the mountain trend; it therefore becomes imperative to seek passages for highways and railroads, which are sometimes blocked in the mountain passes in winter when summits are snowbound. The most famous pass in the range is the Columbia River Gorge, through which the river crosses the mountains at such low levels that the stream has been canalized. Northward from the gorge are passes of moderate altitude, useful through most of the year. Of these, the most easily traveled is Snoqualmie Pass (3,127 feet), connecting the Puget Sound Lowlands on the west and the Columbia Basin on the east by highway and rail.

Lumbering dominates most commercial activity on the western slopes of the Cascades, where a wealth of mid-latitude softwoods of high market value has been exploited for a century or more. Settlements on the western slopes, therefore, are an expression of lumbering, though of late years, with improved highways and trucks, there is less need for the small town and its sawmill to be located very near the timber supply. Today much of the lumber moves by truck downslope to large mills in lowland locations that are convenient for processing the raw material as lumber, paper pulp, or paper, of which the Pacific Northwest is a large producer.

Irrigated agriculture in the bottomlands and dry-farmed grains or grazing operations characterize the dry ground distant from sources of water. This, like so many other sections of the American West, is a region of extractive industry, supplemented by agriculture and grazing. Towns and cities are generally small by U.S. standards and serve mainly as supply centers, being either lumber towns or farm- and ranch-service centers.

In these volcanic mountains are found few valuable minerals, except in northern Washington, where mines produce gold, copper, zinc, and a little coal. The principal factor of the region's economy, however, depends on the vast timber resources and the hydropower supplied by mountain streams. Indeed, such mountain resources support many people who live on nearby lowlands; for example, east of the crest, Bend and Klamath Falls in Oregon are occupied with handling lumber from nearby mountains, and the east-slope valleys of the Yakima and Wenatchee rivers could not produce their profitable fruit crops without the water from the mountain snows. Mountain uplands along the east-facing slopes of the Cascades provide summer sheep pasture; in winter the animals are kept on irrigated lowlands of the valleys.

The scenery and other attractions of the mountains justify the presence of three national parks; the southernmost centers on Lassen Peak, protected because of its distinctive volcanic features, including hot springs and mud geysers. In southern Oregon, catastrophic volcanic eruption formed a huge crater, now occupied by a lake 2,000 feet deep, which has been set aside as Crater Lake National Park. Mt. Hood, Mt. Baker, and other Cascade peaks attract visitors with their scenery and winter sports, and Mt. Rainier National Park includes the most extensive ice fields in the United States outside Alaska. Between Oregon and Washington, in the Columbia River Gorge, the landscape is covered with forests and is distinguished by cascading streams plunging over cliffs to join the Columbia. Set in this gorge is Bonneville Dam, a hydropower project financed by the federal government and tied by transmission lines to Grand Coulee and other power dams both in the Columbia Basin and west of the mountains. Abundant power resources have furthered the expansion of electrochemical and metallurgical industries in the Pacific Northwest.

Sierra Nevada

The Sierra Nevada, a majestic mountain range with Mt. Whitney its highest peak exceeding 14,450 feet, extends from the southern end of the Cascade Range near Lassen Peak southward for a distance of approximately 400 miles, to join the transverse ranges of southern California. The Sierra Nevada Range of California is almost a single granitic block, asymmetrical in cross section, with its steeper and shorter slope facing eastward and overlooking the Great Basin. Its western face is a long and relatively gentle slope descending to the foothills and piedmonts of the Great Valley of California. The higher altitudes, as well as the western slope, have been profoundly affected by glacial scour, for ice and snow accumulated in depths sufficient to supply glaciers of the past as well as many mountain streams of the present. Most ice action has been concentrated on the western face, where deep gorges have been etched by tributaries of the Sacramento and San Joaquin rivers; further entrenchment has usually been due to glacial scour.

The high mountain barrier of the Sierra Nevada greatly interferes with east-west land transport in northern California, since heavy winter snows often make its passes difficult to traverse. Although the northern Sierras are not so high as the southern, their ruggedness prevents easy crossings. The Feather River Canyon is used for rail and highway facilities, but at considerable expense of construction. Near Lake Tahoe are several passes; Donner Pass (7,189 feet), which is also traversed by rail and road, is the most useful, but snows are so deep there that the pass is kept open in winter only with extreme effort. South of Lake Tahoe, most of the passes are higher than 9,000 feet and are useless except in midsummer, and are not traveled greatly even then. For more than 100 miles, the east-facing escarpment is so steep that no wheeled vehicle can traverse this barrier. In the south, Walker Pass (5,248 feet) is used for highway travel, but only at the southern end of the Sierras, at Tehachapi Pass (3,790 feet), do rail lines find a route through the mountains.

Since the western slopes of the Sierra Nevada receive the full force of Pacific storms, precipitation is sufficient for extensive forests. For years softwoods have been produced commercially, but extensive cutting of the accessible timber has limited further expansion of this industry. Several groves of big trees, the giant sequoias, have been set aside in national parks, and much Sierra timberland is protected on a reserve basis in national forests. Yosemite National Park, famed for its waterfalls and canyon, is the most popular of the parks in the region; others are Kings Canyon (including the former General Grant Park) and Sequoia.

Zoning of temperature and precipitation has caused zoning of vegetation on the western mountain slope. In the foothills below 2,000 feet, chaparral mixed with digger pines covers the eroded slopes. From 2,000 to 5,000 feet, because of greater precipitation, there is a zone of western yellow pine (ponderosa pine), sugar pine, and the big trees (sequoias). In the zone from 5,000 to 7,000 feet are several varieties of hardy pine. Up to 9,000 feet juniper and mountain hemlock grow, with unforested mountain meadows found above that elevation. Commercial timber is obtained mainly from altitudes below 5,000 feet.

Midway up the western slope, geologic conditions have favored the formation of a zone about 150 miles long called the Mother Lode, in which gold quartz was deposited as part of the geologic complex. From this source westward-flowing streams carried eroded bits of gold downslope; from these alluvial deposits at lower levels and quartz mines at higher levels, Californians removed quantities of the valuable metal and ore between 1849 and 1855. During this period of the Gold Rush, rapid settlement of the western slope of the Sierra occurred, and many small mining towns were established. Today, with increased costs of mining and the exhaustion of the placers and ore bodies, the remnants of these settlements—"ghost towns"—attract tourists rather than prospectors or mining interests. Except for gold, however, mining has not been an important economic activity in the Sierra Nevada.

The combination of mountain meadows, impressive stands of trees, cool and clear summers, and streams stocked with fish attracts many tourists to these mountains each year, who come mainly in summer, although in Yosemite Valley and some other places they come for winter sports. Peaks of the Sierras take the form of sharp-edged pinnacles, which pose a challenge to experienced mountain climbers.

Many of the canyons cut by stream and glacier on the west slope are spectacular; where they need not be preserved for parkland, these provide fine sites for the installation of dams and reservoirs, from which a steady supply of water is obtained for irrigation and power. Since California generally lacks sources of natural energy (except for some petroleum), the large amounts of available hydropower, developed at relatively moderate cost, provide the state with a highly valuable resource.

Shasta Dam, located in the Central Valley of northern California, impounds the waters of the upper Sacramento River and provides power for valley cities and farms. The dam has other functions, for it controls the flow of water in the Sacramento River so that the valley is less subject to floods and regulates the river flow in summer to aid navigation and provide water for irrigation when rainfall is slight. Oroville Dam on the Feather River was completed in 1968 to impound water for the use of southern California. The western slope of the Sierras also provides water for Millerton Lake at Friant Dam and for power and water-supply installations on the Tuolumne and Mokelumne rivers, which serve San Francisco and Oakland, respectively.

The leading economic activities of this mountain range, then, include a little mining, some lumbering, tourism, production of electric power, and some grazing. Perhaps more important than any one of these, however, is the fact that the abundant water coming from the mountains may be led to lower ground to the west, east, and south, where it may be used for agriculture in the valleys or to supply the growing cities of the neighboring West Coast.

California Valleys and Coast Ranges

The valleys and lowlands of California support more inhabitants than all the rest of the western United States. One of the principal reasons is the Mediterranean-type climate, so attractive to visitors and so favorable for growing fruits, vegetables, and many kinds of farm crops. California has one-third of all the acreage planted in fruits and nuts in the United States and leads in acreage devoted to commercial vegetable farming. Large areas of fertile soil, plentiful water for irrigation, and resources of petroleum, hydropower, and accessible timber have furthered economic growth. In addition, there are transport terminals for the outlets of a vast hinterland and seaports where transcontinental routes meet with ocean shipping. Largest production and the most populous areas are found in the Central Valley, the Los Angeles lowlands, and the San Francisco Bay district. From these centers, smaller valleys penetrate into the mountains or lie within the Coast Ranges, which somewhat parallel the Pacific Ocean and form the western boundary of the Central Valley.

Central Valley

The Central Valley of California, often called the Great Valley, consists of lowlands about 400 miles in length, drained by the Sacramento River from the north and the San Joaquin from the south, except for a small basin of interior drainage at the southern end. The eastern slopes abut against the lower slopes of the Sierras in the form of large alluvial fans or piedmonts, deposited there by streams flowing down from the mountains to the valley. The valley is bordered on the east by the Sierra Nevada and on the west by the California Coast Ranges. At the confluence of the Sacramento and San Joaquin rivers is an extensive delta, formed where the combined streams flow through Carquinez Strait into San Francisco Bay. Shut off from the moderating effects of the Pacific Ocean, the Central Valley is generally warmer than the coastal areas in summer. Temperatures in the Sacramento-Stockton region, however, are somewhat tempered by afternoon breezes from San Francisco Bay.

The broad floodplains and alluvial piedmonts of the Central Valley, with their warm summers, mild winters, and meager winter rains, supported only a small population of primitive Indians before 1849. The arrival of settlers in the wake of the Gold Rush brought a brief period of dry farming of small grains, followed by irrigation. Today this valley is prime farmland, where peaches, pears, asparagus, celery, potatoes, beans, and sugar beets are grown on the delta, and melons, cotton, citrus fruits, alfalfa, and wine and table grapes in the San Joaquin section. Even in the dry

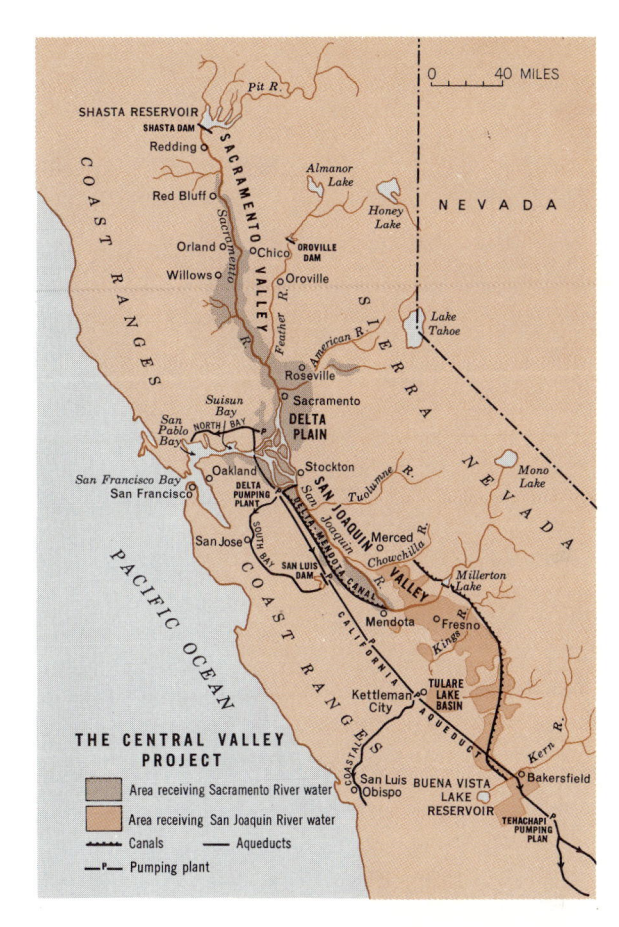

Fig. 3-7 Water stored back of the Shasta Dam in northern California is used to irrigate fields in the south central part of the Great Valley of California. It is necessary that water be moved southward in the San Joaquin Valley by pumps.

southern end of the Great Valley cotton, alfalfa, and potatoes are grown. California is a major dairying area, and it is the leading state in the production of poultry and sugar beets (23 percent of the U.S. total production).

The decline of the large cattle and grain ranches, typical of the early years of the San Joaquin Valley, was accelerated after 1875 by the completion of the Southern Pacific Railroad. Small farm-service centers were estab-

Fig. 3-8 A part of the American Southwest as seen from space. (Courtesy of NASA)

lished, and many communities began specializing in high-value subtropical crops. In Sacramento Valley, Yuba City, and Marysville are food-processing centers for canning peaches, and Willows is surrounded by rice fields. Sacramento owes its origin to agriculture and the needs of the miners, but its later growth came from its importance as state capital; today it is the urban center of Sacramento Valley.

Small communities in the delta process fruit and vegetable crops, and the port of Stockton, connected with the sea by a ship channel, is locally important. Nearby Lodi is noted for wine and table grapes; farther south, Fresno is a center of dried-fruit production, particularly raisins. Other towns ship melons, table grapes, and figs. A common occupation—namely, dairying—runs through the economic fabric, and the valley has many milk condenseries.

The southern part of San Joaquin Valley is dominated by its regional capital Bakersfield, important for rail activity. In the last quarter century the fertility of this area, combined with abundant water from the mountains, has given impetus to the city's growth. In addition, this section of the valley has important petroleum production, and Bakersfield is a center for the sale of oil-well equipment and supplies. Most oil production has been located north and northwest of the city, in the desert on the western side of the valley, where the McKittrick and Coalinga oil fields are situated. Agriculture land near Bakersfield is planted in wheat and irrigated cotton; these developments have come about by nearly draining several large shallow lakes that formerly lay west and north of the city to provide the necessary cropland. Cotton has become the leader in California's agricultural exports.

A major problem in converting the valley to profitable farmland has been solved with much success by means of a vast system of water conservation and redistribution. For decades the northern section (Sacramento Valley) was plagued by floodwaters when melting mountain snows poured their streams down to the valley floor, creating flood hazards along the Sacramento River and its tributaries. At the same time, water was all too scarce in the southern part of the valley (San Joaquin and the interior basins). In the 1930s, under the auspices of the federal government, a gigantic reclamation project was begun, for which dams were built to impound floodwaters of the north and aqueducts conveyed surplus water supplies to the south, crossing the delta of the two rivers through a flume or canal. This water is used by ranchers and fruitgrowers along the San Joaquin River, even though it must be pumped *upgrade* to reach their properties. Power for pumping operations was readily available from hydroelectric plants installed at dams and reservoirs in the mountains. Thus the damaging floods along the lower Sacramento were eliminated, or their threat was reduced.

After the San Joaquin River bottomland residents were supplied with water as described above, they no longer needed supplies from the Sierra Nevada streams; a second aqueduct, parallel to the mountain base, was then built to divert water from these streams (San Joaquin, Merced, Tuolumne, Kings rivers) and carry it southeastward into the interior basins, where it now irrigates large acreages of cotton, potatoes, and alfalfa in the former lake beds. Development of this part of the California water project is completed by the dams, reservoirs, and canals in the arid southeastern section of the valley. This is the San Luis unit, jointly financed by the federal and state governments.

Traffic in the Great Valley flows in two directions: transcontinental routes cross the valley floor from Donner Pass to San Francisco Bay, for example, but local traffic moves mainly north and south, connecting at the northern end with Oregon by way of the Klamath Mountains, and with southern California and Mexico at the south by way of Tejon (4,219 feet), Tehachapi (4,025 feet), and Cajon (3,623 feet) passes.

Southern California

Many visitors, especially winter tourists, think only of southern California when the state is mentioned. Roughly triangular, this area extends from Santa Barbara on the west to San Bernardino and Riverside on the east, and southward to San Diego and Mexico. Inland the region is rimmed by the San Gabriel and San Bernardino ranges on the north and the San Jacinto and other ranges on the east. The Los Angeles lowlands, heavily floored by sediments washed down from the mountains, are divided by low ranges or hills into a number of valleys, including San Fernando, Ventura, Santa Maria, and San Bernardino. Farther south the mountains crowd nearer the coast and leave lowlands but a few miles wide between Los Angeles and San Diego. Offshore the tops of submerged Coast Range peaks project above the Pacific as the Channel Islands, among which only

Santa Catalina Island, a popular resort, is important.

Parallel mountain ridges are obstacles to transportation in the southern Coast Ranges, but a few passes connect the inland desert and the coast. Los Angeles has poor approaches from the north and east; westward from the city the steep faces of the Santa Monica Mountains, extending for 100 miles, descend to the shore. Though a highway has been notched from the sea cliffs, it is expensive to maintain. Southward, two main routes lead to San Diego along the coast and through interior valleys.

Favored by sunny weather, mild winters with little frost threat, fertile alluvial soil, and moderate supplies of water for irrigation, the lowlands and valleys of southern California have undergone an extraordinary agricultural development. Small though the area is compared with other farm regions of the nation, its production of citrus and other fruits, nuts, melons, vegetables, sugar beets, and hay, along with dairy and poultry products, puts Los Angeles County among the leading counties in value of agricultural output.

Southern California's climate is a fundamental resource. During the last quarter of the nineteenth century, out-of-state visitors began to be attracted by the region's warm, sunny winters, and a great influx of settlers and migrants began. Today about 6 million people live in the Los Angeles lowlands and adjacent valleys and coastal areas, and California has become the most populous state, with almost 20 million residents.

The coastal lowlands, originally covered by short grass and scrub trees, were the first part of upper California encountered by the Spaniards in their northward march from Lower California. The mission fathers brought seeds and cuttings—many grains, citrus trees, grapes, figs, and olives. They introduced horses, cows, and sheep. In these imports lay the foundation of California's rise as an agricultural state. As soon as the Americans developed greater sources of irrigation, the grazing economy of the Spanish-Mexicans gave way to land subdivision and settled farming in the late nineteenth century. As the tracts were supplied with water, the land came under cultivation, with large plantings of subtropical fruits. Later, accommodations for winter tourists were expanded, and many visitors became permanent residents. This phase of life in the southern valleys continued until the early 1900s, when discovery of new petroleum fields—the first oil well had been drilled in 1880—provided fuel for industry and helped spur a population growth that eventually became more urban and provided a large consumers' market. The Los Angeles lowlands have been a leading source of petroleum—fortunate, indeed, in view of the fuel needs of industry and the lack of coal in this part of the country.

The city and county of Los Angeles soon took the population lead from surrounding communities, and urban growth eclipsed agricultural activity in southern California. Although inadequate water supply, transport difficulties, and poor location inhibited the city's growth, Los Angeles progressed in spite of such geographical handicaps. Water was imported

Fig. 3-9 Agricultural valley of southern California. The sheltered valleys of southern California are centers for growing citrus fruits, grapes, vegetables, and certain cereals. (Courtesy of Trans World Airlines)

by aqueduct from the east slope of the Sierra Nevada (Owens Valley) and from the Colorado River by way of the Metropolitan Aqueduct. With no good natural harbor, a shallow embayment 18 miles south of the city's center was dredged to serve as an anchorage. Good land connections with the rest of the state and nation were established when the Southern Pacific Railroad reached Los Angeles in 1875, and the Santa Fe Railroad in 1887.

Aggressive advertising did emphasize the few advantages of the city and furthered its growth. To southern California have come people and wealth out of proportion to available resources. California's pressing need for cheap farm labor brought a diversity of workers, including Japanese, Chinese, Spanish-Americans, Negroes, and some East Europeans. Now these and others provide labor, not only for the farms but also for the industries: petroleum refining, tire and chemicals manufacture, fish processing, fruitpacking, motion picture production, aviation, steelmaking, and a long list of other products for which no obvious geographical explanation is always apparent, except that the large population and high level of purchasing power has produced a healthy local market for manufactured goods.

The Los Angeles metropolitan area, the largest in both size and population in the western United States, is faced with many of the same problems as the large eastern urban centers—among them, substandard housing, ghetto areas, water supply, heavily traveled highways, and pollution. The problem of air pollution may well be greater in Los Angeles than in any other city in the nation; the region's air is so polluted that it is killing some of the native vegetation in adjacent hill areas.

Surrounding Los Angeles are many smaller urban agglomerations, originally engaged in agricultural activity but now industrialized. Older and more distant communities such as Ventura, Oxnard, Santa Maria, San Bernardino, and Riverside are pleasant towns, where some subtropical fruits are still processed for shipment, but since 1960 commercial emphasis has been directed away from agricultural occupations. Towns and cities nearer Los Angeles, involved more deeply in industry, include San Fernando, Santa Ana, and Anaheim. Beverly Hills, Pasadena, Van Nuys, and Inglewood are largely dormitory towns, which may, like Beverly Hills, be surrounded by the city of Los Angeles and yet retain separate municipal identities.

West of Los Angeles, facing the sea and backed by the Santa Ynez Mountains, Santa Barbara lies in a secluded location, for it cannot be reached easily from the land side and has no satisfactory harbor. Reflecting its character as one of the earlier mission communities, the city has never been important for industry

Fig. 3-10 The Colorado River is of special importance to southern California. It furnishes water for irrigation in the Imperial Valley as well as for municipal use in the Los Angeles area.

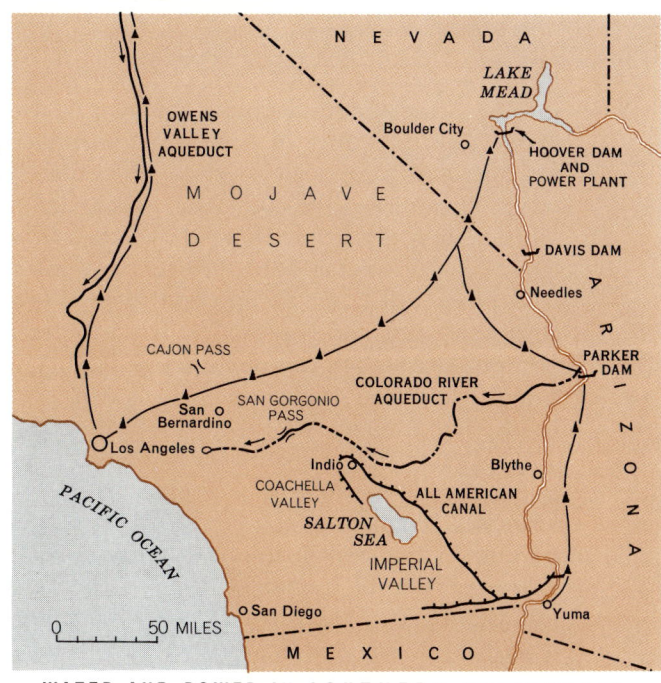

WATER AND POWER IN SOUTHERN CALIFORNIA
—▲— Power transmission ——— Aqueduct ⊢⊢⊢⊢⊢ Canal

or agriculture; like Pasadena, its principal development has been residential, though the character of the latter city has changed. In common with most other southern California urban centers, Pasadena now has a high percentage of minority-group residents—Japanese, Spanish-Americans, and Negro, for the most part.

In the extreme southwest part of the state, situated on an excellent harbor, the city of San Diego occupies a rather remote natural site, somewhat handicapped by inadequate water supply and by a hinterland impeded by mountains and desert to the east and limited by the international boundary to the south. As a port, however, it serves importantly as a base of naval operations, and has also developed light industry (food processing, aircraft, textiles) since local demand is strong, labor supply plentiful, and year-round working conditions are advantageous for specialty industry. It was a center of rapid urban growth in the 1960–1970 decade.

California Coast Ranges and Valleys

The Coast Ranges, forming the western rim of the Central Valley of California, are made up of ridges and valleys whose direction is roughly parallel to the coast itself—a condition associated mainly with faulting of the crust. The mountains present a bold face to the Pacific, and the central Coast Ranges and their valleys all but prevent any penetration from the sea; no first-class harbor can be found between Los Angeles and San Francisco. The submerged northern ends of a few of the troughlike valleys do provide access to the interior and the Central Valley; thus, the submerged block forming San Francisco Bay is a breach in the Coast Ranges leading to the San Joaquin and Sacramento valleys. From the human standpoint, the most significant part of the province is not the mountains but the intervening valleys, in which all the cities and most of the people are situated.

In spite of proximity to the coast, those who live in the Coast Ranges and their valleys today are not directly concerned with the sea as an economic pursuit, though fishing is important in a few places. Most valleys are occupied by farmers—or, as they are known in the west, ranchers. The hills are used by cattlemen. Farming is dependent on the marine climate; where water is available, specialty crops are grown, including artichokes, flower or mustard seeds, flower bulbs, lettuce, sugar beets, apples, grapes, and apricots. Where drought prevails, there are traces of the old-time large-scale ranch economy, with herds of dairy and beef cattle. On some of the mountains there is good grazing, and here beef cattle are raised extensively.

Largest and longest of the intermountain valleys is Salinas Valley, through which the Salinas River flows for more than 100 miles to reach Monterey Bay. The upper valley has scattered ranches, but the lower northern section is intensively farmed in irrigated sugar beets, lettuce, alfalfa, melons, and (near Watsonville) apples. Most of this is large-scale commercial farming, and is highly mechanized, though seasonal labor requirements do present a major problem. The area's urban center, Salinas, derives from earlier ranching activities. Water is pumped from underground sources.

Inland from Salinas is the Santa Clara Valley, where specialty and irrigated fruit crops are also grown, in its lower northern portion. The lower Santa Clara, however, is dominated by large groves of prunes, apricots, and peaches, and dairy herds are important as a source of fresh milk for the San Francisco Bay cities. The Santa Clara urban center is San Jose; like Los Angeles located at the site of one of the early Spanish pueblos, San Jose has become lately a processing center for fruits. This valley is faced with a problem of subsurface seepage of salt water from San Francisco Bay; along with the southward expansion of the San Francisco urban sprawl, this natural condition has almost eliminated the older fruit groves and replaced them with urban settlement that is nearly continuous between San Francisco at the north and San Jose at the south.

South of the 38th parallel, San Francisco Bay is the most conspicuous geographic feature of the coast. Much of the bay is shallow and is seriously polluted by industrial waste; but inside the Golden Gate and a little to the south, depths are sufficient for large vessels. The eastern shore has been dredged to permit ocean vessels to use the port of Oakland. Two peninsulas impinge on the Golden Gate like a pair of tongs: the southern prong is occupied by San Francisco, confined to its peninsula, only 7 miles wide; the northern (Marin) peninsula is so rugged that many parts remain unsettled. Highway connection between the two is maintained over the Golden Gate Bridge.

San Francisco failed to develop under Mexican rule, but upon the discovery of gold in California and the subsequent rush of immigrants, the city became a disembarkation point for settlers entering the state by sea. Thus the early character of the city as a cosmopolitan center of trade and commerce was established, and though San Francisco now has industries such as coffee roasting, metal processing, and sugar refining, its dominant function remains that of a trading and financial center. A century ago, common labor was in such demand in California that immigrant lists included many Latin-Americans, French, Germans, and Chinese. Each of these ethnic stocks left an imprint upon the life of the city, and this variety has given San Francisco a cosmopolitanism unmatched in any other Coast city.

On the eastern side of the bay, the municipality of Oakland is part of the San Francisco metropolitan area, and it is here that industries find cheap and plentiful land, providing for present-day commercial activity in food processing and milling, as well as other industries. North of Oakland, specialized satellite cities line the shore almost solidly to a point beyond Carquinez Strait. Among these are Berkeley, a residential and university center; Richmond and Martinez, with oil refineries; Crockett, with a cane-sugar refinery; and Pittsburg, Hercules, and Giant, with chemical plants. This "Eastbay District" is connected with San Francisco by the Bay Bridge and also by a 3.6-mile underwater transit tunnel.

The total population of the Bay-area towns and cities—that is, of the whole metropolitan area around the bay—is about 3 million. Their economy is widely diversified, with the emphasis being on light and medium industries, even though raw materials for manufacturing are not generally abundant in this part of the West. Fuel, in the form of imported petroleum products and hydropower, is obtained with relatively little difficulty; but mineral resources of use to industry are not at hand, with the exception of salt, which is obtained through solar evaporation from 60,000 acres of salt flats around San Francisco Bay.

The northern California Coast Ranges are unlike the central and southern ranges. They are more rugged, have no good harbors, and lack ready access to the sea since there are few passes; hence they are more isolated. The mountains lack easily accessible intervening valleys, and transport is restricted to routes parallel to the coast. Only in the southern section facing San Francisco Bay are there valley floors suitable for agriculture—those of Santa Rosa, Sonoma, and Napa. Though there is water in these lowlands for irrigating alfalfa and vineyards, the land available for agricultural expansion is limited. The development of dairying is associated with the excellent pasture afforded by the long, narrow coastal terraces that are kept moist by cool, damp air. In the main, these mountains and their small valleys are incapable of supporting many people, and most towns are small. More remote and arid valleys tend to be used for grazing rather than for raising crops. In the north, lowlands at the mouth of the Eel River are dominated by the towns of Eureka and Arcata, where the economy is based on lumbering, especially of the Coast redwoods.

Pacific Northwest Coastal Province

Between the Cascade Range and the Pacific Ocean and from Canada southward to the Klamath Mountains of northern California is a humid region with mild winters and cool summers. Here, in a zone about 100 miles wide from west to east, the Douglas fir, western hemlock, cedar, and spruce attain maximum growth. This marine west-coast climate not only favors tree growth but also encourages dairying and fruit growing in the wider valleys. It is, on the whole, a "green" landscape, with a climate that imposes few hardships on those who live there. Winter is the rainy season, and days are short, with cloudy weather and high relative humidity. The temperature range is small throughout the year. Summers are pleasant, and high temperatures rare. Except in the higher mountains, the snowy ground characteristic of winter in the northeastern United States is lacking.

Most of the residents live in the Puget-Willamette Lowland between the Cascade Range and the Coast Ranges, paralleling the Pacific Ocean. These lowlands are a troughlike basin, whose northern part is submerged to form Puget Sound. The rivers in the north drain into Puget Sound or Grays Harbor on the Pacific, and those in the south flow into the Columbia River or the Willamette, its major tributary.

Coast Ranges of the Pacific Northwest

The mountains bordering the Pacific Northwest Coast adjoin the Pacific shore where coastal uplift has produced a series of marine terraces like giant steps facing the sea. The long, narrow valleys of the California ranges are lacking. This is a rugged and scenic coast, with few corridors leading to the interior. The principal opening through the Oregon and Washington coast mountains is the wide estuary of the Columbia River, which breaches the mountains at right angles at 46° north latitude. This and the opening at San Francisco Bay are two routes by which ocean vessels may penetrate inland along this coast. The third, Juan de Fuca Strait, connects the coast and Puget Sound, and leads far into the interior to provide access to protected ocean ports. A mountain knot, the Olympics, occupies northwest Washington, separated from the Cascades by Puget Sound and from Vancouver Island by Juan de Fuca Strait. The Olympics reach elevations that support glaciers and snowfields, and the central part of their peninsula is set aside as a national park, whose highest peak is Mt. Olympus (7,954 feet).

In Oregon and Washington, the Coast Ranges generally are lower in altitude than those of California. Nevertheless, construction of railways and highways in this terrain is so expensive that some sections remain isolated. The northern mountains are forest-covered, chiefly with coniferous softwoods of great economic value. To some extent the deep forests have been exploited for timber, but much untouched forest serves as a scenic resource. Because of the extensive forests and the expense of clearing them, agricultural land is at a premium, except in sheltered places on the eastern side of the Olympic Peninsula. In the wider valleys, specialized farming has been developed. South of the Olympic Mountains, fishing, lumbering, and paper manufacture are found, but the region is remote and its population density is low.

The Columbia River estuary is so broad and deep that it once presented a major interruption to any north-south travel along the coast; but a bridge now connects Astoria on the southern shore with the Washington side. Astoria is so distant from other modern-day commercial activities in Oregon that its growth has been slow, in spite of its fur-trade fame and its location on deep water. South of Astoria, the Oregon coast is rugged but not high, and residents of the Willamette Valley wishing a seaside vacation use the many coastal resorts here, despite the usual heavy summer fog. Lumbering is a leading industry, and dairying has been profitable. Farther south, flats along the upper Rogue River Valley and the Umpqua River grow deciduous fruits; the former area is famous for its pears. Small population centers such as Medford and Roseburg supply the needs of fruit-growers, farmers, and lumbermen.

Puget-Willamette Lowland

West of the Cascade Range and parallel to it lies a depressed block of the earth's crust whose southern portion is drained by the Willamette River, a tributary of the Columbia. North of the Columbia, small streams provide drainage, but the landscape is confused because of the intensive effects of glacial deposit in some places and extensive submergence of the northern part of the lowlands, forming Puget Sound. In this longitudinal position and west of the Cascade Mountain heights, this depressed area is subjected to the full force of prevailing westerly winds, which cross it at right angles, though some protection against marine influence is provided by the Oregon and Washington Coast Ranges, the Olympic Range, and Vancouver Island.

The marine windward climate of these lowlands provides sufficient rain, mainly in the winter season, and enough high summer temperatures to support excellent growth of trees and grassland, though most of the forests here have been removed for commercial timber. In the Willamette Valley the forest cover is interrupted by occasional open stretches of prairie. Most of the trees are softwoods, Douglas fir and western hemlock. With rainfall of moderate amount, leaching of soils has not been a serious deterrent to farming the lowlands, though occasional flood problems arise when mountain snows melt too rapidly in spring and summer.

Of the two sections of the lowlands, the Willamette Valley is the more productive agriculturally. It has not been subjected to heavy glacial deposits, and its alluvial soils may be farmed with relative ease. Drainage, at least for agricultural purposes, is excellent most of the year. Winter temperatures, though they approach the freezing point for short periods, are seldom so severe that farm animals must be securely protected against cold, and pasturage to maintain the animals is available throughout the winter. Grain crops (especially wheat), forage, small fruits such as berries, the stone fruits of the mid-latitudes, and bulb crops all thrive in this environment. Dairy products, wool, nut crops, apples—these and many other agricultural resources contribute to the varied economic wealth of the valley.

Americans interested in agricultural potentialities came to the Pacific Northwest over the Oregon Trail between 1842 and 1848 and began farming this land. It was in this period that many settlements began. Although land and water transport was inadequate and markets were distant, population growth continued steadily. The Gold Rush to California in 1850 created an excellent market for lumber in that state. Supplies of fruit, grain, and fish were ample, and fortunately the Indians offered little effective opposition to the new settlements.

Wealth was available from forests, but commercial exploitation of timber resources came only with improved transportation after 1900. Since then, lumbering has proceeded so rapidly that the rapid exhaustion of this resource is a matter of concern. Conservation measures look toward cropping northwestern forests on a sustained-yield plan that will make it possible for the Pacific Northwest to continue as a major lumber producer. Forests of the Northwest provide paper, shingles, plywood, building material, telephone poles, and other necessities of modern life.

The lowlands of western Oregon and Washington have been famous for a century for certain crops. Early settlers tended to establish Eastern farm practices in the Northwest; hence parts of this valley may resemble Ohio or Illinois, with their barns, farm animals, and houses repeating the mid-nineteenth-century rural complex of the American Middle West. In addition to staples such as oats and potatoes, Northwest agriculture developed some specialty crops such as berries, mint, spring-flower bulbs, commercial seeds, and flax.

With mild winters, plentiful rain, and a long growing season, agriculture normally is profitable, although occasional summer drought may make it desirable to irrigate in July and August. Though native grasses and introduced forage plants do well and beef cattle and sheep are raised, the concentration is on dairy products. Exports of processed milk, cheese, butter, and poultry are large. The expansion of commercial farming is limited by the extent of valley lands and by the cost of removing stumps from farmland and pasture.

Fishing, especially for salmon and halibut, is a third significant economic factor in the Northwest. Demand for salmon has been so great that sea and streams have been overfished, and production is less than formerly. Commercial fishing is maintained largely through conservation measures such as restocking streams, restrictions on the gear used, and closed seasons. Other economic activities include manufacturing, tourism, and mining.

The Willamette Valley, only 125 miles long and about 35 miles wide, is highly productive. It contains three-fourths of Oregon's people in 5 percent of the state's area. Its grain, wool, lumber, fruit, fish, and other commodities have provided cities with raw material for manufacturing, but the one major resource that is deficient in the Willamette region is mineral fuels. Oil and coal are lacking in any quantity, but abundant waters of streams descending the western slope of the Cascade Range provide conditions suitable for the generation of hydropower, a rich advantage for Oregon's factories.

Urban centers have developed here, especially in and around the city of Portland, near the confluence of the Willamette and Columbia rivers. To secure outlets for the products of the region, the lower Columbia has been dredged, thus making Portland a seaport 100 miles inland from the Pacific. Portland, however, is not wholly dependent upon the products of western Oregon; the city taps eastern Washington and Oregon by a transportation corridor through the Columbia River Gorge. Portland's location, combined with electric power, has led to industrialization, population growth, and market expansion. Smaller cities of the Willamette Valley serve both general and specific functions. Salem is the center of Oregon government; Corvallis and Eugene are service and educational communities. Oregon City has pulp and woolen mills, and other communities prosper from broad agricultural and manufacturing bases.

Northward from the river at Portland, the lowlands extend to Puget Sound. Unlike the Willamette, dense forests once prevailed here, occupying terrain whose features were determined by heavy glacial outwash and deposition. Disturbed drainage patterns and deposits of stony till hinder farming, even after clearing of timber. Forest clearing has left stump-covered land intractable for intensive farming except in the more fertile alluvial valleys. Milk, poultry, vegetables, bulb and seed crops, and small fruits are among the farm products, but most of the lowlands are unsuited to growing grain other than oats.

The northern Puget Lowland is submerged by the waters of Puget Sound. Long, narrow, deep channels make up the sound, whose confused landscape includes irregular peninsulas and many small islands. Originally virgin forests reached the shore, but today the lumber mills have left only second- and third-growth coniferous and deciduous trees.

Seattle is the dominant city of the region, though it competes with neighboring Tacoma. Of the two, Seattle enjoys the superior geographic position, being nearer the open sea and with access enhanced by the Snoqualmie and other passes east of the city. As recently as 1935, Seattle and Tacoma were regional capitals for the Puget Lowland, both meeting the needs of local markets, with Seattle maintaining trade with Alaska and the Orient as well. That trade continues, but since 1940 both cities have emphasized diversified manufacturing; the processing of fish and foods, and lumbering, flour milling, meatpacking, boatbuilding, furnituremaking, copper smelting, aircraft manufacture, and other industries are important. The manufactures are mainly of the lighter types, dependent on hydropower. These cities, after receiving their growth impetus from the establishment of sawmills, became inactive until mining attracted large numbers of people to Alaska and the Yukon at the close of the nineteenth century. Thus, for more than a half century, Seattle has been a leading supply and transportation point for residents of Alaska, for it maintains excellent connections by sea and air with that state. By virtue of its extreme northwesterly location in the conterminous United States, Seattle has also engaged heavily in trade with the Far East.

Hawaii

The fact that Hawaii occupies a position distant from the mainland, with more than 2,000 miles of water intervening, places that state in a location of maritime importance; indeed, insular Hawaii is the only state in the Union whose political boundaries are entirely defined by shorelines. Its climate, too, is unique among the states. Only the extreme southern part of Florida can be regarded as genuinely tropical, but all of Hawaii, located south of the Tropic of Cancer, is in the most tropical environment in the United States, though surrounding seas do modify tropical conditions.

Hawaii lies in the zone of the northeast trade winds most of the year and enjoys a climate free from snow or frost, except on the highest mountains, and steady sea breezes on its northeastern shores. Low islands under the trades normally receive deficient rainfall; but the larger Hawaiian islands are so mountainous, with altitudes exceeding 13,000 feet, that onshore winds sweep against the northeast slopes, chilling as they mount and deluging the mountain faces with heavy rains—as much as 400 to 600 inches in places, though the average is smaller. Downpours on these slopes tend to induce excessive erosion in spite of a protective covering of tropical vegetation, and the island landscapes are scored deeply by canyons and gorges through which many small but swift streams plunge to the sea over impressive cliffs.

On the leeward (southwestern) side, northeast winds produce small amounts of precipitation, for here the warming and drying air has descended mountain slopes. The larger islands, therefore, display two strongly contrasted aspects: tropical torrents, eroded gorges, rugged precipices, and Amazonian foliage on the windward shores; near-desert grasslands and scrub plants, with occasional cacti, on the leeward slopes.

As long as the Pacific Ocean remained relatively uncharted and unused, the Hawaiian Islands made few advances in population or economic life. In the mid-nineteenth century, however, commercial vessels opened the Far East to trade, and then the mid-Pacific became important as a base for naval supplies and as an outfitting point, particularly for whaling vessels. As trans-Pacific trade grew, so did the population of Hawaii, and trade developed between the islands and the mainland, particularly in

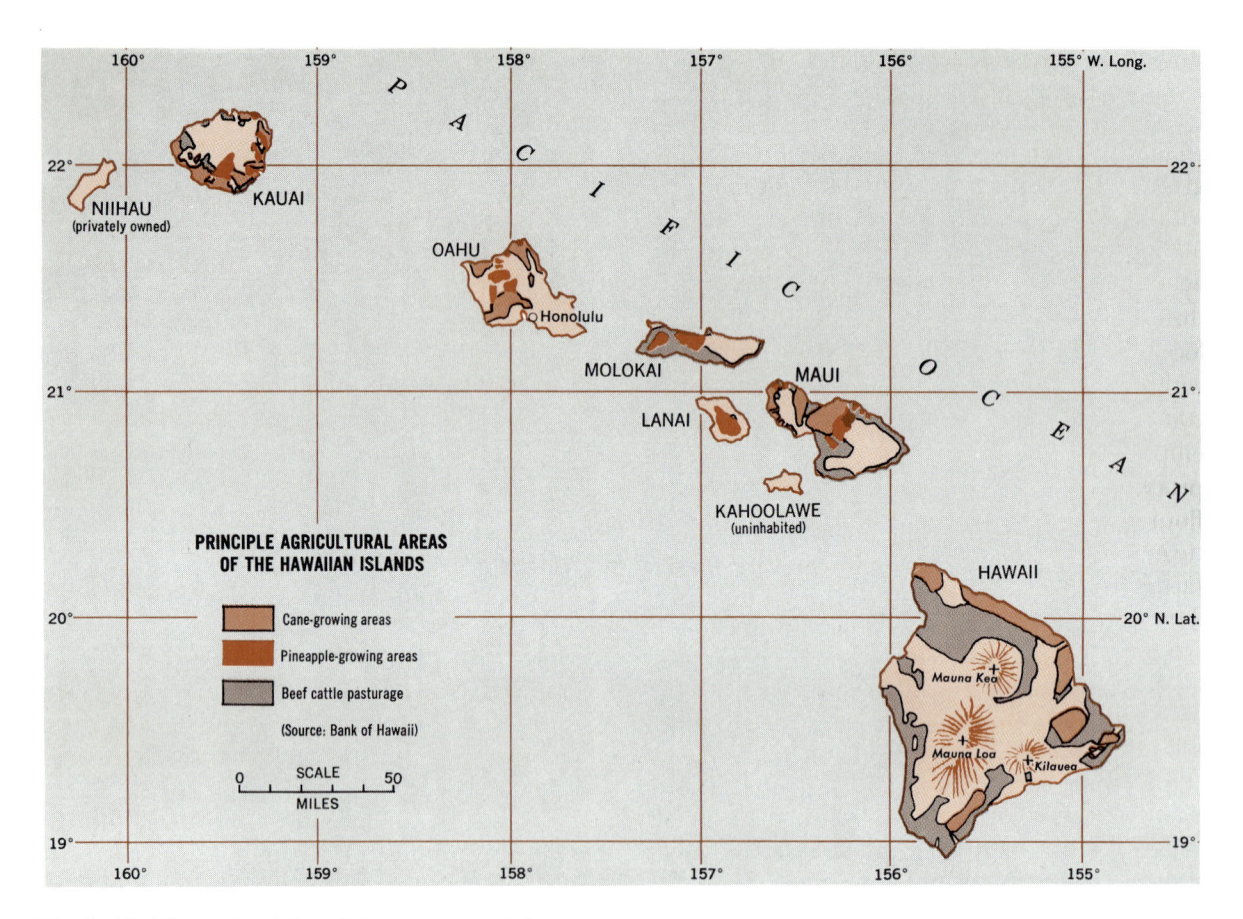

Fig. 3-11 The state of Hawaii is composed of five major islands and several smaller ones. Honolulu, the capital and largest city, is on the island of Oahu.

cargoes of tropical foodstuffs moving eastward and manufactured goods shipped westward.

In the last half century, military operations in the Pacific have led to concentration of defense expenditures in Hawaii—in 1970, to a total of more than $600 million. This income, together with that from a healthy tourist trade, contributes the major share of the support of permanent residents of the islands. Agriculture, the mainstay of former decades, is relatively less important today, and since the agricultural commodities consist mainly of cane sugar and pineapples, the income from that source seems unpromising for future development. Cane sugar from Hawaii must compete against cane sugar from other tropical locations, as well as with the beet sugar of mid-latitudes; but Hawaii, with less than half of the sugar acreage of Louisiana, produces twice as much cane each year. The "sugar" islands are Hawaii, Maui, and Kauai (in about equal value of production) and Oahu. Pineapple cultivation is concentrated mainly on the islands of Oahu, Maui, and Kauai. Relatively minor Hawaiian crops include coffee and tropical flowers, the latter calling for rapid shipment by air. All told, Hawaiian agriculture uses about 500,000 acres of cropland. On some of the less-inhabited islands, grazing of beef cattle has been a principal occupation; but on almost 80 percent of the total area of Hawaii, neither grazing nor cultivation of the soil is profitable. Since the total area of all the islands is little more than 4 million acres, the limited cultivable area promises poorly for future expansion of agriculture.

In common with many other tropical islands, Hawaiian agriculture tends to be the product of the plantation system, with all its advantages and disadvantages. In earlier years, native Hawaiians of Polynesian stock failed to supply the amount or type of agricultural labor needed by the plantations, whereupon owners of cane and pineapple properties introduced a labor force from other parts of the world, including Japan, the Philippines, China, Portugal (Madeira and the Azores), Korea, and Spain. In 1970, in the total population of Hawaii their descendants comprised 25 percent Japanese, 9 percent Filipino, 5 percent Chinese, and 3.6 percent from other homelands; the remainder were Caucasian. Intermarriage was common among the migrants, with the result that the present population of Hawaii is largely of mixed ancestry. It is estimated that only about 12,000 pure-blooded Polynesians remain on the islands today. Perhaps no other state in the Union has experienced such a high degree of racial intermixture, as a direct result of the insistent demand for farm labor. For many years the labor supply in Hawaii remained below the saturation point, but recent mechanization of sugar and pineapple plantations has made Hawaii's farm-labor problem less acute. So far, the installation of defense works has taken up the slack; but a major problem of unemployment is in prospect unless enough work can be found in occupations other than agriculture and military operations.

Tourists bound for the islands are attracted by tropical conditions uncommon in other states: a mild year-round weather condition (temperatures are almost seasonless throughout the year); exotic vegetation; and the prospect of year-long outdoor fishing, swimming, and golfing. These basic attractions are widely advertised, along with luxury hotels (approximately 150 on Waikiki Beach alone) and a type of island life that is foreign to most mainland Americans. In addition, nature has provided many features of active volcanism, such as violent eruptions and lava flows, all available to the sightseer by air or sea transport. Hawaii's Volcanic National Park includes the active volcanoes Kilauea and Mauna Loa. Much tourist income in Hawaii has come from mainland residents who are able to afford a stay, averaging 11 days, in these tropical surroundings—nearly a million and a quarter of them in 1970.

Though Hawaii is the largest island, Oahu is the most densely populated, since it includes the capital city of Honolulu and the defense installations at Pearl Harbor. Large deep embayments formed by lava flows and coastal erosion provide protected anchorages for commercial and naval vessels, and around these sites the urban growth has occurred. Of the total population of the island, approximately 750,000 in 1970, more than one-third live in the Honolulu urban area; the rest live in small villages and outlying centers.

Hawaii is a strongly urban state, dependent mainly on its activities as a center of trade and transport rather than upon any wealth of natural resources. The islands provide no local sources of mineral fuel; coal and petroleum must be imported. Nevertheless, industrial activity has increased rapidly in the last decade, and sales of manufactured goods in 1970 amounted to almost $400 million, compared with a value of $375 million for sugarcane, pineapples, and other agricultural products. The greatest industrial growth has occurred in the areas of food processing (other than sugar and pineapples) apparel and fabrics, printing and publishing, and construction materials.

Fig. 3-12 Honolulu, Hawaii. Thousands of tourists visit Waikiki Beach, which extends from the yacht harbor to Diamond Head, annually. Over 150 hotels and apartment houses are located in the area. (Courtesy of Hawaii Visitors Bureau)

Alaska

The territory of Alaska was purchased by the United States from Russia in 1867. Essentially an isolated peninsula, situated at the northwestern corner of North America, it sprawls over 586,400 square miles, or nearly one-fifth the entire area of the forty-eight conterminous American states. The Aleutian Islands extend toward the Kamchatka Peninsula of the Soviet Union, nearly 1,000 miles west of easternmost Asia, and the coastal panhandle of Alaska reaches to within 400 miles of the state of Washington.

The landforms of Alaska range from the flat marshy plains of the Yukon River delta to the highest mountain peak in North America, Mt. McKinley (20,300 feet). In the southeast, the St. Elias Mountains rise to altitudes of 10,000 to 15,000 feet where Yukon Territory, British Columbia, and Alaska meet. The range breaks into a complex of mountains to the northwest, and then separates into two definite chains, of which the southern spur, the Kenai Mountains, forms the backbone of the peninsula of the same name and reappears offshore as the rugged parts of the Kodiak Islands. The main mountain chain arcs northward as the Alaska Range, which averages 8,000 to 10,000 feet in elevation, and it is here that Mt. McKinley is situated. The mountain system swings southwestward and merges with the sharp volcanic peaks of the Aleutian Range. Elevations decrease westward across the Aleutian Islands, whose peaks are generally between 4,000 and 5,000 feet high. Some mountains of this long island arc are still actively volcanic.

Central Alaska is a region of broad plateaus and plains, quite different in appearance from the mountainous southern section. The land slopes to the west from altitudes of about 2,000 feet at the Yukon border to broad swamplands at the delta. Across this rolling region the Yukon River and its main southern tributary, the Tanana, have cut broad valleys; the streams generally flow over flat-bottomed valley floors.

The northern wall of interior Alaska is formed by the high barren 10,000-foot mountains of the Brooks Range; north of this, a broad foothill belt slopes down to a tundra-covered, lake-dotted coastal plain. Several rivers wind across the plain and have become entrenched slightly into its glacial deposits. Permafrost, which underlies the Arctic lowlands is one of the reasons for the many lakes and swamps on the surface. In this barren wilderness, recent exploration for petroleum indicates the presence of exceptional untapped reserves; and despite the great distance from consuming markets and the difficulties of transport by land or by sea, it now appears that the Alaskan north slope is about to enter a period of rapid economic exploitation. A 470-mile highway already connects western Alaska with this north slope.

The development of the Arctic oil reserves presents a series of cultural as well as physical problems, and the physical problems may be the easier to solve. A pipeline can be built across the mountains and the permafrost zone from the Arctic to the Pacific ports, and there are methods by which the extremely cold temperatures can be compensated. The chief cultural problems are: (1) conservation of wildlife and native vegetation, and (2) effects of such development upon the activities of the Eskimos and Indians living in the area.

Alaska, crossed by the Arctic Circle, is a land of diverse climates of the middle and high latitudes. The interior plains are severely continental, unlike the milder marine climate of the southern coast and the panhandle, where temperatures are moderated by the relatively warm waters of the Gulf of Alaska and summers are generally cool, July temperatures averaging 55°F. Winters are surprisingly mild for the northern latitude. January average mean monthly temperatures are about 30°F, or some 10 degrees cooler than Seattle. Precipitation is heavy, for storms from the Aleutians come in upon the mountainous coast all year round, bringing most of the rain and snow from September to December. The settlements, unless they are sheltered behind island mountains, average about 100 inches of precipitation annually; some stations have recorded as much as 150 inches. Rainfall usually decreases toward the west along the southern coast, with about 60 inches being a common average.

In contrast, central Alaska has a continental climate characterized by cold winters and relatively warm summers. Average January temperatures are slightly above zero near the Bering Sea coast, but are −10 to −15°F in the east-central interior; extreme temperatures of −60 to −70°F may be experienced during most winters. Summers are cool near the Bering Sea, with 50° to 55°F for the July averages, but many days become warm inland; extremes rising into the 80s are common. Because of interior location, annual precipitation is only 10 to 20 inches, with rainfall maximum in summer.

The northern Arctic coast and most of the Brooks Range have a true Arctic climate, in which summer monthly averages remain below 50°F. Although winters are longer than those of interior Alaska, extreme temperatures are not so low because of the modifying influence of the waters of the Arctic Ocean, which, though cold, are warmer than the adjacent land.

Trading for fur seals and sea-otter skins brought Russian ships to the Alaskan coast more than 150 years ago, and these expeditions resulted in several Russian settlements and many Russian place names. The Russians, however, came mainly for exploitation of the resources, not for permanent settlement. Even today the Pribilof Islands are one of the leading sources for fur seals, owing to present regulations governing the hunting of the herds.

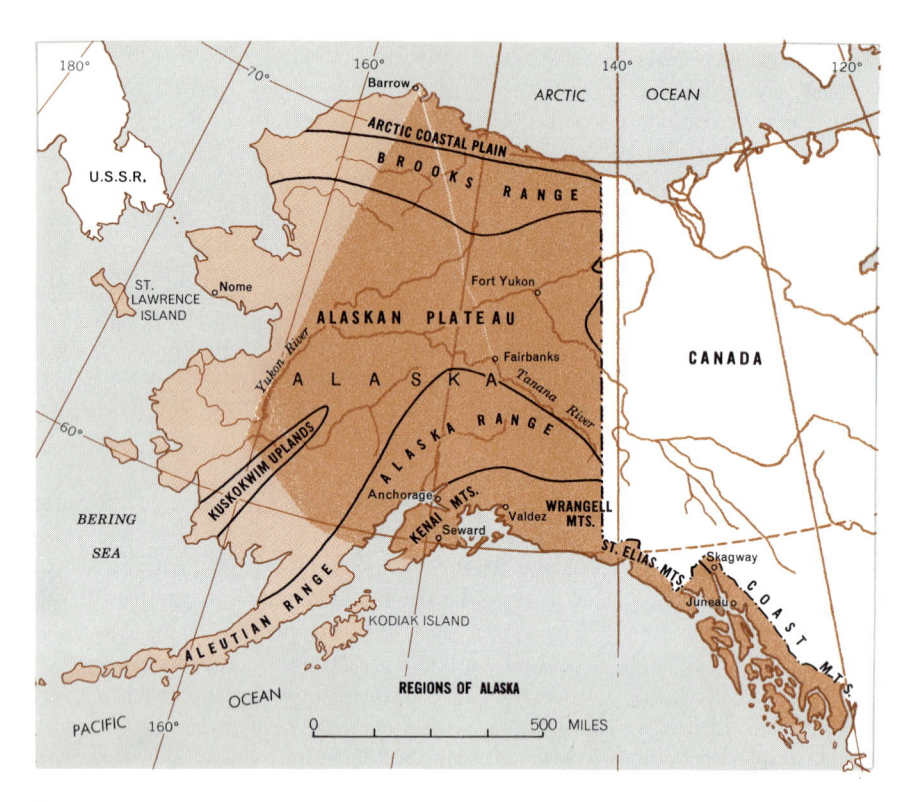

Fig. 3-13 The state of Alaska has only minor areas of plains. Mountains extend to the coast or are near the coast where the state fronts on the Arctic and Pacific oceans. Central Alaska is dominated by a broad, and in some places swampy, plateau.

Placer gold was the attraction that brought many settlers to Alaska at the turn of this century; the gold rushes to Nome and Fairbanks occurred about the same time as those to the famous Klondike region in Yukon Territory. Gold continues to be an important base for the mining industry; the greatest productivity occurs near Juneau and Fairbanks, but mining declined in Alaska during World War II and has not yet regained its former importance. Alaska is a producer of some strategic metallic ores of tin, tungsten, platinum, antimony, and mercury, but deposits have not been large enough or good enough to withstand high production costs. Most mines have been operated near the coast, where accessibility has been an advantage in lowering costs. In the interior the coal of the Nenana area along the railway is used, as is coal from Matanuska Valley.

Salmon packing, which began to expand at the turn of the century also, particularly with improvements in the canning industry, is a chief reason for many of the panhandle settlements. Declining catches since 1950 indicate that the fishery here may be overexploited. Salmon are netted and trapped as they approach the coastal rivers in summer. A large number of seasonal workers usually migrate into southeastern Alaska, especially to work in canneries located at Ketchikan, Wrangell, and Petersburg, though the work has become increasingly mechanized. The salmon catch makes up a large part of the value of the fishing industry, but there are also catches of halibut, herring, cod, smelt, and king crab.

Alaska's forest industry has not yet developed to a large scale, though there are several lumber mills on the coast and some processing of pulp at Ketchikan. The largest and heaviest timber stands are in the south-central and southeastern coastal areas. Central Alaska is part of the Boreal Forest region, which covers Yukon Territory and much of north-central Canada. Trees are smaller than those along the coast, and forest cover is interspersed with swamps, muskegs, and small lakes.

With only 25,000 acres of cropland in the entire state, Alaskan agriculture is a minor economic activity. This cultivated acreage, however, is but a small part of the potential arable land. The leading farming regions are in Matanuska Valley and on Kenai Peninsula. The former area, settled in the mid-1930s, is now Alaska's chief agricultural region; its produce is marketed inland along the Alaska Railway and in nearby coastal cities such as Anchorage. Farming is also practiced, but on a smaller scale, in the Fairbanks area of the interior. Agriculture may attract future settlers, but progress in this commercial sphere is slow. To be sure, the coastal regions, with growing seasons of 140 or more days and ample precipitation, experience cool summers and have only small parcels of level ground. Crops can, and do, grow well in Alaska; but the agricultural sections suffer from such economic problems as high production cost and shortage of local markets.

About two-thirds of Alaska's white population is found in the small cities around the Gulf of Alaska coast. Most of these people are urban dwellers, concerned with fishing, transport, defense, local manufactures, and trade. Anchorage, the largest city, is growing rapidly. About one-third of the state's people live in the Cook Inlet region. In the panhandle, the two largest centers are Juneau, the capital, and Ketchikan, each with less than 10,000 inhabitants. Of the people who live in the central part of the state, about two-thirds are near Fairbanks, which is the important transportation hub for the Alaska Highway, the Alaska Railway, and air travel, as well as the leading center of commerce.

Alaska's population almost doubled in the decade 1940–1950; it increased by 30 percent between 1960 and 1970 and now totals almost 300,000—an increase that has been due almost wholly to the movement of population from the northern states. The military population is not included in this estimate. Native population groups have remained fairly stable; in 1970 there were estimated to be 60,000 residents of Eskimo, Indian, or Aleut stock.

Thus Alaska, the largest of the American states in area, provides the greatest extent of undeveloped and unoccupied land in the entire nation. In many ways, its geography and present stage of development suggests the frontiers of North America over a century ago: a widely scattered population of low density, a wealth of natural resources yet to be tapped, and an environment seemingly hostile. Mountain barriers, only recently breached by air travel, impede normal methods of land transport, and forest barriers are nearly as formidable in some areas. Rugged terrain, flooding streams, precipitous coast, and frozen soil, combined with the severe winters of the northern latitudes, must be overcome before this state can produce a successful living for many people. Alaska still has much to offer to the American pioneer.

In Perspective

During the hundred years that followed the winning of political independence, the people of the United States expanded the Republic

Fig. 3-14 Fairbanks, Alaska. Fairbanks, known as the "Golden Heart of Alaska," has a population of over 15,000. It is the most northern city in the United States. (Courtesy of Alaska Travel Division)

to the Pacific Ocean and then, having acquired Alaska and Hawaii, consolidated political and economic control over the nation's 3,615,211 square miles, and developed its widespread resources. The energy of the people went into clearing and plowing land for farms, founding new cities, logging the great forests, discovering and exploiting minerals, building railroads, establishing and expanding manufacturing industries, and conducting the many other enterprises required by the complexities of modern life.

The nation showed small concern for international relations, even though the Monroe Doctrine did oppose further foreign colonization in the Americas. Two World Wars compelled the United States to discontinue a policy of isolation and to concern itself with foreign problems as well as with those of domestic importance. Without deliberate intent, the country has grown to the status of a world power. In doing so, its energies have been directed along different lines during its second century: development of new sources of energy (nuclear power), invention of labor-saving devices, improvement of machinery, expansion of the highway system, and creation of new modes of travel (aircraft). In a word, the United States has advanced from a youthful agricultural country preoccupied with developing its own resources into a mature industrial nation with a sophisticated economy, deeply concerned with the well-being of its own people as well as of those abroad. In its rapid growth period, the United States made serious mistakes. It permitted needless waste of wildlife, excessive soil erosion, deforestation, and wasteful mineral exploitation. Once it had recognized the evils of these injurious practices, however, remedies began to be devised. Efforts are now made to use land more efficiently than in the past.

The Central Lowland and Great Plains states have the highest proportion of land sown in crops. Here are plains with good to excellent soils, adequate rainfall, and a moderately long growing season. No other part of the nation has as much cropland. Improved farm practices, the substitution of tractors for horses, the use of fertilizers, control of plant pests and diseases, and the development of hybrids or other improved plant species help to account for increasing productivity on American farms. Grazing areas and pastureland are concentrated in the Mountain and Pacific states, where inadequate precipitation limits crops mainly to those produced by irrigation or dry farming. Figures indicate that there is enough forestland in the United States to grow most of its timber requirements if the forests are well managed.

The United States is fortunate in the variety and amount of its minerals. Although the nation is the world's largest producer and consumer of bituminous coal and petroleum, at the present rate of usage the reserves of each will continue to meet the needs of the country for a long time. Iron ore is not so readily procured as in the past, but large quantities of some grades are still available. Lead, zinc, copper, and bauxite are mined in various parts of the country, and nonmetallic mineral reserves such as chemical fertilizers, building stone, and sulfur are more than sufficient to meet national demand.

Since 1900, the United States has become the leading industrial nation of the world. The accessibility of raw materials for manufacturing, the ease with which different types of transport could be developed, the large supply of fuel and hydropower at hand, efficient labor, and plenty of capital have accounted for this rapid economic growth. Great industrial centers such as Chicago, Pittsburgh, Cleveland, Detroit, Philadelphia, Baltimore, Buffalo, and St. Louis, as well as numerous smaller cities and towns, have made the Midwest and the Northeast the leading industrial area. New York, Norfolk, Boston, and other Atlantic ports have aided this growth by providing easy access to the world's markets. Smaller industrial areas have developed in the southeast, along the Gulf Coast of Texas and Louisiana, and around the larger Pacific Coast cities. All told, the value added by the processing of manufactured goods in recent years exceeds $250 billion annually.

With a population greater than 200 million persons, the land and other resources of this nation must be used efficiently if the national needs are to be met. More careful use of natural resources and consideration for the environment will give further evidence of the nation's maturity.

REFERENCES

Booth, Charles W.: *The Northwestern United States*, Van Nostrand Reinhold Company, New York, 1971.
Brown, Ralph C.: "Changing Rural Settlement Patterns in Arctic Alaska," *Professional Geographer*, vol. 21, pp. 324–327, September, 1969.
Cantor, Leonard M.: "The California Water Plan," *Journal of Geography*, vol. 68, pp. 366–371, September, 1969.
Durrenberger, Robert W.: *California*, Van Nostrand Reinhold Company, New York, 1970.
Fenneman, Nevin M.: *Physiography of the Western United States*, McGraw-Hill Book Company, New York, 1931.
Francis, Karl E.: "Outpost Agriculture: The Case of Alaska," *Geographical Review*, vol. 57, pp. 496–505, October, 1967.
Lantis, David W., Rodney Steiner, and Arthur E. Kerinen: *California: Land of Contrasts*, Wadsworth Publishing Company, Inc., Belmont, Calif., 1963.
Lowenthal, David: "The American Scene," *Geographical Review*, vol. 58, pp. 61–88, January, 1968.
Morris, John W.: *The Southwestern United States*, Van Nostrand Reinhold Company, New York, 1970.
Paterson, John H.: *North America*, 4th ed., Oxford University Press, London, 1970.
Quinn, Frank: "Water Transfers: Must the American West Be Won Again?" *Geographical Review*, vol. 58, pp. 108–132, January, 1968.
Rand, Christopher: *Los Angeles, the Ultimate City*, Oxford University Press, New York, 1967.
Watson, J. Wreford: *North America: Its Countries and Regions*, Longmans, Green & Co., Ltd., London, 1964.
Weaver, Glen D.: "Nevada's Federal Lands," *Annals of the Association of American Geographers*, vol. 59, pp. 27–49, March, 1969.

4

Canada and Greenland

Canada is larger than the United States, but it has only about one-tenth as many people. Much of this large country is not densely occupied, for Canadians inhabit only the southern parts in significant numbers. About 70 percent of the Canadians live within 100 miles of their country's political border with the United States.

Much of Canada does not have a favorable physical environment for agricultural settlement; but the country's natural resource potential, in terms of minerals, forests, and power sources, has become more apparent in the past few decades. Although many settlements in eastern North America, including the Canadian sections, were established about the same time in the seventeenth century, the physical conditions of the northern country did not attract as

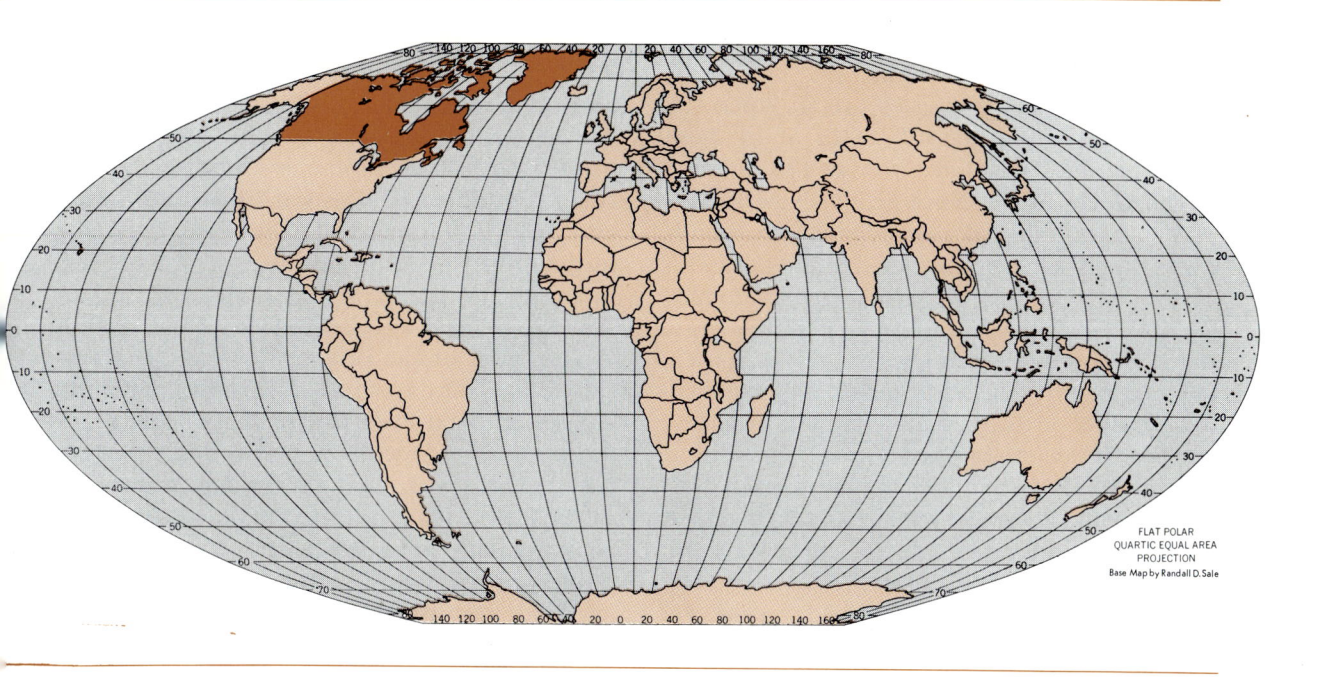

FLAT POLAR
QUARTIC EQUAL AREA
PROJECTION
Base Map by Randall D. Sale

many immigrants as did the areas farther south. Settlement spread across Canada from east to west in the latter part of the nineteenth century, and now resources exploitation is expanding northward. As Canada increased its industrial and manufacturing activities, its population became concentrated in the cities located across the southern sector of the nation. By 1970, more than 75 percent of Canadians lived in urban places—which are defined in Canada as places with more than 1,000 persons.

Canada is a self-governing, independent nation, having a total land area of 3,851,809 square miles, including almost 300,000 square miles of freshwater lakes. It extends northward about 2,800 miles from the tip of southern Ontario, jutting into Lake Erie in latitude 42° north, to the icebound northern coast of Ellesmere Island in latitude 83° north. Canada's width ranges from 3,000 to 3,500 miles, but in longitude its eastern and western tips are about 88° apart, or almost one-quarter the way around the world.

Politically, Canada has a federal government consisting of the union of ten provinces and two territories. The Queen of Britain is recognized as the nominal ruler of Canada; in reality, however, the monarchy (as in the United Kingdom) has no authority. The Queen is represented at ceremonial occasions by a governor-general, who is a Canadian named by the Canadian government. Although the country is a member of the British Commonwealth of Nations, Canada need not agree with the actions or policies of other Commonwealth countries, and each is completely independent in its domestic and foreign relations.

The Canadian provinces are grouped regionally and are usually thought of in these regional contexts within Canada. The Atlantic Provinces consist of Nova Scotia, New Brunswick, Prince Edward Island, and Newfoundland. This last province was added in 1949, having been previously a self-governing dominion in its own right. Central Canada, frequently called "Eastern," includes the more densely populated provinces of Quebec and Ontario. In terms of area, Canada's largest province, Quebec, is twice the size of Texas. The Prairie Provinces of Manitoba, Saskatchewan, and Alberta actually have more forest area than they have prairie grassland; but settlement is mainly across their southern grasslands, and therefore the misnomer has stuck. In the far west, the mountainous province of British Columbia covers as much area as California, Oregon, and Washington together. Northern Canada is sparsely settled; this large area of 1.5 million square miles remains in territorial status, the Yukon and Northwest Territories, governed partly from the federal capital in Ottawa.

Physical Setting

Canada, because of its sizable area, has a great variety of landforms, climate, and native vegetation. Most of the Canadian physical features, especially those south of the 60th parallel, are coextensive with the landforms, climate, and vegetation found in the northern part of the conterminous United States. The physical features of northwestern Canada are continuous with those of Alaska.

Major Landforms

Atlantic Canada, known as the Appalachian-Acadian landform region, is the northern part of the Appalachian Mountains which extend southwest-northeast across eastern North America. The Canadian section has linear hills and low mountains such as the Boundary Ranges of southeastern Quebec and the flat-topped Shickshock Mountains of Gaspé Peninsula, extending northeastward from the Green and White Mountains of New England. East of the linear mountains, a lower region of hills, uplands, and plains is known as the Acadian section. Central New Brunswick is hilly, but the eastern part of the province is a lowland that can be considered a small-scale equivalent of the Atlantic Coastal Plain east of the Appalachians in the eastern United States. Although much of Nova Scotia and northeastern Newfoundland Island is less than 500 feet in altitude, the low coastal areas are rugged and indented with numerous bays and inlets. Areas of level land are small and dispersed. The landform features of the Appalachian-Acadian region are therefore similar to those of New England.

The most important region in Canada is the Great Lakes and St. Lawrence Lowland. Lying between Lakes Ontario, Erie, and Huron, this lowland extends southwestward toward the United States. Although occupying only about one-fiftieth of Canada's area, it has more than half the total population and produces about three-quarters of the total value of Canadian manufactures. The eastern part of the St. Lawrence Lowland, in southern Quebec, is extremely flat. The western part, occupying the peninsulas between the southern Great Lakes, has a rolling surface resulting in part from glacial deposition. This is the "Heartland of Canada," which adjoins a similar core region of importance in the United States.

The Canadian Shield, the largest landform region in Canada, forms a huge horseshoe of almost 2 million square miles of exposed ancient Precambrian rock around Hudson Bay. These worn-down hills and uplands are rough and knobby, but local relief is seldom over 500 to 1,000 feet. There are innumerable lakes of

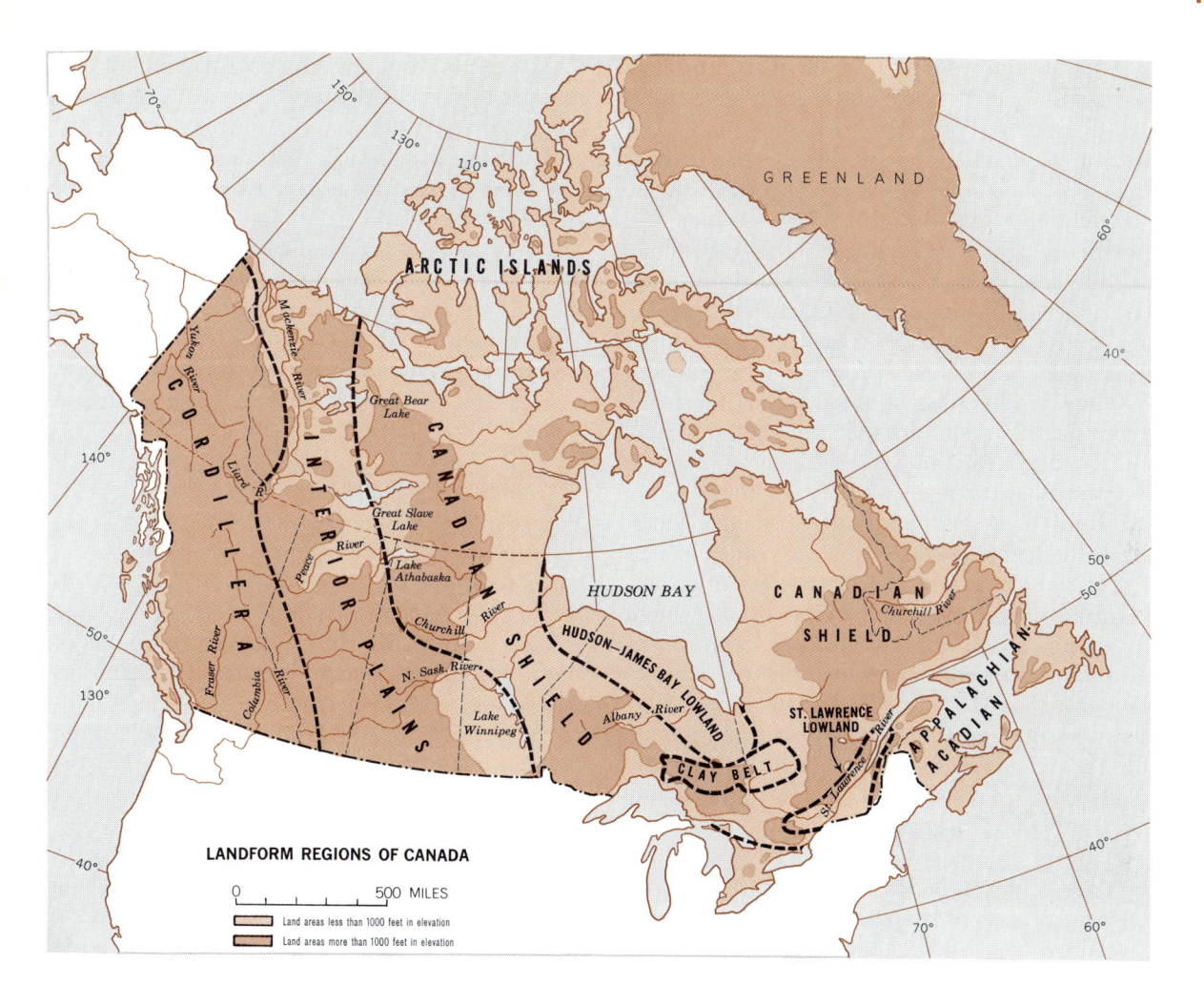

Fig. 4-1 Many of the physiographic regions of Canada are continuous with those of the United States.

glacial origin; the few level areas were the bottoms of former large glacial lakes. Although the Canadian Shield is valuable for its mineral, forest, and power resources, the physical characteristics of rough topography, thin soils, and poor drainage have deterred agricultural settlement and hindered transport.

The Hudson and James Bay Lowland is south of Hudson Bay and west of James Bay. This area of flat-lying, young sedimentary rocks has a different landform character than the old hard-rock hills of the Canadian Shield. The lowland is poorly drained; rivers that have dropped over several rapids while crossing the Shield wander slowly across the boggy surface of the lowland.

The Interior Plains, underlain by gently dipping sedimentary rocks, rise in altitude across the southern part of the Prairie Provinces. The Manitoba lowland is generally under 1,000 feet in altitude, the plains of southeastern Saskatchewan are about 2,000 feet above sea level, and the plateau of southwestern Saskatchewan and southern Alberta slopes upward from 2,500 feet to almost 5,000 feet in the foothills at the base of the Rocky Mountains. The Interior Plains narrow toward the north and are lower in altitude in the level Mackenzie River valley.

The Cordillera region is much narrower and compressed in Canada than in the United States. The linear ranges and valleys are not so wide and the plateaus are smaller than they are to the south. The eastern rampart of the Cordillera is the scenic Rocky Mountains, which terminate in the plain along the Liard River of northeastern British Columbia. In Yukon Territory this eastern wall continues as the Mackenzie Mountains, which curve toward the Brooks Range of Alaska. Along the Pacific Ocean, the Coast Mountains, which are a continuation of the Cascade Range, have some of the most spectacular scenery in Canada; averaging from 6,000 to 9,000 feet, they rise abruptly from sea level above the many twisting fjords that indent the coast. Throughout the interior of the Cordillera, several broad plateaus or basins are incised by deep river valleys.

The Arctic Islands of Canada have very little landform similarity throughout, but they can be grouped together because of their island character. Landform features range in variety from the high, ice-capped alpine peaks of Baffin, Ellesmere, and Axel Heiberg islands to flat, lake-covered lowlands such as those of eastern Victoria, southern Prince of Wales, and King William islands. Many of the Arctic islands are large: Baffin Island is almost as large as the province of Manitoba, or four times the size of New York State; Victoria Island is twice the area of Ohio. The channels between the northern islands are filled with solid or moving sea ice most of the year.

Greenland is a province of Denmark. Although this large island is politically separate from North America, its landform character and ice cap are similar to those of the nearby Canadian Arctic Islands. The east and west coasts of Greenland have high mountains. Filling the space between the mountains, and rising above them in a flattened dome, is the huge mass of ice that constitutes most of the surface of Greenland. Most Greenlanders live on the ice-free strip of fjorded coast in the southwest.

Climate

Canada's large area has a variety of climates. For example, some regions experience continuous cold, whereas others have mild temperatures for most of the year. Similar to the adjoining United States, some sections are very wet and others are dry. There are regions of high heat in summer and, in contrast, ice caps where summer never comes.

The climate of western Canada is influenced by the relatively warm water offshore in the Pacific Ocean. The air masses that pass over the Pacific bring mild temperatures to the west coast throughout the winter, but the water helps maintain cool summers along the coast. During winter, these air masses drop heavy precipitation on the western slopes of the Cordilleran mountains as the moisture-laden air is forced upward to cross them. Climatic conditions in central and eastern Canada are affected by the ice-covered Arctic Ocean, which is a source of cold air masses. In winter the cold air moves southward, spreading over the Prairie Provinces and pushing into the United States. In summer northeastern Canada is cool, owing to the cold water in Hudson Bay and along the Labrador Coast. An Arctic climate, which covers about 1 million square miles of Canada, extends far south of the latitude of the Arctic Circle on both sides of Hudson Bay.

Much of Canada has a continental climate, cold in winter and warm to hot in summer.

Canada and Greenland 107

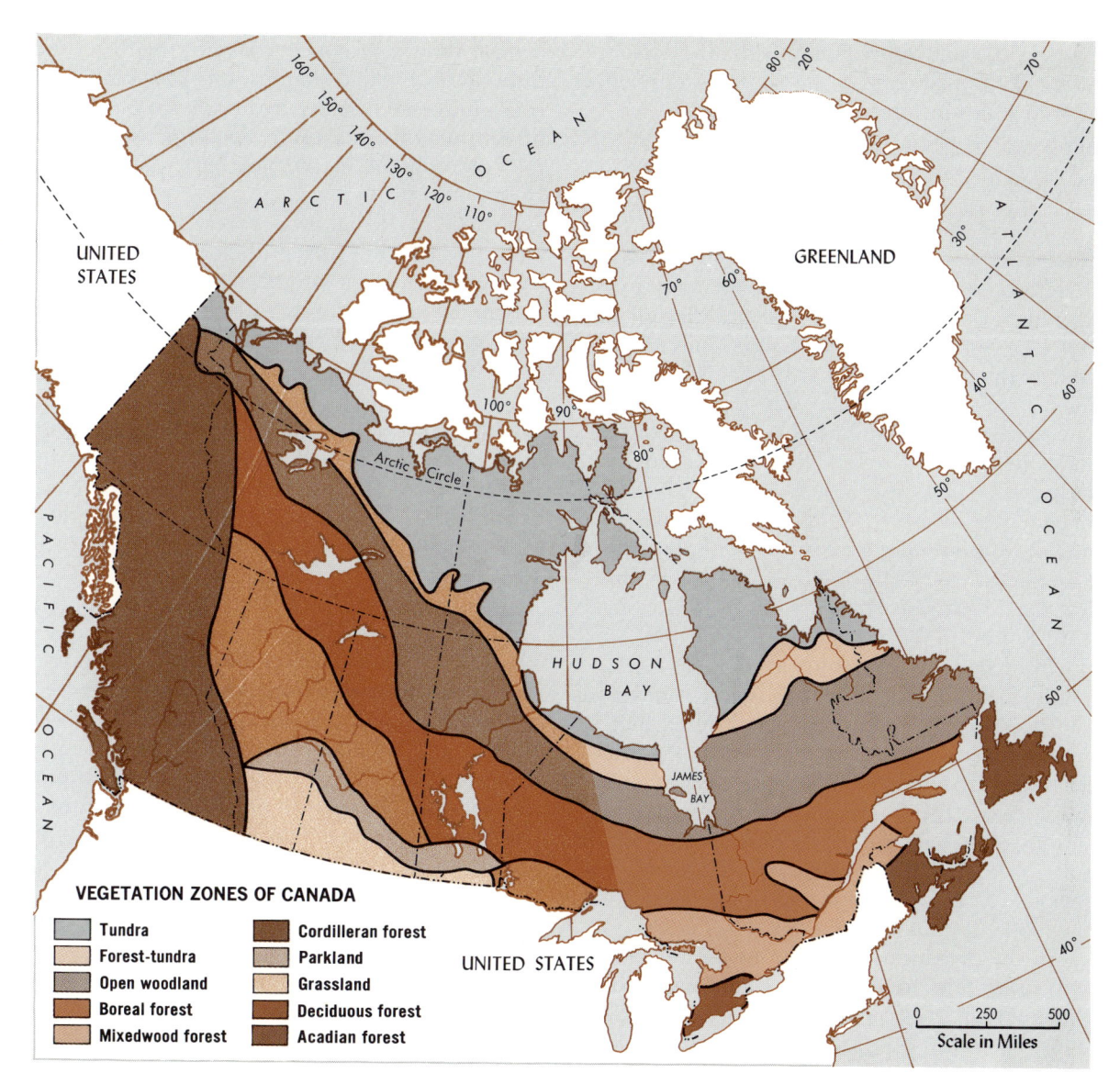

VEGETATION ZONES OF CANADA
- Tundra
- Forest-tundra
- Open woodland
- Boreal forest
- Mixedwood forest
- Cordilleran forest
- Parkland
- Grassland
- Deciduous forest
- Acadian forest

Fig. 4-2 East of the Cordilleran forests the vegetation zones of Canada trend in an east-west direction, as do the climatic belts. As with the landforms, the vegetation zones are continuous with those of the United States.

Since most Canadians live in the southern part of their country, they experience climates quite similar to those of the adjoining United States. The west coast is the mildest region in winter and the wettest part of Canada; the east coast is about half as wet and has colder winters than the west coast. Central Canada is warm in summer; the longest growing season there is found in the most southerly areas of Ontario, which is surrounded by the Great Lakes. Most of

Canada is cold in winter, with the coldest areas of the mainland being those lying northwest of Hudson Bay.

Vegetation

Natural vegetation is a good indicator of climate; thus, a map of vegetation zones is generally suggestive of the climatic regions of Canada. Canada's forested area, although the third largest in the world, covers only 45 percent of the country's total area, that is, about 1.7 million square miles. Because of unfavorable environmental conditions for tree growth, 40 percent of this forested area is covered with trees that grow too slowly to be classed as productive forest. The remainder, about 1 million square miles (or less than 30 percent of Canada's area), is productive forest. This productive forest belt stretches across the central and southern parts of Quebec and Ontario, north of the main agricultural regions.

A small area of deciduous forest, similar to the northern hardwoods of the United States, covered southwestern Ontario when settlers arrived early in the nineteenth century. Because these trees indicated environmental conditions of soil and climate favorable to agriculture, much of the forestland was cleared for farms. Northward, a transitional forest zone has both deciduous and coniferous trees. Because the deciduous trees can be winter-killed, only the more hardy conifers such as spruce, pine, balsam, and tamarack survive in the north. This mixed forest region, close to the industrial centers of Central Canada, supplies much of the nation's woodland wealth.

North of the mixed forest, the Boreal (or Northern) Forest region stretches from the coast of Labrador to northern British Columbia and Yukon Territory. More than 80 percent of Canada's forested area is within this Boreal Forest. It is mainly a coniferous forest, although there are stands of deciduous birch and aspen in the southern sections, notably across the central Prairie Provinces. Within the forest are innumerable lakes, swamps, muskegs, and bare rocky hills—all treeless areas that are the results of continental glaciation. Such open areas are most prominent in the northern parts of the Boreal Forest.

North of the Boreal Forest is the treeless Arctic. The 50° July isotherm, which defines the southern limit of the Arctic as a climatic region, coincides closely with the northern limit of tree growth. North of the tree line, tundra vegetation (mosses, lichens, grasses, low bushes) is seen mainly on the lowlands; most of the rocky hills are barren of vegetation. Another treeless area, a small block of grassland south of the Boreal Forest in the Prairie Provinces, is the result of low annual precipitation. This grassland zone is the northern end of a larger grassland region in the United States.

The vegetation zones of British Columbia are small and complex as a result of that area's variety of climate and landforms. Vegetation there has an altitudinal pattern, with grasslands in the valleys, tall straight conifers on the lower slopes, and smaller conifers on the steeper upper slopes. Many of the mountain ranges, particularly northward in Yukon Territory, rise above the tree line and thus add to the amount of treeless area in Canada.

Soils

Because agriculture has been confined to the southerly parts of Canada, many of the northern soils have not yet been mapped in detail. The grey-brown forest soils in the Great Lakes and St. Lawrence Lowland of Ontario and Quebec are the best agricultural lands of eastern Canada. These soils have developed from a variety of glacial materials, and thus their exact character varies within small areas. Much of the remainder of eastern Canada has poor, acid, and leached soils called "podzols." Soil cover is thin or lacking in parts of the Canadian Shield where there are bare rocky ridges and many lakes, swamps, and muskegs. The northern parts of the Shield and the Mackenzie Valley are underlain by permanently frozen subsoil ("permafrost"), which increases drainage problems and makes the soils cold for root crops. Despite the generally unfavorable environment for agriculture, there are still a few million acres of arable land unoccupied in eastern Canada. The problems of expanding the agricultural use of the soil are related more to economic and cultural conditions than to the natural environment.

The utilized soils of western Canada form a semicircular pattern that coincides closely with the grassland and parkland vegetation regions. Most of the soils here are the same as those extending in a longitudinal arrangement across the Great Plains of the United States. The brown soils are found in the center, in the driest parts of southern Alberta and Saskatchewan; the dark-brown soils lie around them; farther outward are the black soils, which receive more effective precipitation. North of a line running from Edmonton to Prince Albert, the grey forest soils are poorer. Although crops can be grown on these soils in some places, agriculture has penetrated these areas very slowly; at present, it appears more economical to intensify the use of lands closer to the markets of southern Canada.

The soils of the Cordillera are complex, as might be expected in a mountainous region. The useful ones are quite limited in area and are confined to narrow river valleys as alluvial, floodplain, or delta deposits.

Population Distribution

Canada's population of about 22 million is spread thinly across the southern part of the country. More than half of these people are concentrated in the small area in southern Ontario and southern Quebec where most of the original settlements on Canadian territory were located and prospered.

More than 2 million persons live in the four Atlantic Provinces. Because of limited economic opportunities on the eastern edge of Canada, population increases there have been siphoned off to industrial Ontario, the agricultural West, or the Northeastern United States. About 90 percent of the people of Newfoundland live on or near the coast. The island was settled by fishermen, and the inhospitable interior has not attracted other occupations. There is only one large city, the port and capital of St. John's. The rest of the population is scattered along the coast in hundreds of small towns and villages called "outports." Labrador, which politically is part of Newfoundland, is sparsely occupied, with about 15,000 people in an area of 112,000 square miles.

Nova Scotia also has a coastal pattern of settlement, but here a variety of occupations such as manufacturing and agriculture brings a higher level of prosperity than that found in Newfoundland. New Brunswick has a peripheral settlement distribution, with the greatest concentration along the St. John River valley in the western interior. Prince Edward Island, Canada's smallest province, has had a virtually static population of about 100,000 persons throughout this century. Most of the island is fully occupied by agricultural people.

Most of the 6 million inhabitants of Quebec are descendants of the 60,000 original French settlers who lived in the area in 1763. Most live in the St. Lawrence Lowland, with smaller populated pockets to the northeast in the Saguenay Valley and farther north in the Clay Belt. The center of the southern Quebec population core is Montreal, Canada's largest city, having about one-tenth of Canada's total population. Many comparisons can be drawn between the geographic positions and economies of Montreal and New York. As in the rest of Canada, rural population has declined

Fig. 4–3 Politically, Canada is divided into ten provinces and two territories. Most of the country's larger towns and cities are within 100 miles of the Canadian–United States boundary.

steadily in Quebec, where more than 75 percent of the people now live in urban areas.

Ontario, with about 7 million people, has a greater population than any other Canadian province. As in Quebec, the southern one-tenth of the province holds about 85 percent of these people. Whereas the people of Quebec are descendants of French settlers, the forefathers of Ontario residents came mainly from Great Britain or from Western and Central Europe. After most of the good soils were occupied in the last century, people began concentrating in the cities to process the agricultural resources of southern Ontario and the forest and mineral resources of the Canadian Shield to the north. Besides comprising the country's largest urban area, Greater Toronto and a group of adjoining cities around the western end of Lake Ontario have Canada's largest concentration of industrial and commercial occupations. Much of northwestern Ontario, like most of central and northern Quebec, is still virtually unoccupied by white settlement.

Almost 4 million persons occupy the southern parts of the three Prairie Provinces. The settled zone is narrow in Manitoba and broadens toward the west. Most of Manitoba's population is concentrated in the narrow belt between Lake Winnipeg and the United States border. More than one-third of the province's people live in Greater Winnipeg, Canada's fourth largest city. The remainder are farmers who live on large grain or livestock farms to the west, distributed across the fertile black soils. The northern two-thirds of Manitoba, lying in the forested, lake-covered Canadian Shield, is sparsely occupied.

Saskatchewan's population, at one time mainly wheat farmers, increased very little from 1931 to 1951 but began growing in the 1960s following diversification in the province's resource base. The rural population, occupying large, mechanized grain farms, spreads across the black and dark-brown soils, having an average density of about five to eight persons per square mile. The discovery of petroleum, natural gas, and potash and the increased use of irrigation have brought significant changes in the Saskatchewan economy.

The main belt of population concentration in Alberta extends south-north from Lethbridge through Calgary to Edmonton. An outlying pocket of settlement is growing to the northwest in the Peace River area. As a result of oil and natural gas discoveries in the 1950s, Alberta had the highest percentage of population increase among the Canadian provinces in the period 1951–1966. The two largest cities of Alberta, Edmonton and Calgary, have growing industrial and commercial functions, and they promote a friendly rivalry over which shall be the larger.

British Columbia ranks as Canada's third province in population, but most of its land area is not occupied. About 75 percent of its population of more than 2 million persons is concentrated in the southwest sector, in the urban area of Greater Vancouver, the nearby agricultural lowland of the Fraser River, and on the southeastern shores of Vancouver Island. The rest of the province's population is scattered along the southern interior linear valleys. Although the cities of the interior are growing, as a result of expanded natural resource development, the greatest population increases have come to the mild and scenic southwestern corner of British Columbia.

The two northern territories are inhabited by only a few thousand people. Yukon Territory, a vast region of 207,000 square miles, has only about 20,000 persons in all, about one-third of whom live in or near the territorial capital, Whitehorse. In the Northwest Territories, most of the 15,000 non-native inhabitants live in a few small towns in the Mackenzie river valley. Although Canada is sometimes cited for its northward movement of population, the numbers involved in this trend are still small; most of the nation's economic expansion continues to be in the southern parts, or across "middle" Canada. In Northeastern (or Arctic) Canada, there are about 17,000 Eskimos; most of them live within the Territories, but some inhabit the coast of northern Quebec. These migratory people are spread thinly along the coasts of the southern Arctic Islands and the adjoining mainland, with an average density of one Eskimo for about every 100 square miles.

Atlantic Provinces

Physical Features

Although the landforms of Canada's east coast region can be generally characterized as hilly, there is some variety in the landscape. Northwestern New Brunswick, for example, is an upland basin partly surrounded by low mountains; it slopes downward to the northwest, through a gap in the Appalachian chain of the Gaspé Peninsula. This pass is an important part of a transport route between New Brunswick and the densely settled lowlands of southern Quebec. From rugged, forested central New Brunswick, rivers radiate to the flat lowland along the east coast or drop westward into the St. John river valley. Only a few land transport lines penetrate into these central highlands. On Prince Edward Island most of the lowland has been cleared of forest for agriculture; but across Northumberland Strait, in eastern New Brunswick, most of the plain remains forest-covered. This contrast, that is, the different use of the same general environment, suggests that cultural or economic reasons may be the cause of the differences. Relief is not great in the rocky, rough upland of interior Nova Scotia. The indented coastline, with its many sheltered harbors, looks similar to the coast of New England. Newfoundland Island is an elevated plateau, which is highest toward the west, where the Long Ranges have altitudes of over 2,000 feet. Much of central Newfoundland is a barren upland, sloping to an island-fringed, submerged northeast coast. About 40 percent of the island's fishing population is scattered in small villages along this indented northeast coast.

The marine climate of the coasts of the Atlantic Provinces is characterized by mild winters, cool summers, and ample rainfall; but the interior parts have a continental climate, with cold winters, warm summers, and a higher percentage of snowfall. In winter, average monthly temperatures are mildest in southern Nova Scotia, owing to the influence of the warm Gulf Stream offshore. There, harbors do not freeze over; and thus Halifax and Saint John are important winter ports for Eastern Canada. Northwestern Newfoundland and Labrador have cold winters, which are affected more frequently by the cold air masses from interior and northern Canada than by the warm ocean currents. The frost-free season averages about 120 days in some valleys of the Maritime Provinces, but many of the uplands have a marginal climate for agriculture, with less than 100 days without frosts. Daytime temperatures in midsummer are frequently above 80°, particularly in western New Brunswick, but summer days are cool along the coast.

Resources and Economic Development

Although there are broad similarities in the physical environment of the four Atlantic Provinces, the local regional differences produce distinct variations in the major natural resources and occupations of the people in each section. In all of the region, however, economic development suffers from its geographic position on the eastern edge of Canada. Its own population is too small and dispersed to be an adequate market, and it can export little to the industrial centers of Central Canada.

In Newfoundland, fishing has long been the major occupation, but it is steadily declining.

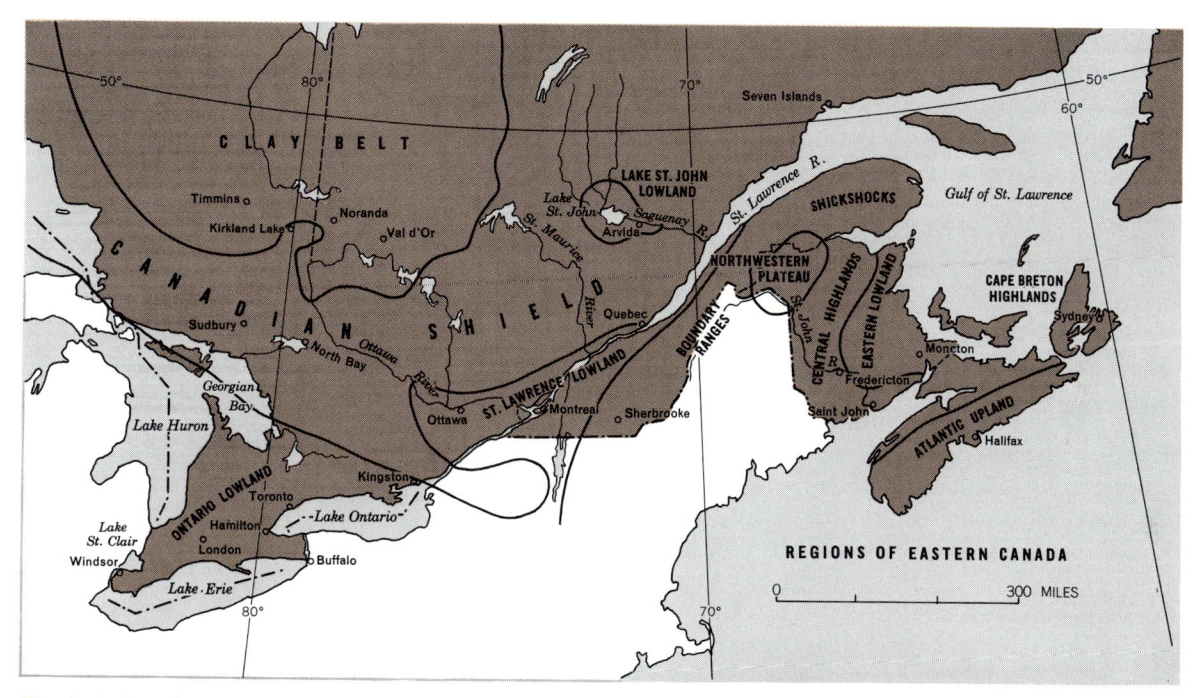

Fig. 4-4 The St. Lawrence Lowland and the Ontario Lowland are the two most intensively developed regions in Canada.

Throughout the history of settlement on the island, its people have had to make a living from the sea, with its plentiful fish. Many sections of the interior are barren, soils are poor, and the summers are too cool for good crop growth. Fishing villages dot the bays of the northeast coast and Avalon Peninsula; nets are set close to shore off the many headlands or small neighboring islands. In the small villages, the catch of cod was cured by traditional family methods, dried or salted, but produced a low-quality product and low incomes for the fishing people. The government is assisting and encouraging the centralization of fishermen around new fish-filleting and -freezing plants, which improve the quality of the export product. Although there are definite economic advantages to larger and centralized fishing settlements, equally important is their provision of social, cultural, and educational amenities that were not available in the scattered, tiny outports. Some fishermen, with new trawlers and draggers, now fish off the south coast of Labrador in summer or go out with the commercial fishing fleets of other nations on the Grand Banks, southeast of Newfoundland. In the past, lack of capital and adequate vessels prevented Newfoundland fishermen from taking a larger share of the catch of the Grand Banks.

The mineral wealth of Newfoundland has been only partly explored. One of the oldest working mines, that which produced iron ore on Bell Isle in Conception Bay, closed in 1966 after more than a half century of operation. Lead and zinc are mined in central Newfoundland, and copper and asbestos come from small mines on the northeast coast. Newfoundland's chief mineral production comes from the rich iron ore deposits of the Ungava-Labrador region, shared with Quebec. These will be discussed below, in the section dealing with the Canadian Shield.

Newfoundland has a coastal population-distribution pattern. The only railway serves the pulpwood areas of the interior but not the coastal people. The first road across the island, completed in the 1960s, also tended to be an inland route, but later branch roads were built to coastal settlements. Many people living on the northeast and south coasts formerly had to rely mainly on summer water transport for supplies. In the last century, more than half of the population of Newfoundland lived in the island's principal port and capital, St. John's. Although still the largest city of the province, and also diversifying its new industry and commercial functions, St. John's now has only about one-quarter of the total provincial population.

Nova Scotia was settled by both fishing and agricultural people in the latter part of the eighteenth century. Many of the farm clearings on the poor soils of the uplands were abandoned in this century. Present agricultural land use is chiefly for hay and pasture, supplying the basis of a livestock and dairy industry. The major agricultural region, the Annapolis-Cornwallis Valley in western Nova Scotia, emphasizes the growing of apples, along with mixed farming. The apple producers, once solely dependent on the export market to Britain, have turned more to processed apple products for the Canadian markets. Agriculture is now becoming more diversified, and there has been considerable increase in such specialties as poultry and dairy farming.

The sheltered harbors of an indented coastline, combined with some of the world's richest fishing banks offshore, were long ago natural attractions to Nova Scotia fishermen. As in Newfoundland, most of the fish catch formerly came from inshore waters along the coast, but now many modern schooners and draggers put out from Nova Scotia to go into the international waters of the Sable Island Banks. An increasing amount of fish is filleted and frozen into blocks for export to the New England fish-stick processing plants. As in present-day Newfoundland, the small outlying fishing villages of Nova Scotia are disappearing. Modern fishermen live in larger towns and cities located near the larger processing plants.

As in the other Atlantic Provinces, mining is not a significant part of the Nova Scotia economy, except for the coalfields of Cape Breton Island. Bituminous coal has been mined in the Sydney–Glace Bay area from seams that slant down under the sea. As in other parts of Anglo-America, coal mining has become a declining activity; a government investigation recommended that many of the subsidized mines be closed. Despite the tidewater location, high production costs have limited the export possibilities of Cape Breton coal. The chief local use of the fuel is in the steel industry at Sydney; however, coal now competes poorly for both home and industrial use with the increasing quantities of imported petroleum.

Nova Scotia has a more balanced economy than does Newfoundland. The province has a broader agricultural base and makes fuller use of its better forest resources for lumber, pulp, and paper. Also, there is a significant development of manufacturing in the large cities such as Halifax. Some manufacturing, based on imported raw materials, is the result of coastal location. Nova Scotia has a higher percentage of urban population than any other of the Atlantic Provinces, and Halifax is the largest city of the region. If the Halifax-Dartmouth area continues to grow and increase its commercial, governmental, and service functions, many of the advantages of a large central market will affect other segments of the Nova Scotia economy in a manner similar to the influence of metropolitan Boston on the New England economy.

The economic development of New Brunswick comes chiefly from its land resources. The long, fertile valley of the St. John River is a major agricultural region of the Atlantic

Provinces. The green of hay and pasture dominates the usual summer landscape, reminding one that dairying or the sale of livestock products is the chief source of agricultural income across southern New Brunswick. The upper St. John Valley specializes in potato production, partly for export, and adjoins the noted Aroostook potato region of Maine. Potato production in New Brunswick was previously hindered by a lack of processing plants; now that these have been built, quantities of potato products are shipped to the urban markets of Central Canada.

About 80 percent of New Brunswick is under forest cover. Forest production, used mainly for pulp and paper, is the leading export from the province. Mills are located at the mouths of rivers emptying into the Gulf of St. Lawrence along the northeast coast, as well as in the upper and lower St. John River Valley. For many of the subsistence farmers of the northeast coast and for part-time fishermen, additional income is obtained by cutting pulpwood in the forest during the winter.

The herring fishery of the Bay of Fundy supports several canneries in southern New Brunswick. The fishing villages here are quite similar to those on the Maine coast. Many of the fishing villages along the east coast are engaged in lobster trapping in Northumberland Strait. Most of the east-coast fishermen are French-speaking Canadians who are descendants of the original Acadian French who settled in the Moncton area.

The port of Saint John is the largest city of New Brunswick. Its manufactures are based partially on local agricultural and forestry resources, but also on imports such as petroleum and sugar. Fredericton, the provincial capital, is an administrative and educational city, and also the distribution center for the central St. John River region.

Prince Edward Island is sometimes called the "Garden Province." Not only is it small in size, but much of it is cleared and fully used as

Fig. 4-5 Potato field in bloom, Prince Edward Island. Over 50 percent of the island, which forms the smallest of the Canadian provinces, is under cultivation. Potatoes are the most important crop. (Courtesy of National Film Board of Canada)

farmland. Farming is a major occupation of the region's people, but there are also a few fishing villages along the coasts. The rolling fields of hay, pasture, and oats are green in summer and contrast with the distinctive red soils. The sale of livestock products, including butter, cheese, and hams, is the chief source of agricultural income. In addition, the island exports its disease-free seed potatoes to the southern states of the United States and to the Caribbean region. Charlottetown, the capital, is the only city on the island.

The Atlantic Provinces have many characteristics in common with the adjoining New England states. In both regions, forests dominate the landscape, agriculture is mainly practiced in narrow strips along the roads, fishing villages are dispersed along the indented coast, and mining is becoming of less significance in the economy. The main difference is the lack of manufacturing in the Atlantic Provinces, in comparison with the industrial centers of New England. A major reason for this difference in economy is the location of New England near the sizable densely populated area of the northeastern United States, whereas the Atlantic Provinces are separated from the consumer markets in the heartland of Canada by appreciable distances.

The Great Lakes and St. Lawrence Lowlands of Quebec and Ontario

Physical Features

The most important lowland in Canada lies between Lakes Huron, Erie, and Ontario and along the St. Lawrence River. Its landforms are similar to those of the Central Lowland of the United States, south of the Great Lakes. There are also many similarities in the cultural landscape on both sides of the international boundary.

The eastern section of the St. Lawrence Lowland, lying mainly in southern Quebec, is flatter than the western part. The Quebec lowland was formerly the bottom of an arm of the sea that extended into North America when the land was depressed at the end of the Ice Age. The chief topographic features are the line of Monteregian hills jutting above the plain east of Mount Royal (Greater Montreal). The part of the lowland in southern Ontario is more rolling in conformation. The Niagara escarpment, for example, rises several hundred feet above the southwest shore of Lake Ontario; the Niagara River tumbles over it at scenic Niagara Falls. North of London, Ontario, this east-facing escarpment is crowned by several hundred feet of glacial debris; the rolling upland is the highest part of southwestern Ontario.

Southwestern Ontario has the highest monthly mean summer temperatures (70 to 75°) of eastern Canada. On many days in the summer, maximum temperatures above 90°F are recorded. The average frost-free period of 160 to 175 days in the flat southwestern peninsula is the second longest in the nation. Because of this long growing season, crops that are not found elsewhere in eastern Canada can be grown there. Winters are mild in southwestern Ontario and resemble those of western New York state. Severe cold, such as below-zero temperatures, occurs only a few times during the usual winter. Average temperatures decrease toward the northeast, where the lowland of Quebec has a January mean of about 10°F. Cold spells, when temperatures drop to −20 or −30°F, are possible during most winters there. Therefore, about half the Canadian population, those living in southern Ontario and Quebec, experiences winter conditions which are little different from those of nearby northeastern United States.

Resources and Economic Development

Although manufacturing is the main basis of the population concentration in the urban centers of the Great Lakes and St. Lawrence Lowlands, behind this important activity is a well-developed, prosperous agriculture. Much of the region was occupied by agricultural settlers by about 1850; most of the forest was cleared and the land in cultivation before 1900. With favorable level to rolling topography, a long frost-free period, and sufficient precipitation, a diversified agriculture developed. As manufacturing and the urban population increased, agriculture gained the further advantage of a large local market.

The average Quebec lowland farm is an example of a mixed farming economy with a dairy emphasis. Most of the land is used for hay, pasture, and oats and some potatoes. Near the back of the long, narrow farms lies the woodlot, from which maple syrup and sugar are obtained each spring and pulpwood may be harvested. Near the large cities, as in other parts of Anglo-America, dairying and truck gardening are the major agricultural occupa-

Fig. 4-6 Montreal, Quebec, located on the St. Lawrence Seaway, has a metropolitan population of almost 2.5 million. It is both an important manufacturing and distribution center. (Courtesy of Montreal Tourist Bureau)

tions. The frugal "habitant," with his large family, describes the characteristic Quebec farmer of forty years ago, but not the modern, mechanized dairy farmer of the Montreal plain.

Southern Ontario has a greater diversity of agricultural production than does Quebec. Over most of the Ontario lowland, dairying is the chief agricultural occupation; large dairy barns and circular silos dot the rural landscape. Because the southwestern part has a climatic advantage over the rest of the lowland, crops ripen there two to four weeks earlier. Corn and soybeans are major crops, similar to the nearby American Corn Belt. A large part of Canada's vegetable crop, grown in this area, is raised for large canning companies. The area near Leamington, for example, has Canada's largest concentration of tomatoes and glass greenhouses.

About 90 percent of Canada's tobacco is grown on the sandy soils of the central north shore of Lake Erie. In this area farms average about 30 acres in tobacco, but they also grow hay, feed grains, and strawberries. Many of the farmers were Central European immigrants who entered Canada after World War I and worked as tenant farmers until they could save enough money to buy their own farms.

Along the north side of the Niagara Peninsula, lying between Lakes Erie and Ontario, is one of Canada's three major fruitgrowing areas. At the base of the Niagara escarpment, miles of well-kept orchards and vineyards cover the lowland east of Hamilton. This area is Canada's main source of peaches and grapes, and it grows more than half of the country's pears, cherries, and plums. Conflicts and choices of land use are common around most of the expanding cities of Anglo-America, and these decisions have become critical factors in the good soil areas of the Niagara Peninsula. Will the urban land use already occupying much of the land around the western end of Lake Ontario soon replace the Niagara orchards and vineyards?

Mining is relatively unimportant in the Ontario and Quebec lowlands. Salt, one of the region's few mineral resources, is mined from the very large reserves found under southwestern Ontario and eastern Michigan. Some of the first petroleum wells in North America came into production near Sarnia, Ontario, and the region still produces some oil and natural gas. Limestone, quarried at several places across the lowlands, is used by the chemical and fertilizer industries as well as for building material.

The economic development of southern Quebec and Ontario has been greatly aided by the Great Lakes and St. Lawrence River system, which serves as a transportation artery and a source of electrical power. The Great Lakes themselves are used by a large fleet of elongated lake freighters designed to fit the locks at the Soo (Sault Ste. Marie) and Welland canals, but these are not suitable for ocean travel. These freighters carry mainly iron ore and wheat cargoes down the lakes during the nine-month navigation season. The return journey to Duluth-Superior or Thunder Bay, formerly known as Port Arthur–Fort William, may bring some coal or manufactured products. To Canada, the chief function of the Great Lakes transport route is to bring wheat from the southern Prairie Provinces to the flour mills of southern Ontario and Quebec, or to the grain elevators along the St. Lawrence River and east coast for export shipment by ocean vessels. Although there was increased overseas traffic to Toronto and Hamilton, on Lake Ontario, after the St. Lawrence Seaway opened in 1958, the bulk of the ocean carriers go to U.S. ports on Lake Erie and Lake Michigan. By making the St. Lawrence River navigable west of Montreal, however, it became possible for the iron ore from Quebec-Labrador to be taken upstream to the steel mills of the Lake Erie region.

The Great Lakes–St. Lawrence system is a major source of electric power for the Canadian heartland and nearby New York state. The large power plants at Niagara Falls and Cornwall are international developments that supply electricity to industries and urban residents on both sides of the border. The great volume of the St. Lawrence River also produces electric power at Beauharnois for Montreal. Each of these power plants produces more than 2 million horsepower; thus their electric power output is comparable with that of Hoover or Grand Coulee dams in the United States.

Cities in the lowlands of southern Ontario and Quebec produce about three-quarters of the total value of manufacturing in Canada, and the largest share of this industry is located in Ontario. The manufacturing concentration of the region is similar, therefore, to that which lies south of the Great Lakes in the United States. The Ontario urban industries produce a wide range of consumer goods, similar to the output of the nearby American Midland cities. In addition, however, some cities are well known for particular manufactures because they make the biggest percentage of certain products. Such examples include iron and steel (Hamilton), automobiles (Windsor and Toronto), nonferrous smelting and refining (Montreal), petroleum refining and chemicals (Sarnia), rubber goods and furniture (Kitchener), cereals and machinery (London), pulp and paper (Trois-Rivières), and book publishing, electrical goods, and agricultural machinery (Toronto). Because many Canadian manufactures are protected by tariffs, American companies have built branch plants in Canada. Much Canadian industry and business is therefore international, and the degree of control over sectors of the Canadian economy by American businessmen has become a matter of serious concern to many Canadians.

In the Great Lakes and St. Lawrence lowlands live more than half the population of Canada. In this region are forty-five of the eighty cities with more than 30,000 population in Canada. Greater Montreal, with some 2.5 million persons, has more inhabitants than all of British Columbia. Greater Toronto, also

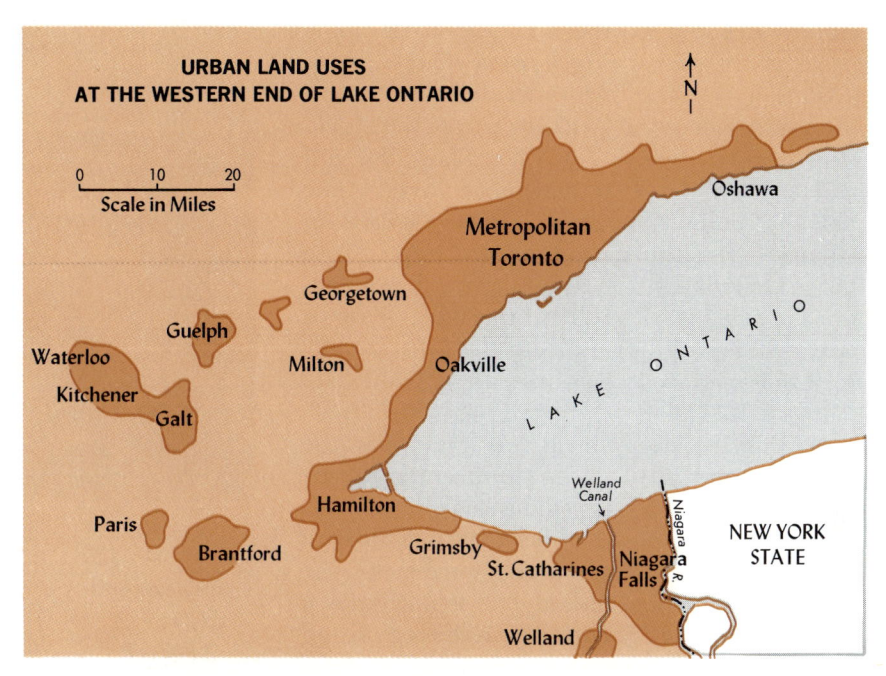

URBAN LAND USES AT THE WESTERN END OF LAKE ONTARIO

Fig. 4–7 That part of Canada lying north of Lake Erie and between Lakes Ontario and Huron is rapidly becoming the industrial heart of the nation. This is the most urbanized area in the nation.

with 2.5 million persons, has more residents than the three Maritime Provinces combined. Although some of the other large cities are located on the Great Lakes or along the St. Lawrence River, owing to original settlement along this route, many Ontario cities have grown up inland, where they have developed excellent road and rail connections. An apparent geometric arrangement of cities can be observed, with the larger cities spaced almost equally apart and the next-lower order of cities set equally between the large ones. This roughly hierarchical pattern of urban distribution suggests the importance of the surrounding agricultural hinterland in relation to urban growth in the past. This geographic pattern of city location is similar to that of the Midwestern United States.

In Ontario and Quebec about 80 percent of the population lives in urban centers. As in the Eastern United States, the large Canadian cities are growing rapidly both in numbers of residents and in area. "Suburban sprawl" is a problem for city planners on both sides of the international boundary. Just as the cities from Boston to Washington, D.C., are gradually merging into one large contiguous urban complex, so is the region at the west end of Lake Ontario becoming a unified large metropolitan concentration, growing outward from Toronto, Hamilton, Brantford, Kitchener, and St. Catharines.

The Canadian Shield

Physical Features

The Canadian Shield extends in a huge semicircle around Hudson Bay. The upwarped eastern section in northern Labrador has alpine peaks and glaciated ridges rising sharply above a fjorded coastline. Altitudes decrease toward the south, where the southern rim of the Shield rises steeply 1,000 to 2,000 feet above the gulf and estuary of the St. Lawrence River. The Shield's southern edge is lower westward in Ontario. Because the drainage divide is close to the Great Lakes, the longest rivers flow northward to shallow, cold Hudson Bay. East of Great Slave and Great Bear lakes, the knobby rock hills are again characteristic, with altitudes of about 1,500 feet.

Within the large Shield two level areas, the Clay Belt and the Lake St. John Lowland, are distinct landform subregions. The former region became known as the Clay Belt early in this century, before soil surveys were made; the name has remained although it is now known that clay soils are not common in the region and that there is much peat. Only part of this level area is potentially arable, owing to a short frost-free period, prevalence of poor soils, and inadequate drainage.

The climate of the Canadian Shield is the climate of almost half of Canada. In winter, cold high-pressure air masses may stagnate over this region for many days at a time, and minimum temperatures may drop below −50°F. Only the southeastern part, from southern Labrador to Lake Superior, has a January average above 0°. Although winters are cold, this does not prevent work from continuing in the region's mines and forests, and comfortable homes in the new resource-based towns are warmly heated. Summer temperatures are warmest in the southern Shield valleys, into which warm air masses from the Great Lakes region may penetrate. The northern sections—both east and west of the cold water of Hudson Bay—have an Arctic climate, with no month averaging above 50°F. Most of the southern Shield has, on the average, 100 days or more without frosts, but local sections north of Lake Superior appear to be frost pockets where the

Fig. 4–8 Canadian Shield, Northwest Territory. This glaciated area is characterized by low rocky hills, many small lakes, and stunted forests. (Courtesy of R.C.A.F.)

average frost-free season is less than two months.

Average annual precipitation decreases to the west and north. The north shore of the Gulf of St. Lawrence receives about 40 inches, much of which is snow. The snowy hills of the southern Shield, just north of the population centers of the Quebec lowland, are popular for winter sports and resorts. The Northwest Territories receive approximately 10 inches of precipitation annually, because the cold air that passes over the northern Shield contains very little moisture, and there are no topographic barriers to cause the air to rise and cool in summer.

Resources and Economic Development

Many of the 2 million people living in the Shield are directly or indirectly employed in the extractive industries of mining or forestry. Others are engaged in transporting the raw materials to the industrial cities to the south.

The Shield is Canada's "storehouse of minerals." As it is explored and mapped geologically, minerals have become one of the bases of the optimistic, expanding economy of Canada. Only a few of the Shield mineral deposits were exploited early in this century, and these were mainly on the outer edges of the south-central section. A wide variety of metallics are now being mined, and roads and railroads have penetrated far into the rugged region. Numerous new place names that have appeared on the map of Canada denote well-planned, modern mining communities, so different from the rough "boomtowns" of the last century.

Iron ore is the chief mineral produced in the eastern section of the Shield. The long belt of iron-bearing rocks extending across the Labrador (Newfoundland)–Quebec boundary became significant because of the approaching depletion of the high-grade Lake Superior ores of the United States. Once the Schefferville iron deposits were explored and a railway built to bring out the ore, it became economical to develop other lower-grade ores in that area. Another rail line, to Gagnon, brought more iron ore to new port facilities on the Gulf of St. Lawrence. As technology changed, it became possible to think of mining lower-grade iron ores and exporting these in concentrated pellet form. The result was another new community on the "mining frontier"—the adjoining Labrador City and Wabush in western Labrador. Ore is shipped from the Gulf of St. Lawrence ports to the Lake Erie steel mills, to the eastern United States, and to northwestern Europe.

The oldest mining district is in the south-central part of the Shield. Copper-nickel ores were being mined at Sudbury in the last century, but the first great "rush," in the years 1904–1912, was to the silver deposits and goldfields of Cobalt, Timmins, and Kirkland Lake in northeastern Ontario. Since that time, dozens of gold mines have been discovered and exploited, and then died; production at the remaining gold mines is subsidized by the Canadian government. These early mining towns developed technology, commercial facilities, and transport—all of which were available for use when other minerals were discovered after 1950. Thus, although the gold mines are closing, base metals such as iron, molybdenum, and other metallics have brought new life to the old mining core. Eastward in Quebec, the gold and copper mines of Noranda and Val d'Or were opened in the 1930s. After 1950 the mining frontier extended to the northeast, where new railroads now bring out base metals from Chibougamau and Matagami. By 1968, of the 120 mines operating in the Canadian Shield, 60 were concentrated in the south-central section, on both sides of the Quebec-Ontario border.

North of Georgian Bay, near Sudbury, several mines account for about half of the world's supply of nickel. These mines are also Canada's main source of copper, a by-product of the nickel ore, and they produce a large share of the world's platinum and palladium. The brown and barren rock hills around Sudbury, devoid of vegetation because of the fumes from the smelters, make this old city one of the less attractive mining centers. Canada's dominance of world nickel production was further increased in the 1950s and 1960s when new deposits were discovered at Lynn Lake and Thompson in northern Manitoba. These towns, and others like them, are examples of the modern, planned communities which now dot the once-empty Canadian Shield. All have modern housing, shopping centers, and curving streets, surrounded by the coniferous forests and set amid the many lakes and rocky hills.

The Lake Superior region in the United States was noted for its iron ore early in this century, but it was not until the 1940s that high-grade iron was mined on the Canadian side of the lake. When iron deposits were discovered beneath Steep Rock Lake, for example, the lake was drained so that the ore could be mined in large open pits.

The northwestern part of the Shield was scantily prospected prior to 1940. The oldest mines, producing copper-gold-zinc, were at

Flin Flon, on the south edge of the Shield and directly on the Manitoba-Saskatchewan boundary. The community was later given new life when several more mines were opened in the region in the 1960s. Flin Flon, like Cobalt fifty years ago, became the commercial and supply center for the expanding mining region. Only the outer edge of the Shield has been carefully explored; thus Canadians can be optimistic that the future will reveal mineral wealth in the Precambrian rocks of the northwestern Shield equal to that now being extracted in the eastern Shield.

Forests are another natural resource of the Canadian Shield. Much of the southern part is densely forested; but in the north, where the climate is more severe in summer, the trees become smaller and less numerous. This Boreal, or Northern Coniferous, Forest is Canada's chief source of pulpwood logs, but it also supplies some lumber to the cities of Ontario and Quebec. Although numerous large and modern pulp and paper mills seem to be scattered across the southern Shield, there are usually specific reasons for their exact location. Many are at the mouths of rivers down which logs are floated, and where they may take advantage of great quantities of available waterpower. The pulp mills turn the forests into newsprint paper, which often ranks as Canada's most valuable export, going mainly to the United States. It has been estimated that three out of every five newspapers in the world are printed on Canadian paper.

The greatest concentration of pulp and paper mills in the world is within an arcing belt extending from Lake St. John and Saguenay Valley along the St. Lawrence River to the Ottawa River. These large mills form a line along the contact zone between the Shield and the St. Lawrence Lowland; they get their raw material, water, and power from the Shield and utilize the transport and markets of the Lowland. Within the Shield, only a few of the northward-flowing rivers that are crossed by rail transport are utilized. The western part of the Shield, although forested along its southern edges, is little utilized for the paper industry because it is farther from markets in Midwest United States. Although the pulp mills themselves employ relatively few people, cutting of pulpwood is a major source of income for many persons throughout the Shield and provides supplemental winter income for some subsistence farmers.

The Canadian Shield has the greatest amount of developed and potential waterpower of any landform region of Canada. Its large size and a number of environmental advantages account for the waterpower resources. Precipitation is adequate throughout the southern sections, with a good percentage being snowfall, which becomes available in spring runoff. The poorly drained surface of the Shield, with its innumerable, interlocking lakes, supplies excellent reservoirs. The southward-flowing rivers have a notable drop, or "fall line," where they spill over the edge of the Shield to the Great Lakes or St. Lawrence Lowland. To these natural advantages must be added the fact of location with regard to the population concentrations of eastern Canada and the power demands of industries of the St. Lawrence Lowland.

Almost half of the developed waterpower of Canada is in Quebec, which produced more than 18 million horsepower in 1970. Two rivers, the Ottawa and the Saguenay, produce more than 4 million horsepower each. The dams and power plants of the Saguenay region, for example, supply the huge aluminum plant at Arvida and several pulp and paper mills, and they export surplus power to Montreal and the Quebec lowland cities. An outward-moving pattern of hydropower production has evolved along the outer edge of the Shield in Quebec. The St. Maurice River, closest to Montreal, was developed for power prior to 1910; the Saguenay Valley, within transmission range of Quebec City, expanded production prior to 1940; to the east, the Bersimis River was generating 2 million horsepower in the 1950s; the harnessing of the Manicouagan-Outardes rivers will produce 6 million more horsepower in the early 1970s. Beyond these enormous power developments lies the power of Churchill Falls on the Churchill (formerly Hamilton) River of Labrador. One of the reasons why it was possible for the economy of Quebec and Ontario to change from an agriculture-dominated economy to a modern industrial base was the continued availability of relatively cheap power from the rivers of the Shield. The Shield sections of Manitoba and Saskatchewan have many potential power sites, but less development because the mining and pulp industries are not found here to the same extent as in Ontario and Quebec. The largest power plants are on the Winnipeg and Nelson rivers. As Canadian exploitive industries and settlement push northward into the Shield, one of the major resource advantages is the almost unlimited supply of hydroelectric power, which is also well dispersed.

Agricultural settlement has been of less significance in the developing economy of the Shield region. The region has few areas of level to rolling topography or of good soils and favorable climate. These physical requirements are found mainly in some of the former glacial lake bottoms in the southern part of the region. The Clay Belt has the largest area of potential agricultural land, only a small part of which is occupied. Subsistence farmers moved into the Clay Belt and Lake St. John region in the early part of this century. In the Ontario section of the Clay Belt, farm abandonment has become common and "pioneering" on the land is no longer done. In Quebec, settlement was more compact, largely centered in parish villages, with community life and facilities that encouraged the hardy French-Canadian pioneers to remain in the region. Because Canada has a surplus of food-producing land closer to the large urban markets, it is probable that many of the favorable lands of the level parts of the

Fig. 4–9 Pulp and paper mill, Kapuskasing, Ontario. Logs may be brought to the mill by water, rail, or highway. Kapuskasing is a planned small city. (Courtesy of Ontario Department of Travel and Publicity)

Shield may be best utilized in the future by the growing of forests.

Population

Many parts of the Shield are completely unoccupied, and some sections are inhabited by only a few hundred migratory Indians who still follow a hunting and trapping way of life. Rural population is sparse and is decreasing, whereas a high percentage of settlers live in urban centers across the Shield. Since most of these cities are new, they have a pleasant, clean appearance; stores are modern, and homes are neat and well-built. There is little of the "frontier" character in Shield cities, although they are the "new north" for eastern Canada. These towns and cities are located along the few major transport lines that function within and across the region. Most of the settlements are supported by the extractive industries of mining and forestry, but there are also towns which are transportation hubs and supply centers or which are located at waterpower sites. For example, there are the mining cities of Sudbury, Timmins, and Kirkland Lake; the transport cities of Thunder

Bay, Sault Ste. Marie, and North Bay; and the diversified cluster in the Saguenay–Lake St. John Lowland, including Chicoutimi, Kenogami, Jonquière, and Arvida.

The southern Shield is "the North" to most eastern Canadians. Much of the construction related to Canada's expanding resource development has been taking place there. New mining towns dot the once-empty land of lakes and rocky hills; pulp mills harvest the almost endless crop of coniferous forests; new power plants and dams have tamed the wild rivers; farm clearing is slowly pushing back the forests near some of the cities; and clear lakes and cool forests attract thousands of summer tourists and visitors from the crowded cities of southern Canada and nearby United States. Just as Canadians moved westward at the turn of the century, now they are looking toward the north —toward the natural riches of the Canadian Shield.

Interior Plains

Physical Features

Although the Interior Plains are frequently dismissed with the word "flat," there are several landform features that break the monotony of the flat areas. Three landform regions—the Manitoba Lowland, the Saskatchewan Plain, and the Alberta Plateau—cross the southern plains and increase in elevation from east to west, from about 800 feet near Lake Winnipeg to above 4,000 feet in the Rocky Mountain foothills.

All of the Manitoba Lowland was covered by ancient glacial Lake Agassiz. The Red River now meanders slowly across the deep clay deposits of the former lake bottom. Much of the northern part of the lowland is still occupied by the three large lakes of southern Manitoba

Fig. 4-10 Winnipeg, Manitoba, is often referred to as "the Chicago of Canada." This marshaling yard is largely controlled by a computer system. (Courtesy of Canadian National)

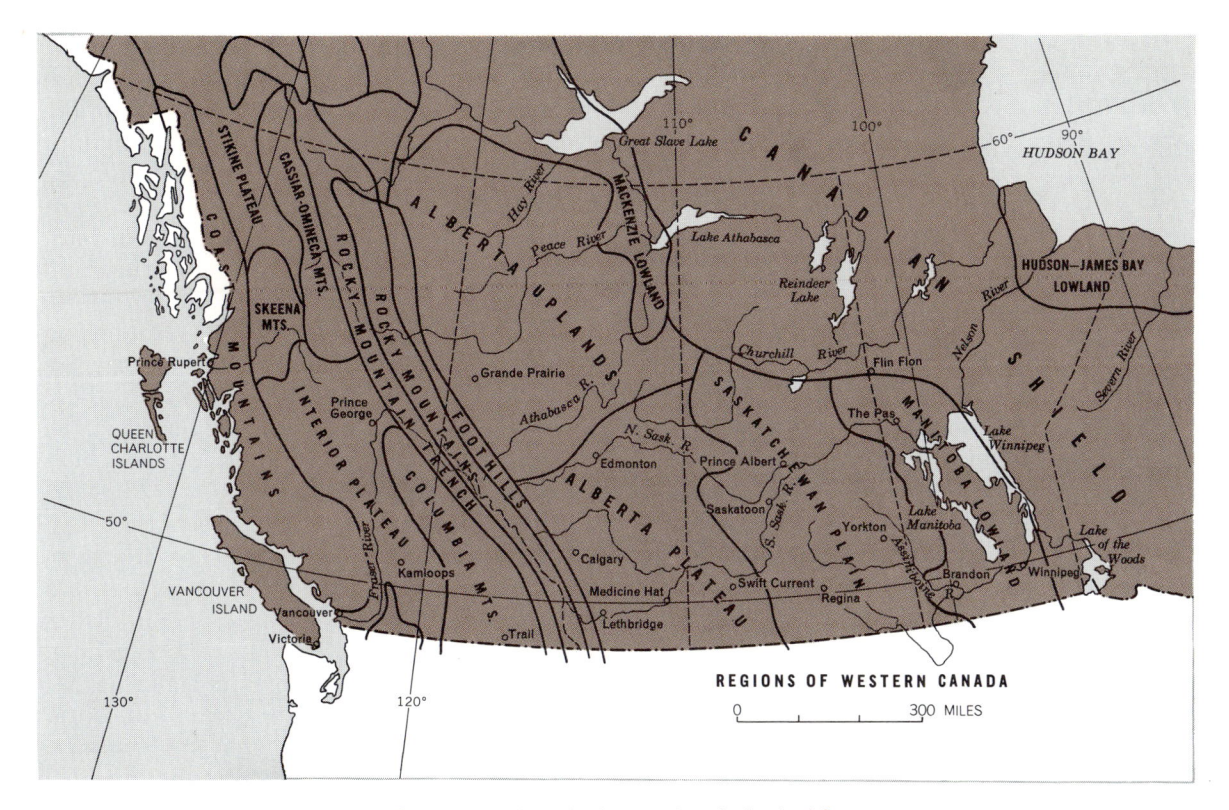

Fig. 4-11 The regions of western Canada are largely the results of physical factors, among which the mountain influence is dominant.

(Winnipeg, Winnipegosis, and Manitoba). The soils between these lakes are thinner and stonier than the fertile farmland of the Red River plain. To the west an abrupt geological escarpment forms the western boundary of the lowland. Above it the gently rolling Saskatchewan Plain has some extremely flat areas, such as the former glacial lake bottom now comprising the Regina Plain. The northern half of this plain is forested and has a much different landscape aspect from that of the southern prairies.

The Alberta Plateau has level to rolling topography similar to the Saskatchewan Plain, but altitudes are higher. There is a minor, low geological escarpment, called the Missouri Coteau southward in the United States, which is used as a landform boundary between the two regions; but in reality, the plain and plateau merge, with the Alberta section having higher elevations. There is greater dissection in Alberta, where the rivers have cut down 100 to 200 feet into the glacial drift and sedimentary bedrock. Above the Alberta Plateau rise flat-topped hills of more-resistant rock, such as the Cypress Hills (4,700 feet). To the north, the Alberta Uplands have more hills than the southern plateau. Drainage is toward the northeast, through the Athabasca, Peace, and Hay rivers toward the Mackenzie Valley.

Summer monthly temperatures, which average 65 to 68°F, have little variation over the large area of the Interior Plains. Although these averages are similar to those of the Great Lakes region, they do not indicate that some searing hot days in the Prairie Provinces are moderated by cool nights. The longest frost-free period, about 120 days, on the Red River plain of Manitoba and in the area along the South Saskatchewan River in Alberta, permits these regions to grow a wider variety of crops than the rest of the Interior Plains. Most of the grassland across the Interior Plains has an average frost-free period of more than 100 days, which is generally sufficient for the maturing of grain.

The distribution pattern of average monthly winter temperatures in the Interior Plains shows a decrease from southwest to northeast. Southwestern Alberta, warmed by occasional Chinook winds, has a January average of 10°F. In general, winter weather conditions of clear, cold days and occasional blizzards are similar to those experienced in the adjoining states of Montana and North Dakota.

The Interior Plains receive very little precipitation, and the amount varies widely from year to year. The highest annual precipitation, about 20 inches, falls in southern Manitoba and along the foothills of southwestern Alberta. Average precipitation decreases toward the central plains, to a minimum of about 12 inches along the South Saskatchewan River. Seasonal distribution is more important to agriculture than annual precipitation: in most areas, about 80 percent of this falls during the six spring and summer months from April to October.

Resources and Economic Development

Agriculture has been the major occupation of people living on the Interior Plains. Although there has been a general relationship between vegetation regions, soil zones, and agricultural economies on the Interior Plains, these relationships are breaking down as a result of changing agricultural techniques and practices. Prairie agriculture has become more diversified and can no longer be classified simply as a grain economy.

Cattle ranching is the characteristic land use across the brown soils of southwestern Saskatchewan and southeastern Alberta. Although these areas have an adequate frost-free period, precipitation of about 12 inches is marginal for agriculture, particularly when below-average rainfall may be received for several years. Wheat is grown by strip-farming methods within the grassland region, more in some years and less in others, thus indicating that the land-use boundary between grain growing and ranching cannot be defined. Although beef cattle are noticeable across the semiarid grasslands, there are larger numbers of cattle supported in the mixed-farming and feed-grain regions near Edmonton and Winnipeg.

Irrigation is making some parts of the grasslands productive. The large dams and associated irrigation aqueducts, mainly in southern Alberta, are extending cultivation into lands that formerly had little use. Sugar beets, alfalfa, and vegetables are the chief irrigated crops, which support increasing densities of rural population. It is estimated that in all about 3 million acres could be irrigated in the southern Prairie Provinces, of which about 1 million acres is now served by irrigation projects. Compared with the 60 million acres now under cultivation (including fallow), irrigation acreage remains a small percentage of the total farmland.

Hard spring wheat has been the dominant crop on the dark-brown soils, sometimes occupying 70 percent of the cultivated acreage. These large, rectangular wheatfields cover much of south-central Saskatchewan. As a result of mechanization, farms are increasing in size; at the same time rural farm population is declining, owing to the decreased need for hired labor. Because neighbors are far apart in this flat lonesome region, where farms may cover a full section of 640 acres, many farmers have moved to the towns to live during the winter. Depending upon weather conditions in a particular summer, the plains wheat region may produce 400 to 700 million bushels annually. Because very little of their crop is consumed

Fig. 4-12 Elgin, Saskatchewan. On those parts of the Great Plains where wheat is dominant, elevators line the railroads of each community, regardless of its size. (Courtesy of Canadian National)

within the sparsely populated Prairie Provinces, the region is one of the world's important food-surplus areas.

The black soil zone grows smaller amounts of wheat, more feed grains for increasing numbers of livestock, and increased acreage of oilseeds. Precipitation is more effective in the black soil zone, as indicated by the parkland vegetation of tall grasses and trees. Farmhouses are not so far apart as in the wheat regions, and barns for the livestock are characteristic of the rural landscape. Numerous small villages, with one or two blocks of wooden stores, are dispersed at fairly regular intervals throughout this mixed-farming area.

The grey soils of the forested parts of the Prairie Provinces are being penetrated slowly. These poorer soils, time and cost of clearing, and lack of transportation have all been factors in delaying settlement in these zones. Land use generally emphasizes feed grains and pasture, as well as some legume or rape seed. The northern fringe of agriculture trends northwestward from Lake Winnipeg to the Peace River area of British Columbia. This demarcation line was fairly well established by about 1940, and only a little new clearing has taken place in the last thirty years. Much of the northern fringe now has well-established farms with permanent buildings, and there is little of the pioneer atmosphere that prevailed forty or fifty years ago.

The sedimentary rocks of the Interior Plains yield different types of minerals from those found in the old hard rocks of the Shield. Petroleum, natural gas, and potash are the most spectacular of the fuel and mineral resources. The first big oil fields were discovered in the Edmonton area in 1947 and 1948. In the next twenty years, an outward-moving pattern of oil-field discoveries extended the producing areas to the northwest, beyond the Peace River area and into northeastern British Columbia. A network of pipelines transports this oil and natural gas to southern Ontario, the U.S. Midwest, southern British Columbia, and the adjoining Pacific Northwest states. Although a small petrochemical industry has developed in Edmonton and a great deal of surplus sulphur became available by extraction from the natural gas, the large reserves of power and fuel in the Prairie Provinces have not yet attracted many industries away from the industrial heartland of southern Ontario. Known reserves of more than 11 billion barrels of petroleum assure production for many years for both a Canadian and an export market. Production is controlled and limited by government regulation; each well has a quota, and there is a maximum permitted for each province. Further reserves are available to the north in the Athabasca Tar Sands. These sands, which cover about 20,000 square miles along the Athabasca River near Fort McMurray, contain a petroleum reserve estimated at about 200 to 300 billion barrels—a quantity about equal to the world's present known reserves of liquid petroleum. Large-scale production from the Athabasca deposits awaits better marketing conditions.

Southern Alberta and Saskatchewan have coal reserves of more than 75 billion tons, chiefly of bituminous grade. Production is on a minor scale, however, because these coal deposits are too far from the markets of southern Ontario and locally can barely compete with petroleum and natural gas as a fuel source. The discovery of large deposits of potash beneath southern Saskatchewan has helped to diversify the economy of that region. There is enough potash available there to supply the world with fertilizer for many years, but Saskatchewan production is curtailed by a quota system agreed upon in order not to close the potash mines of the southwestern United States.

The Interior Plains region is changing from a dominant agricultural economy to one where manufacturing is more significant. As the local urban population increases, there is more manufacturing and processing of consumer goods for these cities. Although industries related to petroleum refining and natural gas purification are important, the most valuable industries are those related to the region's agricultural production. Meatpacking plants, flour and feed mills, and butter factories are major employers in each of the large prairie cities.

Despite an expanding resource base, the Interior Plains remain sparsely populated, and large areas in the northern sector are empty. Most of the agricultural regions average five to eight persons per square mile, and there are only a few large cities. Winnipeg, with one-third of the people of Manitoba, is the largest of the prairie population centers and has a wide range of manufacturing industries. Its strategic transportation position in the center of southern Canada, in the so-called "bottleneck" south of Lake Winnipeg, brings all major east-west transport lines through the city. Edmonton and Calgary are headquarters for the expanding Alberta oil industry, but each city also has a productive agricultural hinterland. Edmonton has an important geographical position as the main supply and distribution center for northwestern Canada. Regina and Saskatoon, the chief cities of Saskatchewan, are supply and service centers for the grain-growing farmlands of that province.

The Cordillera of British Columbia and Yukon

Physical Features

The Cordilleran mountain system is compressed in British Columbia into a width of about 600 miles of spectacular, rugged grandeur. There are high mountains on the east, rising abruptly above the Interior Plains; plateaus lie in the central sections, and another line of mountains runs along the fjorded west coast.

The Rocky Mountains, the eastern range of the Cordillera, form a continuous wall of sharp peaks and ridges broken by only a few passes. Some of the most spectacular mountain scenery in North America is found in Banff and Jasper National Parks in the Canadian Rockies. Several peaks in the range have altitudes of more than 10,000 feet; the highest is Mt. Robson (12,972 feet). The Rocky Mountains decrease in altitude northward and terminate in the broad plain of the Liard River in northeastern British Columbia. Another mountain system, the Mackenzie Mountains, offset to the northeast, arcs along the Yukon–Northwest Territories boundary. The western boundary of the Rocky Mountains is the Rocky Mountain Trench, one of this continent's outstanding topographic features. The trench is a linear, flat-bottomed valley some 1,200 miles in length, extending from south of Flathead Lake, Montana, to the Liard Valley of southern Yukon.

West of the trench lie two mountain systems that are distinct from the Rockies. In the south, the Columbia Mountains consist of four linear, sharp-crested ranges; to the north, the slightly lower, more-rounded Cassiar-Omineca Mountains extend into the Yukon Territory. West of

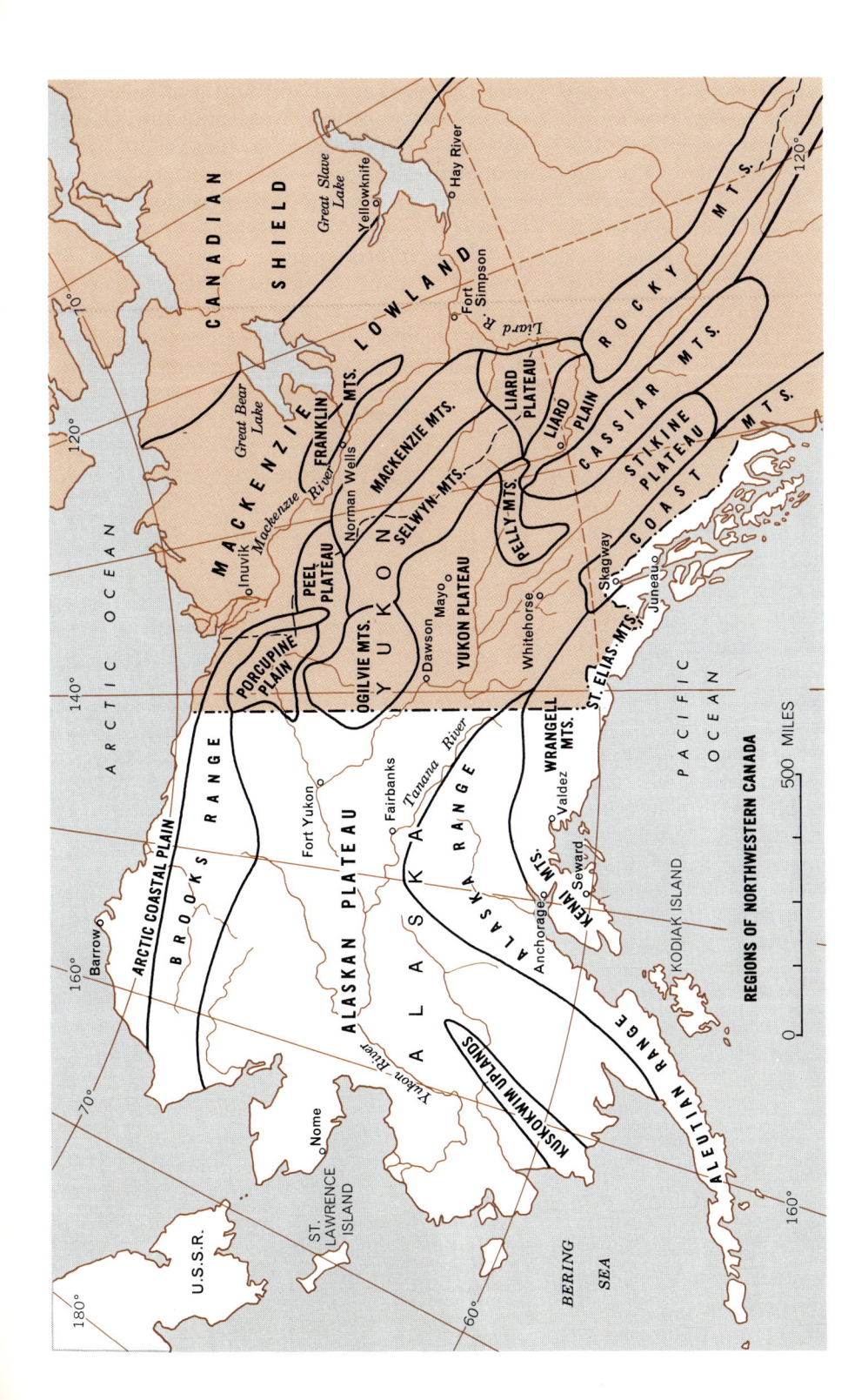

Fig. 4-13 The rough topography of the Yukon Territory tends to make many elongated regions. Such topography is a hindrance to the development of transportation.

these mountains, the Interior Plateau is narrow in the south and entrenched by the Fraser and Thompson rivers; it then broadens to the north into a rolling upland. The Yukon Plateau, about 2,000 feet above sea level, is incised by broad, flat-bottomed valleys and ringed by high mountains. The ice-capped St. Elias Mountains, southwest of the Yukon Plateau, rise above the Alaska coast and are crowned by the highest mountain in Canada, Mt. Logan (19,850 feet).

The Coast Mountains rise in rows of jagged peaks in southwestern British Columbia. Several peaks are over 9,000 feet (Mt. Waddington, 13,260 feet) and tower over an indented, fjorded coast. The Coast Mountains are less of a wall on the north, where they are broken by the Skeena, Nass, and Stikine rivers. There are extensive ice fields on the northern ranges. The Insular Mountains are submerged ranges on Vancouver Island and the Queen Charlotte Islands, off the west coast of British Columbia. This mountain system continues northward in the form of many rugged islands off the Panhandle of Alaska.

Because the Cordillera of British Columbia and Yukon has an area of almost 600,000 square miles, rugged topography, and a latitudinal expanse of 20°, a wide variety of climates exists. Climates range from the mildest part of Canada in the winter, in southwestern British Columbia, to the cold Arctic coast of northern Yukon Territory. The area has the wettest station in Canada, on western Vancouver Island, where an average annual precipitation of 250 inches has been recorded, and the driest station in southern Canada, only 250 miles eastward in a valley of the Interior Plateau.

Winters are mild on the coast, averaging about 35°F in January, but cold temperatures are common inland, where the Interior Plateau and its linear valleys are open to cold air masses from the north. In summer the marine influence maintains cool temperatures in the coastal strip, but the southern interior valleys can become quite hot; summers are slightly cooler in the central and northern valleys. In mountainous British Columbia the climatic statistics usually refer to the valleys where people live. There are greater temperature variations in summer with altitude than with latitude.

Winter precipitation is heavy on the exposed Canadian west coast, which is struck by many storms from the North Pacific. Most coastal stations average 100 inches or more. A dry summer, however, is characteristic of the southern coastal section, which comes under the protective high-pressure air masses that also cover northern California. Because of the rain-shadow effect of the Coast Mountains, interior British Columbia and central Yukon receive little precipitation, except on the western slopes of mountains.

Resources and Economic Development

Forestry is the most valuable primary industry of the Cordilleran region. The mild, wet climate of the coast section has endowed the area with some of the world's largest trees, such as Douglas fir, western cedar, and hemlock. The lumbering economy that characterizes Washington and Oregon extends northward into British Columbia, making this area one of the world's major sources of lumber. Prior to 1940, most of the British Columbia lumber industry and its large sawmills were concentrated on eastern Vancouver Island, the southwestern coast near Vancouver, and in the lower Fraser River valley. By 1970, however, almost half the province's lumber output was being cut from the smaller trees of the interior forests, such as those of Kamloops, around Prince George, and in the Kootenay area of the southeast. The pulp and paper industry has expanded on the Pacific Coast, as part of an integrated forest economy, partially using former waste material from the lumber industry. Since 1960, new pulp and paper mills have been built in the interior, so that the whole provincial forest economy is now integrated. Much of the coast forest production moves by water transport to the eastern United States, whereas interior wood products move by rail to the Prairie Provinces and to eastern Canada and the U.S. Midwest.

Mining is characteristic of most mountainous areas, and the Canadian Cordillera has revealed some of its mineralization. At Kimberley, in southeastern British Columbia, one of the world's large lead-zinc-silver mines has produced as much as 10 percent of the world's annual lead output. Concentrates from this mine are transported by rail to the smelter and refinery at Trail, on the Columbia River, near hydroelectric power. Other mines in the southeast produce a variety of base metals for the Trail smelter. Copper and iron are the major minerals mined in southwestern British Columbia; much of this production is exported to Japan.

The famous Klondike gold rush of 1898 was the reason for the establishment of the Yukon Territory in the Canadian Northwest. Goldseekers moved northward along the Inside Passage steamer route to Skagway, Alaska, and climbed through passes in the Coast Mountains to reach the headwaters of the Yukon River. Within a decade, however, most of the easily accessible gold had been found and removed in the Dawson City region and the boom was over. By 1965 most of the dredges that worked over the alluvial gravels of the Klondike River had ceased operations. Greater mineral wealth now comes from rich silver-lead deposits near Mayo and from base-metal reserves at Faro on Ross River. Improved road transport throughout the southern Yukon permitted several mines to achieve lower production costs and to begin concentrating ores in the late 1960s.

Although fishing was a reason for the many small settlements dispersed along the northwest coast early in this century, most of the canneries are now concentrated near the mouths of the Fraser and Skeena rivers. Owing to increased use of packer vessels, which transfer fish from the specialized fishing boats, the central coast canneries near the fishing grounds are not needed and have been closed; the processing industry has been centralized near the world shipping facilities of Vancouver and Prince Rupert. As in the Panhandle of Alaska, several species of salmon provide the most valuable catches. Starting in midsummer in Alaska, and by late summer in southern British Columbia, the salmon arrive off the river mouths to begin their migration upstream to spawning grounds. The Nass, Skeena, and particularly the Fraser, which has the largest drainage basin, are the chief salmon rivers.

As a mountainous province with heavy rainfall on the coast, it is not surprising that British Columbia has potential waterpower resources larger than the state of Washington. Because rainfall is seasonal and many of the coastal rivers are short, storage of water is a major problem. Although potential power sites are well scattered throughout the province, much of the developed power was first produced in southwestern British Columbia, where most of the people and industries are concentrated. One of the major hydropower developments of the 1950s was built at Kemano-Kitimat on the central west coast. The headwaters of the Nechako, a tributary of the Fraser River, were dammed and diverted through the Coast Mountains by tunnel to Kemano, to produce power for a large aluminum smelter built at tidewater at Kitimat. Similar diversions of the Yukon River headwaters to the heads of coastal fjords are possible in northwestern British Columbia and the Panhandle of Alaska, but such development is stalled because of the political problems created by the international use of this water. Similar political controversies complicated the development of storage, or power, dams on the headwaters of the Columbia River in southeastern British Columbia until 1965. The resulting storage of Columbia River water behind dams in Canada increased power generation at hydro-

Fig. 4-14 Victoria, the capital of the province of British Columbia, is located on Vancouver Island. (Courtesy of Government of British Columbia)

plants downstream on the Columbia in Washington state. As industrialization and urban power demand in Vancouver increased, British Columbia had to turn to the development of the waterpower of the Peace River in the northeast.

Agriculture in this part of Canada is confined to the narrow valleys, and as a result less than 1 percent of British Columbia is under cultivation. Because so much of the province is steep and mountainous, probably less than 5 percent of the land is arable for future agriculture. Half of British Columbia's cultivated land is in the southwest, across the Fraser River delta and along the east coast of Vancouver Island. Most of its crops are for the nearby urban market, and these include dairy products, vegetables, and small fruits. In the Okanagan Valley of the southern interior, irrigation is used to supply water to most of Canada's apricot crop and more than half the apples. This narrow valley ranks second to the Niagara peninsula of Ontario as Canada's other source of cherries, peaches, and pears. The largest areas of level to rolling land and potential arable land are located across the railway belt of central British Columbia, centering on Prince George, and in the well-established grain farms of the Peace River area, east of the Rockies. However, both regions have development problems arising from a short frost-free season and their long distance from large markets.

British Columbia's population increase, on a percentage basis, has been among the highest of the Canadian provinces—similar to the population influx to the Western states in the United States. Most people have concentrated in the southwestern corner of the province where there is a mild climate, some good agricultural land, nearby forest resources, fisheries, and excellent ports. Half of the province's 2 million people are resident in the Greater Vancouver area, which includes New Westminster on the Fraser River. This group of cities produces a wide range of manufactured products, among which those related to wood are most valuable in the local economy. Vancouver has become the financial and commercial center of British Columbia; the head offices of many provincial resource-based companies are located in the many high-rise buildings of the city's downtown core. Victoria, the second city of the province and its administrative capital, located on southern Vancouver Island, is home for many elderly Canadians who retire to its mild climate.

Although the Yukon Territory had 30,000 people at the height of the gold rush at the turn of the century, its population was down to about 5,000 in 1940. Almost one-half of the present population of about 20,000 lives in or near the crossroads town and army base of Whitehorse, situated on the Alaska Highway and at the head of the rail line from Skagway, Alaska. The historic Canadian mining center of Dawson is now almost a ghost town, with less than 1,000 persons.

Northwest Territories

Physical Features

Since the Northwest Territories form approximately one-third of Canada in area, one might expect a variety of landforms. The Mackenzie River lowland is a flat, poorly drained plain of glacial deposition, into which the wide Mackenzie River has cut down about 100 to 200 feet. To the east, bare rocky hills of the Canadian Shield rise steeply above the east sides of Great Bear and Great Slave lakes. West and northwest of Hudson Bay a poorly drained, lake-dotted lowland, with a mixture of water-laid and glacial-drift deposits, extends inland mantling the rocky hills of the Canadian Shield.

The Arctic Islands have no landform uniformity. The upwarped, eroded edge of eastern Baffin Island rises to ice-capped peaks of 7,000 to 8,000 feet; similar altitudes, the highest in eastern North America, are reported in the linear, folded mountains of northern Ellesmere Island. To the west some of the islands are low, barren plateaus that sometimes rise directly from the sea. The flat, lake-covered lowland in the central Arctic is only a few hundred feet above sloping postglacial beach ridges.

Although all of the Northwest Territories is cold in winter, there are wide differences in summer conditions. The Mackenzie River valley has a subarctic climate, with relatively short, warm, summers; the rest of the Territories, in contrast, has an Arctic climate, entirely without summers. The coldest monthly averages occur northwest of Hudson Bay on the mainland and in far-northern islands. The greatest extremes of cold have been recorded in the subarctic Mackenzie Valley, where −60 to −70°F temperatures are recorded in many winters. The coastal weather stations in the Arctic Islands, somewhat "moderated" by cold water around them, have had no recorded temperatures below −63°F.

In summer the subarctic Mackenzie Valley warms, and its July monthly average of 60°F is equal to temperatures experienced much farther south, for example, in the Clay Belt of Ontario and Quebec. Daytime temperatures frequently rise into the 80s. In the Arctic region to the northeast, however, July averages are cool and remain under 50°F. Although these coastal weather stations may be receiving twenty-four hours of daylight in early July, the air is usually cooled by ice floes that are still melting in the channels of the Arctic Islands. Contrary to a popular impression that the Far North is a land of deep and everlasting snow, most of the Northwest Territories receives less than 10 inches of annual precipitation, and about half of this falls as rain during the short summer. Weather stations on the far-northern islands have recorded the lowest annual precipitation figures in Canada, with several stations having yearly averages under 5 inches.

Accessibility in the Northwest Territories is affected greatly by ice conditions. Sea ice freezes along the shores of the mainland and Arctic Islands in early September in the north, and by about early November around Hudson Bay. The Arctic is then closed to sea transport, and planes cannot land on the lakes until the ice thickens. Harbor ice begins to break up in late June in Hudson Bay, and progressively northward until some of the northern islands have open water by early August. Air photos indicate that the sea ice around the northwestern islands remains throughout the entire summer, despite the continuous daylight during June and July. These islands should therefore be thought of as one large "landmass," since they are interconnected throughout the year.

Resources and Economic Activities

Most of the developed and potential resources of the Northwest Territories are in the Mackenzie River valley. Fur-trading posts were established along the Mackenzie River in the last century, and trapping became the basic livelihood of the native Indians. Mining now supports many of the other inhabitants. Yellowknife, on the shore of Great Slave Lake, the capital and largest town in Mackenzie District, is a gold-mining community. Pine Point, south of Great Slave Lake, is a model community housing the workers and management of a large lead-zinc mine. Petroleum is pumped at Norman Wells and supplies the small local market of the Mackenzie Valley.

Other natural resources of the Territories, though of minor significance in the overall Canadian economy, may play a greater role in future local development. There are agricultural lands of limited quality available in many parts of the Mackenzie River valley, but cultivated land there now totals only a few hundred acres. Soils frequently need drainage and are troubled by permafrost near the surface. At present, this land is not being settled since better farmland is available farther south in Canada. Commercial fishing started in Great Slave Lake in 1946, where a large, but controlled, harvest of lake trout and whitefish is taken each year. Fish are caught in both summer and winter, during the latter season through holes in the ice, and shipped in refrigerated transport to southern Canadian cities and the U. S. Midwest.

The Arctic region in northeastern Canada has few developed resources and a very limited resource potential. Since it is a treeless region, there can be no forestry, and all lumber for houses is imported. There is no agriculture. The scour of glacial ice left behind bare rock ridges, and "soils" in the valleys are mainly sands and gravels with little organic matter. Hudson Bay appears to have few fish; some of the Arctic rivers are fished for high-quality Arctic char, a kind of salmon, which is exported by Eskimo cooperative fishing groups.

Mineral resources are the hope of the Arctic for the future. The region has a great deal of bare rock exposed at the surface, and much of it is the same type of old Precambrian rocks that are mineralized in many places in the southern part of the Shield. Although mineralization has been reported from several locations in Arctic Canada, and some small mines have been worked for a few years, there were no mines operating there in 1971. The northwestern Arctic Islands are underlain by sedimentary rock basins that may have an oil potential. If oil reserves similar to those of the north coast of Alaska are discovered, the problem of transporting oil out of the region would then need to be considered.

The main exported resource of the Arctic region has been white fox furs. The Arctic, or white, fox is one of the few furbearers living north of the tree line, and is trapped by Eskimos. Since fur production fluctuates greatly from year to year, the Canadian federal government has encouraged production among the Eskimos of handicrafts, such as stone and bone carvings and paintings; these have become the major source of the Eskimos' income.

Population

All but a few hundred of the 12,000 white inhabitants of the Northwest Territories live in the Mackenzie River valley. The native inhabitants of the Territories are Indians and Eskimos. The tree line, which also is a climatic line dividing Arctic from subarctic, marks a cultural line separating the Indians and the Eskimos. Eskimos, numbering about 17,000 persons, live north of the tree line across the mainland of Arctic Canada, including northern Quebec and Labrador, and inhabit the coasts of the southern Arctic Islands. They are wards of the federal government.

There are wide differences in the level of culture and acceptance of white civilization found among the Eskimos. Those Eskimos living near the delta mouth of the Mackenzie River are modern, well-equipped trappers, and a few of them herd reindeer; other Eskimos are wage earners at air bases and radar stations such as those at Frobisher Bay and Cambridge Bay. A few Eskimos, such as those in the Central Arctic near Boothia Peninsula, still live chiefly off the caribou on the land and the seals of the sea as their forefathers did. A major change in the geographical patterns of the decade 1960–1970 was the centralization of the formerly migratory Eskimos into particular villages, where wooden houses have replaced their traditional snow-block hut, the igloo.

Greenland

The Danish province of Greenland is the largest of the Arctic Islands near the mainland of North America. Its area of about 840,000 square miles is equal to all of the Northeastern and Great Lakes states from New England to Nebraska, or is as large as Mexico. Since 1953, Greenland, a former colony, has been an integral part of Denmark, having equal political status with the other provinces of that country. Greenland is only 12 miles from Ellesmere Island, the northeastern Canadian island, and is inhabited by the same Eskimo people who migrated across northern Canada and reached Greenland more than 1,000 years ago. Greenland has become significant to the world during the modern Air Age. Its geographic position in the north-central Atlantic Ocean placed the ice-covered island on great-circle air routes between Scandinavia and central North America.

About 84 percent of Greenland is covered by a thick ice cap. Alpine ridges and sharp peaks of high mountains rise abruptly above the sea along the fjorded, indented west and east coasts, reaching altitudes of 6,000 feet on the west and 10,000 feet on the east. Behind the old Precambrian rock mountains, and frequently spilling through and around them in beautiful twisting glaciers, lies the Greenland ice cap, which fills an interior basin. The surface of the ice cap consists of two flattened domes arching above the coastal mountains to altitudes of about 10,000 feet. Some soundings indicate that the ice cap is about 10,000 feet thick in the north-central section. It is possible that the land base of northern Greenland is below sea level and may therefore consist of two or more islands. The thickness of the ice cap over southern Greenland is about 6,000 feet. Although icebergs break off from Greenland glaciers at many places around the coast, the greatest number enter the sea in Melville Bay, on the northwest coast.

Greenland has an Arctic climate, owing to its northerly latitude and the cooling influence of the large mass of ice. In addition, the coasts of east and north Greenland are bathed by cold, ice-covered water discharging from the Arctic Ocean. In contrast, however, the southwestern coast has warmer water offshore, coming from the North Atlantic Drift (Gulf Stream), which moderates the climate. Some harbors of the southwest remain ice-free throughout the winter, and in summer the small valleys become warm enough for vegetables to grow in the open and for many sheep and some goats to be grazed. Because of these more favorable climatic conditions, most of Greenland's population lives in the sheltered valleys of the southwest.

The northward-moving ocean current off the island's southwest coast warmed noticeably after about 1930. This increase in temperature is probably related to the general warming of the Arctic regions which became apparent after 1920. To the Greenlanders the warmer waters brought increasing numbers of northward-migrating codfish, which became the basis of a fishing industry.

The livelihood of Greenlanders has changed greatly from that of their Eskimo forefathers, who depended upon the sealife of seals, white whales, walrus, and narwhales for food. Many Greenlanders are now commercial fishermen or work on fishing vessels that come from Denmark. Shrimp grounds were located during World War II in Disko Bay and are harvested during the summer; canned shrimp is now a major export of Greenland. Most fishing is carried on by the Greenlanders fairly close to shore because of their small boats, but newer and larger boats are being built to take part in the offshore fishery in Davis Strait. The traditional Eskimo kayak has almost disappeared and has been replaced by motorboats and fishing cutters.

A thousand years ago, Norse settlers attempted agriculture in the valleys of southwestern Greenland. In recent years agriculture has been revived there, particularly in the Julianehaab district. There are as many sheep as there are people in Greenland, with a surplus of meat and wool for export. Vegetables and grasses grow well, and there are a few head of cattle and many chickens. However, much food is still imported from Denmark.

Greenland's mountains may contain minerals, but little is known about the detailed geology. Although mineralization has been reported from several places, only two natural resources, cryolite and lead, have proved valuable enough for development. The government-controlled mine at Ivigtut, in the southwest, has been one of the world's major sources of cryolite, the strategic mineral used elsewhere in the manufacture of aluminum.

The large land area of Greenland had only about 45,000 people in 1970. Most of the inhabitants are of Eskimo origin but have a strong mixture of Danish blood; they now call themselves Greenlanders. Many people live in small villages, usually near the open southwest coast for winter sealing and summer fishing. As in other parts of the North Atlantic region, centralization into larger towns is now progressing. New apartment houses are being built, and schools and hospitals are more easily supplied to these concentrations of people. With half of the population being under eighteen years of age, the provision of future jobs is a problem to the local and Danish governments. Godthaab, with more than 7,000 persons, is the largest settlement and the main administrative center. Holsteinsborg is a fishing port; on Disko Bay, Egedesminde is a local commercial settlement.

In Perspective

North America, north of the conterminous United States, is a vast extent of land, which is in general only sparsely populated. Its area is over 1.3 times that of the continent of Europe, but the region has only about 4 percent of the population of Europe, although the two landmasses are in approximately the same latitudes.

Canada is a nation of great potential. Vast plains areas, when used more intensively, could produce much larger grain and pasture crops. Rangelands that are now scarcely used may, in the future, become sources of large meat supplies. The mineral wealth of Canada is being developed, but some of its area has not been explored in detail. One of the world's large forest areas extends across the northern part of the provinces and the southern part of the territories. Undeveloped power sites offer great possibilities for producing hydroelectric power. In spite of its northern location and climatic handicaps, Canada is potentially one of the world's great industrial as well as agricultural nations.

The Canadian–United States boundary is the longest common boundary between two nations in the world. These two nations have lived as neighbors for more than 150 years without resorting to war, and today there are no military forts along their boundaries. Many problems have arisen that were not easy to settle, and yet the good judgment of the respective governments has triumphed, and the bonds of peace have not been broken. Future relations between Canada and the United States should be characterized by mutual respect, understanding, and continued cooperation. In the world of today, Canada and the United States can prove that Peace is not an impossible ideal.

All parts of northern North America have been fortunate in their political development, for Canada, Denmark, and the United States have long worked together in harmony for the mutual benefit of all.

REFERENCES

Gentilcore, R. Louis (ed.): *Canada's Changing Geography*, Prentice-Hall, Inc., Toronto, 1967.

———: *Geographical Approaches to Canadian Problems*, Prentice-Hall, Inc., Toronto, 1971.

Greenland, Ministry of Foreign Affairs, Copenhagen, 1969.

Irving, Robert M. (ed.): *Readings in Canadian Geography*, Holt, Rinehart and Winston, Inc., Toronto, 1968.

Krueger, R. R. (ed.): *Urban Problems: A Canadian Reader*, Holt, Rinehart and Winston, Inc., Toronto, 1971.

Putnam, D. F. (ed.): *Canadian Regions*, J. M. Dent & Sons (Canada) Ltd., Don Mills, Ont., 1956.

Robinson, J. Lewis: *Resources of the Canadian Shield*, Methuen, Toronto, 1969.

Somme, Axel (ed.): *A Geography of Norden*, J. W. Cappelens Forlag, Oslo, 1960.

The Canadian Yearbook, Bureau of Statistics, Department of Trade and Commerce, Ottawa, 1970. (Issued yearly.)

Tomkins, G. S., T. L. Hills, and T. R. Weir: *Canada: A Regional Geography*, W. J. Gage Limited, Scarborough, Ont., 1971.

Warkentin, John (ed.): *Canada: A Geographical Interpretation*, Methuen, Toronto, 1968.

Wolforth, John, and Roger Leigh: *Urban Prospects*, McCelland and Stewart, Toronto, 1971.

5

Middle America

Even before the promulgation of the Monroe Doctrine in 1823, the United States had considered Middle America vital to its national security and within its sphere of influence. Consequently, a long history of involvement —political, military, and economic—has characterized American relationships with this area. At times the United States has acted unilaterally to protect its interests in the region; for example, the use of a naval blockade in support of Panama's revolution against Colombia in 1903, the "Bay of Pigs" invasion of Cuba by U.S.-supported Cuban exiles in 1961, and the use of troops in Mexico, Haiti, and the Dominican Republic, both in the early part of the twentieth century and, in the latter nation, again in 1965. Such actions have brought strong criticism from Middle American leaders and perpetuated fears of domination by the "Colossus of the North." In more recent decades, with the advent of the "Good Neighbor Policy" in the 1930s, U.S. foreign policy toward Middle America has mellowed, with diplomatic negotiations and persuasion usually replacing more direct action such as armed intervention. Yet, even today, many Middle American political leaders can often unify dissident groups within their countries by appeals for a united front against the "Northern Imperialist."

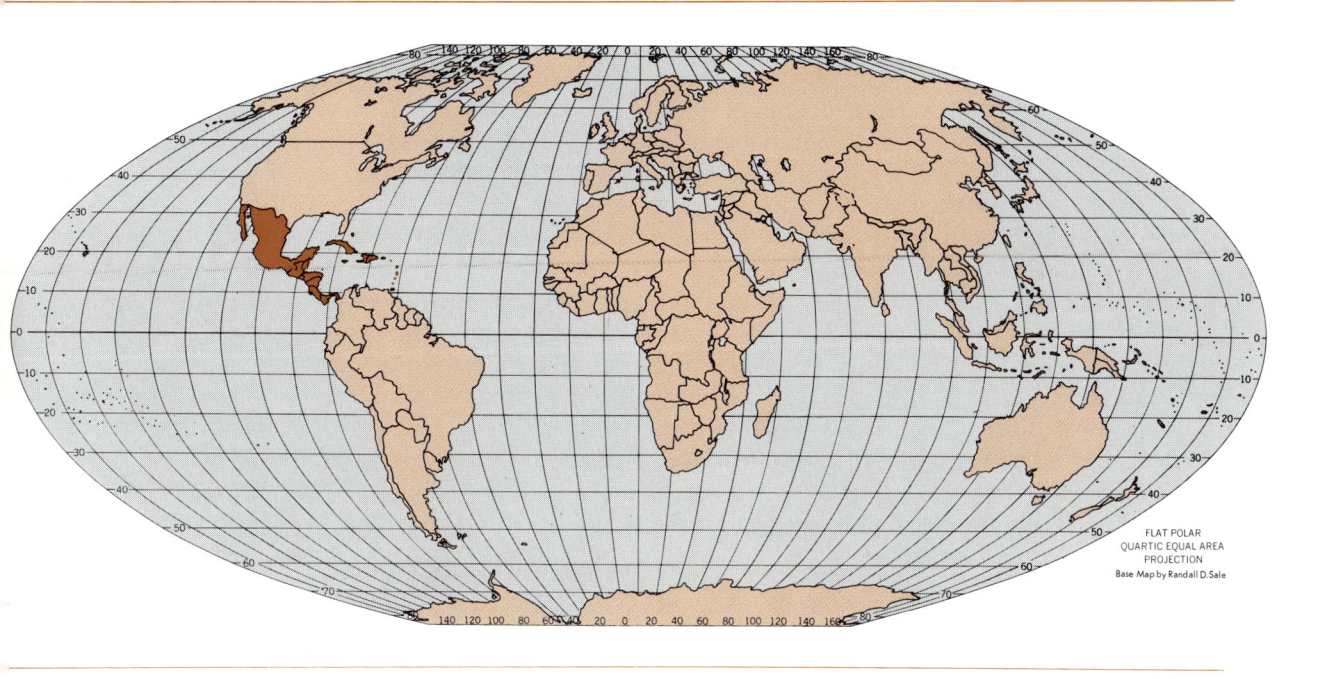

Middle America is that portion of the New World located south and southeast of the United States and north of South America. It includes, as its major subregions, Mexico, Central America, the Greater Antilles, and the Lesser Antillean Archipelago. In reality, Middle America is not considered a distinct region by its inhabitants, but they would recognize a connection with other areas in their subregion. That is, although a Honduran would not perceive that his nation has much in common with Trinidad, he would feel related to Costa Rica. A Cuban, in spite of political differences, would consider his country linked to the Dominican Republic, but not with Nicaragua. Rather, the Middle American region is a concept developed by Anglo-Americans and others who view the world on a broader scale and distinguish characteristics and trends that unify the larger area.

Common Characteristics

One of the most striking characteristics of Middle America is a high rate of population increase. Few other world regions have a comparable rate of growth. Middle America has traditionally had both high birth and death rates; but improved sanitation and medical facilities, coupled with some increase in food supply, have drastically lowered the death rate, particularly infant mortality. The result is increasing life expectancy and a youthful population. In the Antilles, where population densities are already high in relation to the utilized resource base, increasing population pressures are leading to serious discontent in rural areas, urbanization, and a search for new bases of economic livelihood. In Mexico and Central America, increasing population has led to settlement in formerly sparsely occupied tropical lowlands as well as migration to urban areas. Increasing population pressure is also placing heavy burdens upon local governments to supply needed services such as water supply, sewage, transportation facilities, and schools to maintain and upgrade levels of living.

Associated with population growth is a low income level. Increasing national productivity is generally the rule, but often per capita income gains are minimized by increased population. Moreover, distribution of wealth is uneven, since most areas are composed of a two-class population: an upper class which controls the political and economic spheres and a lower class which forms the manual-labor group of

Fig. 5–1 Most of the people of Middle America live in the highland valleys or on the plateaus. The greatest population densities are on the islands of Puerto Rico and Jamaica and around Mexico City.

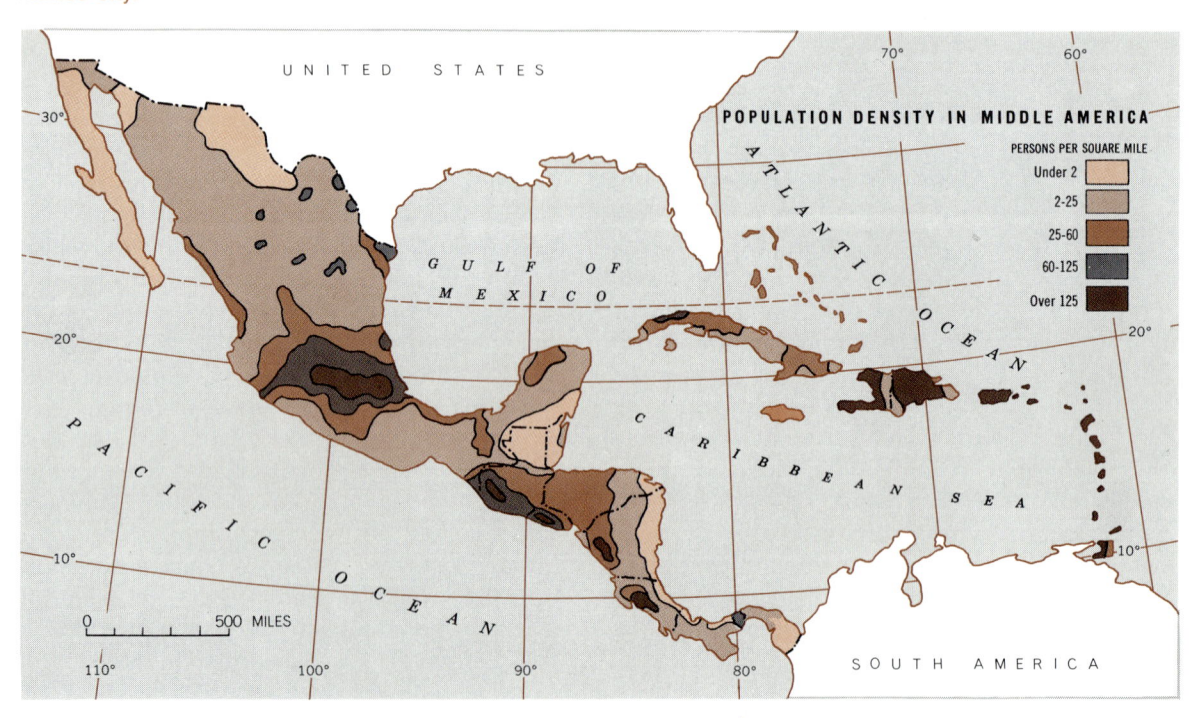

small-plot farmers, rural wage laborers, and urban employees. The average per capita income figures for a nation rarely illustrate the true economic situation, and for most of the population actual income is lower. A middle class is developing and is becoming a significant sector of the economy in Puerto Rico, Mexico, and parts of Central America.

One of the causes of low per capita income throughout Middle America is the heavy reliance on primary activities, particularly agriculture. Agriculture dominates the economic life of all parts of the region. Even in areas such as Puerto Rico, Jamaica, and central and northern Mexico, where manufacturing and tourism are of major importance, agriculture is still the mainstay of most of the population. While types of agriculture practiced vary greatly, nearly all are characterized by low yields per man and per unit of land and a dependence on manual labor in lieu of machines. It is not surprising, then, that income per person is low. Moreover, since the end of World War II, the value of primary products has remained relatively constant, while processed goods have increased significantly in price. Thus, unless the farmer finds ways to increase his output, his purchasing power is substantially decreased, and he must accept a lower standard of living.

Reaction to lower income has taken several courses. In some areas, farmers have become essentially subsistence cultivators, and the commercial structures in the area have suffered accordingly. In addition, removal of the farmer from the commercial economy has resulted in a smaller tax base for government operations and a lower volume of exports, which creates balance-of-payment problems. In other areas where nearby land is available for settlement, farmers have migrated to these sparsely populated zones in the hope of increasing production on better or more land. Almost everywhere in Middle America some farmers have chosen to give up land cultivation and seek employment in the city. To combat the farm problem and stimulate production, agrarian reform programs have been undertaken in some countries.

Still another common feature of Middle America is the character of exports and imports. Almost all areas depend primarily upon exporting only one or two products and sell these goods to only one or two trading partners. The products exported are either a food crop such as bananas, sugar, and coffee or an industrial raw material such as cotton or petroleum. Since the exports are largely unprocessed, their value has not increased a great deal. Because trade is with only one or two countries, the inhabitants of Middle America are highly vulnerable to changing economic conditions in the countries with which they trade. Lastly, the Middle American region imports mainly mid-latitude foodstuffs and manufactured items, goods that have increased in value at a rate greater than the region's exports.

Actually, the export-import problem directly affects only part of Middle America, since the region is further characterized by a dualistic economy that separates goods produced for local and national consumption from those produced for export. Local and national goods are produced by more traditional methods, which usually involve rather inefficient operations. In the industrial section, goods are often produced in small workshops using a minimum of power, local raw materials, and nonstandardized parts and having a limited output. Such manufactures include clothing, furniture, pottery, brick, and tile. In more developed areas such as northern and central Mexico, where a larger and wealthier market exists, some large plants employing modern production techniques common in the United States and Western Europe are gradually replacing the artisan workshops. Agricultural production for local consumption likewise exhibits little-changed, age-old techniques. Here the hoe or machete are the principal, and usually the only, implements of importance; and seed selections, insecticides, herbicides, and fertilizers, except for organic manure, are practically unknown. Farms are small and usually located on poorer land, in terms of soil quality and/or accessibility. Representative crops include corn, beans, and squash in Mexico and northern Central America; corn, manioc, and other starchy tubers in southern Central America; and a variety of garden vegetables such as tomatoes, lettuce, and beans in the Antilles.

In contrast, the export-import sector of the economy uses more modern techniques and is confined to agricultural production and, of lesser importance, to mining. With the exception of Cuba and to a lesser degree Mexico, agriculture is commonly carried out on large land units possessing good physical conditions for crop growth and owned by foreign companies or by members of the local upper class. In many cases, the management and technical staff are foreign, but the manual labor force is of local origin. In these large-scale operations fertilizers, chemical sprays, and quality controls are customary, and mechanized equipment is common in the more level areas. Representative commercial crops include sugarcane and bananas in the Antilles; coffee, cotton, and bananas in Central America; and cotton, vegetables, and sugarcane in Mexico. Actively developed mineral wealth is more restricted in its location, with Mexico by far the most important mining country. There lead, zinc, silver, sulfur, and a myriad of other minerals are produced and exported by companies which are nominally Mexican but which are usually financed by outside funds. Once a leading exporter of petroleum, Mexico nationalized its oil industry in the 1930s, and since then production has been geared mainly to the national economy. In the Antilles, Jamaica is now one of the world's leading exporters of bauxite. Trinidad extracts petroleum from reserves that supply much of the demand in southern Middle America. Elsewhere in the region mineral production is insignificant, although possibilities from petroleum reserves along the Caribbean coast of Central America are strongly suggested and exploration has been active.

Physical Setting

Relief Features

The dominant surface feature of Middle America is the highland backbone and its associated volcanoes. Bordering coastal lowlands are usually narrow and often discontinuous. A major exception is the low, flat limestone plain of the Yucatan Peninsula, which is related geologically to the plains of Cuba, Florida, and the Bahamas.

Mexico north of Tehuantepec is essentially a plateau, bordered on the west by the wide and rugged Sierra Madre Occidental, with peaks over 12,000 feet high, and on the east by the lower, more easily traversed Sierra Madre Oriental. Further west of the Sierra Madre Occidental is a depression filled by the Gulf of California and the rugged, hilly Baja California peninsula. The plateau surface is broken by mesas, mountain ranges, and alluvium-filled basins. It is highest—5,000 to 9,000 feet—in the south, where it terminates along the 19th parallel in a rim of volcanic mountains and ash-filled basins. In this area, lofty volcanic cones such as Orizaba (18,851 feet), Popocatepetl (17,716 feet), and Ixtaccihuatl (17,342 feet) rise above the general plateau level. South of the volcanic rim are the Balsas Valley of Chiapas and the Sierra Madre del Sur, as well as the Central Valley of Chiapas and the Sierra Madre del Chiapas, lying to the east.

South and east of Tehuantepec to Nicaragua, the highland elevations are greatest on the

Pacific side. There, volcanic debris and lava largely obscure the underlying structures and form a plateaulike highland, with ash-filled basins nestling among volcanic cones. With the exception of British Honduras, each of the other Central American countries has mountain peaks reaching about 6,000 feet or higher. The most lofty, in Guatemala, exceed 13,000 feet.

In Costa Rica, highlands and volcanoes again appear and extend in a broken line through Panama and into northwestern Colombia. These highlands attain their maximum height in Costa Rica and diminish gradually to low hills in Panama. Again, volcanoes such as Irazú near San José, Costa Rica, have deposited large quantities of ash masking underlying topography and creating level but dissected upland surfaces; the most important is the Meseta Central centered in San José province.

Three interoceanic passes breach the highlands of Mexico and Central America. One crosses the 130-mile-wide Mexican Isthmus of Tehuantepec at a maximum elevation of 800 feet. A second is part of a structural depression aligned diagonally through Nicaragua from the mouth of the Río San Juan, which forms part of the border between Nicaragua and Costa Rica, to Lake Nicaragua, which has an elevation of a little over 100 feet and is only 17 miles from the Pacific. The third route is that followed by the Panama Canal.

In the Greater Antilles one line of the highlands may be traced from mountainous Puerto Rico westward, through the Cordillera Central of the Dominican Republic, the northern peninsula of Haiti, the Sierra Maestra of southeastern Cuba, and the Cayman Islands, to southern British Honduras. Another group extends through the southern peninsula of Haiti, the Blue Mountains of Jamaica, and the Swan and Bay Islands to the mainland. The highest elevation in the West Indies, 10,417 feet, is located in the Dominican Republic.

The Lesser Antilles, except Barbados, can be divided into high and low islands. The high islands are mountainous, being the tops of volcanic peaks, which in some cases are still actively building or emerging from the sea. Fringing deposits of coral limestone often form reefs and raised terraces. The low islands are relatively level and are generally of coral formation, covering submerged mountain peaks. The island of Barbados was formed by the upturning of sedimentary strata, largely limestone.

Climate, Vegetation, and Soils

Middle American climatic characteristics and their pattern of distribution result from the action of (1) latitudinal location, (2) position of semipermanent pressure and wind systems, (3) location of adjacent large bodies of water, and (4) terrain features. Since most of Middle America is located in a relatively low latitude, the variation in length of daylight throughout the year is slight.

Variation in precipitation results, in part, from four semipermanent pressure and wind systems. Northeast of the Greater Antilles, at about latitude 30° north, is a cell of high pressure—the Bermuda High—from which lower atmospheric winds blow outward in a clockwise direction, creating the Northeast Trades. These winds are remarkably constant in both velocity and direction and prevail throughout the Antilles and along the eastern side of Mexico and Central America southward from about 20° north latitude. The second system is a band of low pressure that moves seasonally with the sun. During the Northern Hemisphere summer, the system is oriented along a northwest-southeast axis paralleling and occasionally overlapping the west coast of Central America and southern Mexico. During the Northern Hemisphere winter, the pressure band moves southward and influences only southern Panama. In both summer and winter, lower atmo-

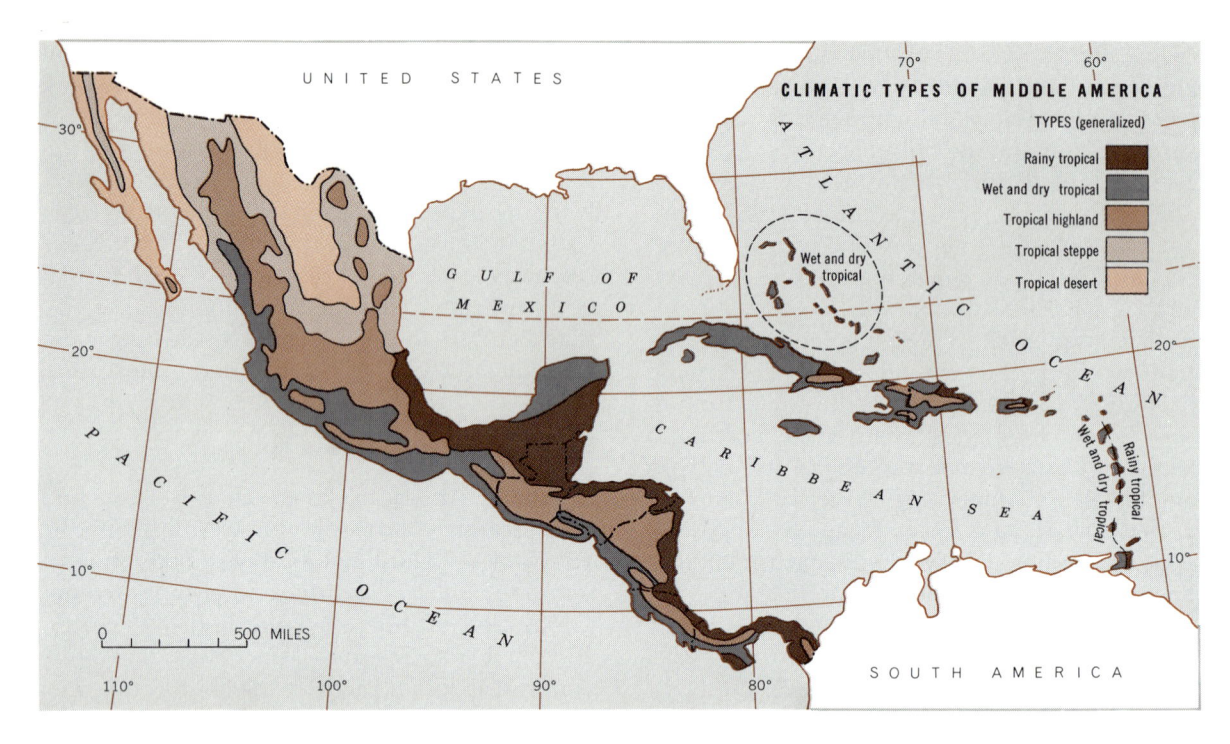

Fig. 5-2 The climatic types of Middle America are greatly influenced by the trade winds and topography. In general, rainy tropical climates are found on the northern and eastern sides of the landmasses; steppes and deserts are in the northern and interior parts of the area.

spheric winds blow toward the low-pressure center. These winds, however, are not so constant as the Trades. The third system is located in central and eastern North Mexico and adjacent areas of the United States, where the Prevailing Westerlies bring alternating high- and low-pressure systems and their associated winds. The fourth system is an area of high pressure similar to the Bermuda High located off the coast of California and Baja California, where winds move outward in a clockwise direction roughly paralleling the coast.

Large bodies of water are the source of practically all moisture taken up in the atmosphere, and their position in relation to prevailing wind direction and land areas strongly influences precipitation potential. Moreover, the temperature of the bodies of water affects the amount of evaporation possible. Where water temperatures are cool, as along the west coast of Baja California and west-central Mexico, air flowing across the water is chilled and consequently is unable to absorb large quantities of water vapor. On the other hand, where water temperatures are warm, as in the Gulf of Mexico, the Caribbean Sea, and along the east coast of Central America and southern Mexico, surface air temperatures are warm, and large amounts of water vapor can be incorporated into the atmosphere. Bodies of water also act as moderating agents of temperature, since they neither heat nor cool as rapidly as land. Along coastal and lakeshore areas, temperatures are less extreme; if winds are onshore and the body of water is large, this influence may penetrate considerable distance inland. Land and water contrasts also can affect the precipitation pattern on a local scale. In areas where semipermanent systems do not completely dominate windflow and where significant differences in temperature exist between land and water, as along the west coast of Mexico and Central America, small localized high- and low-pressure cells form. During the day, surface winds blow from the sea over the land, setting up the possibility of local showers. At night air circulation is reversed, and the chance of showers is minimized.

Terrain features act primarily as disruptors of airflow, and when winds are forced to rise over topographical barriers, the air may be cooled to the point where it is no longer capable of retaining water vapor, and precipitation results. Thus, on the sides of mountain ranges facing the wind ("windward" side), precipitation is common. Examples of this situation in Middle America are the northeast faces of mountains in the Antilles and the eastern side of the upland backbone of southern Mexico and Central America. Conversely, however, on mountainsides that face away from the prevailing wind direction ("leeward" side), the air tends to descend and warm, thereby increasing its capacity to hold moisture. In Middle America this condition is best illustrated on the southeast mountain slopes of Antillean islands and on interior slopes of the highland areas of Mexico and Central America. Additionally, variations in elevation of terrain features affect temperature.

Through the interaction of latitude, pressure and wind systems, bodies of water, and terrain, five general types of climate are represented in Middle America. Correlated with each type are characteristic vegetation and soils.

Rainy Tropical Rainy tropical climate is characterized by constant warm, rainy, and humid conditions. Average monthly temperatures are near 80°F, and daily temperatures range from the low 70s to the mid-90s. Rainfall is abundant during all months, although there may be periods of greater and lesser precipitation. Nevertheless, there is always a supply of water for plant growth. A high amount of humidity in the air and relatively low surface-wind velocities further assist in maintaining a water surplus and cause the sensible temperature—the temperature felt by the human body—to be higher than actual temperature. Areas of rainy tropical climate are confined to low-latitude, low-altitude places where bodies of water and wind and pressure systems combine their influence to bring warm moist air the year around. In Middle America such areas include the windward slopes of the Antillean Mountains, the Caribbean Lowland and the adjacent eastern slopes of Central America, and the extreme

Fig. 5-3 Nogales-Mexico City Highway, west of Guadalajara. The eroded, brush-covered slopes that can be used only for pasture, if used at all, are a result of the dryness of the area. Patches of maize are being grown on alluvial flats. (Courtesy of the Embassy of Mexico)

southern part of Panama. Vegetation in tropical rainy areas is typically selva, a multistoried, heterogenous, broadleaf-evergreen forest, which in Middle America includes some trees noted for their commercial value. Examples include the mahogany and tropical cedar, well known for their use in furniture, and the sapodilla, the source of chicle, which is used in gum manufacture. Soils are generally deep, well-drained, and poor in plant nutrient materials. In areas of recent alluvial deposits, however, soil leaching may be imperfect and nutrients still present; such areas are more favored for agricultural production.

Wet-and-dry Tropical The wet-and-dry tropical climate differs from the rainy tropical in that there is a distinct dry season during the winter months. The difference between the wet and dry seasons is striking. During the wet season rainfall is abundant, and monthly averages may exceed those of rainy tropical areas. During the dry season, however, the climate takes on the characteristics of a steppe or desert area, and chances of significant rainfall are slight. Some observers label this climatic type the "mud-and-dust" climate. Areas of wet-and-dry tropical climate are located on the Pacific fringe of Middle America from central Mexico to southern Panama, in the northern portion of the Yucatan Peninsula, and in the leeward and low-lying areas of the Antilles. Natural vegetation in these areas is mainly a low semideciduous or scrub forest, but in many areas of the Antilles and northern Central America these forests have been removed and the land put to crop and livestock production. Soils vary considerably from area to area. In the Yucatan Peninsula they are thin and poorly developed; in the Antilles they range from the rather deep, fertile sugarcane soils of Cuba to dry, thin soils similar to those of the Yucatan; and along the Pacific they vary from rocky, thin imperfect soils in the hilly sections to deep, fertile alluvial soils in the plains.

Tropical Highland In the mountainous sections of Middle America, where elevations vary greatly and where differences in orientation to the sun and rain-bearing winds likewise change within short distances, many different climates are represented, and several altitudinal temperature zones are commonly recognized in these mountainous areas. The lowest, *tierra caliente*, is below 2,500 feet elevation and has either rainy tropical or wet-and-dry tropical climates. Above this zone and up to altitudes of 6,000 to 7,000 feet is *tierra templada*, with average annual temperatures of 60 to 75°F. In Central America this is the coffee zone, and in Mexico and Central America, except for Nicaragua and Panama, it is also the zone of greatest population density. Higher still is *tierra fría*, which extends upward to the tree line at an elevation of 11,000 to 13,000 feet. Only Mexico and adjacent Guatemala have any sizable areas of *tierra fría*. Above *tierra fría* are the "paramos," alpine grasslands largely limited to Mexico; this country alone has permanently snow-capped peaks extending above the snowline, at approximately 15,000 feet above sea level.

Tropical Steppe and Desert Much of northern Mexico is desert and steppeland; here moisture deficiency characterizes the climate. The area of desert climate is located in the lower, northern part of the Mexican plateau and adjacent Baja California. Its presence is explained by the dry prevailing winds and mountain barriers to the east and west. Higher surrounding lands with steppe climate, especially to the south, commonly receive more rain, or else the rain is more efficient because of lower temperatures. At still higher elevations, more humid conditions support a pine-deciduous forest. What precipitation there is occurs normally in summer, except for the area near the California border, where winter storms bring most of the scant rainfall. This one area is the only exception to the characteristic summer maximum of precipitation throughout Middle America.

Cultural Differences

When Columbus returned to Spain after his first voyage to the Caribbean, he described a land where riches were plentiful and the inhabitants were ignorant of Christianity. These two features of Middle America struck a responsive chord in Spain and set off a wave of exploration, conquest, and exploitation rarely equaled in the history of man. Spain in 1500 was just recovering from the disruption of an 800-year war with Moorish invaders. The nation had a large number of professional soldiers with no one to fight, a religious zeal developed through conflict with the Islamic Moors, and a national treasury almost exhausted from financing the long war. In the New World that Columbus described, troublesome men-at-arms could be put to good use, the fervor of the church utilized to bring religion to the newly discovered peoples, and the royal treasury replenished to rebuild Spain. Soldiers of all ranks, churchmen, and government representatives made up the initial Spanish forces of exploration and conquest. Later, merchants and a small number of artisans and farmers arrived, to assist in the exploitation of the new lands.

These early Spanish groups in the New World contrasted sharply with the colonists who came to Anglo-America. Most Spaniards came in search of quick wealth and fame, with the expectation of returning eventually to Spain. They did not bring their families, nor did they have the skills or desire to develop a permanent utilitization of resources similar to that achieved by the settlers of Anglo-America. Those who did attempt activities such as farming soon found that the tropical setting of Middle America was not suitable for many traditional production techniques and crops of Spain without substantial modifications.

The first attempts at Spanish settlement were made in the Greater Antilles. Here the Arawak Indians, numbering about 1 million, were numerous and practiced crude forms of agriculture based on manioc and corn crops. They supplemented their cultivated food supply by fishing, gathering, and hunting. These people were quickly brought under control, and their meager accumulation of ornamental gold was seized by their conquerors. The Spanish imposed their rule over the Indians and demanded continual tribute from them. Where gold deposits were discovered, the tribute required was gold; in other areas, a specified amount of foodstuffs or labor was demanded. Unable to withstand introduced European diseases, forced labor, and the Spanish conquerors' regimentation of life, the Indian population was quickly decimated. By 1600 the Arawaks had ceased to exist in the Antilles.

The destruction of the Indian population in the Greater Antilles, the limited wealth discovered there, and news from Mexico of still-greater numbers of Indians and gold soon placed the Caribbean islands in the backwash of Spanish settlement. An exodus of Spaniards to the mainland severely depopulated the islands. Some land that formerly had been put to agriculture reverted to natural vegetation or was converted to pastureland. Small plots of land were farmed by Spaniards who no longer desired to follow the wave of conquest, or who were prohibited from doing so by governmental decrees instituted to stop migration. Other land was cultivated with the aid of imported African slaves.

From the viewpoint of the plantation owner, the African slave quickly proved himself an admirable laborer. Most slaves came from a cultural setting where they had developed rather sophisticated agricultural techniques and were accustomed to work regimentation. Moreover, by mixing slaves from many different tribes with diverse languages and backgrounds, control was much easier. Stripped of all worldly possessions, far away from familiar surroundings, and located on islands where the chance of escape was slight, the Negro slaves soon became productive workers.

In the Lesser Antilles the Spaniards were confronted by the fierce Carib Indians, who resisted attempts to subjugate them. The Caribs were an even less suitable labor supply than the Arawaks. Not only were they more warlike, but also their agricultural skills were limited, and they had no knowledge of metals. The Caribs lived mainly by fishing, hunting, and gathering and planted small plots of corn, manioc, and other tubers to supplement their diet. After a few forays into the Lesser Antilles, the Spanish found no accumulation of wealth nor any real attraction for permanent settlement. It was only with the colonization of these islands by the British, French, and Dutch that control over their native inhabitants was firmly established. By 1700 the Caribs were largely exterminated. The Northern European countries began their colonization attempts about 1630 and, like the Spanish, soon introduced African slaves as a labor supply on large sugarcane plantations. Later, French and English influence was extended to parts of the Greater Antilles and the Caribbean Lowland of Central America.

On the mainland of Middle America a densely settled highland area extending from Central Mexico to Honduras was occupied by two indigenous groups of high cultural attainment. Centered in the valleys and basins of south-central Mexico, perhaps as many as 20 million Aztecs had developed a stable and productive sedentary agricultural system based on corn, beans, and squash and used advanced cultivation techniques. So powerful were the Aztecs that tribute and allegiance were paid them by smaller groups whose territories extended from the Gulf Coast of Mexico to the Pacific coast of El Salvador. In the highlands of southern Mexico, Guatemala, and Honduras and extending into the lowlands of Yucatan, the Maya Indians, numbering about 2 million, also had a highly developed culture, based on the corn-bean-squash complex, which supported dense population clusters. These two groups possessed what the Spanish sought: a lavish accumulation of gold used by the Indians as ornamentation, numerous gold and other precious metal deposits to be exploited, and an abundant, well-disciplined, and skilled labor force to work the mines and farms, as well as a sedentary population more susceptible to control and conversion to Christianity. Yet, so large and cohesive were these Indian populations and cultures that, in spite of great decimation from the newly introduced European diseases and from famine in the early years of the Conquest, these Indians were able to maintain themselves, and many of their cultural traditions are still observable today. In some areas, as in Central Mexico, a merging of Spanish and Indian populations and traditions has led to a "mestizo" culture. In others, such as highland Guatemala, many of the Maya Indians have remained distinct groups, clearly separated by language, customs, and their outlook on life.

In northern Mexico and southern Central America, smaller groups of Indians with varying levels of technology prevailed prior to the Spanish Conquest. In contrast to the Antilles, however, some mainland groups were able to find refuge areas where Spanish domination was slight, and many of their descendants have only gradually adopted Europeanized ways of life. Only on the Meseta Central of Costa Rica were the Indians almost completely removed, and a nearly pure Spanish population developed there. Conversely, only in some parts of the Caribbean Lowland did an Indian population persist into the twentieth century.

Disease, probably more than any other factors, contributed most to the decimation of the Indian population. Apparently such diseases as smallpox, measles, yellow fever, malaria, leprosy, and yaws are all of either European or African origin, and against these the Indian had no natural resistance. In the tropical lowlands of Mexico and Central America in particular, malaria is an ever-present danger in some areas.

With the conquest of Middle America essentially completed by the latter portion of the sixteenth century, the influence of Spanish culture became progressively more important. To the New World, Spain introduced crops from Europe, Asia, and Africa, among which the most significant were sugarcane, bananas, coffee, citrus, coconut, breadfruit, wheat, cotton, and rice. Livestock such as cattle, horses, sheep, goats, pigs, and chickens were brought first to the Greater Antilles and then later to the mainland by the Spanish. There the larger grass-eating animals spread quickly into the drier northern portion of Mexico and southward into Central America wherever savanna grasslands and mountain pasture provided forage. Iron and steel implements were also introduced, and most of the cruder stone and wooden tools were gradually discarded. The Spanish constructed new towns laid out on a rectilinear grid pattern, with a central plaza around which the church and government buildings were situated; built trails and harbors to facilitate the movement of goods and people; and organized, albeit imperfectly, the production and sale of goods along European lines.

In the early period of Spanish colonial rule, two policies were implemented that have had fundamental impact on the character of Middle America, even to the present day; these were the system of land distribution and the colonial trade restrictions. The Spanish system of land distribution varied throughout colonial times. At first, *encomiendas* were granted. These *encomiendas* did not involve outright ownership but the right to use the land for a definite period. If Indians were living on the awarded land, the *encomiendero* was officially required to protect and educate them; in practice, however, the Indians became a forced labor supply. Later the *encomienda* system was abolished, and a system of land ownership was established. The division of land was made largely on the basis of service to the Spanish Crown. Individuals of high rank or standing were granted huge tracts of the best lands—good land being determined by accessibility, available Indian labor supply, and physical quality. Individuals of lower rank received lesser amounts of and poorer-quality land. Even today, the effects of this system of land distribution are still evident in some areas. Only in a few cases were Indian rights to land truly considered, and these few instances occurred normally in the high-culture, densely populated areas.

The colonial trade policies of Spain greatly restricted and structured economic activities in Middle America. Spain, as did most other European powers, utilized a mercantilistic relationship with its colonies. In essence, mercantilism meant that the colonies existed primarily for the benefit of the mother country. Under this policy: (1) the colonies could not produce commodities in competition with Spain; (2) trade could only be between a colony and Spain (trade between colonies was discouraged); (3) the transport of goods and people between the colonies and Spain could only be done with Spanish vessels; and (4) manufacturing in the colonies was suppressed. Under this philosophy the colonies were deliberately kept separate from each other; they became raw-material exporters and importers of manufactured goods; and they could not produce such commodities as wheat or wool, which Spain already had in abundance. The mercantilistic policy remained essentially in force until many of the Spanish colonies of the Western Hemisphere obtained independence in the early nineteenth century.

Out of the conflict between Spain and the Indian groups, the introduction of African slaves, and the gradual intrusion of the English, French, and Dutch, a true mosaic of language, cultural attitudes, and population type has been spread over Middle America. In Cuba, Puerto Rico, and the eastern part of Hispaniola (Dominican Republic), the Spanish language and culture have persisted, although some African characteristics are evident in the population. In western Hispaniola (Haiti) and in Jamaica, Trinidad, the Lesser Antilles, parts of Panama, and even some sections of the Caribbean coast of Central America, the English and French languages are dominant, although racially the populations are almost entirely African; in Haiti, moreover, certain African cultural traits have survived. On the Middle American mainland, the Spanish language and culture are found nearly everywhere. In the old Aztec and Maya areas, however, a variety of Indian languages and customs are still strongly represented, and racially the Indian influence is important in all areas but highland Costa Rica and parts of Panama.

Greater Antilles

Diversity in physical setting and in population and culture has created four regions within Middle America. These regions are: (1) Greater Antilles, (2) Lesser Antilles and Trinidad, (3) Mexico, and (4) Central America. The Greater Antilles comprise the islands of Cuba, Puerto Rico, Hispaniola, and Jamaica. The colonial development of Cuba and Puerto Rico was very similar until the Spanish-American War, after which Cuba became an independent nation and Puerto Rico a possession of the United States. On Hispaniola, Spanish control was disputed by the French, who were ceded the western one-third of the island in 1697. Later, with Haitian independence, the Spanish eastern part was incorporated into Haiti. In 1844 a successful revolution led to repartition of the island and the founding of the Dominican Republic. Thereafter, until 1865 the struggling new nation fell for short periods under both Haitian and Spanish control. Jamaica, originally Spanish-occupied territory, became a British colony in 1655 and attained independence in 1962.

Cuba

Cuban development prior to the mid-nineteenth century was characterized by slowly changing social and economic conditions. Population increased gradually in response to improved agricultural production. Large farms were developed on the better lands, with emphasis on cattle in the more outlying areas and on sugarcane nearer Havana. Smaller subsistence land holdings developed in more isolated areas and on old *encomienda* land not effectively occupied by its owners. Negro slaves were introduced early as a labor supply, and their numbers increased as the sugar industry gradually expanded. Urban development was confined largely to Havana, which was the chief port and the island's administrative and trade center. The economy of the island was tied to that of Spain; and since Spain had a limited market, the Cuban economy suffered accordingly.

Beginning with the sporadic wars of independence in 1868 and culminating in the Spanish-American War of 1898–1899, Spanish control over Cuba began to lessen and eventually was completely broken. With independence, Cuba underwent an economic transformation. It became essentially a protectorate of the United States and, in essence, an economic colony. American private capital flowed into the Cuban sugar industry; land prices rose, and many of the old large farms as well as many small farms were bought by U.S. companies and consolidated into still-larger units. Cane production was expanded outward from the tradi-

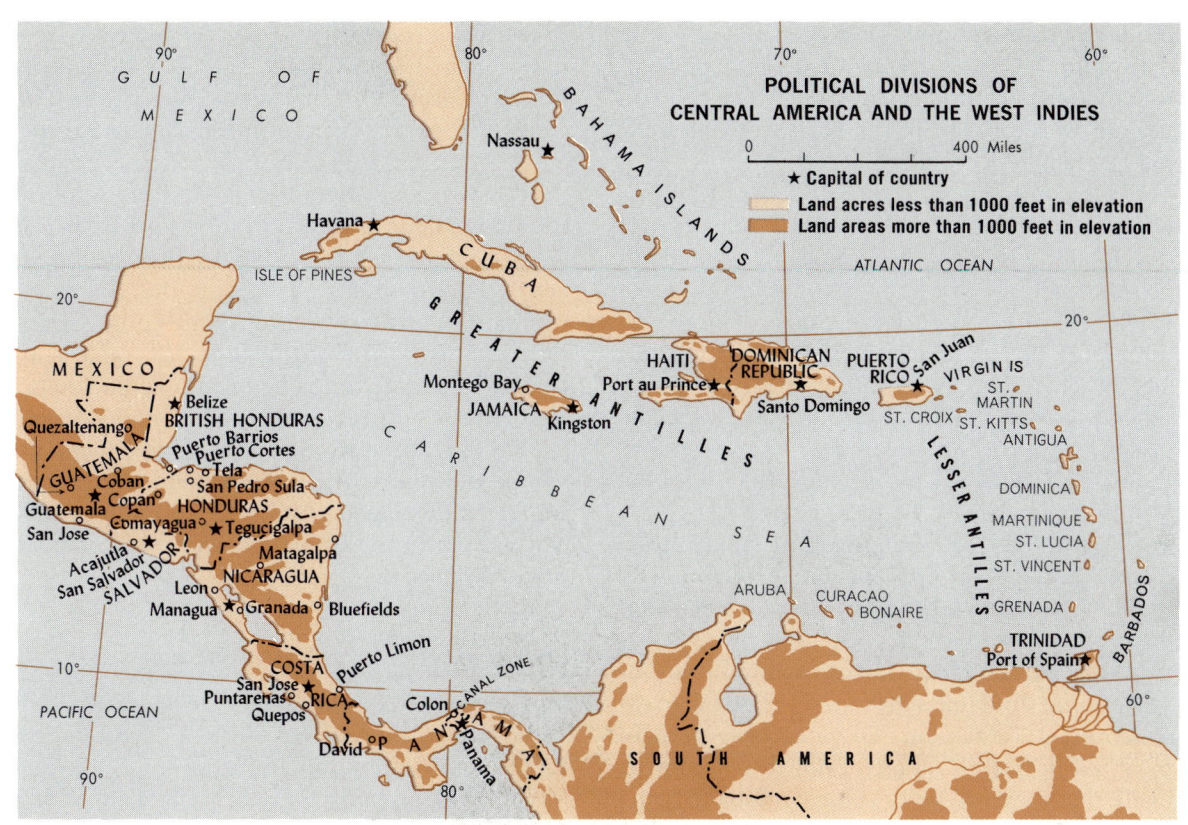

Fig. 5-4 Central America and the West Indies encompass twelve independent nations as well as territories controlled by the United States and several European nations.

tional areas around Havana to the east as far as the Sierra Maestra. Modern sugar centrals were constructed to increase production efficiency, and improved overland transport by rail and roads to new and improved ports further lowered the cost of getting the sugar to world markets, particularly to the United States. Within three decades, "Cuba" and "sugar" had become almost synonymous terms to the American public.

Cuba is an ideal area for sugarcane production. Most of the country is a relatively level plain upon which has developed a deep, fine-grained permeable soil derived from limestone, excellently suited for growing sugarcane. The climate is tropical wet-and-dry, yielding a year-round growing season, a wet season for cane growth and a dry season to "set" the sugar in the cane and facilitate harvesting. Moreover, the proximity of the United States as a major market and the capital inflow for improving production techniques permitted efficient utilization of the physical resources. In addition, when oversupply and the almost worldwide depression of the 1930s caused sugar prices to drop drastically, the United States set up a quota system of sugar imports guaranteeing a fixed market at a price substantially over the world market rate. This, along with receiving a preferential tariff status, carried Cuba's sugar industry through the Depression with little disruption. Cuba's preferred trading position was maintained until relations between the two nations were broken in the early 1960s following the Cuban Revolution.

Along with sugar, U.S. investment interests in Cuba did much to build a tourist industry and developed a small but significant production of minerals. Tourism, principally around the Havana area, brought large sums of money to the island, but today many of the luxury hotels and beach resorts stand empty or are used for other purposes. With travel to Cuba officially discouraged, the American tourist now goes to Puerto Rico, Jamaica, or into the Lesser Antilles. Mining, primarily in the Sierra Maestra of eastern Cuba, has centered around manganese, copper, nickel, and iron ore, but production of these minerals has been comparatively neglected in recent years.

Cuba, under Fidel Castro, is attempting to make some fundamental changes in the life and economy of the nation. To date, these changes have not been fully attained, nor does it appear likely that they will be in the near future. During the period when United States influence was strong, the island had moved more and more toward a one-crop economy. Land consolidation had led to increasing *latifundia*, or huge single holdings, and the traditional small owner-operator became a wage laborer or sharecropper. Wealth produced within the nation was removed as foreign-company profits, or it remained in the hands of a few privileged Cubans. In order to obtain popular support and perhaps to provide a truly better way of life for the majority of the population, Castro instituted an agrarian reform program designed to redistribute agricultural land among the farmers, to diversify the economy away from dependence upon sugar, and to increase the production of local foodstuffs. All these programs have thus far failed; and while effective per capita income rose in the early years, largely because of a more even distribution of wealth, it has more recently dropped below pre-Castro days. Faced with United States trade restrictions, increasing shortages of foodstuffs and consumer goods, a declining balance of trade, and a shortage of machinery and spare parts, the Castro government is again attempting to stimulate sugar production to record levels for export, in order to buy goods on the world market. Today sugarcane is still the dominant commodity and forms the island's only significant export. The sugar industry, once controlled by large companies and foreign capital, has been expropriated and nationalized. Closed off from its former principal market, where premium prices were paid, the Cuban sugar trade has shifted to Communist-bloc countries, where the much-lower world market price prevails.

Puerto Rico and the Virgin Islands

Until 1900 Puerto Rico's development followed similar lines to that of Cuba. Since Puerto Rico is more mountainous, however, sugarcane was not so important an economic commodity as in Cuba, and coffee, tobacco, and local foodstuffs were grown in the more sloping areas. The uplands area in the center portion of the island reaches a maximum elevation of almost 4,400 feet, but averages closer to 2,500–3,000 feet. A coastal plain fringes the center core, ranging in width from only a few hundred feet in the southeast to 5 to 15 miles elsewhere.

The impact of the United States in Puerto Rico was scarcely felt prior to the 1930s. Under the Good Neighbor policy in the 1930s, and later with World War II, Puerto Rican economic development began, the results of which are now apparent. The earlier work of building roads and port facilities, improving sanitation and education, and restructuring the government service formed the foundation for an ambitious plan to transform the Puerto Rican economy in a way similar to the United States model. This effort, known as "Operation Bootstrap," began largely after World War II and continues today. Within "Bootstrap," three segments of the economy are most important: agriculture, manufacturing, and tourism.

In the agricultural sector, sugarcane is the dominant crop and is grown on the coastal plains, on the lower hill slopes, and in some of the interior mountain valleys. In the uplands, tobacco is concentrated in the east-central section and coffee in the west. Elsewhere cropland is devoted to a number of minor export crops and foodstuffs to be consumed on the island. The production of local foodstuffs has augmented under the development program, and with increasing prosperity the island is now able to support truck gardening and dairying in the San Juan area.

One of the basic thrusts of Puerto Rican development has been agrarian reform. Under this program, maximum farm size has been limited to 500 acres, and most farms of greater size have been subdivided and parceled out to landless farmers. Roads, education, housing, sanitation, and medical facilities have been extended into rural areas; and technical assistance in crop production, marketing, and credit have aided in increasing yields per unit of land and per man.

Perhaps the most striking recent change in Puerto Rico is the great increase in manufacturing, which now exceeds agriculture in value of production. The island is handicapped by little mineral wealth, limited energy supplies, and a small local market. On the other hand, there is a relatively plentiful and cheap labor supply and access to the American market without tariff restrictions. These favorable factors have been augmented by government aid to and encouragement of new industry. This aid takes several forms, including: (1) exemption from taxes for a set period, (2) free plant sites and buildings for certain types of industries, (3) labor training programs, and (4) myriad services, ranging from assistance in getting financial credit to advertising finished products.

Today the traditional manufacturing activities of Puerto Rico, those devoted to processing agricultural products such as sugar and coffee, have been joined by a large number of other plants, many of these being subsidiaries of U.S. firms, which import raw materials, process them, and then ship the finished products to the mainland for sale. By and large, these newer industries are labor-intensive and use relatively small amounts of energy and raw materials. The fact that per capita income on the island increased from $115 in 1940 to $600 in 1960 and to over $1,000 in 1970 is due primarily to development in manufacturing.

Tourism, the third principal element in the economic development program, has also played an important role in increasing income. In this industry the natural beauty and tropical setting of Puerto Rico have been enhanced by construction of numerous modern, fully equipped hotels and resorts, some of which were built by the island government and then sold to private companies. In addition low-cost air fares from New York, Miami, and New Orleans have made Puerto Rico as accessible to the American tourist as most parts of the mainland proper. Finally, Puerto Rico has been able to capture much of the former North American clientele of the Cuban tourist industry.

Southeast of Puerto Rico are the Virgin Islands, divided between the United States, which bought its part from Denmark in 1917, and the British. The American Virgin Islands include St. Thomas, St. Croix, and St. John, along with several small, mostly uninhabited islets. Once of some importance as sugarcane producers, these islands now are geared mainly to tourism and retirement homes.

Hispaniola

The two independent nations of Haiti and the Dominican Republic furnish an excellent example of sharp cultural contrasts within a short distance. Haiti is strongly influenced by its African and French history. The population is high in density and is composed mainly of Negroes. A small percentage of mulattoes forms the upper class and dominates the economy and society. The mulattoes speak French and generally are oriented to French traditions. The lower class speaks the "Creole" patois, which, while mainly French, also contains words and expressions of Spanish, English, and African origin. Most of the population is at least nominally Catholic, but African religious beliefs such as voodoo are important among the lower class. When the French controlled the area, the better lands were divided into large sugar plantations; but during the Revolutionary period the French planter fled, never to return. Today about 90 percent of the Haitians are rural dwellers who work thousands of small subsistence farms by hoe and machete methods, and evidence of the old plantations has been largely obliterated. Population pressure has pushed agriculture into the mountains, which make up about 75 percent of the nation's land, leading to serious problems of deforestation, soil erosion, and soil depletion. Coffee is the leading cash crop, and some of it is grown in a semiwild state. Presently, Haiti is one of the few nations in the world where per capita income and living standards are decreasing. Both in the countryside and in the towns, the dominant impression is one of decay.

Differences in language, culture, and race and a lower population density separate the Dominican Republic, which occupies the eastern two-thirds of Hispaniola, from Haiti. The language and culture of the Dominican Republic are Spanish. About 10 percent of the population is of Spanish descent, and an equal number are of African origin; the bulk of the populace is mulatto. The population density is only one-half that of Haiti, and there is a greater percentage of arable land to support the Dominican people.

The country is divided by a sparsely settled hill and mountain area, oriented in an east-west direction. To the north is a heavily settled lowland of small farms. To the south, on the coastal plain, is the core of the nation, centered on the capital city, Santo Domingo. Large farms, associated tenants, and small farmers produce sugarcane, the products of which constitute the principal exports of the nation.

The Dominican Republic was long even more underdeveloped than Haiti. Spanish disinterest and later occupation by Haiti and attempts to reestablish Spanish control created unstable political conditions little suited for progress. In the twentieth century, while Haiti continued to be politically unstable, political stability in the Dominican Republic was established first by U.S. Marine occupation (1916–1924) and later under the long harsh dictatorship of Rafael Trujillo (1930–1961).

Jamaica

Jamaica, like Haiti, was originally settled by the Spanish; but continued incursion by the British led eventually to Spanish withdrawal and British control. Under both Spanish and British rule, settlement was concentrated on the coastal lowlands, where large plantations were directed to raising sugarcane, and on the slopes of the Blue Mountains, where high-quality coffee was grown. The interior, largely karst areas developed on limestone, known as the "cockpit country," remained unused until slavery was abolished in 1833, when emancipated slaves moved inland and began developing small subsistence plots. In the latter part of the nineteenth century, export banana cultivation began, and by 1900 it had surpassed sugarcane in value of exports. Grown on both large and small farms, the banana did much to save Jamaica from the danger of complete economic collapse brought on by the oversupply and increased competition in the sugar market and steadily increasing population pressure on the island. More recently, Jamaica has become noted as a source of bauxite and as a center for tourism.

Bauxite began to be actively mined on the island in the 1950s, and within a few years Jamaica had become one of the leading producers of the mineral in the world. The bauxite deposits, located in the central and northern sections, are overlain by limestone about 25 feet thick. The common method of mineral extraction is by open-pit methods. Since the deposits are located largely in the cockpit country, where only small subsistence plots of farmland were used and transport was little developed, the mining companies, United States and Canadian, have had to build roads, railroads, and port facilities for exporting the ore. Presently, bauxite and alumina (partly processed bauxite) account for over one-half of Jamaica's exports by value, and the monies earned by workers in the industry have done much to support the island's relatively high standard of living. Also important in the local economy are the monies received by the government, which are being used to develop the resource base of the island, to support other industries that are not so competitive, and to provide such basic services to the population as education, medical care, and water, light, and sewage facilities.

Jamaica has also benefited from Cuba's demise as a tourist center and continued North American prosperity. Long known as a resort area for the "jet set," luxurious hotel accommodations and residences are numerous along Montego Bay in the northwest and around Kingston in the south. More recently, the number of tourists has steadily increased, and the industry now ranks with mining and the growing of sugar and bananas as a major economic activity on the island.

The development of these activities has aided in relieving population pressures on the land, but for decades Jamaica has also benefited from migration. Jamaicans, in large numbers, went to Panama to work on the construction of the canal; they worked on the construction of railways in Central America linking the Caribbean coast to the interior; and many remained in Central America as laborers on the banana plantations, development of which followed the railways. These migrations were followed by regular seasonal transfer to the United States to work as harvesters in agriculture, to Cuba to assist in cane cutting, and more recently to the United Kingdom as permanent immigrants. Many of the island's emigrants remained away from Jamaica and now form significant elements in the population of Panama and elsewhere along the Caribbean coast of Central America. In these areas, the English language, place names such as Bluefields (Nicaragua), and the African character of sizable segments of the population are still much in evidence. Other migrants eventually returned to Jamaica, and many of them, with capital saved from their earnings, became small entrepreneurs and invested their money in stores and farms, thereby contributing both to urbanization and to the increasing well-being of the island's population.

Today Jamaica is an independent nation. Agriculture remains the backbone of the economy for most of the population, with large plantations oriented to sugarcane and bananas located on the best and most accessible lands. Small farms are devoted to local foodstuffs, with sugar, bananas, and coffee raised for cash income. While tourism and bauxite mining have greatly increased to the island's economic benefit, opportunities for large-scale migration have largely been closed. Future development in Jamaica will depend upon effective use of local resources relative to population pressure.

Lesser Antilles and Trinidad

The Lesser Antilles form a chain of islands extending from Puerto Rico to Trinidad. The chain is composed of two arcs, which cross about halfway along the chain at Guadeloupe. One arc is made of coral limestone islands that are flat, low-lying, relatively dry, and thinly soiled. Included in this group are parts of the Virgin Islands, Anguilla, St. Martin, Barbuda, Antigua, the eastern half of Guadeloupe, and Marie Galante. Barbados, which is not properly part of this arc, is similar in character. The second arc is composed of volcanic islands that are relatively high, steep-sloping, and rainy on the windward side but semiarid on the leeward side. Among this group are parts of the Virgin Islands, Saba, St. Eustatius, St. Kitts, Monserrat, Western Guadeloupe, Dominica, Martinique, St. Lucia, St. Vincent, the Grenadines, and Grenada.

Settlement of the Lesser Antilles began in the seventeenth century, with the British and French vying for control. Once established, both nations sought to make the islands into sugarcane lands, and large plantations were created using African labor. Until the early nineteenth century, profits realized by these sugar planters were legendary; but then, with the Napoleonic Wars, Western European development of sugar beets, the emancipation of the slaves, and colonial expansion into Africa and Asia, the islands went into a long period of stagnation.

On the larger, flat limestone islands, some plantations were consolidated into still-larger units with newer and larger processing mills and were controlled by companies in the mother countries. These modern plantations, which still form the economic base of the islands, were located in those areas most suitable for cane cultivation. Plantations not so well situated were less able to compete on the sugar market; often these farms were simply abandoned when the planter went bankrupt, and the land was divided among squatters. Where consolidation occurred, there was an intensification of sugar cultivation; where fragmentation took place, subsistence gardens, pasturage, scrub woodland, and sugarcane grown for a cash crop became the dominant land uses.

On the higher, more steeply sloping, and more humid volcanic islands, physical conditions for sugarcane cultivation were never good; but with high sugar prices, even these areas became primarily cane producers, although other crops such as cacao and coffee were also in evidence. When sugar prices dropped, production was greatly reduced; in its place, several other crops became locally important. On the drier leeward parts of some islands, cotton and citrus fruit became ranking crops. In the more humid, sloping areas, crops of coffee, cacao, and spices such as vanilla and nutmeg were expanded on both large and small farms, and on St. Vincent arrowroot became an important export crop. Everywhere in the Lesser Antilles, much land was given over to subsistence crops or abandoned. In the twentieth century, largely since World War II, export banana cultivation has brought some resurgence of economic viability, and on many islands bananas are now the dominant crop, grown on both small and medium-sized farms.

The collapse of the sugarcane market in the Lesser Antilles made the islands less attractive to the European planter, and depopulation of the islands by Europeans created an almost totally black population. In contrast to Haitian blacks, however, the Negro populace of the Lesser Antilles had been thoroughly Europeanized, and their African cultural traits had become insignificant.

The Dutch islands off the coast of Venezuela are extremely arid; consequently, agriculture has never been widespread there. Prior to the twentieth century, these islands were known for their salt production, using seawater as its source. With the development of petroleum production in Venezuela by both U.S. and Dutch interests, refineries were established on the offshore islands, and their economy now depends upon petroleum.

Trinidad was partly settled by the Spanish, but the island became a British possession in 1797. Under the British, sugarcane production

Fig. 5-5 Cacao orchard, Trinidad. Spraying is done regularly to prevent various diseases such as witches' broom. (Courtesy of United Fruit Co.)

expanded in the level area southeast of Port-of-Spain, and cacao plantations were established in the more humid areas. In the twentieth century sugarcane cultivation, sustained by East Indian farmers brought to the island as a labor supply after slavery was abolished, has continued in importance; but cacao cultivation has been almost abandoned in the face of increased competition from Africa and the prevalence of witches'-broom disease, which attacks the trees. In addition, the economy has been diversified by petroleum production, the products of which now form the bulk of the island's exports. More recently, Trinidad and its smaller dependency Tobago have enjoyed an increase in tourist trade, and light manufacturing is developing in the principal towns of Trinidad.

Control of the Lesser Antilles and Trinidad by the British and French has changed since 1950. The French have followed a policy of integration whereby Guadeloupe and Martinique have become politically integral parts of France. The British, on the other hand, have attempted to divest themselves of their Caribbean colonial possessions. In 1958, along with Jamaica and Trinidad, the British colonies in the Lesser Antilles, except for their part of the Virgin Islands, were formed into the West Indies Federation and given independence. Internal conflict arose immediately between Jamaica and Trinidad, and in 1961 the Federation was abolished. Jamaica and Trinidad became independent nations, and the British-administered Lesser Antilles islands reverted to colonial status. In 1966 Barbados attained full independence within the British Commonwealth, and attempts are being made by the British to form a government for their remaining Caribbean colonies.

Mexico

Mexico differs greatly from the Caribbean islands in that pre-Columbian cultures, languages, and ways of life have played and continue to play an important role in its national development. The introduction of Spanish ideas, systems, and agricultural crops, tools, and livestock did not result in the destruction of the indigenous Indian society, although the Indian population was severely reduced through battles, disease, and disruption of its means of production. Indeed, it would have been extremely difficult for the limited number of Europeans who came to Mexico to change fundamentally and lastingly a densely populated, highly organized society such as that of the Aztec and Maya empires.

Actually the impact of the Spanish has varied greatly in Mexico. In the north, where only limited numbers of Indians were found, and in the major urban centers where Spanish society and cultural traits dominate, only vestiges of Indian culture remain. In the region around Mexico City, a mixture of Spanish and Indian systems evolved, and today this area displays a diversity of cultures ranging from almost undisturbed pre-Columbian society to the modern industrial complex characteristic of Western Europe and Anglo-America. South of Mexico City, Spanish influence has been less important, except in the urban centers and along busy lines of transport, and Indian culture remains important in many sections there. The impact of the Indian and Spanish peoples is evidenced, in that probably 30 percent of the nation's 51 million inhabitants are Indian, and over 60 percent are mestizo. Yet, in spite of the Indian presence and influence, the official language, law code, and religion of Mexico are of Spanish origin.

Although industrialization has been rapid, Mexico's prime problem—how to raise the national standard of living—remains essentially agrarian. Fifty-five percent of the gainfully employed are engaged in agriculture that is largely subsistence in character, even though commercial agriculture has expanded to the point that its export totals are more valuable than those for minerals. Village life is the rule in most of the country; fewer than 10 percent of the Mexican population dwells on isolated farmsteads, and only about one-half lives in centers of over 2,500 people.

Actually, the country's natural endowment for agriculture is poor. More than half of its area is too dry to cultivate; and still other parts are too rough, or excessively hot and wet, or possess very poor soils. Only about 10 percent of Mexico's land surface is cultivated—a little over 1 acre per capita. Moreover, because of soil erosion and misuse, primitive tillage methods, poor seed, drought, and frost damage, crop yields per acre are commonly low. A wide variety of crops are grown, but maize, cotton, sugarcane, and wheat account for over one-half of the total value (maize alone accounts for 25 percent). Maize, with beans, chili, and occasionally meat for supplements, is the people's staple diet. Unfortunately, the lands naturally capable of producing the heaviest yields of maize do not coincide with the population centers. Although over half of the cultivated land is planted in this grain, Mexico still must import substantial quantities of it to meet its needs.

As late as 1930, 78 percent of all agricultural land in Mexico was controlled by only 2 percent of the landowners. Today, however, these large

Fig. 5-6 Mexico, like the United States, is divided into political units called states, with each state having a capital. Mexico City, the national capital, is not in any state but is located in the Federal District.

estates are mostly a thing of the past. The government has expropriated such vast holdings and from them has created *ejidos*, in accordance with laws growing out of the Mexican Revolution, which began in 1910. *Ejidos* are agrarian communities whose residents hold the land in common, although the cropland may be worked either by individuals or cooperatively. More recently, government-organized colonization programs have spurred settlement and development in the sparsely occupied southern tropical lowlands, and major land reclamation schemes have expanded irrigated land in the dry north and northwest regions.

Since the land is still not distributed on a wholly equitable basis, further progress must be made in increasing crop yields and the acreage tilled. About 1 acre of cropland in 8 is now irrigated, and further governmental reclamation and conservation projects are under way. An estimated three-fifths or more of the potentially arable land in Mexico is already in use, but almost one-half of this is left fallow each year. The food-supply problem is aggravated because an exceptionally high birthrate increases population by about 1.5 million annually. (In a recent year, there were 43 births and 9 deaths per 1,000 population.)

Like most Latin American countries, Mexico has a central core of dense settlement, with scattered smaller populous districts and sparsely occupied outlying areas. For comparative purposes, the political units of Mexico may be grouped into five divisions: Central Mexico, the Gulf Coast, the North, North Pacific, and South Pacific.

Central Mexico

This region is the heart of the nation. Although the area occupies only one-seventh of the country, about one-half of the population lives here. It is also the region of greatest industrial development. In addition, it possesses important mineral resources; and three of the four top-ranking cities, including Mexico, D.F., the largest, are located on the Central Plateau.

Central Mexico includes the mountains and basins at the higher and more humid southern end of the Mexican plateau, volcanic cones that

Fig. 5-7 The regions of Mexico are the outcome of both physical and cultural influences. The intensity of land utilization varies greatly from region to region.

Fig. 5-8 Mexico City, the capital of the country, has a population of over 6 million. It is located on the Central Plateau at an elevation of approximately 5,000 feet. (Courtesy of the Embassy of Mexico)

rise majestically above the general level, and the deep-cut valleys along the plateau's southern margin. The population is concentrated in the basins and adjacent lower slopes where the land is more level, the soils usually fertile, and the temperatures cool and healthful. In general, winters are mild, with occasional frosts and infrequent snow, and summers are relatively cool. There is usually enough rain, 20 to 40 inches annually, to support agriculture, although irrigation is often practiced. Recognizing that temperatures vary with altitude and rainfall with exposure, the climate of Mexico City, situated 7,486 feet above sea level, can be considered fairly typical of that of Central Mexico.

Although the region is Mexico's leading agricultural area, less than one-fourth of Central Mexico is cultivated. More land is in pasture, and more still is nonproductive, too dry, or on eroded slopes long since deforested to supply firewood and charcoal. Maize and beans are the leading crops. Considerable wheat is grown, and also sugarcane in lower, warmer places such as Morelos. Fruit, vegetables, and flowers in large quantities are supplied to the Mexico City market from the surrounding area. There is also a notable development of dairying tributary to Mexico City.

Mexico City is not only the national capital and the focus of the economic, social, and cultural life of the nation but also one of the largest cities in all Latin America, with nearly 6 million inhabitants. Included within its rectangular grid pattern are numerous parks and plazas, attractive suburbs, and buildings ranging from Spanish Colonial structures to ultramodern, multistoried apartments and office buildings. Its location on the drained bed of part of Lake Texcoco has created a problem for the city. Under the weight of streets and buildings, surface levels have subsided as much as 16 feet in places as ground water was pumped from wells driven into the soft subsoil to provide the city's water supply.

Mexico City is the nation's leading manufacturing center. Mexican manufacturing is generally light industry, producing consumer goods from domestic raw materials. In more recent years, however, there has been a tendency also toward production of goods using imported materials. The bulk of industrial activity in the capital is concerned with textiles, home necessities, paper, chemicals, ceramic wares, and lighter metal products. Handicraft industries, practiced in the home or in small workshops, are common here as elsewhere in the nation.

The government is discouraging the location of new manufacturing plants in the Mexico City area, thereby trying to decentralize industry and to encourage economic growth in other regions. Another reason for decentralization is that air pollution has become an increasing health hazard. The city, located in a sheltered basin, often experiences temperature inversions; under such conditions pollutants released in the air fail to rise and dissipate, thus concentrating in the lower atmosphere. Such a situation is similar to that of the Los Angeles basin in the United States.

The capital and environs are also the focus of Mexico's excellent network of air transportation and its 29,000 miles of all-weather roads. Consequently, the capital is the major center for tourism, which is the republic's largest single source of dollar income. Many tourists travel to Mexico City over the Pan-American Highway, which was opened from Laredo, Texas, in 1936 and has since been completed into Panama. The total length of the highway within Mexico is 1,745 miles.

Gulf Coast

The Gulf Coast region has a level surface and a hot, rainy climate. Much of the land is covered with rain forests. In the state of Veracruz, however, the coastal lowlands, which are often swampy and have considerable areas of savanna, are backed by the forested eastern slopes of the Sierra Madre Oriental, which lie in the *tierra templada*. Still higher towers snow-capped Orizaba, one of North America's highest peaks. Precipitation is especially heavy, and a dense forest covers the lowlands of the Isthmus of Tehuantepec, the state of Tabasco, and southern Campeche. Less rain, a distinct dry season, and rapid drainage into the underlying limestone account for the scrub forest vegetation in the northwestern Yucatan Peninsula.

Although the Gulf Coast region ranks high in output of agricultural products, it still has many undeveloped agricultural resources. Much of the coffee which makes Mexico one of the world's leading producers of this commodity is from this zone. At the same time, however, banana exports, largely from the Gulf Coast area, which exceeded 14 million stems in 1937, have now dropped to less than 3 million stems annually. Maize yields here are higher and more dependable than elsewhere in the nation; and the grasslands, though tropical, support many cattle. From the forests of the Isthmus of Tehuantepec and those extending south into the Yucatan Peninsula is gathered much of the world's chicle. About 15 percent of the world's hard fibers (henequen) is grown in Yucatan. The annual yield of over 125,000 tons of fiber was once largely exported to the United States for the manufacture of binder twine and cordage, but today most of it is fabricated in Mexico and used domestically. Within the Gulf Coast region is located one of the largest reclamation and power projects ever undertaken in Mexico, a $200 million undertaking on the Río Papaloapan. In the 1950s and 1960s, in an attempt to increase production and alleviate overpopulation in other parts of the nation, several ambitious resettlement and colonization projects were carried out in the southern Gulf Coast region.

The principal oil fields of Mexico are located in the Gulf Coast area. Since the first commercial oil discovery was made west of Tampico in 1901, the cumulative production of petroleum within the country has been greater than that in any other nation except the United States, Venezuela, and the Soviet Union. Peak output was reached in 1921, when over 193 million barrels were exported. Fields are largely in three Gulf Coast areas, contiguous with Tampico, south of Tuxpan, and inland from Coatzacoalcos (Puerto Mexico). Present annual production of about 160 million barrels is in approximate balance with yearly national consumption. The petroleum industry is a government monopoly. Tampico is the important petroleum-producing and refining center, as well as one of the principal ports. Production of sulfur in the Isthmus of Tehuantepec area has expanded to over 1 million tons annually since 1954, a quantity placing Mexico second only to the United States in output.

Population is largely centered in small communities. Rural population is densest on the temperate slopes of the Sierra Madre Oriental, focusing on such towns as Jalapa and Orizaba, both centers of cotton manufacturing. Veracruz, the major port on the Gulf of Mexico, is the principal shipping, trading, and processing center in a fertile reclaimed agricultural area. Mérida is the commercial and manufacturing center for the henequen area of Yucatan Peninsula.

The Yucatan area is famous for its Mayan ruins. Many of the former Mayan centers, such as Chichén-Itzá, have been partially reconstructed and now serve as attractions to an increasing number of tourists. These centers, which were built by a people in some ways more advanced than those of the contemporary European civilization, represent the mother culture that contributed much to the peoples of southern Mexico and northern Central America. These vestiges of past grandeur stand in stark contrast to the limited modern-day development characteristic of much of the region.

Fig. 5-9 Open-air market, Mexico City. Open-air markets, common in most towns and cities of Mexico, are especially active on Sundays and holidays. (Courtesy of Les Barry)

The North

In this thinly populated, physically diverse region, winters are cool and summers are warm to hot. The surface is largely plateau, but also included is much of the Sierra Madre Occidental, the northern part of the Sierra Madre Oriental, and a segment of the Gulf Coast plain. Basin floors are lowest and level land most extensive nearer the Rio Grande.

On the plateau, the major land use is grazing; but because of moisture deficiency, the carrying capacity of pastures is low. Crops must almost always be irrigated. The principal commercial crop is cotton, of which one-fourth of the nation's annual production formerly was harvested on *ejido* land in the Laguna area centered on Torréon. New areas such as along the lower Rio Grande, the Gulf Coast, and other sites on the plateau now rival Laguna as producing centers. Much of the expansion is the result of new irrigation projects such as the huge multipurpose Falcon Dam on the Rio Grande, which is part of an international water-conservation project. On unirrigated cropland wheat is dominant, and the North as a whole is the leading wheat-growing section of the republic.

The chief wealth of the region lies in its mineral resources. Chihuahua, Zacatecas, and Durango are three of the nation's five leading mining states. Mexico is the world's leading producer of silver, is commonly second in lead, and is among the first ten in the production of molybdenum, antimony, mercury, zinc, and gold. Lead, zinc, silver, copper, and gold together total some 85 percent of the annual value of minerals mined, exclusive of petroleum.

A good-grade bituminous coal, mined at Sabinas and Monclova in Coahuila, and iron ore from Durango are brought together in Monterrey, making this city the only iron and steel center of consequence in Middle America. Monterrey is second only to Mexico City as an industrial city and is the nation's third largest urban community.

North Pacific

Much of the North Pacific region is too dry, too rough, or too isolated to support more than low population density. In fact, over half of the region is nonproductive, and settlement is focused where water is available for irrigation, in scattered mining camps, and in the border towns of Tijuana, Mexicali, and Nogales.

The region's isolation from Central Mexico was partly remedied in 1948 by the completion of a railway southeast from Mexicali to a junction with the Nogales-to-Mexico City line. More recently, the region has received much government interest and investment. Numerous dams have been built to control floods and supply irrigation water to reclaimed land along the coast. Roads have been constructed to better link the region with the rest of Mexico and to provide access to various parts of the area. Other support in terms of funding and technical assistance has partly overcome the inherent deficiencies of the area and has led to increased agricultural production in favored locations.

Sugarcane, cotton, wheat, and garden vegetables are crops of general consequence. The state of Nayarit is Mexico's primary tobacco-growing area; the Yaqui Valley in southern Sonora is its leading center of rice cultivation; and the Mexican Imperial Valley, one of the places where the United States and Mexico are in dispute over water rights, is its largest long-staple cotton grower. Copper mines, especially at Cananea in Sonora and Santa Rosalía in Baja California, are foremost in output of this metal in the republic.

South Pacific

Separated from Central Mexico by the deep Balsas Valley, the South Pacific area is largely a highland region composed of narrow, flat-topped divides and steep-sided valleys. A strip of coastal plain, the Tehuantepec lowlands, and a flat-floored valley in central Chiapas provide the most extensive areas of level land. Since most of the surface is well watered, there are valuable forests. In places where rainfall is light, however, as in the interior of Oaxaca, irrigation is practiced.

Isolation is the dominant fact in the life of the region. Except for sulfur deposits in the Tehuantepec lowlands, even mining is relatively unimportant, although mineralization is believed to be great. Set apart from the main currents of Mexican life, most of the area's inhabitants lead a subsistence existence; yet with improved transportation, considerable areas now unused may become productive.

The region has no large cities; Oaxaca is the principal urban center. Acapulco, today an internationally renowned resort and in colonial times the chief Pacific port, has a much better harbor than Manzanillo, but few ships now call there because its connection with Mexico City is by motor road only. Exports of the South Pacific region—coffee, bananas, and forest products—leave mostly through the Atlantic port of Coatzacoalcos.

Progress in Mexico

Both the Mexican government and private interests are making strong efforts to improve the economy of the country. Big, new luxury hotels help to attract tourists to Mexico City and other centers, and tourism is a major source of income. Great improvements have been made in education, from village schools to the National University of Mexico, famed for its ultramodern buildings. The government is assisting agriculture by reclamation, land subdivision, and instruction in improved farming methods. Manufacturing has expanded, especially in such light industries as natural and synthetic textiles, tobacco processing, and metal fabrication and assembly. Although iron and steel are produced at Monterrey, where coal and iron ore from local sources are used, the output supplies only part of the national needs and is being increased. Most of the principal Mexican cities have made rapid material gains with the development of industry and trade, along with improvements in transportation, education, and construction. As a close neighbor of the United States, Mexico has more trade with this country than with any other. Gradually even the people of rural villages are learning about developments in the outside world, and throughout Mexico the people are involved in a process of social and economic change.

Central America

Central America is a transition zone. The northern portion, including Guatemala, British Honduras, El Salvador, Honduras, and the northern upland part of Nicaragua, is similar in many respects to southern Mexico. It is here that both Indian and European influence is strong; the basic diet of its population is made up of corn, beans, and squash. The southern part, composed of most of Nicaragua, Costa Rica, and Panama, is more European in culture, with Indian influence limited to small isolated areas; here, although corn and beans are important foods, there is an emphasis on the tuber crops common in adjacent South America.

Central America does have a tradition of unity. In Spanish colonial times, except for Panama, the region was administered as a single entity, and for a short time during the early independence period there was a federation of Central American states. While this federation was soon abandoned and nation-states, formed as we know them today, quickly evolved, there has always been a sentiment—admittedly weak at times—toward reunification. This sentiment has reasserted itself since World War II in the form of economic unity for development, and in the 1950s the Central American Common Market was created; this organization today ties together, at least to some extent, the economies of all Central American nations except for British Honduras and Panama.

In spite of a degree of unity, as exemplified by their growing economic cooperation, the various countries of Central America are individualistic, and relations among nations are often strained at best. Guatemala claims the territory of British Honduras as part of that nation; densely populated El Salvador has quarrelled with Honduras over their boundaries; El Salvadorean migrants in neighboring countries have created difficulties; and rebel groups have launched attacks into their homeland from bases in adjoining countries. Each nation is, therefore, a region itself and merits separate treatment.

Guatemala

In many ways, Guatemala best exemplifies the conflict of Spanish and Indian influence. About half of the nation's population follow the European way of life, speak Spanish, are at least nominally Catholic, and have some national feeling. These people, called *ladinos*, are contrasted with the *indigenas*, who also comprise about 50 percent of the population. The *indigenas*, 400 years after the Spanish Conquest, dress in traditional handwoven fabrics, speak mainly Indian dialects, are devoted to native pagan religion modified by Christian influence, and pay greater allegiance to their local group rather than to the national government.

Settlement patterns in Guatemala reflect the dual cultures. The Indians are concentrated in the western highlands in an area centered on Quezaltenango, just as when the Spanish arrived on the scene. They live on small plots of land and grow corn, beans, and squash and, where roads have penetrated, a variety of vegetables for sale in the larger urban markets of the nation. Proud but extremely poor, these Indians face the prospect of overpopulation. Nearly all the level land was long ago planted with crops, and more recently the steep volcanic mountain slopes have been cleared of forest in search of still more land. Yet, land is scarce and opportunities are limited in response to population increase. Increasing numbers of Indians are being forced to move in search of a livelihood. Some go seasonally to the Pacific coast and fringing mountain slopes, where large farms need labor for harvesting cotton, sugarcane, and coffee. Others move permanently to urban centers in search of any work they can find; and a few drift northward toward the sparsely populated Petén, hoping to establish a way of life similar to that which they knew in the highlands.

South of the highlands and fringing the Pacific coast is a narrow, but level and fertile, coastal plain. Prior to World War II, this plain was sparsely settled, isolated, and disease-ridden, particularly with malaria. After the war, however, an all-weather highway was completed along the coast, and feeder roads were constructed throughout the region, thereby making access relatively easy. At the same time, active and successful campaigns were initiated to eradicate the malarial mosquito. Throughout the 1950s and 1960s, farmers cleared vast tracts of forested land and developed large modern farms devoted to pasturage, cotton, and a number of specialty crops. Today the south coast is the richest agricultural area of the country and grows much of the nation's produce destined for export. Here the land is controlled by *ladinos*, but Indian labor is often used during the harvest. Both lower-class *ladinos* and Indians are flowing into the region and are making it one of the country's most rapidly growing areas.

North of the highlands lies the Petén, the limestone hills near Cobán, and the narrow Caribbean coastal section centered on Puerto Barrios. The Petén, though encompassing about one-third of Guatemala's total land area, is almost uninhabited. Once a seat of lowland Mayan culture, whose majestic ruins dot the landscape and attest a considerable pre-Columbian population, the Petén today has a small population cluster around Flores and a few isolated agricultural settlements along the two or three roads found in the area. Overland transportation between the Petén and the rest of Guatemala is nonexistent. As a consequence, agriculture there is basically of the subsistence type. Only chicle and a few spices such as pepper and allspice are obtained for export via air transport.

To the Guatemalans, the Petén is the frontier

region in which they believe the problems of the nation may find a solution. The optimism generated by the potential of the Petén may be unwarranted, since it is an area of karst topography and lacks the basic ingredients necessary for large-scale agrarian development. To supply these missing elements—namely, roads, people, and marketing facilities—will require very substantial expenditures of capital.

The limestone hills around Cobán rival the upper part of the south coastal plain as a major coffee-growing area. Coffee growing, developed by German immigrants in the mid-nineteenth century, is concentrated in the immediate Cobán area. The remainder of the limestone hills area is largely covered with scrub forest on the sloping lands, with pasturage and subsistence crops being grown in the flat-bottomed depressions.

The Caribbean coast of Guatemala has long been relatively neglected, except as an access route to the highlands. Puerto Barrios and its new sister port, Puerto Galvez, are the major transshipping points of the nation; yet it was not until 1956 that an all-weather road connected them with the rest of the country. Previously, all goods and passengers had to travel by rail. Around Puerto Barrios, banana plantations developed by the United Fruit Company supply nearly all of this crop, which accounts for a substantial part of Guatemala's exports. Away from the Barrios area, the land is sparsely settled and devoted to subsistence cultivation.

Guatemala City is the focal point of the nation, and nearly 500,000 people live in and around the capital. The city is not only the cultural center of the *ladino* population but also the only significant center of manufacturing, wholesaling, and retailing for certain luxury goods.

British Honduras

British Honduras is in a state of political change from colony to independent nation. Once a significant source of commercially valuable tropical hardwoods such as mahogany, tropical cedar, and rosewood, the colony now imports more than it exports. For this reason alone, the United Kingdom prefers to withdraw from the area. Moreover, the gradual withdrawal of Great Britain from British Honduras represents a declining interest in all of Central America and contrasts sharply with conditions in the nineteenth century, when British influence was strong along the Caribbean coast throughout the isthmus.

The slightly more than 100,000 inhabitants of British Honduras live primarily along the coast, with the greatest concentration around the capital, Belize. The topography of the colony is largely a flat coastal lowland, on which agriculture is devoted to citrus, cotton, and local food crops. In recent years, roadbuilding projects have opened new lands to settlement; but much of the colony is still covered in forests, where lumbering and chicle gathering remain the chief forms of livelihood. The population is composed primarily of English-speaking Negroes, descendants of laborers who came to British Honduras from the West Indies in past centuries.

El Salvador

El Salvador is unique among the countries of Central America, in that the nation has no frontage on the Caribbean Sea, and that practically all its land is occupied. Access to the Caribbean is largely by railway through Guatemala to Puerto Barrios. While the Pacific port of Acajutla, improved in the late 1950's, is an important transshipment point for products entering and leaving El Salvador, it is usually cheaper to move goods traded with European countries by way of Puerto Barrios. Thus, El Salvador depends to a large degree upon maintaining good relations with Guatemala. Relations with neighboring countries are often strained, however, since population pressure upon the land is extremely high in El Salvador, and Salvadorean migrants, both legal and illegal, have crossed the national borders and settled in surrounding countries. The lack of unoccupied land in El Salvador, the high population density, and a rapid rate of population increase are the fundamental problems facing the nation today, and these have resulted in occasional brief armed conflicts with neighboring nations. Such conflicts have most often been with Honduras, which has a relatively low population density and large amounts of unoccupied land.

The bulk of El Salvador is made up of two types of landforms. The southern part is a continuation of the low coastal plain extending from Guatemala. The northern part is composed of hilly to mountainous terrain, which has a few active volcanoes rising above the general elevation of 2,000 to 3,000 feet. Agriculture dominates life in El Salvador, with coffee in the foothills and cotton on the coastal plain being the two principal export crops. Food cultivation for local consumption occupies most of the land, with corn the principal crop. While food crops are grown primarily on small and medium-sized farms, many of the larger holdings are devoted to pasturage for cattle. Attempts at alleviating the problem of population pressure have led to increased manufacturing in the San Salvador area. However, industrialization is limited by lack of power resources and raw materials, except for agricultural products produced in the country.

Honduras

In the ratio of land area to population, Honduras differs radically from El Salvador. Large areas in the eastern part of the country are inhabited only by a few primitive Indians; and even in the more densely settled west, population pressures are not great. In the west, there are two basic population nodes. In the highlands, the population is concentrated in mountain basins and valleys in the vicinity of the capital Tegucigalpa and Comayagua. Here are found the older towns of the country and the greatest concentration of Indian population. It is an area that gives an impression of stagnation. Magnificent buildings of the past stand side by side with poorly constructed, more recent structures. Means of production are representative of nineteenth-century and earlier techniques. Even the major cities such as the capital Tegucigalpa give an impression of poverty, and of having been bypassed by the modern era.

In contrast, the second population node, in the Caribbean Lowland, presents a picture of change, movement, and progress. San Pedro Sula, near the Guatemalan border, is the principal city in the lowlands and now rivals the capital in importance. While the highlands are largely oriented to subsistence agriculture, the lowlands are devoted to commercial crop production, mainly bananas, and manufacturing is of increasing importance.

Until 1969, overland connection between the two population nodes was by dirt road, impassable for much of the year. Now a two-lane all-weather road links the areas and provides for greater interchange of goods and people. Nevertheless, transportation throughout the nation is still underdeveloped and hinders efficient utilization of resources.

Nicaragua

Nicaragua is the most sparsely populated Central American country and competes with Honduras in having the least foreign trade in the area. Only a small percentage of its surface is cultivated, and only a little more is used for pasture. Except for a few mining camps and small seaside settlements along the flat, swampy Caribbean shore, the forested, inaccessible eastern half of the country has few people. Banana production, important along the coast in most Central American countries, is focused on Puerto Cabezas and Bluefields but is of small consequence to the economy.

The country is unusual in that many of its

people live in *tierra caliente*. Population is densest in the lakes region, especially between Managua and Granada. Here, much of the cotton, the country's most valuable export, is grown. Settlement is also fairly dense in the highlands northeast of the lakes, as in the coffee-growing area of Matagalpa. Most coffee, however, is grown southwest of Managua, the capital. Cattle, raised on grasslands, especially those east of Lake Nicaragua, enter foreign trade both as live animals and in the form of hides.

Costa Rica

Costa Rica differs from other Central American countries in that its population is largely of European origin. A land of small farms, it is the most democratic and literate nation of the isthmus. Population pressures are increasing rapidly, although much of the tropical lowland is still sparsely settled. Foreign trade is based largely on three agricultural products: coffee, bananas, and cacao.

Population is concentrated in the fertile and temperate Meseta Central, where most of the people are white or near-white. In this area is located the capital, San José, and three other of the nation's largest cities. The Meseta Central is the chief producer of Costa Rica's fine-flavored coffee. Large amounts of maize, sugarcane, beans, potatoes, fruit, and vegetables are grown for local consumption. There are also substantial grazing and dairy industries.

In the Caribbean Lowlands, settlement is of consequence only along the main line and branches of the Puerto Limón–San José railroad. The world's first large-scale commercial production of bananas was established in these lowlands. Peak exports, in 1913, totaled over 11 million stems. More recently, because of Panama disease, production has greatly diminished, and the center of banana farming has shifted to the country's west coast. Meanwhile, much of Costa Rica's 40,000 acres of cacao have been established in the Caribbean Lowlands. In addition, Negroes, who constitute most of the lowland population, grow considerable quantities of food for domestic use.

Fig. 5-10 Banana plantation, Panama. Cableways carry the bananas many miles from the farms to the packing station. (Courtesy of United Fruit Co.)

Banana production is now concentrated along the Pacific, with Golfito and Quepos the leading ports. The fields must be irrigated during the dry season and are sprayed to prevent sigatoka (leaf-spot) damage. It is significant, however, that over 12,000 cultivated acres of bananas in the Quepos division have already been abandoned because of Panama disease and have been replaced by African oil palm, cacao, pasture, and trial plantings of mahogany and teak. Banana land controlled by major foreign concerns is now largely subdivided into small owner-operated plots. Many former plantation workers are from Guanacaste, a northwestern province that is a scrub forest and grassland area of cattle ranches and cereal production, with a moderately dense mestizo population.

Panama and the Canal Zone

First traversed by Balboa, who discovered the Pacific Ocean in 1513, the Isthmus of Panama has since been most significant as an interoceanic pass route. Over the "Old Gold Road" traveled much of colonial South America's wealth. United States interest in this interocean crossing increased during the California Gold Rush of 1849. In 1855, U.S. capital completed the transisthmian railroad, but thereafter interest waned. French attempts to dig a canal, between 1880 and 1889, failed because of yellow fever and mismanagement. When the Spanish-American War started in 1898, the U.S. battleship *Oregon* required two months to travel the 13,000 miles from the Pacific to the Atlantic, and the necessity for a Central American canal became evident to the United States government.

In 1903, the United States encouraged Panama, then a part of Colombia, to revolt. With tacit assistance from the United States, the rebellion was soon successful, and by treaty with the new nation the United States gained the right to build, operate, and defend an interoceanic canal and to have jurisdiction in perpetuity over the 10-mile-wide Canal Zone. The Canal Zone has a land area of 362 square miles and a population, including those in the ports of Balboa and Cristobal, of approximately 50,000. The Panama Canal, completed in 1914, is 50 miles long and has three sets of lift locks on each side of the summit elevation of 85 feet in Gatun Lake. On the Atlantic side, three locks are located at Gatun; on the Pacific side, one set is at Pedro Miguel and two at Miraflores. The Canal's value to the United States as a wartime facility is, of course, inestimable. In addition, it has greatly aided the expansion of American intercoastal commerce and of trade between the eastern coast of the United States and the west coast of South America and Eastern Asia. Bulky raw materials constitute the chief Pacific-to-Atlantic cargo, and mostly manufactured goods move in the opposite direction.

The Panama Canal cannot accommodate some of the giant tankers and other ships built in the post–World War II period, and the pressure of traffic on the Canal has spurred plans for construction of a second waterway, with suggested passages including a sea-level route in the Darien area east of the present canal. Actual selection of such a site has been hampered by growing conflict between Panama and the United States over control of the present canal.

The Republic of Panama itself is divided into two parts by the Canal Zone. To the east, much of the rough, rainy, malarial, and selva-covered land, with maximum elevations of about 4,000 feet, is almost unoccupied, except for the San Blas Indians living along the Caribbean littoral. To the west, on the Caribbean side of the mountain backbone, precipitation is very heavy, and the rain forest is thinly peopled. On the Pacific side, where rainfall is less abundant and more seasonal, there is considerable savanna vegetation. In this area live most of Panama's population. The completed Panamanian section of the Pan-American Highway loosely ties this population together. A transisthmian highway between Colón and Panama City was constructed during World War II.

Over three-fourths of Panama's land area remains clothed in forests, the source of fine cabinet woods and other tree products. Cattle, yielding hides for export, graze over much of the remaining surface. Only about 6 percent of the country's land is cultivated. Rice is the leading subsistence crop, and bananas, of which 10 to 12 million stems are shipped annually, are mostly grown on the Pacific side near Costa Rica. The old banana area on the Caribbean shore is now significant for plantations of cacao, coconuts, and abaca, a fiber source.

In Perspective

Middle America is in the process of fundamental change. Some developments, such as growing industrialization, may well lead the region into the European economic mode. Others, such as increasing population pressure, may negate economic advances and lead toward more subsistence producers. These conflicting trends may well change the geography of Middle America in the near future.

Perhaps the most important trend in Middle America is that of the increasing population pressure upon the resource base. Everywhere in the area the annual rate of population increase is high—nearly twice that for the world as a whole. In the Antilles, where practically all the potentially arable land is already in use, population densities per square mile have reached alarming proportions. Further population increases cannot be supported solely on present systems of agricultural land use. While some improvement in farming techniques (e.g., more intensive cultivation by means of irrigation or fertilization) is possible, other economic resources must be developed. To this end, tourism is being encouraged, and on the larger islands some light industry has developed. Manufacturing opportunities, however, are limited by a lack of local raw materials, available energy supplies, and local markets.

In Mexico and Central America, natural advantages and opportunities for supporting an increasing population are greater. Here two significant trends combine to lessen the population problem. One is the development of heretofore sparsely settled tropical lowlands. Long considered as unhealthful sites with a limited resource base, the low coastal zones of Mexico and Central America have been avoided. Now with effective medical action against diseases such as malaria and yellow fever, and with the

Fig. 5–11 The population of Middle America is increasing more rapidly than in most other parts of the world.

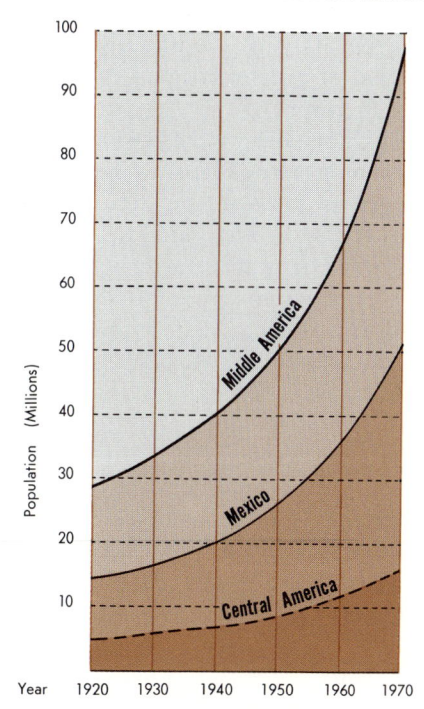

POPULATION GROWTH IN MIDDLE AMERICA

Source: 1970 World Population Data Sheet,
Population Reference Bureau, Washington, 1970;
Center of Latin American Studies,
Statistical Abstract of Latin America, 1968,
University of California Press, Los Angeles, 1969.

construction of roads and other supporting facilities, these tropical lowlands are the scene of new settlement and colonization. The second favorable trend is the development of manufacturing in the traditional centers of population. In this respect, Mexico is the economic leader in Middle America, for it now possesses many modern factories producing complex products such as petrochemicals, synthetic fibers, automobiles, and washing machines. Many of these products are fabricated completely within Mexico by Mexican labor for the growing domestic market. In Central America, manufacturing is less advanced but of increasing importance. Industrialization may be a partial answer to population growth, since it offers increased employment opportunities and assists in more efficient utilization of resources.

Another trend that may have great impact on the geography of Middle America is the development of greater cooperation among the countries to improve their common standard of living. The nations of Central America, except Panama and British Honduras, have joined together to form the Central American Common Market. The purpose of the Common Market is to reduce tariff barriers among member countries in order to create greater trade and to provide a large market for industries within the area. Mexico belongs to the Latin American Free Trade Association (LAFTA), which includes most of the larger nations of Latin America. To date, however, LAFTA exists more as a concept than a reality. Finally, in 1967, the Caribbean Free Trade Association (CARIFTA) was formed to link together economically the British colonial islands of the Lesser Antilles and the Commonwealth countries of Trinidad, Jamaica, Barbados, and Guyana. The basic purpose of both LAFTA and CARIFTA is the stimulation of greater internal trade in order to promote economic growth. All three of these economic unions have followed the sample of the highly successful European Common Market.

Migration of people constitutes still another trend. Within Middle America, two main directions of population movement are most apparent. The first is the flow of rural peoples from more densely settled areas to less populated zones; this movement is occurring largely in Mexico and Central America. The second is the urban-directed movement, from rural areas and smaller towns to the major cities of the region. Increasing urbanization results from two factors working together. One factor is the "push" of people generated in the countryside, where employment opportunities have not kept pace with population increase. People have literally been forced to leave these rural areas because there is no means of livelihood. The second factor is the "pull" or attraction of the larger cities, where, because of increasing industrialization and the existence of better services and facilities such as schools, electricity, and indoor water supply and sewage disposal, the cities are considered desirable places to live. For these reasons, urban centers are growing rapidly, at rates of 5 to 7 percent per year. Such rapid increases create problems of absorption for the cities. With limited budgets, it is increasingly difficult for them to provide adequate housing and public utilities. Furthermore, many of the new city dwellers do not have the skills needed by urban employers.

To resolve some of the problems created by increasing population and urban migration, most nations are instituting agrarian reforms, through which attempts are being made to improve the rural scene by land redistribution and better systems of land use. In Mexico, the reform movement began with the revolution of 1910 and has had a fundamental impact on all aspects of Mexican life. Much of the nation's land, which was then held by a small percentage of the population in very large units, was expropriated and transferred into *ejidos*, or communal land holdings controlled by village councils. In more recent years, the movement has been more oriented to improving land use by means of cooperatives, credit facilities, irrigation projects, and the creation of small privately owned farms in sparsely populated lowlands. In Central America, agrarian reform is largely of post–World War II origin, with emphasis on land redistribution and settling of unoccupied lowlands. In the Greater Antilles, Puerto Rican agrarian reform began in the late 1940s and has progressed to a second phase. Its focus was at first one of land redistribution and improvement of small-farm production; today reform continues but now emphasizes social benefits such as electrification, water supply, transportation, and housing. Cuba has had a fundamental land reform initiated by the Castro regime. There, land units have been nationalized and the government has taken over the role of farm management, with the workers sharing in the profits. As agrarian reform continues in Middle America, it is expected that the numbers of *latifundia* will be reduced, but it is questionable if significant advances in production for export will be made.

Finally, the general characterization of Middle America as a region of a small upper class and a large lower class may need revision in the near future. By gradually improving standards of living, increasing economic opportunity, industrialization, urbanization, and agrarian reform, a middle class is developing. Already Mexico and Puerto Rico possess a sizable middle class, and in Jamaica, the Dominican Republic, and parts of Central America, the middle class is increasing. The formation of a substantial middle class will have a profound influence on the political, social, and economic structures of Middle America.

REFERENCES

Bounds, John H.: "The Bahamas," *Focus,* vol. 19, pp. 1–7, May, 1969.
Fox, David J.: "Man-Water Relationships in Metropolitan Mexico," *Geographical Review,* vol. 55, pp. 523–545, October, 1965.
Fuson, Robert H.: "The Orientation of Mayan Ceremonial Centers," *Annals of the Association of American Geographers,* vol. 59, pp. 494–511, September, 1969.
Horst, Oscar H.: "The Specter of Death in a Guatemalan Highland Community," *Geographical Review,* vol. 57, pp. 151–167, April, 1967.
Hoy, Don R.: "Changing Agricultural Land Use on Guadeloupe, French West Indies," *Annals of the Association of American Geographers,* vol. 52, pp. 441–454, December, 1962.
James, Preston E.: *Latin America* (4th ed.), The Odyssey Press, New York, 1969.
MacPhail, Donald: "Puerto Rican Dairying: A Revolution in Tropical Agriculture," *Geographical Review,* vol. 53, pp. 224–246, April, 1963.
Minkel, Clarence W.: "Problems of Agricultural Colonization and Settlement in Central America," *Revista Geografica,* vol. 66, pp. 19–54, June, 1967.
Pearcy, G. Etzel: *The West Indian Scene,* D. Van Nostrand Company, Inc., Princeton, N.J., 1965.
Sauer, Carl O.: *The Early Spanish Main,* University of California Press, Berkeley, 1966.
West, Robert C., and John P. Augelli: *Middle America: Its Lands and Peoples,* Prentice-Hall, Inc., Englewood Cliffs, N.J., 1966.
Wilgus, A. Curtis (ed.): *The Caribbean: Its Hemispheric Role,* University of Florida Press, Gainesville, 1967.
Young, Bruce: "Jamaica's Bauxite and Alumina Industries," *Annals of the Association of American Geographers,* vol. 55, pp. 449–464, September, 1965.

6

South America

During the 1960s, South Americans by the millions became aware of their material poverty. In that decade, demands to remedy economic and social conditions and to alleviate the suffering caused by malnutrition, lack of medical assistance, or absence of opportunity became irrepressible at many social levels and in several Latin American nations. Programs and doctrines have been and are being pursued that will alter or replace traditional systems of the past not suitable to modern-day needs. At times a man, such as Juan Peron, became the embodiment of a doctrine. In some countries, however, even though leaders or programs won the allegiance of a majority of that nation's population—rural peasantry, miners or plantation workers, urban slum dwellers—if the leader or program was unable to enlist the support of the ruling socioeconomic hierarchy, the efforts toward change usually failed.

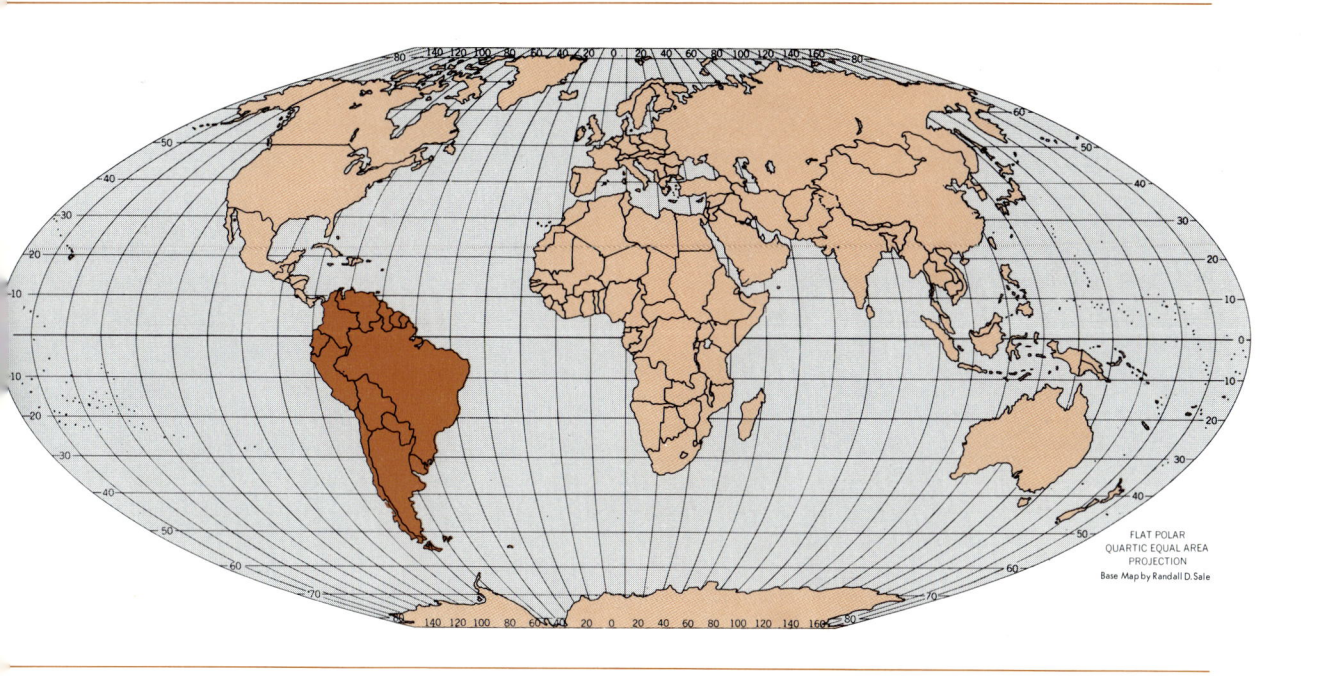

When the desire for change resulted in chaotic conditions or adversely affected vested interests of the ruling hierarchy, suppression by police and military establishments largely serving the interests of the traditional hierarchy often followed. Brazil, Peru, and Argentina were examples of this pattern in the 1960s. At the onset of the 1970s, more than two-thirds of the South American people were ruled by basically nondemocratic "caretaker" governments that are usually no more than the traditional Latin junta in disguise. Recently the guerrilla terrorism that has plagued Colombia for more than a decade appeared in several other nations, including Bolivia, Brazil, and Uruguay.

During recent decades, high rates of population growth coupled with a declining death rate have added large increments to the populations of many South American nations. These increases are shaping the pattern of economic development for several nations for decades to come. The task for many such countries—to which Argentina and Uruguay, with their stable population levels, are striking exceptions—is to produce or obtain food supplies, to provide adequate minimum levels of public health, welfare, and education, and, finally, to open doors to useful lives for added millions of people.

Among the many avenues open for attacking this problem effectively, two approaches

175

Fig. 6-1 South America is divided into eleven independent countries plus French Guiana and Surinam, which is controlled by the Netherlands.

stand out. One that government planners and private groups are using to alleviate the population crisis is the development of pioneer communities. The pioneer frontier may bring into rational use nearly empty or underutilized portions of national territories or may recover abandoned lands. Examples of such pioneer occupancy, marked by varying degrees of success, include settlements in the Venezuelan and Colombian llanos and the Orinoco lowlands, the Peruvian Montaña, the Bolivian Yungas and plains, and Brazil's vast West Central region as well as Greater Amazonia. A second solution being utilized is to enlist underemployed urban and rural populations in the development of such industrial technology as road and dam construction. Construction industries in cities, for example, are capable of absorbing many of the unskilled; on the other hand, most modern manufacturing establishments require some degree of literacy or skill from their labor, aptitudes in which the peasant or slum dweller may be lacking. Nevertheless, from the industrialization of the lower Orinoco Valley in Venezuela and the ambitious construction of the new national capital at Brasília during the 1950s and 1960s, it is now evident that successful urbanization and skills upgrading of hundreds of thousands of people can be achieved in less than a generation with large-scale, careful government planning.

For centuries, and still today, a host of primary agricultural, forest, and mineral commodities have been bases for commercial interaction between Middle American and South American nations and industrialized neighbors around the North Atlantic, and now with the Far East. Commercial resource exploitation often required construction of means of surface transportation over sometimes difficult terrain. Sometimes these travel routes penetrated former no-man's-lands, as in the Chilean nitrate pampas or in Amazonia's selva. Often, however, such routes did not fulfill the needs of traditional settlements or complement the economic institutions that had evolved around these communities. Today, the building of fast, inexpensive surface transportation into resource regions, especially along secondary routes, is essential to foster interaction between outlying regions and the urban industrial centers that require the resources. Possible investors willing to venture into new activities, whether the pioneer agriculturalist or the small merchant, need the supportive incentives offered by competition and market growth. How swiftly barriers of distance, time, cost, and politics recede will very likely decide the future for continued commodity exports and for industrial progress inside each nation.

For most South American nations, the ultimate economic goal is away from dependence on nations outside the Latin American region. Implementation of the Latin American Free Trade Association in 1970, with a 1985 regional-integration goal, hopefully is destined to overcome petty differences and deep-rooted suspicions. One of the most critical problems, however, is the readiness of all the nations to participate. Since there has been a differential and incomplete assimilation of the technologies and institutions needed to support twentieth-century commerce, some nations in the trade association will have advantages over others.

As a world region, South America is characterized by spectacular conditions and contrasts: hot, humid tropical forests, deserts where rainfall is all but unknown, mountains higher than anywhere but Asia, vast river plains, and a poverty of material goods for societies rich in tradition. South America's continental area is smaller than Asia, Africa, or North America; considerably smaller, fortunately, is its population, totaling about 180 million at the onset of the 1970s. Brazil alone occupies approximately half the continent's territory and has more than half of the people. Argentina accounts for another third of the continent, and the balance is shared by nine other nations and two European-related territories. Metallic ore deposits abound in Peru, Bolivia, and Chile in Andean South America, as well as in Brazil; petroleum is concentrated in Venezuela and Colombia in the north, and along the Andean foothills of Colombia, Ecuador, Peru, and Argentina, which has sizable reserves. Developed plantations for several commodities are still largely in Brazil and Colombia; the middle-latitude granary and ranches are largely in Argentina. During nearly two centuries of independence, many of these people have been unable to evolve modern societies or to organize their economies by their own efforts. What the future holds remains to be seen; but there can be no question that change is on the horizon.

Physical Setting

South America has an area of nearly 7 million square miles and is nearly twice the size of the United States. The continent is situated southeast of North America and reaches beyond the equator far into the Southern Hemisphere. The meridians of New York or Miami intersect the extreme western margin of Peru. The eastern continental cities of Rio de Janeiro and Buenos Aires are actually much closer to the capitals of Western Europe and West Africa than to much of the United States.

Relief Features

The Andean Cordillera is the longest continuous mountain chain in the world. The outstanding landform feature of South America, this mountain system is second only to the Himalayas in altitude. The Andean Cordillera extends north from Tierra del Fuego to the Caribbean. Approaching the Caribbean, the ranges turn eastward as the backbone of the north coast, only to disappear in the Atlantic beyond Trinidad.

In the south the elevation of the Andean Cordillera is lower, with crests averaging 5,000 feet. The Cordillera rises to peaks of 20,000 feet or more in middle Chile and western Argentina, maintaining this altitude as far as northern Peru. Mt. Aconcagua (22,834 feet), the Western Hemisphere's highest peak, stands amid its solitary snowfields and glaciers in western Argentina, northeast of Santiago, Chile. The high Cordillera divides into two ranges in northern Chile, supporting plateaus (altiplanos) with altitudes of 12,000 to 14,000 feet as they traverse Bolivia and southern Peru. In Bolivia and Peru, the ranges merge into a broad system that is incised deeply on the east by Amazon tributaries. The system narrows across Ecuador, before it thrusts northward into Colombia. In southern Colombia, the ranges separate and extend north as three smaller cordillera.

Fig. 6-2 Andes Mountains near Cuzco, Peru. Over much of their length, the Andes are high, rugged, and barren. The Peruvian Indians shown here are dressed in traditional garb. (Courtesy of Pan American Airways)

The Caribbean littoral north of the Andean Cordillera is dominated by the Magdalena River basin, the Maracaibo lowlands, and outliers of the mountains. Flanked on the west by narrow coastal lowlands and plains on the north, the western slopes of the Cordillera drop precipitously to the Pacific shore in its middle section. To the south, the Cordillera is paralleled by a lower coastal range and longitudinal valley. In the extreme south of Chile, this valley has been invaded by the ocean and forms a lengthy archipelago. Along its eastern flank, the Andean Cordillera faces the three great river basins of South America: the Orinoco, Amazon, and Paraguay-Paraná.

The Orinoco River system is fed by tributaries rising in the Andes and the Guiana Plateau. It flows in a great semicircle across Colombia and Venezuela, from south to northeast, before its muddy waters enter the sea. The Orinoco plain, or llanos, is low and gently rolling, with large areas that are nearly flat. During the rainy seasons, these flat areas are covered by floodwaters spreading over hundreds of square miles.

The Amazon is the grandest natural spectacle in South America. With its tributaries, it drains at least 40 percent of the continent. The Amazon rises within 150 miles of the Pacific coast in Peru and flows 4,000 miles to the Atlantic, through a wilderness of rain forest. With its multitude of tributaries, many of them rivers of consequence, the Amazon passes through hot, steamy lowlands that seldom rise more than 1,000 feet above sea level. The river basin is 800 miles wide at its widest but narrows to only 20 miles where it passes between the Guiana Plateau and the Brazilian highlands on its journey to the ocean.

A low tableland extends westward from the Brazilian highlands to divide the watersheds of the Amazon to the north from the basin of the Paraguay-Paraná system to the south. The Paraguay-Paraná system drains the southern portion of the Brazilian highlands and the eastern slopes of the Andean Cordillera as it crosses the Gran Chaco. The Chaco region merges imperceptibly with the Pampas of Argentina and Uruguay as the system enters the estuary of the Río de la Plata.

The Plateau of Patagonia occupies the extreme southern segment of the continent, south of the Argentine Pampa. The plateau rises to altitudes of more than 2,500 feet where it stands between the sub-Andean depression at the foot of the Andean Cordillera and the Atlantic. It has a rolling surface, through which mountain-fed streams have cut canyons to the cliffed, rugged coast.

The eastern margin of South America is marked by the Guiana Plateau, north of the Amazon channel, and the vast Brazilian highlands occupying almost one-quarter of the continent. The Guiana Plateau has a rolling surface composed of rounded hills with broad tablelands, and is intersected by several short mountain ranges.

The Brazilian highlands are the largest and most extensive of the eastern plateaus. A few miles inland from the Atlantic, they rise along a steep escarpment to altitudes of 1,000 to 3,000 feet and then descend by steplike terraces to the valley of the Amazon in the north and the Paraguay-Paraná River basins in the south and the west. The edge of the escarpment, called the Serra do Mar, overlooks narrow, discontinuous lowlands along the coast. The plateau surface is rolling, with numerous areas of steep hills and several isolated ranges of low mountains. In the west the surface is moderately elevated and rolling, interrupted by areas of swamp and marshes.

Climate and Vegetation

South America has the largest area of tropical climate found on any continent. A majority of the continent's people who dwell within the

CLIMATES OF SOUTH AMERICA

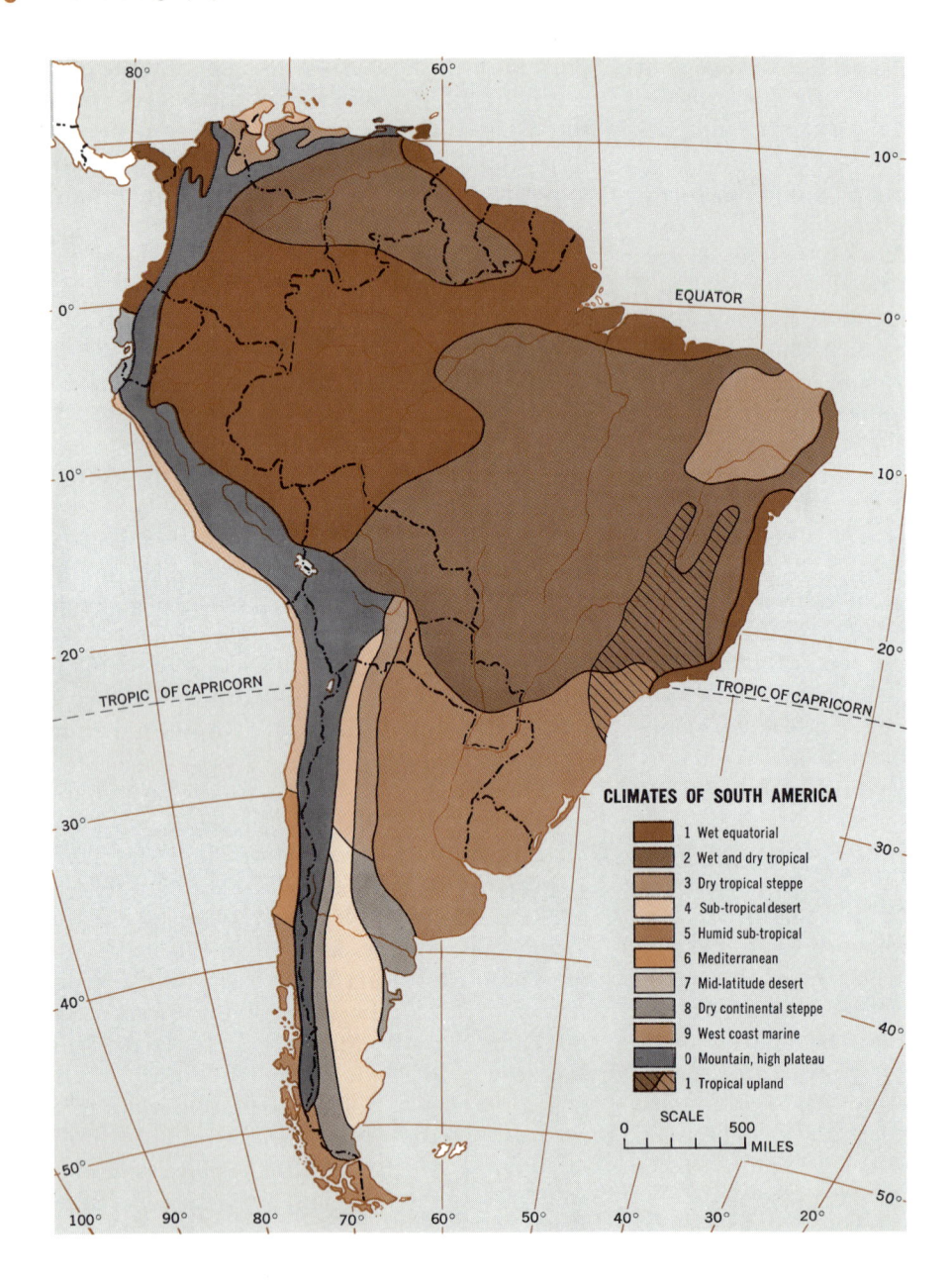

1. Wet equatorial
2. Wet and dry tropical
3. Dry tropical steppe
4. Sub-tropical desert
5. Humid sub-tropical
6. Mediterranean
7. Mid-latitude desert
8. Dry continental steppe
9. West coast marine
0. Mountain, high plateau
1. Tropical upland

Fig. 6-3 Much of South America lies within the tropical climate zone. Southern South America, where the continent narrows, has the greatest variety of climates. The most highly developed area of the continent, from Rio de Janeiro south to Buenos Aires, is in the transition belt between the tropical and continental climates.

tropics live in the highlands or in subtropical zones around the periphery. The altitude of the Andes and the Brazilian plateau offset significantly the constant hot temperatures that would prevail across lowlands, and the great latitudinal tapering in the south affords additional areas of mid-latitude climate favoring settlement.

Rainy Tropical The rainy tropical climate of South America dominates the Amazon Basin and the trade-wind coasts to the northeast and southeast and in Pacific Colombia. In this hot and humid region, rainfall is possible every day of the year. Temperatures average 80°F annually, but may range less than 10° from season to season or from day to night. Rainfall averages 60 inches yearly, increasing to 80 inches or more upslope on the highlands and along the coastal littorals.

Although the lateritic topsoils of the rainy tropical climate are usually badly leached of plant nutrients, the deeper soils support a thick selva of luxuriant broadleaf, evergreen forest. The forest consists of tall hardwoods and some softwoods, with a tremendous number of species. Only low shrubs and bushes occupy the open spaces between the trees. Along the rivers that penetrate the thick forest, the sunlight seeks out the forest floor, and a riot of jungle palms and shrubs fills in between the trees.

Savanna On the poleward margins of the rainy tropical climate, rainfall declines in amount and is distributed during the year so that a distinct dry season occurs when the sun is low in the heavens. North of the Amazon region, the Orinoco lowlands and the northern third of the Guiana highlands experience this climate type. It has a distinct low-sun drought that lasts from November through March, followed by rainfall of 30 to 50 inches during the high-sun rainy season. South of the Amazon lowlands, in the Southern Hemisphere, the extensive areas of west-central Brazil and the Gran Chaco have high-sun precipitation during December to April, followed by low-sun drought from May to October. Since the Chaco region extends farther poleward than the Orinoco basin, its southern margin has cooler temperatures during the low-sun period. The better-watered savannas close to the Amazonian rain forests have large numbers of trees, interrupted by grassland openings in the drier or poorly drained areas. As rainfall amounts decrease and the dry season becomes prolonged, broken woodland with tall grasses gives way to shorter grasses mixed with trees limited to stream courses or to low areas where sufficient moisture is present. Over many of the savannas, rainfall is scant and evaporation so effective that little moisture remains even to support short grasses.

Semiarid Grasslands Away from the wet-and-dry tropical climates, seasonal rainfall finally may become so meager, less than 18 inches, and so variable from year to year that the environment can support little more than a cover of short grasses, with occasional clumps of xerophytic shrubs. The semiarid grasslands are generally narrow belts of transition between the savannas and the desertlike regions of South America. Although small in area, they are significant zones. The foreland of the Cordillera Mérida in Venezuela, the western slopes of the Andean Cordillera in the *tierra templada* (except in western Colombia), and the *sertões* of northeast Brazil are dominated by this climate type.

The semiarid steppe also is found where temperatures are cooler, either because of altitude or seasonal variations. The high altiplanos of Peru, Bolivia, and extreme northwestern Argentina are semiarid temperate steppes, with generally cooler temperatures and even less precipitation during the year. From the Puna de Atacama in northern Argentina, skirting the western deserts at the foot of the mountains, a tremendous area of semiarid grassland with intermediate climate dominates the tableland of Patagonia.

None of the steppelands of South America are so crucial or important to large population concentrations as is that in northeast Brazil. Covering 250,000 square miles and harboring 30 million people, this vast area occupies the eastern bulge of Brazil. In few other areas is the rainfall more erratic from year to year or the vegetation more nearly desertlike in appearance. Overgrazing by livestock has cleared much of the natural grass cover, which has been replaced with thorny shrubs and other useless xerophytic types.

Desert Western South America, from northern Peru to La Serena in Chile, embraces a narrow desert backed by the Andean Cordillera. The Peruvian coastal desert and the Atacama Desert in northern Chile are among the driest areas on earth. Iquique, in northern Chile, has received little rainfall since the time of its first European settlement. Some areas in Peru have heavy coastal fog that moistens the land and permits some shrubs to exist. For the most part, however, the desert is barren, dominated by extensive rock surfaces. Water from melting snows and from rainfall on the higher Cordilleran slopes finds its way to the Pacific in alternately wet and dry braided channels. In Peru the streams have been successfully exploited for irrigation agriculture. The desert extends up the mountain slopes to about 5,000 feet in Peru, and virtually to the Cordilleran crests in Chile.

In western Argentina a narrow desert parallels the foot of the Andes south to Patagonia. Lying in a rain shadow where the Andes shelter the east slope from Pacific storms, this barren region has no settlements except near streams that rise in the snowfields of the Andes or where there is water from artesian wells.

Mediterranean Middle Chile occupies a small segment of western South America from 30° to 38° south latitude and enjoys a climate similar to that of southern California. During the rainy winters the temperatures are usually mild, but vary with the southern frontal storms that sweep in from the Pacific. On the other hand, the summers are quite dry, with temperatures more like those of the desert areas to the north. Temperatures range from 68°F for the warmest months to 48°F during the winter, from May to September. Usually, total rainfall is greater in the subhumid south, averaging about 40 inches, than in the semiarid north, nearer the Atacama, with 10 to 12 inches. In addition to the frontal storms from the Pacific, air masses from the ocean are forced to rise up and over the Andes, thereby causing orographic rainfall. The vegetation of middle Chile closely resembles the vegetation types of California, whence many of the trees, shrubs, and grasses have been introduced.

Marine West-coast In southern Chile the dry season becomes gradually shorter and finally terminates as rainfall increases in amount and is distributed more evenly during the year. The frontal storms carried along in the "roaring forties" persist all year and carry large amounts of precipitation inland. Pushing inland, the air masses are forced to rise and are cooled, releasing nearly continuous and heavy precipitation. Summers are chilled by the persistent cover of cloud and mist, and winters are quite cold, with snow blanketing the mountain slopes down to the shore. A veil of fog hangs continually over the island archipelagoes. Precipitation averages 80 inches, but may be over 150 inches in windward headlands or on exposed mountain slopes. The islands and fjorded coastline are covered with thick stands of Chilean pine in the north and beech, aspen, fir, and spruce in the south.

Humid Subtropical The humid subtropical climate is second only to the highland climate in the numbers of people who dwell and gain a

South America 183

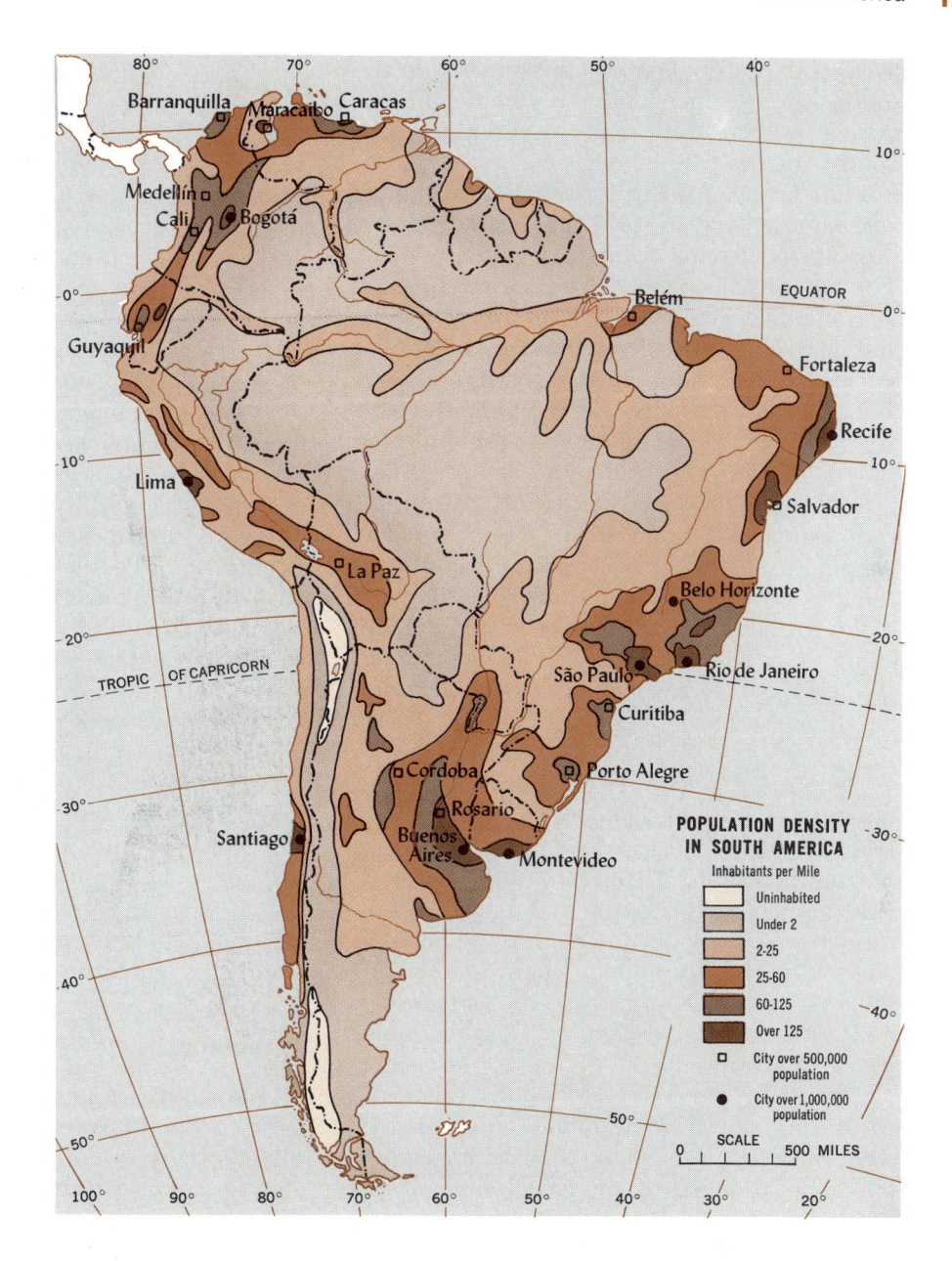

Fig. 6-4 The centers of dense population are largely clusters around the larger cities. In tropical areas, the most populous districts are in the mountain valleys.

livelihood in a single area. It is situated in the southeastern segment of the continent, approximately between 25° and 40° south latitude. In South America, the humid subtropical climate extends farther equatorward than the comparable climate type in the United States and is shut off from frontal storms during the summer. The humid subtropical climate is more strongly marine-influenced; thus its winters are milder and marked by frequent invasions of frontal storms from the south. The climate conditions and fine soils that prevail over the land in this region support natural grasslands and pine forests. The grasslands, or pampas, have proved unusually suited to the introduction of European agricultural systems. In the summer temperatures average 68°F, and in the winters about 55°F, varying somewhat from southern Brazil to Bahía Blanca in Argentina. Precipitation is heavier in winter, ranging between 40 inches in the eastern pampas to 22 inches on the drier margins in the interior.

Tropical Highland The variety of climates on the mountainsides and valleys of the highlands are quite similar to those present throughout the highlands of tropical Middle America. But the major climate zones occur at lower altitudes on the mountains nearer the equator, and the greater latitudinal length of the Cordillera causes the lower zones—the *tierra caliente* and *tierra templada*—to disappear entirely south of 35° south latitude.

Prevailing easterly trade winds cause the east-facing slopes to be quite rainy; whereas western slopes, in general, lie in a rain shadow and are considerably drier. Exposed headlands receive large amounts of precipitation, but slopes facing them across valleys may lie in a local rain shadow and be quite dry. Each valley and slope may thus have its own distinct microclimate, which is perhaps very different from that of the surrounding slopes.

Economic Development

The majority of the people of South America are still tied closely to the land. Although the growth of urban-industrial centers proceeds unabated, and even medium-sized cities are experiencing the tide of migration from the countryside that has already flooded the great cities, it is the farmer, tenant, farm laborer, and rancher who provide the food staples for the cities and the major export commodities on which the national economies of several nations depend. South America is a provisioner of tropical agricultural products and mid-latitude grains and livestock commodities of worldwide significance. To such more or less luxury exports of the colonial era as gold, silver, indigo, brazilwood, natural rubber, sugar, and coffee have been added cargoes of bananas, wheat, corn, wool, mutton, and beef. The climatic and seasonal differences of South America are an implicit factor in the development of commerce with Northern Hemisphere markets and the foundation of commercial agriculture today. The roles of the individual countries differ strikingly; but the output of export commodities should increase substantially in the decades ahead as domestic and international markets for these essential agricultural products expand.

Agriculture

Agriculture is in a critical state throughout most of South America because of the widespread occurrence of two systems of land tenure and agricultural technology that exist side by side. The one harvests the commercial export commodities; the other provides the day-to-day needs of millions of families who dwell on the land and eke out a marginal existence. These divergent systems prevail and must be understood clearly to appreciate the severe technological, social, and institutional challenges facing the South American nations.

Traditional commercial landholding is in the form of a plantation, referred to in various places as the hacienda, fazenda, estancia, or fundo. Whatever its name locally, older plantations were often handed down from one generation to another, with possession sometimes dating from the original royal land grants. Newer types of plantations have been purchased by individuals or corporate enterprises. The owners may either hire laborers to work their land or rent the land to tenant farmers by sharing the income from the crops they produce. These lands have been used rather efficiently, especially in recent decades, to produce export commodities. The introduction of modern agricultural techniques and machinery has been necessary to increase output and quality.

In sharp contrast, tiny plots of land long since depleted by unwise agricultural methods or overexploitation, often situated on hills or mountainsides, are tilled by peasant farmers, who may or may not own the land. In accordance with the practice of their ancestral societies, land assigned to Indian communities in Spanish South America was supposed to be protected and developed in the interest and to the benefit of the Indian peoples. The roles of guardian and guide were assigned at first to the Catholic Church. After political independence, this supervisory responsibility was shifted to the new national states. Land theft and Indian labor exploitation were practices not uncommon as private hacienda owners attempted to acquire much of the communal lands.

Using farming methods often handed down from previous generations, and simple though sometimes effective tools such as the machete and hoe, these subsistence farmers provide food for their families and graze small flocks on the poor pastures of the mountains, valleys, basins, or plains. Today, more often, the peasant works on plantations because of the cash wage incentive; here he is guided in the cultivation and harvest of commercial crops and in livestock care. While the best, most modern techniques are not invariably used on the plantations, those practiced by the small farmer on his own land are generally even more destructive of the soil.

The great inequality in land distribution is a major challenge to the economic and social structures of every South American country. The problem is that if the land were to be distributed equally under some land-reform system, commercial crop production indispensable to the existence of some countries would decline or all but cease. In the hands of the untutored and ill-equipped small farmer, destructive farming practices would spread and ruin much more land. In a number of countries —Venezuela, Colombia, Peru, and Brazil, for example—there are sizable empty areas into which pioneer families, equipped with modern farm training and aided by government supervision and support, are opening new frontiers.

Meanwhile, more of the commercial farmland is being shifted to domestic food production for the rapidly growing cities; thus, commodity output for export is sustained only with effort. The major challenge, then, is how to sustain expanded production of export commodities and also to provide the cities with enough food, while at the same time land-distribution and rural education programs progress.

Manufacturing

In the cities, the development of manufacturing industries to supply the domestic market with consumer items and the growth of heavy industry in countries whose markets can support them keynote a major change in the South American economy. Manufacturing is largely a development of the past forty years. As recently as 1950, every South American nation relied on revenue earned by the export abroad of mineral and agricultural commodities to purchase the larger share of the needs of its people. Since then, Brazil, Chile, and Colombia have joined Argentina in managing to supply from their own factories a wide variety of consumer needs such as clothing, shoes, foodstuffs, small appliances, and simple chemical products. Brazil and Chile pioneered in integrated iron and steel industries, and were followed by Colombia, Argentina, Peru, and Venezuela. Output and by-products of basic manufacturing industries have multiplied job opportunities for those skilled in various subsidiary industries and have increased the number and variety of goods available.

To protect the home industries it has been necessary to raise tariff walls, to license corporations, and to restrict certain classes of items. At the same time, large factories producing goods on a mass-production basis do not employ the numbers of unskilled people who are swarming to the cities, in response to the good news of a better life elsewhere.

Transportation

Within each country and across South America, land transportation is still more a feat of accomplishment than a fact of everyday life. Only the Argentine and Uruguayan pampas, eastern Brazil, and central Chile have anything approaching an integrated network of highways and railroads. In most other places, railroad construction has always faced the nearly insuperable hurdles of the mountains and sparsely settled tropical lowlands, and was thus limited to service lines leading to commercial agricultural areas and mining camps.

The past three decades have witnessed the construction of thousands of miles of all-weather, hard-surface highways in South America. Today it is the highway, traveled by trucks, buses, and automobiles, that gives promise of better use and assembly of raw materials, circulation of people, distribution of manufactures, and the spread of education. Of the older means of inland transportation, only the navigable Amazon, Río de la Plata, lower Paraná, and lower Orinoco show improved prospects for the future. Tributaries of the chief rivers, including those of the Amazon and Magdalena, are slowly declining in importance as faster or more efficient avenues of transport are developed.

The Caribbean Countries

The independent countries of Venezuela, Colombia, and Guyana and the dependencies of Surinam and French Guiana comprise the Caribbean region of South America. Except for a small part of southeastern Colombia, this entire area lies north of the equator. Large tracts of land in each country are sparsely settled and generally underdeveloped. Each also has highlands and lowlands that create a topographical pattern which greatly influences the activities of the people. In Colombia and Venezuela the highlands offset, to some extent, the climatic disadvantages of the low-latitude tropics. The Caribbean countries are well situated for world trade, because their northward-facing coast is near and pointed toward the principal world-trade lanes, those leading to the United States and northwestern Europe. The Caribbean countries are areas of great economic potential.

Venezuela

Petroleum has flowed from beneath Lake Maracaibo since the oil fields of Venezuela were opened in the 1920s. Before the advent of

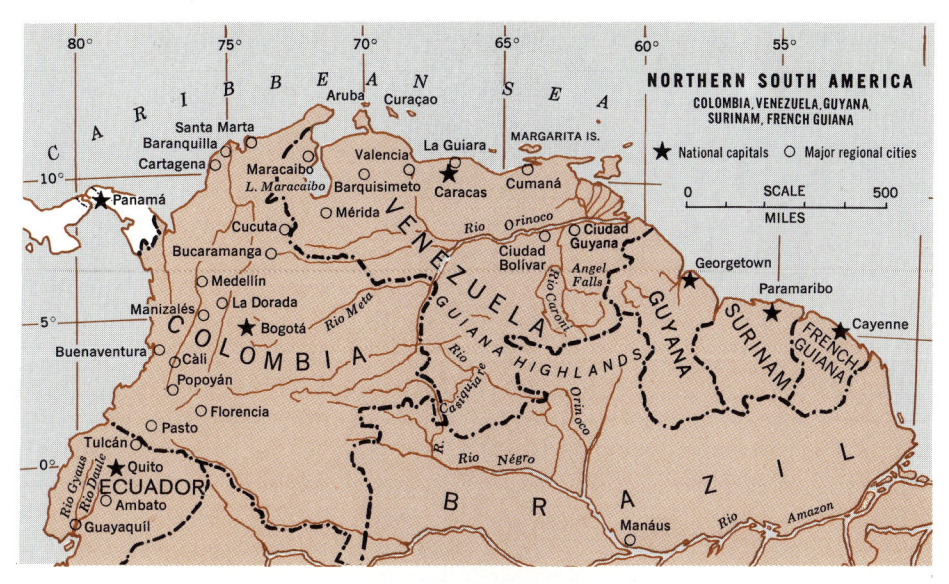

Fig. 6-5 Northern South America is dominated topographically be the Andes Mountains and the Guiana Highlands, with the llanos separating the two areas.

petroleum, coffee and cacao, along with cattle and hides, provided the economic base of Venezuela, which was then a poor Caribbean nation. Today petroleum and iron ore exports provide the largest share of the revenue essential to the life and vitality of the nation. The farms and ranches of Venezuela employ about 40 percent of the working force of the country. During the 1960s, with the support of the government, agriculture made great strides in development. The country has shifted from one having food deficits to that of an exporter of food products, especially sugar and rice.

Thousands of Venezuelans seeking better opportunity have been drawn to the oil fields of Maracaibo, to the cities, and to the iron district of the Caroní Valley. Often they left their fields and herds untended, with no one to take their places. Many small farms have been abandoned, and large haciendas find it increasingly difficult to recruit dependable, skilled field hands. Many rural migrants arrive in the cities and mining centers only to discover that they lack the technical skills necessary for employment.

The Maracaibo lowlands, with their large petroleum production, are the greatest source of Venezuela's wealth. Surrounded on three sides by highlands, the region is nearly shut off from the moderating influence of the Caribbean trade winds. With their high temperatures and constant humidity, the lowlands have one of the most unpleasant climates in South America. The petroleum fields are mainly along the east side of Lake Maracaibo or on the shallow lake bottom. To facilitate entrance, once blocked to deep-draft vessels, a channel has been dredged, and larger oceangoing tankers now have direct access to the oil terminals on the eastern shore. Maracaibo, a modern tropical city, is the center of the nation's petroleum industry.

The coastal oil fields produce more than 1 million barrels of petroleum daily, or slightly more than two-thirds of the total national production. Over one-half of the 4,400 producing wells are situated offshore in the shallow waters of the lake. Venezuela ranks among the four leading world producers and exporters of petroleum. Petroleum and its derivatives provide about 90 percent of the country's foreign-exchange credits.

The Cordillera Mérida rises like a great arch from the Maracaibo lowlands and Orinoco basin to the only snow-covered peaks in Venezuela. The people of the Cordillera live in settlements ranging between 2,500 and 6,000 feet in elevation, in the upper *tierra caliente* and lower *tierra templada*. Coffee is the major crop from the tree-shaded slopes and valleys. The Cordillera is the chief coffee region and a major cacao region of Venezuela. Between 7,500 and 10,000 feet above sea level, the basins and mountain slopes are covered with fields of wheat, barley, and potatoes, and cattle forage on the temperate grasses. Sheep graze on the higher pastures of the paramos. In Venezuela, as elsewhere in Caribbean South America, lowland diets emphasize maize (corn), a basic staple, along with beans, yucca, rice, bananas, and other subsistence crops. The diet in this region is notably short of animal or vegetable protein foods.

The coastal ranges and eastern highlands include the densely settled Valencia basin and the Caracas Valley, with their urban centers and smaller valleys and basins in between. Caracas, the capital, is a modern city with tall buildings and tree-lined boulevards; it is the financial and industrial center of Venezuela.

A series of basins and valleys lie in a trough between the two paralleling mountain ranges that comprise the coastal ranges. From the Caribbean, the coastal ranges rise abruptly to altitudes of 7,000 to 9,000 feet. As the most important agricultural district of the country, the Valencia basin grows sugarcane, cotton, rice, corn, and beans. Earlier in the century the Caracas Valley was equally important, but agriculture there has declined with the great expansion of suburban Caracas. To replace the food supply lost from the Caracas Valley, farming emphasis has shifted to the Tuy River valley in the *tierra caliente*, south of the city. On the mountain slopes in the *tierra templada* around the basins and valleys, coffee and cacao remain important.

The elevations of the coastal ranges, particularly in the Caracas Valley and the Valencia basin, are great enough to alleviate the extreme heat of the Caribbean shore. Valencia is located at an elevation of 1,500 feet, and Caracas is 3,000 feet above sea level. Situated only 8 airline miles from its port at La Guaira, Caracas was for centuries nearly cut off from the outside world by the steep slopes of the north coastal range. Today, the modern *autopista* running inland from the coast covers the distance and

Fig. 6-6 With a population of almost 800,000, Venezuela's capital, Caracas, is said to have more miles of concrete expressway than any city of comparable size in the world. (Courtesy of Delta Air Lines)

altitude in only 20 miles, a trip of no more than thirty minutes.

The Orinoco lowlands include a large river watershed that drains much of eastern Colombia. Nearly 1,000 miles long and averaging 200 miles in width, the lowlands stretch from the Andean Cordillera on the west and north to the Guiana Plateau in the south. Even though they have witnessed remarkable recent industrial development, pioneer settlement, and improved animal husbandry, the lowlands remain very sparsely settled.

The Orinoco River plain is flat and interrupted only by occasional rolling hills until it rises against the Cordillera Mérida and the coastal ranges as a series of piedmont terraces. The wet-and-dry climate brings heavy rainfall during the hot, high-sun season; the downpours are so great that vast areas of the plain are flooded. Grasses grow in thick luxuriance, but are unavailable to cattle that have been moved by ranchers to protected uplands. With the onset of the warm, dry season, the grasses ripen and dry out, becoming almost inedible and of little value as cattle forage. Insect pests and diseases plague the animals in the lowlands at all times. The cattle seldom fatten properly, but nevertheless walk to markets at river towns or cities along the coast. They produce low-quality beef and usually are of value only for their hides.

Two factors are altering the character of the Orinoco lowlands. Agricultural pioneering under government leadership has opened new lands to field crops along the Andean piedmonts and in alluvial soil zones in the plains. With irrigation waters from the mountains available, plus good soils, the warm climate favors commercial production of basic food staples in short supply in the highland cities. The settlers have been selected carefully and trained in modern farming methods. Another type of settlement, near Calabozo in the central lowlands, depends for its water supply on a large reservoir that collects water during the wet season for irrigating pastures and field crops. The objective, beside self-sufficiency in food, is a modern cattle-finishing district. Both of these agricultural experiments have proved so successful that others are being implemented.

The second factor of economic importance has been the exploitation of minerals in the Orinoco lowlands and the neighboring Guiana Plateau. Almost one-third of Venezuela's oil comes from the eastern fields that extend some 50 miles north of the river and westward from the Orinoco delta for over 300 miles. Moreover, the lower Orinoco plain is experiencing the impact of iron-ore exploitation in the Caroní River district just south of the Orinoco. Below Ciudad Bolívar, river ports and several industrial communities have developed and are changing the outlook of that region.

Iron ore is the most important mineral mined and is the second export in total value of Venezuela. Very large deposits of high-grade hematite, with an iron content of 50 to 70 percent, are being mined. Subsidiaries of two U.S. corporations are exploiting the huge reserves at the rate of over 25 million tons a year. To aid in shipping the ore, the muddy Orinoco has been dredged to San Félix, where ocean vessels may enter to take on cargoes of the ore. Mining is done in open cuts, and the ore is transported to the river ports, from which it is shipped in ore carriers to tidewater steel mills near Baltimore and Philadelphia.

Ciudad Guayana is the site of a major integrated steel mill developed by the Venezuelan government. Coal is brought in on returning ore carriers. Upstream on the Caroní River a large hydroelectric station has been installed in the Guri Dam, and abundant power is supplied to the evolving industrial region in the lower Orinoco Valley. New activities include oil refining, metals processing, and a petrochemical industry.

Venezuela's per capita foreign trade is the

highest of the Latin American countries, and in dollar value it accounts for more than one-fourth of the whole Caribbean region's export trade. Petroleum and derivatives account for 92 percent of the total exports, in monetary value more than 1.25 billion dollars annually. Iron ore now accounts for almost 6 percent of total exports. From foreign-exchange credits earned by these sales, Venezuela obtains the money to purchase goods in foreign countries. With foreign-exchange credits earned, Venezuela also purchases foreign "know-how" and technology to upgrade its productive capacity. The United States remains the country's chief trade partner, sending in return for exports investment capital, transport equipment, chemicals, and food products. Demand from the mining areas and the cities, especially Caracas, accounts for large imports of foodstuffs, which help to keep the cost of living high.

Colombia

Having coasts on both the Pacific and the Caribbean, Colombia enjoys a unique position in South America. Rugged arms of the Andean Cordillera cut across the country and divide it into several regions, separating the lowlands of the Pacific coast, the Caribbean river plains, and the eastern rain forests and llanos. Conditions in several highland areas in the tropical zone favored settlement, but the nearly insurmountable terrain acted as a barrier to surface communication and isolated almost completely one population cluster from another. Only with the advent of air transport and the construction of modern highways have the major Colombian population centers been tied together effectively. The shadow of this early isolation, however, still hovers over the political and economic thinking of the country.

The tremendous diversity of landforms and climates in Colombia results in several distinct major regions. The Caribbean Lowland in the north incorporates the river systems of the Magdalena and its tributaries and the islandlike outlier, the Santa Marta highlands. The Andean Cordillera that divides the country is bordered by lowlands and low mountains on the Pacific and is sliced by the valleys of rivers that traverse the country flowing from their headwaters in the south to the Caribbean Lowland. The eastern arm of the Andean Cordillera thrusts across Colombia into Venezuela, where it carries the system eastward to the Atlantic. Thus the Cordillera stands as a bulwark overlooking the hot, tropical rain forest of the Amazon to the south and the llanos and the Orinoco tributaries to the east.

The Caribbean Lowland of Colombia encompasses several of the most highly developed tropical grazing districts of South America. The wet-and-dry savanna climate affords adequate rainfall in the high-sun period to support good tropical grasses, many of which have been introduced and are more nourishing to cattle than the native grasses. Here, improved livestock such as the zebu breed of India has supplanted the older breeds still common in the llanos and the Orinoco lowlands.

Close to Cartagena and Barranquilla and in the Sinú Valley, available water for irrigation combined with good alluvial soils has aided in the development of plantations yielding cotton, oilseeds, and food staples. The west piedmont along the Santa Marta highlands was among the earliest commercial banana districts in the Caribbean; plant diseases have now reduced the output there. Of the cities of colonial heritage, Cartagena and Santa Marta have survived since the days of the Spanish Main. Cartagena is the control point for pipelines from the interior Magdalena Valley oil fields. Santa Marta is a thriving city because it became the center for the movement of heavy goods between landlocked Bogotá and the Caribbean after the completion of a rail link in 1961. Barranquilla, a functional port city situated slightly inland on the Magdalena floodplain, dominates the regional economy, is the center for air traffic, and handles the residue of river commerce through a channel deepened by dredging for large vessels.

The middle Magdalena basin holds Colombia's older petroleum fields. In the late 1960s new discoveries were made east of the Andes near the Ecuadorean border. The new fields are linked to the coast by a trans-Andean pipeline to the Pacific port of Tumaco. Originally developed by foreign capital, the industry is now shared with a government-owned enterprise.

Hot, steamy, and cloud-cloaked much of the time, the Pacific slope and lowlands are blanketed by humid forests that challenge settlement and economic activity. Nonetheless, Buenaventura has attained prominence as a port since it became accessible for exportation of coffee when a highway and a rail line were pushed to the coast from Cali, in the Cauca Valley. As much as half of the annual coffee crop moves to overseas markets through Buenaventura; it is also a major entrepôt for imports. Elsewhere, Pacific Colombia remains thinly settled.

The Cauca Valley rests between the arms of the Andes. Shut off by the mountains from the moisture-bearing air masses of the Pacific, the Cauca Valley and Antioquia to the north stand intermediate between the *tierra caliente* of the Pacific and the Caribbean Lowland and the *tierra templada*. The Cauca Valley Corporation (CVC) was established to assure scientific progress for farm and livestock industries. With its economic hub at Cali, the intensively developed irrigated valley supplies basic staples and industrial raw materials. The entire valley benefits from the multipurpose development of the Cauca watershed, including the regulated flow of river waters, a plentiful supply of hydroelectricity, and the wise utilization of soils. The mountain slopes of the western and central Cordillera have many small *fincas*, or family-sized farms, growing coffee and cacao.

North of the Cauca Valley is the Antioquia department, with its prosperous modern metropolis, Medellín. Few population concentrations in Latin America are cores from which pioneer settlement has migrated voluntarily into unoccupied lands. From Antioquia, however, colonies have pushed forward vigorously, south into the Cauca Valley, across the central Cordillera into the Magdalena Valley, and north into the margins of the Caribbean Lowland. As independent farmers, they have carried with them the coffee culture that today serves as the economic base of Antioquia. More than coffee is produced here; cacao, sugarcane, potatoes, dairy products, and many other crops help to support the people of the region. Medellín is an important export center for textiles and apparel.

The Magdalena Valley, east of the Cauca Valley and Antioquia, is flanked by the central and eastern Cordillera. The volcanic soils of the mountain slopes and the climate of the *tierra templada* that so abundantly favor coffee and cacao in Antioquia are also present. The valley floor of the Magdalena is closer to sea level and drier than its neighbor to the west. Dry tropical grasslands of the valley floor support lower-grade cattle. However, by channeling the river waters to irrigation projects, major yields of rice, cotton, sugarcane, beans, and corn supply many of the needs of the country.

The eastern Cordillera is much broader than the other ranges and includes numerous highland basins and valleys. The largest and most important of these is the Sabana de Bogotá, where the nation's capital is located. Situated in *tierra fría*, these extensive basins are producers of numerous food staples. The excellent grasses of the basins support fine beef herds and the most important dairying region in northern South America. Though much of the land is still farmed below capacity and small farmers must cling to tiny patches on hillsides, important progress has been made to raise the crop yields and to mechanize production on large haciendas in the basins and valleys.

Bogotá is both the political and intellectual

capital of Colombia. In recent years, however, with the establishment of the iron and steel industry near supplies of iron ore and coal at Paz de Rio in the highlands northeast of the capital, Bogotá has developed an industrial base worthy of note.

The eastern lowlands, or llanos, of Venezuela continue south into Colombia, where *criollo* cattle roam the tropical grasslands and encounter the same hazards of flooding and disease that plague grazing animals to the north. Southward, across a low divide formed by outliers of the Guiana Plateau, wooded areas appear and gradually merge into tropical forests, as rainfall increases and the dry season present in the llanos disappears completely.

The creation of an adequate surface transportation system across seemingly insurmountable barriers has wrought a change in the political and economic intercourse of Colombia's otherwise isolated population centers. Modern roads, passable for most of the year, connect with the Inter-American Highway System, which links Colombia with Venezuela and Ecuador. Arterial links also extend to the Caribbean and Pacific coasts and to the principal provincial capitals. Secondary road systems serve the various plantation districts. Faster rail transport, bypassing the sand-clogged Magdalena River below La Dorada, ended a century-long era for paddlewheel steamers downriver to Barranquilla. Air travel, however, has solved most of the passenger circulation problem. Colombia was the first South American nation in which passenger aviation was used.

As in Middle America and neighboring Venezuela, a one-item economy places a nation in an extremely vulnerable position; for Colombia, coffee makes up three-fifths of its annual exports by value. Petroleum, production of which is likely to increase, amounts to another 10 to 12 percent. Colombia's mountain-grown coffees, chiefly blending grades like those from Central America, are harvested and processed with greater selectivity than are base grades from Brazil. Earning 3 to 5 cents more per pound, they are eagerly sought in world markets. Other noteworthy Colombian exports are gold, bananas, and livestock.

Colombia's imports still emphasize capital goods and basic staples for its city dwellers. As in Venezuela, the ever-growing tourist industry is also important as a source of "invisible income," especially when winter Caribbean cruises stop over at the country's beautiful colonial ports along the Caribbean.

Guyana, Surinam, and French Guiana

The former Guiana colonies have each attained political status involving either full independence or new relations with the mother countries. Guyana, formerly British Guiana, became independent in 1966; Surinam is now considered an integral part of the Dutch Kingdom; and French Guiana is now an overseas department of France. Although these states have been associated with European nations for over 400 years, they remain thinly populated overall and economically underdeveloped. Of the approximately 1 million people who inhabit the Guianas, about 677,000 are in Guyana; and a third of these live in and around that nation's chief port and capital, Georgetown. Surinam has about 250,000 inhabitants, mostly living along the coast or in Paramaribo, the capital. Of the 45,000 dwelling in French Guiana, half live in its port-capital, Cayenne.

The narrow coastal plain, covered with an almost impassable mangrove barrier, made much of the area inaccessible. Spanish and Portuguese ships had skirted the coast, but it was the Dutch who made the first successful settlement. When drained, the coastal plain proved satisfactory for plantation agriculture. A few miles inland, the Guiana Plateau rises to elevations as great as 8,000 feet. Important and navigable rivers traverse the plain from the hot, rainy interior uplands. Alluvial deposition along these rivers has built up vast deposits of bauxite, the wellspring of wealth and the economic future of these states.

The absence of any concentration of native peoples caused a labor scarcity when the area was first settled. As in the Antilles, black Africans were first brought in to fill this manpower gap, but the experiment failed since many of the slaves soon fled into the refuge of the forests. The labor deficiency was finally resolved by introducing laborers from India and the East Indies. Although descendants of the early escapees are now involved in the modern economies in greater numbers than ever before, Bush Negroes, people of the backcountry, still subsist by patch farming in clan settlements within the wilderness.

Striking similarities prevail in settlement and resources between Guyana and Surinam. Overall, population densities are among the lowest in South America; yet in the settled districts, densities are very high. In each of these states, the leading agricultural commodities are plantation-produced sugar and rice. Sugar and its by-products go to European markets; surplus rice, second in crop acreage, goes to Trinidad's Asian population or to Europe. Local farms supply cacao, coconuts, yucca, yams, and fruit. In both of these states, Europeans are few in number, and mostly temporarily resident technicians, owners and managers of business enterprises, and foreign government representatives; only a handful of the indigenous Indians remain.

Bauxite and its derivatives have constituted the boon sparking hope for these otherwise economically undiversified states. Guyana and Surinam are the world's two leading suppliers of this vital ore, principal source of aluminum. Some of the bauxite is washed and concentrated locally, but all is exported; North American markets are the major importers. Other minerals exported include gold and diamonds obtained from placers along stream courses.

Until the aerospace research center sponsored by the French government, for atmospheric and other space investigation, begins operation, French Guiana's small population and weak economy will remain more of a burden than a benefit to the mother country. The notorious penal colony, Devil's Island, that once supported this state was closed in 1946. Total land under cultivation is only a few thousand acres.

West-Central South America

In the Andes of Ecuador, Peru, and Bolivia, on gentle slopes at high elevations, Indians have dwelt for thousands of years. It was on the high, cool plateaus of the central Andes that the Spanish first became established in South America, selecting this region because of its mineral wealth and the considerable supply of native labor. Despite more than four centuries of occupation, the Spanish methods of developing the land and other resources have been only partially successful. The modern prosperity of Colombia and Venezuela has not been attained by most of the Indians living on the high plateaus of the central Andes.

Semiaridity and the chilled air of the 10,000- to 14,000-foot altiplanos favored the evolution of the hacienda; and with more than four and a half centuries of conservative aristocratic landowner management, technical progress has been slow. Nowhere have the descendants of the ancient Inca civilization realized the prosperity of mestizo-European communities elsewhere. The High Andes are uniquely the domain of the Quechua, Aymara, and other Indians, where they have evolved the physical adaptation that permits them to live and work in the rarefied atmosphere and brilliant sunlight.

The Pacific lowland economy has gone to the highland for such items as minerals, wool, and labor. Avoiding the remoteness, heat, and humidity of the Amazonian selva east of the

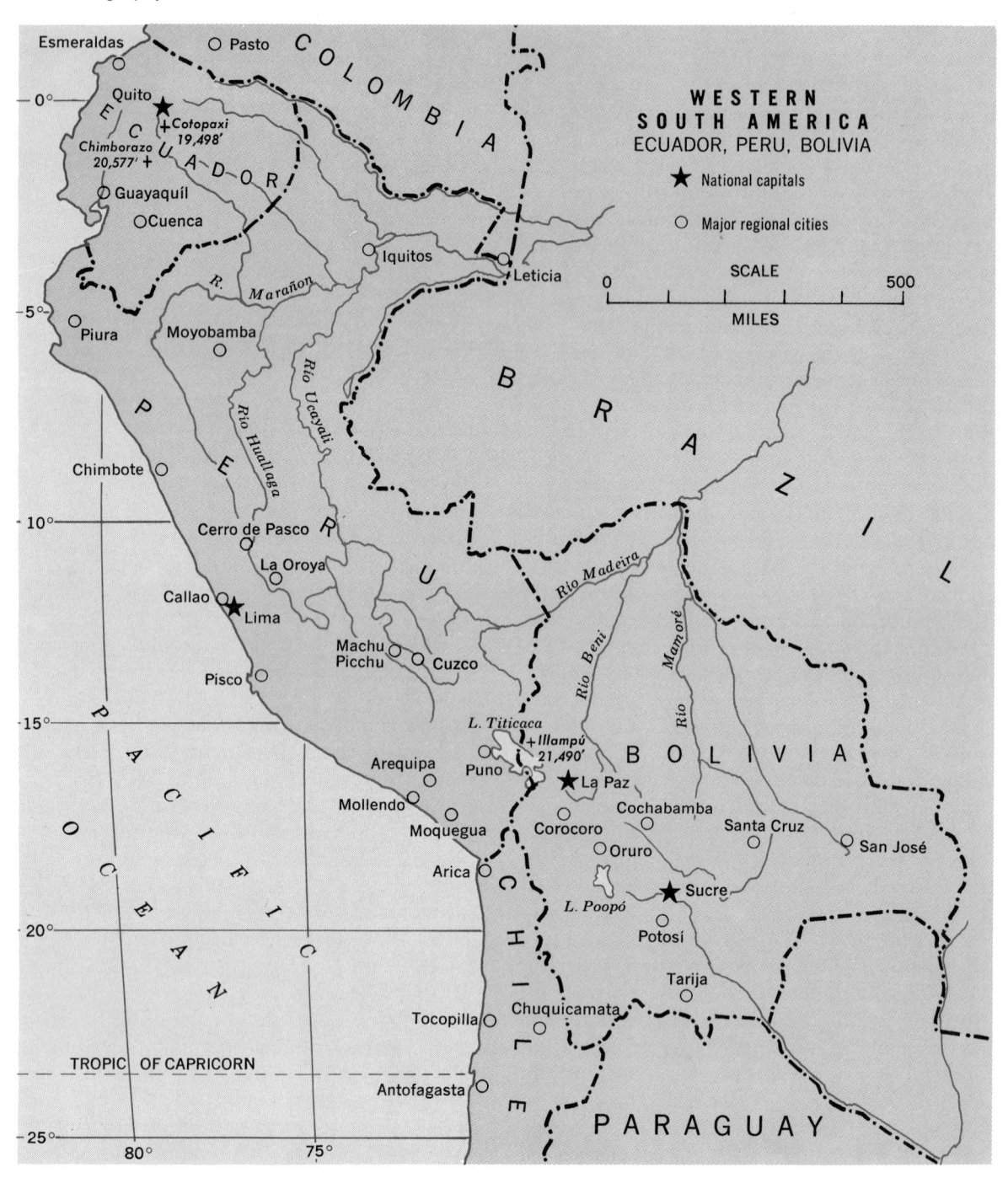

Fig. 6-7 Activities in the three western countries of South America are greatly influenced by the Andes. Here the only large plains areas are in eastern Peru and Bolivia.

Andes, the mestizo-European communities have turned westward, looking to the Pacific and utilizing resources from the western Andean slopes. At the beginning of the 1970s, some 22 million people live on about 1 million square miles of land in Ecuador, Peru, and Bolivia. Two-thirds of them, however, dwell above the 10,000-foot level in the cold, dry Andean refuge.

Ecuador

Ecuador, the third smallest nation in South America, lies on both sides of the equator. Its territory also includes the famed Galapagos Islands of Charles Darwin, situated on the equator several hundred miles to the west. Overshadowed by its larger neighbors, Ecuador is backed up against the Andes, with only a token east Amazonian slope. Two contrasting areas of the country favored human occupancy: the high, cool valleys walled in between the Andean crests, and the alluvial watershed of the Guayas-Daule river system, set between west slopes of the Andes and coastal hills, opening on the Gulf of Guayaquil at the port city of that name.

Across Ecuador, from north to south, the Andean Cordillera is pocketed with several basins and a multitude of eroded valleys at elevations between 5,000 and 12,000 feet. Highlands above the *tierra templada* contain almost two-thirds of Ecuador's people. Unlike the hot, humid Pacific plains and valleys where the mestizo, Negro, European, and Indian newcomers from the highlands have begun to convert the *tierra caliente* into a productive garden, the highland remains an Indian realm. Only cities such as Quito, Cuenca, and other provincial capitals have any numbers of mestizos or Europeans.

Livelihoods differ from north to south in the valleys, but upland grazing of sheep is universal. The north basins are lower in elevation, sometimes lying within the *tierra templada*, and are the most highly developed for field grains and gardening. Family villages supply their own needs for various foods such as wheat, potatoes, and poultry. Throughout the north and central basins, soils have developed from volcanic deposits dispersed in the distant past from high, snow-crested volcanoes. Towering majestically are several snow-clad volcanic cones, having elevations of 19,000 to 20,000 feet. The symmetry and grandeur of Cotopaxi and Chimborazo, among others, are justly famous.

The irrigated basin of Quito and other central basins yield temperate grains, vegetables, and fruit to sustain day-to-day needs of the peasant farmer, who delivers any surplus to nearby towns. Although numerous large haciendas still emphasize cattle ranching, many

Fig. 6-8 Ecuador's capital Quito, with a population of about 500,000, is located in a high mountain basin almost on the equator. (Courtesy of Pan American Airways)

have been divided into farm units and rented to tenants. Some land is held communally by the Indians, as was common among their forebears. Field technology ranges from sophisticated farm equipment to the far more common wooden hoe. Since the basins south of Quito are more rugged and often drier, cattle and sheep grazing is more widespread. These basins are linked by the Pan-American Highway, entering in the north from Ipiales (Colombia) and winding its way south to enter north Peru. A railway is the principal surface link with the ocean gateway, chief port, and largest city on the Ecuadorean coast, Guayaquil.

The hot, Pacific lowlands afford a distinct contrast with the highlands and differ in themselves as they span the coast from north to south. From the Ecuador-Colombia border southward, rainy tropics give way to savanna and, in turn, to semiarid steppe into Peru. From the Gulf of Guayaquil northward, tropical forest envelopes lowlands, coastal hills, and lower Andes slopes in the *tierra caliente*. Settlement is scant, with towns of some size at the ports of Esmeraldas and San Lorenzo. Peasants subsist by family farming in clearings or by occasional wage work on banana or cacao plantations. Upslope, pioneer settlement in the *tierra templada* sends out commercial quantities of coffee and fruit to the coast.

Projecting into the waters of the Peruvian Current west of Guayaquil is the dry Santa Elena Peninsula. Offshore waters along the entire Pacific coast of the continent are a storehouse rich in commercial fish. Salinas is one of numerous fishing ports. Santa Elena produces oil, which is sent by pipeline to refineries at Guayaquil. Traditionally, Santa Elena is a delightful resort for those escaping the rigors of heat and humidity in the Guayas Valley or the chill of the highlands. Guayaquil, once it was able to achieve proper health and sanitation conditions, emerged as the chief trade and manufacturing center and main seaport of Ecuador. As a gateway, Guayaquil commands surface movement toward Quito and the highlands, as well as river and land traffic in its valley hinterland. Hills to the west and the gentle rising piedmonts of the Andes support coffee, citrus, and other *tierra templada* agriculture. In the humid lowlands in between, plantations on fertile alluvial soils that once supplied cacao to the world now yield harvests of rice, sugarcane, and other staples. In the Guayas-Daule lowland and around the fringe of the Gulf of Guayaquil, modern banana plantations send out the world's largest exports of that commodity from any single nation. South of the gulf, the dry lowland is given over to pasturing cattle.

Small quantities of gold, silver, copper, and wool come out of the highlands, but the economic destiny of Ecuador appears to be as a supplier of tropical agriculture. Besides being world leader in banana exports, mostly sent to North American markets via the Panama Canal, Ecuador exports such commodities as cacao, coffee, and rice.

Peru

The great empire of the Incas, whose political seat was at Cuzco and whose mantle of authority encompassed Indian communities as far north as Colombia, westward across Bolivia, and south into Chile, occupied the southern highlands of Peru, the center of this impressive pre-Columbian civilization. Well-disciplined by their culture, the Incas developed remarkable agricultural and irrigation systems and constructed massive roadways, temples, and dwellings that represent feats of engineering and architecture. The Spanish conquistadores were able to substitute themselves in the system of totalitarian rule by the Incas. Seeking gold, the Spanish directed the labors of the Incas away from agriculture, which fell into disuse and decay, and turned instead to the search for and accumulation of precious metals from the mines of southern Peru and Bolivia. Today, however, as traditional landholdings and newer plantations held by domestic and foreign firms are directed into the nation's economic mainstream, increased responsibility for successful management is again resting with the long-disinherited Indian peoples.

Though the indigenous culture of the Indians was long submerged, it was not obliterated. Today the highland Indians usually speak the Quechua and Aymara tongues of their ancestors, rather than Spanish. Only along the Pacific coast and in the Montaña pioneer communities to the east is Spanish widely used. A knowledge of their background is essential as we approach the problems, paradoxes, and cultural provincialism of the Indian communities in the highlands of Peru and Bolivia. The significance of the Indian population today becomes quite apparent when we remember that in Peru and Ecuador about two-thirds of each nation's population live in the highlands, with an even higher percentage in Bolivia, and that they are Indian peoples.

In 1970, a series of earthquakes caused massive devastation throughout the Sierra and Pacific coastal region of Peru north of Lima. Scores of towns were totally destroyed, and hundreds of thousands of people were displaced. Dams and power stations were destroyed, transportation disrupted, mines collapsed, and the important industrial city of Chimbote suffered a setback that will require years for recovery.

The Pacific lowlands and west Andean flanks are dominated by an extremely arid desert. In a few places the desert attains a width of 100 miles, but for most of its length it is less than 10 miles wide or disappears entirely where cliffed headlands and mountain spurs plunge directly to the Pacific. Fed by melting ice and snow and upslope precipitation, some fifty streams maintain at least a seasonal flow across the desert. These waters are the lifeblood that sustains valley oases. Several of these oases are first-rank among agricultural districts on the continent and produce sugar, cotton, and rice on modern industrialized plantations for export. These and other Pacific oases send food staples to Lima, other coastal towns, mining camps, and highland communities deficient in food. Having the best water supplies and modern technology, the northern oases lead in the growing of sugarcane, rice, and cotton. The middle oases, from the Pativilca River south to Ica, including the productive Rimac Valley behind Lima, provide sugarcane, truck crops, and fruit. Irrigated forage sustains a modern dairy industry near the capital. The smaller, drier south oases grow subsistence crops and grapes for a famous brandy.

Offshore, the cold waters of the Peruvian (Humboldt) Current are so rich in marine life that Peru soon came to the forefront among fishing nations after the development of a modern fishing fleet under Japanese guidance. Millions of birds live off the sea and nest on tiny islands and peninsulas projecting from the coast. Their droppings form guano, a phosphate-rich natural fertilizer used by the ancients to enrich their fields. Today, guano is mined as a government monopoly.

At the north end of Peru, near Talara, are three notable oil fields. These fields and tidewater refineries presently are government-owned. With rising domestic oil consumption, most of the output now goes to home markets.

Lima, the largest metropolis in western South America, and its nearby port at Callao are rich in the flavor of Spanish colonial history, but also reflect the quickened pace of twentieth-century commerce and industry. Their location midway along Peru's coast is advantageous, since the ocean is still the most important means of surface commerce from north to south. They also control the chief gateway to the highlands, where they fit into a general north-south alignment of upland routes. Since 1968, three arterial routes have been opened from the coastal Pan-American Highway across the Andes ranges to pioneer districts in the Montaña. The northern route reaches the navigable headwaters of the Amazon; a second terminates at Pucallpa, a river town and springboard to the interior lowlands; and the third enters the lowlands below Cuzco.

The Andean System is only some 150 miles wide near the Ecuador border but widens to almost 450 miles farther south between Peru and Bolivia. The western Cordillera forms a high barricade, with towering peaks and volcano-crested uplands extending to 18,000 and 19,000 feet and traversed by low passes at 12,000 feet. The eastern Cordillera slope is scarred, dissected, and incised by Amazon headstreams. In between, a series of high plateaus are suspended within a series of ranges that cross and recross the highlands. These plateau basins lie in the *tierra fría* and paramos, and extend higher as snowfields and glaciers. Here dwell two-thirds of Peru's people.

As in Ecuador, two contrasting and sometimes contradictory systems have existed side by side for hundreds of years. Haciendas encompass thousands of square miles, while smaller, communally held Indian communities are wedged between. Highland agriculture clings to steep slopes and occupies valleys and basins up to elevations of 14,000 feet. Since the early years of the Spanish presence, the conservative Indian workers of the land have added little to their knowledge of farm technology. Many still use little more than a simple hoe or machete to plant, cultivate, and harvest food staples. Soils have been overworked for centuries. Tiny properties, communal Indian lands, or parts of haciendas may be grazed when in fallow. Under cultivation, they grow moderate yields of wheat and barley at low altitudes and quinoa (Indian barley) on higher slopes. The potato and related native tubers will grow everywhere. Livestock raising adds to a pitiful livelihood for the Indians. Llamas and alpacas, in their wild state native to the Andes, and sheep provide wool to make blankets and clothing for trading at local markets. Cattle are raised on the lower slopes and in the warmer valleys. The 12 to 40 inches of annual precipitation that barely sustains sparse highland steppe grasses must also support field crops. Valley canyons set deep in the highlands isolate huge segments of the Andes. Cuzco, a regional center, is a tourist attraction and today is one of the four major settlement clusters in the Peruvian Andes. Arequipa, midway between coast and highlands, has been a traditional marketplace for items coming out of the Andes.

The Montaña, east of the Andean Cordillera on the slopes and in the lowlands, covers one-half of the land area of Peru. Several tributaries of the Amazon have sheared deep clefts into the eastern Sierra and afford valleys and piedmonts suitable for settlement. Pioneer settlements are gradually advancing into the wilderness. Several resources have spurred their progress; one is the fine virgin timber awaiting harvest for markets in the highlands and along the Pacific coast. Petroleum has been discovered near Pucallpa. Otherwise, most of the region's people live along the Amazon tributaries. Iquitos is the major city in this inhospitable environment. Situated at the head of Amazon River navigation, some 2,000 miles from the Atlantic, this city of 75,000 serves as a collecting center for products of the rain forest.

The legendary mines of Peru, worked before the Spaniard set foot in the New World, still yield some silver and gold; but today industrial metals such as copper, lead, and zinc from such mines as the 14,200-foot Cerro de Pasco are more important mineral ores. Larger copper mines have been developed to the south, on the Pacific slope near Toquepala. Iron ore mined at Marcona is shipped overseas, as well as to the national steel mill at Chimbote, north of Lima. Bituminous coal comes from inland deposits near the mill.

As a phenomenon of the 1960s, the fishing

industry gained prime importance. Fish meal and fish products rank first in the export commerce of Peru and make up nearly a third of the total trade by value. Metals, chiefly copper and iron ore, comprise another third of the exports. Agricultural commodities, cotton and sugar, make up almost 18 percent of the country's exports by value. The United States is both Peru's biggest customer and supplier. Relative to West Germany and Japan, however, its position has declined in recent years. Peru needs industrial machinery, transportation equipment, and foodstuffs, all of which relate to urbanization and industrialization of the economy.

Bolivia

The population core of Bolivia, living in the cold, dry altiplano, is surrounded by the walls of the Andean Cordillera. The people are cut off, as is the nation, from access to the Pacific by topographic and political boundaries. In comparison with area, Bolivia's population appears small, since the country is almost as large as Colombia. The Indian community and the pattern of land use in the highlands of Bolivia differ little from those found in highland Peru. At least two-thirds of the Bolivian people are wholly or partly of Indian heritage and live at high elevations in plateaus and basins among the Andes. The high plateau, or altiplano, lies between the western Cordillera that continues south from Peru into Chile and the central and eastern Cordillera that continues south into Argentina. To the east, the deeply dissected slopes and valleys of the Cordillera give way to the river plains of Amazonian tributaries in the north. A low divide of hill lands separates the Amazon watershed from the Paraná-Paraguay lowlands occupied by the Gran Chaco and the Pilcomayo River.

The western Cordillera and the two merged ranges of the central and eastern Cordillera support the more than 12,000-foot-high plateau. Above this altiplano, the snow-crested Cordillera rises beyond 20,000 feet. The western range is very dry, with little snow on its crests, as it extends from the dry Pacific lowlands of Chile to the altiplano. The Cordillera to the east blocks the southeast trade winds and thereby captures a considerable amount of moisture on its north- and east-facing slopes. Thus the altiplano, which stands at about the tree line in the cold paramos, is a dry, bleak grassland. In the fields and mines at these high elevations, oxygen is so sparse that only the barrel-chested highland Indian is capable of exerting himself without experiencing shortness of breath. The altiplano is about 60 miles wide and 500 miles long, with spurs of the Cordillera cutting the plateau into basins. The southern part of the altiplano is a desert, with salt plains and the salt lake Poopó, which receives overflow from Lake Titicaca to the north via the Río Desaguadero. The northern part of the plateau, near the capital at La Paz and around Lake Titicaca, is the most heavily populated area of the highlands. Since rainfall is somewhat more plentiful, coming during the high-sun season, an agricultural community similar to that of highland Peru has been able to subsist. Tilled fields and hill-terrace farms grow potatoes, barley, and quinoa. Herders tend flocks of llamas, alpacas, sheep, and burros on the slopes.

La Paz supports nearly 450,000 people at an altitude of 12,000 feet. The city is located in a steep-walled canyon, situated in the shadows of the glacier-covered Illampu and Illimani peaks. The *de jure* capital is at Sucre, a valley town in the east Cordillera in the pleasant *tierra templada*. Linked to the Pacific by rail, La Paz has "free port" privileges at Antofagasta, in northern Chile. Rail and lake steamers across Titicaca connect it to Mollendo-Matarani, Peru's chief southern port, and rails also tie it with Chile's port of Arica. Both road and rail arteries penetrate the rugged east Cordillera to the important crossroads city of Cochabamba, gateway to the east. The wealth of the highlands comes not from poverty-stricken peasant agriculture but from the rich deposits of metals in the surrounding Andes. South of La Paz, important mining camps are located at Potosí (elevation 13,780 feet), where Cerro Rico's wealth of silver lured the early Spaniards, at Orurú in the foothills, and at Unica on the plateau floor. These mines yield tin and by-products of lead, zinc, and silver. Corocoro is an important copper mining camp near the capital. To these cold deserts, every life-support item and every piece of equipment must be brought in at great expense. Only since 1968 has an ore smelter been operating in the country, at Orurú.

Challenges confronting the Bolivian nation are manifold. One task is completing the destiny of the 1952 revolution that ended the era of haciendas and brought nationalization to the tin industry. A second is coping with the emptiness of three-fourths of its territory, while nearly two-thirds of its people live on a tenth of the land area in the altiplano. Highland agriculture has changed little since the early 1950s. People dwelling in the altiplano are unable to supply their own everyday requirements of food, fuel, or manufactured goods; yet, they are the majority of the nation. Mining companies, now mostly government-owned, are heavily subsidized. Though modernization and renovations have improved output, Bolivia now supplies approximately 11 percent of the world's tin.

Beyond the east Cordillera crests, the Andean flanks slope into Amazonia and the tropical savannas in the heart of South America. River valleys and slopes in the *tierra templada* zone have high-sun rainfall and good soils. Fertile valleys indent the forested slopes of Yungas into the *tierra caliente,* and from them come sugar, cacao, staple foods, and the narcotic coca leaf, chewed by highland Indians to ward off cold and hunger in their rigorous realm. South from Cochabama and its important farming valley, crops are subtropical citrus, maize, coffee, and mid-latitude wheat, which are sold to highland markets. Lumber and firewood, essential for construction and fuel in the treeless highlands, come from the eastern valleys and the Santa Cruz pioneer districts for altiplano cities. Along the rail link between Santa Cruz and Brazil, pioneer settlements send back cotton, cattle, and sugar to Cochabamba. One outstanding factor in causing settlement along the eastern front of the Andes has been the sizable petroleum discoveries south of Santa Cruz. Oil moves by pipeline across the Andes through Cochabamba to the Chilean port of Arica.

Aside from eroding internal barriers between the Indian and mestizo-European cultures, Bolivia's major economic drawback is the government's inability to foster interaction between its underdeveloped, but complementary regions. Effective surface-transport systems are lacking. Railways, costly to build and maintain, seldom run through trade-generating hinterlands; their destiny is to transport minerals. Except for the Santa Cruz–Cochabamba–La Paz roads, modern highways are lacking. Even air transport has failed to offer the degree of flexibility present in Colombia or Venezuela.

Southern South America

The four nations forming the southernmost part of South America—Argentina, Chile, Paraguay, and Uruguay—for the most part occupy that part of the continent south of the Tropic of Capricorn. Extending into mid-latitude and subpolar climates from subtropical northern borders, Chile, along the western flank of the Andes, and Argentina fill out the southern end of the continent, with Uruguay and Paraguay situated as buffer states between the Spanish and Portuguese heritages in South America.

The agricultural riches of the Argentine and Uruguayan pampas, the industries of cities

Fig. 6-9 Southern South America is the best-developed part of the continent agriculturally. The pampa is a world leader in the production of wheat and meat.

fringing the Río de la Plata, and the affluence and cosmopolitan atmosphere of Buenos Aires and Montevideo present an astonishing contrast with the more provincial nations and capitals of Andean South America. The mild, mid-latitude climate, well-watered during the year, and the large stretches of flatland with good soils have favored both livestock raising and farming. Growing markets in Europe for food staples during the late nineteenth and early twentieth centuries and the immigration of tens of thousands of Europeans from Spain, Italy, and other lands, who came equipped with capital and modern farm methods, spurred the emergence of commercial agriculture. Rail lines and, later, highways spread the benefits of European civilization across the land. Without an avenue to the sea, isolated Paraguay still retains its provincial character and Guarani Indian culture, modified by contact with Europeans.

Urbanization has progressed rapidly, and generally high literacy attainments are keynotes for these societies. The agricultural riches of Chile's Central Valley and the Argentine and Uruguayan pampas complement the manufacturing complexes in middle Chile and those arrayed along the Río de la Plata. When coupled with the affluence and cosmopolitan atmosphere of Buenos Aires, Montevideo, and Santiago, these metropolitan areas and their support hinterlands offer great contrasts to most core regions elsewhere on the continent.

Chile

Chile extends southward from Peru along the Pacific shore for over 2,600 miles, over half the total length of the continent. Although its length is great, the country attains a maximum width of only 250 miles from the Pacific across the low coastal range into a segment of the altiplano; seldom is its width greater than 100 miles. The exceptional length of the country, from the subtropical deserts of the Atacama to the chilled waters and grasslands of Tierra del Fuego, creates striking climatic contrasts. In much the same fashion, the coastal ranges, the longitudinal valley, and the Andean Cordillera which extend the length of each region add a second interesting dimension to the climate divisions arising in accordance with the increasing distance of the country from the equator.

The Andean Cordillera, extending the length of the country and demarcating Chile's political boundaries with Bolivia and Argentina, forms one of that nation's major regions. The other main regions, according to latitudinal position from north to south are: the Atacama Desert, Mediterranean Chile, and the marine west coast of South Chile.

The Atacama continues the dry, desolate, and barren landscape of Peru southward to the vicinity of La Serena, where rainfall amounts to about 10 inches yearly. The desert of North Chile is reputedly the driest in the world. The extreme dryness has preserved deposits of sodium nitrate, the chief product of the Atacama. In the longitudinal valley, between the coastal ranges and the Andes, this soluble salt is found in layers near the surface, where it is mined and then processed for export. Chile is the world's leading producer of natural nitrate and is also the outstanding producer of iodine and borax as by-products. All necessities to sustain the workers and the mining facilities must be shipped in, including water. The water is carried by pipelines from reservoirs in the high Cordillera. Near Coquimbo, high-grade iron ore is mined from rich deposits at El Tofo. The iron ore supplies Chile's iron and steel mill at Huachipato, near Concepción, and is also exported to the eastern United States through the Panama Canal.

The international boundary follows the Andean Cordillera. Mighty peaks, including Mt. Aconcagua, rise to nearly 23,000 feet; others descend to elevations of 5,000 feet or less. Although sparsely settled, the mountains provide the leading export commodities by value for Chile. Rich in copper, iron, cobalt, gold, zinc, and other industrial metals, the great deposits developed by several U.S. companies have gradually been "Chileanized." In 1969, the vast Anaconda interests were purchased by Chile for $250 million in government bonds. Usually Chile ranks among the top three nations in copper production. The three principal mining districts, all situated on the western slopes of the Cordillera, are Chuquicamata, El Salvador, and El Teniente. Chuquicamata, which has a fabulous history, has the distinction of being the largest single copper mining camp in the world. To develop these rich deposits in the remote, barren wastes of the mountain desert, millions of dollars have been invested in equipment, housing, electric power, transportation, and planning. Hydroelectric stations constructed in the high mountain valleys provide electricity, and this power is supplemented by diesel electric stations using imported fuel oil. Every year thousands of tons of food staples, expensive machinery, and other necessities are shipped in at great cost to support the workers and to maintain the operation of the mines and reduction facilities.

With the free election of the first leftist government in Chile in 1970, the atmosphere favoring foreign investment changed dramatically. Under the Allende government, nationalization proceeded apace in the basic metals industries and spread into other service and manufacturing activities.

Mediterranean Chile is located from La Serena south to Valdivia, in the area where winter (May to October) concentrated precipitation increases appreciably, from 10 inches to over 40 inches. The seasonality of temperature becomes quite evident, especially in the central valley, some 1,500 feet above sea level.

During the summer dry season (November to April), melting snowfields and glaciers in the Cordillera supply water for irrigation. Over the uncultivated mountain slopes and hills that divide the valley into smaller basins, there is dry shrub vegetation similar to that of southern California. In this delightful climate live nearly 90 percent of the Chilean people. They pursue farming activities on the good soils of the central valley or are engaged in manufacturing activities in the large number of medium-sized cities in the valley and along the coast. Cultivated land held in large estates is underutilized. Land-reform laws, however, are aimed at abolishing the long rule of the Chilean hacienda, or fundo. Agricultural self-sufficiency for the nation is the goal. Farm technology has lagged, and inefficiency has been perpetuated. Where Chile was once an exporter, in recent years grain imports have been necessary. Estates

Fig. 6–10 Valley in Central Chile. Many wheat fields in the Andean valleys are separated by stone walls. (Courtesy of Hamilton Wright)

produce winter wheat, but much cultivable land is kept out of food production to support dairy and beef cattle. In the drier, sunnier valleys around Santiago, vineyards on small farms supply fresh table grapes, and Chilean wines have an international reputation. Citrus and basic staples such as corn, beans, and potatoes, together with green truck crops, are grown in abundance.

Santiago, the capital and largest city, with its gateway port of Valparaíso and resort-industrial center of Viña del Mar, forms the largest population concentration in Chile. About 45 miles inland from the Pacific, Santiago is about centrally located between the country's lateral extremities. It is also the commercial entrepôt for the central valley and the cultural and manufacturing center for all of Chile. The capital is the terminal for trans-Andean routes through the Uspallata Pass to Buenos Aires. As in Peru and Ecuador, Chilean fishing enterprises working out from ports between Antofagasta in the north through Valparaíso and Talcahuano are exploiting the rich resources of the Pacific.

South of the Bío-Bío River at Concepción to Puerto Montt, a marine, mid-latitude climate prevails. Temperatures are generally cooler, and winds are more pronounced. Descendants of European immigrants and indigenous Araucanian Indians are engaged in forest clearing that supports important lumber and by-products industries and in opening new land for agriculture. Along the coast, around Valdivia and in the central valley, nearly a million people are involved in agricultural occupations. Heartier crops such as spring wheat, barley, rye, and potatoes are common, as are apple and pear orchards. Cattle, grazed on lush pastures around deep-blue lakes, and sheep and goats pastured on hills and slopes furnish milk for dairy and cheese industries. Here and in Mediterranean Chile, government and private agencies are striving for an effective, economically sensible resolution to the land-distribution challenge.

South Chile extends to the subpolar margins of cold, bleak Tierra del Fuego. The annual precipitation exceeds 100 inches; winters are rigorous, and summers are merely cool interludes of less rainfall. The grand archipelago, extending in an arc southward from Chiloé Island, encloses an inland passage for much of its 1,500-mile length. Tierra del Fuego has become an important petroleum exporter for Chile. Rain-soaked and fog-cloaked, the coast is pocketed with fjords, some of them holding glaciers down to the sea's brink. Far-south Chile, east of the low Andes and facing the Atlantic, is protected from torrential precipitation. Cool, grassy pampas support quality sheep herds numbering in the thousands. Sheep thrive and produce fleece of superior length and quality. The wool clip is shipped out through Punta Arenas, facing the Strait of Magellan.

With the opening of the Panama Canal, Chile's reserves of metallic ores began to enter the channels of world commerce in volume for the first time. Earlier, only the unique nitrate deposits of the Atacama had provided sizable exports.

Copper, nitrates, iodine, iron ore, and other industrial minerals constitute over four-fifths of Chile's exports by value; petroleum, wool, mutton, lumber, and even some manufactures are significant. Much of Chile's foreign trade is with West European nations, though the United States has continually been the most important single trade partner. Japan is a newcomer on the scene of Chilean trade. Imports consist of consumer and industrial goods, such as machinery, transport equipment, and agricultural commodities.

Argentina

Since the middle of the last century, Argentina has become one of the important nations of the world and one of the most prosperous in Latin America. The country's wealth springs from the resources of the Pampa, an area that possesses a remarkable combination of excellent climate, fine soils, and extensive cultivable surfaces, all exploited by a willing and technologically equipped people. Foreign initiative, capital, and know-how have played leading roles in its economic development. Though foreign investments are still welcome, the Argentines themselves have assumed the major role in management, design, and expansion of transport and industry.

Within its latitudinal range, from the Tropic of Capricorn to the Straits of Magellan, and its breadth, from desert foothills of the Andes to humid plains and river basins along the Atlantic, Argentina offers numerous possibilities for regional identity. Of its several subregions, the Pampa is the vital heart of the country. In an area of 250,000 square miles, spreading outward from Buenos Aires at the Río de la Plata, live two-thirds of the Argentine people. In little more than one-fifth of the national territory, the Pampa has most of the surface transportation and accounts for three-fifths of the cattle, 90 percent of the grain lands, and virtually all of the manufacturing.

The humid Pampa stretches west and south from Buenos Aires for over 300 miles. Few areas of its size have been so favorably endowed for agriculture. Among its assets are a nearly level surface, interrupted only by two low hilly ridges in the south, covered with fine, mellow, nearly stone-free soils, and a humid climate hardly excelled for mid-latitude agriculture. Summers are warm to hot; rainfall is plentiful and usually reliable, though occasional droughts do occur. In contrast, winters are mild to cool at Buenos Aires; snowfall, rare in most sectors, is common in the south at the port of Bahía Blanca.

The wetter eastern third of the Pampa is the center of the livestock estancia; the drier west encompasses the grain-growing fertile crescent. Nowhere in the greater Pampa region is less than half the land in pasture. South and east of Buenos Aires the flattish, poorly drained surface supports excellent alfalfa and clover grasses and sleek herds of cattle and purebred sheep. Since the marketing of Argentine beef in Europe developed with the adaptation of refrigeration to ocean freighters, herds of prime stock have expanded rapidly. Much of the east Pampa is subdivided into 100- to 5,000-acre pastures fenced with barbed wire. Year-round grazing finishes stock for market a year younger than do the ranges elsewhere. Mutton and wool breeds use the southern hilly belts as range, which supports nearly a third of Argentina's sheep population in the Pampa. Westward, field crops appear, though livestock retains importance. Corn, wheat, barley, and flaxseed are notable cash crops cultivated in rotation with alfalfa. Favored by the long, hot summers of the district in the northwest Pampa, a corn-hog economy has grown up to surround the city of Rosario. Here corn (maize) may be either a cash crop for export or be used to fatten hogs, depending upon which promises the better income in that year. The wheatlands of the western Pampa extend in a vast crescent for nearly 600 miles, south from Santa Fe to Bahía Blanca in southern Buenos Aires province. Wheat is the single most valuable export commodity of Argentina. The wheat crescent is situated along a climatic transition zone where precipitation becomes too marginal to support many crops. Grain farming is pursued generally by prosperous tenant farmers on 150- to 300-acre sites. Most of the grain is cultivated with modern machinery, in contrast to more backward farming technology found elsewhere on the continent.

Buenos Aires is a splendid and attractive city, the second largest metropolis, after Mexico City, in Latin America. In the heart of the city, near the port and commercial houses, tree-lined boulevards fan out between government buildings, including the Casa Rosada—residence of the government—banking institutions, fashionable shopping districts, theaters, and waterfront parks. Buenos Aires and other cities facing the Río de la Plata and the Paraná River comprise one of the three key manufacturing centers in Latin America. Until recently, limited national resources for sustaining heavy manufacturing restricted output to consumer goods. Today, barriers of distance and cost have been so reduced that iron ore from Sierra Grande and coal from the United States support an integrated steel plant at San Nicolás, north of Buenos Aires on the Paraná. From the oil fields at Comodoro Rivadavia in Patagonia, petroleum is delivered by tanker fleets to tidewater petrochemical complexes to supply one of the largest consumer markets in Latin America. Rosario is a leading processor and exporter of farm and livestock commodities; La Plata is especially important for livestock exports. Córdoba, located inland at the edge of the Pampa, has emerged as an automotive and machinery center.

Lodged between the Paraná and Uruguay rivers north of the capital is the "Argentine Mesopotamia." Here, temperatures are warm and rainfall is plentiful. Bordered by floodplains, the interior is a series of rolling hilly belts, sometimes covered with grasslands favored for grazing. A pioneer version of Pampa agriculture spills across the southern provinces, and grain farming and livestock grazing predominate. The cattle are of lower quality than in the Pampa. Sheep are also found there in large numbers. Where a small tongue of the Paraná plateau crosses into a tiny panhandle in the extreme north, fine stands of Paraná pine support a lumbering industry. Where Brazil, Paraguay, and Argentina share a common border, the magnificent Iguassu Falls plummet over several cataracts descending the edge of the Brazilian plateau. A tourist attraction of the first order, they are broader and higher than Niagara Falls.

Where the Gran Chaco extends southward from Bolivia and west Brazil across Paraguay and into subtropical northern Argentina, the terrain is flat and the climate hot much of the year, complemented by a high-sun wet and low-sun drought seasonality. Scrub woodland, low-nutrient grasses, and a flood-drought cycle provide support for at best poor-grade cattle. Efforts to upgrade utilization of this tropical savanna, as in the Orinoco llanos, remain to the future. Since early in the century, the Chaco region has been Argentina's cotton-growing area. Good-grade cotton from pioneer Chaco farms is shipped south by truck or rail to Pampa textile factories.

The western oases and deserts of Argentina are located between the humid Pampa and the Andes. Nearly a quarter of the Argentine people live in this area, pinpointed mostly at nodes where water from mountain streams or well aquifers is adequate for irrigation. Although these population nodes are separated by barriers, each is linked by modern rail and road transport to the important urban-manufacturing core they serve in the Pampa. The northern oases, including those of Jujuy, Salta, and Tucumán, contain the principal sugar districts, which send out more than half of Argentina's annual output. The middle oases, focusing on Mendoza, are Argentina's vineyards, but they are small and short of water. At Mendoza, the east gateway to the Uspallata Pass leading to Chile, wine grapes are harvested, and grains and irrigated pastures finish cattle raised on the steppe ranges for eastern markets.

The vast, windswept Patagonian plateau spans southern South America from the Andean foothills to jagged, cliffed headlands along its sometimes inaccessible coast. Deep valleys and west-east canyons split the plateau surface into segments. Covering a third of the country, Patagonia is at the same time one of the most thinly peopled areas on the continent. Sheep ranching there is declining because of low-quality wool produced by sheep that must survive the harsh, semiarid environment. Far to the south, Argentine sheep ranchers share the grassy pampas along the Strait of Magellan with their Chilean counterparts. Here the large estancias send fine fleece and wool of the best quality to small coastal stations. Over a quarter of the country's sheep are maintained in this cool realm, and these produce about half of the export wool. Argentina ranks with Australia, New Zealand, and the Soviet Union as a major wool producer.

Argentina's network of railways and highways radiating from Buenos Aires and the Paraná River tidewater to adjacent regions is among the most modern and extensive in South America. During the late 1800s, modern farming and livestock industries pioneered inland along the frontier, and were followed by a wave of railway construction. Unimpeded by the natural barriers present in western South America, these modern transport arteries have facilitated maximum surface circulation and swift and inexpensive assembly of a wide range of commodities. Beyond the Pampa network, links to provincial towns have fostered regional interaction in securing supplementary or complementary goods for eastern tidewater cities. Each region harvests agricultural commodities and sells livestock and livestock products that are moved quickly to processing points in the eastern port and market cities.

Uruguay

Uruguay, the smallest independent nation of South America, is wedged into the southeast corner of the continent between the two largest nations. The pampas extend from Argentina across Uruguay and into southern Brazil. Except for a small, rugged upland in the extreme north, Uruguay has an undulating to rolling surface, subjected to a uniform land use throughout. Quite unlike Argentina or Brazil, in which large areas separate population centers and vast segments of the national territory lie unused, virtually every square mile of Uruguay is productive and contributes to the livelihood of the nation. In a country that has few resources other than its land, farming and grazing are the foremost commercial activities.

Sheep range over the natural grasses fenced into areas of 100 to 5,000 acres on large haciendas, and from their clip comes the most important export of the country. The sheep ranges are usually found north of Montevideo and away from the lowlands of the Uruguay River and the Río de la Plata, where farming is more important. Together with the sheep, fine herds of cattle are fattened on the pampas grasses before they are shipped south to packing plants in Montevideo. In recent years the number of cattle herded has declined, but the overseas markets for wool and mutton have held firm.

A typical Uruguayan landscape resembles the Argentine Pampa, except that natural grasses rather than alfalfa provide the forage for sheep and cattle. Small frame houses, the homes of the cattlemen, are surrounded by trees to protect them against the winds that sweep inland from the Atlantic throughout the year. Since grazing can proceed all year in this mild climate, no shelter or stored feed is required, and there are only a few outbuildings.

In the south and west, crop farming is especially important to Uruguay's economy, though production may be small by comparison with Argentina. Every endeavor is being made to stimulate increased harvests of grain and flax, for the breadgrains are needed at home and flax is a valuable export commodity. The expanding farming is encroaching on lands formerly used to fatten and finish cattle. The continued growth of farming has contributed to the diminishing of cattle herds and the reduction of a major export commodity.

Montevideo is the hub of Uruguay's universe; in the capital city live nearly one-third of all Uruguayans, and it is the commercial, social, economic, cultural, and political center of the nation. Moreover, Montevideo is fast becoming a manufacturing center. The metropolis is the seaport and chief meatpacking center for the leading export industry. The climate is mild in the summer, surpassed in its healthfulness only by the delightful climate of Punta del Este, the nearby resort center. Winters are equally mild, though cooler and marked by frontal storms that push in from the South Atlantic.

Though its population is increasing rather slowly for Latin America, Uruguay is nonetheless already feeling the pinch in the drive for exports and the need also to supply food grains to its people. Still, Uruguayans enjoy a remarkably high standard of living in comparison with other Latin-American nations. To sustain both a high level of exports and provide an adequate food supply, without sacrificing the high living standards already achieved, is the nation's continuing challenge.

Paraguay

Hundreds of miles from the Atlantic, with no direct route to the sea, Paraguay shares with Bolivia the handicaps of isolation and an undeveloped economy. Yet the rather mild climate and fertile soils of well-watered eastern Paraguay indicate good natural endowment for this nation of the Guarani Indian and the mestizo. The dual culture of the full-blooded Indians and city-dwelling mestizos recognizes the Guarani tongue as well as Spanish as official languages of the country.

Flowing from north to south, the Paraguay River divides the country into two major regions, the Paraguayan Chaco to the west and the eastern river plains and uplands. Most of the western Chaco is flat, with a distinct wet-and-dry climate during the year. Temperatures are so hot that they may become almost unbearable during the summer, and near desertlike conditions prevail in some districts. When the rainy season begins, temperatures moderate, but torrents of rain inundate tremendous segments of the land. Herds of very low-grade cattle, valuable only for their hides, graze over the scattered patches of grass and through the dry scrub forest. With scant population and few resources, only the quebracho tree, a source of tannin for dyes, produces any additional income for the people of this region.

The eastern plains are likewise rather sparsely settled, except for the triangle of population clusters extending from Asunción, the capital, through Villarrica to Encarnación on the Paraná river and the floodplains of the Paraguay and Paraná rivers that form the south and west boundaries of the country. Forests have been leveled, and fields of cotton, rice, and wheat have been planted wherever the slope or climate favors the crops. In the settled area and near the Paraguay River north of Asunción are numerous large cattle ranches. A good number of the herds consist of improved cattle breeds or the zebu, which have been imported and acclimated to the warm, subtropical climate. Though meat and hides are the chief exports, cotton also holds a primary position in foreign trade. Yerba maté, the national beverage, is brewed from the powdered leaves of the maté tree (an American holly), which grows in many parts of the eastern upland forests. Timber and some coffee are sent to markets from the eastern uplands.

Asunción is both the capital and the chief river port of Paraguay. When water is high, riverboats loaded with cargoes of commodities make the long and dangerous journey to Argentine ports, which are still prime outlets for Paraguayan exports. In recent years a railroad has bridged the Paraná River and extends to the Brazilian port of Paranaguá, thereby offering an economic alternative to the former Argentine trade monopoly.

Brazil

Brazil covers about half of South America, and, at the beginning of the 1970s, its more than 90 million people comprise half of the continent's entire population. The Treaty of Tordesillas, in 1494, which divided the New World between Spain and Portugal, assigned the eastern portions of South America to the Portuguese. Although its European heritage is primarily Portuguese, as is its language, the indigenous Indians, a major admixture of black Africans, and numerous other Europeans all comprise significant segments of the population.

Despite its area, Brazil is still largely empty, and only about 2 percent of the land is under cultivation. Most Brazilians earn a livelihood in modern agriculture, manufacturing, or mining activities localized within 300 miles of Rio de Janeiro and São Paulo in Brazil's East-Central region. These Brazilians possess the highest literacy levels and the best living standards.

Fig. 6-11 Brazil is the largest country in area in South America, and one of the five largest nations in the world. Eastern Brazil has no high mountains but does have a great variety of landforms.

Arrayed around the triangle formed by Rio de Janeiro, São Paulo, and Belo Horizonte are the nation's other subregions, sometimes separated from the urban-industrial core by hundreds of miles of thinly occupied territory or jungle wilderness. Over a decade after the political capital was relocated from Rio de Janeiro to Brasilia, in the sparsely settled interior, the forces driving the nation remain tied to the coast.

Little of the Brazilian national realm extends south into subtropical areas. Only the vast, rolling Brazilian highlands spreading over the eastern third of the country, with their 3,000- to 7,000-foot elevations, offset the hot temperatures that would prevail at this latitude. The Brazilian highlands have also proved to be a storehouse of minerals vital to the nation. Strangely, the luxuriant forest of the wet Amazon region, which once directed the course of the nation when rubber was "king," and the Northeast, the "New England" of Brazil, today remain more of a burden than a blessing.

East-Central Brazil

São Paulo and Rio de Janeiro, two of the largest metropolitan areas of the continent, the great coffee fazendas and sugar plantations, the largest iron and steel industry, and the largest diversified manufacturing base of South America are all concentrated in East-Central Brazil and situated within 300 miles of the Atlantic Ocean.

The federal state of São Paulo is notable for its location where the great coffee fazendas begin, the plantations that extend south into Paraná state, west across the Paraná River into eastern Matto Grosso state, and even into southern Minas Gerais state. Coffee growing and processing is the biggest industry of the country, and Brazil is the world's largest coffee producer. In an average year, Brazil will export some 40 to 45 percent of the world's coffee. Brazil's coffee is sold principally to the United States, and it helps forge a strong community of interest between the two nations.

The coffee fazenda is a large plantationlike unit that usually has been divided into separate farms, each placed in the care of a tenant farmer. Although each farm has space set aside for food crops and a homestead, most of the land is planted in coffee trees by the owner. Once each year the crop is gathered, and the berries, after being hulled to recover the coffee beans, are dried, sorted, selected, sacked, and shipped in bags to the seaports. For almost a century Santos, situated below the Serra do Mar about 45 miles from São Paulo, was the world's largest coffee export center. During the past two decades, however, as the coffee frontier moved south and west into the state of Paraná, the city of Paranaguá gradually superseded Santos. For each city, the construction of transportation links across the escarpments that separated these ports from the inland plateau was a major challenge.

A rare combination of terrain, soils, and climate has remarkably favored the coffee culture and the diversified agricultural economy of the entire region. The altitude of the highlands, broken along the east by the high escarpment of the Serra do Mar, which rises suddenly and precipitously from the ocean or from narrow coastal lowlands, offsets significantly the otherwise hot temperatures. The southeast trade winds carry moisture-bearing winds inland during the year, giving only a less rainy season to the region. During much of the year, cloud cover protects the coffee trees from the harsh tropical sun. Without the deep, rich terra-rossa soils, derived from the lava base that spreads over the highland in the south, coffee cultivation and agriculture in general would be limited. Growing of coffee is confined largely to the low plateau—mostly between 1,500 and 3,000 feet—where elevation keeps the air temperatures mild, at the same time that the tropical location precludes killing frosts.

Coffee is still the reigning "king" in Brazil.

Fig. 6-12 Rio de Janeiro, Brazil, with a population of over 4 million, has one of the best natural harbors in the world. It is dominated by Sugar Loaf Mountain, an international tourist attraction. (Courtesy of Pan American Airways)

But São Paulo state, the Paraíba Valley near Rio de Janeiro, and southern Minas Gerais are also national leaders in the production of sugarcane, cotton, rice, and other food staples. Modern farming methods, the increasing use of agricultural machinery, and better fertilizers have hastened growth and pushed the region into leadership in commodity production. Dairy and beef herds graze in planted pastures on adequate natural grasses. That farming and livestock industries of the first order are necessary is quite clear, since they must support the millions of people who live and work in the metropolitan areas of São Paulo, Rio de Janeiro, Belo Horizonte, and the large number of smaller communities.

Throughout East-Central Brazil, the principal cities are not isolated urban nodes as is encountered elsewhere in South America. Sizable industrial and service cities are dispersed within the resource hinterlands, to which they are linked by surface transport arteries. From them, first-class rail lines and highways speed commodities to the prime urban-industrial centers they support, and on which they depend. Commercial interaction thus fostered sustains São Paulo.

Although its initial prosperity is derived from coffee, alert "Paulistas" are only too well aware of the "boom-and-bust" cycles recurrent in Brazil's history. The fast-paced, spirited growth of São Paulo City stems from its role as coffee "capital," as well as being political center for the richest state in Brazil. The nation's industrial revolution has its roots here, relying upon a support hinterland market that possesses wide-

Fig. 6-13 Coffee plantation near São Paulo, Brazil. (Courtesy of Brazilian Government Trade Bureau)

ranging industrial resources and peoples having above-average education and incomes. Heavy industries include steel, petrochemicals, tires, and machinery. The city is the center of electronics and consumer goods fabrication and also is usually referred to as the "Detroit of Brazil" because of its large automobile industry.

The former capital, Rio de Janeiro, is also a leading manufacturing center. But it is better known as the cultural and political center of Brazil, the leading port, and a recreation center of international fame.

The state of Minas Gerais has been important for centuries for the exploitation of several rich mineral deposits. The state is the source of almost all the gold mined in Brazil, as well as of diamonds of value for industrial stones. However, the much-larger and more-abundant deposits of iron ore, manganese, and base metals have now eclipsed the original mining activities. In the rolling hill land and low mountains east of Belo Horizonte, near Itabira, are located substantial iron ore deposits of high-grade hematite without impurities in harmful amounts, which rank among the finest in the world. Much of the ore is utilized principally in iron and steel industries at Itabira or Volta Redonda, and relatively little of the ore enters the export market. The ore for Volta Redonda, the national steel-industry center located in the Paraíba Valley north of Rio de Janeiro, is sent by rail to the coastal port of Vitória, where it is carried by ore ships to the south. There it is transshipped once again inland to the steel plant. Most of the coal used comes as return cargo on carriers delivering goods in Western Europe or the United States. Some domestic coals from the Tubarão mines in south Brazil are mixed with the imports.

Northeast Brazil

The Portuguese colony in the New World was first settled along the northeast coast of modern Brazil. Captaincies, which were large land grants or patents, extended inland along the coast and were important from the outset. They were the first centers of sugarcane culture in the Americas. The several small states extending inland, with ports that are now state capitals, attest to the lingering impact of the early settlement pattern. The narrow coastal lowlands, backed by an escarpment as in the south, were planted and harvested first by the few Indians, and then by Negro slaves, who were needed to work in the canefields. Few crops have responded more rapidly to the rise and fall in prices or to the expansion and contraction of plantations than has the sugar culture of the Northeast. In an effort to decrease the dependence of the region on sugar, the growing of plantation cotton, cacao, and bananas was introduced into the coastal lowlands.

In the lowlands and foothills of the highlands behind the bay at Salvador (Bahia), the largest and most important cacao district in the Western Hemisphere has been developed. The Brazilian cacao crop usually provides about three-quarters of the South American crop and 15 percent of the total world supply.

The critical problem faced by the agricultural economy of the Northeast is that other regions in Brazil, or in other nations, have been able to produce the same commodities in large quantities of better quality or at lower prices. The older plantations, using outmoded methods, cannot compete effectively, and much of their land falls into disuse. Thousands of laborers are left without jobs and migrate into the dry interior, where they barely eke out a living.

The unpredictable droughts that befall the interior of Brazil's Northeast drive families, bearing what possessions they can carry, out of the region to the coastal cities, where government relief keeps them alive. Nevertheless, the population of the Northeast has increased steadily even in the face of adversity. Hardships and poverty are seldom dispelled even by abandoning the barren, bleak interior. The Northeast of Brazil is the poorhouse of the nation, potentially a political powderkeg, and one of the most economically depressed regions in the entire Western Hemisphere.

Drought calamities are matched only by flood disasters when torrential downpours inundate vast flatlands. An entire year's rain may occur as heavy showers within a few days or weeks, and the highly variable annual average seldom exceeds 25 inches. Only when the intertropical front shifts south and eastward into Brazil's Atlantic bulge may the moisture-bearing air from the equatorial North Atlantic invade the region.

The tropical rain forests of the coast disappear quickly beyond the escarpment, and grasslands, thornbush, and cacti become common. In rainy years—about two out of seven—herds of cattle do well and crops flourish; in the years of rainfall deficiency, crops wither and the herds of livestock waste away. Large reservoirs constructed to catch and hold water for irrigation usually stand dry and empty. When there is water, the crops give fairly high yields, and in wet years they satisfy the very small needs of the subsistence farmer and rancher. But in dry years, the land is parched, crops die, and people and livestock starve.

South Brazil

South Brazil embraces the northern margin of the Pampa, the southern edge of the Brazilian highlands, narrow river plains in the west, and an equally narrow coastal plain along the Atlantic. Climatically, south Brazil resembles Uruguay; but in the highlands and in the interior, temperatures are cooler, and frost, or even snow, may occur with some regularity. Most of the highlands and much of the lowlands were originally forested. During the period of first settlement, the Paulistas moved southward, following the grasslands as far as Uruguay and using them as breeding grounds for cattle and mules. Later European settlement, beginning in the 1800s, was composed largely of Germans, Italians, and Slavs who pushed inland along river valleys from coastal bridgeheads. Settlement by a dominant European population has progressed slowly but steadily. The small, family-sized farm is the principal economic unit, quite in contrast to the giant fazendas and plantations of the regions to the north. Farming has proved to be prosperous and is pursued with vigor. The small farms grow wheat, barley, corn, and other grains, together with potatoes. In a few places along the western rivers and the eastern coastal lowlands, rice cultivation is important, and in warmer areas cotton is a leading crop. On the open pampas, pastures occupy most of the land, and cattle and sheep provide most of the income. Porto Alegre, the region's chief port and manufacturing center, lies near the northern shore of Lagôa dos Patos. Inland, population clusters quickly become dispersed, and densities decline as the frontier is reached. The three principal cities in south Brazil—Porto Alegre, Florianópolis, and Curitiba—are local service and commercial centers, supporting some industry as well as serving as state capitals. The rich forests of paraná pine are of prime importance in providing lumber for building materials and pulp and cellulose for paper and chemical industries; unfortunately, the wood is used widely for charcoal and firewood in a region and nation short of fuels. In spite of the danger of frost, the coffee frontier has been pushed south into northern Paraná state, and coffee is being grown in a marginal climate as older coffee lands become exhausted or are shifted to other crops in São Paulo state.

Interior Plateau and Plains

The interior uplands of Brazil are a westward extension of the coastal highlands. Sheared on the north and south by rivers and their tributaries that flow north to the Amazon or south to the Paraguay and Paraná rivers, the low plateau surface is a rolling hill land with numerous cliff-bordered tablelands. Farther to the west are extensive plains. More than one-third of Brazil's national territory is encompassed by the plateau and plains, which contain only a fraction of the country's population.

To Brasília, the capital of Brazil, which was transferred from Rio de Janeiro to a sparsely populated frontier region in 1960, Brazilians from the poverty corner of the Northeast have migrated by the thousands, but the residents of Rio de Janeiro and São Paulo have been less enthusiastic. After more than a decade, civil servants, politicians, and businessmen still demand special subsidies to make a permanent transfer to the interior. Situated on a high plateau, very near the headwaters of two rivers —one flowing north, the other south—Brasília, a magnificent city of beautiful symmetry and daring and of boldly symbolic contemporary architecture, has risen in the hope that the Brazilian spirit will look inland from the coast to a future of occupying the country's vast and virtually uninhabited interior. The elevation of the upland moderates the tropical temperatures and gives relief from the otherwise very great heat. Though warm throughout the year, the higher eastern segment of the plateau, north of São Paulo state, is marked with daily and seasonal weather variations.

Brazilians are reluctant to leave the coast and move inland; thus, after centuries of pioneering efforts, the greatest part of the land is still in the early stages of frontier development. Ranging through the open woodlands and tropical grasslands, cattle herds provide most of the income. From remote and isolated corners of the interior, hides, skins, and sun-dried beef are still shipped to eastern markets. West of São Paulo state, in southern Matto Grosso, herds of cattle move quickly to urban markets across new roads and railroads. The use of zebu cattle and improved breeds developed in Brazil has raised the quality of beef animals. West of Minas Gerais, toward Goiás and the new capital at Brasília, commercial farming and livestock ranching have made substantial progress in a short time. The most important changes that have occurred is the construction of modern highways, the bridging of the Paraná River, and the construction of hydroelectric stations to provide power along the pioneer fringe.

Amazonia

Over a million square miles, constituting one-third of Brazil's area, are blanketed by the forests and grasslands of the vast Amazon Basin, the largest single rain forest in the world. Yet Amazonia is the most sparsely settled region of Brazil. The basin is flat to gently undulating throughout, and only in a few places stands higher than 1,000 feet above sea level. The climate is monotonous, constantly warm, humid, and rainy. No moderation of temperature occurs during the day or significantly from season to season. The trade winds sweep across the highlands from the east and carry moisture that gives between 70 and 100 inches of rainfall during the year. Violent thunderstorms break the quiet of the forests and dump tremendous volumes of water on the land. The Amazon Basin, however, does for the most part experience a less rainy season during the low-sun period. At Manaus, one of the older population clusters located on the Río Negro, the average temperature for the warmest month, October, is 83°F; for the "coldest" month, April, it is 80°F.

Of the hundreds of species of trees that may be found in every square mile of the Amazon forest, only the rubber tree has been widely exploited. The Amazon Basin is the original home of *Hevea brasiliensis*, the tree from which natural rubber is obtained. The rubber industry has long languished, and today Brazil obtains most of its rubber from synthetics fabricated at São Paulo.

Because rivers are the principal means of surface transportation into the rain forests, settlements are scattered along the Amazon and its tributaries. Forest gatherers still bring gums, nuts, and woods to the river settlements for trade, but the rubber trade has ceased to be of any significance. As elsewhere in Brazil, a long era of quiet is coming to an end. A highway from northeast Brazil already reaches to Belém, the once-isolated seaport of the Amazon. Manaus, although it lies 1,500 miles up the Amazon, may be reached by oceangoing vessels because of the great depth of the river. Already, a road under construction from Brasília is pushing slowly through the tropical forest wilderness toward the south bank of the river.

The great undeveloped resource of the Amazon is its forest, capable of yielding a considerable amount of lumber and possibly pulpwood. The valuable species are rather few and are scattered among a larger forest of more or less useless types. The timber must be dried in the forest before it will float, and another problem involves getting the logs out through the dense forest growth. The Amazon Basin remains an enigma to modern man, for it probably will not bloom, as some believe, into a rich agricultural region that will support millions.

In Perspective

The thrust of challenges to South America's peoples is to evolve new systems for interaction and communication on a scale inconceivable even a decade ago. Fortunately, the continent is adequately endowed with cultivable land and with areas suitable for settlement in tropical highlands. By and large, South America is more evenly and reliably watered than tropical lands elsewhere, though water management and conservation are vital concerns. Still a comparatively moderate population for its land area, South America's 200 millions at the onset of the 1970s possess significant agricultural and mineral reserves that are still untouched. Furthermore, great numbers of South American peoples enjoy a common tradition of Iberian history and culture and a dichotomy of but two major modern languages—factors that should work strongly to favor unity. Long and favorable trade, technical, and cultural ties with Western Europe, North America, and now Japan have brought benefits to the lives of millions.

For the future, agriculture may advance into now thinly populated or empty savanna regions. Moreover, few lands already being exploited would fail to respond with bigger yields if better agricultural methods were applied. Successful Japanese and native colonization in Brazil and Peru dispels the notion that tropical rainy zones of Amazonia cannot be harnessed productively. National manufacturing of impressive range and sophistication is already located around the margins of the continent. That manufacturing can thrive and prosper in South America is no longer questioned.

A two-century-long epoch of massive trans-Atlantic migration from Italy, Germany, and East European nations, joining earlier Spanish, Portuguese, and African Negro arrivals, has come to a close. Peopling of unoccupied lands and adaptation and diffusion of modern technology now rests squarely on the shoulders of the inhabitants themselves. If the inexperienced and irresponsible carry the day and forward destructive practices, great natural resources will be wasted.

Perhaps the most critical challenge clouding South America's future is the matter of property. Will ownership systems of the past be perpetuated: will land be divided into commercial, family-size farms, or will it be developed as commercial cooperative ventures? Perhaps still another, as yet unrecognized, solution will be forthcoming. Unresolved socioeconomic problems of critical proportions could spark dangerous unrest in such places as Brazil's Northeast and the Indian highlands of Peru and Bolivia. In every instance, one great weakness threading through nearly all of South America's nations and peoples is their inability to provide sufficient food for burgeoning populations.

REFERENCES

Brundage, Burr C.: *Lords of Cuzco*, University of Oklahoma Press, Norman, 1967.
Castro, Josue: *Death in the Northeast*, Random House, Inc., New York, 1966.
Friedmann, John: *Regional Development Policy: A Case Study of Venezuela*, The M.I.T. Press, Cambridge, Mass., 1966.
Glassner, Martin I.: "Feeding a Desert City: Antofagasta, Chile," *Economic Geography*, vol. 45, pp. 339–348, October, 1969.
Hanke, Lewis: *Contempory Latin America*, Princeton University Press, Princeton, N.J., 1968.

James, Preston E.: *Latin America* (4th ed.), The Odyssey Press, New York, 1969.
McGann, Thomas F.: *Argentina: The Divided Land,* D. Van Nostrand Company, Inc., Princeton, N.J., 1966.
Momsen, Jr., Richard P.: *Brazil: A Giant Stirs,* D. Van Nostrand Company, Inc., Princeton, N.J., 1968.
Pendle, George: *Paraguay: A Riverside Nation* (3d ed.), Oxford University Press, London, 1967.
Preston, David A.: "The Revolutionary Landscape of Highland Bolivia," *Geographical Journal,* vol. 135, pp. 1–16, March, 1969.
Smith, T. Lynn: *Brazil: People and Institutions,* Louisiana State University Press, Baton Rouge, 1963.
Toynbee, Arnold J.: *Between Maule and Amazon,* Oxford University Press, New York, 1967.

7

Northwestern and Central Europe

Western Europe rose miraculously from the depths of World War II to a state of economic health, productive power, and social concern and a political stature that make these nations collectively a major power bloc in today's world. Individually, the countries of Northwestern and Central Europe have declined relatively in world significance, partly as a result of their loss of direct political control over extensive overseas territories and partly because of the economic growth of countries in other areas, particularly the United States and the U.S.S.R. However, the world impact of this small geographical area, both past and present, has been of immeasurable importance. Its location, as well as the diversity, number, and ability of its people, its resources, advanced culture, contributions to science and technology, tremendous agricultural and industrial productiveness, and dominant position in world trade make the understanding of this small continent of prime concern to people everywhere in the world. Exclusive of the U.S.S.R., Europe contains 13 percent of the world's people on only 3.6 percent of the planet's total land area. For its size, Northwestern and Central Europe is the most urbanized area in the world.

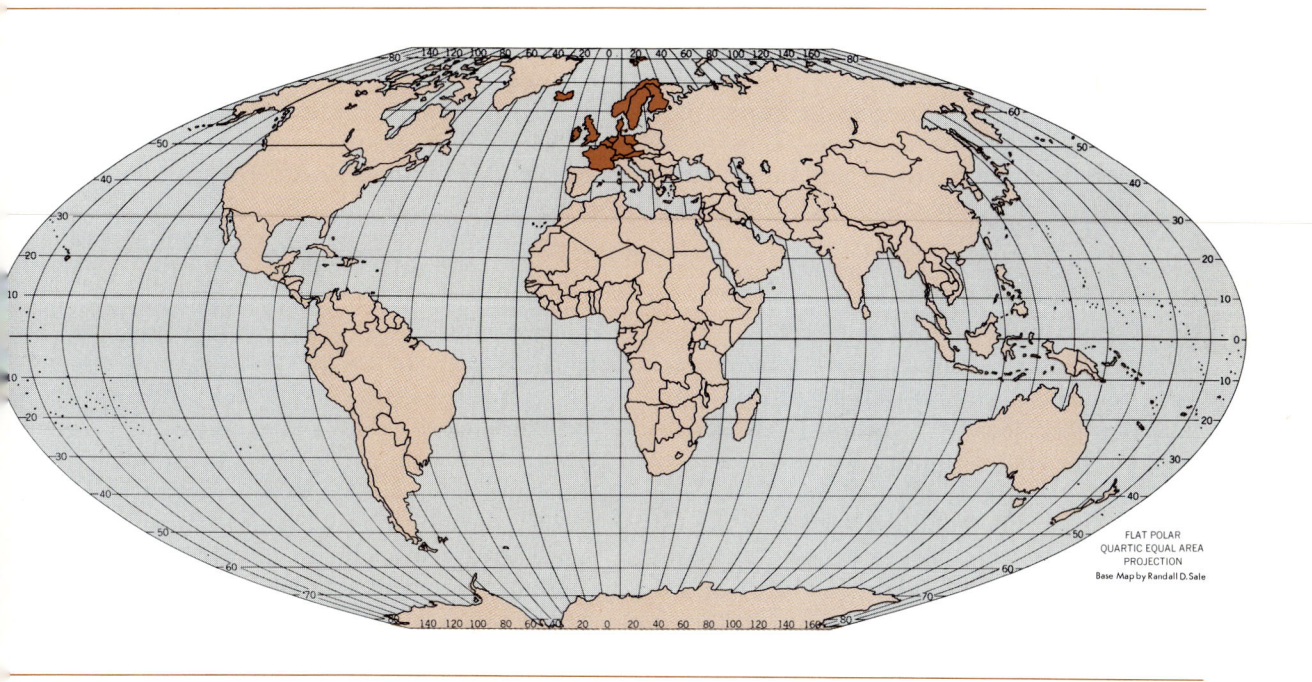

Marked political fragmentation, a dominant characteristic of Northwestern and Central Europe for the past century, indicated the deep desire of diverse peoples to have sovereignty over their own affairs. The disadvantages of small size have been partially offset by regional groupings of nations for economic, trade, defense, and political reasons. The ultimate aim may be the formation of a federated Europe, but the area currently possesses political units ranging in size from France (212,737 square miles) to tiny Liechtenstein (61 square miles).

Europe's proportion of the world's industrial production decreased from 68 percent in 1870 to 25 percent in 1948. Pre-World War II European industrial output was one-third greater than that of the United States; by 1948, as a result of World War II and rapid economic growth in the United States, European productivity was more than one-fourth less. Today the highly industrialized continent, excluding the U.S.S.R., produces about 40 percent of the world's motor vehicles, 35 percent of the steel, and 30 percent of the electricity.

The densely populated, coal-rich, industrial core of England, the Benelux countries (Belgium, Netherlands, Luxembourg), northern France, and West Germany is surrounded by areas where agriculture, forestry, fishing, and mining produce raw materials of importance. Nevertheless, the area is heavily deficient in food, petroleum, and many other essential items and must rely on imports to supplement local production. Europe, excluding the Soviet Union and other Eastern European countries, accounts for about 46 percent of the total world imports and 42 percent of the exports. Thus, Western Europe is the focus of major world trade routes, as well as of developed land transportation facilities vital to commercial relations with neighboring countries. Trade deficits of these countries are offset by such revenue sources as earnings of their merchant vessels, foreign investments, insurance, and spending by numerous tourists. The Common or European Economic Community Market, (EEC), and European Free Trade Association (EFTA) are examples of attempts to cooperate to gain trading and tariff advantages for their members.

Despite political and other differences, the area is a well-integrated spatial system favoring communications and commerce. The economic and transportation linkages are strong between the industrialized core of England, Benelux, West Germany, and Northern France and the less intensely developed surrounding areas. This spatial interaction within and between countries is central to the geography of Western Europe. Nevertheless, sharp cultural differences do exist and often disrupt the unity of the total area, as well as of individual countries such as Belgium.

Northwestern and Central Europe, as here defined, includes 1,055,215 square miles, or 56 percent of Europe (excluding the U.S.S.R.). The fifteen countries and principalities comprising these regions would cover less than one-third of the United States. The distance from northern Norway to southern France measures 2,125 miles—equivalent to the distance from Chicago to Los Angeles. Small countries have the advantages of unity, compactness, and ease of exchange of goods and ideas with other countries but usually lack sufficient domestic resources on which to base an economy without heavy reliance upon imports.

The area's strategic buffer-zone location between the two world powers, the United States and the U.S.S.R., is fraught with political problems, despite the economic advantages of proximity to major consumption areas. The central location in the land hemisphere facilitates world trade, especially since Western Europe faces three major bodies of water —the Arctic Ocean, the Atlantic Ocean, and the Mediterranean Sea. No other comparable world area has such disrupted coastlines and irregularly shaped countries. The North Sea, the Baltic Sea, the English Channel, and the Bay of Biscay cause major geographical separations and indentations. The peninsulas of Scandinavia, Jutland, and Brittany, as well as several smaller ones, add to the irregularity of the landmass, as do the numerous islands associated with Western Europe, ranging in size from Great Britain and Iceland to tiny skerries along the Norwegian coast. Norway and Scotland exhibit highly fjorded coasts.

Physical Setting

Relief Features

Northwestern and Central Europe has a complex physical structure. East of the Carpathian Mountains and the Vistula River is the low, stable Russian geologic platform. West of this platform, repeated elevation and depression of the land, along with folding and erosion, have left mountains, plateaus, hills, and plains all in close proximity.

Highlands The highlands found in Scandinavia, northern Scotland, and Ireland are composed of very old, hard rocks that have been repeatedly uplifted and eroded. Moreover, during the Glacial Period huge ice sheets sculptured, scoured, and smoothed the topography. The fjorded coastline of Norway, with some fjords reaching 120 miles inland, and the rounded highlands of northern Scotland demonstrate the great erosive power of the ice. Most of Norway is a high, glaciated plateau averaging between 4,000 and 5,000 feet in altitude. The Scandinavian highlands drop sharply to the west, but more gradually toward the Baltic. Soils are generally infertile; valleys and coastal lowlands possess limited arable land. Timber and waterpower are abundant, especially in Scandinavia.

Mountains and plateaus in West-Central Europe, although geologically younger than the Scandinavian highlands, have been eroded for so long that their relief is one of rounded maturity. Such remnants, called the Central Highlands, extend from the southern uplands of Ireland to the highlands of central Germany and include southwest England, Brittany, the Central Plateau of France, the Ardennes Plateau, and the Vosges Mountains. These uplands consist of hard, resistant rock with considerable relief and poor soil. Adjacent basins and lowlands contain more recent water-deposited soils of higher agricultural value. The metallic ores and coal measures which are associated with these highlands form the basis of major industrial regions. Agriculture is limited in the uplands, but grazing and forestry are locally significant.

The Alpine System, a complex combination of young folded mountains with associated forelands and intermontane structural basins and plains, dominates Switzerland, Austria, and the eastern borderlands of France. The Alps average 12,000 feet, with higher individual sharp peaks such as Mt. Blanc (15,781 feet); intermontane valleys are narrow, steep, and rocky. The south slope of the Alps is abrupt, but the north slope is a gradual incline. Despite the high altitudes, numerous passes and tunnels allow remarkably easy north-south communication between Northwestern and Central Europe and the Mediterranean Basin.

The Pyrenees Mountains extend east-west from the Bay of Biscay to the Mediterranean Sea. Like the Alps, they are young, folded mountains and form a considerable barrier between France and Spain.

Lowlands Fortunately, Northwestern and Central Europe contains considerable areas of lowland. The great European Plain extends from southeastern England through western and northern France, Belgium, Netherlands, Denmark, north Germany, Poland, and on into the Soviet Union. It is narrow in the west and widens in the east. Generally, it is level; rolling to hilly topography occurs, but elevations rarely rise to 500 feet. Much of the existing relief is a result of continental glaciation, and ridges of glacial debris (moraines) border the Baltic in north Germany and Denmark and occur in the remainder of northern Europe. Outcrops of slightly folded beds of shale, chalk, and limestone, called "downs" or "scarp and vale," give topographic relief to southern England and northern France. Sand, gravel, and silt of glacial origin covers much of the remaining lowland area; in central Germany, there are fertile wind-deposited glacial soils called "loess."

Lowlands also border the Baltic Sea and the Gulf of Bothnia in Sweden and Finland. Southern Sweden and southern Finland are mostly less than 500 feet above sea level and are dotted with numerous lakes, swamps, and glacial ridges. The Baltic lowlands are underlain by very old crystalline rocks, except for the fertile sedimentary plains of Skåne in the extreme southern tip of Sweden and the islands of Öland and Gotland off the Swedish coast.

Soils are usually thin, with scattered deposits of fertile clay.

Rivers The Alps and associated mountains are the source of the major rivers of Central Europe. Large north-flowing rivers—such as the Oder, draining into the Baltic, and the Elbe and Rhine, draining into the North Sea—have swift currents and are of value for waterpower near their sources, but they become broad and sluggish in crossing the European Plain. France has three major west- or northwest-flowing rivers, the Seine, Loire, and Garonne; the Rhone empties into the Mediterranean. A system of interconnecting canals and rivers gives Western Europe the best inland water-transportation network in the world. In the British Isles, which have numerous short rivers, the longest are the Thames and the Shannon. Scandinavian rivers draining into the Atlantic are short and rapid; those entering the Baltic are longer and flow more slowly. Finland, the Netherlands, and northern Germany, are all handicapped by considerable amounts of poorly drained land.

Climate

Reliability of climate, the keynote of physical conditions in Northwestern and Central Europe, fosters agricultural development and other human activities. Temperatures are far more moderate than those of similar latitudes in eastern North America or in the interior of Eurasia. Only the higher mountain regions with rigorous temperatures and the exposed coastal areas with heavy rainfall are climatically unsuitable to crops. Barley and rye, for example, are grown beyond 70° north latitude in Norway—farther north than on any other continent.

Mild winters, cool to warm summers, small temperature ranges for the latitude, and well-distributed precipitation typify the overall climate of this geographical zone. Situation on the western side of a large landmass, prevailing westerly winds off the warm North Atlantic, moving inland without land barriers, and the prevalence of cyclonic influences are largely responsible for these conditions. The movement of storms from west to east results in frequent weather changes affecting temperature, winds, and precipitation. The cyclonic influence is especially strong in the winter. Drought rarely occurs.

Natural Vegetation and Soils

Natural vegetation and soils in Europe are related closely to the varied relief, subsurface rock structures, climate, and drainage conditions. Man's intensive utilization of the land has greatly disrupted the original natural patterns of Western Europe, especially the pattern of vegetation. Rapid soil changes, because of glaciation or type of parent rock, result in much less uniform belts of soil here than are found in Eastern Europe. Nevertheless, certain broad associations of native vegetation and soils can be identified.

Significance of Diversity of Cultures

The thirteen political units (besides Luxembourg and Liechtenstein) it includes indicate the tremendous diversity of peoples and cultures found in this area. Each country, no matter how small it is, has developed a national identity and traditions peculiar to its individual culture. In most cases, the country has its own language, within which several dialects may have evolved because of separation by bodies of water or land barriers. Switzerland has four major language and ethnic groups, and Belgium and Finland each have two.

Northwestern and Central Europe

CLIMATES OF NORTHWESTERN AND CENTRAL EUROPE

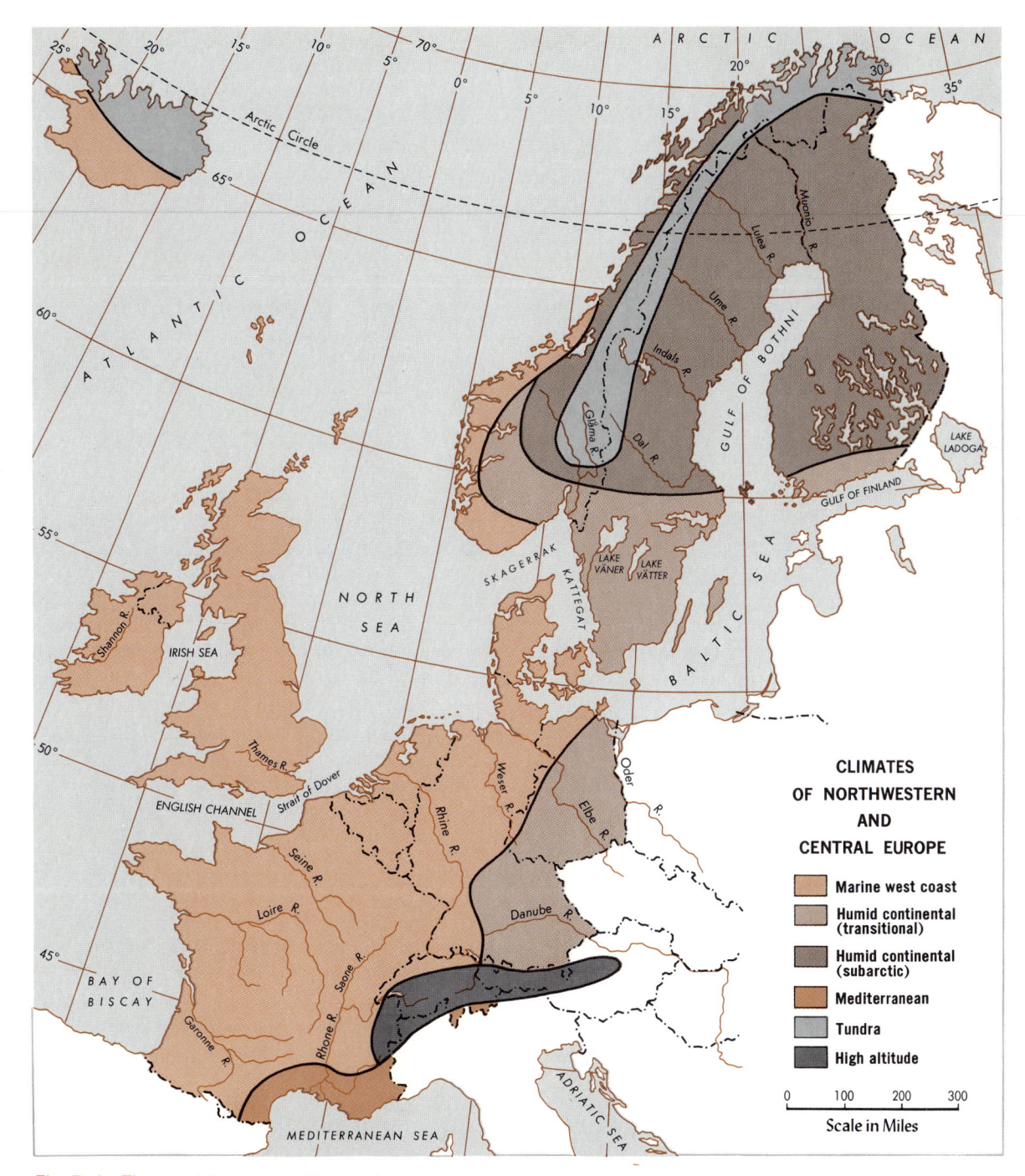

- Marine west coast
- Humid continental (transitional)
- Humid continental (subarctic)
- Mediterranean
- Tundra
- High altitude

Fig. 7-1 The most important climates in northwestern and central Europe are classed as marine west coast. The climates of the area, however, range from Mediterranean in southern France to tundra in northern Finland.

Various races of people are found in Northwestern and Central Europe; but cultural, particularly linguistic, rather than racial, differences form the basis of nationalistic desires and territorial problems. Germanic, Celtic, and Romance languages predominate; the Asiatic-derived Finnish and Lapp languages are found in the north. This diversity of languages, dialects, customs, and traditions is a major barrier to federation in Western Europe.

Economic difficulties brought on by two world wars, common trade problems, the loss of colonies by Western European powers, and joint defense efforts have resulted in the development of various supranational economic, political, and defense unions. Belgium, the Netherlands, and Luxembourg entered into an economic (customs) agreement called the "Benelux Union" immediately following World War II. In 1953, the so-called Schuman Plan, creating the European Coal and Steel Community for pooling steel and coal in six Western European nations, was initiated in an attempt to offset by international cooperation the raw-material deficiencies and market disadvantages of small nations. This evolved into the Common Market Treaty (European Economic Community), signed in Rome in 1957, which created a six-country customs union of Benelux, France, West Germany, and Italy. The objective was the formation of a single market of 182 million consumers, in which trade barriers between member nations were to be gradually removed over a 15-year period, along with encouraging the free movement of goods, labor, capital, and services. The treaty also directs that the economic policy of the six countries be coordinated and that common policies be applied over broad sectors of the economy. Objectives of the ECC have not yet been completely achieved, owing to economic and political difficulties. Agricultural policy differences and conditions are proving a major stumbling block, as are such other factors as the stability of currency and differences in labor costs among member nations. The United Kingdom, after its initial decision not to join, so far has been blocked in its subsequent attempts at ECC membership. It is likely that Great Britain will be included and that many other European countries will associate with the organization in some way. The Euratom Treaty, aimed at the peaceful uses of nuclear energy, was also signed by the six Common Market nations in 1957; this treaty was later ratified by Switzerland as well.

To counteract the commercial effects of the Common Market, seven other European nations formed the European Free Trade Association. An even broader grouping of nations belonging to the Atlantic Community, including the United States, is being considered as a commercial union to be called the North Atlantic Free Trade Area (NAFTA). Various organizations with political and collective security implications have also been formed, such as the Council of Europe and the North Atlantic Treaty Organization (NATO). The success of such attempts at international cooperation would be an important step toward the unification of Western Europe. Any federative movement faces the tremendous obstacles of strong nationalistic and self-sufficiency movements. Geographical (spatial) differences, both physical and cultural, play dominant roles in the success of any regrouping efforts.

Population Distribution and Present Stage of Economic Development

A belt of extremely dense population extends from central and southern England across the English Channel to include northern France, Benelux, the Rhine Valley, and the contact zone between the highlands and lowlands in Germany. As one of the most intensely urbanized and industrialized regions in the world,

Northwestern and Central Europe 225

POPULATION DENSITY IN NORTHWESTERN AND WEST CENTRAL EUROPE

Persons per square mile

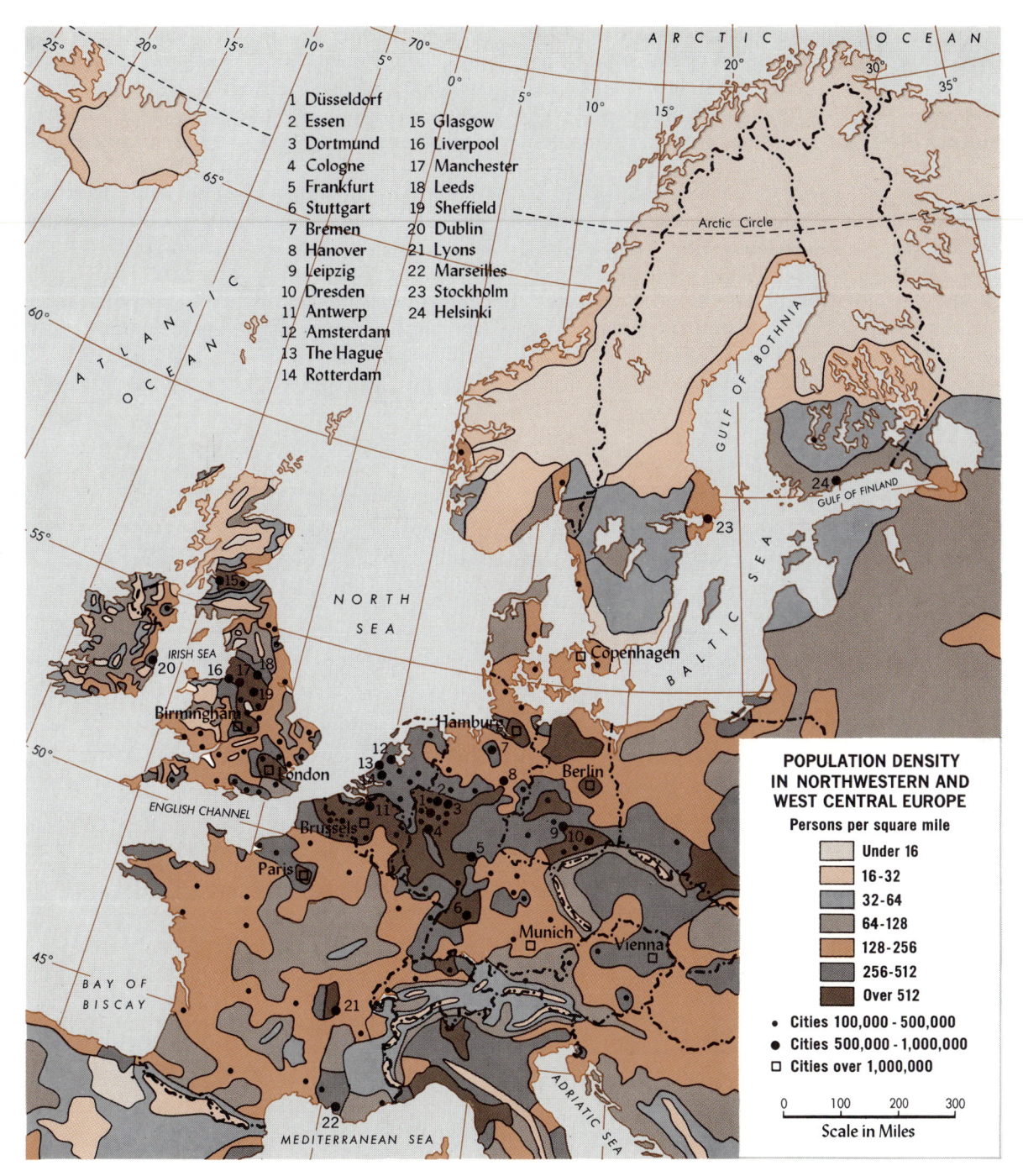

- Under 16
- 16–32
- 32–64
- 64–128
- 128–256
- 256–512
- Over 512

- • Cities 100,000–500,000
- ● Cities 500,000–1,000,000
- □ Cities over 1,000,000

Scale in Miles: 0 – 100 – 200 – 300

City key:
1. Düsseldorf
2. Essen
3. Dortmund
4. Cologne
5. Frankfurt
6. Stuttgart
7. Bremen
8. Hanover
9. Leipzig
10. Dresden
11. Antwerp
12. Amsterdam
13. The Hague
14. Rotterdam
15. Glasgow
16. Liverpool
17. Manchester
18. Leeds
19. Sheffield
20. Dublin
21. Lyons
22. Marseilles
23. Stockholm
24. Helsinki

Fig. 7–2 The central part of the region has a high population density; the northern areas are sparsely populated. Belgium and the Netherlands are two of the most densely populated nations in the world.

this is the core area of Europe. More than 70 percent of the population of Great Britain live in cities of over 10,000. The Netherlands has the highest average population density of any industrialized nation in the world: 1,068 persons per square mile. If Australia and New Zealand were as densely inhabited, for instance, they would have populations equal to the entire present-day world's population. The Scottish Lowland, South Wales, eastern Denmark, the northern tip of Ireland, the Swiss Plateau, the Danube Valley of Austria, and the Rhone Valley of France are other densely populated areas of Northwestern and Central Europe.

Surrounding the industrial and commercial cities are thickly populated and intensively developed agricultural areas. Despite intensive agriculture, the large urban populations mean that most of these areas are characterized by deficiencies of food and surpluses of industrial products. This situation along with the need for considerable imports of fuel and industrial raw materials, makes Northwestern Europe foremost in world trade. The area's commercial development is aided by excellent ports, large merchant marines, and good inland transportation facilities.

Access to mineral reserves present in the

TABLE 7-1
MEMBERSHIP OF EUROPEAN ORGANIZATIONS

	OECD[3]	Council of Europe	Common Market (EEC)[4]	Euratom	European Coal and Steel Community[5]	NATO[6]	EFTA[7]
Austria[2]	X[1]	X	[8]				X
Belgium[2]	X	X	X	X	X	X	
Denmark[2]	X	X	[9]			X	X
Eire[2] (Ireland)	X	X	[9]				X
Finland[2]	[10]						X
France[2]	X	X	X	X	X	X	
Germany, West[2]	X	X	X	X	X	X	
Greece	X	X				X	
Iceland[2]	X	X				X	
Italy	X	X	X	X	X	X	
Luxembourg[2]	X	X	X	X	X	X	
Netherlands[2]	X	X	X	X	X	X	
Norway[2]	X	X	[9]			X	X
Portugal	X					X	X
Spain	X		[8]				
Sweden[2]	X	X	[8]				X
Switzerland[2]	X	X	[8]	X	[10]		X
Turkey	X	X				X	
United Kingdom[2]	X	X	[9]	[10]	[10]	X	X

[1]X indicates membership or proposed membership
[2]Political units in Northwestern and Central Europe
[3]Organization for Economic Cooperation and Development
[4]European Economic Community
[5]Also known as the Schuman Plan
[6]North Atlantic Treaty Organization
[7]European Free Trade Association
[8]Seeking Associate status
[9]Will become a member in 1973
[10]Associate

highlands of Northwestern and Central Europe, especially coal and iron, favors industry. Although the United Kingdom, Belgium, France, and Germany have large coal supplies, each imports some coal as well as most of the petroleum consumed. The Netherlands has large reserves of petroleum and natural gas, and the fuel potential of the North Sea is being explored and developed. Low-quality iron ore is available in France, Luxembourg, and Great Britain; Sweden is an abundant producer of high-quality iron ore. France has a large output of bauxite for aluminum production, and France and Germany also have potash deposits. Waterpower development or potential is significant in mountainous Northern and Central Europe.

Many peripheral areas of Northwestern and Central Europe are sparsely populated but are significant as sources of certain raw materials. They are generally regions of poor soils, rough topography, and severe climate. Fish, forest, and mineral resources, notably iron ore in Sweden, are vital to Europe. Much of this area's agriculture is subsistence in nature, but grazing and dairying are locally highly developed.

The differences accentuate the need for economic integration of the whole area and the development of linkages. Western Europe is a focus for products being exchanged both within the area and around the world. The skills of the people, combined with their concentration largley in cities, make this region of major economic importance to the world.

Scandinavian Countries and Finland (Norden)

The Northern European countries (Norden), loosely referred to as Scandinavia, consist of the five nations of Norway, Sweden, Denmark, Finland, and Iceland. They have a great deal in common; yet each shows a degree of individuality. As a result of past association, their cultures and traditions are in part similar. Except for the Finns and Lapps, language similarities exist throughout this area, and racially the majority of the people stem from the characteristically blond Nordic stock. All the countries are socialist democracies, although monarchs still reign in Denmark, Norway, and Sweden. The Lutheran religion is strongly predominant. Illiteracy is practically nonexistent.

Similarities are many. With the exception of Denmark and the southernmost tip of Sweden, old, resistant rocks are characteristic features, and the entire area has been glaciated. Iceland and Norway still possess sizable permanent glaciers. Soils, with a few exceptions, are generally infertile. With long and indented coastlines, fishing and commercial development rank high in importance. The Atlantic slopes of the Scandinavian Peninsula are rugged, abrupt, and fjorded; the Baltic slopes are gradual, and hence southern Sweden, southern Finland, and all of Denmark are relatively flat. The population is concentrated in the southern and seaward peripheries of these countries.

The severe climate to be expected in these high latitudes is offset by prevailing westerly winds off the warm North Atlantic. This oceanic influence is strongest during the winter, as is evidenced by the 32°F isotherm paralleling the entire Norwegian coast, which is ice-free the year round. Annual temperature ranges increase as one moves eastward. Summer days are rather cool, but the long days offset in part the short growing season. Precipitation varies from over 60 inches in western Norway to 25 in Finland, with less than 20 inches in the extreme north.

In the past, agricultural and maritime interests have been foremost; but recently industry

228 World Geography

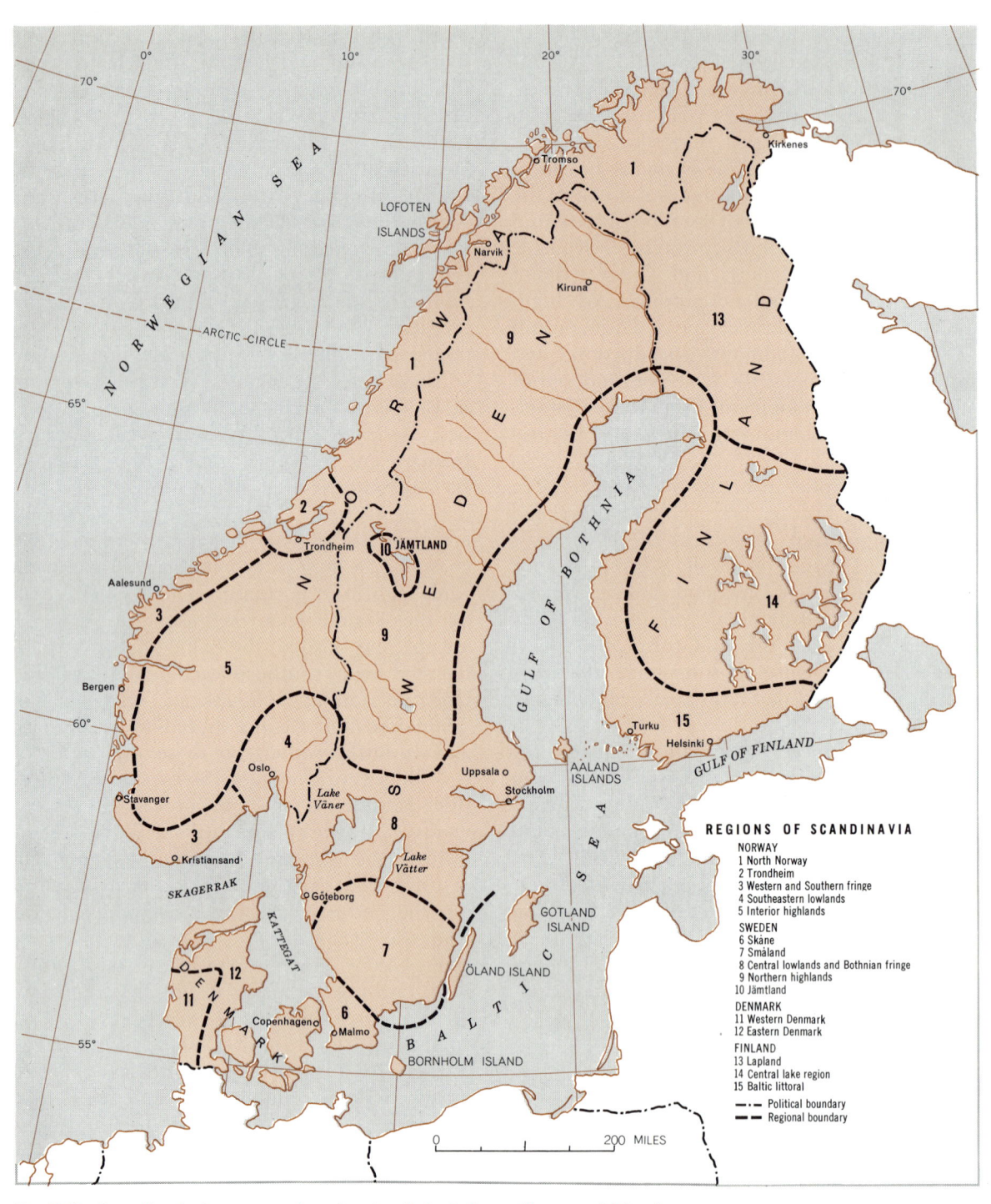

Fig. 7-3 Scandinavia is an area of peninsulas, indented coastlines, and islands.

REGIONS OF SCANDINAVIA

NORWAY
1 North Norway
2 Trondheim
3 Western and Southern fringe
4 Southeastern lowlands
5 Interior highlands

SWEDEN
6 Skåne
7 Småland
8 Central lowlands and Bothnian fringe
9 Northern highlands
10 Jämtland

DENMARK
11 Western Denmark
12 Eastern Denmark

FINLAND
13 Lapland
14 Central lake region
15 Baltic littoral

—·—·— Political boundary
— — — Regional boundary

(manufacturing) has emerged as the principal sector of the respective economies, and a major portion of the population is now urban. All the Northern countries except Norway lack coal, and most of the iron ore mined in Sweden is exported. Local raw materials supply forest-products and food-processing industries. Waterpower is abundant throughout these countries except in Denmark. Their combined merchant marines are significant carriers of world trade.

NORWAY

Norway extends for more than 13° of latitude—a distance of 1,000 miles—in Northwestern Continental Europe. Its total area of 124,587 square miles includes thousands of offshore islands and skerries (rocky islets). The fjorded coast and rugged rocky highlands dominate the topography of Norway. Much of the highland surface is plateaulike in character, averaging between 2,000 and 5,000 feet in elevation. Glaciation has rounded the landforms and has cut the numerous fjords along the west coast, some extending over 100 miles inland. The 3 percent of the country's area that is arable is located primarily along the coast, in inland valleys, and in the flatter southeast section.

Forests, which cover one-fourth of Norway's land area, occur at elevations below 3,300 feet in the south and below 1,000 feet in the north. The best forests are on the eastern slopes of the highlands. About two-thirds of the commercial forest is coniferous Norway spruce and fir, and one-third consists of deciduous trees, with birch dominant. Forests provide the raw materials for one of Norway's major industries.

Norway's deficiency in industrial fuels is partially offset by plentiful hydroelectric power. Abundant precipitation, melting glaciers, negligible freezing, and high stream gradients give Norway more available waterpower than any other European country. Pyrites (source of sulfur and sulfuric acid) and iron ore, the principal minerals, normally represent three-fourths of the total value of ore production. Molybdenum and silver are also mined. Coal is mined on Spitsbergen (Svalbard Archipelago), a large group of Arctic islands belonging to Norway. Oil and natural gas potential exists on the adjoining continental shelf, where a promising area has been discovered off the southwest coast.

Agriculture must contend with handicaps of relief, climate, and poor soil in most of Norway; dairying and livestock production predominate. Farming is often combined with fishing in the coastal districts and with forestry in the interior. Farms, mainly small and 90 percent owner-operated, are now increasing in size because of labor shortages and increased mechanization. Hay and other pasturage, cereal grains, and root crops are grown.

Industrial activity, intensified in the past decade and largely based on abundance of cheap waterpower, employs over one-third of the labor force and contributes nearly 40 percent of the national production. Processing fish, forest, and agricultural products and manufacturing electrochemical and electrometallurgical goods are leading industries. A large integrated state iron-and-steel works, located at Mo-i-Rana, is based on available waterpower, limited local iron ore reserves, and cheap water transportation.

Many Norwegians look to the sea for their livelihood. The large merchant marine is a major factor in the economy of the country. From the waters adjacent to Norway and certain foreign territorial waters are landed nearly 3 million metric tons of fish per year. Norway ranks first in Europe and fifth in the world as a fishing nation. Fish and fish products normally account for 7 percent of the total value of the country's exports. Herring are caught predominantly south of Trondheim Fjord and

Fig. 7-4 Lake Olden and Nord Fjord, Norway. Considering the topography and latitudinal location, the fjords of Norway are used rather intensively. (Courtesy of the Norwegian National Travel Office)

cod to the north, especially in the Lofoten Islands area. Undependable catches of herring and, to a certain extent, cod are causing shifts in the dependence upon Norwegian coastal fishing and are also forcing more emphasis on larger vessels and distant fishing banks, such as those off West Greenland and Iceland.

The urbanized southeastern lowlands dominate Norway and its economy. Nearly 50 percent of Norway's 3.9 million people live in Oslo and seven surrounding counties, an area comprising the center of economic intensity in the country. A larger proportion of rolling but relatively arable land exists here than elsewhere. As usual in Scandinavia, dairying predominates, although cash crop farming is increasing. Excellent forests are found on the slopes, and many farmers work in lumbering and related activities during the long, cold winters. Coastal sawmills and paper and pulp mills are common. Oslo, Norway's capital and largest city, is the industrial, transportation, commercial, and cultural center of the whole country. It has shipbuilding and a variety of manufacturing industries. Electrochemical and electrometallurgical industries, located in smaller cities, utilize large amounts of hydroelectric power.

The country's western and southern fringes, south of the Trondheim depression, are composed of a fjorded, precipitous, island-dotted coast, where fishing and farming are important activities, along with scattered industry in the fjord and coastal cities. The population is concentrated along the coast, on the offshore islands, on the fringes of the mainland fjords, and in interior valleys. Farming is often combined with fishing to provide sufficient income. The heavy precipitation, cool temperatures, and rough topography favor hardy fodder crops and dairying. Aalesund, an island city off the west coast, is the leading fishing port and herring capital; Kristiansund, a port on the southeast coast, the klipfish (dried, salted cod) center; Stavanger, a fish-canning and shipbuilding center; and Bergen, the major port and industrial and cultural center of western Norway. Here, too, the growth of industry is based on waterpower. Kristiansund is also an industrial center and port.

The Trondheim Depression, surrounding Trondheim Fjord, is a fertile dairying area, where hay, oats, barley, and potatoes are the principal crops. Its well-forested slopes form the basis of the paper, pulp, and lumber industry. Trondheim is the major commercial center and port of west-central Norway. North Norway, the area north of the 65th parallel, is largely a narrow, barren, thinly populated mountainous plateau region with a highly dissected coastline. Small-scale fishing is often combined with marginal agriculture. The Lapps graze reindeer on the tundra vegetation of the far north. In recent years, several freezing and fishpacking plants and other commercial ventures have been established, based on local waterpower and raw materials, as part of a plan to increase economic opportunities in the resource-poor north. Narvik, is an important port and railhead, shipping iron ore brought by rail from north Sweden. Tromsö, another island city, is a regional capital of North Norway that, owing to major government and private investment and a new university, has become a growth center. Iron ore is mined near Kirkenes, within a few miles of the U.S.S.R. boundary.

The interior highlands are sparsely populated. Some forestry is found on the lower slopes, and cattle are summer-pastured on the grasslands of the upland plateaus. Communications have been constructed with difficulty in the interior.

Norway is normally an exporter of processed raw materials and an importer of finished goods, and the many resources she lacks. Its exports consist of electrochemicals, forest products, fishing and whaling products, and ores and electrometals. Imports are dominated by foodstuffs, coal, petroleum products, textile raw materials, iron and steel, and machinery. Earnings from the nation's merchant marine and tourism help offset a trade deficit. Norway is belatedly enjoying the economic boom and high postwar standard of living of Western Europe. Other Western European countries and the United States dominate Norway's foreign trade. Both economically and politically, Norway is oriented toward the Western world.

A major economic problem exists in the uneven development of the various parts of the country. The north has had difficulty as many of its people have migrated to the better opportunities of the south, and the central government has attempted to aid such hardship areas through loans and various development schemes.

SWEDEN

Sweden, the largest and economically most significant of the Scandinavian countries, is slightly larger than California and ten times the size of Denmark. Nearly 90 percent of its 8 million people live in the increasingly urban southern half of Sweden, which is industrialized and also has a humid continental climate more favorable for crops than the north.

The country's land slopes gradually from the mountainous heights along the Norwegian border to the Baltic Sea. The many large rivers that follow this gradient, flowing northwest to southeast, are vital for hydroelectric power development and transportation of logs to coastal sawmills, although trucks have taken over much of the latter function. Sedimentary rocks occur in the southern peninsula of Skåne and the islands of Öland and Gotland, but old crystalline rock prevails elsewhere. Soils are generally thin and infertile, except for those in Skåne and certain marine clay soils in the central lowlands.

Only 9 percent of Sweden's total land area is arable, 55 percent is forested, 3 percent is permanent pasture, and 33 percent is wasteland or uncultivated. Inland lakes and rivers cover about 9 percent of the country's total area. Despite the northerly latitude, agriculture is aided by a climate moderated by Atlantic and Baltic influences.

The principal natural resources of Sweden are forests, iron ore, waterpower, and the fertile soils in certain areas. About 35 percent of its exports are forest products. Pine, spruce, and birch are the main commercial species. About one-half of the forests are state-owned, and an efficient conservation program is in effect.

Sweden mines about 40 percent of the iron ore of Europe (excluding the output of the Soviet Union). The Lapland ores average over 60 percent metallic content but are high in phosphorus. A large part of the iron ore is exported either via Narvik, Norway, or Luleå, Sweden, and the remainder is utilized by the steel plant in the latter city. The central Swedish (Bergslagen) ore is lower in metallic content but contains less sulfer and phosphorus; some is exported, and some utilized in Sweden's high-quality iron and steel industry. Gold, low-grade coal, and copper are among other minerals found.

Sweden's relief and climate favor waterpower development. Just one-fourth of the country's 176-billion-kilowatt-hour potential is now utilized. Although approximately 80 percent of the hydroelectric potential is in the northern half of the country, distant from the southern urban and industrial centers, improved technology allows it to be utilized increasingly both in the north and, by transporting it, in the south.

Sweden is nearly self-sufficient in food production. Infertile soils are offset by hard work, application of fertilizer, and improved plant breeding. Farming is marginal and allied with forestry in the north, but intensive and mechanized in the favored soils of the south. Over half of the country's farms contain less than 25 acres.

Industry which has increased rapidly since the 1880s, now considerably outranks agriculture as a supporter of Sweden's population. Industrialization is based on the availability of high-quality raw materials, abundant waterpower, capital, and technical skill, as well as the contributions of Swedish inventors. Urban-industrial regions are found around Stockholm and Göteborg and, to a lesser extent, in the areas in between these cities.

Sweden is divided roughly into two parts by the 60th parallel. The sparsely populated northern zone is a source of raw materials and power from its forests, mines, and rivers; the south accounts for most of the country's agriculture, commerce, industry, and population. Skåne Peninsula, the most intensely cultivated area of Sweden, contains about 12 percent of the people on only 2.5 percent of the total land area. Here, 90 percent of the land is cultivated. Wheat and sugar beets are grown, in addition to the common Scandinavian crops; dairying is highly developed. Malmö, with ferry connections to Copenhagen, is the main urban center in the south. The islands of Öland and Gotland, which are geologically similar to Malmö, are mostly sheep-grazing and resort areas. The Småland highlands are significant for handicraft industries such as woodworking and glass.

The central lowlands and the Bothnian fringes are the agricultural, industrial, and commercial heart of Sweden. The marine clay deposits favor agriculture with emphasis on dairying. Large lakes such as Lake Vänern and Lake Vättern cover much of the area. Rolling glacial topography is characteristic, with rougher and poorer soils left to forests. Metal and wood products are leading manufactures. Iron ore for the high-quality steel comes from the Bergslagen district. Industrial specialties are found in many smaller centers. Numerous wood-processing and export centers are located along the Bothnian coast. Luleå exports Lapland ore and has an iron-and steelworks. Umeå is a growth point, having a university and associated laboratories and research institutes. Stockholm, the major city of Sweden, is the nation's capital, cultural center, and chief Baltic port and also has a variety of manufacturing. Göteborg, with a favored westcoast position, is the principal foreign trade port, and its shipyards are a chief factor in placing Sweden fourth in world shipbuilding. Textiles and metal goods are also manufactured there. Water plays a major role in the site, beauty, and human activity of both Stockholm and Göteborg.

The northern highlands, covering approximately two-thirds of Sweden, are characterized by glaciated topography, with numerous large, fast-flowing rivers and long, narrow glacial lakes in the upland valleys. Forestry is the chief inland occupation. Minerals, especially highgrade iron ore, provide a second mainstay of the

Fig. 7–5 Lulea, Sweden. Lulea is the principal port on the Gulf of Bothnia for the transshipment of Kiruna iron ore. (Courtesy of the Swedish National Travel Office)

northern economy. The Lapland and Bergslagen districts are found, respectively in the extreme northern and southern portions of the highlands. Isolation, darkness for nearly six months, and the severe winter cold hamper workers and production in the area centering around Kiruna and Gällivare. Iron ore can be shipped only during the warm season from Luleå, but ore cargoes leave the year round from ice-free Narvik.

Climate, soils, and topographic features create many handicaps to agriculture. Limited development is found in the valleys and around the edges of the lakes. Jämtland is the Swedish counterpart of the Norwegian Trondheim area. In the far north, some 10,000 Lapps make a living by grazing 300,000 reindeer.

Finished steel, paper and paperboard, passenger cars, lumber, and iron ore dominate Swedish exports. Coal, petroleum, metals and machines, and raw materials for industry are major imports. England, Germany, the United States, and Scandinavian neighbors are leading trading partners. Invisible exports such as merchant marine earnings and tourism generally offset import excesses.

A major question facing neutral Sweden is how to continue the "middle way" economically and politically. Sweden faces trade and economic problems, especially since her neutrality hampers association in any way with such organizations as the Common Market. Her comparative economic strength makes the functioning of a Nordic Union difficult.

The Swedish life-style and standard of living are among the highest in Europe.

DENMARK

Denmark consists of the peninsula of Jutland and about 500 islands, lying mostly in the Baltic Sea; its land area is less than one-eighth the size of Norway. Denmark, however, has one million more people, with about 29 percent of the country's 4.9 million inhabitants living in metropolitan Copenhagen. The average population density reaches 400 per square mile in the fertile eastern islands but drops to 125 in sandy, infertile western Denmark. Approximately 70 percent of the gently rolling terrain is used for agriculture. The surrounding waters and moderating westerly winds result in a mild marine climate, with temperatures averaging 60° in summer and 32°F in winter.

Physically, Denmark is similar to the north German glaciated plains, but culturally it is linked with Scandinavia. The position of Denmark, controlling the narrow water outlets of the Baltic, is highly strategic. Germany occupied this tiny nation during World War II in order to command the Baltic. Russian maritime development gives Denmark's position added significance today.

Denmark lacks the mineral, waterpower, and forest resources of her neighbors; even the soil is not exceptionally fertile. Nevertheless, intensive agriculture is vital to present-day Denmark. Small, carefully tended holdings, when supplemented by considerable imports of protein concentrates and grains, support large numbers of cattle, swine, and poultry. Dairy products, meat, and eggs are leading Danish exports. High-quality agricultural production is largely a result of the effort of people aided by the highly developed cooperative movement. Postwar economic problems have stemmed from the austerity program in Britain, Denmark's major foreign market, and from the difficulty of finding trade outlets in other Western European countries and the United States.

Agriculture and fishing employ 13 percent of the working population; industry, which employs 37 percent, is increasing. Much of Danish industry consists of processing domestic agricultural commodities. Most other industrial raw materials and all power sources must be imported.

Western Denmark, occupying the western two-thirds of the Jutland Peninsula, is a sandy, infertile outwash plain interspersed with subdued old moraine hills. Sandy heaths, made productive by great effort, support grasslands and hardy crops. Christmas trees and lumber are harvested from pine trees planted on poorer soils. Sand dunes, marshes, lagoons, and shallow water are found along the unindented coast. The port of Esbjerg, constructed to carry on trade with Great Britain and Western Europe, together with Lim Fjord and Frederikshavn to the north, are fishing centers.

With better agricultural land, many islands, and an indented coastline that favors harbor development, as well as the presence of the Copenhagen metropolitan region, eastern Denmark has the bulk of the country's population. Despite a Baltic location, most of its trade and other relations are with Atlantic nations.

Rolling morainic hills and intervening valleys with fertile clay soil replace the flat, sandy terrain to the west. Intensive methods of cultivation, favorable temperature and rainfall conditions, and careful plant breeding result in some of the highest crop yields per acre in the world. Most of the agricultural exports, butter, bacon, cheese, and eggs, originate in this section. The intensive forced feeding of animals, accomplished in large part with imported feeds, has been termed "factory farming."

Copenhagen dominates the industrial and commercial development not only of eastern Denmark but of the entire country. It is the national capital, as well as the cultural and

Northwestern and Central Europe 235

REGIONS OF WEST CENTRAL EUROPE

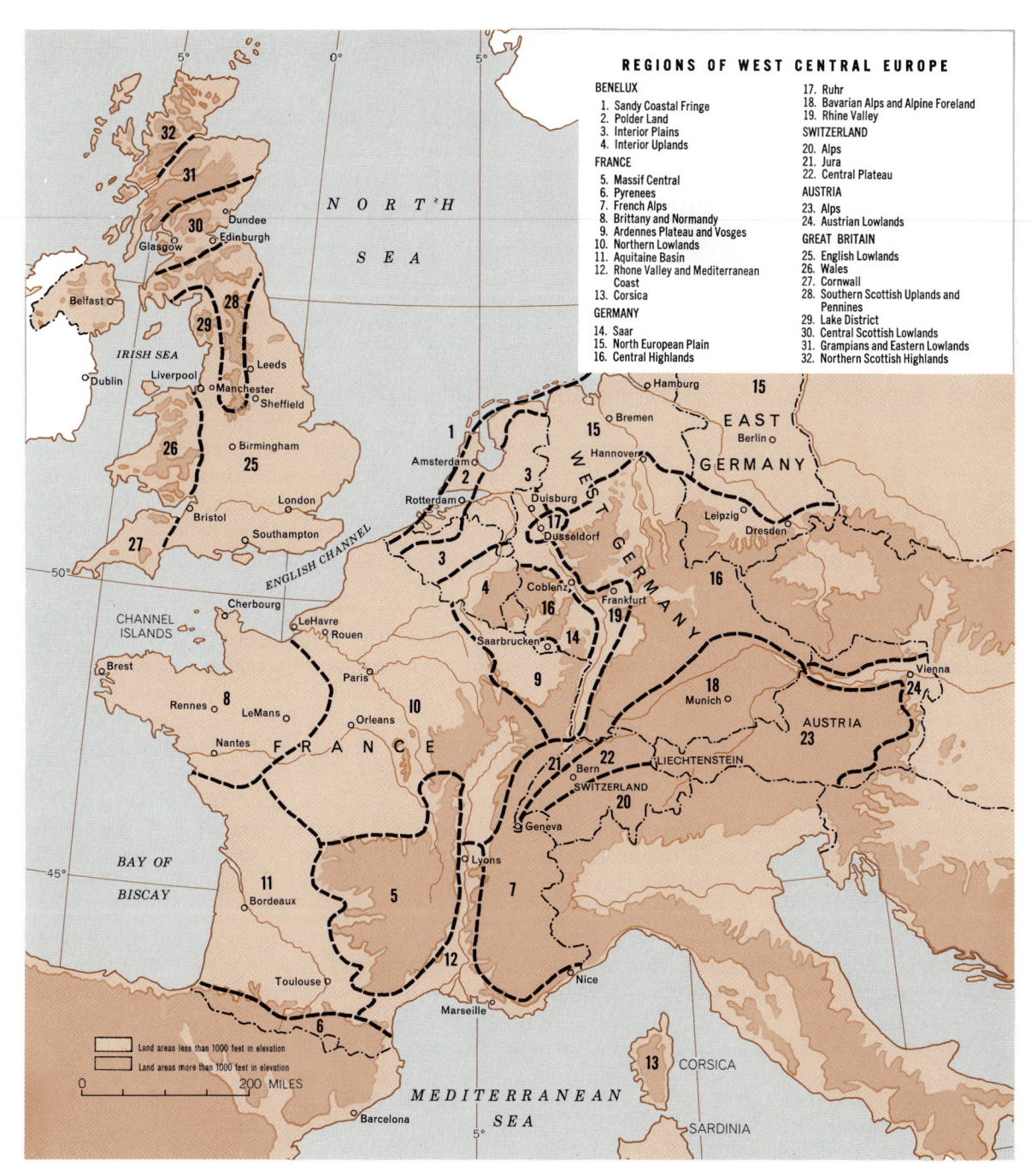

BENELUX
1. Sandy Coastal Fringe
2. Polder Land
3. Interior Plains
4. Interior Uplands

FRANCE
5. Massif Central
6. Pyrenees
7. French Alps
8. Brittany and Normandy
9. Ardennes Plateau and Vosges
10. Northern Lowlands
11. Aquitaine Basin
12. Rhone Valley and Mediterranean Coast
13. Corsica

GERMANY
14. Saar
15. North European Plain
16. Central Highlands
17. Ruhr
18. Bavarian Alps and Alpine Foreland
19. Rhine Valley

SWITZERLAND
20. Alps
21. Jura
22. Central Plateau

AUSTRIA
23. Alps
24. Austrian Lowlands

GREAT BRITAIN
25. English Lowlands
26. Wales
27. Cornwall
28. Southern Scottish Uplands and Pennines
29. Lake District
30. Central Scottish Lowlands
31. Grampians and Eastern Lowlands
32. Northern Scottish Highlands

Fig. 7-6 The regions of west-central Europe are the result of cultural activities as well as the physical environment. Away from the mountain areas, most of the land is intensively used.

intellectual center, and rivals Stockholm for beauty and charm among North European cities. Shipbuilding, food processing, brewing beer, and manufacture of margarine (so that the fine Danish butter can be exported), textiles, world-famous silverware, porcelain, and china are the chief industrial activities. The excellent port of Copenhagen is the headquarters of Denmark's maritime interests and also has an international free port. Small fishing ports are common along the Danish coast. The island of Bornholm, located 95 miles east of Copenhagen in the Baltic, provides kaolin for the pottery industries and is also a major tourist attraction.

ICELAND

Iceland, located just south of the Arctic Circle in the North Atlantic, has the lowest average population density per square mile (5.0) of any European nation. Over 80 percent of its area is uninhabitable, and less than 1 percent is arable. Almost half of the island's 200,000 inhabitants live in the capital Reykjavík, a bustling, crowded city with the usual rush-hour traffic.

Iceland is a mountainous island with active volcanoes and hot springs. Over 13 percent of its surface is permanently covered with snow and ice. Limited sections of the coastal lowlands have grass vegetation, but trees are rare on the island. The marine location and the North Atlantic Drift combine to produce mild winters and cool summers. The average January temperature of Reykjavík is 33°F; the summer mean is 50°.

The waters off Iceland yield abundant cod, herring, coalfish, and ling; the current average annual catch is over 1 million tons, or about 5 tons per capita. Fish normally provide over 90 percent of the island's exports, and Iceland's prosperity fluctuates with the fish catch and world-market prices for fish. Fishermen from other nations also exploit Icelandic territorial waters. Grazing of sheep and cattle and limited crops of hay, potatoes, turnips, and greenhouse-grown fruits and vegetables are found on the coastal lowlands. Iceland must import many of its everyday food and raw-material needs. The rapid population increase and accompanying urbanization has meant a marked increase in industry, which now supports over one-fourth of the people.

FINLAND

One-fifth of the area of Finland is north of the Arctic Circle. Despite this geographical position, the climate is tempered by warm and westerly winds. The summers are short and changeable, and the winters are long, cold, and snowy. Hardy crops, aided by long, sunlit days, mature in the three-month growing season.

War destruction and peace-treaty concessions resulting from World War II were costly to Finland's economy. The 12 percent (17,780 square miles) of its territory on the east and north that Finland lost to the Soviet Union included its Arctic outlet, 13 percent of its forest resources, one-third of the installed hydroelectric power, 30 percent of its fisheries, 11 percent of the gross value of industrial production, the Petsamo nickel mines, and some of its most fertile land. The U.S.S.R. also demanded a staggering $226.5 million in reparations, to be paid in wood products, cables and ships, and metal goods. To meet these demands, the Finnish shipbuilding industry increased six times, and the metal industry doubled in size; these changes virtually revolutionized Finland's industrial structure. The debt was paid in full in September, 1952, despite an annual drain of more than 10 percent of the national income. In order to find outlets for the expanded shipping, metals, and other industries, Finland had to look for certain markets in the Eastern-bloc nations, since she could not compete in Western markets

—an impact that is still seen in her trade structure. Another problem, resulting from the territorial losses, was the strain of resettling 420,000 people, or one-tenth of the total population, in a country with limited resources.

Finland is a glaciated, old crystalline-rock platform. Only in the northeast and in Lapland does the generally low elevation of the land reach over 650 feet. Glacial scour and deposition left a large number of swamps and lakes, which cover 10 percent of the country's area. Coastal lowlands and lake borders have more fertile recent sand and clay deposits but comprise only 9 percent of the soils.

Besides the skilled, able people of this nation, Finland's most valuable resource is the 56 percent of its area that is productive forestland, which provides over 80 percent of the gross value of exports. There is a variety of minerals, among which copper is the most important. The complete lack of coal and petroleum reserves is partially offset by abundant waterpower for industrial use.

The population numbers approximately 4.8 million, over 90 percent of whom are found in the southern half; they live mainly on the south and west coastal fringes, with about three-fifths residing in cities and towns. A majority of Finns, thus, now live in urban areas and are supported by urban-oriented activities. Swedish-speaking peoples, composing about 8 percent of Finland's population, are located primarily in the coastal districts nearest Sweden and in the Åland Islands.

Finland's chief livelihoods are agriculture and forestry. The Finns produce approximately 98 percent of their food needs; they normally have dairy surpluses and shortages of cereals and sugar. About one-third of the nation's gainfully employed obtain their livelihood from manufacturing and mining and another third from commerce, transportation, and other sources. Forests and hydroelectric power are the two major industrial assets.

The Baltic littoral, having more favorable clay and silt soils, better climate, and greater accessibility than other areas, is the principal agricultural region of Finland. Hay, grains, and potatoes are grown; but two-thirds of Finland's arable area is used for fodder production, indicating the predominance of dairying. Commercial fishing is found off the southwest coast and among the Åland Islands.

The coastal regions are also the chief industrial areas. Helsinki, Turku, and Tampere, which are the chief manufacturing cities, have access to transportation facilities, forest resources, labor, and imported raw materials. Waterpower is abundant where streams break through the Salpauselka, a double moraine located south of the lakes district. Manufactures include paper and other wood products, metals and metal items, food, drink, textiles, glassware, chemicals, and tobacco products. About one-fifth of the country's industry is located in Helsinki, the capital and the cultural, financial, and university center. Helsinki is also the center of a nodal region that, in a spatial organizational sense, is the focus of the entire country.

The central lakes region is a swampy and lake-dotted area, where forests are the primary resource. Lumbering is a winter occupation; the huge accumulations of logs are floated out in summer. Dairying is practiced around the edges of some lakes.

Finland's postwar trade was dislocated by the burdensome war reparations. Even after completion of payments, Finland has had to continue looking eastward for markets for its ships, metals, and other products. Forest products dominate Finnish exports, though ships, machinery, and dairy products are also important. Metals and metal products, cereal grains, raw textiles, coal, and petroleum are leading imports. Important among Finland's trading partners are the United Kingdom, Sweden, other countries of Western Europe, the U.S.S.R., and the United States.

Economically and culturally, Finland leans toward the West; but politically, though independent, it is strongly within the Soviet sphere of influence. Trade and defense agreements with the Soviet Union restrict development of relations with the West. Finland's future depends largely on being able to reconcile these conflicting interests.

British Isles

Over 57 million people inhabit the British Isles—about 5,000 islands and islets in all, of which Great Britain and Ireland are the largest. This relatively small area has played an extremely prominent role in the development of Europe and the world. The Industrial Revolution began here, and urban activity now supports a large share of Britain's people.

Before the fifteenth century the British Isles were situated on the periphery of the known world, but since then they have occupied a central location among the populated landmasses. Their inhabitants became maritime-minded, and the power of England increased as the country secured extensive colonies abroad, developed its manufactures, and expanded its merchant fleet and overseas commerce. Excellent harbors along the indented coastline and the inventions of the Industrial Revolution were significant factors in the country's expansion. The islands, part of the European platform, are separated from the Continental mainland by the narrow Strait of Dover. This geographical separation allowed the English to spend much of their energy on overseas trade and colonies rather than on Continental wars. With the loss of most of her remaining colonies in the twentieth century and the trade threat of international economic unions on the Continent, the United Kingdom has gradually abandoned its aloofness to developments on the neighboring European mainland.

Physical Setting

Hilly and mountainous terrain dominates the north, west, and southwest of Great Britain, and undulating downs and plains the south and southeast. The highlands average between 500 and 2,000 feet in elevation. Rounded peaks and deepend valleys are a result of glacial erosion; rocks composing the highlands are mostly old granites and schists. Flanking some of the highlands are valuable coal deposits. Young sediments in the south and east are largely limestone, sandstone, and chalk, which form resistant ridges called "downs." Weak chalk and clay underlie the valleys and plains. Recent glacial drift covers all the British Isles except the Cornwall Peninsula and extreme southern England. Most of Ireland is a low, poorly drained glaciated plain rimmed by low mountains. The central plain is drained by the Shannon River.

The marine climate is free from extremes except in the highlands. Summer temperatures range from a July average of 63° in southeastern England to 55° in northern Scotland and 59°F on the west coast of Ireland. The moderating influence of the sea in winter is shown in the January averages of 44° in southwest Ireland, 39° in southeast England, and 37°F in northern Scotland.

The annual precipitation ranges from 24 to 55 inches, with generally a slight winter maximum. The exposed and higher western portions have the heaviest precipitation; Valencia in western Ireland averages 55.6 inches per year, as compared with 24.5 in London. Eastern districts, situated in the rain shadow of the highlands, have a summer maximum, largely of convectional origin. The cyclonic influence causes dominantly cloudy, humid, rainy, and often foggy weather, and southern England has a cloud cover about 70 percent of the time.

Natural Resources

Soils range from a leached, moderately fertile type in the rainy western districts to a less leached, more fertile soil in the drier east. The thin, infertile soils of the highlands and limestone escarpments support grasslands used for grazing. Glacial soils and those in the clay vales and chalk areas of the southeast are generally good for cultivated crops.

Minerals are the principal natural resources of Great Britain. Coal, the outstanding mineral, accounts for 80 percent of the mineral output by value. The 175-million-ton annual domestic production must be augmented by coal imports and large quantities of petroleum and domestic waterpower, as well as by nuclear energy, to supply the steadily increasing power needs. Principal coalfields include those on the flanks of the Pennines, the Northumberland-Durham (Newcastle) fields, and those in South Wales, Cumberland, and the Scottish Lowlands. All these areas are leading industrial districts, although recent manufacturing growth has occurred more rapidly along the coast and in the London area rather than near the coalfields.

Some current British economic problems can be traced to difficulties in the coalfields. Production and exports of this mineral resource steadily decreased, and the present output is less than 180 million tons annually. In 1970 coal provided less than 1 percent of the total value of exports, as compared with 33 percent in 1913. Handicaps facing the British coal industry are thin seams, deep mines, dipping and faulted beds, lack of mechanization and modernization, and labor shortages. The coal-mining industry has been state-operated since 1947.

Iron ore, the second mineral in importance, has the low average metallic content of 29 percent in the British Isles. It is a low phosphorus- and high sulfur-content ore. The principal iron deposits are east of the Pennines. The substantial British iron and steel industry imports over two-thirds of its raw-material needs. Excluding the U.S.S.R., Great Britain ranks third in European iron-ore production.

Fish abound in the North Sea and adjacent bodies of water. Easy access to these nearby banks on the Continental Shelf and the maritime-minded people of the region have led to the development of a significant fishing industry. Although the catch averages slightly over a million metric tons annually, fish exports have decreased, and increasing amounts are now imported from various sources. The British commercial fishing fleet operates from such ports as Aberdeen, Grimsby, Hull, and Yarmouth.

Economic Development

The marked predominance of industry over agriculture is the outstanding feature of the British economy. Manufacturing accounts for employment of over 38 percent of the civilian work force. Transport and trade activities are another important sector of the economy. Industrial concentrations show marked correlation with the location of coalfields, with the exception of those which have grown up around London and coastal ports. Industries are often localized, such as the cotton textile industry in Lancashire. In manufacturing, there has been a definite shift from basic to lighter metal products, and from cotton and wool to rayon, nylon, and other synthetic fibers. In general, the trend has been to utilize Britain's technical skill and ingenuity in making quality consumer goods. The English industrial system has Continental and worldwide linkages of major importance both for raw materials and for its finished products.

The large, quite dense population places considerable pressure upon the country's land area to produce food. Nevertheless, prior to 1940, much agricultural land was extensively rather than intensively used. The United Kingdom imported 70 percent of its prewar food needs, including grains and meat products to supplement home production. Land use was intensified during World War II; and immediately after World War II, Great Britain produced nearly one-half of its own food requirements. Despite the wartime and postwar increases in domestic food production, the British are again back to about the 1940 level of dependency upon imports of this kind.

The World War II "plow-up" campaign increased the acreage devoted to grain crops and intensive dairy farming in the British Isles. Scientific grassland cultivation and mechanization also increased crop and dairy production; British agriculture is now among the most highly mechanized in the world. Agriculture employs nearly 800,000 people and utilizes 48 of the 60 million acres of land in the United Kingdom. In Eire approximately 35 percent of the total workers are engaged in agriculture. In Ireland, the economy still centers primarily upon agriculture, although industry is rapidly increasing, especially in Northern Ireland and the Dublin area.

Divisions of Great Britain

Great Britain, the largest island of the British Isles, includes England, Scotland, and Wales. The principal divisions of the island of Ireland are Ulster (Northern Ireland) and Eire (Ireland); Eire, occupying the larger southern part of the island, is an independent nation. The United Kingdom includes Great Britain and Northern Ireland.

ENGLAND

The densely populated English lowlands are the agricultural, industrial, and commercial heart of England and the British Isles. Less than 500 feet above sea level, they make up approximately three-fourths of England. The bulk of the regional topography is undulating to rolling, but rough hills and downs reach nearly 1,000 feet in a few places.

Cropland is predominant in the fertile eastern sections of England, which have less than 25 inches of rainfall yearly and warm summers. Wheat, barley, oats, and root crops are utilized in the winter fattening of beef cattle. Certain areas specialize in supplying fruits, potatoes and other vegetables, and dairy products for the large urban market. The western districts, with wet, cool summers, have extensive rotational grass pastures. Dairying is common, and oats is the major cereal crop. Some districts ship cattle eastward for fattening, and others fatten them locally. The rougher summits of the chalk downs and limestone escarpments in the south and central portions are generally sheep-grazing areas; the lowlands are intensively cropped.

England's industrial concentrations occur in or near the coalfields and iron deposits or the contact zones between highlands and lowlands. London is the principal exception in this general distribution. The industrial northeast, which is dominated by the iron and steel industry, normally produces about one-fifth of Great Britain's steel. Coking coal is obtained locally, but much of the iron ore must be imported to supplement Cleveland Hills and Northampton ore. Iron and steel manufacturing is concentrated in Middlesbrough and the surrounding Tees River towns, and shipbuilding is chiefly on the Tyne and Tees River estuaries. Oil refining, the fabrication of iron and steel products, and chemical industries based partly on local salt, gypsum, and lime deposits are also important. Exports of coal from Newcastle have declined to relative insignificance.

East of the Pennines, industry is based primarily on Yorkshire coal. This coalfield, the largest producer in Great Britain, extends over 70 miles in a north-south and 15 to 20 miles in an east-west direction. The northern part of this district specializes in woolen manufacturing, and the southern part in high-grade steel products. The woolen industry was originally based on available local raw materials and waterpower; but now a large portion of the wool is imported, and coal supplies most of the power. The presence of soft water for washing the wool is another local advantage for the industry. Leeds and Bradford are the major woolen centers. Sheffield is the well-known center for high-grade cutlery, hardware, machinery, and other steel products. Some steel is imported from other parts of England or from abroad. Sheffield usually accounts for nearly 15 percent of British steel production.

At the southern end of the Pennine Chain in central England is the leading English industrial complex, often referred to as the Midlands, or the "Black Country." Though these terms characterize the cities proper, there is much green agricultural land between them. As local iron supplies declined, basic iron and steel production tended to move to coastal areas because of reliance on imports of foreign ore. Some iron and steel is still produced in the Midlands region, but industries using steel made elsewhere, for such manufactures as metal goods, motors, automobiles, and locomotives are now more important. Birmingham, the major industrial center of the Midlands, manufactures a variety of materials, especially brass, other nonferrous metal products, leather goods, clothing, and pottery.

Lancashire leads the British Isles in cotton spinning and weaving. Access to nearby coal, early development of waterpower, plentiful pure water, a mild and moist climate that lessens thread breakage, and an advantageous position for the import of foreign cotton supplies were the principal reasons for this localization of the industry. Though cotton still dominates the region's output, fibers, textile machinery, engines, and a variety of other products are manufactured in this district's numerous cities, with a population totaling 2.5 million people. Situated on the Irish Sea, Liverpool is the third-ranking British port; it is connected to the major textile-manufacturing center of Manchester by a deepwater canal, the Manchester Ship Canal, opened in 1894.

London, long the world's largest metropolitan district, is rivaled only by New York City and Tokyo in this respect. Greater London contains over 8.5 million people; and some 10 million, or nearly one-fifth of the total population of Britain, live in the city and surrounding area. Although located 65 miles inland on the Thames, London is a major world seaport, which collects, processes, and distributes goods from numerous distant points. The single leading industrial center of Great Britain, London has a large variety of processing and light industries. It is also a cultural, transportation, financial, political, and commercial center of both British and world importance. The city is the organizational hub of the country, and all routes of transportation focus on it, resulting in congestion within the city and in much of southeast England. London and the other large cities of the British Isles, like the large cities of the United States, are faced with problems of air pollution, water pollution, poor housing conditions, and increasing crime rates.

The Cornwall Peninsula of southwest England is dominated by several high moors on which sheep are grazed, and its luxuriant lowlands furnish grass for beef and, especially, dairy cattle. The mild climate of the southern coast attracts tourists and, in protected valleys, permits the growth of early flowers and vegetables for the London market. Small fishing villages dot the rugged coast. Considerable china clay is extracted both for a local pottery industry and for export.

The topographical backbone of England is

the Pennine Chain, a rolling to flat-topped upland averaging between 600 and 2,000 feet in elevation. Most of the sparsely populated Pennine Chain is heathland, moorland, and rough pasture. It is bordered by the Lake District and the Cumberland coal and industrial center.

SCOTLAND

The Central Scottish Lowland is one of the major agricultural and industrial regions of the British Isles. On 20 percent of the land area live 80 percent of the people of Scotland. These lowlands have recent alluvial and glacial soils and a moderate climate. Approximately two-thirds of the eastern portion is planted in cereal grains; sheep graze the pastures and rougher areas. Grassfarming, with emphasis on breeding dairy cattle, dominates the agricultural economy of the rainier western portions.

Scotland produces about one-tenth of the British coal output, much of which is of excellent bituminous grade, in the central and east portions of the basin, with some also coming from the Ayrshire field in the western part of the country. The iron and steel industry, localized in and around Glasgow, produces about 10 percent of the total United Kingdom tonnage and depends mainly on imported iron ore. Glasgow, Edinburgh, and Dundee are the three major urban centers of the Central Scottish Lowland. Glasgow, located on the Clyde River, is the sixth-ranking port in the British Isles, as well as a heavy-industry center, specializing in iron and steel, machinery, and chemicals, and one of the world's leading shipbuilding centers. Edinburgh, Scotland's capital, is a cultural, university, printing and publishing, and light-industry center. Dundee is noted for its manufactures of imported jute fiber.

The Southern Scottish Uplands, although smaller and less rugged than the Northern Highlands, form a communications barrier between the Scottish Lowland and England. The uplands are dissected by numerous small river valleys. The hill sheep industry is found here, and a high-quality woolen tweed industry is centered in the Tweed River valley. Cattle grazing, dairying, and crop agriculture have been expanded in the coastal districts, in river valleys, and on the lower slopes.

The Northern Highlands, rising 1,000 to 3,000 feet above sea level, is a plateau strongly eroded by the action of water and glaciers. The western portion is high and rugged, with a fjorded coast. Agriculture here is handicapped by poor soils, excessive precipitation, and rather severe winters. Approximately 95 percent is sparsely populated moorland utilized for crofting (limited sheep grazing).

The eastern coastal lowlands have better soil, less rainfall, and higher summer temperatures. Grass, oats, turnips, and barley are the principal crops. Scottish arable farming is closely associated with raising beef cattle for the English market. Aberdeen, the largest city of this northeast region, is a manufacturing center, cattle market, and major fishing port on the North Sea.

WALES

Wales consists of a highly dissected central plateau surrounded by narrow coastal lowlands; the most densely populated portion is the south. The highlands culminate in Mt. Snowdon, which rises 3,560 feet. The rugged, isolated character of the geography of Wales has lessened interchange with outside areas, as indicated by the persistence of the Welsh language and intense national feelings. The grasslands of the excessively rainy highland interior favor sheep grazing. Dairy farming, accompanied by oats as a major crop, is found on the broader southern coastal lowlands. Small resort towns are located along the northern and western coasts of Wales.

The South Wales coalfield supports a densely populated mining, industrial, and exporting region. The extensive coal measures vary in grade from bituminous in the east to anthracite in the west. They are found in narrow valleys, with accompanying tightly spaced villages in the bottoms and often extending up hillsides. The changeover to oil-burning vessels and the general decline in the British coal trade resulted in large-scale prewar unemployment and poverty in these congested mining valleys. Much of their population has now found employment in the expanded and modernized steel, food, and chemical industries.

Cardiff is the administrative and diversified industry center of Wales. The smelters of Swansea process tin, zinc, nickel, and copper, and the city is also a shipbuilding center. These and other nearby cities comprise an urban area of 1.5 million that is now much less dependent upon basic industries than was true prior to 1950.

Ireland

Since 1921 Ireland has been divided into Northern Ireland and the Republic of Ireland (Eire); the latter area withdrew from and became completely independent of the British Commonwealth on April 18, 1949. Eire occupies about five-sixths of the island's 31,840 square miles and has a population of 2,885,000, as compared with 1,480,000 in Northern Ireland. The average population density, however, is 108 people per square mile in agricultural Eire, compared with 300 per square mile in more industrialized Northern Ireland. Eire is predominantly Roman Catholic, and Northern Ireland is largely Protestant—their religions difference being the basis of much turmoil between the two areas, as well as between these two religious groups within Northern Ireland.

The long English domination has geared the trade of both sections to that nation, with a general pattern of supplying agricultural products and buying manufactured goods. The protectionist policy of independent Eire and the continued close political and economic ties of Northern Ireland with the British government, however, are causing their economies to become gradually more diverse.

EIRE

Since 1921 Eire has become economically more self-sufficient by breaking up many large estates, placing greater emphasis upon raising wheat and sugar beets, and increasing industrialization. About one-fifth of the labor force are employed in agriculture, and one-third in industry. Most industries must be based on agricultural or imported raw materials, for Eire lacks coal and iron. Developments along the Shannon River system provide some waterpower. One-fourth of the country's exports are live animals shipped to England for fattening, but manufactured products are increasing. Eggs, dressed poultry, bacon, butter, ham, tobacco, beer, leather goods, china and glassware, and linens are also exported. About 75 percent of Eire's export trade and 50 percent of its import trade are with the United Kingdom. The steadily declining rural population (a 60 percent decrease in the last century) continues to be a problem facing Eire.

ULSTER

Northern Ireland, or Ulster, the principal area of Irish industrial development, is dominated by linen and rayon manufacturing and shipbuilding; both activities are centered in Belfast and vicinity. The linen industry was a natural outcome of local flax growing, a water supply and climate favorable for bleaching, and the large supply of female labor available in families of shipyard workers. Electronics and aircraft manufacturing are recent additions to the area's industrial production. The shipbuilding operations of Belfast, based on imported coal, iron, and steel supplies from Great Britain, are currently experiencing problems in world competition.

The value of exports from Northern Ireland is far greater than the total of those from Eire. Ulster trade is almost entirely with Great Britain. Imports of food, drink, tobacco, raw materials, and manufactured products usually are greater than Northern Ireland's exports; this trade deficit is offset by income from tourists, investments abroad, and pensions paid from Britain. Industrial output such as ships and textiles dominates the exports.

Regions of Ireland

The central plain of Ireland, comprising largely the poorly drained basin of the Shannon River, covers most of central Eire and extends into Ulster. Small farms, with considerable area left to pasture and meadow, are characteristic; but corresponding with increasing annual rainfall, the amount of cropland decreases as one moves westward. Livestock and occasional cash crops provide the chief source of income. Sheep raising and fattening, along with limited cattle raising, is the chief type of farming in the western central plain; mixed crops and livestock are predominant in the center, and dairying around Dublin and in Northern Ireland. Dublin, the chief industrial city, is the administrative, cultural, and transportation focus and marketing outlet of the central plain.

A highland rim and associated valleys surround the central plain, except on the east. The higher lands are rough, sheep-grazing areas. Dairying is predominant in the ridge-and-valley section of the south and around Belfast. Crop farming is dominant in the southeast and in the valleys of Northern Ireland. Livestock supplies most of the cash income. Commercial fishing is of local importance in the southwest. Cork and Cobh have excellent natural harbors that give access to the Atlantic for oceangoing vessels. Shannon is an international airport that has also begun to promote industrial development in the area.

Fig. 7-7 Connemara country, County Galaway, Ireland. Fields divided by stone walls and thatch-roofed buildings are characteristics of the Irish landscape. (Courtesy of the Irish Travel Board)

PROBLEMS AND TRENDS IN THE BRITISH ISLES

The high degree of urbanization in the British Isles, especially in London and environs and in southeast England, presents a modern-day congestion problem of major proportions. The growth of the internal transportation network has been insufficient to meet the need of the population. The new so-called new towns created near London have helped provide additional urban centers for more people, but have also added to the overall degree of urbanization and crowdedness in southeast England.

The shift in part from basic industries to quality manufactures based on skilled labor has shifted the emphasis of location away from coal districts to coastal ports and the London area. New power sources such as nuclear energy and North Sea natural gas reserves undoubtedly will further accentuate these geographical location trends.

The densely populated, heavily industrialized British Isles are highly dependent upon foreign trade to supply much of their food and considerable quantities of industrial raw materials. British exports are completely dominated by manufactured products, which find worldwide markets. Imports exceed exports in value by about 4 billion dollars annually in recent times, and this foreign-trade deficit must be covered by profits from foreign investments, world insurance services, merchant marine earnings, tourism, and other forms of "invisible" income. Foreign markets are vital to British survival; yet the nation's commercial situation is made difficult by increasing competition, loss of direct control over sizable overseas possessions, newly independent countries which were formerly good customers but which now favor domestic products, protective regional tariff accords such as the European Common Market, and decreased dominance of trade with Commonwealth countries.

Food, raw materials for industry such as metals and textile fibers, mineral fuels (particularly petroleum), and semifinished manufactures are Great Britain's leading imports. Principal exports include machinery, vehicles, other iron and steel products, finished textiles, and chemicals. The exports represent mainly the British quality items for which technical capability and skill give the British a competitive edge. The Commonwealth, the Common Market, EFTA countries, the United States, the Near East, and Argentina are foremost among British trade partners. Commonwealth trade is decreasing; about 25 percent of current-day British commerce is with Western Europe, mostly with Common Market countries. Tariffs among the seven member countries of the European Free Trade Association are gradually being decreased, and British trade thus benefits from this affiliation. The increasing likelihood of eventual membership in the Common Market would tie the economic future of the British increasingly with that of Western Europe, after several centuries of primarily worldwide and secondarily European interests.

Benelux Countries

Belgium, the Netherlands, and Luxembourg —the Benelux countries—have a combined area of nearly 26,000 square miles, or only slightly larger than that of the state of West Virginia. Within this area live approximately 23 million people; this is nearly half the population of France, the area of which is eight times larger. The Netherlands (1,068 people per square mile) and Belgium (803 per square mile) are unrivaled in Europe in population density. Benelux is a microcosm of urbanized Western Europe and its many accompanying problems.

The triangle of land occupied by the three Benelux countries is wedged between Germany and France. This buffer position is an advantage for peacetime trade but a disadvantage in time of war. The natural funneling of Rhine and West German traffic through the Netherlands and Belgium is of tremendous value to their respective economies. The proximity of the North Sea, whose coastal regions are traffic-generating areas, also aids Benelux commercial development. Despite limited natural resources, the industrious inhabitants have made these countries relatively significant in both agricultural and industrial production, thereby offsetting in part Dutch and Belgian loss of control over large overseas territories.

These three countries are often referred to as the Low Countries or as Benelux. The first usage is partially a misnomer, for certain interior areas, especially in Belgium and Luxembourg, are quite rugged and reach altitudes of 2,000 feet. The Benelux designation developed in the late 1940's when the governments of these countries decided to form a customs union with the ultimate goal of a free-trade area and a larger market for Benelux products. Subsequently the Benelux countries became key members in the Common Market.

Cultural Differences

The people of the Netherlands have a common culture and language; they speak the Dutch language, which developed from ancient Frankish or Germanic dialects. Thrifty, industrious, and strongly nationalistic, the Dutch have high educational standards and a literacy rate that has produced exceptional reading habits.

The Walloons and Flemings, two highly diverse peoples, comprise the inhabitants of Belgium. Generally short, dark, and French-speaking, the Walloons populate the southern industrial portions of the country. The Flemings are of a dominantly tall, fair Nordic type; they speak Flemish, a Germanic language, and are largely in agricultural and commercial occupations. The linguistic and cultural line separating these two groups is an important factor affecting most political issues in Belgium. The people of Luxembourg show both French and German influences in their language and culture.

Natural Resources

The natural resources of the Benelux countries consist largely of fertile soil and some mineral deposits, but the major economic asset is the skill of their people. Coal is the most prominent mineral reserve in Belgium and the Netherlands. The Sambre-Meuse coalfields of Belgium and the Limburg province of southern Netherlands are deep and faulted and have thin seams; mining here is difficult and expensive. The Campine field in northern Belgium, though 1,500 to 3,000 feet deep, has thicker seams and greater reserves. The Netherlands produces some 8 million tons of coal, and Belgium about 16 million tons of coal and 9.7 million tons of lignite a year. Luxembourg, which shares the Lorraine iron-ore field with France, mines about 1.7 million tons of ore annually and produces 4.1 million tons of steel.

Other Benelux mineral resources include petroleum, natural gas, and salt deposits in the Netherlands. A petroleum field in northeast Netherlands produces 2.3 million metric tons a year, or about 25 percent of the home consumption. Over one-third of Europe's natural gas is also derived from Dutch reserves.

The soil, much of it reclaimed from the sea, is a major resource in all three countries. Infertile, sandy soils have been made productive through great effort and ingenuity. Limited forests are found on rougher and sandy soils of the interior.

Regions

The sandy coastal fringe area, a narrow belt of beach backed by sand dunes, separates the interior lowlands from the sea in Belgium and the Netherlands. In northwestern Netherlands this topographical zone exists as the Frisian Islands. Some of the higher dunes are forested, and sheep graze on the grassy portions. These landforms are commonly utilized for town sites such as that of The Hague, the Dutch royal residence, de facto capital, and seat of government. The coast has generally poor harbors, but Ostende in Belgium and Hoek van Holland (the Hook of Holland) are crossing points to the British Isles. The Frisian and Zeeland islands are important for commercial fishing. The extensive excellent sandy beaches foster

Fig. 7-8 Polder area, Netherlands. In some polder areas sand covers the fertile clay soil. If not too deep, the clay is brought to the surface by special plows. (Courtesy of the Netherlands Information Service)

numerous summer resorts along the Dutch-Belgian coast.

The diked and drained polder lands, which form approximately 40 percent of the area of the Netherlands and nearly 10 percent of Belgium, would be subject to flooding at storm- or spring-tide levels if they were not protected by massive dikes along the sea and rivers. These lowlands must be constantly drained by the power pumps which have replaced the picturesque windmills formerly used for the purpose. In the Netherlands, 1.3 million acres have been reclaimed from the sea since the thirteenth century, and an additional 549,000 acres are currently being reclaimed from the shallow Zuider Zee, cut off from the sea by an enormous barrier dam erected in 1932. The completed project will add 7 percent to the country's total land area. Much of the recent reclamation, since it lies near Amsterdam and other cities, will be used for urban rather than agricultural purposes. The Delta Project, which will control the river outlets on the southwest coast, will give further protection from the sea as well as a needed source of fresh water for industry, homes, and recreation purposes.

The polders are often left as grassland, since it is uneconomical to lower the water table sufficiently to grow crops. Dairying is predominant; but on higher ground, wheat and fodder are grown. Cash crops such as seed potatoes, flax, sugar beets, vegetables, and horticultural crops are raised in the north and southwest. Flower bulbs and vegetables are found on clay and peat soils inland from the coastal sand region, especially between Haarlem and Leiden. South of The Hague, horticultural crops are extensively grown in greenhouses.

Amsterdam, Rotterdam, and Antwerp, major commercial and industrial cities of the Low Countries, are connected by smaller cities forming an almost continuous urbanized strip. The Dutch are much concerned about the growing population concentrations in what is called

"Randstadt" (or "Rim City") Holland, which extends from Utrecht to Dordrecht. Despite inland locations, rivers and canals provide various major cities with ocean connections. Antwerp and Rotterdam handle Rhine hinterland trade and transshipments; grain, iron ore, forest products, petroleum, cotton, and oilseeds are transported upstream, and iron and steel products, coal, and building stones are sent downstream. A large port facility to serve Common Market members, called Europoort, is situated at the mouth of the Rotterdam waterway. Amsterdam is the capital, the financial hub, and a diversified manufacturing center of the Netherlands.

The interior plains of eastern Netherlands and central Belgium are slightly rolling and sandy. Their infertile soils have been improved and, when heavily fertilized, nourish a variety of crops. Crops of rye and buckwheat, forest, and heath are found on the poorer soils. Dairy cattle are common livestock. Brussels, the capital of Belgium, is also the country's commercial, transportation, and industrial hub.

The interior uplands in the extreme southeast of the Netherlands, southeast Belgium, and Luxembourg vary from 300 to 2,000 feet in elevation. The Ardennes proper is a sparsely populated, dissected, infertile plateau with forests and moors, suitable for dairying; but the fertile loam soils of the foreland are densely populated and intensively utilized. The Limburg region has many orchards and intensive dairy farming.

The industry of the Sambre-Meuse Valley and of the adjacent province of Limburg in the Netherlands is based on local coal reserves. Iron and steel machinery, zinc smelting, textiles, chemicals, glass, and leather products are the principal manufactures and activities in such cities as Liège, Mons, and Charleroi.

Economic Development

The Netherlands The Netherlands is an agricultural and commercial nation whose industry is increasing in importance. About 8 percent of the labor force are employed in agriculture, 41 percent in industry and construction, 25 percent in trade and transport, and 25 percent in services. The highest per acre use of fertilizer and the highest average wheat yield per acre in the world indicate how intensive is Netherlands agriculture. Farms are small; 42 percent are less than 12.5 acres. The land use percentages are as follows: pasture, 36; arable, 30; built-up, 15; woodland, 7; horticulture, 3; and wasteland, 9 percent. Agricultural produce such as vegetables, flowers, bulbs, butter, and cheese is a major export. Cooperatives control a large percentage of the dairy output and maintain high-quality products. Leading industries include shipbuilding, refining of tin and other metals, and manufacture of textiles, chemicals, and leather goods.

Geographical location, extensive trade and a long mercantile tradition, skill, experience, and necessity have all greatly aided the commercial development of the Netherlands, including the important transit trade to and from Germany and adjoining areas. Besides agricultural commodities, Dutch exports include a wide variety of finished and semifinished goods. Imports include petroleum, coal, timber, iron and steel, and textiles. The Netherlands merchant marine, totaling over 5 million tons, ranks eleventh in the world. Lack of natural resources, rapidly increasing population, a rational development of the extensive natural gas field, and increased competition within the Common Market are major economic problems.

Belgium The economy of Belgium places greater emphasis upon industry and less upon agriculture and commerce than does that of the Netherlands. Heavy industries such as basic iron and steel manufacture, contrast with the lighter types prevalent in the Netherlands. Using iron ore from Luxembourg, France, and elsewhere, Belgium and Luxembourg together produce about 14 million tons of steel annually, about 65 percent of which is exported. The refining of zinc and copper is another leading industry, along with production of textiles, machinery, chemicals, and glass. Foodstuffs, iron ore, petroleum, and other raw materials for industry are prominent imports; metal products, chemicals, textiles, and glass are exported.

Nearly 40 percent of the industrial production must be exported to maintain a prosperous economy. German transit trade is also important to Belgium. Participation in the Common Market, the Belgian-Luxembourg Economic Union, and the Benelux Customs Union is vital to Belgium's economy.

Luxembourg The smallest of the Benelux countries, Luxembourg is a tiny principality. The majority of its people are supported by the iron and steel industry; a few inhabitants engage in forestry and agriculture. Luxembourg produces 4.5 million tons of steel annually, a total that gives it the highest per capita steel production in the world.

Economically, all three Benelux countries are heavily dependent upon the Common Market, which they hope will offset their individual small areas, lack of natural resources and substantial domestic markets, and the tariffs and trade restrictions imposed by other nations.

France

The location of France gives it a dual continental and maritime orientation. Though 1,870 miles of French coastline face the Atlantic Ocean and the Mediterranean Sea, the nation's strongest links have long been with Continental Europe. At one time, France possessed the world's second largest overseas empire, but since 1954 most of this has been lost or relinquished. To the east and south, common land boundaries exist with eight nations. Much of the international border follows the natural barrier-forming ridges of the Pyrenees, the French Alps, and the Juras, but northeast France is open to neighboring areas of dissimilar language and culture.

Population

France's population, which remained stabilized at 40 million since the middle of the nineteenth century, has now increased to 50 million. It thus leads most European countries in current rate of population growth. Immigrants from Algeria after the Algerian crisis and independence and laborers from Italy, as a result of free movement within the Common Market, helped to solve the country's post–World War II labor problem and maintain the rising level of productivity. The French population, which increased 4 million during a ten-year postwar period, gives every indication of continuing this rate of increase. Although 68 percent of the total population is urban, a rural character still persists in many areas. Only the Central Plateau and the Pyrenees, Alps, Juras, and highland areas of the northeast are sparsely populated; nevertheless, with 235 people per square mile, France has the lowest average population density among the industrial nations of Western Europe.

The French population is remarkably homogeneous, although minor differences exist in regional dialects and other cultural traits, particularly between the north and the south. Small minority groups are found along some French borders. For example, Basques are found toward the western end of the Pyrenees; and a rather large population group in Alsace-Lorraine, along the German border, is Germanic in language and culture. All but about one million of the country's people are classified at least nominally as Roman Catholics, but for many formal religious ties are not strong.

Physical Setting

The varied topography of France is relatively simple in structure. Three major landform types exist: (1) areas of old rock such as the Massif Central (Central Plateau), the hills of Brittany and Normandy, the Ardennes, and the Vosges Mountains; (2) the rugged, young, folded mountains represented by the Alps, Juras, and Pyrenees; and (3) the recent lowland deposits in river basins and valleys. Strategic corridors such as the Belfort Gap between the Rhine and Saône valleys and that of Verdun between the Rhine and the Paris Basin are among the connecting lowlands.

The varied climates of France are the result of several factors, including topographical relief, latitude, and location between the Atlantic Ocean and the Mediterranean Sea. The country's position of 42° to 51° north latitude and its location on the west coast of Europe place it in the path of the tempering prevailing westerlies and cyclonic storms. The daily and seasonal temperature ranges and the amount of summer rainfall increase toward the east, with the lessening sea influences. Paris has a blend of maritime and transitional climates. The higher mountain regions generally have heavier precipitation and lower temperatures. These diverse climates give France growing conditions suited to a variety of crops.

Natural Resources

Among the world's iron ore producers, France ranks fourth, following the United States, the Soviet Union, and Canada. Over 60 million tons are mined yearly in France: 90 percent in the Lorraine district and 10 percent in the Normandy and East Pyrenees regions. The Lorraine ore averages 30 to 35 percent metallic content but is self-fluxing. It was not usable, however, until development of the Thomas-Gilchrist process for reducing high-phosphorus-content ore. The Lorraine ore, occurring on or near the surface in an area of about 70 by 12 miles, is easily mined. Much smelting is done near the mines, since it is uneconomical to ship the low-quality ore long distances. Considerable quantities, however, move freely into Belgium, Luxembourg, and Germany as a result of the European Coal and Steel Community agreement. The Normandy ore is of higher quality, but it is less abundant and less accessible.

The coalfields of France normally supply only two-thirds of the country's needs; especially significant is the lack of coking-quality coal. Over one-half of the 54 million tons mined comes from the Sambre-Meuse field near the Belgian border; scattered fields on the fringes of the Massif Central and in the lower Loire River valley provide the remainder. The Saar coalfield was transferred to West Germany in 1957. Thin and broken seams, deep beds, and coal of poorer quality than German deposits caused difficulty in competition for the French industry. Modernization has taken place, however, to the extent that the French coal output per manshift is now among the highest in Europe.

In bauxite production, France ranks fifth in the world and first in Europe, except for the Soviet Union. Bauxite is mined on the southern flanks of the French Alps, near the Spanish border in the Pyrenees, and just west of the Rhone River near the Mediterranean coast. With hydroelectric power available nearby, much of the ore is processed into aluminum; some also is exported. Potash deposits are located near Mulhouse in Alsace. Potash is used in the commercial fertilizers vital to the intensive agriculture of Northwestern Europe. Extensive salt deposits in Lorraine are used along with potash in the chemical industry.

Petroleum discoveries in southwestern France have greatly increased domestic production. Natural gas from the Lacq field in

Northwestern and Central Europe 251

PRINCIPAL MINERAL RESOURCES
OF NORTHWESTERN AND
CENTRAL EUROPE

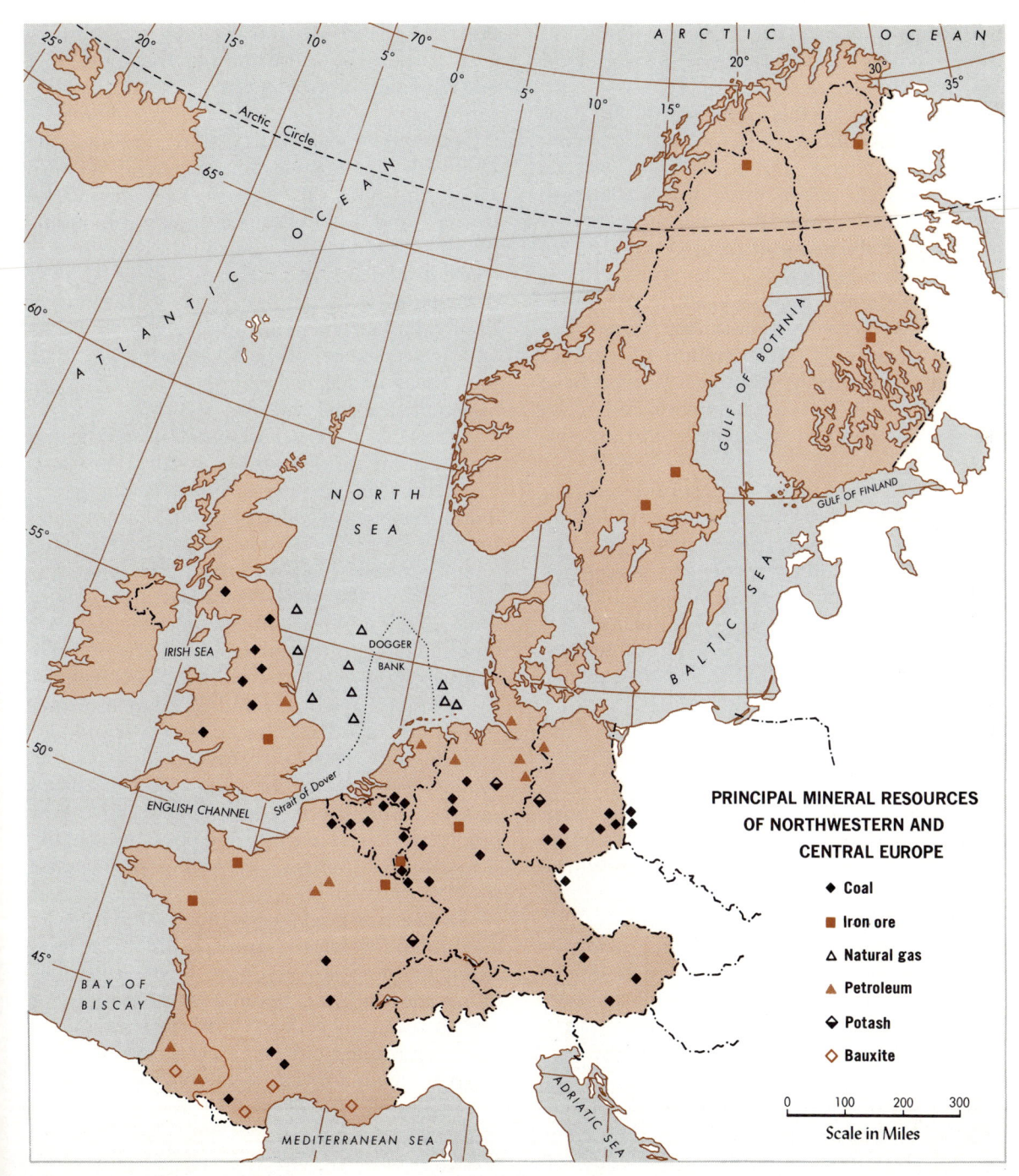

- ♦ Coal
- ■ Iron ore
- △ Natural gas
- ▲ Petroleum
- ◇ Potash
- ◇ Bauxite

Fig. 7-9 Although a variety of minerals are mined in Europe, most of the nations in west-central Europe are importers.

southwestern France has become significant. Atomic power is under active development. The waterpower resources of France are limited primarily to the highlands of the east, central, and southern districts. Of the approximately 105 billion kilowatt-hours of electric energy output, about three-fifths is hydroelectricity and two-fifths thermoelectricity.

Forest products are of primary importance in the highland economy. The Alps, Vosges, and Massif Central regions produce furniture, pit props, and construction timber. Pine trees planted to prevent the migration of sand in the Landes district of southwest France have become a major source of lumber and naval stores. Lumber and wood pulp also are imported.

The soils of France vary considerably in quality, but the country is fortunate in having large areas of arable, productive lowlands. Postwar land-use figures show 39 percent in arable cropland, 24 percent in meadow and pasture, 21 percent in forest and woodland, and 16 percent relegated to other uses. Only one-tenth of the area of France is classified as unproductive.

Economic Development

The significance of France in the European economy is illustrated by the fact that it accounts for 21 percent of the total value of agricultural output and 18 percent of the gross national product of Western Europe. Agriculture produces 7 percent of national income; manufacturing and construction, 47 percent; trade, 14 percent; transportation and communications, 5 percent; and other activities, 27 percent. Industry is handicapped by domestic raw material deficiencies (especially power), insufficient skilled labor to man and manage new industries, and slow modernization of equipment and industrial procedures. National-level economic planning is placing emphasis on modernization and decentralization of industry away from Paris, where housing, transportation, and distribution problems are critical.

Industry France ranks third, behind the United Kingdom and Germany, as a Western European industrial nation. The presence of iron ore and coal deposits is largely responsible for this position; thus, the country's industry is concentrated in the north and northeast, near the raw materials sources. This localization in French industry depends upon access to stream and hydroelectric power, transportation facilities, cheap and skilled labor, and raw materials, expecially the coal and iron ore.

The bulk of French manufacturing is found east of a curving line drawn south from Calais to Orleans to the mouth of the Rhone River. The following individual industrial areas stand out: (1) the northern area around Lille and the Sambre-Meuse coalfield—wool and linen textiles and iron and steel; (2) the northeast (Alsace and Lorraine)—metallurgy, cotton textiles, and chemicals; (3) metropolitan Paris—diverse manufactured products, with emphasis on quality items; (4) the Lyon and St-Étienne district—silk and rayon textiles and metal products; and (5) the lower Loire River valley and the vicinity of Bordeaux—shipbuilding, metal products, and diversified manufacturing. Areas other than Paris have received impetus from the dispersion of industry occurring during and since World War II.

France has a well-developed rail, highway, and inland-waterway transportation system, although there is congestion in Paris because of overdependence upon it as a focal point and distribution center. The Rhone, Saône, Loire, Garonne, and Seine rivers are all navigable and connected by canals.

Agriculture Agriculture, a vital sector of the French economy, employs about 15 percent of the labor force. Diversified agriculture predominates, but specialties based on long tradition and highly skilled farm labor have developed. Examples of specialty products are wine, cheese, fruit, and vegetables. Cash income comes largely from the sale of livestock, animal products, and certain crops. Wheat is the country's single most important crop. Olives, cork oak, irrigated fruits, and vegetables are found in the Mediterranean region; corn is peculiar to southern France. Vineyards and winemaking are widespread over the southern two-thirds of the country, with certain areas of world-renowned specialization. France is second to Italy in world production of wine, but first in consumption. The areas of cool, moist climates and rugged topography encourage livestock raising and dairying. Crop yields per acre are increasing but are generally lower than in the United Kingdom or Belgium, because of poorer soils and less-advanced agricultural methods. Farms average less than 40 acres, with about one-fourth being less than 12 acres as a result of the inheritance system. Government planning has done a great deal to increase individual farm size and further modernization, especially from the Loire Valley north.

Regions

The northern lowlands, including the Paris Basin, the Loire and Saône valleys, and the industrial north, form the heart of the French nation and economy. The Paris Basin, the most productive agricultural as well as industrial region of France, is formed by saucerlike sedimentary rock layers. The outer layers, consisting of a resistant limestone and chalk, form rugged and wooded outcrops. Transportation routes pass through breaks or gaps in the escarpment faces. Agriculture is intensive here; grains, vegetables, and fruits are the principal crops. Cattle predominate in the more humid west, and sheep in the drier, rougher east portion of the northern lowlands. Two important wine areas are found in the north: white wine in the middle Loire Valley and champagne made from grapes grown on the warm, south-facing slopes east of Paris.

Greater Paris is the cultural, artistic, and leading manufacturing center, the hub of water and rail traffic, and the leading inland or river port of France. Nearly one-seventh of France's total population lives in this metropolitan district. Industry consists largely of finishing types of manufacture. The luxury goods for which Paris is famous play a relatively minor role in the suburbs, as compared with automobile assembly and the electrical, chemical, food-processing, and printing industries. Paris thus is an excellent example of a "primate city," or a city that clearly dominates a country in size and importance in relation to the nation's other urban centers.

Lille is the principal industrial center of northern France. Its industry is powered by coal from the Sambre-Meuse field; iron- and steelmaking dominate along with manufacture of woolen and linen textiles. Rouen, near the mouth of the Seine, is a cotton-milling center and the deepwater port for Paris. Le Havre, a busy port for both transatlantic and trans-channel trade, is also important for refining imported crude oil as well as for being the starting point of a pipeline to Paris. Tourist resorts are common along the French Channel coast.

The Aquitaine Basin is drained primarily by the Garonne River and its tributaries. The Garonne Valley and the central parts of the basin grow wheat, corn, vineyards, and pastures for cattle grazing. The area has one-fourth of all the French vineyards; production of red wines is predominant here, but poor farming methods and unreliable precipitation result in low yields. Sheep grazing is a leading activity in the infertile sub-Pyrenean, northeast, and sandy Landes districts. The deepwater port of Bordeaux is the outlet for wine and many other products of the region, including those from the forests of the Landes district. The city also has refineries for petroleum, some of it from nearby areas. Toulouse, which commands the narrow gap between the Aquitaine and Mediterranean lowlands, is an important industrial center of southwestern France. There is a natural gas field nearby.

The climate of the Mediterranean lowlands in the Rhone Valley results in a distinctive agriculture. West of the Rhone River is the leading vineyard and wine-producing district of France. East of the Rhone, where the topography is rougher and soils are poorer, olive groves, vineyards, and grazing of sheep and goats are common agrarian features. Wheat and irrigated crops such as citrus fruits, rice, and vegetables are raised on the better soils. Flowers for perfumemaking are grown. The Rhone Valley has vineyards on its slopes, and crops and cattle grazing are found in the valley bottom. Mulberry trees of use to the silk industry are found near Lyon.

The large bauxite deposits form the basis of the French aluminum industry and profitable bauxite exports. Marseilles, the second most populous city in France and the first in ocean shipping, is the nation's major Mediterranean port. It also has a wide range of manufacturing activities, including the processing of tropical materials and petroleum refining. Lyon, strategically located at the confluence of the Rhone and the Saône, utilizes nearby coal deposits and waterpower for silk and rayon production as well as for automobile and other manufacturing. The Riviera coast, protected from cold north winds by the Alps, has long been a famous resort area, with extensive beaches and numerous hotels.

The Massif Central, dominating south-central France, averages 3,000 feet above sea level. This region is mostly a dissected crystalline rock plateau with some volcanic rock; sedimentary rock is found on the edges and in the lowlands. Except for a few favored valleys, generally poor soil and rigorous climate restrict agriculture to hardy crops and grazing. This is the chief rye-growing area of France and is important for cattle and sheep. The Massif is a barrier to communications and transportation within France. St-Étienne is an iron and steel and textile center.

The crest of the Pyrenees, rising to over 9,000 feet, forms the French-Spanish boundary. Their inaccessible rugged and infertile slopes are sparsely populated. The climate of the western portion is marine; that of the eastern sector is Mediterranean. Grazing of the transhumance type is dominant in the higher grasslands, with some crop agriculture practiced on the lower slopes and in the valleys. Availability of hydroelectric power, iron ore, and bauxite provides the basis of growing electrochemical and electrometallurgical industries.

The French Alps and Jura Mountains regions are thinly populated and play a minor economic role in France. Cattle and sheep are pastured on alpine grasslands. Small chemical and metallurgical centers utilize part of the large hydroelectric potential. Dairying, forestry, and small workshop industries, especially watchmaking, are predominant.

Brittany and Normandy differ from the Massif Central in that their outcrops of crystalline rocks have lower relief, no volcanic material, and a peninsular shape with numerous offshore islands. The hedgerow landscape is characteristic. Cattle and some sheep are grazed on the interior uplands. There is little land suitable for wheat; thus buckwheat is the major cereal grown on the poor soils. Apple orchards are common, and cider is the usual beverage. More than 2 million tons of iron ore are mined annually, and smelting works are located in Caen. Rennes is the regional capital.

The narrow Brittany-Normandy coastal strip is densely populated. Rich soil and a mild marine climate favor vegetable gardening.

Nantes and St-Nazaire are metallurgical and shipbuilding centers, with their industries based on Swedish and Spanish iron ore and British coal. Ports and naval bases are located also at Cherbourg, Brest, and Lorient. Several smaller villages are fishing and resort centers.

The Lorraine and Ardennes plateaus and the Vosges Mountains, which have a strategic location bordering on Germany, contain nearly one-half of the iron ore resources of Europe.

Fig. 7-10 Fishing fleet, Brittany, France. In those areas bordering or near the shallow waters of the Bay of Biscay, English Channel and North Sea fishing is of prime economic importance. (Courtesy of the French Government Tourist Office)

The iron and steel industry, concentrated in Nancy, Metz, and other smaller centers, obtains coking coal from the Ruhr. The Common Market facilitates the iron ore–coal trade between France and West Germany. The chemical industry is based on the by-products of iron and steel manufacturing and on local salt and potash deposits. The textile industry also is noteworthy. The Vosges region supports lumbering, dairying, and paper and cotton-textile manufacturing, based on available waterpower.

In the fertile Rhine Valley, a down-dropped block, are grown grapes, tobacco, hops, sugar beets, and wheat. Strasbourg, a key rail and water-transportation center on the Rhine, has served as Council of Europe headquarters since 1949.

Trade and Problems

The position of France is favorable for trade with Europe and the world. Although reserves of some minerals and other raw materials are scarce, the great technical skill of the people is reflected in the making of commodities such as clothing and fine art goods, for which the country is famous. Raw cotton, mineral fuels, hides and skins, and wool are raw material imports. Machinery, various manufactured products of agriculture, including wine and cheese, are the chief exports. Western European countries and the United States are France's major trading partners; 41 percent of French imports and 41 percent of the exports are in exchange with Common Market countries.

France faces many economic, political, and social problems. During the De Gaulle era, political stability was achieved and an intense effort was made to bring France back to world-power status. The country has regained a position of leadership in Europe, particularly within the Common Market. In fact, France has for years prevented the United Kingdom from achieving membership in that trade group.

There are significant economic weaknesses in lack of modernization of both agriculture and industry. Characteristic French individualism has slowed the pace of introducing mass-production techniques. Possession of the atom bomb has helped France in achieving world power in the postwar era. The large number of able young people who are giving France a dynamic character reflected in its industrial and commercial growth indicates that the nation will continue to play a key role in Europe. Tourism is another major factor in the economy, with such major attractions as Paris, the Riviera, the Alps, and the chateaux country.

Germany

United in 1871 after a long period of disunity, Germany has twice risen to the status of a major world power in this century and has twice lost that position in the world wars. The countries that defeated her in World War II, especially the United States, have also helped to revitalize West Germany. Recovery was remarkably rapid after 1948, and this part of Germany emerged as a major agricultural, industrial, and trading nation and a key state in postwar Europe. West Germany is cooperating in NATO, the Common Market, and other Western European economic and political plans but still faces the perils of being a major instrument in the cold war between the Communist and Western world. The status of West Berlin, for instance, and its isolated and vulnerable position within East German territory continue as unsolved issues.

The location of Germany in Europe is generally advantageous in peacetime and a liability during war. The central location, transitional between Eastern and Western Europe, offers the following advantages: (1) ease of development of trade relations with nine bordering countries; (2) facilitating of water, rail, and highway routes for east-west and north-south transit traffic; (3) easy development of waterborne trade by access to both the North and Baltic seas as well as the Rhine River outlet; (4) a transitional climate for agricultural development; (5) a central location in relation to European raw material sources, particularly for iron ore and coal; (6) access to Southeastern Europe via the Danube River and rail routes. Encirclement by possible enemies in wartime is a major disadvantage of Germany's position. Other liabilities are its boundaries with Poland and France, which have been shifted repeatedly in modern times; the fact that the Rhine outlet, vital to the industrial regions of western Germany, is controlled by the Netherlands; and the vulnerable position of the Ruhr industrial district and Berlin.

Post-World War II Changes and Problems

The postwar boundaries, though not yet official by treaty, considerably alter Germany's size, shape, and available natural resources. Major changes are the loss of all territory east of the Oder-Neisse River line. Included in that territory are some of the best agricultural land, the important Silesian coalfields, and associated industrial districts. Minor shifts were also made at the Belgian and Netherlands borders. The Saar and its important coalfields and industrial development were obtained from France by plebiscite in 1957 and were fully integrated into West Germany by 1959. The resultant postwar Germany has a more compact shape. The nation's economy is much less self-sufficient in food but still has surpluses of industrial goods, since the Germans retained over 90 percent of their prewar industrial output potential.

Germany, now separated into eastern and

western parts, has approximately 75 million people in an area about equal to that of Minnesota and Wisconsin combined. The population density in West Germany averages over 600 people per square mile; if one considers only arable area, the density reaches nearly 1,300. The densest zone of concentration is found near the contact between the highlands and plains extending from the Ruhr Valley to Dresden. About 70 percent of Germany's population is urban and 30 percent rural.

The German ethnic group is numerically the largest in Europe, excluding the Russians. About 15 million ethnic Germans and German nationals have been forced out of Eastern European countries and former German territory now occupied by Czechoslovakia, Poland, and the Soviet Union. The net result is that East and West Germany together have 5 million more people than Germany's total prewar population living on three-fourths of the country's former area. Regional differences in dialect, religion, and group character sometimes lead to antagonisms, like that between Prussians and Bavarians.

Divided Germany

After World War II, Germany was divided into four zones of occupation. The British, Americans, and French occupied about 94,700 square miles of western Germany, which became the Federal Republic of Germany in 1949. In 1955, it became a sovereign state with free and equal partnership in the Western community, including membership in such organizations as NATO and the European Common Market. Because a World War II peace treaty has not been signed, Allied forces remain in West Germany and West Berlin.

Dominantly industrial West Germany has a population of 59 million. It contains a little less than half of prewar Germany's arable land and must import one-third of its food requirements, much of which previously came from East Germany. The Ruhr Valley district mines most of Germany's hard coal and produces a large share of the 37-million-ton output of steel as well as other industrial products. Potash, salt, and limited quantities of iron ore and petroleum are found in West Germany. Located here also are the two leading German ports, Hamburg and Bremen, giving access to the North Sea and the Atlantic. Industrial production and over 70 percent of the country's trade are oriented toward the West.

Following World War II, the Soviet Union occupied about 41,700 square miles of eastern Germany. In 1949 a Soviet-sponsored puppet Communist government, called the German Democratic Republic, was established. Berlin, situated 100 miles within Soviet-controlled East Germany, has been a continuing focal point of the cold-war struggle between the Eastern European Communist bloc and the West. After millions of refugees had crossed over to the West through Berlin, this escape hatch was closed by the East Germans in 1961 by erecting a concrete wall to separate East and West Berlin.

East Germany, with over 16 million people, is nearly self-sufficient in food production. It has valuable lignite, potash, and salt deposits and large chemical and electrical industries. About 10 percent of the national income is received from agriculture and 65 percent from industry. Its production and trade are oriented toward Communist Eastern Europe. A unified Germany would benefit from a mutually complementary domestic exchange of agriculture, fuel, and industrial products.

Physical Setting

The two general German landform types, in both East and West Germany, are the glaciated lowlands in the north and the highland complex in the south. The northern lowlands, covered with recent unconsolidated glacial deposits, are generally below 600 feet in elevation. The central zone is characterized by the dissected remains of Hercynian mountains, reaching nearly 5,000 feet at their highest point. In the extreme south, the rugged Bavarian Alps and their forelands dominate. The German Alps average between 6,000 and 7,000 feet; the forelands are lower and are strewn with rock debris brought down by the Alpine streams.

Western, and especially northwestern, Germany (considered as a whole) has a marine west-coast climate. Eastern Germany has a modified continental type, and that of the southern portions reflects the altitude and varied relief of the highland areas. The major characteristics of climate for the whole country are cool summers, variable but relatively mild winters (except in highland Germany), and summer maximum precipitation. The marine climate influence lessens eastward and southward. Most of the crops raised are suited to cool, moist conditions, except the grapes, corn, and tobacco found in sheltered valleys of the south and west.

The original vegetation cover of both East and West Germany was largely a mixed coniferous and deciduous forest. Forests still cover more than one-fourth of the country, with a greater percentage in the central and southern highland regions and the poorly drained areas of the lowland north. An efficient conservation program, including replantings of cut timber, has been developed. The forests are an important resource for timber, for pulp and paper, and locally for wood carvings and toys. The acid, leached soils that prevail in Germany are generally mediocre to poor, except for the loess soils at the southern edge of the North German Plain. Most of Germany's soils are highly productive only through careful management and fertilization.

Germany has eight river systems, among which the Rhine, Elbe, Oder, and Danube are the most important. These rivers, connected by a system of canals, give Germany one of the best waterway networks in the world. Evenly distributed precipitation plus the melting snows of the Alps support a fairly steady stream flow, although at times traffic is handicapped by low water level or ice. Normally, 30 percent of the country's total freight traffic is carried on the inland waterways, and much of this is concentrated on the Rhine. The Oder is now shared by the East Germans and the Poles, with the latter controlling the outlet; the Elbe, cut by the Iron Curtain, is of limited navigational use. All the large rivers of Germany, especially those flowing through industrial districts, are greatly polluted. The Rhine has been referred to as "the sewer of Europe."

Natural Resources

Germany's greatest mineral resource is coal, of which all grades from anthracite to lignite are found. The Ruhr, or Westphalian, coalfields are the principal producers of anthracite and bituminous and of 102 million tons of lignite annually, as compared with the nearly 3 million tons of bituminous coal and 250 million tons of lignite mined in East Germany. In addition, West Germany imports quantities of coal and exports coal and coke to European neighbors deficient in that resource.

Germany has limited, widely scattered iron ore deposits and must import approximately 60 percent of its consumption, primarily from France and Sweden. The Siegenland iron deposits south of the Ruhr are the most valuable. Germany benefits from the free flow of iron ore from the French Lorraine fields within the Common Market.

On the flanks of the Harz Mountains, between the Weser and Elbe rivers, is a 100-square-mile area estimated to contain 20 billion metric tons of potash, essential to the commercial fertilizer and chemical industries.

Potash mining areas are divided by the Iron Curtain, with West Germany producing 2.7 million tons of K_2O content annually. Quantities of copper, lead, zinc, pyrites, petroleum, and salt insufficient for domestic needs are mined. Petroleum production is increasing, however, and about one-fifth of the West German needs are supplied from fields within the country. Hydroelectric power is important in the south.

Economic Development

German agriculture is favored by a relatively large amount of arable land. Much of the North German Plain, a number of river valleys and basins in the central and southern regions, and considerable areas on the lower slopes of the highlands can be cultivated.

The West German economy is one of the most prosperous and stable in Western Europe. Industrial production levels reflect the fruits of postwar recovery aid and the technical ability and industriousness of the German work force. The skills of the numerous Eastern refugees have also aided considerably. West Germany, a key to trade exchanges in Western Europe, is a cornerstone in the stability and success of the Common Market. Germany depends on quality output, competitive wage rates, and efficiency rather than mass-production techniques to meet foreign competition, although the internationally bestselling Volkswagen automobile is a product of the modern assembly line. German industry is diversified; both the heavy (basic) and light (finished and consumer goods) sectors are represented.

In West Germany, 53 percent of gross national product is provided by industry, 43 percent by services, and only 4 percent by agriculture and forestry. With an annual production of about 37 million tons of steel, West Germany ranks third in the world. Industry in East Germany is geared to the needs of the Soviet sphere, and greater emphasis is placed on agriculture in the overall economy.

German agriculture is characterized by mixed farming, small farms, low income per farm, large families providing labor, and increasing use of machinery. The inherently poor soils necessitate scientific farming and heavy fertilization for high production. Prewar Germany produced a little more than three-fourths of its food requirements; now, two-thirds of its own needs are grown in West Germany, and East Germany no longer produces and exchanges the surpluses formerly depended upon by the industrial West.

Regions

The German portion of the Central European lowlands extends from the Netherlands to the Polish border, with tongues, or lowland bays, protruding southward into the highlands in both West and East Germany. The maximum east-west extent is about 300 miles. Surface materials and relief are largely a result of continental glaciation. The previous drainage system was disrupted, thereby leaving lakes, large areas of bog and marsh, and east-west-oriented glacial valleys that facilitated the building of canals connecting the navigable rivers. The topographical differences can be summarized as follows: (1) a poorly drained outwash plain extending inland from the North Sea; (2) a sandy, heath-covered outwash area between the Elbe and Weser rivers; (3) a sandy, lake, and forest district east of the Elbe, with greater relief deriving from morainic ridges; and (4) the southern transitional zone, largely loess-covered, bordering on the highlands. Most of the plain is less than 300 feet above sea level and only rarely reaches over 600 feet.

Agricultural development reflects the regional differences mentioned above. Dairying and livestock grazing are the predominant uses of the bogs, marsh, and sandy heaths of the northwest, with crop agriculture, especially vegetables and sugar beets, practiced in artificially drained areas. To the east, crops are increasingly important, with potatoes and rye grown on the poorer soils and sugar beets and wheat on the better soils. The rougher grasslands are grazing areas. The more sparsely populated northeast normally produces food surpluses. Forestry and resort activity are significant economic contributors. The transitional zone to the south is one of the best farming areas in Germany; its comparatively dense rural population grows wheat, barley, sugar beets, and fodder crops for stall-fed cattle.

The North Sea coast is much more favorable than Germany's Baltic shores for ports and trade. Both coasts are sandy and have shallow water offshore. The North Sea ports, improved by deepened narrow estuaries, have better access to the Atlantic and world trade routes. Hamburg and Bremen, the leading German ports, are located considerable distances up the Elbe and Weser rivers respectively, and enhanced by access to rich agricultural and industrial hinterlands. Hamburg is the first port and second most populous city of Germany. Although its free port attracts considerable trade, Hamburg is handicapped by the proximity of the Iron Curtain, which bisects the productive Elbe River hinterland 20 miles farther upstream. Hamburg has engineering, shipbuilding, and other industries. Cuxhaven, Hamburg's outport, is used by large passenger vessels. Bremen, Germany's second port, does not have such an extensive hinterland but normally accounts for as much ocean traffic. Bremerhaven, Bremen's outport, is also important for commercial fishing.

Berlin, the prewar governmental, cultural, and artistic center of Germany, continues to be handicapped by its location deep in Soviet-controlled East Germany and its unsettled position. Its present population of about 3.3 million is over 1 million less than in 1939; of the total, about 2 million live in West Berlin and 1 million in East Berlin. Varied manufacturing establishments made prewar Berlin the leading German industrial and commercial center and the hub of a vast rail and waterway network. Reunification seems unlikely, and the status of isolated West Berlin remains precarious. The wall constructed between East and West Berlin in 1961 has closed off much of the movement of refugees and makes everyday relations between the two sectors of the city difficult. West Berlin produces electric machinery, clothing, and chemicals. The printing industry is much less important than it was in prewar Berlin. East Berlin, the capital of Communist East Germany, has made major economic strides in recent years.

Ruhr Industrial The huge industrial complex based on Ruhr resources is one of the outstanding manufacturing concentrations in the world and is the major single region of economic intensity in Germany and Western Europe. Over 10 million people are found in an area about the size of Delaware (which has a population of about 500,000). The Ruhr region accounts for over 90 percent of the anthracite and bituminous coal production of West Germany and for over one-half of that mined in all six Common Market countries. It contributes 70 percent of the 37-million-ton steel output of West Germany; its steel production equals that of the United Kingdom. The Ruhr Valley is a leading manufacturing district for metal products, coke and coal derivatives, machinery, chemicals, and textiles. It was the arsenal of both the Kaiser and Hitler, and is now a pivotal production area within the Common Market as well as in all Western Europe.

The following factors favor the development of the Ruhr industrial potential: (1) the most extensive deposits of excellent coking and other grades of coal in Europe (130 seams,

Fig. 7-11 Dusseldorf, Germany. Dusseldorf is one of the manufacturing centers of the Ruhr area. The Rhine River, the most heavily used river in Europe, gives easy access to the sea. (Courtesy of the German National Travel Office)

of which 57 are used, occurring in succession, with the lowest at a depth of 9,000 feet); (2) good water and rail connections for access to raw materials, expecially Lorraine and Swedish iron ore, as well as for export of finished products to foreign markets; (3) a location at the junction of the Rhine Valley and east-west rail and water routes along the northern edge of the central highlands; (4) nearby Sieg Valley iron ore deposits; and (5) abundance of skilled labor.

The Ruhr region in the south is hilly and forested, but the relatively level and fertile agricultural north provides about half of the area's food needs. Three Ruhr cities—Essen, Dortmund, and Düsseldorf—have populations of more than 650,000; 15 other cities have more than 100,000 people. Urban districts in the north specialize in coal mining and heavy industry such as iron- and steelmaking.

High-quality metal industries requiring skilled labor and only small amounts of raw materials, producing such goods as machine tools, motors, springs, and locks, are found in the hilly southern district. Located here and in cities west of the Rhine are textile and clothing industries; chemical, metal, and glass industries are scattered throughout the Ruhr region.

Central Highlands A complex of worn-down mountains and dissected plateaus, the central highlands, extends southward from the northern lowlands to the Alpine forelands. The forested or grass-covered highlands are infertile, but the basins and valleys of the region are generally intensively cultivated. Livestock is grazed on the poor soils of the rough grasslands.

The populous contact zone between the highlands and the northern lowlands is one of the most intensely developed industrial and agricultural regions of Germany. Ruhr coal and the lignite fields near Cologne and Leipzig are the major sources of power for industry. From Hannover to Dresden are a number of prominent industrial towns. Lignite, potash, and salt provide raw materials for the heavy chemical and fertilizer industries in cities on the flanks of the Harz Mountains such as Halle, Stassfurt, and Magdeburg, and others in the Leipzig area. Lignite is also used in synthetics manufacture for rubber plastics and for dyestuffs, explosives, and other materials. Textiles, scientific instruments, and watches are among the other products made in this area. Leipzig is a great commercial, rail, and publishing and printing center. Dresden, long a focus of music and art, supports most of its inhabitants through manufacturing a variety of specialty products such as scientific instruments, chinaware, and optical goods. The latter two industrial cities are in East Germany.

Uranium and other minerals are mined in the Erzgebirge (Ore Mountains) of Saxony.

Lumbering, wood carving, and toy- and clock-making are commercial activities developed in the Black and Thuringian Forests.

Bavarian Alps The northern limestone fringe of the Alpine system, the Bavarian Alps, extends through southern Germany from Lake Constance to the Austrian border. Though not so high as the main Alps, the Bavarian peaks average from 6,000 to 7,000 feet in elevation and are snow-covered for most of the year. Tourism is a major industry; the wooded slopes and mountain meadows form the basis of lumbering and pastoral pursuits.

Alpine Forelands Averaging 1,000 to 3,000 feet above sea level, the Alpine forelands slope northward to the Danube River, forming a relatively level plateau covered with coarse glacial debris. Their generally poor soils and swampy areas foster an important dairying industry. The richer soils of the Danube and tributary valleys are most intensively utilized for wheat, barley, oats, and hops and for pasturing cattle. A lumbering industry has developed on the forested slopes.

Munich, the third largest city of Germany, is a manufacturing, art, and music center located at the intersection of the Berlin-Rome and Vienna-Paris routes. Its industries include breweries and factories making textiles, electrical items, and mechanical goods. Augsburg is of regional importance; Ulm and Regensburg are Danube River ports. Considerable waterpower is available for industry.

Rhine Valley The Rhine Valley is divided into three sections; the upper rift valley, the central gorge, and the lower plains. The rift valley is a down-dropped block of rock about 20 miles wide and 185 miles long, part of which is in France. Except for swampy meadows near the river, the valley floor produces abundant crops of wheat, tobacco, and hops and has extensive fruit orchards. Vineyards and pastures occupy adjoining terraced lower slopes; the upper slopes are forested. Karlsruhe and Mannheim are strategic transportation centers in this region. Petroleum was discovered in the vicinity in 1952. The Main and Neckar valleys are tributary to the rift valley. Fertile, terraced fields and a moderate climate result in agriculture similar to that of the rift valley. Frankfurt (or Frankfurt on Main), situated on the Main River, and Nürnberg, on the banks of the Pegnitz, a tributary of the Main, are the chief cities of this region. Frankfurt, which has a major airport and extensive rail connections, is an important industrial city. Stuttgart, a noted publishing center, dominates the Neckar industrial region.

Running from just west of Mainz to Bonn is the picturesque Rhine gorge. Cut into resistant rock, the valley is barely wide enough in places for a railway and highway on each side of the river, and terraced vineyard slopes are prominent along its length. Bonn, the capital of West Germany, is the most important city here.

North of Bonn, the Rhine enters the flat European Plain. Cologne, which has access to large lignite deposits, is a diversified industrial center and an important Rhine port. Duisburg (incorporating Ruhrort) is the chief German Rhine port and the principal river port in Europe. The Rhine, an international artery, carries more traffic than any other European waterway.

Trade and Problems

About 40 percent of West Germany's trade is with Common Market countries, and another 20 percent is with the European Free Trade Association nations. West Germany exports chiefly manufactured goods: iron and steel, machinery, vehicles, chemicals, textiles, and precision equipment. Principal imports are foodstuffs, machinery, and industrial raw materials such as petroleum, iron and other ores, coal, cotton, wool, wood pulp, and paper.

East German imports include primarily agricultural products, coal, iron and steel, petroleum products, and other industrial raw materials. Chief exports are brown coal, potash, chemicals, metallic ores, machinery, and various consumer goods. Other Eastern European countries account for about three-fourths of East Germany's foreign trade. Its trade with West Berlin and West Germany is increasing but is handicapped by political considerations.

Western Europe and the United States dominate West Germany's trade; only 4 percent of its trade is with Eastern Europe and the U.S.S.R. West Germany accounts for about 10 percent of total world trade and is second only to the United States as a trading nation. With aid funds, hard work, and exceptional technical ability, the Germans reconstructed their war-torn country and again established one of the great industrial powers of the world. Large numbers of refugees and foreign workers added materially to their labor force. The West German economy is still expanding, but at a slower rate. This picture of prosperity has contrasted markedly with the state of East Germany, controlled by the Communists, where postwar economic conditions were long stagnant and political conditions caused thousands to flee to West Germany before the avenues of escape were largely closed in 1961. East Germany is now slowly emerging as an industrial power among the satellite countries, however.

Land-locked Countries

SWITZERLAND

Switzerland is one of the smallest but most consistently economically successful of all European countries. Since 1815 the landlocked Swiss have maintained a neutrality policy, and thus have avoided the war destruction and disruption that has at various times been so disastrous in much of Europe. Switzerland has 6.1 million people living within 15,944 square miles, an area slightly less than that of Denmark. Its population is concentrated largely in the Central Plateau areas, since much of Switzerland is too mountainous for settlement.

Switzerland has for centuries exemplified how peoples diverse in ethnic background, language, and religion can be welded into one nation. It has four official languages: German dialects are spoken by 70 percent of the population, who live in the central and northern districts; French is spoken by 19 percent, in the southwest; Italian is spoken by 10 percent, on the southern slopes of the Alps; and Romansh, a Latin-derived tongue, is spoken in a few southeastern districts by about 1 percent of the people. In religion, 53 percent of the Swiss are Protestant, 46 percent are Catholic, and about 1 percent belong to other faiths.

The physical resources of Switzerland are limited. Severe climate, poor soil, and much mountainous terrain all limit agricultural development to favored valleys and the Central Plateau. Small, low-quality and poorly located coal deposits, building stone, sand, and clay are the only minerals. Mountains are the dominant feature and occupy almost three-fourths of the country's total area. A prime factor in Switzerland's long-maintained neutrality, they are of major significance for defense; for grazing and dairying, partially based on Alpine grasslands; for waterpower, the nation's greatest natural resource; for channeling profit-earning traffic through Switzerland, since several major arteries of communication use the low passes in the Alps and Juras; and for tourism, attracted by their spectacular beauty and winter sports facilities.

Since about 1880, industry has emerged as

the major segment of the Swiss economy, despite an almost complete lack of raw materials. The Swiss economy is highly industrialized, with over 90 percent of the labor force employed in nonagricultural pursuits. Only Belgium and England rival Switzerland in emphasis on manufacturing. Many of the small and scattered enterprises specialize in high-quality goods such as watches and precision instruments. Swiss industries of major importance are engineering and metals, electronics, chemicals, watchmaking, and textiles. Despite handicaps of soil and terrain, however, the Swiss have developed a rather intensive agriculture. Dairying is dominant, supplemented by the growing of hay and cereal crops. Sugar beets and tobacco, along with orchard and vineyard produce, are among the cash crops.

The Alps, the Jura Mountains, and the Central Plateau are the major physiographic subdivisions of Switzerland. The rugged Alps cover 60 percent of the country, with the famed Matterhorn on the Swiss-Italian border reaching 14,780 feet. Forestry, dairying, and tourism are the major Alpine economic activities. Transhumance, the seasonal transfer of animals and herdsmen to summer mountain pastures, is practiced.

The Jura Mountains, located along the French boundary, are composed of folded parallel valleys and ridges reaching heights of 5,000 feet. Dairying and forestry are common occupations, with vineyards planted on south-facing slopes. La Chaux-de-Fonds is a center of the watchmaking industry. Basel, a Rhine port and railway hub, is known for chemical, machinery, and silk manufacturing.

The glaciated and stream-dissected Central Plateau covers 29 percent of Switzerland's area but contains 70 percent of its people. This region is the heart of Swiss economic activity. Elevation varies from 1,300 to 4,600 feet, and the climate is damp and rather severe. Agriculture is intensive, with specialization in dairying. Industry is based primarily upon available hydroelectric power. The silk and cotton textile industries are located in the northern cantons, around Zurich and St. Gallen, and manufacture of machinery, watches, and textiles is found in Geneva and Bern, the national capital.

The prosperity of Switzerland depends upon the export of much of its industrial production, as evidenced by foreign purchase of 97 percent of its watches and over 70 percent of its machinery output. Machinery, watches, chemicals, pharmaceuticals, precision instruments, and high-grade textiles are the chief exports. Watches and watch movements comprise 14 percent of the total value of exports. Leading imports are iron and steel, chemical products, heavy industrial machinery, coal, foodstuffs, automobiles, petroleum, and other industrial raw materials. Neighboring industrial countries of the Common Market and EFTA, as well as the United States, are the chief trading partners. Quality production is essential to Swiss success in the keen world competition. The nation's adverse balance of trade is offset by transit traffic on Rhine barges and Swiss railroads, tourism, foreign insurance and investments, and a strong currency.

LIECHTENSTEIN

Liechtenstein, an independent principality of 20,000 inhabitants and some 60 square miles situated on the eastern Swiss border, has an economic union with Switzerland. The majority of its people are of German origin and are Roman Catholics. About one-third of the population is foreign, and nearly half of the labor force is employed in handicrafts and light industry. The capital and chief city is Vaduz. The people pay no taxes; the ruling prince secures his income from fees of foreign corporations that are licensed by Liechtenstein and maintain headquarters here for tax advantages elsewhere, as well as from the substantial sale of postage stamps to collectors.

AUSTRIA

Neutral Austria, strategically located in the heart of Europe astride vital modern rail, water, and air routes, forms a link between the West and the East. Its 1,648-mile boundary line is shared with seven countries; along two-thirds of its borders Austria touches Communist territory. The Austrian population numbers 7,500,000; about 23 percent live in Vienna, and most of the remainder occupy lowlands such as the Danube Valley. The sparsely populated Alps occupy three-fourths of the total area. Though about 95 percent of the inhabitants speak German, cultural and other differences from the Germans are apparent. About 95 percent of the Austrians are Roman Catholics.

Austria is dominated by the Alps; only one-fourth of its area consists of plains and low hills. The climate is transitional, with temperature extremes increasing to the east and at higher altitudes. Soils are varied; the most fertile are found in the Danube Valley and in favored basins. Approximately 21 percent of the total area is arable; 28 percent is in meadow and pasture, fostering important development of grazing and dairying; and 37 percent is in forests, making Austria one of the few European exporters of timber products and paper. It ranks high in production of hydroelectric power. Important petroleum fields are located in eastern Austria. The Styrian low-quality iron ore fields are, in area, among Europe's largest. Other significant minerals are coal and magnesite.

Approximately half of Austria's gross national product comes from manufacturing and only 12 percent from agriculture. Consumer goods, machinery, chemicals, and metallurgy are the main types of manufacturing. Vienna, Linz, Graz, and Salzburg are the principal industrial centers. Industrial levels have advanced rapidly since the withdrawal of occupation forces in 1955.

Domestic agricultural production supplies about 85 percent of Austrian food requirements. Cereals, potatoes, and sugar beets are the chief crops. Dairying and pig raising are very important to the agrarian economy. There are some broader areas of fertile valley land, but the average farmer wrests a living from small, rugged farms through hard work and with limited machinery.

The severely folded Alps reach their maximum width in Austria. The highest peak of the Tirolese Alps is Grossglockner, (12,461 feet). Grazing and forestry are dominant activities on the rougher slopes, and crops such as rye, oats, potatoes, and hay are concentrated in the valleys and small basins. The Arlberg and Brenner passes accommodate important arteries of trade and traffic through Austria. Tourism is of growing importance in the Austrian economy; Salzburg and Innsbruck are the major cities for the tourist trade. Northern Austria is a forested plateau.

Limited areas in eastern and southern Austria and the narrow Danube Valley compose the lowlands. Their more favorable soil and climate encourage the growth of wheat, corn, vineyards, and garden crops, along with dairying. Lumbering is commercially important on the lower mountain slopes. The eastern lowlands contain valuable oil deposits, producing 2.7 million tons annually. Linz, Graz, and Vienna are the leading and most populous cities. Vienna presents a major problem, for it is too large a capital city for current-day Austria; this is because it developed as the seat of the sprawling Austro-Hungarian Empire. Its industry is largely finishing and processing manufactures, but it contains important service and commercial facilities as well.

Austria is one of Europe's most important

producers of electricity, and much of its output is exported to West Germany. The physical geography of the country, with its rugged mountain system and high, swift-running Alpine streams in the west, plus the energy of the Danube in the east, gives it an outstanding waterpower potential. It has been estimated that by 1975 Switzerland, Germany, and northern Italy will have developed their waterpower potential fully, but that Austria will have reached only about 50 percent of its possible capacity. With the building of storage dams to compensate seasonal variation in flow, Austria can be a major producer throughout the year. In western Austria, during the summer, water from melting glaciers can be used for the generation of power and then stored to be used again at lower elevations during the winter. Eastern Austria and adjacent countries can be supplied by hydroelectric developments along the Danube.

Austria, with its limited natural resources, must trade or die. Foodstuffs, coal and coke, raw cotton and wool, and heavy machinery are Austria's principal imports. Forest products and paper are the leading exports, along with iron and steel products, textiles, magnesite, and machinery. Trade with Eastern European countries and the U.S.S.R. decreased from over 40 percent of the total value in 1938 to about 12 percent at present. About half of Austria's trade is carried on with Common Market countries and one-fifth with those belonging to the EFTA, of which it is a member. Austria's painstakingly achieved neutrality, proclaimed permanently in 1955, also may preclude a much-desired membership or association with the European Common Market. Increased utilization of existing resources and expansion of industry, agriculture, transportation, and trade are necessary to overcome the lack of natural resources. The nation's unfavorable balance of trade must be offset by earnings from tourism and foreign investments.

In Perspective

The political expansion and economic growth of Western Europe that followed the discovery of America and the sailing routes to the Far East was the outstanding geographic development after the fifteenth century. It was then that many European governments emerged from feudalism. Discoveries of new lands and peoples led to the establishment of colonies and trade overseas. Inventions and the increased use of minerals and other resources, especially the application of power to manufacturing, later resulted in the Industrial Revolution, which caused a great growth in population density, particularly in urban areas. For centuries after the great age of exploration, the world was largely dominated by the countries of Western Europe; until World War II, most of Africa and Oceania, much of Asia, and parts of the Americas were governed from European capitals. Since 1950, however, most former European overseas possessions have become independent nations.

During these centuries the culture of Europe spread to distant lands; European languages were used around the world in trade and education; the Christian religion gained adherents abroad; and many backward peoples learned the principles of self-government. The contributions of Western Europe to the advancement of world civilization have been great. Today most European countries are prosperous, and their citizens enjoy a high standard of living. The combined productivity of the nations of Western Europe is tremendous.

Not only is Western Europe highly urbanized and the site of huge and varied manufacturing enterprises, but the area also has a noteworthy agricultural output. Wheat and other cereal grains, potatoes, sugar beets, fruits of many sorts, vineyard products, and various crop vegetables, along with dairy products and livestock such as pigs, beef cattle, and sheep, are among the substantial produce of Western European farms. Soils, rainfall, length of growing season, experienced farmers, and closeness of markets are among the factors that favor Western European agriculture. Coal, iron, and other minerals are present in abundance, and forest and fishery resources are locally in large supply.

Most important of all the factors affecting the growth of Western European industries are the technical skill and experience of the factory workmen, the efficiency of management, and the availability of investment capital. Although exports of manufacturers help to provide income for the countries of Western Europe, the large local populations constitute a big market in themselves. Organizations such as the Common Market represent attempts to take advantage of these large markets and of the diversity of physical and human resources in member countries. Europe has much that is of historic and cultural interest to visitors from abroad, and the several billion dollars that tourists spend there annually is in turn available for the purchase of foodstuffs, minerals, and other materials and goods not found or manufactured in sufficient quantity on the Continent itself. Although the Western European countries have lost most of their overseas colonial empires and other nations have developed into commercial and industrial rivals, the region as a whole remains, for its size, the most urbanized and potentially the most influential in the world. Western Europe, as a key node in the world spatial system, has links and influence throughout earth-space.

REFERENCES

Buchanan, Ronald H.: "Toward Netherlands 2000: The Dutch National Plan," *Economic Geography*, vol. 45, pp. 258–274, July, 1959.

Day, E. E. D.: "The British Sea Fishing Industry," *Geography*, vol. 54, pp. 165–180, April, 1969.

Dollfus, Jean: *Atlas of Western Europe*, Rand McNally & Company, Chicago, 1963.

Elkins, T. H.: *Germany: An Introductory Geography* (2d ed.), Frederick A. Praeger, Inc., New York, 1966.

Evans, E. Estyn: *France: A Geographical Introduction*, Frederick A. Praeger, Inc., New York, 1966.

Fleming, Douglas K.: "Coastal Steelworks in the Common Market Countries," *Geographical Review*, vol. 60, pp. 48–72, January, 1967.

Gottmann, Jean: *A Geography of Europe* (4th ed.), Holt, Rinehart and Winston, Inc., New York, 1969.

Hoffman, George, W.: *Geography of Europe* (3d ed.), The Ronald Press Company, New York, 1969.

Mayhew, Alan: "Structural Reform and the Future of West Germany," *Geographical Review*, vol. 60, pp. 54–68, January, 1970.

Rees, Henry: *The British Isles, A Regional Geography*, George G. Harrap & Co. Ltd., London, 1966.

Shackleton, Margaret R.: *Europe: A Regional Geography*, Frederick A. Praeger, Inc., New York, 1969.

Somme, Axel (ed.): *A Geography of Norden*, J. W. Cappelens Forlag, Oslo, 1960.

8

Southern Peninsular Europe

Europe is bathed along its southern shore by the Mediterranean, a long narrow sea. It was historically known as the "Middle Sea" because of its significant situation, amid the lands that were first known to Western man. In fact, its very name, Mediterranean (*medi-terra,* or *mid-terrace*), means "in the midst of the land" (or earth).

Projecting into this sea, or making the terminus along some of its edges, are the four peninsulas that make up Southern Peninsular Europe: the Iberian, Italian, Greek, and Turkish. Italy and Greece jut far out into the sea; and together with Sicily, Italy divides the Mediterranean into two great basins, eastern and western. The Iberian Peninsula and the peninsula containing Turkey in Europe mark the

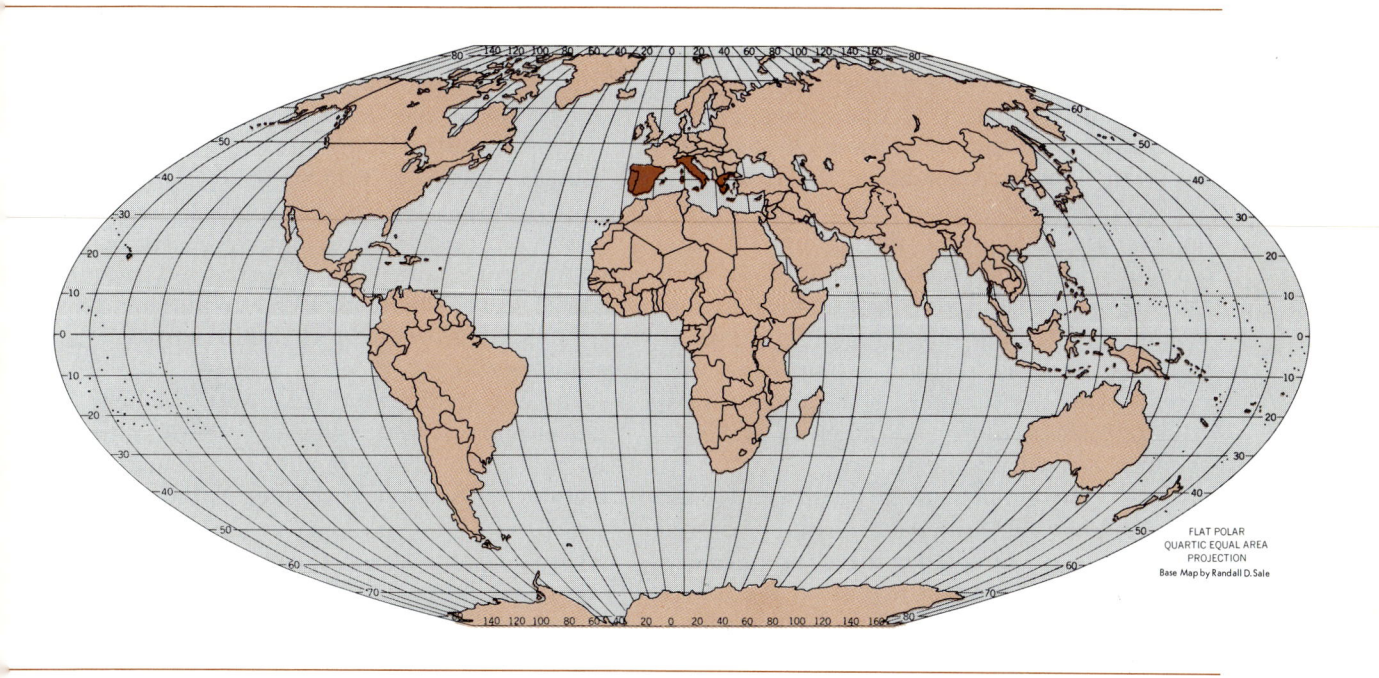

FLAT POLAR
QUARTIC EQUAL AREA
PROJECTION
Base Map by Randall D. Sale

termini of the vast "Middle Sea": the blocky bulk of Iberia all but shuts off the Mediterranean from the Atlantic Ocean, and the peninsula of European Turkey, reaching toward Anatolia, delimits the narrow water connection with the Black Sea which marks the separation of the two continental landmasses.

The four peninsulas not only separate the Mediterranean from the Atlantic Ocean and the Black Sea, and divide the "Middle Sea" into two large basins, but they also subdivide the vast land-bound sea that is the Mediterranean into several smaller basins that, in some instances, are also known as seas—limited seas within the greater one. Thus, within the western basin, the waters that fill the deep between Italy, Sicily, and Corsica-Sardina are called the Tyrrhenian Sea; eastward, between Italy and Yugoslavia-Albania, is the long narrow arm of the Adriatic; and, just south beyond the Strait of Otranto lies the Ionian; the island-strewn waters of the basin that separates Greece and Anatolia are identified as the Aegean Sea.

The protected Mediterranean basin fostered communication among its inhabitants. Continuous intercourse back and forth across the long span of the Mediterranean, from Gibraltar to Suez and the Dardanelles, and penetration into the smaller seas that lie among the land projections of Southern Europe bound the peoples of the basin together economically and led to an interchange of knowledge and ideas that gave a unified character to the lands bordering the basin—a unity evident not only physically and climatically but historically and culturally as well.

As a result, aided by likeness of landscape and climatic conditions, Mediterranean culture took on an essentially homogeneous character, and something that might be described as a Mediterranean civilization arose. In particular, the pattern of land use throughout the area developed along nearly identical lines. At once challenging man and assisting him to adapt and invent, the Mediterranean region was like an incubator in which culture was bred and nurtured. Many of man's earliest advances unfolded here: in agriculture, art, literature, religion, commerce, navigation, and law.

Progress began first at the eastern end of the basin and filtered westward as group borrowed from group. Thus, from the river valleys of Western Asia and Egypt and from the coastal lands of the eastern Mediterranean, military and commercial contacts gradually disseminated Eastern civilizations toward Gibralter. Over a period of centuries, a series of cultural and power centers arose along the inland sea progressively from east to west. All of Southern Peninsular Europe participated in the spread of ideas and cultural influence. Through several millennia, Crete (Minoan civilization), Phoenicia, Greece, Rome, Spain, and Portugal rose to eminence in succession. As one region receded, another succeeded to first place. Thus Rome overthrew the power of Greece and, at its height, reached out to conquer North Africa, Western Asia, and the barbarians of Northern Europe and Britain. The statement "All roads lead to Rome" became literally true, for Rome ranged east, north, and south to extend its political and economic influence. But eventually Rome declined; and in the Dark Ages that followed, much that had been learned in the past almost disappeared.

For nearly 2,000 years Mediterranean culture was dominant. During all that time, European civilization faced south. The ebb of this brilliant period of Mediterranean supremacy came at the beginning of the sixteenth century as control of the Oriental spice trade slipped from the hands of the Turks. By 1503, practically no spices were arriving overland at the ports of the eastern Mediterranean. They were now being carried in Portuguese vessels, by way of the ocean route around southern Africa and thence to Lisbon, to be picked up there and distributed in Europe by the ships of the Hanseatic League.

So the Mediterranean faded as the heart of power and progress, and the center of influence moved north. For a time, the bypassed Mediterranean became a backwater in the affairs of the world. Though there was a revival of Mediterranean commerce in the first half of the nineteenth century, only with the opening of the Suez Canal did that sea return to great prominence as a trade route. The nations that bordered the sea, however, remained in relative obscurity. Not until the last half century have there been indications of recovery. Nevertheless, the innovations in economics, politics, navigation, and culture originating here carried Western man from barbarism to an advanced level of civilization, and furnished the base for many of his present-day achievements.

The Mediterranean landscape presents a scene of alternating fragmented, littoral plains and projecting headlands, backed by lofty mountain ranges toward Europe and fronted by the waters of the inland sea. Add to this a prevalence of sunny, cloudless skies, with a pattern of precipitation that shows a winter maximum and comes in brief, heavy showers; paint in the vine, the olive, the citrus, the cork oak, wheat, barley, and some goats; set in proud, inventive, and industrious people with a great past, and you have the landscape, and the scene, almost in its entirety. The limits of the Mediterranean zone are quite clearly defined, for

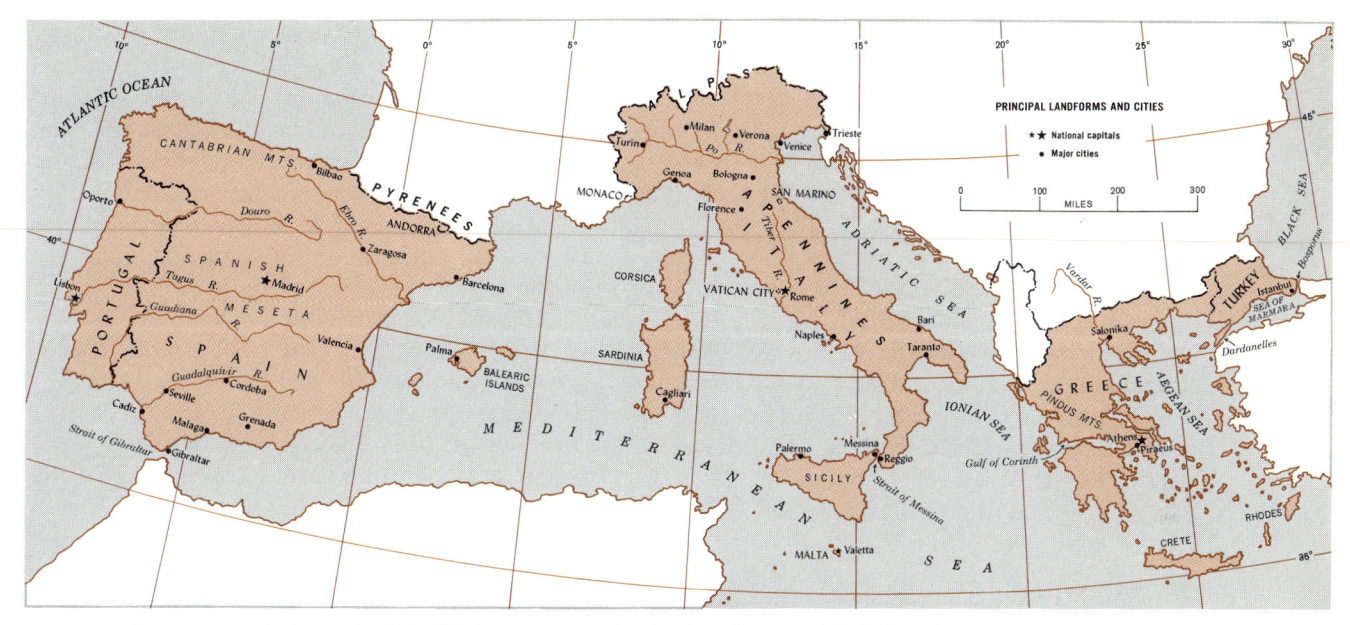

Fig. 8-1 The largest city in each of the Peninsular countries is also the capital. Italy has the greatest number of large cities among these countries, largely owing to the development of sizable industries centered in the Po Valley.

both the climate and adjustment change as the coast is left behind and distance from the sea increases.

Physical Setting

Relief Features

The Mediterranean topographic relief is characteristically one of mountains, hills, and constricted valleys; plains, and especially large plains, are a rarity. The Po Valley and narrow coastal plains in Italy, portions of coastal Portugal and Spain, and the coastal region of southern France are the most important Mediterranean lowlands. Greece, dissected by mountains and fringed with multitudes of islands, is a jumble of ranges and valleys; the region of Thessaly is the largest of the Greek lowlands. Promontories thrust into the sea, so that the coastline is an alternation of forbidding cliffs and deep indentations. Offshore islands are the tops of submerged ranges that are a continuation of the mountains of the mainland.

Climate and Vegetation

Over much of Southern Peninsular Europe, the so-called Mediterranean climate prevails. The principal areas of Southern Peninsular Europe where climates other than the Mediterranean predominate include northwestern Spain, which has a marine west-coast type; interior Spain, where semiarid steppes occur; and northern Italy, with its considerable summer rainfall and greater seasonal variation of temperature than is typical in Mediterranean lands. The climate of northern Italy is of modified continental type, rather than Mediterranean. In winter, cold winds from Northeastern Europe sometimes blow through the gaps that pierce the mountain barrier north of the Mediterranean. The minstral of the Rhone Valley and the bora of the Adriatic and Aegean seas are examples of such northerly winds.

The Mediterranean climate has two seasons: a cool winter with rain, which results from the in-movement of the westerly wind belt and its associated cyclonic storms; and a hot, dry summer. Rarely does the temperature drop to freezing, for mountain barriers shut out most of the cold winds from the continental interior. The proportion of sunny days is high, even in winter. Summer is a period of drought, broken only occasionally by convectional showers, as well as of brilliant sunshine, hot south winds, and, in some places, of dust storms that blow across the sea from the Sahara. The dry season parches the lands so that grass and other vegetation becomes brown and scorched-looking. Fields are bare and often are plowed in readiness for the planting that takes place with the onset of the wet season.

Precipitation, averaging between 15 and 35 inches, varies considerably from place to place and from year to year but is dependable for the growing of winter crops. Only by careful water conservation, however, are the people of the Mediterranean able to maintain a supply that satisfies even the minimum requirements. Irrigation supplements rainfall in many vineyards, orchards, and gardens and is essential where rice is grown.

Mediterranean plant life reflects a high degree of adaptation to these climatic conditions. It is drought-resistant, and species such as maqui and thorned flora—cacti and thorny shrubs—are found in great number and variety. Deep-rooted plants, like the grape, fig, and olive, penetrate their taproots to the level of permanent groundwater for moisture sustenance. In fact, the olive is a classic example of the type of vegetation known as Mediterranean and is found only within Mediterranean regions. Another plant indigenous to and found only in Mediterranean regions is the cork oak. Its thick, pulpy bark acts as an insulator to reduce evaporation of moisture from its surface.

Because of the extended dry season, pasture is scarce; hence the goat, sheep, and donkey are the characteristic animals of the Mediterranean. Only nimble-footed creatures that thrive on poor forage can scramble among the rough scrubby lands and survive. Goats, the best users of poor forage of any of the world's domesticated animals (except perhaps the camel of the deserts), even climb trees to secure enough to eat. They are, therefore, extremely destructive of the natural vegetation.

Cultural Landscape

The two physical factors of topography and climate have acted to shape Mediterranean life throughout the ages. Here man has more completely exhausted the natural environmental possibilities offered him than in most regions. Three types of agriculture are carried on side by side: the growing of winter grains using natural precipitation, terracing and the raising of tree crops whose deep roots reach down to subsoil water and survive through the drought period, and irrigation agriculture.

Wheat and barley are the two Mediterranean grains par excellence. Wheat takes up the better-watered and more fertile soil; barley is planted on the marginal lands where wheat will not do so well. Grains are planted in the fall, grow under the moisture of the winter rains, and are harvested in the spring. Even then, in this land of high evaporation, dry farming must be employed to make the most of the uncertain and scanty precipitation. Under this method, fields are planted in alternate years; during the fallow year they are plowed and frequently reworked to keep the soil loose and free from cracks and weeds, so that the rains can penetrate and moisture can be stored. By this technique, two years of precipitation is used to grow one crop of grain.

Since the amount of plains area in the Mediterranean Basin is definitely limited, even intensive use of the lowlands is insufficient to provide a livelihood for the large population; therefore, hillsides are cultivated. To combat erosion, the slopes are terraced, often from bottom to top. A succession of embankments, made on the hillsides with broken stones and filled with soil, frequently carried up from the lowlands, are planted with vines; fig trees and gray-green olive trees fill other terraced slopes. These crops grow without irrigation.

In a land that is characteristically a nondairy region, the oil of the olive replaces butter and other cooking fats in the diet of the people, and also provides a skin lotion against the intense summer heat and sunshine. The vine supplies not only fruit for eating but also the major drink. Vineyards frequently occupy lowlands as well as hill slopes. On the plains, in order to keep the plants off the ground, the grapevines are generally trained on trellises or supports that extend between trees; between the widely spaced rows of trees and trellis, there are frequently plantings of other crops. Land is intensively used, and no amount of hand labor is considered too great to assure a good crop yield.

The grains, wheat and barley, depend on the winter rains. The orchard crops of olives, figs, grapes, and even citrus fruits at times exist through the drought of summer. Other crops such as rice, vegetables, small fruits, and, in times of great and prolonged drought, the citrus fruits require irrigation. Wet rice, a somewhat unusual crop in Mediterranean lands, is grown in a few selected places, for example, the river valleys of the Po, Ebro, and Tagus (Tejo), the Spanish coastal plain near Valencia, and on the reclaimed salt flats of Greece. Vercelli, Italy, is the major rice market of Europe.

Because the environmental conditions are meager for the large population that is found everywhere in the Mediterranean Basin, and also because the neighboring sea invites, fishing and commerce are principal occupations. Fish ranks high in the diet of Mediterranean peoples. Large quantities of dried fish are imported to this area from the North Atlantic fishing countries, because the amount caught locally is too small to meet the demand.

Fig. 8-2 Southern Apennines in Calabria, Italy. The rugged Apennines extend lengthwise through the Italian Peninsula. The demand for land in this densely populated country makes it necessary to terrace all slopes where cultivation is practical. (Courtesy of Italian Information Center)

Iberian Peninsula: Spain and Portugal

No explanation of the political partition of the Iberian Peninsula into the two nations of Spain and Portugal can be found in the ethnographic composition of these two countries. Their peoples are closely similar, for all the populace of Iberia is of very old Mediterranean stock; and though this unity has been modified as a result of invasion and colonization by Vandals, Greeks, Goths, Moors from North Africa, and other aliens, the basic character of the people has been little changed.

Because the peninsula lies between Europe and Africa, it has long been a roadway between the two continents. The Iberian Peninsula has been, and seems physically, almost as much a part of Africa as of Europe. The Pyrenees are a formidable barrier, in later centuries even more formidable than are the water narrows that separate the continents. Iberia was long the battleground of Islam and Christianity, with Europeans and Moors seesawing back and forth across it for many centuries—all a part of the "turbulence of ethnic commingling" with which all Europe seethed for a millennium. The wars subsided earlier in Portugal than in most other parts of the continent. Welded together by more than six centuries of struggle against the foreign Moorish civilization inflicted upon them by the Moslem invaders from North Africa, the people of this compact little land were early bound into a social and political unity. Two hundred and fifty years before neighboring Spain was liberated and unified, the African Moors had been ejected from Portugal, and the area had emerged as a national state.

While the Portuguese navigators were slowly following the outline of coastal Africa in a search for routes to the Orient, the Spanish were undergoing experiences like those which their western neighbor had already had in creating a nation. The original success of the Moorish invasion lay less in the organization or strength of the invaders than in the social disorganization and regional conflicts among the people of Spain.

The early division of Iberia into two sovereign and distinct nations can be explained, then, only in historical terms. But once the division had been made, geographic factors tended to perpetuate it. Separate action by Portugal and its eventual independence arose, in part, as a result of its physical seclusion on the coastal edge of the tableland, for it is separated from the rest of the peninsula by the rough mountain country that features the plateau margin of western Spain. Even within Spain itself, according to Isaiah Bowman, "patriotism is a local thing—reflecting the geographical division of the country; a man says he is a Galician, an Asturian, a Castilian, an Andalusian; he rarely thinks of himself as a Spaniard." Dialects, and even the languages, differ from one locale to another, and Spaniard is separated from Spaniard by physical barriers, by language, by custom, and by social class. Only the forced collaboration against the Moorish invaders of the half-dozen little Iberian realms, of which Castile and Aragon were the most important, brought about Spain's coalescence into a nation, a unity that has endured precariously to the present.

Portugal slopes to the sea. The value of its coastal situation is enhanced by the fact that the lower reaches of the Douro (Duero) and Tagus rivers are navigable. The Portuguese are, as a result, naturally sea-minded; historically, they were among the first great navigators of Europe, exploring the coastline of Africa in their search for a sea route to the "lands of spice." Fishing currently holds an important place in the economy of the country; regardless of its size, every coastal village along the Portuguese shores is a fishing port. However, less than 1 percent of the world's deep-sea fish are caught by Portugal, and Portugal's present interest in oceanic trade is negligible.

Spain instead, largely plateau, is oriented landward. Though it formerly had a brilliant career as a seafaring, exploring, and colonizing nation, the period of conquest was episodic and relatively short-lived in its history.

After the crushing defeat by the English of her "invincible" Armada (1588), which meant near-financial ruin, and, even more, after the falling away of most of her overseas empire in the nineteenth century, Spain practically withdrew from the sea and turned inward on herself. The potential of modern-day Spain as a sea power is weak. Although half of the country's population live near the coasts, the fine harbors of northwest Spain on the Bay of Biscay are not easily accessible from the interior, and much of the rugged coastline elsewhere offers meager opportunity for modern harbor development. The Castilians, plateau dwellers, show little aptitude for the sea. Recently, however, there has been some revival of interest in sea trade; at present, the Spanish merchant marine accounts for about 1.9 percent of the world's total tonnage.

Relief Features

The Iberian Peninsula is a bulky block of land, the greater part of which is made up of an ancient massif, an eroded stump of old folded mountains whose margins, formed by faulting, are steep and straight. It is tipped higher in the east than in the west, so that the longest rivers—the Douro, the Tagus, the Guadiana, and the Guadalquivir—rise near the eastern margin of the block but cross the plateau westward and empty into the Atlantic Ocean. Faulting also occurred on the interior of the tableland, where huge masses of rock were upthrust along the fault lines to form mountain ranges. The greatest of these, the Sierra de Guadarrama, extend centrally across the plateau from Portugal to Aragon. Faulting also explains the occurrence of the Cantabrian Mountains near the northern edge and the Toledo Range to the south of the Sierra de Guadarrama.

The plateau is frequently referred to as the Spanish Meseta. It stands high, over 2,000 feet above sea level, and its surface has something of the character of a plain with mountains standing upon it. Although great areas are extremely flat, the Meseta is not featureless topographically. North and south of the central range are two large basins known, respectively, as Old and New Castile. The surfaces of these basins are covered with sedimentary deposits that are younger than are the rocks of the massif itself; these sediments evidently represent the deposits of interior drainage. The pattern of interior drainage later changed as rivers eroded through to the coast, cutting deep valleys into the extensive upland and dissecting the plateau into a series of broad level tablelands alternating with wide valleys, terraced and steplike. Among the sediments of the central basins, not only gravel and clay are common but also gypsum and salt. Their presence indicates the arid to semiarid character of the central Meseta in ancient times; even now, it is extremely dry.

Young, folded mountains were crushed against the northern, northeastern, and southern edges of the plateau to form impressive highlands. In the south, the highest of these, the Sierra Nevada, mount to 11,660 feet and tower abruptly above the plains of the Guadalquivir. In the north, the plateau is partially rimmed by the Cantabrians and the Pyrenees, which rise almost as high as the southern ranges. Though not so lofty as the Alps, these northern mountains are more difficult barriers to communication than are the mountains that separate Switzerland and Italy.

Between the folded structures and the plateau are wedgelike lowlands. One of these, the valley of the Ebro River, is separated from the Mediterranean Sea by the Catalonian coastal ranges; the other, the plain of Andalusia, drained by the Rio Guadalquivir, opens into the Atlantic through the Gulf of Cádiz. Narrow plains are found along the coasts of the peninsula and are widest to the west in Portugal and along the Mediterranean shore of eastern Spain.

Most of the fertile land of Iberia lies in these lowlands.

Climate and Vegetation

Three types of climate are found in Iberia: marine, Mediterranean, and continental. Although the peninsula is relatively small and its geographical situation seemingly favorable, maritime influences are singularly lacking in the interior, thus making climatic conditions unexpectedly severe for the region's latitude and location. Because of the latitude, the Mediterranean pattern of a dry summer and a winter with precipitation holds true. During the winter months, however, the Meseta is dry as compared with most other parts of Spain and Portugal, because of local high pressures that form over the plateau. Madrid, most centrally located, averages less than 17 inches of precipitation annually; other large area, covering about one-third of Spain and including portions of the Ebro Valley in the northeast, receive only 12 to 16 inches annually. Temperatures likewise are more extreme on the plateau than elsewhere; Madrid, at an elevation of 2,000 feet, averages 40°F in January, the coldest month, and 77°F in July.

The drought that prevails over the interior is reflected in salt flats, treeless steppes, and grasslands suited to the raising of sheep. In both Old and New Castile, large areas are adapted to the growing of winter wheat; whereas the valleys in the south support Mediterranean vegetation. In the western portion of the plateau, on the borderlands between Spain and Portugal, greater humidity and milder temperatures favor the growth of the cork oak, and the region is an important producer of this forest product.

There are two sets of climatic contrast in Iberia: one between the interior and coastal areas, and the other between the east and west coasts. The most marked distinction is found between the northwest (Galicia) and the southeast. In the former area, as well as in the mountainous north, a typically marine climate exists, with year-round precipitation showing a winter maximum and with mild temperatures throughout the year. Mean temperatures range from a low of 45°F for the coldest month to not above 70°F for the warmest month. Precipitation varies from 30 to over 60 inches annually. The results are a rich forest vegetation, interspersed with grasslands, and a prosperous agriculture, producing fruits as well as grains. Climatically this is the best part of Spain, and though topographically less favorable than the flatter plateaus and lowlands, it is economically one of the most progressive. Unfortunately, the areal extent of this marine region is small.

Southward along the Atlantic coast from Galicia, the westerly winds continue to moderate temperatures and bring sufficient precipitation, so that an overview of Portugal presents a land with green and lush vegetation. The dry summer season prevails, but the period of drought is short; rainfall reaches nearly the amounts that are typical in the lowlands of northwest Spain. For example, Lisbon has more than 29 inches of rain annually.

In southern and eastern Spain, a true Mediterranean climate prevails. Bare hills, maqui-like vegetation, and dry, brown fields characterize these lowlands in summer. Only the gray-green of the drought-resistant olive and almond and the green of irrigated crops such as rice, citrus, and sugarcane breaks the barren monotony produced by lack of rain.

Agricultural Land Use

Though 45 percent of the Iberian Peninsula is cultivated and another 40 percent is grazed, for several reasons neither Spain nor Portugal is self-sufficient in food. Methods of cultivation continue to be largely outmoded. Crude wooden plows prepare the soil for planting; seeds are hand-sown; and grains are still harvested with hand sickles. The threshing floor, where grains are trodden from the stalks by animals, recalls the methods used in Medi-

LAND USE IN SPAIN AND PORTUGAL

- Commercial livestock and crop farming
- Subsistence livestock and crop farming
- Non-agricultural area
- Mediterranean agriculture
- Dairy farming

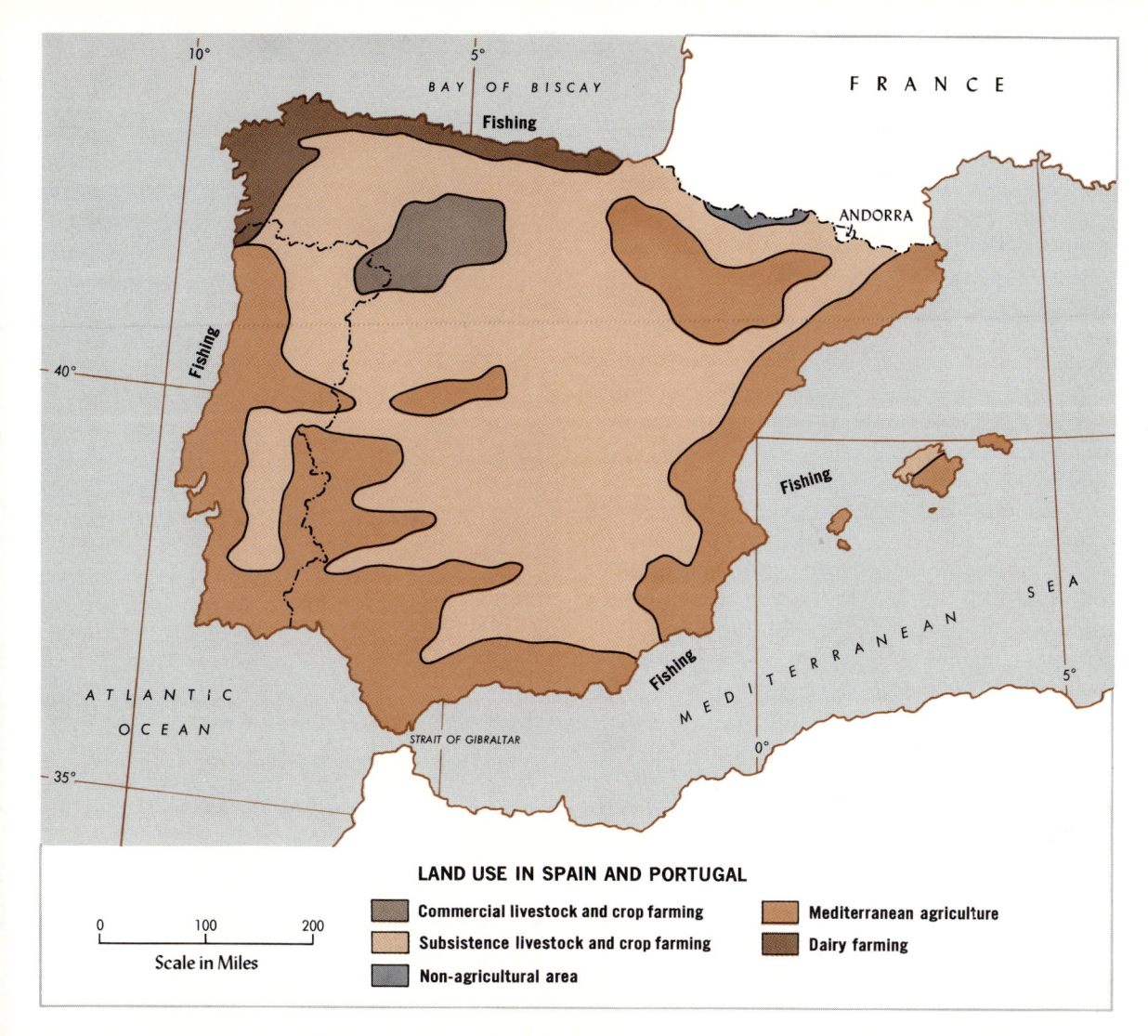

Fig. 8-3 Mediterranean-type agriculture is found along much of the coastal area. Subsistence livestock and crop farming dominate the higher interior lands.

terranean lands during Biblical times. Most of the farmers of Iberia till infertile, semiarid soil, moistened by rainfall which is erratic in amount and which at times may vary as much as 50 percent from one year to the next. Inability to depend upon precipitation means that anticipated yields can never be counted upon. Also, inadequacy of fertilizer keeps the gross output unnecessarily low. The pressure of people upon the land, not apparent from statistics, is in reality great. Population density per square mile averages from 165 in Spain to 265 in Portugal; but for the present level of Iberian agricultural technology, population pressure is excessive.

Livestock raising is important. In the humid marine north, dairy and beef cattle feed in the mountain pastures; bulls for the arena are carefully bred; besides those kept in pens, droves of pigs roam about the oak forests of the western plateau, fattening on the acorns; several million goats scramble among the maqui in the rough and scrubby landscape of the Mediterranean sections; oxen, and even cows, as well as horses, donkeys, and mules, draw carts and work in the fields. As in all Mediterranean lands, the donkey, always seemingly burdened beyond his size and strength, is one of the most familiar sights. Although cultivation is infringing on their grazing grounds, millions of sheep are raised throughout the peninsula. These are mainly of the Merino type, and their wool is of superior quality.

Northern and northwestern Iberia is a wheat-corn country. The humid climate makes possible the growing of maize, not found elsewhere on the peninsula. Apple trees are scattered about the pastures, and potatoes and rye are grown. The Meseta is a sheep and wheat country, with a scattering of other crops. Wheat is grown in all parts of the plateau, but most extensively on the Meseta to the north of the dividing ranges. Although it is the principal agricultural crop of both Spain and Portugal, yields per acre are low, compared with those of most other European countries. Barley is grown in the more marginal areas of the upland to the east of the major wheatlands and in the south, where better-than-average grazing conditions exist but where soil and rainfall are not suited to growing the preferred grain, wheat.

The cultivation of the Mediterranean sections of the Iberian Peninsula approaches horticulture. Rice, grown under irrigation on some of the rich river lands, is a specialized crop. Though the rice fields are limited in area, the yields per acre are high. The olive tree is almost coextensive with Mediterranean Iberia. Only in the north, northwest, and parts of the Meseta where rainfall and altitude make the climate unsuitable does the olive tree disappear. Most of the fruit is pressed for its oil. In Spain, however, a considerable quantity of olives are pickled; table olives are the leading Spanish export to the United States. In Portugal a large share of the olive oil produced is used by the sardine-canning industry.

Fig. 8-4 Olive orchard, Portugal. The olive is a product common to all Mediterranean lands. Olives are an important export item, as well as a source of oil and fat for local diets. (Courtesy of Sni-Yan)

Although grapes are grown over much of Spain and Portugal, the cultivation of the vine is especially important along the sheltered sides of the river valleys, where the vineyards are on terraces that cover the slopes. Vine cultivation is the most important of all farming activities in Portugal. Certain Iberian localities have become famous for the products of their vines. Among these are the port wines of the Portugal's Douro Valley, which are named after Oporto, the city located at the mouth of the river; the sherry of Andalusia; and the table grapes of Almería. Wine is a leading export of both Spain and Portugal.

Portugal, the world's foremost producer of cork, normally supplies about 50 percent of the cork exports moving into international trade. The major cork-producing area in Iberia is along the western edge of the plateau, in the border region between the two countries.

Fisheries

Fishing nets wreathe the Atlantic coast of Iberia, for the coastal waters off Western Europe are alive with fish. One of the most important sources of national wealth, the fishing industry of Portugal takes in three main fields of activity: (1) the coastal fishery, from which sardines are by far the most important variety taken, also includes catches of tunny, anchovies, mackeral, and chinchards; (2) trawl fishing on the high seas for such species as whiting, pargo, and sea bream is engaged in mainly off the west coast of Africa; and (3) cod fishing, involving a large modern fleet of Portuguese schooners and trawlers, is conducted on the Newfoundland Grand Banks and off the west coast of Greenland. Of lesser importance is the whaling carried on off the coast south of Lisbon and near the Azores.

Fishing is vital to Portugal as a source both of food and of raw material for a large canning industry. Tinned sardines and tunny are important exports from the country. The little fishing villages that send their men out to sea are no less picturesque than the hardy fishermen themselves, colorful in their traditional wool caps and striped sweaters. Their sailing galleys resemble those of the early Phoenicians.

The Basques and Galicians of northern Spain, famous for their fishing ability, are the only real seafarers of that country. A spare, mountainous land with many sheltering harbors adjoined by a sea rich in fish encouraged these northerners to take to the water. In fair weather or foul, Basque and Galician fishermen set out in their ketch-rigged boats to take the sardines, tunny, and other fish in the coastal waters; oceangoing vessels, trawlers, cod-fishing craft, and even whalers work the more distant fishing grounds.

Despite the relatively greater importance of fishing to the Portuguese economy than to that of Spain, the Spanish fishing fleet is larger, and the Spanish catch is nearly twice that of Portugal: 1,430,000 as against 506,000 metric tons. The Spanish commercial fishing fleet comprises some 11,500 boats operating in coastal waters; over 1,325 oceangoing vessels and 130 trawlers and cod-fishing ships go out beyond the territorial waters of Spain into the deep-sea fishing grounds and to the major Atlantic fishing banks. Renovation of the Spanish fleet was undertaken as early as 1961, with major effort being put into building the boats stronger and capable of heavier operations. Portugal also is investing in the improvement of her fishing fleet. Under the current investment-development plan, over 64 million dollars is expected to be devoted to the project.

Minerals

Iberia has notable mineral resources, most of them concentrated in Spain. The famous Río Tinto copper deposits of the southwest (near Huelva) are known to have been mined since 1240 B.C. Between Santander and Bilbao is located one of Europe's important iron ore reserves; exported in considerable quantity, this ore is also the basis for a small but growing iron and steel industry in Spain. Mineral exploration in 1967 led to the discovery of another significant deposit of iron ore in the southwest, in the Huelva-Sentla-Badajoz region. Mercury is secured from the Almadén mines of south-central Spain; smaller deposits occur in several other regions. The Almadén ores are the richest in the world, and Spain is the world's leading producer of this rarely occurring mineral.

Coal deposits occur in northwest Spain. Unfortunately, the principal fields mine only a low-grade coal unsuitable for coking; thus the steel industry of Spain depends on imported fuel. Gradually, however, Spain is attempting to replace coal with oil and electricity as a source of power and energy. Some petroleum is found in Burgos Province, but reserves are small, estimated at about 20 million tons, with an annual production capacity of about 1 million tons. Spain needs to import some 20 million additional tons of oil annually. Most of the domestic oil production is consumed by oil refineries, because most of the imported petroleum is crude or only partly refined.

Mining is also a traditional activity in Portugal, although the exploitation of mineral resources is relatively unimportant to the nation's economy and exploration and mapping of deposits is far from complete. Portugal does, however, mine small quantities of coal, iron, kaolin, wolframite, tin, and manganese.

Industry

Neither Spain nor Portugal can be termed industrial countries. In Portugal, manufacturing is almost entirely concerned with processing farm crops or canning fish. In Spain, despite the variety and substantial quantity of mineral reserves, until the last decade industry was still a minor activity. After the Civil War of the 1930s Spanish industry was largely reorganized, with the National Institute of Industry (INI) taking dominant control in many areas. Although production rose during the 1950s, most investment was put into state-controlled projects; private industry had difficulty obtaining money for modernization and expansion. After 1959, official policy toward private business became more tolerant. Foreign investment was encouraged, and Spanish firms were able to obtain not only needed financial but also technological assistance.

With the inauguration of the first of Spain's "Social Development Plans" in 1964, the government began a restructuring of the industrial complex of the country, with the result that manufacturing output rose by 28 percent between 1965 and 1967. This restructuring represented a regional reorganization of industry into what are called "development poles." A few carefully selected regional "poles" were designated, and types of developments were decided upon on the basis of the resources that the geographical area offered, what was already developed, and the needs of the region and the country as a whole. It is, in other words, an enormous nationwide program of regional planning. In the first year of the plan, seven development poles were designated, with the intent that emphasis in industrial development would be centered on those areas for the next five years, especially in and around the cities chosen as the focal points of the individual regions. The plan and the designation of poles was made flexible so that, as the program proceeds, certain poles (having attained the objectives set up) are dropped or are subdivided and organized into smaller regions that present problems requiring separate attention or reveal themselves to be of sufficient importance to be treated separately as "poles" of development. Thus, a special pole was set up for Campo de Gibraltar in late summer of 1969 to take care of conditions that resulted from the closing of the frontier between the British colony of Gibraltar and Spain, with the concurrent withdrawal by Spain of nearly 5,000 Spanish workers who had till then crossed this frontier daily to work for the British.

The first seven poles of industrial development designated were: (1) Madrid, the capital, serving central and western Spain; (2) Barcelona, major industrial city and port in eastern Spain, for the northeast provinces comprising the region of Catalonia (Cataluña); (3) Zaragoza (Saragossa), also in the northeast but more centrally situated west of Catalonia; (4) Valencia, serving as nodal center of the central Mediterranean coastal area, the major region of Spanish citrus production; (5) the northern coastal region of heavy industry, with Bilbao as the focal point; (6) Andalusia, embracing the southern provinces, with Seville as the center of activity (this region of development also includes the basin of the navigable Guadalquivir River); (7) Galicia in the northwest, with La Coruña and Vigo serving as the principal centers and ports within the development pole. Secondary centers of development ("city centers") within some of the poles were also designated. These were Valladolid and Badajoz in the northwest; Córdoba, Granada, Málaga, Cadiz, and Huelva in the southern provinces; and Burgos, Oviedo (Asturias), and Pamplona in the north-central region.

Early in 1969, four additional development poles were set up: Granada, Córdoba, Oviedo, and Logrono. Two of these are in the southern provinces, and two are in the northern district of heavy industry. Each has a scheduled development period of five years, which can be extended another five years if conditions demand it.

Implementation of the plan moved slowly at first, particularly in the private sector of the economy, which was actually characterized by a decline in investment during the two years preceding 1968. Since late 1968, however, there has been an acceleration and expansion of investment and an upswing in the general level of economic activity.

There are three areas of industrial concentration in Spain: Barcelona, the center of a textile industry originally based on wool from the Meseta; a belt of heavy industry centers in and around Bilbao; and Madrid, the center of a region of diversified manufacturing. Compared with the manufactures of Northwest Europe, Spanish industrial production is still small. Copper refining, food processing, some shipbuilding at Bilbao and Barcelona, cotton textiles, a little silk manufacturing, leatherwork, paper, cement, wine, and handicrafts more or less complete the list of Spanish industry and manufactures. A recent development has been that of the chemical industries, particularly the production of nitrogenous fertilizer and superphosphates. Spain has latent possibilities for manufacturing that, under the regional plans, the country is attempting to develop. These include not only the natural resources noted but also waterpower, for which the country is estimated to have a greater potential than either Switzerland or Sweden.

Like Spain, Portugal inaugurated a series of development plans aimed at attacking a backwardness that has marked and still characterizes the Portuguese economy and social order. The first plan, for the 1953–1958 period, called for the outlay of $350 million for industrial improvement, mainly for development of electric power, transportation, and communications. The second five-year plan (1959–1964) raised the investment figure to nearly 1 billion dollars and set a goal for the establishment of new industries. An interim plan, for the years 1965–1967, aimed at and succeeded in increasing the GNP by 6.5 percent. Although the increase in GNP was not maintained in the following two years (4.5 and 4.0 percent, respectively), Portugal continued to channel about 18 percent of its GNP into new industries.

Infrastructure and industrial investments have been given priority in Portugal. Power installations at Gondomar (just outside Oporto) and along the Douro River helped to increase the production of electric power by 300 percent between 1955 and 1965; an 80-million-dollar bridge was built across the Tagus River to connect Lisbon with the industrial area on Setúbal Peninsula; and the irrigation system was extended. Between 1958 and 1965, private industry also increased, by about 80 percent.

Nearly one-fourth of the investment al-

Fig. 8-5 Praia da Rocha, Portugal. Located about 200 miles south of Lisbon, this area is becoming a center for the tourist industry. Tourism is one of the chief sources of income for all Mediterranean countries. (Courtesy of Trans World Airlines)

lotted to agriculture is being diverted into the livestock and dairy industry. Portugal imports an average of 12 to 15 million dollars worth of meat and dairy products a year. The same lack is true of feed grains; hence, the emphasis upon livestock and forage-crop improvement. High-quality breeding stock—dairy cattle and meat producers (beef cattle and swine)—have been imported since 1967, when 2,000 head of Hereford beef cattle were purchased from the United States. The improvement of feed crops is being accorded the same high priority that improvement of animal strains is given because, to improve animal strains and keep the quality high, proper forage must be available.

Although more than one-third of the Portuguese population are farmers, agriculture produces only 19 percent of the GNP. The imbalance in the Portuguese economy is further indicated in the fact that 21 percent of the total export and 33 percent of the import value is agricultural. The successive governmental economic plans expect to even out such imbalances. However, with the rising competition of industry, agriculture faces an increasing loss of farmworkers, increasing wages, and mounting farm costs. It is felt that the only solution is to assure greater farm productivity per acre and per man, animal, and machine.

Part of the agricultural policy of Portugal is designed to "encourage and protect" production in the Overseas Provinces in Africa (Angola, Mozambique, and Portuguese Guinea), which supply the mother country with large quantities of cotton fiber, tobacco, corn, sugar, and edible oils. They also comprise the largest market for the wines, canned fish, and some of the textiles manufactured or processed in Portugal.

Food processing, including fish canning, sugar refining, flour milling, and production of olive oil, is an important part of Portuguese manufacture, although textiles are the dominant sector, employing 40 percent of the people engaged in manufacturing. Lisbon is the center of this industry, which produces almost entirely for the domestic market. Handicrafts, such as embroidery, and winemaking are widespread commercial activities, the latter being especially notable in the Douro River valley.

Trade and Transportation

Iberia imports more than it exports. Foodstuffs, raw materials, machinery, and consumer goods are the major imports of both Spain and Portugal. Spain also brings in large quantities of nitrogenous fertilizer and phosphate rock for agricultural purposes. Among the exports of Portugal, cork, in both raw and manufactured form, heads the list and normally accounts for more than one-eighth of the country's total export value. Fish and wine rank next. Portugal is also an important producer of naval stores and ranks second—though far behind the United States—in the export of rosin and turpentine. The economy of Spain relies even more heavily on foreign trade than does that of Portugal; Spain's exports are three and one-half times greater than those of Portugal.

Though the railroad systems of Spain and Portugal connect with each other, transportation is inadequate and not up to modern standards in either country. Roads have been improved and extended, however, and commercial air transport has steadily increased. Because of their geographical position, Lisbon and the Azores are of great importance in international air commerce, with Lisbon serving as a major European air terminal and transit point; many foreign airlines make regularly scheduled stops in that city. The Azores international airport on Santa Maria Island is a stopover on mid-Atlantic crossings.

Towns and Cities

Iberia is a land of many small agglomerations. Along the coasts of Portugal and the Bay of Biscay, these take the form of picturesque fishing villages, where nets stretched for drying and repairing, fish-drying racks, and many small fishing craft dominate the scene. On the Meseta, as well as elsewhere in the interior, rural villages are scattered; often dull in appearance and frequently located some distance from good highways, these reflect the depressed economic condition of their inhabitants.

Madrid and Barcelona are the two most populous cities of Spain and of the whole peninsula; Madrid has a population of about 3 million, and Barcelona almost 2 million. Portugal's two largest cities, Lisbon and Oporto, are considerably smaller, with populations of about 1 million and 350,000, respectively. Spain has a number of other large cities, including Valencia and Seville, each with more than 600,000 inhabitants.

Landlocked Madrid owes its size and importance to the fact that it is the political center of the nation. It is situated centrally on the plateau, nearly equidistant from the Mediterranean Sea, the Bay of Biscay, and the Atlantic Ocean. It occupies a relatively barren area, and the city is handicapped by an inadequate water supply. Madrid is young when compared with some other major capitals such as Rome, London, and Paris. It was not until King Philip II made it his capital in 1560 that the city began to attain importance. Previously, it had stood as an isolated, lonely village in the midst of the treeless, sun-baked tableland. Today Madrid is a modern city with many industries and also the hub of the Spanish transportation system. Barcelona is Spain's most important economic

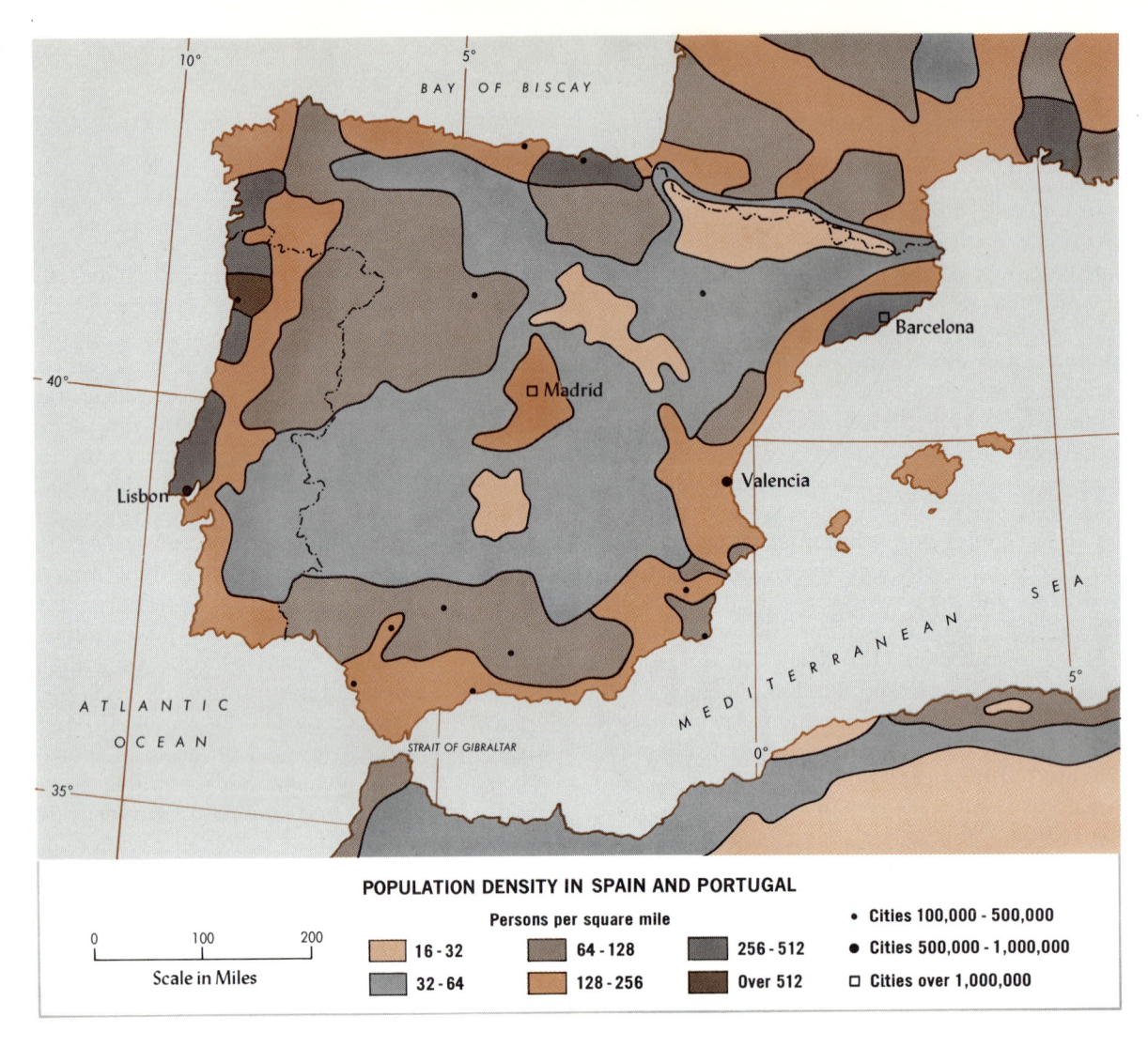

Fig. 8-6 The greatest population density in Spain and Portugal is, in general, concentrated along the coast or in the larger river valleys.

center and great port, the gateway on the northeast coast for the export of cork, wine, olives, and citrus. Cosmopolitan and exuding an atmosphere of world importance, it is Spain's most important industrial city. Situated on a good harbor and in one of the best agricultural areas of Europe, the city has a favorable location for the import-export trade. Large textile manufactures—cotton, linen, and wool—have been developed.

Valencia and Seville are historic centers on and near the coast, located in irrigated areas and surrounded by citrus orchards and gardens. Here notable churches, narrow streets, and ancient walls and buildings attract tourists. Various industries have also been established, and handicrafts are prevalent. Bilbao, the principal northern port on the Bay of Biscay, lies in a valley surrounded by high and fairly steep hills. Railroads enter the city through tunnels; iron mines not far from the city supply the chief item of export. Modern steel mills, fueled with imported coal, are among the heavy industries that have developed here and in nearby Santander.

Lisbon and Oporto, the chief urban centers in Portugal, are both ports situated on the Atlantic coast at the mouth of a great river: Lisbon on the Tagus, and Oporto on the Douro. Lisbon, the national capital, is one of the principal commercial centers of the world and has the best harbor on the Iberian Peninsula and one of the finest in Europe. It ranks as an important center for transshipment and entrepôt activities and has one of the leading airports on the globe. Industry has developed; it is also an important tourist center. Oporto, the second largest city and second port, is noted chiefly for its wines, although the city manufactures other items.

Overseas Territories

Portugal is the last of the European colonial powers to retain most of its large overseas empire. Twenty-three times the size of Portugal, its colonial territories once constituted a domain that was the fourth largest overseas empire in the world. Only Britain, France, and Belgium had greater holdings within the twentieth century. The strategic Azores Archipelago, consisting of nine islands, lies in the Atlantic some 800 miles due west of Portugal. The Madeira Islands are approximately 600 miles southwest of the continent. Politically and administratively, these island groups are part of continental Portugal.

The status of the other Portuguese colonies has been changed to that of Overseas Provinces; the largest and most important of these are Mozambique and Angola in Africa. Of Portugal's once vast holdings in the Far East, only part of the East Indian island of Timor and Macao, near China, remain. The total population of the overseas territories, approximately 13 million, exceeds that of Portugal. Millions of people besides the Portuguese still speak the Portuguese tongue, including Brazilians.

The present-day colonies of Spain are inconsiderable compared with those of her neighbor. What remains of a worldwide empire is found in small holdings scattered along the coast of Africa from Gilbraltar to the Gulf of Guinea. Their total population is small, and the colonial trade is insignificant. All of Spanish Morocco, except Ceuta and Melilla, became a part of the Kingdom of Morocco upon the independence of that French territory. Ifni, a Spanish enclave within Morocco along the Atlantic coast, was acceded to Morocco in July, 1969; the Spanish Sahara remains Spanish, however, although claimed by Morocco.

Gibraltar

The control points of the Straight of Gibraltar are the southern Pillar of Hercules on the North African coast, the Spanish-owned port of Ceuta, and the famed "Rock" across the strait, a peninsular boulder off southern Spain. The Rock of Gibraltar fell to the English in 1704 when a combined Dutch and British fleet captured it from Spain. From the moment it became British, Gibraltar has played a mighty role in the fortunes of Europe; however, its history is strictly military.

Gibraltar surges up out of the Mediterranean in sheer cliff walls, from a base 3 miles long by three-fourths of a mile wide, to a crooked ridge running north-south and reaching a maximum elevation of 1,396 feet. Seemingly impregnable to military attack, it is very vulnerable from another viewpoint—namely, the water supply, fresh fruit and vegetables, and the workers have always come from Iberia. The harbor at Gibraltar is a major naval base, capable of bunkering twelve ships at a time without using any dockspace for that purpose. Its drydocks can care for all but the largest aircraft carriers; and within the solid rock are miles of tunnels used for various purposes, including electronic detection.

Gibraltar may have lost some of its long-recognized strategic importance as a result of modern methods of warfare. Artillery, firing from Spain, could doubtless render the harbor and the airstrip impotent; nuclear bombs could close the water passage. Nevertheless, there is no doubting its continuing value as a base for the North Atlantic Treaty Organization, as long as the Spanish mainland remains friendly, or at least neutral.

Italy

Italy has a past of brilliant accomplishment, a history that is perhaps unique in majesty. From Italy came a progression of notable accomplishments: the reach of imperial Rome, the might and power of the Roman Catholic Church, and the genius of the Renaissance, among others. Three times Italy has ruled the world—once in military might and government, once in religion, and once in art.

For many decades, overpopulation and underproduction have been the foremost economic problems of the Italians, representing the pressure generated by too many people on too little land. The land itself is overworked and rocky and is too dry during much of the year; rainfall and available water decrease steadily in amount with distance southward in Italy. Resources are meager or have long awaited an advanced technology for development. As long as population was not large, and while industry was on a handicraft scale and did not make the huge demands for raw materials and fuels that modern industry does, Italy did not feel this pressure too much; moreover, such pressure has always been partially relieved through emigration. Migration will continue to offer an outlet for some of Italy's people, because large numbers emigrate every year. More people leave Italy for other lands than depart from any other West European nation. The majority of emigrants from Mediterranean lands go to Latin America, but a considerable stream, especially of Italians, flows into France, Switzerland, Belgium, and other North European industrial states to enter the work force either temporarily or permanently.

During the first fifty years of the twentieth century, the economy of Italy merely limped along. Under Mussolini's rule, the government attempted to spur production and inject vigor into the country's economic activities, but the Fascist philosophy and a course of military aggression eventually turned the economy away from trends that might have benefited the people.

Beginning about 1950, however, Italy entered a period of economic resurgence. So strong has its recovery been that Italy has already entered the front ranks of industrial nations. This record-breaking industrial development and economic growth has been assisted by loans to Italian industry from the United States and by membership in the Common Market. Living standards have risen markedly, and Italian labor productivity has increased.

Despite this progress, industry has not freed Italy of all her population problems. There are two Italys, a North and a South—the first well-endowed and developed, and the second underdeveloped and poverty-plagued. Most of Italy's industry is located in the North, much of it within the small triangle of land that lies between Milan, Turin, and Genoa. In the Po Valley of the North is found the bulk of the fertile farmland of the country. The gap in wealth and development between the two areas is vast and striking. The farther south one goes along the peninsula, the less industrialized the country becomes; fertile plains are fragmental along the coasts toward the south, and the interior is rugged and difficult. Economically speaking, the Italian South, or "Mezzogiorno," as it is often called, is almost a country by itself—an underdeveloped country that, since Italy's unification more than a century ago, has lived side by side with the far more advanced North.

On the basis of economic, historical, and cultural characteristics, the South includes all of Italy below a line drawn from just south of Rome to Ascoli Piceno near the eastern coast; Naples lies within this zone, as well as the islands of Sicily, Sardinia, Elba, and others. Southern Italy's economy has long been characterized by a low level of income and saving. Although 38 percent of the population of Italy live in the South, only 23.7 percent of the GNP comes from there. Characteristically, the level of consumption, relative to GNP, is much higher in the South than in other regions, though as an effect of growth it is declining—a sign of growing maturity. A large majority of the nation's unemployed are found here, and intranational migration shows a continuous flow from the South to the North. Between 1961 and 1965, nearly 1 million persons left the South for the northern areas of industry.

The Italian five-year plan (1965–1970) was committed to curbing this outflow of southerners, especially of the South's energetic young people, through a process of industrialization concentrated on labor-intensive industries. It set out to increase the labor force in the Mezzogiorno from 6 to 6.3 million workers during the duration of the program, despite the fact that agricultural jobs have been markedly declining. In the initial stages of development, the Cassa per Il Mezzogiorno (Southern Italy Development Fund) placed emphasis upon bettering and expanding the infrastructure of the South, that is, roads, railroads, harbor installations, communications, and the like. Later, emphasis shifted to the stimulation of private investment and the establishment of government-controlled industries in the South. By law, 60 percent of the new state-controlled enterprises were required to locate in the Mezzogiorno. Notwithstanding this impressive program of aid and the infrastructure developments and improvements, as one Italian economic publication has said, "Southern Italy is not yet at the take-off stage of economic development." The level of local savings is still too low to sustain a tolerable growth rate, and the area lacks the energetic entrepreneurial class that exists elsewhere in the country. The southward savings and investment flow is thus likely to continue for many years to come; the Italian government is thinking in terms of 15 to 20 more years, though the proportions in which the various instruments for stimulation of growth will be applied are likely to change markedly. Although the system of subsidized loans and contributions to individual projects will be continued, the time is judged ripe for more decisive action, particularly in the labor-intensive sections of the engineering and modern food-processing industries and in tourism. The idea is to create, practically from scratch, an industrial environment in which external economies are possible, including those derived from expansion of the market which the labor-intensive industries are likely to induce.

This imbalance between the North and the South was recognized as serious at least fifty years ago, and it probably had its origins far back in history. The land problem was due in large part to the prevalence of landed estates on which extensive holdings were not even worked, while all around was a landless peasantry that could not even get land to cultivate. This was the first problem attacked. By the 1950s, the demand for economic betterment and for equilibrium between North and South, became so insistent that the government had to act quickly. Land reforms have been a large part of this effort; land has been redistributed, and irrigation has been developed. One of the greatest public works undertaken in Italy's program in the South was the Campano Aqueduct, which brings fresh water from the Apennines to the large plain of Campania, where Naples and Salerno are located, and supplies the entire area with water for drinking, irrigation, and power. Included in the government's program in the South also are attempts to reduce unemployment by two-thirds, to do away with underemployment, and to open up opportunities for youth as they enter the labor market. Available labor supply is one of Italy's greatest attractions to investment. Since this is one of the unique resources of the South, the labor market and the favorable climate are expected to attract newly expanding industries into the Mezzogiorno.

Another government project has involved the building of a system of modern highways to and in the South (including the spectacular north-south "Autostrada del Sole"), the hope being that good roads will transform the area into a tourist mecca. There is much in the South to attract tourists, including the natural beauty and beaches, volcanic wonders, and the ruins of the ancient cities of Magna Graecia, remnants of Greek colonies established in southern Italy. Some of these ruins are now being excavated, along with other sites from Roman times.

Another immense project which facilitates movement between Italy and Western Europe and which makes travel by car into Italy most attractive is the joint construction by Italy and France of the Mont Blanc Tunnel. Some 7½ miles long, it pierces a hole through one of Europe's highest mountains and shortens the former driving distance between Paris and Rome by 125 miles. The Italians began tunneling on May 14, 1959, and the French on January 8 of the same year; they met at the "hole-through" point on August 14, 1962, about 3⅛ miles from either end at a depth of 7,400 feet under the mountain's rugged peak. The tunnel was ready for traffic in the spring of 1964.

The Italian state-owned railroads connect all major population centers, and ferryboats are employed for the railway crossing of the Strait of Messina between the mainland peninsula and the island of Sicily. Since these boats transport up to 1,500 railroad cars daily, the water barrier between the island and mainland is no longer a bottleneck to traffic. (A bridge across the Straits, a formidable undertaking, has long been discussed.) Highways between the larger cities are paved and in good condition. Passenger bus service, especially in the North, is excellent, even when compared with that of the United States. Both trucking and air transport have shown a marked increase.

According to tonnage, Italy ranks among the first ten maritime nations of the world. Although its commercial shipping was almost totally destroyed during World War II, it has made a remarkable recovery in the postwar period. Italians raised sunken vessels from the bottom of the sea, reconditioned ships returned by the Allies, and built and bought new tonnage. As a result, the Italian merchant marine has been brought up to its prewar level, doubling the merchant tonnage handled between 1953 and 1967. Income from shipping partly supplied Italy with funds to overcome its unfavorable balance of trade. These vessels are also of great importance in moving passengers and cargo between the mainland and the island regions of Sardinia and Sicily.

Relief Features

Italy falls into four natural regions. To the north is the Alpine border, where the boundary in general follows the crest line. Though the Alps Mountains seem to isolate Italy from the rest of the continent, the Italian ranges actually provide the natural transit zone between continental Europe and the Mediterranean Basin. In early days, passes like the Simplon, Great and Little St. Bernard, St. Gotthard, and Brenner, situated at the heads of valleys permitted ranges to be crossed in spite of ice and snow. Today, airplanes, railroads, and tunnels have almost eliminated the Alps as a travel and communications barrier.

A number of long narrow lakes are interspersed along the frontal zone where the mountains drop to the plains. Occupying valleys deepened by the action of glacial ice and dammed by terminal moraines that impounded the waters, the lakes are of great economic significance to Italy for the development of power and irrigation, especially for the agricultural and industrial Po Valley. The Italian lakes district of the North is milder than the plains below, for it lies on the warm, sunny, south-facing slopes. Foehn winds that descend from the higher Alps have a tempering effect. This region is famous as a winter and summer tourist resort.

The Po Valley is the northern end of the Adriatic depression that has been filled with alluvium carried down the mountainsides by tributaries of the Po and other streams. An actively building delta is extending the plain still farther toward the east, noticeably encroaching upon the quiet waters of the sea. Fertile soil and level land, combined with a favorable climate and ample water for irrigation, make this one of the great agricultural plains of the world. Though the political center lies to the south in Rome, the economic core of the nation is focused on these plains in the north.

The Apennine Mountains, extending the length of the narrow peninsula, constitute, with Sicily, the third natural region. The surface configuration is one of a central folded mountain backbone bordered by marginal plains. A zone of active and quiescent volcanoes extends along the western coast from Rome southward through Sicily. In the north, the Apennines are considerably lower than the Maritime Alps with which they merge. Farther south, however, the peninsular ranges become higher and more rugged, with elevations in the central portions reaching over 9,500 feet. As a whole, the Apennine country has difficult access. Railroads here are few and have been costly to construct. Paved highways cross the mountains to connect major cities; though excellent, they are narrow and have many curves.

Slowly and transitionally, geology and geography change the mountain scene from north to south. Where the northern Apennines meet the Mediterranean along the Ligurian coast, steep slopes rise high and shut out the cold winds from the north. Here a narrow strip of land, known as the Italian Riviera, edges the sea. As a resort center, it is a close competitor with the contiguous Riviera in France.

The Apennines, with their jutting spurs and foothills, occupy most of the peninsula and leave few areas of plains. Of importance, however, are the three fertile but isolated lowlands along the western shore, each centered about a commanding city. First, there is Tuscany in the valley of the Arno, with Florence, the art center of all Italy, in a strategic position at a major pass through the Apennines. Then there is Lazio in the Tiber Valley, focusing on Rome. Third is Campania, to the south, spread around the beautiful Bay of Naples, on which the great port city of the same name is situated. The lowlands of the east are of lesser importance than those of the west, though the largest of the peninsular plains is found here, occupying the heel of the boot of Italy and extending northward along the shore to the Gargano Peninsula.

Agriculture

Italy is a cereal-growing land, with the greatest proportion of its cultivated land planted in

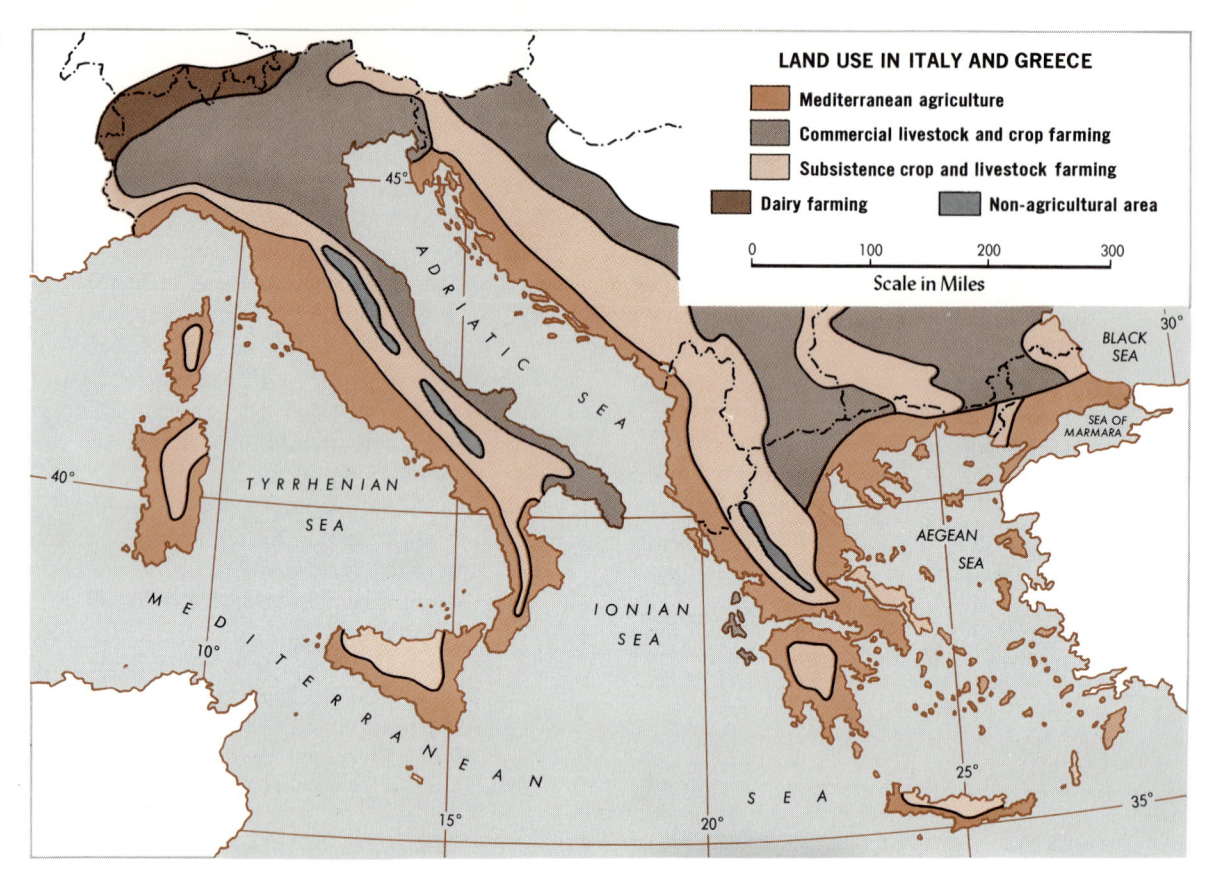

Fig. 8-7 The Po Valley, in eastern Italy, and northeastern Greece are highly developed areas of commercial farming. Some mountain areas are too rugged for utilization as agricultural land.

that type of crop; 25 percent of the cultivated land is planted in wheat. Although grown throughout all of Italy, this typically Mediterranean grain is found to be most productive in the Po Basin. This region is dominated by cereals, with wheat leading and corn ranking second in importance. Despite this predominance of cereals, Italy cannot feed itself and is obliged to import large quantities of these and other foodstuffs. Rice is another cereal of importance, and Italy is the leading producer of this crop in Europe. The upper valley of the Po, where alpine streams provide water for irrigation, is the center of rice growing. Grapes for winemaking are prominent, as well as the two industrial crops of sugar beets and hemp. The olive, scattered throughout much of Italy, is not found in the North, where cold winters inhibit its growth; but Italy is second only to Spain in olive oil production, and second to France in the production of wine. The beet-sugar industry has expanded to a point where Italy is now self-sufficient in sugar.

Livestock raising and dairying are better developed in the North than elsewhere, and a high proportion of the land is maintained in improved pasture and fodder crops. These extend into the mountains, where agriculture takes on an alpine character. In favorable parts of the northern foothills wheat is grown;

but, in general, rye replaces the wheat, corn, and rice of the lowlands.

Mediterranean Italy is characterized by two types of agriculture that might be classed as horticultural and nonhorticultural, depending on the intensity of the cultivation and the variety of crops grown. In the former type, intensive methods of land use, including irrigation, terracing, high fertilization, and triple-cropping produce fruits and vegetables in a garden type of agriculture. Here grapes attain their highest perfection. Olives are especially important in the heel of the boot in southern Italy. Lemons, oranges, and peaches, along with some vegetables, move from these lands of intensive husbandry into the markets of Northern Europe. Sicily is the center of citrus production, especially lemons, though lemon culture extends north along the western side of the peninsula to Naples, and oranges grow almost as far north as Rome. In the Mediterranean regions where the less intensive type of agriculture is practiced, five crops predominate: wheat, oats, grapes, olives, and beans. Few cattle are reared in the Mediterranean lands; because of sparse and poor grasses, sheep and goats typically replace the larger animals.

A system of five-year planning, to facilitate the social and economic development of agriculture and rural areas, was initiated by the Italian Government in 1961; this is known as the *Plano Verde* ("Green Plan"). The first terminated in 1965, as planned; and although many difficulties were encountered, some of the economic goals were achieved. The second Green Plan became law in 1966, differing from the first in that intervention was concentrated in certain economic sectors rather than generally. Emphasis is placed on increasing profitable forms of production and productivity and on raising the living standard of rural workers through the provision of loans, land improvement, extension of electricity supply and irrigation, and the like.

The island of Sardinia, though a part of Italy, is remote from the rest of Italian life. Rugged and rocky, deforested, and eroded, it is a spare land. Although Sardinia produces famous wines and cork and has some areas of intensive cultivation, it is principally a pastoral land. Animals, especially sheep, are grazed on the plateau and mountain country that occupies seven-tenths of the island. The inhabitants are known for their hardy endurance and their interest in livestock. Petroleum refineries have recently been built at the northern tip of the island.

Industry

Industry has not replaced agriculture as the dominant sector of the Italian economy. Industry in Italy is of two general kinds, handicrafts and modern machine production, and both occupy an important place in the economy of the country. In Italy, where art is a way of life, the handicrafts earn their keep; they are "big little businesses." Local handicrafts are a part of Italian life. They are carried on not only for tourist trade. Italian households harbor and treasure them, using them interior and exterior: carved chairs from the Trentino, a Venetian blown-glass vase, an elaborate handwrought iron lamp from Lombardy. Exteriors of houses are often ornately decorated with the work of local artisans. There is much variety to the country's artisan crafts, which range from primitive to sophisticated, but all correctly reflecting the mood of the country and the particular region, heritage, and color. It is rare for an Italian to turn to foreign design in furnishings, regardless of how wealthy he may be.

If there seems to be a paradox in a country whose labor force supplies workers to build supertankers, machine tools, exquisite pottery, hydroelectric dams, marble statuary, nuclear reactors and fine lace, it is made the more puzzling by the fact that handicrafts are by no means the products of the less-industrialized regions only. Sicily, Sardinia, and Calabria specialize in ceramics, woodworking and tapestries, among other artifacts. But then the Piedmont, home of mighty Fiat (producer of automobiles) is also home to artisans who can work minor miracles with wrought iron, precious metals, ceramics, rush and cane. Lombardy is one of the richest regions of Italy. It boasts Milan—heavy industry—and at last count 153,569 artisan concerns producing jewelry, ceramics, fine furniture, and even lace and tapestry.[1] Some 8 million Italians are engaged in the handicrafts as artisans, one out of every seven men, women, and children in Italy; 13 percent of the country's exports are accounted for by the handicraft industries.

The other side of Italy's industry is basic manufacture. The country produces little pig iron and steel, because it does not have the raw materials (iron and coal) for their production. Iron and steel are instead imported, and become the raw materials for the products of heavy industry that Italians produce. Despite the handicaps of lacking iron ore and coking coal, Italy is nonetheless attempting to increase its production of pig iron and alloy steel. In recent years, production of pig iron, ferroalloys, and crude steel has doubled. Italy has been the main recipient of Common Market aid in this field. Most of the new steel plants are situated along the coast, because both the coal and ores for the industry must be imported. Genoa in the north and Taranto, on the inner side of the heel of the boot of Italy, are among the major centers of iron and steel production. The manufacturing industry is mainly concentrated in the northwestern part of the country, especially in and around Milan, Turin, and Genoa. This centralization is especially pronounced in the steel, mechanical, chemical, electrical, and textile industries, although several coastal centers—namely, Taranto, Piombino (near Livorno), and Bagnoli (near Naples)—also produce steel and/or steel products. It took years of large investments and plant construction to close the gap between steel production and domestic demand; but finally, in 1965, this was accomplished. The completion of a second blast furnace at Taranto and of a blast furnace and three oxygen converters at Bagnoli was largely responsible for the 57 percent increase in production. Thereafter, Italy accounted for 15 percent of total EEC steel production.

Leading the industrial products of Italy are machines, which range from machine tools and nuclear reactors to automobiles, locomotives, tankers, and ships to sewing machines and precision instruments. Among the products that have had a recent appreciable development are chemicals, steel, and machines. In 1968, out of a total of 23,860,000 automotive vehicles manufactured worldwide, 663,600 came from Italian plants. Turin occupies first place in automobile manufacture, followed by Milan. The Fiat and Lancia companies are located in Turin. The making of textiles, one of Italy's traditional manufactures, also has expanded, but to a lesser degree than other manufactures.

Although most Italian factories are small, often being family enterprises, Italy also has some of the largest industrial complexes in Europe: for example, Fiat (automobiles), Montecantini (chemicals), SNIA Viscosa (synthetic fibers), Pirelli (tires), and Olivetti (office machines). The government is an important stockholder in industry; IRI, the government's giant holding company, controls about 30 percent of Italy's industrial capital, including much of the steel, machinery, shipping, and electrical power. ENI is a huge, government-

---

[1] "The Craftsmen," *Italian Trade Topics*, Vol. XIII, No. 6, p. 7, June 1969.

owned oil and gas complex that ramifies into all aspects of the petroleum industry—petroleum refining, petrochemicals, diverse overseas oil operations, service stations, and even motels and nuclear developments.

Industry works under handicaps in Italy, in that the industrial resources are meager. Most of the raw materials and fuels (aside from electricity generated by waterpower in the North and volcanic power in South and Central Italy) must be imported. In late 1968, valuable finds of natural gas were discovered in the Po Valley and under the Adriatic Sea. It is believed that these fuel resources will go far toward assuring, from domestic sources, the large supply of energy needed for Italy's expanding industries. ENI, in consortium with Shell Italiana, is carrying out the necessary explorations. When the field has been fully evaluated, foreign companies will be permitted to apply for concessions to develop. "Until the discovery of natural gas deposits in the Po Valley, Italy was scarcely endowed with energy sources, with the exception of abundant water supplies. Out of sheer expediency, therefore, Italy was bound to develop through the years unsurpassed techniques in the exploitation of water resources, so that now it has become the major producer of hydro-electric power in Europe. As things stand now, Italy can avail itself of a highly specialized engineering industry that looks eagerly at foreign markets where its particular skills can be put to profitable use."[2]

Genoa is Italy's great port of the North, through which the articles of trade flow to and from the industrial Po Valley across the densest network of roads and railroads in Italy. Italy's long coastline favors shipping, although not all portions of the coast are blessed with protected harbors. The east coast, on the Adriatic, is remarkably straight and practically devoid of good harbors.

Foodstuffs and live animals are, by value, the country's largest import, with cereals and meat being the largest single items; crude materials (inedible, excluding fuels) rank next, with textile fibers and waste having the highest value among these. Each of the above import categories (food, meat, crude materials) make up over 20 percent of the total value of Italian imports. Mineral fuels, lubricants, and the like are third, comprising about one-sixth of the import value; almost as large a category comprises machinery and transport equipment, with chemicals ranking next.

Among Italian exports, machinery and transport equipment (including electrical

Fig. 8-8 Fiat assembly line in Milan, Italy. Northern Italy has become one of the leading industrial centers of the world. (Courtesy of Italian Government Travel Office)

---

[2]"Italian Projects Around the World," *ibid.*, no. 12. pp. 1-2.

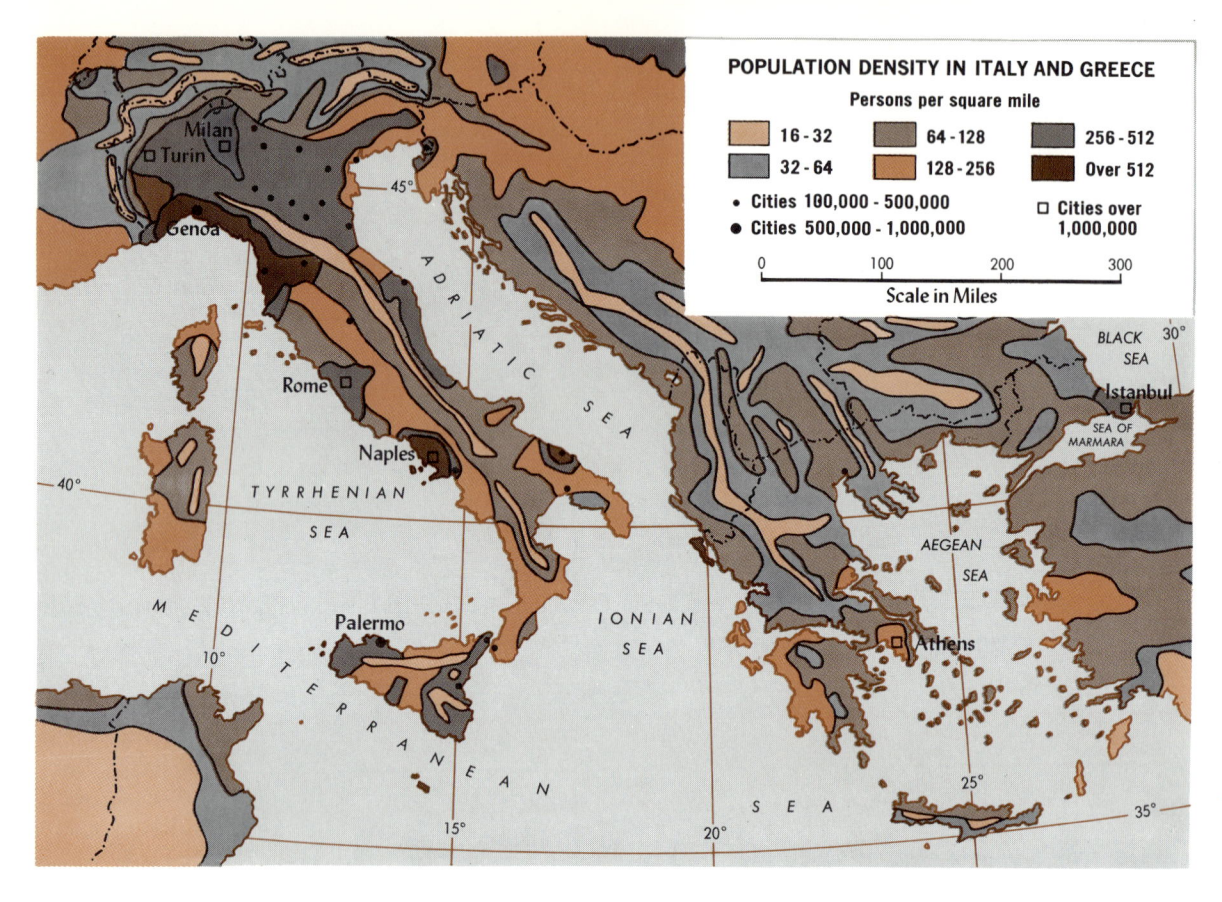

Fig. 8-9 The Po Valley and the area around Naples are the most densely populated parts of Italy; the higher and rougher areas, the least. Greece has a lower population density than other major countries of Europe.

equipment and appliances, motor cars, railway vehicles, and aircraft) lead, making up over 30 percent of the total export value and going to all parts of the world. Recognition of Italy's engineering capabilities is partially illustrated in the winning of a much-contested contract by Fiat to manufacture the first automobiles from a "Western" country in the Soviet Union. Chemicals, fresh fruits and nuts, and textiles also rank high among exports. Although Italy competes with France for first place as a wine producer, the total value of the wine exported from Italy is small and, by weight, equals only a little less than half the wine exports of France.

The value of Italy's imports generally exceeds that of its exports, but the difference is compensated by invisible earnings, mainly deriving from tourism and remittances from abroad.

Cities

The world has no cities more famed than those of Italy. Every year millions of dollars of income flow into the country as tourists and religious pilgrims come to enjoy the beauty of the Italian countryside and seacoast, the delight of her agreeable climate, and, perhaps most of all, the splendor of her historic cities.

No city in the world puts on a better spectacle than Rome. From ancient days it has had a reputation for circuses, celebrations, coronations, fetes and fairs, torchlight parades, and the corteges of emperors and popes. Because the Vatican, the residence of the popes, is located here, the millions of the world's Roman Catholics look upon Rome as the "Eternal City." Rome is well known for its historic and religious associations, its art treasures and ruins, and its educational institutions. It is first of all a cultural city; but as the capital of Italy and seat of the Catholic Church, it is also political. Rome is also an important manufacturing center for such goods as precision instruments, electronics, pharmaceuticals, food products, printing, and communications equipment. It is situated along the Tiber River on its celebrated seven hills, occupying a central position on the western side of the peninsula and dominating Latium.

About 125 miles south of Rome lies Naples, the only large and industrial city in the Mezzogiorno, as well as Italy's second port. The city circles around the bay that gives Naples one of the most beautiful sites in the world. As a striking backdrop, Mt. Vesuvius stands across the bay from the city. The lower slopes of the volcanic mountain are green, and cultivated plots and houses nestle at the base and even climb some distance toward the crater. Although temporarily driven away by threatened or actual eruption, the inhabitants of its slopes soon move back to Vesuvius, seemingly oblivious of the potential danger, as the sun-bleached and deserted ruins of Pompeii testify. Although destructive, this prevalence of volcanic activity has several beneficial effects: volcanic ash and disintegrated lava produce very fertile soils, and modern technology has permitted the volcanic gases to be harnessed for power production; sulfur occurs in the vicinity and is one of Italy's valuable minerals. An important tourist center and passenger port, Naples also handles several million tons of cargo goods annually, and has developed such industries as chemicals and food-processing plants.

Around Naples spreads the fertile Campania countryside with its intensively farmed soils which produce citrus fruits, vegetables, and some cereals, and which would produce even more if the land could be irrigated. Throughout most of the Campania, which marks the beginning of the Mezzogiorno, water must be carried to the plants. The region spreads eastward into the mountains and blends southward with Calabria. Population densities of up to 910 persons per square mile here are greater than those of Lombardy, with its great industrial cities of Milan and Turin. The Campania does not have the industry of these two northern regions, and as the people from the surrounding country have moved into Naples seeking employment, they have contributed more to its misery than to its upbuilding. Naples shows greater physical and social contrasts than possibly any other city in the Western world, contrasts between wealth and beauty on the one hand, and poverty and misery on the other.

Genoa is Italy's greatest port and the second-ranking port of the entire Mediterranean, after Marseilles. Just as it challenged Venice for maritime leadership in the Middle Ages, so it has attempted to rival Marseilles in the modern era. It serves as outlet not only for the important industrial triangle of the Italian North but also for some of the trade of Switzerland; the entire Po Valley and much of the Alps are its hinterland. There is little space on which industry can develop, however, because Genoa clings to the coast on a narrow strip of plain that actually plays out in places as the Italian Alps approach the shoreline. Houses and other buildings, both older and modern, climb the hills and ascend to the top of mountains that rise within the city limits. The heights are attained by funicular cabs, which are pulled by cable up the seemingly perpendicular slopes. It is a colorful city on a spectacular site. The stone home of Columbus's boyhood still stands at the foot of the ramp leading to a great and ancient gateway; in contrast to this small dwelling, a modern skyscraper rises just opposite. The atmosphere of the whole city seems to be of the sea and of the mountains.

Milan, Italy's second city in size, is the financial and economic center of Italy. Its stock exchange surpasses that of Rome in importance, and the head offices and branches of most of the large Italian and foreign industrial firms are found here. The products of its factories are varied and include silk and other textiles, machinery that ranges from agricultural devices to airplane engines, electrical equipment and precision instruments, basic steel and copper, pottery, and pharmaceuticals. Industrial though it is, Milan is also a city with great cultural traditions and reputation. The original version of Leonardo da Vinci's celebrated fresco *The Last Supper* is found on the walls of a Milan abbey; the Duomo of Milan, its cathedral, is one of the masterpieces of world architecture; La Scala Opera House is of world repute for its productions. In the city's immediate vicinity are many industrial satellites of Milan.

Turin, capital of the Piedmont, is heavily industrial. Although it lies at the convergence of several valley routes from the western Alps, Turin does not have the central location in the plain that Milan does, hence, Milan is regarded as the actual "capital" of the plain. Nevertheless, the manufactures of Turin are notable: automobiles, electrical equipment, and a large proportion of the cotton and rayon textiles. Historically, it was the center from which the House of Savoy set out to unify Italy.

Venice, dominating the eastern sector of the Po plain, is a city built on many man-made, pile-supported islands; its arteries of traffic are canals, rather than roadways. The Grand Canal, the principal throughway of the city, is the broad "street" along which "water buses" run (that is, motor launches which zigzag from side to side of the canal to pick up and discharge passengers); gondolas are the "taxis" of this remarkable city. The city had its beginnings on the islands of a lagoon when refugees from barbarian invasions on the mainland chose the site for their village in the fifth century A.D. because it was protected by the surrounding water.

Venice, today, remains a world-renowned tourist center, but it is also an important Adriatic port, at times rivaling Naples for second place in Italy. Industry, especially glass manufacture, and handicrafts have grown up in the outskirts to supply the tourist trade. There is much to attract travelers to the city. Foremost perhaps, are the unique floating quality of the city and the canals themselves, which so intricately interlace the city that all movement, except foot travel, occurs on the water. The grandiose buildings fronting the canals drop directly into the water; gondolas, launches, and other small craft tie up along the edges; graceful bridges arch and cross over the channels, the most famous of them being the Rialto and the Bridge of Sighs. The Byzantinizing Basilica of St. Mark, the Doge's Palace, and the huge Campanile, all reminders of the city's past glories, surround the stupendous piazza that is Venice's focal point and gathering place. So famed is Italian Venice for its beauty and its canals that other canal cities, especially of Northern Europe (e.g., Amsterdam), one after another have characterized themselves as the "Venice of the North."

However, these same picturesque canals of Venice are slowly eating into the heart of this unique city. The salt waterways that divide Venice into about 180 little islands are corroding the foundations of hundreds of priceless palaces and churches. Lapping incessantly against the bases of the buildings that flank the canals, the water wears tiny furrows in the building stone, which gradually widen into fissures and then become dangerous cracks. In some buildings the decay is already too advanced to save the crumbling structures. Strengthening the foundations is a slow, difficult process, but it has been begun, along with some more ambitious and ingenious schemes for saving the city sponsored by UNESCO and other international aid groups.

Florence on the Arno, Pisa, Siena, and Trieste, the outlet for Central Europe and a port for the Balkans, are among the many other significant and treasured cities of the Italian past and present. Florence was preeminent in the Italian Renaissance, which in turn spurred the Renaissance of all Europe—that transition from medieval ways and times into the modern which marked the Humanistic revival of the arts and literature and the beginning of modern science.

Greece

Physical Setting

Greece is situated at the tip of the Balkan Peninsula. Washed on three sides by seas and indented by many inlets and bays, it has one of the longest coastlines in Europe. A rugged mountain country, with four-fifths of its area made up of mountain chains and spurs, Greece is probably the most barren and sterile among the Mediterranean lands. The Pindus Mountains, a series of long, continuous ranges, extend southeast through central Greece. East of the Pindus, the topography is one of basins separated by southeast- and eastward- trending spurs. The principal plains are along the east side of the peninsula and in central and western Macedonia. Northern Greece is extremely mountainous, the central section less so. Here, in central Greece, are located Athens and Piraeus, which is the main seaport, contiguous with and serving Athens. To the south lies the Peloponnesus, separated from the mainland by the Corinth Canal, a breathtaking sheer-walled waterway cut through a narrow isthmus between 1881 and 1893 to sever the former peninsula from the mainland. The many islands in the bordering seas make up nearly one-sixth of the country's total area.

A Mediterranean climate predominates throughout Greece, though prevailing winds, altitude, and continental location cause local differences in temperature and precipitation. The south and east sections are generally drier and warmer than the north and west. The lowlands are warmer and have less rain than the uplands. The Greek island of Corfu, for example, off the extreme northwestern tip of Greece, has an average of 47.6 inches of well-distributed annual rainfall; whereas Athens, in the southeast, averages 15.4 inches and undergoes a marked dry period in summer. Salonika, in the northeast, receives 21.5 inches of precipitation, fairly well distributed throughout the year, though the three summer months get somewhat less than the others. Large parts of Greece have rainfall averages that correspond to those of Athens.

Problems of Land and Population

On an area about the size of North Carolina live nearly 9 million people trying to wrest a living from a largely harsh, rugged land. Although only about 28 percent of Greece's land is cultivated, about one-half of its population are classed as rural. As in Italy, overpopulation and underemployment have been prime domestic problems. The per capita income, although it has risen somewhat in recent years, is one of the lowest in Europe.

Other problems burden Greece. Mountains isolate one part of the country from another; a chain of rocky mountains divides the nation north-south into what are commonly referred to as Western and Eastern Greece. These mountains, like a backbone of the country, form the watershed that divides Ionian from Aegean Sea drainage. The country as a whole is generally mountainous, and mountain barriers separate one small plains region from another, particularly in southern Greece. Hence, there are only a few large and fertile plains; of these, Thessaly in west-central Greece (490 square miles) and central and western Macedonia (875 square miles) are the largest. Of the cultivated area, more than 60 percent is over 750 feet above sea level, with the result that most of the land is steep, rocky, and cut by gullies. Although a few rivers flow all year round, most streams are dry from June to September.

About one-sixth of the Greek domain is insular, comprising the many islands of the Ionian and Aegean Seas, including the Dodecanese, just off the west coast of Anatolia, and the large islands of Crete and Corfu. Insular Greece has a surface area of 9,981 square miles, much of which is rocky and difficult terrain though picturesque.

The holdings of Greek farmers are generally small and fragmented: 85 percent are under 12.5 acres. Moreover, these small holdings tend to be fragmented even further through inheritance. On well-drained plains, the land is frequently divided into many small units, with each peasant holding several of these plots scattered throughout the cultivated area. Although fertile, the lower lands are generally poorly drained.

American economic aid has done much to help the Greeks with problems of land reclamation, drainage, erosion, flood control, irrigation, and soil conservation. Spectacular results have been attained in the reclamation of alkali lands through rice production. Thousands of acres of such areas, considered useless since before the time of Christ, are being sweetened by the steady flow of fresh river water poured in to irrigate the grain. It is expected that within a few years these soils will be sufficiently cleansed to permit the raising of such other crops as wheat and cotton. Greece is an importer of rice, but it is presumed that the rice produced on these reclamation projects will eventually meet its needs. Many swamps and lakes have been drained, and large areas have been provided, for the first time, with irrigation by drilling new wells and channeling rivers that for centuries have drained off, unused, into the surrounding seas.

Agriculture

Agriculture remains the backbone of the Greek economy. Of the approximately 9.2 million acres of land cultivated, about one-half are planted in cereals; over one-fourth in pulses, tobacco, and cotton; and about one-fifth in grapes and other fruits. The main crops, produced chiefly in valleys nestled among mountains, are cereals, fruits and vegetables, olives, cotton, and tobacco. Cereals, including wheat, corn, barley, oats, and rice, are the most basic and widespread of all crops. Potatoes, pulses, olives, fruits (including citrus and grapes), nuts, and cotton also rank high. Tobacco is outstandingly important because, although making up only 10 percent of total agricultural income, it accounts for about two-fifths of all export earnings. Greek tobacco is classed as Oriental. Interestingly enough, Macedonia and Thrace, the major areas of tobacco production, are those regions which were settled by Greeks repatriated, between 1907 and 1928, from Turkey, where they had learned the skill of growing this demanding crop. The Oriental tobacco produced in Greece is considered the highest quality grown in the world.

As in other Mediterranean lands, vineyards are extensively cultivated and provide a money crop, since much of the Greek grape output is dried and exported as currants and raisins, which together make up nearly 15 percent of the total export value. Greece ranks third among nations in the production of olive oil. Though slow in growth, requiring 15 to 20 years to mature, the olive is a small sturdy tree that lives and produces for centuries. Spreading from Albania southward through the Peloponnesus and north to the Macedonian coast, olive culture extends eastward to within 50 miles of the Dardanelles. Greece is a net exporter of products of the field; but, in certain items, notably meat and livestock, dairy products, and feedstuff, the country is heavily dependent upon imports.

Food processing is the second most important industry in Greece, surpassed only by textile manufacture. From the fields and sea come the numerous commodities that are canned. The output of this industry has steadily increased since 1950, owing in part to modernization of the factories.

Despite the important place that agriculture and the commodities it produces have in the Greek economy, serious problems still remain in the agricultural sector. Farms are so small as to prevent the significant economies that large holdings would permit. Support from the government has not been sufficient to allow anything but a production geared to physical capacity rather than to market demand. Extensive capital should be invested for the extension and improvement of irrigation; animal husbandry industries must be improved and extended; cultivation of feed grains must be expanded; and the dairy industry should be extended. Efforts must be made to transfer some of the labor force out of agriculture and into other economic sectors. Expenditures must be put into modernizing farm technology, expanding irrigation and land reclamation, and improving marketing techniques. As noted, agricultural production needs to be keyed to demand, and the balance within the agricultural sector must be bettered; in other words, there should be a greater emphasis on livestock and dairy production vis-à-vis crop production.

Resources and Industry

Less than 2 percent of the GNP of Greece is accounted for by mining; and the industry employs only some 22,000 persons. Yet, beneath the bare and rocky hills of Greece are minerals which, if properly developed, could contribute much toward national solvency and prosperity. Although not rich in natural resources as compared with many other countries, Greece can claim minerals as one of her few assets. Some deposits are too small for profitable development, but many could be commercially significant if worked by modern techniques. About twenty basic minerals and fuels, in quantities and grades worth mining, are known to exist. These include, among others, iron, lead, zinc, lignite, chromite, and bauxite. Official estimates of bauxite, lignite, and chromiferous iron reserves are in hundreds of millions of tons for each.

One important resource virtually untouched until recent years is lignite. Greece lacks high-grade coal. Thus, the expansion of lignite mining and use is the single most important development in the industrial field. The lignite mines, which have been termed the "fuel bin" of the nation, furnish practically all the country's powerhouses, and 90 percent of the total output is used for generating electricity. Another significant mine undertaking has been the revival of chromite extraction at Skoumtsa in central Greece. Near Lavrion, at the tip of the peninsula of Attica, important deposits of zinc and lead, yielding by-products of silver and iron pyrites, are being worked. North and south of Athens are the two mountains Pentelikon (Pentelicus) and Hymettus, from which were quarried the exquisite marble used for the buildings and statues of ancient classical Greece.

Private ownership dominates Greek industry, although foreign private investors control some. The government owns and exploits some mineral deposits, such as the Aliveri lignite and the Naxos emery mines. Greek postwar rehabilitation of industry has been slow; however, production has increased steadily in recent years. The Athens-Piraeus district is the major center of industrial production; approximately 57 percent of the industrial labor force and 45 percent of the industry are located here. A large consumer market, excellent port installations, adequate transport and electric facilities, and a large labor reservoir draw industry to metropolitan Athens. Thessalonika is the second largest focus of Greek manufacturing. Whereas most Greek industries must relie on imports for some or most of their raw materials, the tobacco-related industry is self-sufficient, because Greece is a large producer of Oriental tobacco; cigarette manufacture is by far the most important sector of this industry. Large investments of capital are needed, however, to bring about any major improvements in the industrial field.

Trade and Transportation

Invisible earnings make up an unusually large part of the balance of payments in Greek trade. In recent years the largest part of these invisible items has come from wages to sailors, shipping receipts, remittances from Greeks living abroad, and earnings from tourism. However, ability to better the balance of payments in the future will be determined not by the above items but by the ability to increase exports and to expand consumer-goods production. This depends on stimulating foreign investment in Greece, even though the effects might not be felt for several years. Foreign trade receipts have increased greatly in recent years. Much of this increase was due to an increase in grain sales; but, even without this, exports increased in value by 13 percent in 1970, resulting for the first year in a favorable balance of trade for Greece. The earnings of exported industrial products, making up 16 percent of the foreign-exchange revenues, increased by 80 percent that year. The countries of the European Economic Community account for over 40 percent of Greek imports and 35 percent of the exports. This trade should expand as Greece becomes a full member of the Community. Greece's second greatest trading partner is the United States, supplying 11 percent of Greek imports and taking about 12 percent, by value, of the exports.

The Greek merchant marine ranks high among the commercial fleets of the nations of the world, ranking seventh or eighth by tonnage; however, if the over 1,000 Greek-owned ships flying flags of other nations are considered, the fleet ranks third in the world. Of the "flags of convenience" flown by Greek merchant vessels, the Liberian is by far the major one, with the flag of Cyprus being second. Most of the new ships in the Greek fleet are small; if they are large vessels, they are almost invariably registered under foreign flags. Piraeus, outlet to the sea for Athens, is the principal port of Greece and one of the major Mediterranean ports. It handles well over half of all Greek shipping. Because the interior topography of the country is so rugged, coastwise shipping constitutes a major form of domestic transport. Also, the many islands belonging to Greece make a numerous fleet essential.

Greece has approximately 1,700 miles of railroads; but because of the increasing importance of trucking, the railways now transport about one-fourth less tonnage than previously. Air transport is gaining in importance, although the international carrier Olympic Airways (privately owned) is also the only domestic airline of Greece; it has exclusive rights to operate until 1986. However, a number of foreign airlines make scheduled stops in Greece.

Ellenikon Airport outside Athens is a busy focal point for planes flying from elsewhere in Europe and the Eastern United States, on the one hand, and from Africa and Asia, on the other. Out of the eighteen commercial airports of the country, seven are international.

Fisheries

Greece is probably one of the most ancient fishing nations in the world, despite the fact that the Mediterranean Sea falls short of some of the requirements of a great fishing ground. One specialty of Greek fishermen in the warm waters of the Aegean and along the North African coast of the eastern Mediterranean is sponge gathering. Greeks bring up most of the Mediterranean sponge take; nearly two-thirds of the total value of fish exports from Greece derives from sponge sales.

The sponging boats go out in fleets of ten or twelve vessels, operate for the first few weeks in Greek waters, and then move on to fishing grounds along the shores of Libya and Tunisia, where fine sponges are found on underwater rocky ledges at depths ranging from 120 to 150 feet. Several factors have begun to cut into the success of this dangerous but traditional occupation of the Greeks. First, artificial sponges are capturing the world markets and displacing the costlier natural product. (The United States is still the largest buyer of Greek sponges, taking nearly half of the total catch.) Second, the sponge-fishing grounds are being depleted, and despite greater efforts by the energetic Greek divers, the catch is growing smaller. Third, the governments of the North African nations along whose shores some of the finest sponges are taken have begun to impose severe restrictions on foreign fishermen. Egypt asks 18 to 20 percent of each catch for permission to operate in her waters; Libya imposes a high tax on each diving team. Beyond this, African divers now share in what was formerly a Greek monopoly. Kalymnos, a bleak mass of rocks among the hundreds of small islands in the Aegean Sea, is the main sponging base in Greece.

Fig. 8-10 Athens, Greece, including its port of Piraeus, has a population in excess of 1.8 million. The city is the center of important archaeological and historical studies as well as a modern trade and tourist center. (Courtesy of Trans World Airlines)

Cities

Greece is not a country of large cities. Rough topography, poor highways, and adverse economic conditions make for small and sometimes isolated settlements. Athens is the undisputed political, economic, and cultural center of the country. One of the oldest cities in the world, established before 1000 B.C., Athens has long been noted for its culture—its great architecture, sculpture, literature, philosophy, and drama—and for the early development of democratic forms of government. Modern Athens and Piraeus form a sprawling, continuously built-up urban area. The port of Piraeus is also very ancient, built more than 2,500 years ago. It has suffered many catastrophes, including great destruction during World War II. Reconstruction, however, has made it one of the most modern ports of the Mediterranean. Besides new port facilities in the old harbor, a new harbor was built close to the old one. The Athens-Piraeus urban center is growing rapidly. Athens itself increased in population by one-third over the last decade. This is causing concern in Greece, and it has been suggested that the capital be moved from Athens to Pella, in northern Greece, a city that was once the capital of Macedonia in the time of Alexander the Great. Thessalonika, situated at the mouth of the Vardar River, is the second port of Greece. This port is the leading outlet and commercial center for the Vardar-Morava Valley. The city sprawls over a slope, rising from the edge of a large and sheltered harbor. Salonika (ancient Thessalonike) was founded in 315 B.C. by Cassander, King of Macedonia. Its earliest role was that of port for Macedonia. As the Romans came in, they built the Ignatian Way, which led from the Adriatic Sea through Thessalonike to the Bosporus; this road made the city a center of east-west as well as north-south traffic. In population, modern Salonika

Fig. 8-11 European Turkey is separated from Asiatic Turkey by the Dardanelles, the Sea of Marmara, and the Bosporus. These narrow waterways are strategic and effective in controlling ship movement between the Black Sea and the Mediterranean Sea.

ranks second to Athens and, as an industrial center, is surpassed only by Athens. It is still the "hub of Macedonia," an ancient land that is now divided among Greece, Yugoslavia, and Bulgaria.

European Turkey

All that remains of the European portion of the vast empire of the Ottoman Turks is a small sector of land, 9,254 square miles in area, at the eastern extremity of the Balkan Peninsula. It reaches toward Asian Turkey in a double-pronged land bridge, the points of which are separated from Anatolia by the Bosporus on the north and the Dardanelles on the south; the two straits are joined by the Sea of Marmara, which lies between. This bit of land is the only political division of the Balkans that consists, in the main, of lowlands, the only one in which there are no mountains of any great height. Almost 3 million people live here, or approximately 8.5 percent of the population of Turkey, on a little more than 3 percent of that nation's land.

Most of European Turkey is fertile and could support a larger population. The holding of this small area in Europe places all the strategic and historic eastern Mediterranean "straits" within Turkish territory. The Dardanelles-Marmara-Bosporus waterway has had and still has a quite extraordinary importance. Control of the Straits by Turkey has been a thorn in the side of the Russians ever since Empress Catherine the Great succeeded in securing a foothold along the Black Sea. Russian efforts to gain some control or supervision over the water passage, or to secure a land bypass that would bring Russian influence into the Mediterranean itself, have increased. In peacetime, ships of diverse types of all nations pass freely through these waters; in wartime, the Straits can become a bottleneck readily controlled by the Turks.

Guardian city of this waterway is Istanbul, situated on the Golden Horn at the southern end of the Bosporus; overlooking the Straits from the European side, it is built on one of the most commanding sites in the Mediterranean. It is the fabled Byzantium and Constantinople of the past, a city that served as the capital of a succession of empires from the fourth century to the twentieth—until the modern country's founder Ataturk transferred the Turkish capital to the interior plateau at Ankara.

There is perhaps no better example in the world of a cultural and commercial crossroads than the area of the Turkish straits. People and goods have moved not only by water across this sector of land but also east and west by land, negotiating the narrow straits of the Dardanelles and the Bosporus. Armies and goods and migrants have swept from continent to continent via these two land bridges. The

Fig. 8-12 The Galata Bridge connects the two sectors of Istanbul, Turkey, which have developed around the shipping and industrial area known as the Golden Horn. (Courtesy of Turkish Tourism Office)

Fertile Crescent route between Baghdad and Aleppo continued on, westward and northward, across the Turkish peninsula and the land bridge of Asia Minor; and the Berlin-to-Baghdad railroad was built to follow the same course, passing through Istanbul. Strategically and centrally located as it is, Istanbul stands alongside the continuous flow and, so to speak, merely observes it. Despite this, Istanbul is Turkey's leading port and largest city. Across the Bosporus from Istanbul stands Usküdar, the Asian twin of the historic European city. However, two zones of civilization are represented in these two centers: notwithstanding its ancient past, Istanbul is a metropolis of today; Usküdar is of the long ago—or is the difference simply Europe versus Asia? The contrast is startling, sharp, and impressive. Istanbul, though only about half the size it was a century ago, has kept up with the times, modernizing and industrializing, and is still a city of great significance. Annually, more and more tourists are drawn here, to view the magnificence related to the city's past and to look upon those celebrated waters, so strategic historically—the Bosporus and the "Straits."

Although in our age of nuclear power and space exploration a narrow passageway like the Bosporus could be effectively closed with relative ease, Istanbul and the straits are still important. Without this outlet, the Soviet Union would have no access to the ocean lanes from the south. Although terms of international treaties specify the role of Turkey relative to the Straits, Turkey is the gatekeeper of the Hellespont. The strategic character of this bit of European territory overshadows all other aspects of its importance. It will be further enhanced when the contemplated bridge across the Bosporus, connecting from Straitside with Istanbul by superhighway, is completed. On July 1, 1969, the European Investment Bank announced it was loaning Turkey 79.5 million dollars for this project, the total cost of which is expected to be about 185 million dollars.

Minor Mediterranean States

Five tiny independent nations—Andorra, Monaco, San Marino, Malta, and Vatican City—lie in the western Mediterranean area. Andorra, Monaco, and San Marino are relic countries, remnants of state's established in mountainous areas centuries ago; the papal state of Vatican City, as it is characterized today, was established by the Lateran Treaty of 1929, drawn up between the Pope and the Italian government. Vatican City has existed for many centuries, however, for the popes have held temporal authority over certain areas in Italy since the establishment of the Roman Catholic Church. Malta, a former British colony, became independent in 1964.

ANDORRA

A country of narrow mountain valleys and a few high peaks situated in the heart of the Pyrenees, Andorra is about 190 square miles in area and has a population of slightly less than 20,000. It is an area of cold winters and cool to mild summers, because no part of it lies under 3,500 feet in altitude. For several months of the year, most of the mountain passes are snowbound. In the Andorran valleys during the short growing season, subsistence crops are cultivated. Many of the inhabitants are engaged in sheep raising. The principal natural resources of Andorra include extensive growths of pine, iron ore deposits, and marble. Although some iron ore is mined and a small amount of marble is quarried, the chief industries are related to the nation's agricultural production. Small-scale manufacturing activities include spinning, wool combing, and tobacco processing and are centered primarily around the town of San Julian de Loria.

Politically the nation is semifeudal, governed by a twenty-four-member elective council that rules under a joint suzerainty of the president of France and the Spanish bishop of Urgel. Andorra la Vella, the largest city and the national capital, is a busy mart for tourists.

MONACO

Near the Italian-French border is Monaco, an independent principality joined in a customs union with France. This bit of a country, with an area slightly greater than ½ square mile (370 acres) and a population of about 23,000, is, after Vatican City, the smallest sovereign state in the world. On a rocky promontory facing the Mediterranean is the old capital, Monaco. More noted is Monte Carlo, famous for its gambling casino and luxurious hotels, which form a primary attraction for Riviera tourists. The country receives its income from the casino, taxes on hotels, tourists, and the sale of postage stamps.

Monaco is a remnant of a formerly large and powerful feudal estate. As long as the ruling line has a male heir, Monaco will remain sovereign; if and when the ruling family should cease to produce such an heir, the little principality will automatically become a part of France.

SAN MARINO

San Marino is said to be the oldest independent nation in Europe and the smallest republic in the world. Located only 12 miles from the Adriatic coast in the Etruscan Apennines and situated on an isolated hill, the nation has a total area of some 23 square miles, enclosed within an irregular frontier that is completely surrounded by Italian territory. Population is about 19,000. The country is rugged; Mt. Titano, the highest peak, has an elevation of about 2,400 feet. Most of San Marino lies in the drainage basin of the Marecchia River.

Agriculture is the chief occupation, and grapes and other fruits, wheat, and corn are the principal crops grown. Cattle and swine are common domesticated animals. Industry caters to local needs or to tourists, for whom souvenirs are produced. Transportation is provided by highway and rail connections made with the systems of Italy. The city of San Marino, situated on the western slope of Mt. Titano, is the national capital. Government revenues are derived largely from the tourist trade and from the sale of postage stamps issued for collectors.

VATICAN CITY

The independent papal state of Vatican City, in area the smallest nation in the world, covers only 108.7 acres of land. It was formally created on February 11, 1929, with the signing of the Lateran Pact. The ecclesiastical city-state lies along the west side of the Tiber River, within the commune of Rome, entirely surrounded by the "Eternal City." Inside the small but well-defined bounds of Vatican City stands the Basilica of St. Peter, the largest church in Christendom; the Vatican, the collection of palaces that serves as papal headquarters; the famed Sistine Chapel and extensive museums and libraries; and the papal gardens. Extraterritorial status also extends to a number of palaces and churches within Rome proper, including the Basilica of St. John Lateran, designated the cathedral church of Rome. About 1,000 persons live within the papal walls, including the Pope, several hundred priests and other religious, the renowned Swiss Guard, and laity who perform a variety of essential tasks.

As the home of the Pope, this small nation is an ecclesiastical state with worldwide influence out of all proportion to its physical size. Over 617 million people, the world's Roman Catholics, give ear to the edicts of their theocratic ruler. Only mainland China commands authority over a greater number of subjects.

MALTA

Malta is situated in the narrows between the eastern and western basins of the Mediterranean, just east of what is known as the "waist" of that sea. The strategy of the "waist" is based upon naval strength, and Malta has been considered the traditional key to the marine passageway. This small, strategically situated island was held by the British since 1800. Valletta, the capital and main port, was a naval dockyard where British Mediterranean fleet units were refitted and repaired. About 75 percent of the Maltese depended directly or indirectly upon the British armed forces on the island for their livelihood. Important as Malta has been in the military strategy of the Mediterranean, however, it is vulnerable to nuclear attack.

Malta became an independent nation in 1964, thus ending more than a century and a half of British rule; simultaneously, it became a member of the British Commonwealth of Nations. The road to independence was not an easy one, and the road after independence has not been easy either, because, aside from its value from a military viewpoint, there is little to support an autonomous economy except the island's scenery as a lure for tourists. In the past, British military spending and money spent by service personnel and their families stationed on the island kept the Maltese economy going. The Malta naval dockyard had been the island's main source of employment. As the Mediterranean headquarters of the North Atlantic Treaty Organization, some money will continue to flow into Malta. The former naval dockyard has been converted to a commercial ship repair dockyard, and a number of factories producing a wide variety of consumer goods have risen nearby. All these are providing new means of employment for the Maltese. It is likely, however, that tourism will be Malta's most important industry, and it has developed rapidly in recent years with the building of luxury hotels and a gambling casino.

Malta is made up of three islands: the largest is Malta, next is Gozo, and lying between the other two is Comino, the smallest. The total area of the islands is 122 square miles; their population is 329,030.

In Perspective

Southern Europe is a land of peninsulas and islands, a land in which the regional unifying characteristics include mountains that dominate the landscape, the Mediterranean-type climate, closeness to the sea or ocean (except in interior Spain), similarities in crops, and large areas of drought-resistant natural vegetation. This is also the land in which the remarkable cultures of Greece and Rome developed. Isolation, caused by difficulties of communication and transportation, resulted in the early development of numerous small independent states and nations. Eventually, most united to form the larger modern nations of Portugal, Spain, Italy, and Greece, but a few of these tiny anachronistic states continue to exist.

During the past quarter century, the development of manufacturing in northern Italy and areas of Spain near Barcelona, Madrid, and along the north coast has made great strides. During the past decade, manufacturing has also increased its foothold in Greece. Manufactured products, especially machinery, automobiles, and textiles, are now moving from this region to world markets. Fishing has long been an important factor in the economy of each Mediterranean nation. Since World War II, with the establishment of more stable governments, tourism has increased greatly, and the "invisible" income from this source has added strength to the area's economy.

In general, the most common Mediterranean occupation is agriculture, which employs almost a third of the total working force. Most farming is still done on small land holdings with simple implements, often by hand methods. The same grains—wheat, barley, and rice—are grown throughout the region; and vineyards and groves of fruit, nut, and olive trees are commercially important. The chief domesticated animals are the goat, sheep, and donkey; even cattle and pigs are scarce in some sections. Most Mediterranean farmers live in small settlements, in houses built of stone and adobe, rather than on individual land holdings.

This region, like many other parts of the world, is confronted with the dual problems of increasing population and urbanization. In most parts of Southern Peninsular Europe, the standard of living is generally lower than that of Northwestern and Central Europe. The continuing general movement from rural to urban areas places still a greater burden upon city utilities, tax structures, educational systems, and general development. Even though the ordinary European automobiles are small, the ancient streets of the historic cities are not capable of handling the traffic. Air pollution and noise, the need for a sufficient supply of good water, and the continued hunt for new industries are the common problems of city administrators and national governments here as in other parts of the world.

REFERENCES

Cole, J. F.: *Italy: An Introductory Geography*, Frederick A. Praeger, Inc., New York, 1966.
Enggass, Peter M.: "Land Reclamation and Resettlement in the Guadalquivir Delta—Las Marisinas," *Economic Geography*, vol. 44, pp. 125–143, April, 1968.
Gentilcore, R. Louis: "Reclamation in the Argo Pontino, Italy," *Geographical Review*, vol. 50, pp. 301–327, July, 1970.
Guzzardi, Jr., Walter: "Boom Italian Style," *Fortune*, vol. 78, pp. 136–145, May, 1968.
Houston, J. M.: *The Western Mediterranean World: An Introduction to Regional Landscapes*, Frederick A. Praeger, Inc., New York, 1967.
Kayser, Bernard, and Kenneth Thompson: *Economic and Social Atlas of Greece*, Center of Economic Research, Social Sciences Center, Athens, 1964.
Owen, Charles: "Island in Transition," *Geographical Magazine*, vol. 39, pp. 203–213, July, 1966.
Rodgers, Allan L.: *The Industrial Geography of the Port of Genova*, Department of Geography Research Paper 66, University of Chicago, 1960.
Stanislawski, Dan: *The Individuality of Portugal: A Study in Historical-Political Geography*, University of Texas Press, Austin, 1959.
Walker, D. S.: *A Geography of Italy* (2d ed.), Methuen & Co., Ltd., London, 1967.
Walker, D. S.: *The Mediterranean Lands*, John Wiley & Sons, Inc., New York, 1960.
Way, Ruth: *A Geography of Spain and Portugal*, Methuen & Co., Ltd., London, 1962.
Yearbook of International Trade Statistics, 1968, United Nations, New York, 1970.

9

Eastern Europe

Eastern Europe, sometimes referred to as the Shatter Belt, extends from the Baltic a thousand miles southward to the Ionian Sea, and 800 miles eastward from the Adriatic to the Black Sea. The seven countries of Albania, Bulgaria, Czechoslovakia, Hungary, Poland, Romania, and Yugoslavia are within this broad belt. Together, they embrace an area of some 450,000 square miles, which in size is equivalent to a little more than the combined areas of Texas and California. The appellatives used to designate this area conform more to common usage than to actual conditions. The expression "Shatter Belt," on the one hand, implies political instability; but long history has proved the indigenous cultures more shatter-proof than the political boundaries enclosing them.

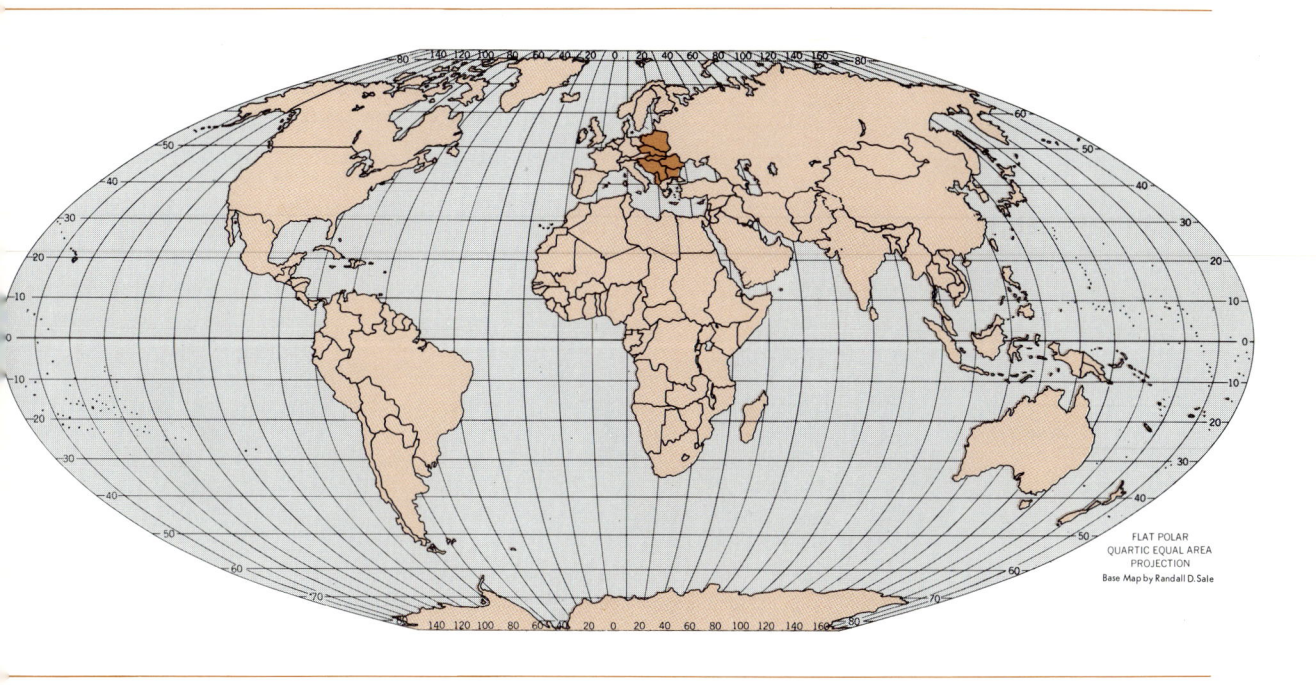

On the other hand, the locational term "Eastern Europe" signifies relative position. Budapest, located in the midst of Eastern Europe, is as far west of Moscow as it is east of London. Geographically, then, this area is the central core of the European Peninsula where intra-Shatter Belt conflicts have been far fewer than is popularly supposed.

Located between the more populous, unified, and powerful Germanic and Russian realms, Eastern Europe was long the chosen war arena of its expansion-bent greater neighbors. The times of tranquility in this area were few and short. The last was at the turn of the century, when Austria-Hungary, Turkey, and Russia dominated the scene. The nearsighted jealousies of the fragmentary Succession States after World War I were a continuing threat and an insurmountable obstacle to enduring peace in Europe. Individually small and weak, the Eastern European states, victor and vanquished alike, looked to powerful allies for security and redress. The great powers, taking selfish advantage of the situation, used these nations as pawns in their game of power politics. The last two great global wars were sparked in Eastern Europe. In the aftermath of World War II, all Eastern European nations were brought under Communist control. The ensuing close ideological, economic, and military ties with the Soviet Union resulted from forced Soviet military imposition rather than from free popular preference. Dissatisfaction with this forcible arrangement, so explosively expressed by the

Fig. 9-1 Eastern Europe includes seven nations, all of which are communist. All except Yugoslavia and Albania are aligned with the U.S.S.R. Albania is aligned with China, but Yugoslavia is independent of either of these political blocs.

Hungarian uprising of 1956 and by the 1968 civil disturbances in Czechoslovakia, may yet see Eastern Europe spark another war. Although local chauvinism contributed its share to feelings of political impermanence, the area's repeated boundary adjustments most often came prefabricated from distant world capitals. Apparently no quick and ready redress in boundaries and political ideologies is possible; moreover, totalitarian policy confines the mass populations of Eastern Europe within the prearranged boundaries. With the exception of higher governmental and party officials and technicians, few others are permitted to cross, or take the risk of crossing over, the physical and psychological boundaries now established between Eastern and Western Europe. Were the people not literally fenced in, it would be interesting to see where and in what numbers the inhabitants of the East would go of their own free will.

Over 115 million people inhabit the area called Eastern Europe, most of whom are Slavs. Nevertheless, a strong core of Magyars, Romanians, and Germans also exists in their midst, and lesser concentrations of Albanians and Turks are found along the area's southern borders. The Poles, Czechs, Slovaks, and Ruthenians make up the Northern Slav group; the Slovenes, Croats, Serbs, Bulgarians, and Macedonians represent the Southern Slavs. The racial kinship of Slavic and Orthodox Christians linked together the Balkans, and more especially the Southern Slavs, who were closer to imperial Russia; but the Northern Slavs were oriented westward. Although Christianity—Catholic, Orthodox, and Protestant—as well as Islam and Judaism are the traditional religious preferences, in recent years public declarations of atheism have been popular.

Eastern Europe is noted for its physical and cultural diversity. Highly developed industrial areas give way, especially as one travels eastward and southward, to some of the poorest farmlands on the Continent. Brilliant cosmopolitan centers of art and refinement are offset by hamlets of backward peasants. Ancient structures exist in the shadow of ultramodern edifices. Grandiose post-World War II plans for industrialization and agricultural collectivization have yielded some spectacular local results, but the planned opulence has not yet pervaded the area as a whole. As with other systems the world over and in times past, the present "omnicompetent" Eastern European welfare states have yet to create a new type of happy citizen and technological paradise. If heterogeneity is a typical characteristic of Europe, then the Shatter Belt is most typically European.

Physical Setting

Relief Features

In its most simplified aspect, the topography of the Shatter Belt consists of five major lowland and four major mountain units. Enumerated, the lowlands are: European Plain (I), Hungarian Plain (II), Walachian Plain (III), Maritsa Valley (IV), and Bohemian Basin (V). The mountain chains are: Carpathian-Balkan System (*A*), Rhodope (*B*), Dinaric Alps (*C*), and Bohemian Rim System (*D*). Many lesser lowlands and uplands, having local importance and local names, are also associated with these major topographic units. The topographic complexity of the region's central and southern two-thirds is in contrast to the simplicity of its remaining northern sector.

The Mountains (*A*) The Carpathian and Balkan Mountains are the longest, highest, and most rugged mountains in Eastern Europe. Together, they form a great double horseshoe arc that cradles the Hungarian and the Walachian plains. Evidences of earlier glacial erosion and deposition are common, and deep snowfields last most of the year at higher elevations. While their great length and height make them the prime watershed of the region, the deep glacial valleys offer excellent hydroelectric damsites. The Carpathians, especially, stood as definite barriers and provided stable political boundaries for about a thousand years—this despite the easily negotiable passes through them. Their three notable passes, the Uzhok, the Veretski, and the Jablonica, offer access from the Ukrainian S.S.R. to the Hungarian Plain. After World War II the Soviet Union gained a highly strategic foothold by incorporating this segment of the Carpathians, and Ruthenia as well, into the Ukraine.

(B) Alternating crests and valleys form the Rhodope Mountains. Running water, rather than glaciers, cut deep valleys into the upstanding igneous mass, but broad areas of high meadows remain above the forest-covered slopes. Racial diversity compensates for population sparsity, especially in the western parts of the Rhodopes. Here, side by side yet almost completely separated, live Bulgars, Serbs, Pomaks (Bulgars converted to Islam), Greeks, Turks, Jews, and Macedonians, speaking different languages, using different scripts, having different social organizations and mores, and often ready to fight each other.

(C) The Dinaric Alps and associated ranges of Yugoslavia form a corrugated arch with one limb rising from the Adriatic and the other from the Hungarian Plain. The folds are neither so steep nor so intense as those in the mountains elsewhere. Aside from the parallel valleys and ranges, the key to topographic development is the thick limestone and dolomite strata present nearly throughout. These calcium carbonate rocks were subjected to solution, especially after deforestation, with the result that inumerable sinkholes, caverns, poljes, and dolines now dot, if they do not dominate, much of the landscape. The features produced by rainwater on soluble strata are termed, collectively, karst topography, after their excellent development in the Karst (Carso) highlands and plateaus that represent the northernmost extension of these mountains.

The narrow Adriatic littoral, with its many elongated offshore islands, is wedged between the sea and the Dinaric Alps. The coastal strip is as accessible from Italy as from the interior. Once forested, the backing Dinaric Alps appear white because of the dominant light-colored limestone and the absence of vegetation and topsoil. Along the Yugoslav-Albanian border the geological complexity of the mountains is augmented by rugged relief, solution, lack of surface water, and lack of any broad flatland with fertile soils. The many poljes, representing inset windows dissolved from limestone, are local steep-sided basins with tillable flat floors. These are oases of good agricultural and spring-watered lands. Unfortunately, their possessors frequently fought over the poljes and developed an inbred, isolated, clan social system therein. Empires came and vanished, each leaving some imprints on the people of this area, but no empire was able to bring them completely into its own cultural community.

(D) The Bohemian Rim actually consists of three ranges; the Bohemian Forest and the Ore and the Sudeten Mountains, which together form a quadrilateral diamond-shaped pattern enclosing, or "rimming," the Bohemian Basin. These mountains attain elevations of over 4,000 feet above sea level, but generally they rise only about a thousand feet above their immediate surroundings. Both their igneous summits and their sediment-overlapped flanks have long been subdued by erosion. Although they are old worn-down mountains—again, neither so high nor so rough as the Carpathians or the Alps—they are nevertheless densely forested, and therefore form significant natural barriers between the Germanic and the Slavic peoples. Traditionally, the Czech people occupied the basin, and the Germans inhabited both flanks of the mountains.

Fig. 9-2 Eastern Europe is divided into several physical regions by the arrangement of various mountain chains. Three major plains areas separated from each other by the mountains are the European Plain, the Great Hungarian Plain, and the Walachian Plain.

The Lowlands (I) Extending from the Urals as an ever-westward-narrowing wedge whose apex in Belgium points to the Atlantic, the European Plain is by far the largest flatland on the Continent. The portion of this vast plain stretching from the Oder-Neisse river system to the Pripyat Marshes and a 300-mile breadth from the Baltic Sea to the Carpathian foothills together compose the northern one-third of the Shatter Belt. Besides the low surface sloping toward the Baltic, a key factor in the European Plain is glaciation. While the Continental ice sheet extended to the foothills of the Carpathians, the Sudeten, and other mountains farther west, it did not advance or retreat at a uniform pace. Nor was the glacier clean; great quantities of ground-up rock were incorporated within the ice body. Where the rate of melting exceeded forward motion, the residual rock debris was spread as a flat, low, and waterlogged till or "drift plain." Where the rate of melting and the forward motion were in balance, long and higher ridges, known as "moraines," accumulated in front of the ice sheet. The Southern Morainic Hills and the Baltic Moraine are parallel belts of morainal hills sandwiching the broad glacial valley between them. The accumulating moraines, especially the Baltic Morainic Hills, dammed the northward-flowing rivers into glacial lakes. Addition of glacial meltwaters raised the lake levels until the waters coalesced and began draining westward in great, shallow glacial spillways. After glacial recession, the spillways were drained and the present drainage pattern was established. In places the east-west spillways were easily canalized to provide interconnecting waterways between the present rivers. This network of inland waterways facilitates barge traffic in bulky commodities.

The Oder and Vistula rivers break through the Baltic Moraine to flow sluggishly across the low marshy Baltic Sea coast. The coastline is made irregular by the deep indentations of the Pomeranian and the Danzig bays, which in turn are separated from the inner lagoons by offshore islands consisting of sand dunes. Backed inland by the higher lake-dotted and poorly drained Baltic Moraine and fronted by the Baltic Sea, this coastal strip has seen the eastward march (*Drang nach Osten*) of Germanic settlers since medieval times. Although the coastal regions became German, the hinterlands beyond the moraine remained Slavic. The Germans established large port cities at the mouths of rivers and through them, often in confederation with the Hanseatic League, controlled the commerce of the hinterlands and the Baltic. The expanding German and Slavic populations, together with the accompanying growth of agriculture and trade, caused these ports to become centers of international intrigue and strife. Slavic fortunes increased as the outcome of the two World Wars finally deprived the Germans of their centuries-old Baltic coastal settlements. Even farther inland, the Oder-Neisse rivers now form the de facto boundary of the westward-marching Slavs.

The other four major lowlands, sometimes designated as basins or plains, are diastrophic downfolds. They contain most of the people, most of the arable land, the densest transportation network, and the most intensive economic activities. If the lowlands are the most lively area in terms of human activity, then the Danube is the central nerve of these mountain-segmented basins.

(II) The Danube has some 300 tributaries, nearly all in the Shatter Belt, draining over 300,000 square miles. It is neither "blue" nor known anywhere along its course by the name Danube. Called the Donau in Germany and Austria, the Duna in Hungary, the Dunav in Slavic lands, and the Dunărea in Romania, its seasonally green and mud-yellowed waters run through all but two of the Eastern European nations; Poland and Albania are the exceptions. The Danube enters the Hungarian Plain only to shift in braided channels across deposits of its own making. The constricting line of hills known as the Bakony Forest stands as the final obstacle to its southward flow; here the Hungarians plan large hydroelectric installations. Aside from the Bakony Forest, the vastness of the Hungarian Plain is interrupted by the domal igneous but subdued Bihar (Bihor) Mountains to the east and by the large freshwater Lake Balaton to the west. The immense plain, called the Alföld in Hungarian, is actually a diastrophic basin partly filled with alluvium. While flowing over the Alföld, the Danube is joined by its important Drava and Sava (Száva) affluents, as well as by the flood-prone Tisza. It leaves the Alföld through a spectacular canyon, the Kazan Pass. The impressive rapids, just below the canyon, called the "Iron Gate," were cleared for navigation by blasting the channel during the 1860s. As a joint venture of Yugoslavia and Romania, a high dam is under construction in the canyon for purposes of supplying hydroelectric power to the two nations. Navigable throughout its entire Shatter Belt course, the Danube is, indeed, an international stream.

(III) Genetically and topographically similar to the Hungarian Plain, the Walachian Plain itself consists of an irregular amphitheater-shaped lowland open to the Black Sea. Here, the lands north of the Danube tend to be low, marshy, and prone to floods. By contrast, the land south of the Danube rises to form higher bench levels, or "platforms," and therefore is sometimes called the Danubian Foreland. Thus, while one bank of the Danube is low and marshy, the other (the south bank) is marked by escarpments. The topographic break causing these two distinct levels has long served as a stable political boundary between Romania and Bulgaria. The yellow, accreting muds of the Danube delta caused the Walachian Plain to protrude into the Black Sea. The boundary of the Soviet Union is now contiguous with the lowest segment of the Danube.

(IV) The Maritsa Valley, confined between the Balkan and the Rhodope Mountains, is similar in several ways to the Walachian Plain but is much smaller. The Maritsa River itself parallels the valley for about half its length, only to turn southward and then leave it at a common boundary between Bulgaria, Greece, and Turkey. The Bulgars were long desirous of turning south along the Maritsa to gain access to the Aegean. The floor of the valley is relatively flat, densely populated, and highly productive agriculturally. These conditions have prevailed since classical historic times.

(V) The hilly and well-watered Bohemian Basin is diverse geologically and topographically. It is densely inhabited, highly mineralized, and highly industrialized. Partly because of its central location and partly because of several natural gateways, it is also highly strategic. The Elbe and its tributaries, including the Moldau River, drain the Bohemian Basin through a wide water gap in the Ore Mountains. This gap, known as the Elbe Gate, affords easy access to the German lands of the north. Farther to the east another gateway, the Moravian Corridor (or Gap), is a broad structural and topographic depression between the Sudeten and the Carpathian Mountains. Drained northward by the Oder and southward by the Danube tributaries, the Moravian Corridor is a historic avenue of communication between the Danubian basins and the European Plain.

Climate

Because of latitudinal position and extent, central position, and exposure, the Shatter Belt falls under the influence of four distinct climatic realms. The Mediterranean and humid subtropical climates are restricted to narrow southern and southeastern sectors. One-third of the Shatter Belt, especially the basins of the Danube, lies under the warm, or long-summer phase, influence of the humid continental climate. Other than the main climatic controls of latitude, the Atlantic Ocean, the Eurasian continent, and the Mediterranean Sea, the mountains provide secondary controls that cause intricate local diversity within the highlands. Although in winter cyclonic storms are prevalent throughout, they are rare occurrences along the southern periphery during summer. Hurricanes and tornadoes are unknown.

The mean annual temperatures are expectedly higher in the south and the lowlands than in the north and the uplands. However, the generally higher elevations of the south partly offset the expected warming effect of lower latitudes. Thus the January temperature of Warsaw, averaging 24°F, is only 10° lower than that of the Albanian lowlands. The average July temperature along the Baltic coast increases from 65° to nearly 80°F in interior valleys of the south, but is between 73° and 75°F in the southerly latitudes along the Adriatic. Despite the vast size of the Eastern European area, the summer and winter temperatures in any two given localities do not vary more than 20°F.

Precipitation generally decreases with distance inland from the main Atlantic moisture source. Thus, the 24-inch maximums of the west taper off to less than 15 inches on the Dobrogea (Dobruja) Platform in the east, in Bulgaria and Romania. The eastward-decreasing precipitation is associated with a tendency for summer concentration and lessening reliability. Moisture decrease and variability are controlled by the intensified continentality of the eastward-widening landmass. With the notable exception of the southern coastal strips, the Eastern European summers are everywhere wet and warm to hot. Only along the Balkan littorals are summers dry. Winter snows occur throughout Eastern Europe, but snowfalls are heavier, more frequent, and longer-lasting in the north and the uplands. Precipitation is largely convectional and cyclonic in origin; nevertheless, the mountains, especially along the Adriatic, induce orographic rainfall. Occasional summer hailstorms pelt the lowlands of the southeast but are more frequent in the mountains.

Humid Continental The short- and long-summer phases of the humid continental climate dominate nearly the entire area of the Shatter Belt. The long summer phase, sometimes referred to as the "Central European climate," extends over the Alföld of Hungary, interior Yugoslavia, most of Romania, and northern Bulgaria. Akin to the climate of Iowa, it is also often called the "Corn Belt climate" or that of the oak-hickory deciduous forest. Although the four seasons are well developed, the longer summer and milder spring and fall produce a decidedly longer growing season than in climates of adjacent northern lands. The winters (25°F for January) are, on the average, about 50° colder than the summers (74°F for July). Less than one-fifth of the precipitation, mostly in the form of snow, falls during winter, but nearly three-fifths of the total annual precipitation of 20 to 25 inches comes in the warmer half of the year. The summers are wet and humid, but the winters tend to be relatively dry and crisp. A large number of thundershowers account for the May-June rainfall maximums, especially on the Alföld.

The short-summer phase of the humid continental climate, much like the climate of Minnesota and Wisconsin, is distinguished by longer winters, cooler summers, a shorter growing season, and by somewhat less precipitation, particularly along the eastward margins of the Shatter Belt. Poland, Czechoslovakia, and north-central Romania underlie the short-summer-phase climate. Winters are not only long, but decidedly cold and relatively dry despite frequent snowfalls. Summers are short, cool, and wet. The average July temperature of 65°F is some 40° higher than the January average of 25°F. The short growing season and the less-reliable rainfall limit agricultural productivity to levels below those of the adjacent climate of the southlands.

Humid Subtropical The sheltered Maritsa Valley of Bulgaria and the Yugoslavia highlands facing the Adriatic experience mild winters (January average, 35°F) and year-round precipitation in excess of 26 inches. Though the Maritsa Valley is densely inhabited and intensively tilled, the calcareous uplands of Yugoslavia are severely leached and sparsely populated. Rice and cotton fields in the Maritsa Valley attest the subtropical character of that region's climate. These restricted portions of the Shatter Belt find a climatic, if not a topographic, analogy in the United States in the state of Tennessee.

Mediterranean California-type climate in the Shatter Belt is confined to the Adriatic islands and the coastal strip of Yugoslavia and Albania, as well as to the lesser southward-facing valleys of the Rhodopes of Bulgaria. While winter temperatures are generally above freezing, occasional frosts and snowfalls do occur. Also, the disagreeably strong and cold north winter wind, the bora, at times drains into valleys with devastating results to plant life. Nearly all precipitation falls during late autumn and winter, leaving the summer months hot (July, 73 to 80°F) and dry. Depending upon local altitude and exposure conditions, the average yearly rainfall ranges from 20 to 60 inches. The comparatively high winter temperatures, together with a scenic location along marginal seas backed by mountains, make the Mediterranean portions of Eastern Europe one of the most popular "playground climates" of Europe.

Mountain An endless variety of climates prevail in the mountains. Although elevation and slope facings produce myriad local departures, in the main the regimes of the four principal climatic realms are discernible in the Eastern European mountains. Generally, mountains have the coldest temperatures and receive the heaviest summer rains and deepest winter snows. While many mountain slopes are high enough to extend above the timberline into the tundra-type climate, only a few peaks attain the zone of eternal snow.

Vegetation and Soils

Occupation of Eastern Europe has been so long and continuous that only isolated marshes and mountains have escaped radical alteration. Man seems to be always in competition with natural vegetation for the use of the land. After long association and a continuing need for wood and pasture, however, man has learned conditional competition. In most areas he now practices tree farming rather than forestry, regulates tree cutting and meadowland grazing, and elsewhere reforests cutover lands.

Originally at least two-thirds of the Shatter Belt was forest-covered. Today, after two millennia, less than one-third of its area remains in forest. Over one-fourth of the total areas of Romania, Bulgaria, Yugoslavia, Albania, and Czechoslovakia is classed as forestland. Generally, the densely forested lands are coextensive with the mountain slopes otherwise unattractive for agriculture. The principal zone of forest lies between the upper limit of tree growth, at 6,500 to 7,000 feet, down to the less-steep foothills. Valuable stands of softwood conifers, mainly spruce and pine, yield to beech and oak hardwoods in the lower lands or along stream courses. The countries least- and most-forested are Hungary (13 percent) and Albania (46 percent). Lumbering is a leading activity in the highlands.

Distinctive low, Mediterranean-type xerophytic brush dominates the vegetation along the Adriatic strip and atop the physiologically dry calcareous rocks of Yugoslavia. This dense, thorny vegetation bursts into green leaf and brilliant flower after winter rains, only to be withered to a dusty gray by summer drought. Herds of goats browse amidst these low brushy plants for lack of more palatable vegetation.

Originally, coniferous forests covered much of the Polish sector of the European Plain, but only isolated wood plots remain there today. The ever encroaching hygrophytic vegetation of the glacial marshes has filled many lakes to a stage of extinction, but provides a source of peat for the present-day inhabitants of these areas. By contrast, the landward-drifting dunes along the Baltic fringe are barren or are precariously anchored by low, drought-resistant plants.

The eastern part of the Walachian Plain and some lesser open parks on the Dobrogea Platform were originally grass-covered. But the most extensive natural grasslands of Eastern Europe have always been the "puszta" of the Alföld. This treeless steppe was the choice grassland, and subsequently the choice plowland, of the Shatter Belt. The early home-seeking Magyars, guided by their grassland economy, deliberately chose it as their permanent homeland.

Through extensive drainage in the northlands and local irrigation in the south, and through long use over the entire area, the soils of Eastern Europe have been greatly altered and generally improved. Although the gray, podzolic, glacial soils of Poland tend to be waterlogged and leached of soluble plant nutrients, they respond readily to drainage and fertilization. But the reddish terra-rossa soils prevalent within the Mediterranean strip and upon limestone surfaces, representing the accumulated insoluble residues of chert and clay, are seldom fertile or easy to alter. Large-scale transfer and mixing of the limestone-derived pedocals of the south with the glacially derived pedalfers of the north would greatly benefit both soil groups. But possibilities of radical change in soil characteristics are far less likely than those for altering the vegetation ground cover.

The most naturally fertile soils of Eastern Europe are the chernozems developed under a grass cover in the Alföld and the Walachian Plain. Slightly less fertile, but nevertheless highly productive, are the loessal soils along the northern forelands of the Carpathian and Sudeten Mountains of southern Poland. Generally, the glacial soils of Poland are naturally the least fertile and productive within the area. Although wide variations in soil type and fertility are present along alluvial or diluvial strips and within poljes, the swifter rivers, steeper slopes, and calcareous rocks permit only rubble-strewn skeletal soil development.

Human Relationships

Population

A hypothetical spreading of 100 million Eastern Europeans over 450,000 square miles would result in an average of some 222 persons to each square mile. Such uniform distribution, of course, is not found; rather, limited areas of intense concentration alternate with larger areas inhabited by few people. Hungary's overall population density of 292 to the square mile stands in contrast to Albania's density of 196 per square mile. The population densities of the other five Eastern European nations lie between these extremes. Although density figures are indicative of areal carrying power, they are not altogether reliable either with reference to present land capabilities, to potential capabilities, or to future capacity. While most of the land of Hungary is arable and productive, most of Albania is high, rugged, dry,

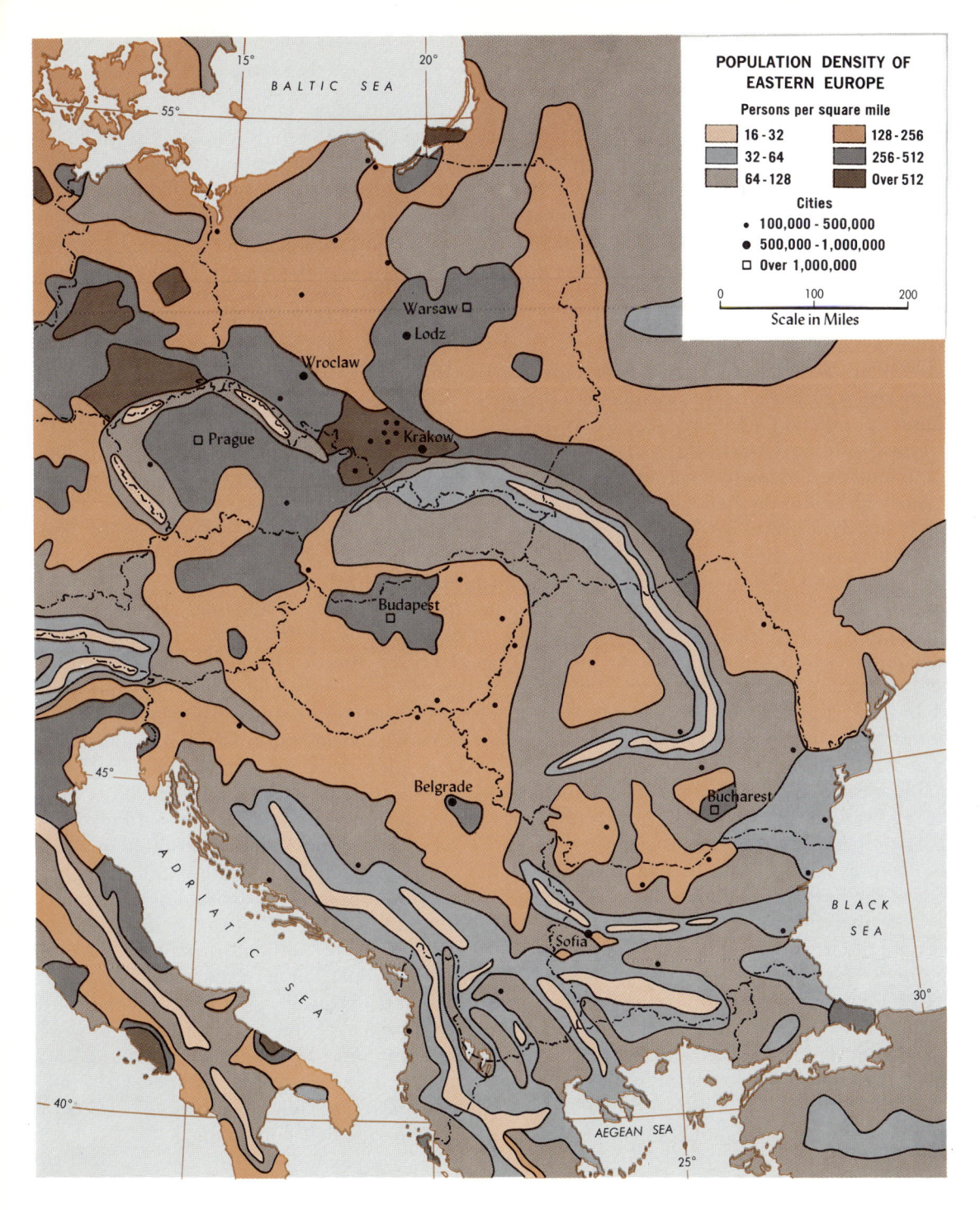

Fig. 9-3 The population pattern of Eastern Europe also clearly indicates the location of the mountainous regions, for the higher and more rugged areas are those having the lowest population density.

malarial, or otherwise limited agriculturally. Selected portions of Eastern Europe have been inhabited since prehistoric times, and population increase has been numerically steady since then; nevertheless, not every part of the region has been occupied with equal thoroughness. Disadvantageous topographic conditions, soil infertility, distribution of minerals, and stages of economic development have all contributed to the unequal dispersal of people.

In the main, Eastern Europe has been agricultural in regions where the best farmlands and the densest population stood in direct correlation. This condition is still largely true. The densest population, namely, over 250 persons per square mile along the best soils of southern Poland, is matched by similar high densities in the Danubian basins. By contrast, the high and rugged Carpathians, Balkans, Rhodopes, and Dinaric Alps, as well as the Baltic Moraine zones, exhibit the low average densities of 61 to 125 persons to the square mile. Densities in the central Yugoslav-Albanian highlands fall to record lows of less than 60 per square mile.

Continuously improving transportation facilities and new discoveries of minerals in the mountains have provided the bases for urban concentrations in the midst of agricultural lands, as well as for huge industrial clusters in mountainous lands. As an example, the minerals of the Sudeten Mountains have been a magnetic attraction for settlement, until today they are the world's most densely populated mountains. Many major contemporary economic developments are predicated upon mineral deposits in mountains, particularly in Bulgaria, Yugoslavia, and to a lesser extent Romania. What minerals did for population density in the Sudeten Mountains may well be repeated in other mountainous zones of the Shatter Belt.

Good roadways enable the farmer to transport his agricultural products to focal points where they are prepared for the table or used as industrial raw materials for other products. As transportation facilities improved, large cities grew up in the midst of agricultural lands. Budapest is the largest metropolitan city in Eastern Europe; it is approached in size by Bucharest. Warsaw and Prague are the other cities with populations in excess of 1 million. Poland, with 23 cities of over 100,000 inhabitants, has the greatest number of big cities; Albania has only one city in this class, its capital Tiranë.

Generally, urbanism in Eastern Europe does not attain the high levels common in Western Europe. Indeed, the Shatter Belt is historically a land of villages and smaller rural settlements, where even towns of more than 25,000 people retain the aspects of overgrown villages. However, recent agricultural collectivization and, more especially, planned industrialization are causing rapid urban growth.

Again in general, while the Eastern European birth rate is highest in Albania and lowest in Hungary, the death rate is highest in Hungary and lowest in Poland. The infant mortality rate is highest in Albania (also the highest in all Europe) and lowest in Czechoslovakia. Statistics are less accurate in recording the losses from wars or from racial and political persecutions. Changing boundaries have often meant the wholesale expulsion of unwanted minorities. Neither higher education nor a higher living standard seems to have lessened man's inhumanity to man. This proclivity, it should be stressed, is not confined to Eastern Europe. Large-scale repatriation of minority groups was a deliberate postwar policy of Poland and Czechoslovakia. Forced recruitment for work in the Communist-bloc nations, political deportations, and surreptitious escapes to other lands are less carefully documented aspects of shifting Shatter Belt population.

Cultural Diversity

Although all Shatter Belt countries are superficially united under the aegis of a Communist utopian social system, deep-seated cultural forces preclude both uniformity, or standardization, and everlasting unity. In lands where traditional cultures were badges of identity, mutual respect for the other man's cultural heritage was readily obtained, simply because it was so fiercely retained. Cultural identity is perhaps the one thing for which Shatter Belt peoples are prepared to fight. The Western individualistic, free, and democratic attitude and the contrasting Eastern collectivist, fatalistic, and autocratic viewpoint represent two differing ideological streams flowing through Eastern European life. Shatter Belt developments may be understood in the light of this contrast and conflict between "Occidental" and "Oriental" ideologies.

Languages

If language is a basic tool for human communication, then there is neither a scarcity nor a sparsity of it in the Shatter Belt, where no less than seventeen languages are spoken. The Slavic languages, all closely related, are spoken by the largest number of people here. The Poles, Czechs, Slovaks, and Ruthenians belong to the Northern Slav group and prefer the Latin alphabet for written communication. The Serbs, Croats, Slovenes, Montenegrins, Bulgars, and Macedonians belong to the Southern Slav category. The Croats and Slovenes use Latin script, but the other Southern Slavs learn the Cyrillic alphabet.

Major non-Slavic languages used in the area are Hungarian, Romanian, and Albanian. German, Yiddish, Italian, Turkish, and Gypsy are minor languages spoken by numerically decreasing groups of communicants. The Hungarian language is unrelated to any other in the area. Although, by virtue of prolonged contact, modern Hungarian employs some Turkish, Latin, and German expressions, the main language body is Finno-Ugric, with strong Sumerian affinities. The Albanian language is an amalgam of ancient Illyric with modern Slavic and Turkish words. These major non-Slavic ethnic groups employ Latin script, but the older Turkish (Arabic) script has not yet faded from use among some small Muslim groups.

Religions

Roman Catholicism, Greek Orthodoxy, and Protestantism are the dominant religions of Eastern Europe. Poland is almost entirely Roman Catholic; Czechoslovakia is about three-fourths so. More than 2 million Hungarians represent the largest Protestant minority in Eastern Europe, but two-thirds of Hungary is Roman Catholic. The Slovenes and Croats are largely Roman Catholic, but elsewhere Catholicism finds fewer adherents. The

Fig. 9-4 The Alföld, a part of the Great Hungarian Lowland, has fertile chernozem soils and is a prime agricultural area in Eastern Europe. (Courtesy of Hungarian People's Republic)

main body of Greek Orthodox adherents are found among the Bulgars, Serbs, and Romanians; but again, fewer numbers are present in Poland, Albania, and Czechoslovakia.

With two-thirds of its population adhering to the teachings of the Koran, Albania is the only Muslim country in Europe, although large Muslim groups still live along the Yugoslav-Bulgarian border.

Perhaps the world's largest concentration of Jews was found in prewar southeastern Poland; to this area, many Americans of the Jewish faith are able to retrace their national origins. Although all Shatter Belt countries have Jewish minorities, the number of Jews in the Balkans was comparatively small. The tragic events of World War II saw many of these Jews killed or expelled; but about a half million Jews still remain in the Shatter Belt.

Economic Development

Eastern Europe is richly endowed agriculturally and mineralogically. Southwestern Poland, Bohemia, Moravia, and western Hungary boast a high order of economic development, but the eastern sectors of the area are just now planting the seeds of an industrial revolution. Through recent plans, each country ambitiously projects a "great leap forward," thereby hoping to reap the fruits of planned industrialization. International cooperation for purposes of economic, technical, and scientific development functions through the Council for Mutual Economic Assistance (CMEA). Although Albania ceased to participate in CMEA, and Yugoslavia does so selectively, the other nations of the Shatter Belt consummated various long-term multilateral trade agreements, planned further development of mutual resources, and planned the increase of production, all in relation to the broader needs of the Eastern European bloc. More recently, Western nations also have been invited to contribute technological know-how, as exemplified by the proposed building of fertilizer, glass, and plastics plants by American companies in Romania and Hungary.

Agriculture

Agricultural output is sensitive to both peace and war. During normal times, Eastern Europe produces a significant food surplus. Wheat, concentrated in the Danubian lands, is the principal crop; it is trailed in importance by maize in the southeast and by rye and potatoes in the northern glaciated sectors. Oat fields are common in central Poland and in the highlands of the south. Rice, a cereal crop recently introduced, is restricted to eastern Hungary and to several Balkan valleys. Barley is grown nearly throughout the Shatter Belt. The sugar beet is intensively cultivated in Poland, Czechoslovakia, and Hungary for domestic use and for export. Hemp, flax, and hops are raised in the north; corn and tobacco are important crops in the south. Fruit orchards and vineyards, although common throughout the area, become specialties on sandy soils and slope facings in the foothills of Hungary, Romania, Czechoslovakia, Yugoslavia, and Bulgaria. Soybeans, cotton, sunflowers for seed, roses for oil, and opium poppies are other specialty crops of local significance.

Animals have always been important in the agricultural economy of Eastern Europe. Beef cattle, especially in Hungary, are most numerous in the northern areas; but large dairy herds have replaced the horned cattle in western Poland and Bohemia. Sheep and goats are the dominant animals in the Balkans. Poultry, including millions of ducks and geese, is ubiquitous but most numerous in the western sectors. Horses and oxen are the traditional draft animals on better farmlands; mules and donkeys are still used on poorer southern farmlands. Collectivization has provided for cen-

LARGER TERRITORIAL CHANGES SINCE W.W. II

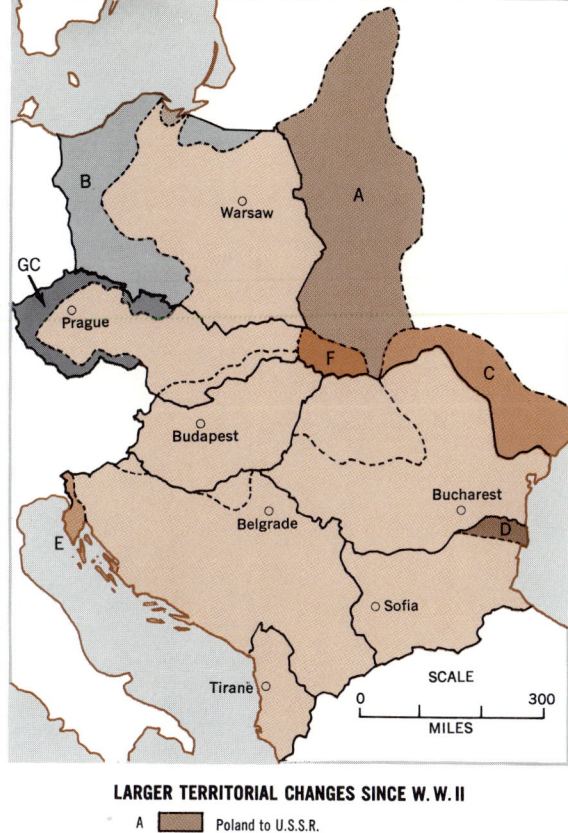

A	Poland to U.S.S.R.
B	Germany to Poland
C	Romania to U.S.S.R.
D	Romania to Bulgaria
E	Italy to Yugoslavia
GC	Alternately German and Czech
F	Hungary & Czechoslovakia to U.S.S.R.

Hungarian boundaries changed alternately with neighboring countries during war.

Fig. 9-5 Seventeen different languages and most major religions are represented in the different areas of Eastern Europe.

tral tractor stations where agricultural mechanization has displaced a large number of draft animals.

At present, most of Eastern Europe's employable labor force is used in agriculture. In Romania and Bulgaria over 50 percent of this force are farmers; only in Czechoslovakia, Hungary, and Poland is less than one-half of the labor force engaged in agriculture. Agricultural development was retarded throughout the area by centuries of autocratic Russian rule and wars over the control of Poland, as well as by long Turkish occupation of the Danube basins and the Balkans. Wars not only drained off needed farm workers, but ensuing boundary changes reoriented, if they did not disrupt, agricultural practices and markets. The problem of landless peasants and absentee aristocratic landowners became a sociopolitical storm center for which everybody, within and outside the Shatter Belt, offered fool-proof panaceas. The large landed estates—often smaller than an average 250-acre farm in Iowa and never as large as a Western ranch in North America—were eventually broken up and parceled out among the peasants. These small parcels were promptly collectivized into much-larger collective farms ("kolkhozes") or state farms ("sovkhozes") administered by absentee agropolitical officials. The questions of who and how many own the land and the livestock, and whether or not new owners and new social and political systems can encourage cows to give more milk, sheep to have more lambs, and hectares to yield more corn all remain to be answered.

Agricultural output seems to have remained on a plateau during the past decades. In the case of Hungary, a traditional wheat exporter, the nation found itself unable to meet domestic wheat requirements. Aside from man-induced difficulties, the soils of the northern glacial lands and the southern rugged lands are weak or outright infertile. Fertilization and other modern practices would greatly increase the farm output, but heretofore the individual farmer has had little available capital for soil improvement. Over much of the area climate too, if not actually marginal, is capricious enough to limit agricultural production. Crop yields are therefore lower than in Northwestern Europe, and farm labor productivity is below that of the American farmer. Farm mechanization is not likely to increase productivity markedly; agricultural machinery produces more efficiently, not more abundantly.

Mining and Manufacturing

The mountain-and-basin warped rocky crust of Eastern Europe affected not only the topography, climate, hydrography, soils, and vegetation of the area, but it also brought many different minerals within minable depths. Coal and lignite, petroleum and natural gas, bauxite, copper, zinc, lead, iron ore, potash, salt, and several other natural resources are extracted or mined on a large scale, and increased production quotas are strongly promoted by the governments of all seven Shatter Belt nations. Under the Soviet system, all earth resources, like other aspects of the economy, are exploited through nationalized industries. Except for those in the U.S.S.R., the Romanian petroleum fields are the largest reserves in Europe. The production of petroleum in Hungary and Yugoslavia is also significant; Bulgaria, Poland, Albania, and Czechoslovakia are lesser producers. Czechoslovakia, Hungary, Yugoslavia, and Poland work the leading lignite deposits of the area, but Bulgaria has recently matched the output of the last three countries. Altogether, the seven Eastern European countries mine over 150 million tons of coal, 170 million tons of lignite, and 19 million tons of petroleum annually. In addition to the power fuels, all these nations except flatland Hungary have great hydroelectric potential, as yet only slightly developed even by those countries most industrially advanced. Nevertheless, hydroelectric installations are being constructed at rapid rates throughout the region.

Eastern Europe mines some 10 million tons of iron ore annually, mainly from mines in Yugoslavia, Poland, Romania, and Czechoslovakia. Bauxite from Hungary and Yugoslavia and lead and zinc from Yugoslavia and Poland are nonferrous metals mined in sizable amounts. Although some minerals have been mined in Eastern Europe since prehistoric times, wars and frequent changes in management and markets have retarded large-scale, protracted development. Many nations are concerned over the problem of industrial demand exceeding the supply of mineral resources. In Eastern Europe this concern is reversed: development or exploitation lags behind known mineral resources.

Fig. 9-6 Lignite mine at Turoszow, Poland. The development of this mine and power plant was one of the biggest investments made by the Polish government in the late 1960s. (Courtesy of Polish Embassy)

The lifeblood of Eastern Europe's industry flows from its mines. Czechoslovakia and Poland are the leading mining nations and, correlatively, are the most industrialized nations of Eastern Europe. Czechoslovakia alone has perhaps as many industrial establishments as the other six countries together. Poland, however, is in possession of Silesia, one of the principal industrial zones of Europe. Other Shatter

Belt nations, hoping that an industrial economy will find firm support around the newly formed iron-steel backbone, built steel plants at Zenica, Yugoslavia; Dunaujvaros, Hungary; Pernik, Bulgaria; and Hunedoara, Romania. Without exception, industrial production over the past decade has risen at an accelerated pace in all Shatter Belt countries.

Transportation

Over one-half of the more than 42,000 miles of railroads serving Eastern Europe are concentrated in the western sectors of Poland and Czechoslovakia. Railway mileage elsewhere reflects the relative size as well as the geographic disadvantages of the other countries; small and mountainous Albania, for example, has less than 100 miles of railway. The fine early railway system of Hungary radiated from Budapest, but the changed World War I boundaries disrupted the system by placing international boundaries between the radial and peripheral connecting trackage. Indeed, in one instance a town was left in Hungary, but its connecting tracks and railroad station were turned over to Czechoslovakia. Such matters not only impaired transportation efficiency but also embittered the participants, thus increasing the neighbors' willingness to go to war. Ignoring regional cultural differences, the Communist leadership of the Shatter Belt has encouraged an all-region "internationalization" of railway traffic that is oriented toward the Soviet Union's needs.

Nearly 250,000 miles of public roads traverse Eastern Europe. As with railroads, the denser and better highway pattern of the west gives way to sparse and poor roads—often country dirt roads—farther east and south. Nearly all 6,000 miles of the area's navigable inland waterways are in Poland, Yugoslavia, and Hungary. The Baltic and Adriatic seas afford the large Polish and Yugoslav merchant marine deepwater access to major international trade routes; the Black Sea gives the Romanian and Bulgarian fleets only local coastwise contacts. Even though all major Shatter Belt cities are served by airlines, Prague, as the westernmost terminus of the "Prague-to-Moscow axis," has the greatest international air traffic.

Countries of Eastern Europe

It is unrealistic to consider the seven countries of the Shatter Belt as forming a homogeneous geographic or an indissoluble socioeconomic unit. Although all are united by Communist governments, the light of long national histories and traditions illuminates their broad cultural and physical differences.

In a cultural and religious sense Poland, Czechoslovakia, and Hungary have for centuries been oriented to the West. On the other hand, the Balkan states of Yugoslavia, Romania, Bulgaria, and Albania have long looked to Greek Orthodoxy and the Byzantine autocratic political systems of the East. Moreover, Poland, Hungary, and Bulgaria have maintained historic friendships, whereas the Slovaks, Croats, and the Transylvanian peoples were for centuries an integral part of the Magyar domain. The Southern Slavs, united by the common misfortune of long Turkish rule, found themselves in transitory entente with Romania and Czechoslovakia. It is unfortunate that the unconstructed political conditions after World War I and again after World War II were not appreciated by the decision-making great powers. It is unfortunate also that the spirit of "The enemy of my enemy is my friend" prevailed over the democratic process of "Let the people most intimately affected decide for themselves." Prolonged happiness and mutual satisfaction with political compartmentalization have not been the outstanding experiences of Shatter Belt peoples.

POLAND

Poland is the largest and the most populous Shatter Belt state. The nation's area is larger than that of Michigan and Wisconsin combined; its population of approximately 33 million exceeds that of the four U.S. southwestern states from California to Texas. Poland slopes from the Carpathian-Sudeten crests of neighboring Czechoslovakia across the glaciated European Plain to the Baltic Sea. Its western Oder-Neisse riverine boundary is contiguous with restive East Germany; the nation's longer eastern boundary, partly along the Bug River but generally unmarked by other natural features, is shared with the Soviet Union. The contemporary eastern and western boundaries of Poland are thus as vulnerable as they are new. Indeed, the resilient Poles have long lived within historically more transitory than stable boundaries and have always known the fate of a buffer state where many new starts were made in organizing political independence.

The cultural heart of Poland lies around the ancient capital of Kraków in the upper Vistula Valley. From here Poland once ruled the lands between the Baltic and the Black seas and from the Oder to the Dnepr rivers. Her powerful legions not only checked the eastward-marching Tuetonic Knights but also entered Moscow and defeated the westward-marching Turks at Vienna (1683). After three partitions between Russia, Prussia, and Austria, Poland disappeared from the scene in 1795, only to be resuscitated by Napolean in 1812. A short-lived three-year period of independence was ended by repeated Russian occupation that lasted to the end of World War I. Following the peace treaties of Versailles and Riga (1918–1921), an expanded Poland included German, Russian, and Lithuanian minorities, and this change at once embroiled newly independent Poland with its neighbors. Nazi Germany and Communist Russia jointly invaded Poland in 1939 ostensibly to free German and Russian minorities, thereby precipitating World War II and the fourth partition of Poland. Post-World War II diplomatic accords provided for a revived Poland. The eastern Polish territories were transferred to the Soviet Union, but Pomerania, Silesia (Śląsk), and the southern part of East Prussia were transferred from Germany to Poland. Thus, geographically, the Polish nation was shifted westward.

The Polish state appears to have come out of

Fig. 9-7 With the exception of Albania, the boundaries of all the countries of Eastern Europe were changed as a result of World War II.

the war richer and with better boundaries than it had before. Whereas the ceded eastern strip, as large as Missouri, was a poor and undeveloped region, the newly occupied western part, as large as Kentucky, was rich in resources and highly developed, despite the heavy damage it had suffered during World War II. Poland fell heir to (1) Europe's largest and second-best coalfield in exchange for most of its oil fields and potash deposits; (2) the great industrial capacity of all of Silesia; (3) the navigable Oder River; and (4) the Baltic seaports of Szczecin (Stettin) and Gdańsk (Danzig). Pending a final peace treaty, the United Kingdom and the United States consented to this arrangement as a de facto solution at the Potsdam Conference (1945).

Contemporary Poland is almost free of ethnic minorities; but the price for this cultural homogeneity was paid in millions of human lives lost or displaced. Some 3 million Galician Jews were killed or dispersed during the Nazi occupation; the Soviet Union claimed the country's Ukrainian, Belorussian, and Lithuanian minorities; and the Poles and Russians expelled or annihilated some 8 to 9 million Germans. Several million Polish workers and peasants from Poland proper joined the stream of Polish migrants from the east to new homes on former German farms and in cities of the western sector. Here, the expression "Shatter Belt" is not without an ironic implication of vengeance.

The northern two-thirds of Poland is a low plain ribboned by parallel east-west morainal ridges and intervening valleys. Although the sluggish Vistula and Oder rivers drain the nation's waters to the Baltic, innumerable shallow relict mud-basin lakes and marshes remain scattered over the surface. The podzolic soils are frequently stony, acidic, infertile, and heather-covered when not in cultivation or under wood plots. The soils, moraines, spillways, lakes, and rivers are all of glacial origin.

The southern one-third of the nation consists of a discontinuous plateau surface blanketed by loessal soils of unusually high fertility. This marginally dissected plateau merges southward with the rapidly ascending, densely forested slopes of the Carpathian and Sudeten Mountains.

The Polish climate is characteristically continental. Since higher altitude cancels out the advantage of southern latitude, the Polish average temperatures are remarkably uniform nationwide. The 24-inch yearly precipitation is fairly evenly distributed throughout the year; the western sector, nevertheless, is slightly wetter and warmer than the east.

The observation that the best farmer gravitates toward the best lands and that the inefficient farmer ruins the best farms may be proved in western Poland. Many less skilled Polish farmers from agriculturally backward eastern sectors replaced the traditionally careful and productive German farmers on the newly acquired western lands. Although the western and southern sections of Poland are agriculturally the most productive, over one-half of the nation's land area is tillable; however, slightly less than one-half of the Poles are agriculturists. The main cereals grown are rye, wheat, oats, and barley. In addition to potatoes, of which Poland is one of the world's principal growers, large quantities of sugar beets and lesser amounts of hemp, flax, and hops are grown. Approximately 10 million cattle and larger numbers of hogs, together with fewer sheep and horses, are kept on Polish farms. Fish from inland and coastal waters and poultry augment the nation's food supply.

The Silesian coalfield is the cornerstone of the Polish industrial economy. This highly developed industrial area, acquired virtually unimpaired from Germany following World War II, is the "Polish Ruhr." Katowice, Zabrze, Chorzów, Sosnowiec, Gliwice, and Bytom are the larger industrial and mining cities clustered

Fig. 9–8 The breeding of geese, valued for their meat and feathers, is a commercial activity in many parts of Poland. Both products are exported to the Americas. (P. A. Interpress Photo)

around the Silesian coal deposits. More than 100 million tons of bituminous coal and 20 million tons of lignite are mined yearly. The bituminous coal is used in the coking plants, blast furnaces, and steel mills of the area itself; the lignite is used in electric generating plants. Some coal, however, is exported in exchange for foreign iron ores, which when added to the domestic ores from Silesian, Radom, and Kielce mines yield about 10 million tons of crude steel annually. The manufacturing of heavy machinery, coke, and chemicals makes Silesia the center of Polish heavy industry. Zinc (of which Poland is one of the world's leading producers), nickel, lead, low-grade copper, arsenite, pitchblende, and smaller quantities of petroleum and natural gas are mined in areas along the southern foothills.

Warsaw, located on the Vistula River, is the principal political, commercial, cultural, and scientific center of the country. Heavily damaged during the war, the capital city has been largely rebuilt. Many imposing and modernistic buildings now stand in place of the rubble, along with meticulous reconstructions of many of the city's historical landmarks and entire old quarters. The city's outstanding manufacturing activities and goods include metalworking industries, electrical appliances, and tractors and automobiles.

Łódź, sometimes termed the "Polish Manchester," is a well-known textile-processing center, but metallurgical, chemical, and food-processing plants also raise the city's industrial rank. Wrocław and Poznań, the fourth and fifth largest Polish cities, are regional trade and manufacturing centers. Gdańsk and Szczecin are the nation's two chief ports. Gdańsk is a transshipment point for places along the Vistula; Szczecin serves the Oder and the canals connected with it. Each has numerous industrial establishments, particularly shipbuilding yards. Kraków, the capital of the country until 1595, is also an important trading and commercial center of southern Poland; Wałbrzych (Waldenburg) and Lublin serve similar regional functions; in addition to its industrial functions, the old city of Częstochowa is the nation's foremost religious shrine and the venerable objective of Polish Catholic pilgrimages.

In addition to ensuing agricultural and industrial improvements, the westward-shifted Polish boundaries improved transportation and communication facilities as well. In terms of value, Polish imports generally exceed exports.

Coal and its products, meat, sugar, semi-manufactures, and textiles are leading exports. Iron ore, petroleum, chemicals, cotton and wool, and machinery are the principal imports. The bulk of Polish foreign trade is carried on with other members of the Soviet bloc.

For the future, as in the past, Poland faces great problems. The status of the western postwar-occupied lands is not clear or permanently fixed by treaty. Recrudescent Germany, dissatisfied with its heavy territorial losses, will no doubt press for a more favorable settlement at an opportune time. However, if left in its present form without another partition, Poland may yet promote an affluent society, for it now has the geographic resources to do so.

CZECHOSLOVAKIA

Entirely landlocked Czechoslovakia is slightly larger (49,366 square miles) and more populous (14.7 million) than Pennsylvania. The Czechs (*Böhmer* in German; Bohemians in English) themselves represent the old Slavic peoples within the basin of the Vltava (Moldau), where they had experienced earlier independence and prolonged political and cultural associations with the neighboring Germanic peoples, especially the Austrians. The modern Czechoslovak state was carved out of the old Austro-Hungarian Empire in 1918. Prominent Czech emigrés, armed with a strong passion for independence and an astute appreciation of profitable political connections, were successful in convincing the victorious Allied powers of the righteousness of their cause. Pan-Slavism was a correlative of Czech independence; the new state synthesized the Moravians, Slovaks, and Ruthenians with the Czechs themselves. Czech national aspirations were supported by France's desire for a weakened Austria, Germany, and Hungary. Thus abetted, Czech ambitions were permitted to extend beyond Slavic-inhabited lands, and the new state included about 4 million non-Slavic peoples of German and Hungarian ancestry. Launched with the blessings of some diplomats as a "shining example of self-determination and democracy," a reputation appreciated more in distant lands than among its neighbors, Czechoslovakia was drawn into immediate difficulty with Germany, Austria, Hungary, and Poland. The relentless pressures applied by neighboring states having territorial claims on Czechoslovakia were significant factors leading to World War II.

While the predominantly German fringes were integrally joined to Germany, the Bohemian and Moravian core land became a subdivision of the Reich. Shorn of her Magyar areas, which were returned to Hungary, Slovakia became an independent, pro-German republic. At the war's end the Allies reestablished the status quo, with the exception of Carpatho-Ruthenia, which was incorporated into the Soviet Union. As partial compensation for this loss, Czechoslovakia received about 26 square miles of additional Hungarian territory near Bratislava and, like Poland, gained the right to expel the resident German and Hungarian minorities. Today about 90 percent of the Czechoslovak state's nearly 15 million people are Czech and Slovak.

The Bohemian Forest and the Ore and Sudeten Mountains rim the Bohemian Basin on three sides, with the low Moravian Hills forming the less-prominent fourth side. Chief egress from this fortresslike topographic basin is northward through the Elbe Gate and eastward through the Moravian Corridor. The Carpathian arc effectively separates Slovakia from the Bohemian Basin and its Moravian foreland. All natural passageways through Slovakia follow the river valleys and lead to the Alföld and Budapest, not to Prague. The forest-covered and mineralized mountains of food-deficient Slovakia are naturally complementary to the treeless food basket or granary of Hungary. The international boundary between these two adjacent regions has made economic integration impossible; instead, to the detriment of both countries, rivalry has prevailed.

Generally, the climate of Czechoslovakia is the humid continental, cool-summer type. The mountainous character of the country, however, produces a variety of local conditions dependent on altitude and exposure.

While the Czech portion of the country is mainly industrial, the Slovakian sector is primarily agricultural. Nevertheless, agriculture in the Bohemian Basin is intensive and highly productive. In Slovakia agriculture is extensive, with lower yields. Wheat, barley, potatoes, and sugar beets are dominant crops; but rye, oats, tobacco, hay, hops, and deciduous fruits are grown on smaller acreages restricted by soil and climate. The cold climate precludes corn growing outside the lowlands adjacent to the Hungarian border.

Dairy cattle, pigs, and poultry are numerous, but fodder crops must be intensively cultivated to feed the large stocks of domesticated animals. Production of both cereals and meat falls short of domestic demand.

As a part of the Austro-Hungarian Empire, the Bohemian Basin was deliberately assigned an industrial function. Hungary was to be agricultural, and Austria was the administrative center of the extensive realm. This apportionment was in keeping with the characteristic resources of the empire; as a result, development along these lines was rapid, and prosperity was at a high level. On gaining independence, the Czechs fell heir to one of Europe's most highly developed industrial regions. With the exception of East Germany, Czechoslovakia today is the most thoroughly industrialized member of the Soviet bloc.

Besides owing much to the skill of its workers, Czechoslovakian industrial might is based on the nation's abundant mineral resources, especially coal and iron ore. Some 26 million tons of high-grade coal is mined annually from the southern extension of the Silesian fields at Moravská Ostrava; another 70 million tons of lignite is mined from scattered fields elsewhere. Although Czechoslovakia is the second-ranking coal producer in the Shatter Belt, the small amount of petroleum taken from the southern Moravian fields is insufficient to meet domestic needs. Czechoslovakia is usually the leading pig-iron and crude-steel producer in Eastern Europe; however, most of the iron ores used in its blast furnaces are imported. Domestic iron ore, as well as some wolframite, pitchblende, and manganese, is derived mainly from the Ore Mountains. The mines of Slovakia also yield iron ore, but their output of magnesite, antimony, copper, lead, zinc, and mercury is more notable. The manufacture of fine chinaware and several chemical industries depend on kaolin and rock salt mined within the nation's borders.

Czechoslovakian industry is centered around Moravská Ostrava; secondly, in a belt from Prague to Plzeň (Pilsen); and thirdly, in the Ore Mountains region. Brno (Brünn) and Bratislava are smaller detached centers of industry that are also significant. The Prague-Plzeň belt produces armaments, locomotives, electric motors, airplanes, automobiles, machinery and machine tools, textiles, and many other types of products. Brno, the second largest city of the country, is the chief textile center of the nation; but armaments factories, chemical plants, machine shops, and breweries augment the city's industrial economy.

Bratislava, besides being a bridge city, a rail and road center, and a provincial capital with many food-processing and light industrial establishments, is also the principal Danubian port of Czechoslovakia. The nation's largest historic and crossroads city is Prague, the political, cultural, and commercial capital of Czechoslovakia. Košice is a rapidly growing industrial center in eastern Slovakia.

Czechoslovakia is Eastern Europe's principal

Fig. 9-9 Prague, the historic capital of the Czechs, has a population of more than a million and is the leading industrial and trade center of the nation. (Courtesy of Czechoslovakian Embassy)

trader. Although it has no direct outlet to the sea or a merchant marine, access is available to the Danubian, Elbe, and Oder inland waterways. In addition to an excellent network of national transportation lines, it is also served by international rail and road connections and airlines. Exports normally include munitions and armaments, iron and steel products, machinery, textiles, glass and chinaware, shoes, and sugar. Principal imports are foodstuffs, cotton, and iron ore. In former times, the country's trade was oriented westward, but at present Czechoslovakia's trade alignments are largely with the Soviet Union and its satellites. The desire and attempts of Czechoslovakia to re-establish traditional Western trade lines and a degree of individual incentive was an important contributor to recent Czech-Soviet difficulties.

HUNGARY

Roughly equivalent in area (35,919 square miles) to the state of Indiana but twice as populous (10.3 million), Hungary is one of the smaller and densely inhabited European states, and at the same time it is one of the oldest. Preceded by earlier kindred peoples, the Magyars—the only name by which the Hungarians ever call themselves—entered and settled the Alföld more than a thousand years ago. Their inflammable relations with the Germanic peoples to the west, the Tartar invaders from the east, Turks from the south, and more recently with Slavs from north and south have repeatedly consumed their energy and outweighed domestic difficulties. With the loss of the country's main industrial areas to Czechoslovakia, the richest farmlands to Yugoslavia, and the best mines and timberlands to Romania, World War I deprived Hungary of much of its territory, natural wealth, and population. Their uncontrollable passion to regain these lost lands and the Hungarians living therein led Hungary in 1940 to enter World War II on the Axis side, and to an eventual repetition of the previous disaster, with the appended tragedy of Russian occupation and sovietization. Late in 1956, the desperation of the Hungarian people resulted in a popular uprising, which was crushed within a fortnight by Soviet arms. Beaten in their wrecked cities and demoralized on their farms by foreign "forces of liberation" supplied first by the Nazi Germans and then by the Communist Slavs during World War II, the Magyars remain in dumbfounded stoicism, unable to regain their nostalgic past or to see hope in a Soviet-oriented Communist future.

The two World Wars left truncated Hungary almost entirely homogeneous in language and culture; such a condition is unduplicated to this degree in any other country of Eastern Europe. The Magyars, speaking a language unrelated to any other in Europe, are regarded as a characteristically proud, meditative, and industrious people.

Hungary is landlocked within the downwarped Carpathian Basin, where only the Central Hungarian Hills stand interposed between the generally low, flat, and deeply alluviated floor of the great structural and topographic basin. The Danube River, navigable for its entire 255-mile length within Hungary, further divides the basin into Transdanubia, or the lands west of the Danube, and the Alföld, the great plain east of the Danube. Although most Hungarian soils are chernozems developed under steppe-grass cover, lesser areas of alluvial, loessal, and volcanic soils produce local variations. The overall climate is typically humid continental.

Nearly two-thirds of Hungary is arable land, and nearly one-half of the population are engaged in agriculture. Hungary has long been among the world's leading wheat and corn producers, and these grains are still the dominant crops. Barley, oats, rye, and the recently introduced rice are other cereals grown on a smaller scale. Sugar beets, potatoes, and tobacco are cultivated intensively, but flax, hemp, and cotton occupy smaller acreages. Two Hungarian specialties are Tokay grapes, produced from extensive vineyards along the volcanic slopes of the north, and the paprika grown around Szeged in the south. Truck crops, some under irrigation, as well as deciduous fruits, are grown for domestic markets and for limited export. Cattle, pigs, and poultry are the principal meat animals; smaller herds of horses and sheep are kept in the eastern part of the country.

The nation's ambitions for an increased industrial economy are frustrated by insufficient reserves of power fuels and metalliferous ores. Hungary's waterpower potential is correlatively low with its dominantly level topographic relief. The 25 million tons of lignite mined annually and the 4 million added tons of bituminous coal mined around Pécs do not meet national needs; thus, the nation's industry depends on coal imports. Although the better-grade bituminous coal is used by blast furnaces, the lignite is consumed by thermoelectric installations, by chemical industries, and by individuals for heating homes. Hungarian petroleum production, amounting to some 1.7 million tons annually, also falls short of meeting domestic needs.

About 15 percent of the world's bauxite reserves lie along Lake Balaton, where annual production of this ore amounts to some 1.7 million tons. Although most of the ore is exported, domestic concentrators account for 62,000 tons of alumina. Hungary's only iron ore mine, near Miskolc, yields less than 500,000 tons of ore annually. Lesser quantities of uranium, manganese, lead, zinc, and copper are mined from the nearby Matra hills.

Hungary's industrial capacity has far greater local than international significance. Except for the three main centers—Miskolc, Györ, and Budapest—industrial establishments elsewhere in the country are insignificant. The rapidly expanded steel center around Miskolc produces heavy machinery and armaments and depends on imported ores and coal from Soviet-bloc neighbors. Györ, the second center in the northwest, has large cement, chemical, and metallurgical plants. The size and dominance of Budapest overshadows all other Hungarian cities. Located on both banks of the Danube, it is an important bridge city, river port, and the main cultural, political, commercial, and industrial center of Hungary. A cluster of light industrial establishments process domestic agricultural products. Textile mills, shipyards, refineries, tractor plants, and factories manufacturing electrical goods further augment the city's industrial economy. Szeged and Debrecen are regional commercial centers in the south and east, respectively.

Cut off from many of its former natural resources, from much of its integrated transportation systems, and without ready access to earlier industrial centers, residual Hungary found itself underdeveloped. World War II interrupted and retarded planned industrial and social adjustments. Less from internal choice than from outside coercion, Hungary today is in total alignment with the Soviet Union and its other Shatter Belt satellites. Foreign trade is conducted largely within this Eastern Communist bloc. The nation's main exports are bauxite, electrical goods and machinery, textiles, pharmaceuticals, and small quantities of agricultural products. Wood products, cotton, coke, metal ores, and petroleum are the principal imports.

ROMANIA

Territorially, Romania is slightly smaller (91,700 square miles) than Oregon, but it is ten times as populous (20.3 million). Historically, the nucleus of modern Romania evolved in 1861 with the unification of Walachia and Moldavia. With the acquisition of Transylvania, part of the Alföld of Hungary, and smaller segments of adjacent Russia and Bulgaria, Romania doubled in size after World War I. Boundaries, however, are changed more easily than human personalities. The heterogeneous character of the added peoples created wide cultural discrepancies, if not outright resentments, that led to mutual liabilities and a checkered political development. Indeed, Romania grew more rapidly territorially in the twentieth century than any other European country; but its growth was attributable more to the grants of the victors of World War I than to her own

Fig. 9–10 Budapest, Hungary, was formed by the consolidation of the two cities Buda and Pest, which were separated by the Danube. The city, with a population in excess of 2 million, is the largest in Eastern Europe. (Courtesy of Hungarian People's Republic)

military or administrative prowess. Although the indigenous trained technological and political talent among the newly acquired peoples was unacceptable to the new Romanian rulers, the low-level educational attainment and technological and administrative training of the Romanian himself proved an unequal replacement. Romania thus became a prime example of an underdeveloped country large in area, diversified geographically and culturally, rich in resources, but low in trained manpower.

Economic stagnation, restive minorities, and the territorial claims of neighboring nations involved Romania in World War II, which the country entered in 1941 on the Axis side. On the basis of its dominant Hungarian population, part of Transylvania (Székelyland, 16,650 square miles) was returned to Hungary in 1940 (Second Vienna Award). Today, however, this is again part of Romania, with the status of an "Autonomous Hungarian Territory," where most of the Magyars residents are Protestants. The Soviet Union reoccupied Bessarabia and incorporated it into the Moldavian S.S.R., and Bulgaria regained southern Dobrogea.

Romania forms the easternmost wedge separating the Northern from the Southern Slavs. Some two-thirds of its 20 million people speak Romanian, a language derived from Latin, Slavic, Magyar, and Turkic; about 1.8 million speak the Magyar tongue. A lesser number of Germans and Bulgarians and smaller groups of Serbs, Jews, Turks, Greeks, and Gypsies are also present within the confines of the nation's contemporary boundaries. Pointed toward the West by language and toward the East by its Greek Orthodoxy and other cultural characteristics, Romania is a transitional link between the two cultures.

There are striking similarities between Romania and Czechoslovakia. Both are comparatively recently created countries; both are subdivided by the Carpathian Mountains; both harbor large ethnic minorities; both have rich mineral and forest resources and the greatest waterpower potential in the Shatter Belt. But where resemblances end, contrasts begin. The Czechs were the most highly developed peoples in their country, whereas the Romanians of the Old Kingdom were generally on a low level of industrial and cultural development. Transylvania showed a higher order of progress in all fields of human endeavor than did Old Romania. From its condition of prolonged underdevelopment, Romania had a long way to go; and if for no other reason than this, evidence now seemingly points to remarkable economic progress.

Romania is dominated by the densely wooded, well-watered, and snowcapped Carpathian Mountains. The great Walachian Plain, lying between the Carpathians and the Danube River, merges northward into the somewhat higher dissected Moldavian lands again lying between the Carpathians and the Prut River. The eastern third of the country is thus bounded by the two rivers and the mountains. To the west, the domal Bihar Mountains are cradled within the great westward-arcing Carpathians. Between the subdued and forested Bihar and the higher Carpathians themselves, the Transylvanian Basin, another flatland, opens westward onto the Alföld. All physiographic regions are structurally related to the Carpathians; only the small Dobrogea Platform, lying between the Bulgarian border and the Danubian deltas, stands unrelated. The Walachian Plain, like the Alföld, has fertile chernozem-like soils which furnish the basic asset for an agricultural economy. Elsewhere farming is limited to alluvial soils along the many riverine strips. The Bihar and Carpathian Mountains contain dense stands of coniferous evergreen forests; the Transylvanian Basin and lowlands support large oak and beech groves. The climate is generally of the humid continental type.

Two-fifths of Romania is arable land, on which four-fifths of the total working population are employed. Corn and wheat, the principal crops, occupy three-fourths of the arable lands. Despite the low yields associated with the extensive type of farming, the cultivation of large acreages enables Romania to rank as the leading corn and wheat grower of the Shatter Belt. Since corn is the main staple of the Romanian peasantry, the bulk of the wheat remains for export. Barley, oats, rye, and potatoes are cultivated largely on farms operated by Magyar and German peoples in Transylvania. The sugar beet is restricted to the Banat region, and soybeans, flax, tobacco, rice, and cotton are grown on limited acreages elsewhere, but peas, beans, and sunflowers are widely distributed. Fruit orchards and vineyards are most common in Transylvania. The principal handicaps of Romanian agriculture have been the traditionally poor farming methods used by the peasantry; and collectivization does not seem to have increased farm output.

The Romanian was traditionally a sheepherder, keeping some 14 million multipurpose breeds of sheep in the mountains, where a strong transhumance movement has developed on the drier eastern lands. Cattle and horses as well as pigs are most numerous on the Danubian plains, but the water buffalo is popular only in Transylvania. Although fish abound in the inland streams and the Black Sea, commercial fishing is underdeveloped.

Romania's mineral wealth equals its surface riches. With the exception of petroleum and natural gas, the country's mineral resources are as underdeveloped as its surface potentialities. The Ploesti oil fields contain some 1,000 million barrels of reserves and yield about 13 million tons of petroleum annually. In addition to its foremost position as a petroleum producer, the natural gas fields of Ploesti and the Banat districts enable Romania to rank second only to the U.S.S.R. as a gas producer. Most Romanian petroleum is exported through Danubian and Black Sea ports to the Soviet Union and other members of the Communist bloc.

The nation's 5 million tons of coal and its somewhat larger production of lignite are utilized largely by the rapidly expanding blast furnace capacities in the Banat district. The less than a million tons of iron ore extracted enable Romania to produce over 4 million tons of crude steel annually, thus becoming the third largest steel producer in the Soviet bloc. Mountainous Transylvania has numerous mines producing salt, phosphate, silver, lead, manganese, and bauxite. The highlands and heavy precipitation provide great hydroelectric potential, which remains largely underdeveloped.

Until recently, urban and industrial life in Romania was confined mainly to the capital city of Bucharest, to the cities of the former Hungarian territories, and to Ploesti, Galați, and Brăila. Today, these towns are growing rapidly; likewise, rapid urban and industrial development has overtaken the cities of Oradea, Brașov, Constanța, and others. Thus, in a land of numerous villages and small towns, larger-scale urbanism is gradually becoming prevalent. The modernity of Bucharest's central section gives way to sprawling tracts of small houses and unpaved streets. Many of its 1.5 million inhabitants are engaged in service occupations, general trading, or work in heavy industrial establishments, machine shops, flour mills, textile factories, and distilling and chemical plants. Cluj and Timișoara, with their large Hungarian minorities are regional capitals of Transylvania and the Banat, respectively. Brăila and Galați are principal Danubian ports; Constanța is Romania's chief port on the Black Sea. Apart from the rapidly developing Banat mining and metallurgical complex, the diversified industries of Bucharest, and the petroleum refining of Ploesti, other manufacturing centers of Romania are

Fig. 9-11 Bucharest, Romania, the second largest city in Eastern Europe, is the center for expanding oil and agricultural activities. (Courtesy of Socialist Republic of Romania)

Galați, with its steel mills, and Craiova; and in smaller woodworking, wool-processing and nonferrous metalliferous plants mainly within Transylvanian towns.

Romania's principal exports are petroleum, timber, wheat and flour, and animals. Machinery, iron ore, coke, and chemicals are leading import items. Most of Romania's foreign trade is carried on within the Soviet-bloc nations.

YUGOSLAVIA

Yugoslavia, with an area of 98,766 square miles and a population of 20.6 million, is as large and as populous as the states of Pennsylvania, Virginia, and Maryland combined. A recently founded country, Yugoslavia was created in 1918 out of a patchwork of Balkan and peripheral peoples living in one of Europe's most politically sensitive sectors. Present-day Yugoslavia lies between the Adriatic Sea and seven inland neighbors.

Largely through the aid of the victorious Triple Entente, Serbia, liberated after four centuries of Turkish occupation and serving as the nucleus for the ensuing condominium, succeeded in uniting the various Slavs formerly under the control of the disintegrating Austro-Hungarian and Turkish empires. The Croatians, formerly united with the Hungarian Kingdom for eight centuries, and the Slovenes formed with the Serbs a post-World War I union known as the "Kingdom of the Serbs, Croats, and Slovenes," which subsequently became Yugoslavia. The Serbs, however, considered the new state an enlarged Serbia—an assumption that immediately created difficulties with the related Slavs, especially with the Croats. Here again, as in Poland, Czechoslovakia, and Romania, large minorities from peripheral nations were included within the newly formed state. In 1941, beset by domestic Slavic troubles and by outside pressures, Yugoslavia became the object of Italian and German invasion. Italy, Hungary, Bulgaria, and Germany temporarily annexed those parts of Yugoslavia inhabited by kindred peoples. Guerrilla warfare raged for four years among the Slavs themselves over unresolved ethnic, political, and religious issues. Backed by the United Kingdom and the Soviet Union, the political-military movement headed by Marshal Tito succeeded in reestablishing former boundaries and later in acquiring former Italian islands in the Adriatic and part of the Trieste zone. Since Soviet armies at no time entered Yugoslavia, the domestic Communist regime chose to follow an independent course based on conditional cooperation and the practice of utmost opportunism in its relations with the Western democracies and the Communist bloc. The neutralist intent of Yugoslav leadership, together with its geographically sensitive location, qualified the troubled state for heavy United States aid. American aid, in turn, contributed to Yugoslav economic development and abetted the Yugoslav one-party, one-leader domestic arrangement.

The members of the Socialist Federal Republic of Yugoslavia consist of the Serbs, Croats, Slovenes, Montenegrins, and Macedonians who, on the basis of close ethnic relationship, formed a separate federal republic. Ethnically mixed Bosnia and Herzegovina were together added as the sixth member of the federation. Voyvodina (Bačka and Banat), with a large Hungarian minority, was made a semiautonomous oblast, as was Kosmet (Kosovo-Metohija), with its Albanian minority.

The Adriatic Sea laps against the narrow littoral at the base of the Dinaric Alps, whose parallel ranges and intervening valleys descend, one by one, toward the Alföld. The topographic dominance of the Dinaric Alps is reflected in the hydrographic pattern: three-fifths of the Yugoslavian rivers empty into the Danube. Except for the Vardar, the other rivers flowing to the Adriatic and the Aegean seas are few, short, and swift. Two-thirds of the country is mountainous or otherwise rugged. Moreover, the honey combed calcareous strata are filled with sinkholes, caverns, dolines, and poljes. Karsting thus intensified the natural inhospitableness of mountains and, at the same time, encouraged isolation and the development of provincialism.

Aside from the alluvial soils of the narrow valley bottoms, the residual soils of the dolines and poljes represent the only tillable lands in the mountains. Yugoslavia's most extensive flatlands, containing the most fertile alluvial soils, lie north of the Sava River toward the Hungarian and Romanian borders, especially in the Banat-Bačka district sometimes known as the Voyvodina. About one-third of the country is forested; although the limestone and dolomitic mountain surfaces and the low Danubian lands are only sparsely wooded, the mountains of the south, especially near the Albanian border, are densely forested. Yugoslavia is mainly under humid continental climatic influences. Nevertheless, the southward-opening Vardar Valley and the Adriatic littoral exhibit Mediterranean tendencies. Exposure and altitude produce wide local temperature and precipitation variations. The windward Adriatic side of the mountains receives over 180 inches of annual precipitation, but this amount decreases rapidly on the leeward side, until not more than 20 inches falls on the lowlands of the Voyvodina. The Adriatic Sea itself, although it abounds in valuable marine life, scenery, and historic tradition, remains largely unexploited. Recently, however, tourists have discovered the amenities of the Adriatic shores and have begun to arrive there in large numbers.

Agricultural and industrial development vary widely, if not spectacularly, from one section of the country to another. While only one-third of Yugoslavia is arable, two-thirds of the working population are farmers. The best plowlands by far are in the northeastern parts of the country, where, paradoxically, the Slavic population is sparse in the midst of the traditionally small Magyar land holdings and dense farm communities. Without these former Magyar lands the country's agricultural production would be almost insignificant. Agriculture in the mountains, particularly in Bosnia and Herzegovina, is reduced to subsistence levels.

Corn, wheat, barley, and oats are the leading cereals raised in Yugoslavia. In normal years the corn yield per acre in the Voyvodina is one of the highest in Europe and is often better than that of the United States. Sugar beets and hemp are also widely grown. A quick-maturing variety of cotton, although locally significant, is a new agricultural venture. Yugoslavia is one of the main tobacco growers of the Shatter Belt. Rice, like cotton, is grown for local consumption. Within sheltered valleys and on favorably exposed slopes, orchards and vineyards produce a variety of fruits and grapes. Plums are especially abundant in Slovenia and Bosnia. When not exported as fresh fruit, the plums are desiccated in dry weather and exported as prunes; or more frequently, they are distilled as fruit brandies.

Sheep are the country's dominant domesticated animals, partly because of rugged topography but more particularly because of tradition and the Muslim population's penchant for mutton. Both hogs and cattle are most numerous on the best agricultural lands of the northeast.

Geographical isolation breeds conservatism and an attitude of "What was good for grandfather is good enough for me." Thus, after forced agricultural collectivization, the Yugoslav peasantry reacted with traditional hostility, if not open sabotage, and agricultural output declined. In the wake of relenting government decrees and increased farm mechanization, production has again attained prewar levels. Agriculturally, Yugoslavia is currently the least-collectivized country in the Shatter Belt.

The unfavorable agricultural environment of the mountains is offset by the abundant mineral resources existing in them. The high value set on a planned program of industrial expansion encouraged extensive mining operations. Some 2.2 million tons of iron ore are taken from mines at Ljubljana and in Bosnia. Moreover, Yugoslavia is one of the ten leading world producers of copper, lead, antimony, chromite, and bauxite. Mercury, zinc, manganese, asbestos, salt, and gypsum are also mined on an increasingly larger scale. Yugoslavia is the principal gold producer within the Shatter Belt, and mines producing silver, magnesite, and tungsten are also active within its borders. While most of the ores are exported as raw ores or concentrates, several notable ore-dressing and refining centers have appeared on the new industrial skyline.

Yugoslav lignite production, amounting to 26 million tons yearly, ranks next to that of Hungary, which country it is likely to displace as the third producer of the Shatter Belt. Petroleum production from the northern foothills has increased rapidly, amounting to 2.4 million tons annually. Nevertheless, neither the domestic coal nor petroleum is sufficient to meet the fast-growing industrial demands.

The northern foothills from Ljubljana to Belgrade are the "Ruhr of Yugoslavia." Nearness to raw materials, a dense railway network, the navigable Sava, and the proximity of the Adriatic all contribute their share to industrialization along this axis. The new Zenica steel complex and the metalliferous industries of the foothills belt are surrounded by plants for the manufacture or processing of building materials, timber, paper, leather, and chemicals.

Yugoslavia, like Romania and Hungary, is a land of small cities and numerous villages. Located near the confluence of the Sava, Danube, Tisza, Temeš (Timis), and Morava rivers, Belgrade has been historically a nodal and gateway city. Despite its strategic site, ancient origin, and function as the capital, it is still comparatively small, but is growing rapidly. The other large Yugoslav cities were developed and are located in former Austro-Hungarian territories. Zagreb, Ljubljana, Subotica, Novi Sad, and Rijeka are regional capitals, with additional functions of local trade and recent industrialization. Sarajevo in Bosnia-Herzegovina and Skoplje in Macedonia are other large communities carrying the undeniable imprint of their earlier and, to a lesser extent, that of their remaining, Muslim occupants.

Rijeka, developed as the chief port of the Austro-Hungarian monarchy, was inherited by Italy (Fiume) and more recently by Yugoslavia. Although the outlet from the Danubian lands is difficult, nevertheless Rijeka is Yugoslavia's best port and handles annually some 6 million tons of inbound and outgoing sea trade. Dubrovnik, Sibenik, Split, Zadar, and Pula (Pola) are other small but important ports, naval bases, and resort centers along the rocky but climatically pleasant Adriatic coast.

Yugoslavia's principal exports of nonfer-

Fig. 9-12 Zenica Metalworks, about 100 miles northwest of Sarajevo, Yugoslavia. The plant, which includes a steel-rolling mill with a capacity of 450,000 tons per year, is the largest in Yugoslavia. (Courtesy of Consulate of Yugoslavia)

rous ores and metals, chemicals, fruits and vegetables, tobacco, hides, timber and pulpwood are exchanged for the main imports of machinery, coal, petroleum, cotton, and various foodstuffs. West Germany, Italy, the United States, and the United Kingdom, followed by the U.S.S.R., are Yugoslavia's principal trade partners.

BULGARIA

Bulgaria, occupying 42,823 square miles, is as large as Tennessee but more than twice as populous (8.5 million). The land now known as Bulgaria was already occupied in Paleolithic times, and vestiges of the historic Thracians, Greeks, Romans, and Byzantines remain mute evidences of other early civilizations in the area. Indeed, the land was well known to Alexander the Great and to his renowned teacher, Aristotle. The early Bulgars themselves were Finno-Ugric peoples, related to the Magyars. They appeared, horse-mounted, on the Balkan scene in the seventh century, turned to the Greek Orthodox form of Christianity in 865, and quickly extended their sway over the entire Balkan Peninsula. The Bulgars were eventually absorbed by the probably indigenous if not the southward-migrating Slav population, until today only their name remains as a memento of their non-Slav past. The modern Bulgarian alphabet, language, and religion, together with present political commitment, are closely akin to those of Russia.

The national independence gained in 1878 at the end of the Russo-Turkish Wars did not efface five centuries of Turkish occupation; today some 10 percent of the Bulgarian population still speak Turkish. Nearly 90 percent of the population speak Bulgarian, with only a few enclaves of Greeks, Romanians, and scattered Gypsies. As an ally of Germany in both World Wars, Bulgaria had as her objective the winning of territory inhabited by kindred peoples. With the exception of southern Dobrogea, all such acquisitions were in the end temporary.

Bulgaria is dominated topographically by the rounded ranges of the Balkan Mountains and the nearly parallel serrated ranges of the Rhodope Mountains. Between the two, the wedgelike Maritsa Valley opens to the Black Sea and, laterally, toward the Greek and Turkish shores of the Aegean and the Sea of Marmara. This is the nation's principal lowland. The Balkan Mountains descend northward to form a discontinuous and dissected plateau surface (Danubian Foreland), which terminates as an escarpment overlooking the Danube River. The country is predominantly mountainous.

Forest, confined largely to the mountains, covers nearly one-third of Bulgaria's surface and represents a great source of national wealth. The mountains and the northern plateaus are under humid continental climatic influences, but the sheltered Maritsa Valley has the milder humid subtropical climate. A Mediterranean climate penetrates the southward-oriented small valleys of the Rhodopes. The broad interfluves of the northern forelands contain fertile residual limestone and transported loessal soils, but the soils of the Maritsa Valley owe their fertility to alluvial and diluvial origins. Despite the country's broad frontage on the Black Sea, the Bulgarians have not turned to seafaring.

Although only about two-fifths of the country's surface is arable, over one-half of the country's working population are engaged in agricultural pursuits. Agricultural lands are confined largely to the forelands north of the Balkan Mountains and to the lowlands of the Maritsa. Wheat and corn are the main crops cultivated, but rye, barley, oats, and rice are also produced. Sugar beets, potatoes, flax, alfalfa, sunflowers, and cotton, together with lesser acreages of soybeans and opium poppies, are significant local crops. Bulgarian tobaccos and wine grapes are famous internationally; but even more renowned is the attar of roses obtained from roses grown in the Tundzha Valley.

Numerous pigs, cattle, and horses are raised on lowlands; but Bulgaria, like neighboring Yugoslavia, is best known for its large number of sheep and goats. Bulgarian cheese and yogurt made from ewe's and goat's milk are nearly as famous as its rose gardens. Bulgarian agriculture is the most highly collectivized in the Shatter Belt. The government's eager, if not unrealistic, program to increase agricultural production twofold in the next decade remains to be accomplished.

Bulgaria has made a successful leap forward in developing her mineral resources. In the past decade, lignite output doubled to the present level of 27 million tons. This rapid growth was occasioned by the opening of new coal mines. Some 500,000 tons of petroleum are pumped yearly from recently discovered fields, and new hydroelectric installations utilize part of the country's great waterpower potential. Some 800,000 tons of domestic iron ore are smelted in the new steel plants at Pernik and Kremitovtsy, where more than 1.25 million tons of crude steel are produced annually. Lead, zinc, copper, manganese, chromite, rock salt, and gypsum are mined in increasing quantities.

Previously, the modest Bulgarian industries processed food products and tobacco leaf and made fertilizers and textiles. The more recent emphasis on heavy industry now enables Bulgaria to produce its own metals and machinery. Nevertheless, machinery, iron ore, and petroleum are still the most significant import items. The principal new industries are centered about the lignite fields of the Maritsa Valley and near Sofiya.

Sofiya, aside from its important nodal position in the Balkan Peninsula, is the nation's capital and largest city, with outstanding trading and manufacturing functions. Plovdiv (the ancient Philippopolis), the second largest city, is the main commercial center of the Maritsa Valley. Ruse is the country's leading Danubian port city, and Varna and Burgas are Bulgaria's main Black Sea ports. Pernik, Sliven, and Pleven are regional trade centers, but more recently are assuming mining and manufacturing functions as well. Tobacco leaf is an export item of leading value, and attar of roses, fruits and vegetables, cereals, timber, and hides are also exported. Nearly all Bulgarian foreign trade is within the context of the Soviet-bloc economy.

ALBANIA

Albania (11,100 square miles) slightly larger than Maryland, is only slightly more than half as populous (2.2 million). Size, however, is not the only criterion for importance among nations. Albania lies strategically a mere 50 miles from the Italian heel, across the narrowed southern Adriatic strait. The Albanians have the unenviable distinction of being Europe's most foreign-ruled people. After 2,000 years of rule by Romans, Byzantines, Goths, Serbs, and Turks, the country obtained independence in 1912, mostly to forestall the territorial claims of its rivaling neighbors. During World War I the country was prey to Italian, Greek, Serbian, Austrian, and Bulgarian forces; during the interwar years it became a political dependency of Italy. Following the Axis powers' occupation in World War II and the establishment of a Communist regime at the end of 1944, Albania nearly became a political and economic satellite of Yugoslavia, her largest neighbor.

The ancient Illyrian tongue, antedating all other languages used on the Balkan Peninsula, is preserved in the Gheg dialect in the northern and by Tosk in the southern parts of Albania. Five centuries of Turkish occupation bequeathed Islam, the religion professed by three-fourths of the population. Roman Catholicism and Greek Orthodoxy are minority religions; Turkish and Greek are lesser languages.

In most places Albania looms from the Adriatic shore to elevations of 7,000 feet within a 50-mile distance inland to the Yugoslavian border. The short, swift streams emanating, if not leaping, from the mountains unloaded their rocky burden and thereby provided a flat coastal plain. Its marshy and malarial, mosquito-infested character has been partly overcome by modern drainage and reclamation projects. The Adriatic littoral is the population, agricultural, and economic center of the country.

The country's agriculture, if not its entire economy, is based upon subsistence farming and sheep raising. Although the climate permits the cultivation of citrus and olive groves, the predominant crops are corn and wheat; tobacco, cotton, and rice are also cultivated.

The magnificent stands of timber in the mountainous interior, as well as the known copper, chromite, bauxite, bitumen, asphalt, and lignite resources, remain practically unworked. Petroleum production of some 1 million tons annually is by far the nation's current leading resource. While most of this is exported, a smaller amount is refined at Cerrik. Like agriculture, manufacturing is also mainly on a modest scale, in establishments devoted to processing agricultural products. With exception of Tiranë (Tirana), the capital, Albania has no towns with populations of over 50,000; Shkodër (Scutari), Durrës (Durazzo), and Vlona are the largest communities. Except for a short railway, modern means of transportation are not developed. Modest exports of petroleum, chrome, and copper ore are traded for imports of manufactured goods and machinery.

In Perspective

All peoples of the Shatter Belt were deeply involved in the two World Wars and suffered the total consequences of physical and spiritual devastation and degradation. After wars, there is only one way to go: accept the misfortunes, then work and rebuild. Without exception, all Eastern European countries made remarkable strides in their postwar recovery and technological advances. Production from mines and factories has consistently increased in successive years. However, technological production is perhaps less the function of political ideology, or of ethnic or religious affinity, than of human ingenuity and willingness to work. The Shatter Belt peoples are traditionally hard workers. That their territories should draw the attention of most, if not all, of the great world powers is not altogether by local choice. What draws the mighty "Mastodon of Nations," China, to the side of tiny Albania, or the Great Russia to little Serbia, or the distant United States to the side of Yugoslavia? The answers to these relationships often appear learned and articulate and perhaps, nearly as often, replete with self-righteous rationalizations on the part of the nations involved.

Diastrophism, glaciation, and the workings of a blast furnace are easier by far to understand then nationalism, tradition, and the workings of political ideologies. Moreover, the human geographic problems posed by a 1,300-year uninterrupted record of settlement in Bulgaria are on a different level from the same problems involved in little Albania, less than a century old as a nation. While there are many similarities in the physical geography of the two areas, the historical geography of the Shatter Belt is quite unlike that of the United States. The history of the United States since the coming of the Europeans is comparatively brief; the history of the Shatter Belt is long and varied. Since the present and, more emphatically, the future are understood in terms of the past, a brief evaluative account of the Shatter Belt's past has been deliberately presented. But history without land is meaningless; the two factors are transmuted into the human geography of the Shatter Belt, which at times may seem superficially tortuous, if not negative.

The world's largest land power, the Soviet Union, and secondarily the great powers west of the Shatter Belt appear driven insatiably and relentlessly toward territorial expansion. When not passionately defending their heritage, the small intermediate nations, not altogether innocent of dreams of aggrandizement themselves, have become willing allies of the great powers. The resulting wars have left a human aftermath of ethnic minorities, expellees, and pent-up resentments that wait for powerful allies to set right that which went wrong for them. But who or what is to set things right? Not a desire for vindication or to "cast the first stone," but a concerted effort to cast ballots; not a warlike frenzy to count the dead for the fatherland, but a peaceful counting of votes for every man's freedom. These have been for centuries the sorely needed adjustments in the Shatter Belt. Unlike forced expulsions and exterminations of minorities, self-determination has never been tried on a grand scale in the Shatter Belt. The outcome of widespread free elections, if allowed, would undoubtedly bring about a popular settlement of long-standing disputes more acceptable than the resentful, if not vengeful, adjustments comprising the present *modus vivendi*.

Large and diverse in area, rich agriculturally and mineralogically, populous and rich in human ingenuity, the Shatter Belt is well endowed geographically. Indeed, the natural geographic environment has been far kinder than the area's domestic or foreign human neighbors. The long-standing Shatter Belt problems are evidently less the result of geography than of human psychology.

REFERENCES

Burck, Gilbert: "East Europe's Struggle for Economic Freedom," *Fortune*, vol. 75, pp. 124–127, May, 1967.

Hamilton, F. E. I.: *Yugoslavia: Patterns of Economic Activity*, Frederick A. Praeger, Inc., New York, 1968.

Harrington, Richard: "Albania, Europe's Least Known Country," *Canadian Geographical Journal*, vol. 74, pp. 132–143, April, 1967.

Kosinski, Leszek A.: "Changes in the Ethnic Structure in East-Central Europe, 1930–1960," *Geographical Review*, vol. 59, pp. 388–402, July, 1969.

London, Kurt (ed.): *Eastern Europe in Transition*, The Johns Hopkins Press, Baltimore, 1966.

Matley, Ian M.: "Transhumance in Bosnia and Herzegovina," *Geographical Review*, vol. 58, pp. 231–261, April, 1968.

Montias, John M.: *Economic Development in Communist Rumania*, The M.I.T. Press, Cambridge, Mass., 1967.

Osborne, R. H.: *East-Central Europe: An Introductory Geography*, Frederick A. Praeger, Inc., New York, 1967.

Singleton, F. B.: *Background to Eastern Europe*, Pergamon Press, New York, 1965.

Williamson, David: "The New Warsaw," *Geographical Journal*, vol. 38, pp. 596–607, December, 1965.

Zotschew, Theodor D.: "Social Changes in the Communist Danubian States," *Central Eastern Europe Journal*, vol. 18, pp. 48–49, February, 1970.

10

Union of Soviet Socialist Republics

In less than half a century, the Union of Soviet Socialist Republics has risen to a position of world industrial importance second only to that of the United States. Its industrial strength, too, is matched by its military might and its successes in space exploration and advanced technology.

Though the USSR competes with the People's Republic of China for leadership of the Communist movement, its attention nevertheless has not been deflected from traditional ideological goals: the promotion of economic, political, and social revolution in the countries outside its bloc. Because of these aims, we must seek greater understanding of the country and what it stands for if we are to be in a better position to defend and strengthen our own freedom and diverse way of life.

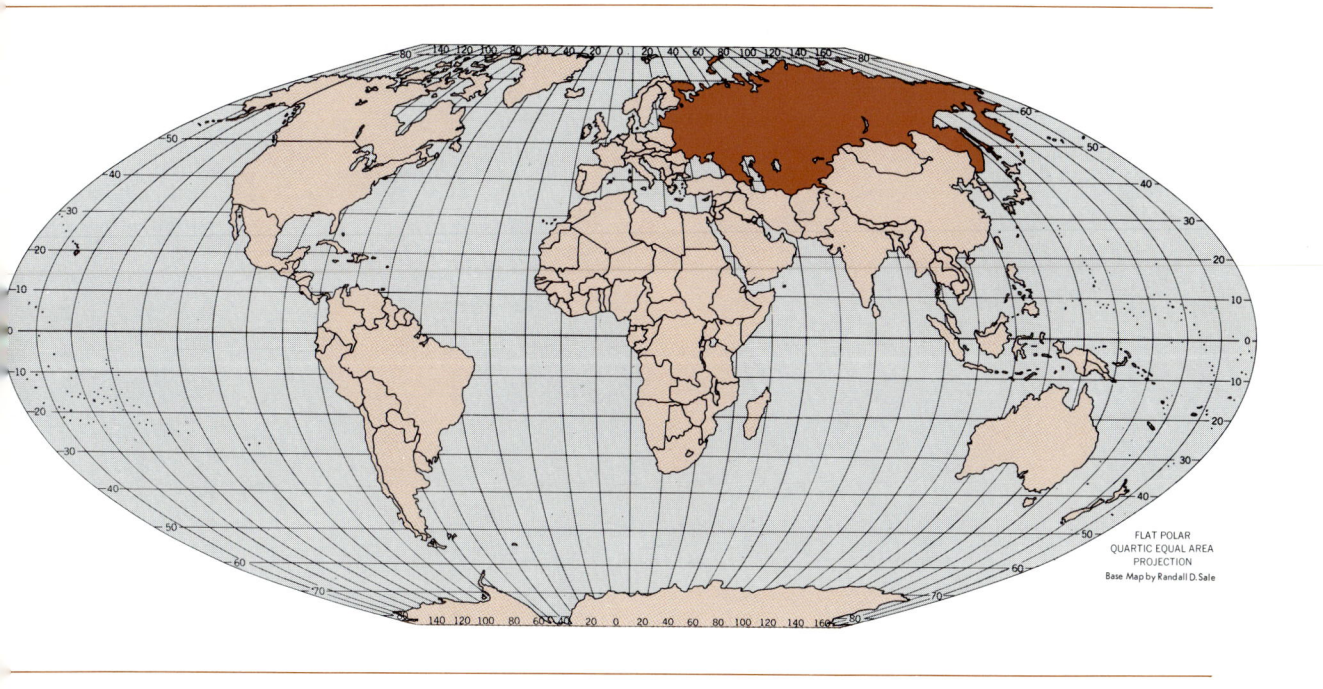

FLAT POLAR
QUARTIC EQUAL AREA
PROJECTION
Base Map by Randall D. Sale

The Union of Soviet Socialist Republics, ranking first in size among the sovereign states of the world, covers an area of 8.5 million square miles, or roughly one-sixth of the earth's total land surface. Essentially continental in dimensions, it is somewhat larger than South America and nearly as large as North America. Moreover, its size exceeds by about two and one-half times that of some of the largest countries of the world, notably the United States, China, and Brazil.

Extending from 19°30′ east longitude, just west of Kaliningrad on the Baltic Sea, to 169°30′ west longitude in the Bering Strait off Alaska, the Soviet Union spans a distance from west to east of approximately 7,000 miles. This vast longitudinal extent marks a difference of eleven hours, or nearly four times the time difference in hours between New York and California. The latitudinal spread, while much shorter, is nevertheless impressive. From Cape Chelyuskin at 77°35′ north latitude in Asiatic Russia to Kushka at 35°15′ north latitude, on the southern border with Afghanistan, the distance is nearly 3,000 miles. Relative to North America, the U.S.S.R.'s southern border lies at approximately the same latitude as Memphis or Chattanooga, Tennessee, while the latitude of the northernmost mainland point is the same as that of the Queen Elizabeth Islands of the Canadian Arctic. While it extends into subtropical latitudes, the Soviet Union is essentially a northern land. It has physical characteristics that, in many respects, are similar to those of Canada. Three-fourths of its territory lies north of the 49th parallel, which in turn forms much of the

345

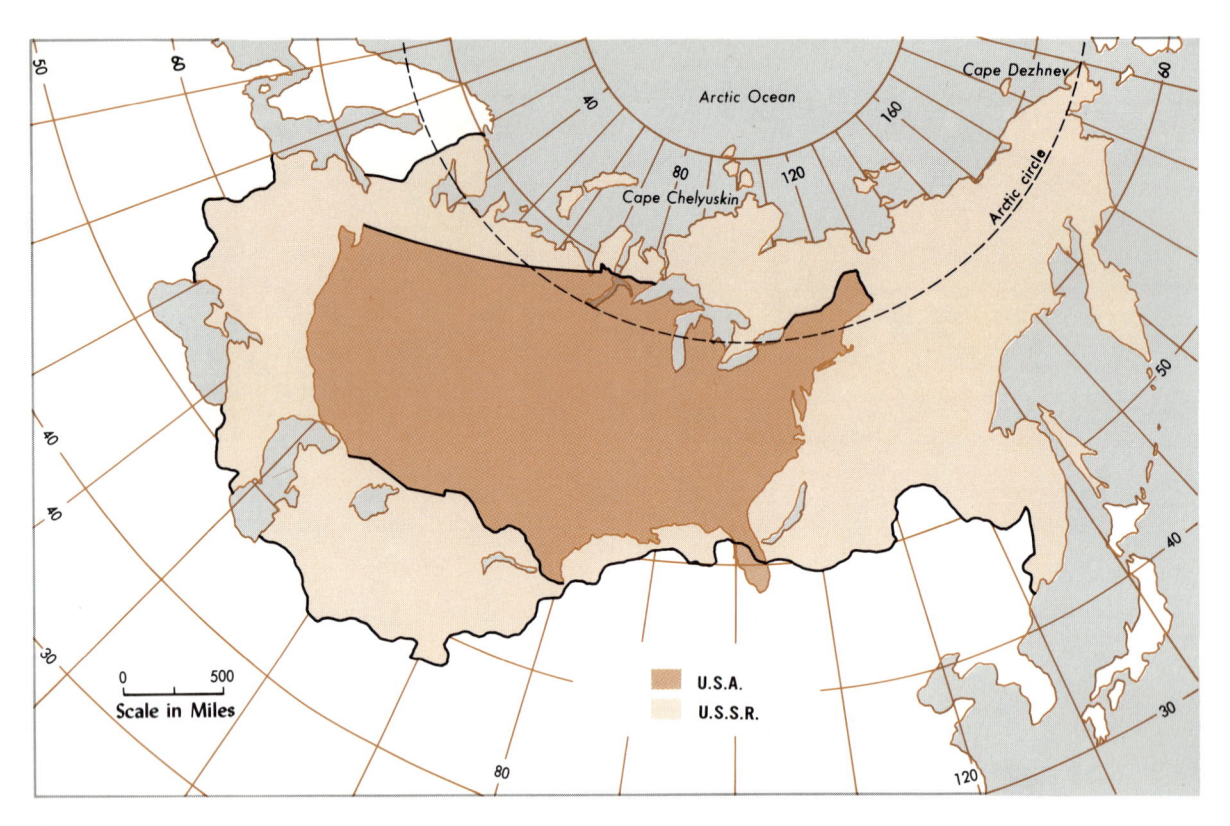

Fig. 10-1 About three-fourths of the Soviet Union lies north of the 49th parallel. In area the Soviet Union is more than 2.3 times the size of the United States.

southern boundary of Canada. In addition to this, one-fourth of its area lies north of the Arctic Circle.

The vast size of the Soviet Union is reflected in its extensive boundaries. The combined length of its coastline and land borders totals more than 37,000 miles. Of this, coastlines account for about two-thirds, or 27,000 miles. Yet, in spite of its extensive coasts, the Soviet Union exhibits climatic characteristics that are, on the whole, continental rather than marine. The county appears locked up in its enormous size and territorial spread; one-half of the Soviet Union, in fact, is situated more than 400 miles from the sea. For the most part, the coastline is regular, and there are few good harbors. The coastal waters of the Arctic and Pacific oceans are cold and, except for a stretch of the Barents Sea coast near Murmansk, are ice bound in winter. The northern part of the Caspian Sea is also frozen in winter, as is much of the Baltic coast. Ice may also form along the north shore of the Black Sea.

Many Soviet ports are located on seas of which the entrances are controlled by other countries. The approaches from the Atlantic Ocean to the Soviet Baltic Sea ports are flanked by territories of Western European countries. The Black Sea exit to the Mediterranean and Aegean seas is controlled by Turkey. Even the southern and central exits from the Sea of Japan, off Vladivostok, are bordered by Japan and South Korea.

The Soviet Union has land boundaries in common with twelve other countries, both in Europe and Asia. From the Black Sea to the Pacific Ocean, the Soviet Union borders Turkey, Iran, Afghanistan, the People's Republic of China, the Mongolian People's Republic, and the Democratic People's Republic of Korea (North Korea). Over most of its 8,000-mile extent, the Asian boundary passes through sparsely populated desert and mountainous areas. In fact, throughout its length, this boundary is crossed by only four railroad trunk lines.

The 1,500-mile European boundary, extending from the Black Sea to the Barents Sea, borders the People's Republics of Romania, Hungary, Czechoslovakia, and Poland, as well as Finland and Norway. Unlike the Asian boundary, that on the west passes through lowlands which, with the exception of Finland and Norway, are well populated. No less than sixteen railroads link the Soviet Union to the countries of Eastern Europe.

Whereas the Asian boundary of the U.S.S.R. has been altered little in recent times,[1] the western boundary has fluctuated greatly. Since 1939 the latter has been pushed progressively farther to the west. The Soviet Union regained in 1940 the provinces of Northern Bukovina and Bessarabia, held by Romania since 1918. Bessarabia was subsequently added to the Moldavian Autonomous Soviet Socialist Republic to form a new and enlarged constituent republic. From Czechoslovakia after World War II the Soviet Union gained Ruthenia, which, including a portion of the Carpathian Mountains and the upper Tisza River valley, was added to the Ukrainian S.S.R. This territorial cession not only gave the Soviet Union access to the middle Danube but also afforded a common boundary with Hungary. The present boundary with Poland lies as much as 200 miles west of its position prior to World War II. The Soviet incorporation of the eastern territories of Poland was accompanied by an expansion of Poland westward, at the expense of Germany, to the Oder-Neisse rivers. Though as yet unrecognized as a final settlement by the Western powers, the Oder-Neisse line remains the *de facto* western boundary of Poland. Similarly the German exclave of East Prussia was divided between Poland and the Soviet Union, with the latter gaining the city of Königsberg and its surrounding territory, now comprising Kaliningrad Oblast. Finally, in 1940 the Soviet Union seized the independent Baltic republics of Lithuania, Latvia, and Estonia, which resulted in expanded Soviet frontage on the Baltic Sea.

The remainder of the western boundary, extending from the Gulf of Finland to the Barents Sea, was altered also as a result of the Russo-Finnish War of 1939 and World War II. Consequently, the Karelian Isthmus, formerly one of the richest districts of Finland, was ceded to the Soviet Union, along with other borderlands to the north, including the Pechenga District (Petsamo), containing rich nickel mines. The loss of Pechenga deprived Finland of its Arctic coast and established a common boundary between Norway and the Soviet Union.

While the territorial extent of the Soviet Union is very large, only a relatively small part of it, frequently called the "fertile triangle," is suitable for intensive economic development. In that triangle live 80 percent of the U.S.S.R.'s population, which is nearing 242 million. Yet the area in which the bulk of economic activity is presently concentrated, the core area, is an even smaller one within this fertile triangle. The core lies within the west-central part of the country, extending from Leningrad east-

---

[1] The exceptions are the annexation of the People's Republic of Tannu Tuva during World War II and of the Kurile Islands and southern Sakhalin at war's end.

ward to and including the Middle Urals, thence through Volgograd (formerly Stalingrad) and Rostov-on-Don to the southwestern boundary of the country.

Enclosed within the delimited area are many of the large cities of the Soviet Union, including, besides the above-mentioned, Moscow, Kiev, Gorki, Kharkov, Kuybyshev, Sverdlovsk, Donetsk, Dnepropetrovsk, and Odessa. These and other cities of the core area account for 70 percent of the country's manufacturing. There, too, the transportation network, especially rail lines, is most dense. Finally, much of the richest agricultural land of the Soviet Union lies within the core, mainly in the black-earth steppe of the southwest.

This area of concentration of economic activity has its roots in development beginning in the Czarist period. On the eve of the Bolshevik Revolution of 1917, the main centers of manufacturing were St. Petersburg, now Leningrad (shipbuilding, textiles, specialty items); Moscow and nearby cities (textiles, pig iron, machinery); and the southern Ukraine, where extensive deposits of coal and iron ore favored a metallurgical industry. In the early eighteenth century an iron industry had grown up mainly in the Urals, which turned out much of the Russian pig-iron output, but by the end of the nineteenth century it had lost its prominent place in Russian metallurgy to the Donets Basin (Donbas).

Most of the population of Czarist Russia, too, was located in the European part of the country; especially dense was the rural population settled in the south and southwest. Outside the core, however, the land was only sparsely inhabited. Settlement had taken place in the steppe region of western Siberia, stimulated by the emancipation of the peasantry in 1861, construction of the Trans-Siberian Railway in 1894, and governmental policy, but much of eastern and, indeed, all of northern Siberia remained virtually uninhabited, save for small bands of native tribesmen. In Transcaucasia and in Central Asia, which the Russians had annexed in the 19th century, the native peoples followed a way of life that modern economic development had scarcely touched.

Though the Industrial Revolution did not reach Russia until the 1880's, large towns and cities—though well under the size of St. Petersburg and Moscow—had appeared much earlier. Linked mainly by an intricate system of rivers, these population centers functioned as seats of provincial administration as well as centers of trade and artisan industries.

After the middle of the nineteenth century the rail network began to take shape. The first major line was built between the imperial capital, St. Petersburg, and Moscow in 1851. This was followed by lines linking the grain-producing regions of the steppe to the cities to the north. Later, to facilitate grain exports to Western European countries, railways were built to the ports on the Baltic, Black, and Azov seas. The growth of heavy industry in the Ukraine resulted in the construction of additional lines between the mines and mills and between the southern centers of production and the northern markets. Finally, the need for raw materials and a governmental policy of territorial expansion compelled the construction of long rail lines into the Caucasus and Central Asia, as well as across Siberia to the Pacific Ocean.

The economic development of the core area before the Bolshevik Revolution was enhanced, too, by its accessibility to Western Europe. Here, at the time of Russian industrial expansion lay major sources of capital and technology, as well as of machinery and equipment and some of the raw materials for Russia's important textile and leather industries. Western Europe, too, was an important market for Russian grains and other agricultural products, as well as for the timber, furs, and

Union of Soviet Socialist Republics 349

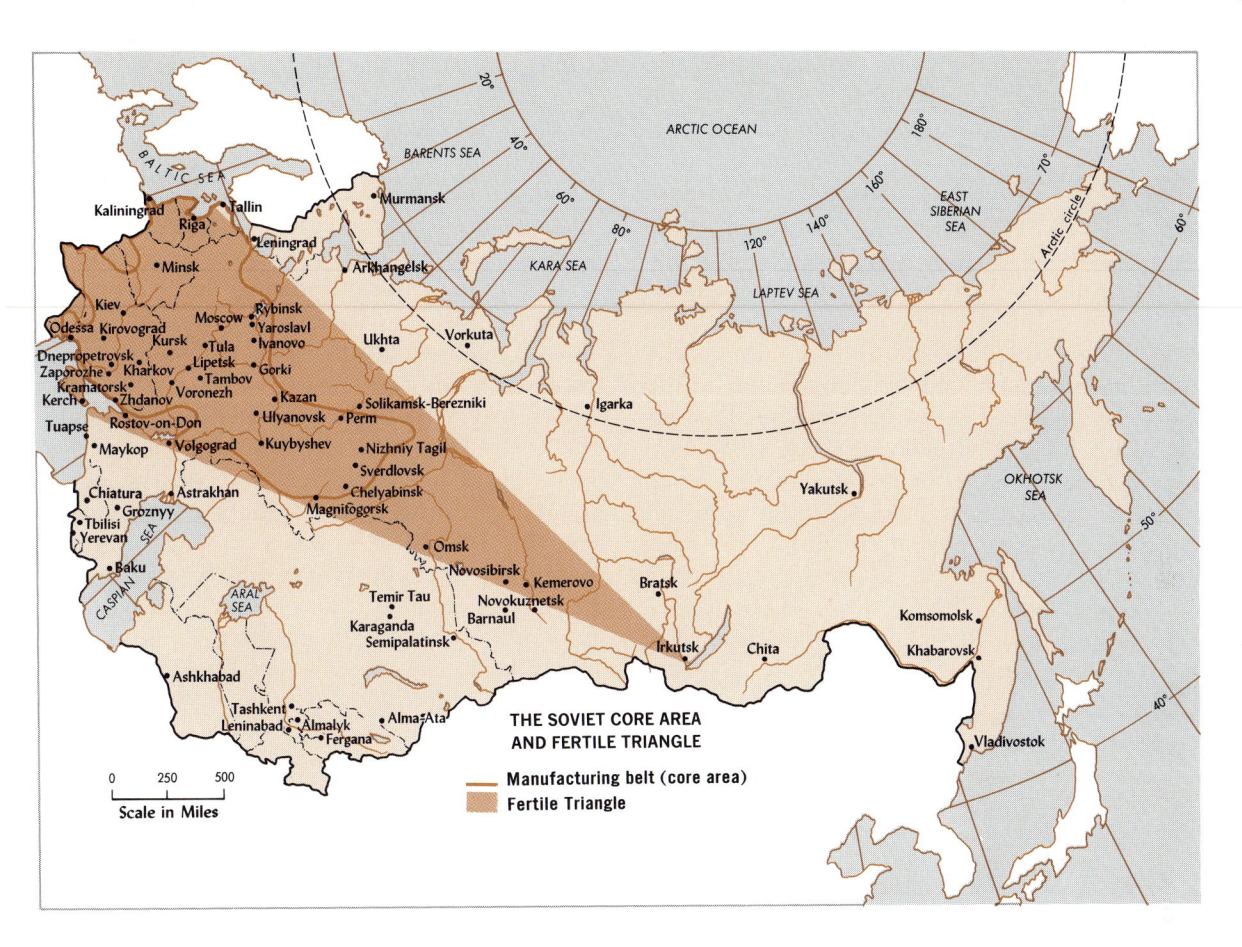

Fig. 10-2 The core area and the fertile triangle are both best developed in the European part of the Soviet Union.

petroleum that were important earners of foreign currency.

Under Soviet rule, a program of rapid industrial development, beginning in 1928-1929, accompanied by an enforced collectivization of agriculture that released a large surplus rural population for industry, contributed substantially to the further growth of the core area. Old centers of manufacturing were rebuilt and expanded. New industries appeared in ancient towns and new manufacturing centers were built elsewhere. Additional rail lines were laid to support the new industry, and waterways were improved through the building of canals and by increasing the navigability of rivers. Dams were built on the rivers to provide for the generation of electricity, and an extensive search for and exploitation of mineral reserves and power resources began throughout the country. The Soviet drive to expand its industrial base, and particularly that of heavy industry, led above all to the emergence of new industrial regions, first in the Urals and then beyond. Though rich in minerals, the Ural Mountains lack adequate supplies of coking coal; but this deficiency was met by an immediate expansion of coal production in the rich Kuznetsk Basin, about 1,000 miles to the east. The marriage of Kuznetsk coal to Urals iron ore not only led to the establishment of large new metallurgical and machine industries in the Urals, which included such centers of production as Magnitogorsk, Sverdlovsk, and Chelyabinsk, but also to the founding of a smaller producing center in the Kuznetsk region. Finally, new centers of manufacturing have emerged in more peripheral areas of the country such as in Transcaucasia, Central Asia, eastern Siberia, and the Soviet Far East. Yet, in spite of these developments and a shift of industry eastward both during and after World War II, the traditional core area in European Russia remains the economic heart of the country.

Soviet Goals

The main objective of the Soviet regime has been the erection of a citadel of socialism in the U.S.S.R. from which the revolution, preserved and strengthened, might be exported to other parts of the world. At the time of the Bolshevik Revolution, Russia was still essentially a backward agricultural country. Lenin and his fellow revolutionaries originally believed that the revolution in Russia could not succeed unless it had the support of an international revolutionary movement, originating particularly in the industrial countries of Western Europe, where Marx had predicted the Communist revolution would occur first anyway. In the years immediately following the Bolshevik seizure of power in Russia, a Communist revolution did not sweep over the Western countries. Before his death in 1924, Lenin came to realize that the rebuilding of Russia along Marxist socialist or Communist lines would not have significant support from abroad; his successor Stalin, on the other hand, proclaimed the unorthodox Marxist idea that socialism could be built in one country. But, if socialism were to survive in Russia, that nation and its political and economic systems must be strengthened, said Stalin, to guard against attack from its enemies, the capitalist countries.

In order to build a strong socialist state, the Soviet regime launched in 1928 a series of Five-Year Plans designed to speed the development of industry under tight governmental control. Initiated also was a complete reorganization of the countryside, which involved the merger of small private peasant land holdings into numerous collective farms, or kolkhozes, and the creation of large specialized state farms, or sovkhozes.

Until the Nazi invasion of the Soviet Union in June, 1941, two Five-Year Plans had been completed and a third initiated. Industrial output had risen substantially over 1928, but at great cost and suffering to the Soviet peasant or farmworker and the city dweller. As a result of peasant hostility to collectivization, even by the outbreak of the war agricultural production had not regained its pre-collectivization level. After the war a fourth Five-Year Plan begun in 1946 aimed at rebuilding the national economy, severely damaged by the war. A fifth plan, introduced in 1950, was followed by a sixth in 1956, only to be discarded two years later in favor of a longer-range plan, a Seven-Year Plan, which was concluded in 1965. Another Five-Year Plan was in operation until 1970.

Soviet planning, according to Soviet political economists, operates under the socialist law of planned proportional development. This means theoretically that, under socialist planning, all parts of the country (as well as all

Union of Soviet Socialist Republics   351

branches of the economy) undergo "even" development, as opposed to capitalism, where development, said to be chaotic, results in the creation of a core area, which in turn exploits the peripheral areas, just as capital is said to exploit labor. Hence, under socialism, it is claimed, economic development should occur everywhere. Differences between advanced regions and backward regions and between town and countryside thereby would be eliminated.

An even or proportional development of all parts of the Soviet Union, however, would necessitate an even distribution of natural resources. Yet because the Soviet Union covers such a huge territory, resources are not evenly spread and the distances between consuming areas and producing areas may be very great. Moreover, the Soviet Union is a northern land with an intemperate climate. Both climate and soils impose severe limitations on crop production in many parts of the country. The vast size of the country and its varied and unequal natural conditions create major difficul-

Fig. 10-3 For purposes of development, the Soviet Union has been divided into Economic Planning Regions. The size of such a region depends on physical factors as well as cultural development.

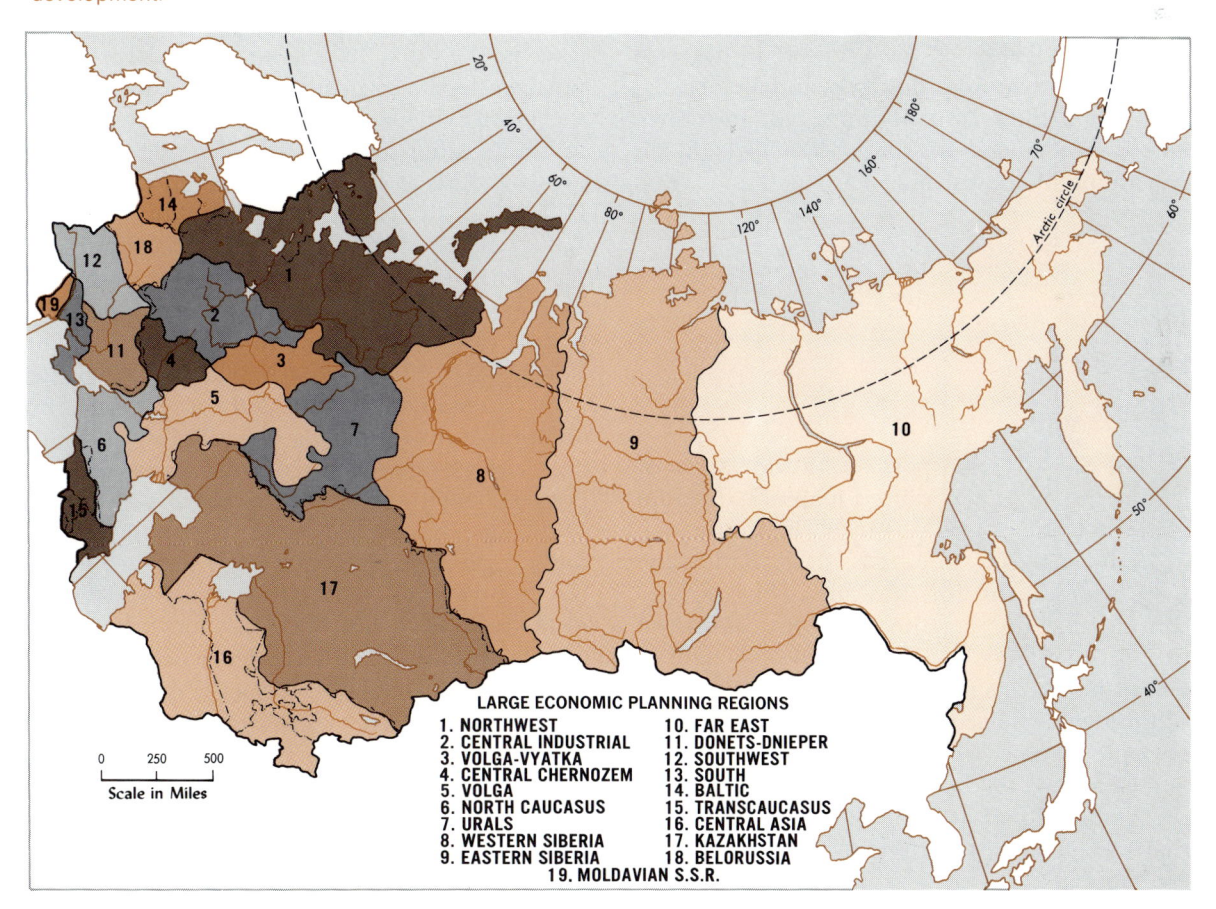

LARGE ECONOMIC PLANNING REGIONS
1. NORTHWEST
2. CENTRAL INDUSTRIAL
3. VOLGA-VYATKA
4. CENTRAL CHERNOZEM
5. VOLGA
6. NORTH CAUCASUS
7. URALS
8. WESTERN SIBERIA
9. EASTERN SIBERIA
10. FAR EAST
11. DONETS-DNIEPER
12. SOUTHWEST
13. SOUTH
14. BALTIC
15. TRANSCAUCASUS
16. CENTRAL ASIA
17. KAZAKHSTAN
18. BELORUSSIA
19. MOLDAVIAN S.S.R.

ties for Soviet economic planners attempting to work within the framework of an ideology that refuses to recognize the limitations imposed by nature.

In order to achieve correct socialist planning, the regime has sought, largely through the agency of the State Planning Commission, to subdivide the Soviet Union into large economic regions. These efforts have not always proved satisfactory, either because the regions proposed were not ideologically correct or ran counter to nationality requirements, or regional boundaries proved too arbitrary, or the needs of the state changed. At any rate, in the development following World War II, the number of large economic regions into which the Soviet Union was divided changed from thirteen to sixteen in 1956–1957, was raised in 1961 to seventeen (excluding Moldavia and Belorussia, which had their own planning commissions), and then changed in 1962–1963 to eighteen (excluding Moldavia).

Physical Setting

Relief Features

The Soviet Union possesses a wide array of landforms, but a generalized map or a diagram of these features would convey the impression of uniformity over vast areas. It is a landscape for a wide canvas, for the scale is huge. To a considerable degree, the U.S.S.R. consists of extensive lowlands lying poleward of a complex belt of mountains and plateaus that form much of the southern borderlands. On the west, these lowlands extend without a break into Poland and Germany; on the east, they are bounded by the Yenisey River and the plateaus and ranges that predominate throughout central and eastern Siberia.

European Part The Russian Plain, which occupies the greater part of European Russia, is a broad structural basin, composed mainly of horizontal layers of sedimentary rocks. Its average elevation is 300 feet, but uplands reaching to 1,000 feet appear in places; these give the land a gently rolling character. Some upland areas—notably the Central Russian Upland, between the Dnepr and Don rivers, and the Pre-Volga Heights, between the Don and Volga rivers—result from a warping of the earth's crust. Other uplands, such as the Azov-Podolian Shield west of the Dnepr River, are formed where older and more resistant crystalline rocks appear near the surface. The lowland plain has been affected, too, by glaciation. Of three glacial advances from northern Scandinavia, the third (or Würm), which covered the northwestern and northern parts of Russia, left the most evident traces, especially in extensive morainic ridges. One of these is the Smolensk-Moscow Ridge; others, making up the Valdai Hills, form the divide for rivers flowing to the Baltic and those flowing to the Black and Caspian seas. Lakes, bogs, and marshes are found in many places. The most extensive area of poorly drained land is the Polesye region in Belorussia. The southern part of the plain, which remained unglaciated, is covered by accumulations of loess, a fine silt derived from clay and sandy glacial outwash material deposited near the edge of the glaciers. Erosion of the loess has caused the formation of numerous ravines and gullies, though the loess itself has contributed substantially to the fertility of the soils.

The Russian Plain is bordered on the northwest by the Karelian Shield, a region of ancient hard rock scoured by glaciation, which in the Kola Peninsula rises to heights of over 4,000 feet. On the east the Urals, stretching for 1,600 miles in a north-south direction, form the boundary. These mountains are not a formidable divide, however, since the elevation of their

Fig. 10-4 The lowlands of the Soviet Union are principally in the western regions; the highest mountain areas are along the border with China and Afghanistan. The most highly developed part of the nation is in the European sector.

middle section is low and easily traversed. Between the industrial cities of Chelyabinsk and Nizhni Tagil, elevation does not exceed 2,500 feet, and at Sverdlovsk it is less than 1,000 feet. The Northern and Southern Urals, on the other hand, are higher, with elevations of 5,000 to 6,000 feet. In the north a spur of the Urals, the Timan Ridge, extends in a northwesterly direction, interrupting the northern portion of the Russian Plain.

In southern European Russia, apart from the Black and Caspian seas, the Russian Plain ends at the base of the high Caucasus Mountains, which have peaks of 18,000 feet, and at the lower Yaila Mountains in the Crimean Peninsula. In the southwest, the Carpathian Mountains form the border. On the west, the Russian Plain remains unbroken, except for the Polesye, and it sweeps into Poland and beyond as part of the North European Plain.

Asiatic Part The Asiatic part of the Soviet Union may be divided, for convenience of description, into western Siberia and Kazakhstan, central Siberia, eastern Siberia, the Soviet Far East, and Soviet Central Asia.

Western Siberia, extending between the Ural Mountains and the Yenisey River, is composed of a vast and extremely flat lowland. Across the lowland flow the Ob River and its major tributary, the Irtysh. The flatness of this lowland is apparent in the fact that the Ob River is less than 300 feet above sea level at a distance of some 1,800 miles from its mouth. The lowland has been affected by glaciation in its northern part, but evidences of this have been largely removed by stream erosion. Vast floodplains have been formed as streams meander across the lowland. Indeed, the distinction between floodplain and interfluve is almost imperceptible. The interfluve between the Ob and Irtysh rivers is, in fact, an extensive waterlogged area called the Vasyugan Swamp. The southern part of the lowland, though for the most part flat, is better drained.

On the south, the lowland terminates in the Kazakh Upland and in the broadened base of the Southern Urals. Between the two is the Turgay Gap, which leads into the Aral Sea Basin in Soviet Central Asia. To the southeast, the West Siberian Lowland is bordered by the Altai and Sayan Mountain systems. Within these uplands, but open on the north to the West Siberian Lowland, is the Kuznetsk Basin. To the east, along the Upper Yenisey, lies the Minusinsk Basin.

Central Siberia comprises the territory lying between the Yenisey and Lena rivers. An upland region dissected by the major tributaries of the Yenisey and Lena, it has the appearance, over much of its area, of a broad horizontal tableland. Elevations here ordinarily do not exceed 3,000 feet. On the north, the plateau descends to a lowland that stretches along the Arctic coast. On the south, it is terminated by the Sayan and Baikal Mountains, with the deeply set Lake Baikal.

Eastern Siberia is situated between the Lena River and the Pacific coast ranges. The region is composed of an array of mountain systems, enclosing rolling plateaus and swampy lowlands. The major ranges are the Verkhoyansk, the Cherski (Cherskogo), and in the extreme northeast the Kolyma. To the east of Lake Baikal stretch the relatively low, rounded Yablonovoi (Yablonoi) and Stanovoi Mountains.

The region termed the Soviet Far East comprises the Pacific coastal portions of Siberia and includes the Kamchatka Peninsula, the Kurile Islands, the basin of the lower Amur River, and Sakhalin Island. This region is predominantly mountainous, with only narrow coastal lowlands. However, west of the Sikhote Alin Range, a broad and poorly drained valley has been formed by the Amur River and its major tributary, the Ussuri. Mountains prevail in southern Sakhalin, but the northern part of the island is relatively low and flat. The Kuriles form a chain of volcanic islands extending from Kamchatka Peninsula to Hokkaido, the northernmost island of Japan.

The last of the Asiatic regions, Soviet Central Asia, consists mainly of an interior drainage basin centering on the Aral Sea. Sandy deserts, such as the Kara Kum and Kyzyl Kum south of the Aral Sea, are extensive features of Soviet Central Asia. The lowland is bordered almost entirely by plateaus and mountains. Between the Caspian and Aral seas lies the Ust Urt Plateau, which is virtually lacking in surface water. On the Soviet border, to the south, are a series of mountain systems including the Kopet Dagh, the high Pamirs, and the Tien Shan, and to the east the Dzungarian Ala Tau, the Tarbagatay Range and the Altai Mountains. The permanent snows of the mountainous borderlands, particularly in the Pamirs and Tien Shan, form the sources of rivers such as the Amu Darya and Syr Darya, which flow across the sandy desert to the Aral Sea. Smaller streams such as the Zeravshan, descend into the lowland but eventually disappear in the sands. At the base of the mountains, irregular zones of loess foothills occur and mountain spurs enclose fertile valleys, the most important of which, the Fergana Valley, is drained by the Syr Darya.

Climate

The Soviet Union lies, for the most part, in the higher mid-latitude of the Northern Hemisphere, where the landmass reaches its greatest extent, and much of the country is far removed from the moderating climatic influences of the surrounding oceans. Thus, a northerly location on the largest of the continents gives the Soviet Union an extremely continental climate. Its winters are long and cold, and its summers short and hot. Precipitation is moderate to low. These characteristics of the country's climate gradually intensify from west to east.

The moist, mild air from the Atlantic and Pacific oceans does not penetrate deeply into the country. Such air, entering from the Atlantic, noticeably affects only the western and northwestern parts of European Russia. The moderating effect of the Pacific Ocean is limited principally to Pacific coastal areas. The latitudinal position of the Soviet Union, on the other hand, exposes it to strongly developed cold air masses of Arctic and polar origin which, in the absence of any east-west trending moun-

Fig. 10-5 In general, climatic and agricultural zones extend from west to east across the U.S.S.R. Note the relationship of these belts to latitude.

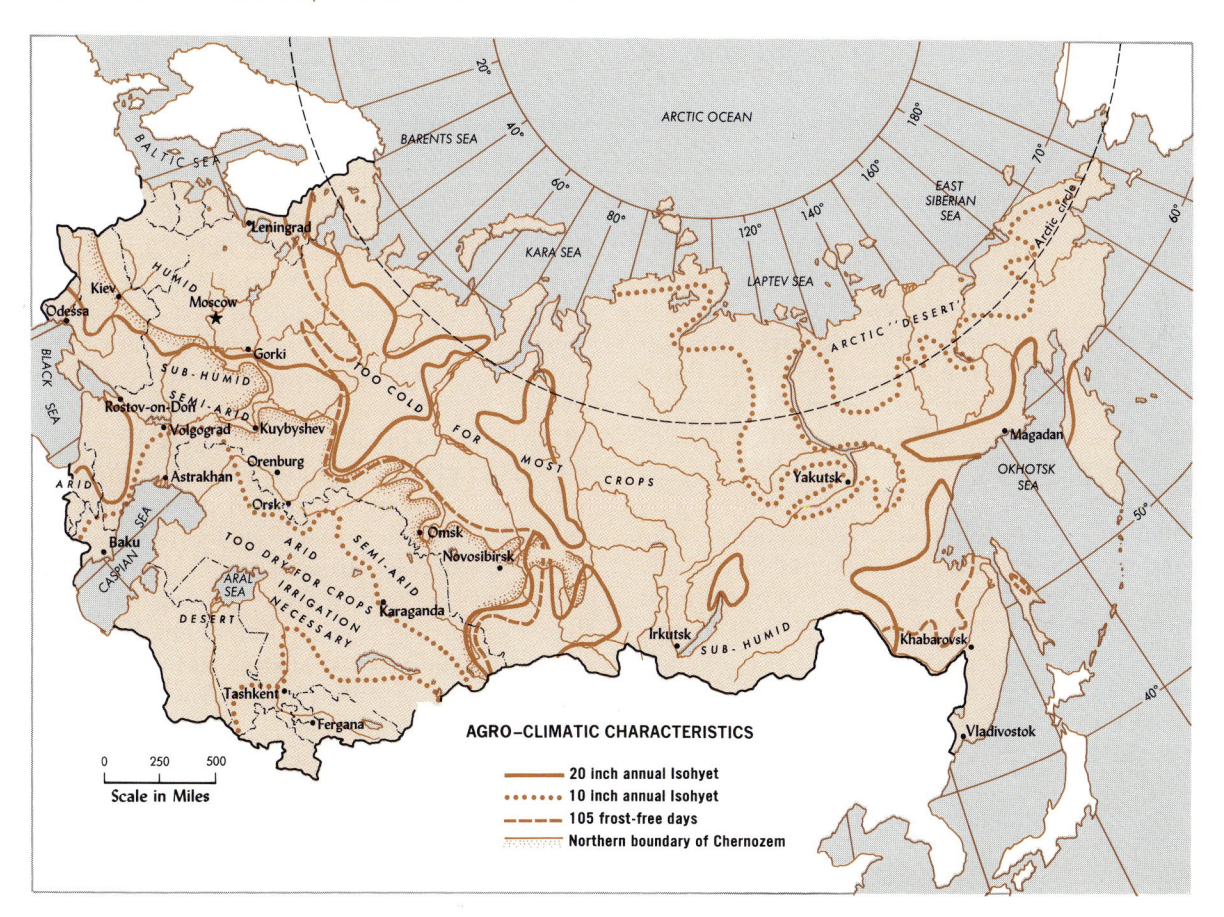

tains, often move far into the southern reaches of the country. But tropical air from the southern oceans is prevented from entering because of the high mountains and plateaus along the southern border of the country.

The effect of maritime air is limited further by the development in winter of a zone of high pressure centering over eastern Siberia, with a narrow ridge of high pressure extending into European Russia approximately along the 50th parallel. Outward from these high-pressure areas flows cold, dry air. Thus, all of Siberia, as well as Central Asia, the southwestern parts of European Russia, and even the Pacific coast, receive the impact of this outblowing air in winter. Northwestern European Russia, subject to some marine influence, is not as dry or as steadily cold in winter as the country as a whole. The average January temperature in Leningrad is 18°F; whereas farther into the interior, at Omsk in western Siberia, it is −3°F, and at Verkhoyansk in eastern Siberia it averages −58°F.

Very limited areas on the southern coast of the Crimea, which is bordered by mountains on the north, as well as Transcaucasia and the valleys of the extreme southern portions of Central Asia, because of their low latitudes and protected positions, have relatively mild winter temperatures. Sochi, on the Black Sea coast, has a January average temperature of 43°F; Baku on the Caspian Sea averages 37°F.

In summer westerly and northwesterly Atlantic air predominates over much of the country, owing to a much-weakened or virtually absent high-pressure system. Air moves inward from the cooler Pacific Ocean as well during this season. Summer is the period of maximum precipitation for the country, except for the Black Sea coast. From northwestern Russia, precipitation totals decrease toward the interior, except in upland and mountain areas, where the orographic influence is pronounced. Summer temperatures are high, except along the Pacific and Arctic coasts. The hottest parts are in Soviet Central Asia, where the July averages range from 80° to 90°F.

For purposes of agriculture, the length of the winter season, relieved by short, though generally warm summers and only brief spring and autumn periods, is a definite handicap. A short frost-free period (less than 105 days) places the northern part of European Russia and most of Siberia beyond the limits of crop production. The short growing season, too, narrows the variety of crops that may be grown. This helps to account for the large acreages of winter rye and spring wheat, grains that mature relatively rapidly.

Deficiency of moisture exerts even greater limitations on agriculture than the length of the frost-free season. Only the western parts of the country, together with the eastern Black Sea and Pacific coastal areas, have total precipitation exceeding 20 inches a year. Only in these regions is there adequate, though not in all cases abundant, precipitation for crop production. Indeed, where poorly drained land prevails, the problem is often one of too much, rather than too little, moisture. The black-earth country to the south and west of Moscow, the most important agricultural region of the country, receives between 15 and 20 inches yearly. To the southeast, precipitation drops off sharply. The annual average is 12 inches at Volgograd on the middle Volga, but the total at Astrakhan near the mouth of the Volga is only 6 inches. Less than 6 inches per year are found in extensive areas of the Aral Sea Basin. Almost all lowland areas of Siberia have less than 16 inches of precipitation annually. Much of central and eastern Siberia, in fact, has less than 12 inches per year, but the lower temperatures reduce evaporation.

It is important to note, then, that the larger part of the Soviet cropland has less than 20 inches of precipitation per year. Moreover, in these subhumid and semiarid regions, precipi-

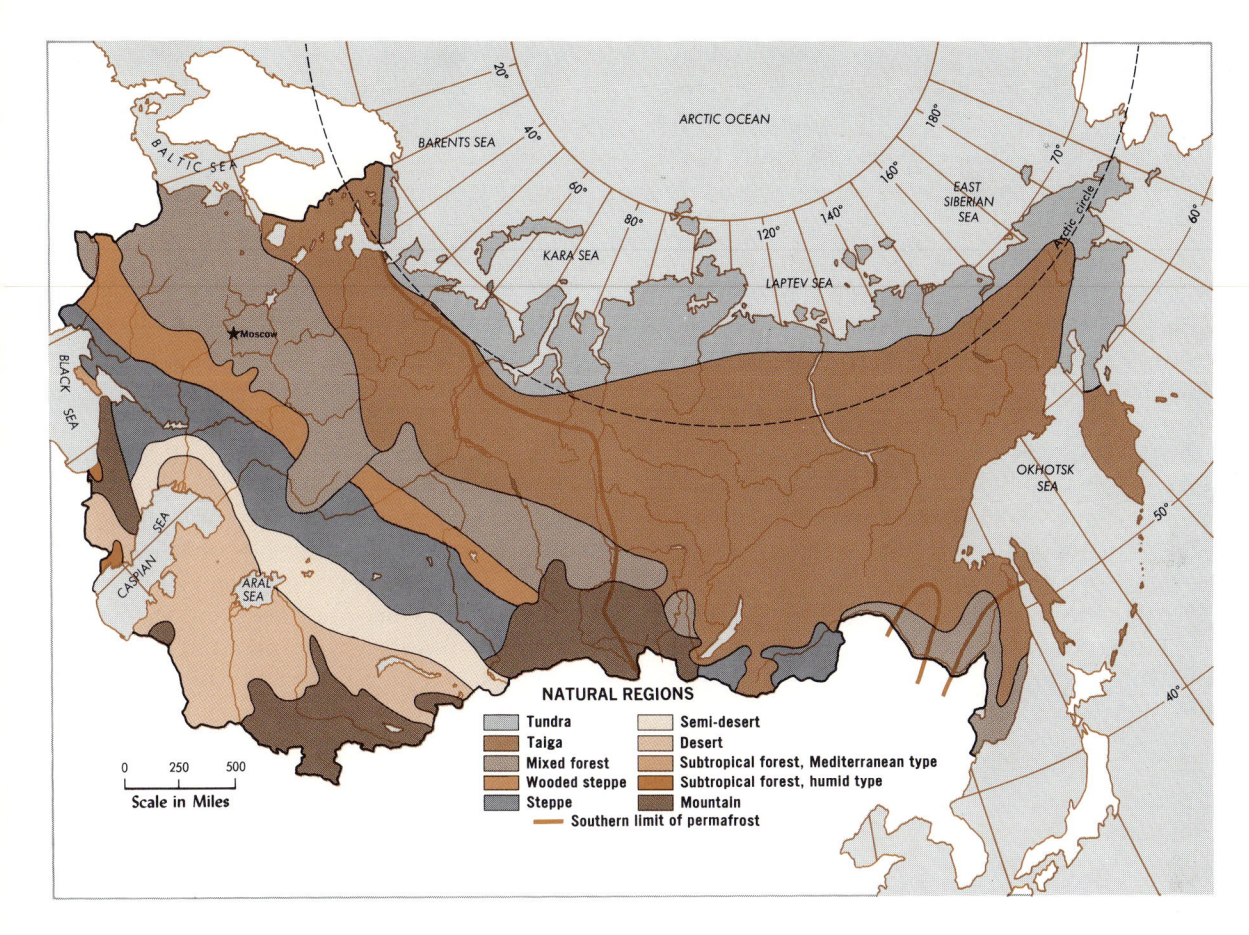

Fig. 10-6 The vegetation zones follow the climatic zones closely. The Soviet Union is so large that its vegetation zones range the gamut from desert to tundra.

tation is unreliable; they are therefore subject to frequent droughts. Years of adequate precipitation are followed by years of very little moisture. Indeed, there are in the Soviet Union extensive areas that, because of moisture deficiency, are unsuited to agriculture without irrigation.

Natural Regions

Even though the Soviet Union is of vast extent, the number of distinctly different natural features therein is limited. There are relatively few distinct climatic regions, for example. Climate tends to be broadly similar over wide areas, and changes in climate occur only gradually. These gradual changes are explained, in considerable measure, by the broad expanses of lowland and the lack of any appreciable interruption of them by mountains. As with climate, there are broad zones of vegetation and soils, which reflect to a large extent the dominant influence of climate. The Soviet geographer L. S. Berg has grouped the complex of climate, vegetation, and soils, as well as native animal life, into eight wide latitudinal zones called "natural regions," with natural features that are more or less similar throughout the extent of each region.

Tundra The zone of Arctic tundra is an unforested region with moss and lichen vegetation predominating. Where valleys offer protection, and particularly toward the south, low, dwarfed birches, spruces, and larches appear. Bogs and marshes are widespread in this region. The long, severe winters, with strong winds, make it impossible for trees to grow. Summers are cool and short. Their short duration is offset by long hours of daylight, however, which does permit small flowers to bloom. Annual precipitation is light, averaging 8 to 12 inches.

Tundra soils are thin and contain little decaying organic matter. Moreover, the poorly developed soils rest upon permanently frozen subsoil. Wet ground is common in summer when the surface layer undergoes a brief period of thaw.

Taiga South of the tundra lies a broad zone of coniferous forest, or "taiga," extending from the western borders of the country to the Pacific coast. In European Russia the taiga lies, for the most part, north of the core area. The cities of Leningrad, Yaroslavl, and Gorki lie on or close to the southern boundary of the taiga.

The taiga is composed principally of coniferous trees, notably spruce, larch, fir, and pine. The transition from the tundra zone is gradual. At first, trees are dwarfed; but with increasing moisture, higher temperatures, and more gentle winds, the trees become taller and the stands are denser. Bogs are widespread in the Northern European and western Siberian parts of the taiga. In parts of western Siberia, waterlogged

Fig. 10-7 Taiga in central Siberia, U.S.S.R. In central Siberia the taiga is an upland area crossed by many major rivers. Here the Lena River has cut a broad valley through dense forests. (Courtesy of Tass)

areas occupy entire interstream areas. Bogs are few, however, in central and eastern Siberia.

The climate of this vast forested region may be classed as subpolar continental. The features of the climate vary considerably within the region; but, in general, the winters are cold and long and the summers warm. January average temperatures range from 20°F in the west, where Atlantic influences are felt, to below −50°F in eastern Siberia. Average July temperatures range from 50 to 68°F, but daytime temperatures during the short summer in Siberia may exceed 85°F. Precipitation is moderate to light. More than 24 inches a year fall at Leningrad. In winter, in the west, days are often cloudy, and the snowfall forms a thick cover over the earth. Toward the east, snowfall is lighter, but prolonged low temperatures cause the snow to remain on the ground throughout the winter.

The soils of the coniferous forest are known as podzols. Highly leached, with a strongly acidic reaction, the podzol is infertile. Large amounts of lime and organic matter must be applied to the soil before the podzol can be productive. Moreover, the short growing season prevents the cultivation of anything except a few hardy vegetables and grains.

Permafrost, or permanently frozen ground, underlies much of the Siberian taiga. Its thickness varies from 3 feet to many hundreds of feet. During the short summer, thawing occurs only in the surface layers.

Mixed Forest South of the taiga in European Russia, deciduous broad-leafed species become mixed with conifers. The mixed forest region forms a wedgelike zone extending from the western border (between Leningrad and the western Ukraine, at approximately the latitude of Kiev) eastward to the Middle Urals. In the valley of the Amur River in the Soviet Far East, there is another, less extensive region of mixed forest, but its species differ from those of European Russia. The mixed-forest zone of European Russia, first settled centuries ago, was the heart of the old principality of Muscovy and today forms the northern half of the Soviet core area.

Among the deciduous species are oak, elm, maple, and ash. Oak is especially prominent; this is an indication of more favorable soil and climate conditions. Bogs are generally less prevalent than in the north, despite the existence of the Polesye, a poorly drained forested lowland in the basin of the Pripyat River.

The climate of much of the region may be considered humid continental. Winters remain long and cold, but summers are warmer and wetter than in the taiga. At Moscow, the average January and July temperatures are 12 and 66°F, respectively. Precipitation varies from over 30 inches in parts of the Baltic republics to less than 20 inches in the east. Summer rain predominates and is, thus, an aid to agriculture. Snowfall is heavy and remains on the ground from October to March. Upon thawing in spring, snow and ice provide a valuable reserve of moisture for agriculture.

Wooded Steppe The wooded steppe forms a zone of transition between the mixed forests to the north and the grasslands to the south. Its vegetation consists of both trees and grasses, with the latter becoming increasingly predominant southward toward the true steppe, also called the grassland zone. Beginning at the western borders of the country, the forest steppe extends to the Urals; beyond the Urals it stretches in a narrow zone across western Siberia as far as the foothills of the Altai Mountains. In the west, the wooded areas consist almost entirely of oak, which is replaced in Siberia by birch and aspen. Small isolated areas of wooded steppe may be found in the southern valleys of central Siberia.

Since it is transitional, the zone has, in the north, a climate similar to that of the mixed forest and, in the south, a climate more nearly like that of the steppe. The southern border, in fact, coincides with the axis of high pressure that extends across the Ukraine in winters. As a result, the northern part is affected by moisture-bearing winds from the Atlantic. In the south, on the other hand, the winds coming from the north and east are dry. The mildest and wettest conditions are found in the extreme western part, in the southwestern Ukraine and in the Moldavian S.S.R.

January temperatures decrease to the east, from an average of 12°F in the western Ukraine to 3°F near the Urals. July temperatures are more nearly uniform, ranging from 69 to 72°F. In Siberia, conditions are more extreme. While the July temperatures are comparable to those in the European part, January temperatures are lower. Throughout the wooded steppe, precipitation is moderate. Annual totals range from 18 inches in the north to 12 inches in the south.

The soils of the wooded steppe are described generally as leached black-earths, that is, degraded chernozems. The leaching of the surface horizon, however, is not excessive. On the other hand, owing to the decay of grasses, the soils are rich in humus. In fact, humus comprises up to 10 percent of the content of the soil. Beneath the horizon of humus accumulation is a horizon of lime accumulation. The soil near the surface has a loose, crumbly texture. Easily eroded, it often results in deep gullies and ravines on exposed surfaces. In the western part of the wooded steppe are extensive loess deposits, which further enhance the fertility of the soils.

On the whole, the wooded-steppe zone, with its more favorable climatic conditions and fertile soils, forms one of the more productive agricultural regions of the Soviet Union.

Steppe The steppe or grassland zone comprises a vast area extending from the southwestern borders of the country as far east as the foothills of the Altai Mountains in western Siberia. Except for the river valleys, it is entirely unforested. Its soils are predominantly black (chernozem) and chestnut-brown.

The steppe climate is warmer and drier than the climate of the forest zone to the north. Light to moderate precipitation prevails, but totals, ranging from 16 inches to about 10 inches annually, decrease eastward. Most of the rainfall occurs in early summer; however, its effectiveness for agricultural purposes is reduced by the fact that high summer temperatures induce a high degree of evaporation. Moreover, precipitation fluctuates in amount from year to year. Adding further to agricultural difficulties are the frequent dry and unusually hot winds, the "sukhovey," arriving from Central Asia.

In the northern half of the steppe, the soils are black and rich in humus, having developed under a tall grass cover. In the Ukraine, where a heavy loess mantle prevails, the soils are especially productive. Southeastward, however, as precipitation decreases, the humus content of the soils falls, and they become lighter and browner in color. They, too, are subject to erosion by the strong winds that sweep unobstructed across the steppe. On the whole, Siberian chernozems are neither so extensive nor so rich as those in the European part of the U.S.S.R. Moreover, east of the Volga, the incidence of alkalinity increases, making the soils unsuitable for wheat, the predominant crop. Such soils are referred to as "solonets-chernozems."

In the southern steppe, where precipitation is both light and highly irregular, the grasses are shorter and sparser. Here the soils are decidedly chestnut-brown in color. Low in humus, they are lumpier and more tightly packed than the crumbly, friable chernozems to the north. Alkalinity and salinity are more extensive and pronounced than in the chernozem region. However, in years when the warm-season precipitation is sufficient, the chestnut-brown soils may yield good harvests of grain.

Desert The desert region lies south of the steppe. It extends through the lower Volga and Aral Sea basins as far as the foothills of the southern mountain region. It comprises extensive flats of stony, clayey, and sandy desert, in many places virtually devoid of vegetation.

The desert has meager precipitation; everywhere in this zone the annual total is less than 10 inches. Most of the precipitation appears in summer, in the form of short and heavy showers. Summer downpours, coupled with sparse vegetation, result in excessive runoff. Winters are cold, and especially so in the northern parts, which are strongly affected by winds from central Siberia. At Novokazalinsk, the average January temperature is 11°F; whereas at Tashkent, to the south, it is 32°F. Summers are very hot throughout the region, with absolute maximum temperatures reaching as high as 122°F.

Gray desert soils predominate; they contain little humus and are highly alkaline and saline. On the loessial piedmont plains along the base of the mountains forming the southern border of the country, the soils, while low in humus content, are high in lime and are little salinized. These soils are extremely fertile and, when irrigated, give substantial yields of cotton.

Subtropical Forest Along the southern coast of the Crimea and the southern slopes of the Caucasus Mountains, from Novorossisk to Tuapse, a Mediterranean-type climate prevails. Winters are mild and rainy, and summers are hot and dry. January average temperatures are about 38°F; in July, the average reaches 76°F.

These mild regions have a rich and varied vegetation. Open groves of oak and juniper, characteristic of the shores of the Mediterranean Sea, are prominent. These appear on the coastal lowland and reach up to 1,000 feet on the mountain slopes. There is also a variety of evergreen woody plants. Less evident, however, are the shrub thickets ("maquis") so typical of true Mediterranean lands. Many exotic plants from the regions of mild climate of other countries have been introduced; among these are the Italian cypress, palm, magnolia, and wisteria.

Farther south, in the western part of Transcaucasia, the climate is more humid. Precipitation, which occurs throughout the year, totals between 90 and 100 inches. The high rainfall is combined with high relative humidity. Under these conditions, there is a luxuriant and rapidly growing vegetation, both in the Colchis Lowland and on the lower mountain slopes. Mixed broadleafed and coniferous types appear together with many vines and ferns. Much of the Colchis, however, is swampy and remains unreclaimed.

The Talysh (or Lenkoran) Lowland on the eastern coast of Transcaucasia is also subtropical in character, but continental influences are somewhat stronger here than in western Transcaucasia. Summers are hotter, and winters are occasionally severe. Rainfall, though abundant, is somewhat less, and a definite dry season prevails in late summer. The vegetation, though less luxuriant than that of the Black Sea coast, is of similar type. Much of the original forest cover in the lowland is gone and has been replaced by crops. Alluvial soils are widespread on the lowland but, with increasing elevation, give way to lateritic types.

Mountain Regions The mountains often display on their slopes the same vegetation zones that the Russian lowlands possess latitudinally. In some cases, the vegetation may change from desert types at the base of the mountains into grasslands, and thence to forests, alpine meadows, and tundra. On the loftier peaks of the Caucasus and the mountains of Central Asia, there are permanent snowcaps. In summer, the alpine meadows, with abundant grasses and cool temperatures, provide a favorable base for livestock grazing. However, such alpine meadows are lacking in the mountains of eastern Siberia and the Soviet Far East.

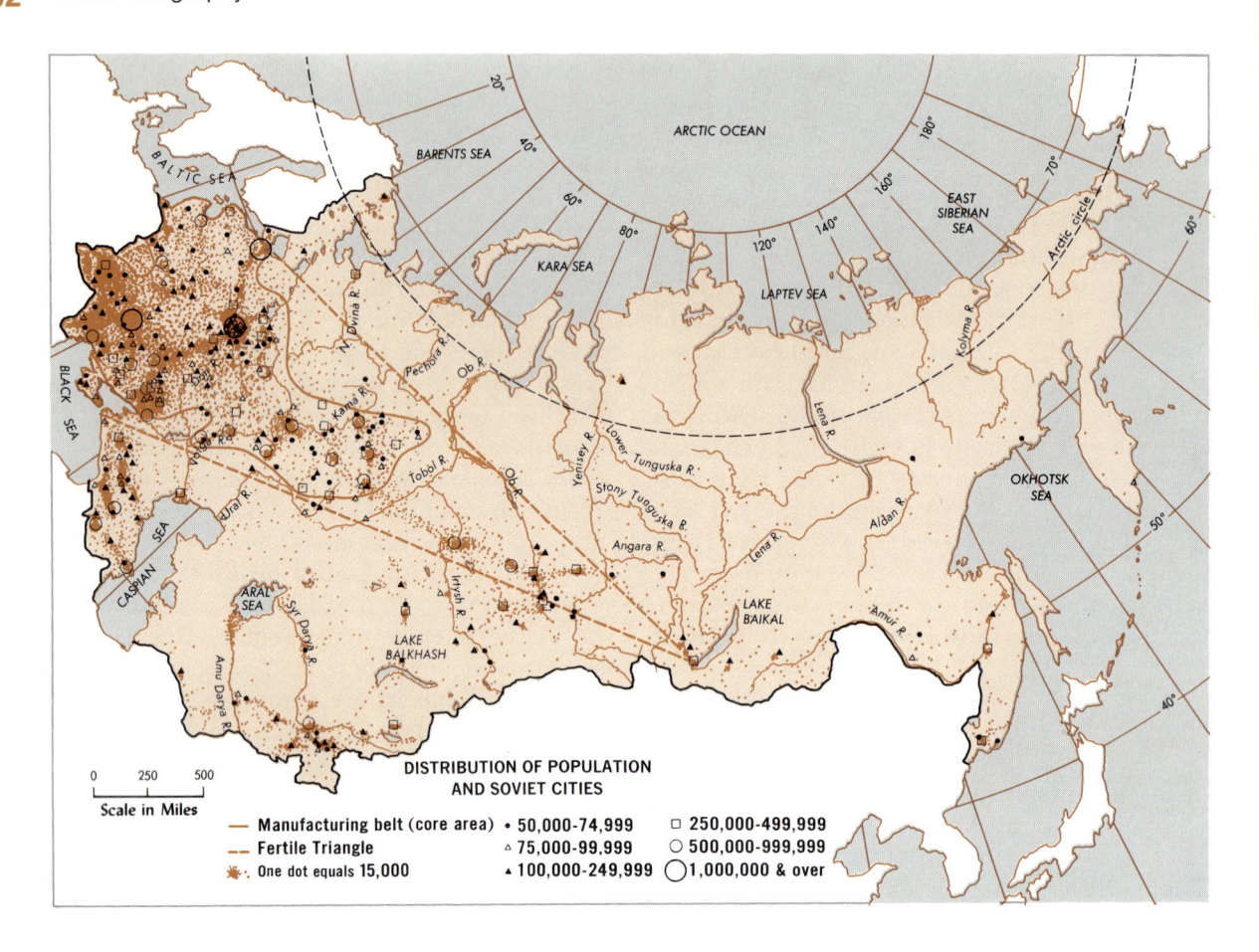

Fig. 10-8 Most of the population and most of the large cities of the Soviet Union are located in the European part of the country.

Population

On the basis of population, the U.S.S.R. ranks third among the countries of the world, after China and India. In January 1959, according to the first official postwar Soviet census, the population of the U.S.S.R. totaled 208.8 million. This estimate represented an increase of nearly 70 million since 1913, or nearly 40 million since 1939 (prewar boundaries). Not all of the increase, however, was due to natural causes. As a result of the extension of Soviet boundaries during and after World War II, principally westward into Europe, more than 20 million additional persons were brought within the Soviet fold. On the other hand, had the war not taken so many lives, the Soviet population today would have reached well over 250 million —assuming, of course, that prewar demographic trends had prevailed. As it is, the most recent Soviet estimate (1970) gives a total population of 241.7 million.

Until the early 1960s, the Soviet population

increased at an annual rate close to 1.75 percent, representing about 3.5 million persons more per year. Such a growth rate was high compared with that of Western industrial countries. It seems high, too, if one recalls the severe losses sustained by the people of the U.S.S.R. during World War II, particularly among men of marriageable age. Indeed, it is often said that the U.S.S.R. has the highest proportion of widows of any country in the world. This fact helps to explain the large numbers of older women in the labor force. Any visitor to the U.S.S.R. will find women not only driving taxis and sweeping streets but also undertaking heavy construction work.

The high natural rate of increase that prevailed in the postwar period was due primarily to a sharp drop in the death rate: from 18.0 per 1,000 in 1940 to 7.1 per 1,000 in 1960. This represented a major decline in infant mortality, thereby causing the Soviet death rate to fall below that of the United States. The Soviet birthrate, which stood at 24.9 per 1,000 in 1960, had declined also, from 31.2 per 1,000 in 1940. In recent years, however, the trends have changed. The number of births has continued to fall even more substantially (17.4 per 1,000 in 1967), while the number of deaths has tended to increase (7.6 per 1,000), thus leading to a sharp reduction in the natural rate of increase (9.8 per 1,000). The latter statistic is now lower than at any previous time in the history of the Soviet Union, with the exception of such "crisis periods" as the Revolution, the famines of 1921–1922 and 1932–1933, and World War II. Urban migration and other factors associated with modern industrial societies probably account for the lowering Soviet birthrate; and together with a decline in urban fertility, there is some lowering of the birthrate in the more backward, non-Russian areas of the U.S.S.R., notably in Soviet Central Asia and Transcaucasia.

Of the total Soviet population, over 171 million (or 70 percent) live in the European sector, including the Urals. The bulk of these, roughly 160 million, are found in the European core area. Between the Urals and the Pacific Ocean (i.e., in Siberia and the Soviet Far East) live 25.4 million, and in Soviet Central Asia and Kazakhstan live another 32.8 million. An additional 12.3 million live in the republics of Transcaucasia. In short, most of the Soviet people live in the west-central part of the Union, with smaller but important concentrations in Soviet Central Asia, Transcaucasia, and the Soviet Far East.

While average population density for the Soviet Union as a whole is about 24 persons per square mile, in the core area, outside the cities and industrial areas, it rises to 130 persons or more. The greatest rural densities are found in the rich chernozem zone in the southwestern Ukraine, where they reach over 260 persons per square mile.

In the steppe region of Western Siberia and northern Kazakhstan, the population distribution forms a narrow wedge about 300 miles wide astride the route of the Trans-Siberian Railway. Here densities range from 25 to 65 persons per square mile. To the north, owing to the presence of the Vasyugan Swamp, and to the south, because of aridity, the population dwindles rapidly. To the east, beyond the Yenisey River, a lower density also prevails, but most of the inhabitants are found within a short distance of the railway. The only sizable concentration in the Soviet Far East is north of Vladivostok in the Lake Khanka Plain. Throughout the whole of northeastern Siberia, the population density ranges from only 2 to 4 persons per square mile. In much of Transcaucasia, densities reach European Russian proportions, and in some irrigated areas in Central Asia they climb to over a thousand.

Unlike the United States, which until recent times received large numbers of immigrants from the countries of Western Europe, the population of Czarist Russia, in addition to natural increase, grew mainly through the conquest or annexation of lands and peoples that lay in the path of imperial ambitions. As the empire pushed out from European Russia to the shores of the Pacific and to the lofty mountains of Central Asia, there was an accompanying movement of Great Russian and, to a lesser extent, of Ukrainian settlers. During the latter decades of the nineteenth century, apart from organized military settlements (Cossacks) along portions of the Asian frontier, much of the population movement into the new territories was attracted by opportunities for agricultural development. This was especially true of the migration into the steppelands of Siberia.

The Bolshevik Revolution, if anything, initiated an even greater movement of people out into the peripheral areas of the Soviet Union. Since much of the good land had already been occupied, these Soviet migrants were sent into new mining and lumbering areas and to cities that were expanding under the industrialization drive. From 1913 to 1939, well over 3 million persons moved eastward to the Urals and Siberia. During World War II, and especially in the years following, additional millions moved east. According to the official 1959 census, it is evident that a number of older regions in the western part of the Soviet Union suffered net losses, while various eastern regions grew significantly. Between 1939 and 1959 the population of the Urals increased by over 29 percent, of western Siberia by over 19 percent, of eastern Siberia by over 24 percent, and of the Soviet Far East by 69 percent. Since 1953–1954, the expansion of crop cultivation in northern Kazakhstan, under the so-called "virgin and idle land" program, resulted in an influx of several hundred thousand persons and contributed to the increase of over 39 percent in that republic's population. Important population gains have also been reported in the Central Asian republics and in Armenia, in Transcaucasia.

Along with the general migration eastward or into peripheral areas, there has occurred in the Soviet Union, since the beginning of the First Five-Year Plan, a pronounced movement to the cities. Whereas in 1928 about 18 percent of the population was listed as urban, this ratio had increased to over 56 percent by 1970. Indeed, apart from the migrants to the virgin lands, much of the eastward movement of population under the Soviets has been into the cities. Collectivization of agriculture released millions of workers for industry, not only because they were no longer needed on the farms, but also because living conditions in the countryside became so unsatisfactory that life in the cities seemed preferable. In short, from 1926 to the late 1960s, well over 40 million persons migrated to the cities. Consequently, the number of cities grew substantially; those with a population of 100,000 or more increased in number from 28 to 188. At present, 33 Soviet cities have 500,000 or more inhabitants. Ten cities total over a million. Moscow is by far the largest city, with 7.1 million persons, followed by Leningrad (3.9 million) and Kiev (1.6). Seven additional cities that have in recent years reached the one-million mark include: Tashkent (1.4), Baku (1.3), Kharkov (1.2), Gorki (1.2), Novosibirsk (1.2), Kuybyshev (1.0), and Sverdlovsk (1.0).

On the other hand, it should be noted that nearly half of the Soviet population still lives in the countryside or is associated with rural life. Increasing mechanization of agriculture, if accompanied by greater efficiency per farmworker, could release many additional workers for industry, as has happened in the United States. Nevertheless, the U.S.S.R., despite a drive for autarky, is still unable to provide food in sufficient quantity for its needs, owing to inadequate planning, mismanagement, inefficiency, and a lack of strong individual incentive. However, as we have seen, climate sharply limits the range of crops that can be grown successfully.

Minority Groups and Soviet National Policy

The expansion of Muscovy from the fifteenth century on created a multinational empire. Presided over by the Great Russians, the empire included, besides related Slavs, peoples of widely differing race and culture, such as the fair-haired Finnic peoples of the northern forests, the dark-haired Turko-Tatars and Mongols of the grasslands and desert, the mixed peoples of the Caucasus, and the primitive tribes of northeastern Siberia.

The early expansion had been into lands sparsely inhabited by peoples less culturally

Fig. 10-9 The U.S.S.R. is divided into fifteen large subdivisions, each known as a Soviet Socialist Republic. The largest of these republics is Russia. Of the ethnic groups within the nation, the Great Russians are by far the most numerous.

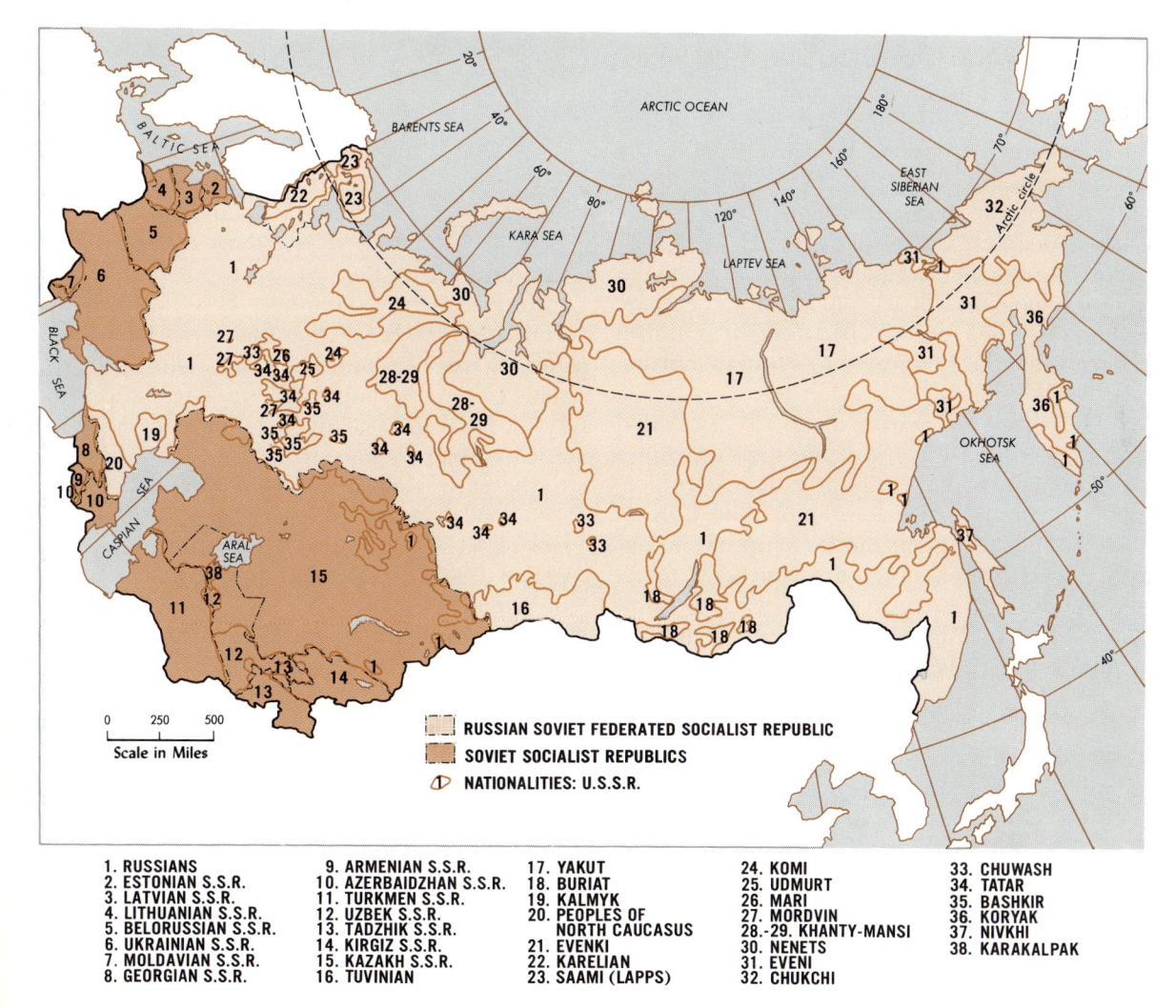

RUSSIAN SOVIET FEDERATED SOCIALIST REPUBLIC
SOVIET SOCIALIST REPUBLICS
NATIONALITIES: U.S.S.R.

1. RUSSIANS
2. ESTONIAN S.S.R.
3. LATVIAN S.S.R.
4. LITHUANIAN S.S.R.
5. BELORUSSIAN S.S.R.
6. UKRAINIAN S.S.R.
7. MOLDAVIAN S.S.R.
8. GEORGIAN S.S.R.
9. ARMENIAN S.S.R.
10. AZERBAIDZHAN S.S.R.
11. TURKMEN S.S.R.
12. UZBEK S.S.R.
13. TADZHIK S.S.R.
14. KIRGIZ S.S.R.
15. KAZAKH S.S.R.
16. TUVINIAN
17. YAKUT
18. BURIAT
19. KALMYK
20. PEOPLES OF NORTH CAUCASUS
21. EVENKI
22. KARELIAN
23. SAAMI (LAPPS)
24. KOMI
25. UDMURT
26. MARI
27. MORDVIN
28.-29. KHANTY-MANSI
30. NENETS
31. EVENI
32. CHUKCHI
33. CHUWASH
34. TATAR
35. BASHKIR
36. KORYAK
37. NIVKHI
38. KARAKALPAK

advanced than the Russians. In those areas, assimilation had come easily, largely through Russian occupancy. The conquests and annexations of the eighteenth and nineteenth centuries, however, had led to the imposition of Russian control over peoples who had an older history than the Russians themselves and who were, in many instances, more culturally advanced. The Czarist government was faced, therefore, with a difficult problem of control.

An uprising among the subject Poles in 1863 aimed at establishing an independent Polish state, led Czar Alexander II to adopt an official policy of "Russification." This policy took the form of tighter central control and the imposition of Russian as the universal language of the empire. These measures were accompanied, wherever possible, by conversion to the Orthodox faith. In time, Russification was directed at the Finns, Ukrainians, and other subject peoples; those who resisted were persecuted and exiled. Finally, at the turn of the century, pogroms broke out against the Jews, mainly in the western part of the empire. Those who could migrated to Western Europe and North America.

The collapse of czarist rule in 1917 provided an opportunity for some of the subject peoples to break away and establish independent republics. Prior to the Revolution, the Bolsheviks, hoping to hasten the breakup of the old order, had promised self-determination to all the minority groups. In reality, what the Bolsheviks had in mind was not the formation of a number of independent states on the ruins of the Russian empire, but something quite different. They hoped that the minorities would throw off the autocratic yoke of the czars, establish working-class states, and then reunite with the Great Russians to form a new socialist republic. Later, after having seized power, the new Bolshevik, or Communist, regime did its utmost to ensure that the minorities remained part of the larger Soviet state. Though

Fig. 10-10 Moscow is the largest city, with a population of approximately 7 million, and the political center of the Soviet Union. (Courtesy of Tass)

the Finns, Latvians, Estonians, Lithuanians, and the Poles managed to break away, independence movements in the Ukraine, the Caucasus, Central Asia, and the Far East collapsed with the advance of Red Army troops or the conspiracies of the Communist Party. It was not until 1940, however, that the Soviet Union was able to regain control of the Baltic peoples; the Poles were reduced to satellite status after World War II, leaving the Finns to enjoy an uneasy independence thereafter. Nevertheless, in the early years of Soviet power, nationalism remained a powerful force among the minorities of Russia, and the new regime was compelled to make concessions to national feeling—on paper at least.

The diversity in ethnic structure led the Soviet regime to proclaim, in the early days of its power, a "Declaration of the Rights of the People of Russia." Embodied later in the constitutions of 1924 and 1936, this principle theoretically guaranteed the sovereignty and equality of all the associated peoples in the Soviet Union. In an effort to maintain the semblance of a local national autonomy, a hierarchy of political-territorial units was created corresponding to the size of the respective minority groups and their degree of cultural development.

At the highest level, constituent national republics were established. These are the Soviet Socialist Republics, of which there are fifteen at present. Only peoples who represent a distinct, stable nationality totaling a million or more, and who live on the periphery of the country, may enjoy such political status. According to the constitution, location somewhere along the border is a necessary condition, so that the national group may exercise its right of secession—a right not yet implemented by any of the republics.

The largest of the Soviet republics is the Russian Soviet Federated Socialist Republic, which, in addition to its Great Russian majority, includes numerous smaller national or ethnic groups. On the west the Russian Republic is bounded by the Soviet republics of Estonia, Latvia, and Lithuania, established in 1940; the sister Slavic republics of Belorussia and the Ukraine, both of which obtained representation as "sovereign states" in the United Nations in 1945; and Moldavia, created in 1940 when Bessarabia was annexed from Romania. Along the southern border are the three Transcaucasian republics, Georgia, Armenia, and Azerbaidzhan; and the five Central Asian republics, the Turkmen, Tadzhik, Uzbek, Kirgiz, and Kazakh S.S.R.'s. All of these republics had been established by 1936.

The Autonomous Soviet Socialist Republics, which represent a lower order in the political-administrative structure, are comprised of national minorities that do not meet the prerequisites for the status of Soviet Socialist Republic. Moreover, they are directly subordinate to the Union republic in which they are located. Altogether there are twenty A.S.S.R.'s, sixteen of which are in the Russian S.F.S.R. The autonomous oblasts and the lowest-ranking of all, the national okrugs, have even less of the trappings of autonomy.

Stalin described the national administrative system as one that was "national in form, socialist in content." In reality, this credo allowed the minority groups to preserve their folksongs, costumes, dances, and literature, as long as these traits did not conflict with the overall interests of the Soviet state. At the same time, it meant that in every important aspect of life, direction would come from the Communist Party of the Soviet Union. All power would emanate from Moscow, and administration would remain in the hands of Great Russians or of individuals such as Stalin, the Georgian, who had assumed the outlook of the Russian or the new Soviet man. Finally, in the planning of the country's economy, the national units were in no way given preference over the non-national units, the oblasts and krays into which the rest of the country is subdivided.

According to the official 1959 census, the U.S.S.R. is composed of more than 100 separate nationalities, the largest of which is the Russian. Closely related in language and culture are the Ukrainians and Belorussians. Altogether, the East Slavs account for three-fourths of the total population.

Most of the national variety within the Soviet Union is found among non-Slavic groups:

TABLE 10-1
SOVIET POLITICAL-ADMINISTRATIVE STRUCTURE

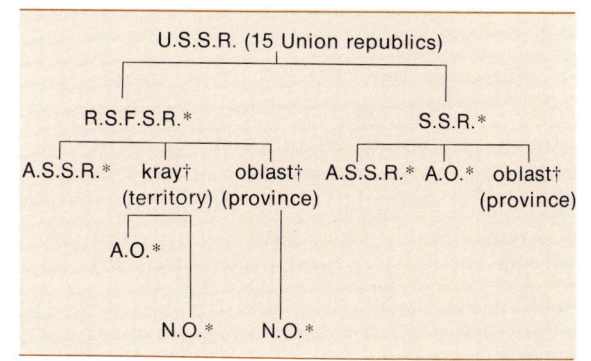

```
                    U.S.S.R. (15 Union republics)
                              |
              ┌───────────────┴───────────────┐
           R.S.F.S.R.*                      S.S.R.*
        ┌─────┬─────┬─────┐              ┌─────┬─────┬─────┐
     A.S.S.R.* kray† oblast†          A.S.S.R.* A.O.* oblast†
          (territory)(province)                        (province)
              |
            A.O.*
              |
            N.O.*          N.O.*
```

*National units
†Non-national units

the Turko-Tartars, the Mongols, and the Tungus-Manchurians who belong to the Altaic family. Distantly related is the Uralian family, which includes the Finns and Karelians, the Estonians, the Mari, the Udmurts (Votyaks), and the Samoyeds. The Japhetic peoples, located entirely within the Caucasus region, include among others the Georgians and the Dagestanis. Finally, there are significant minorities of Jews, other European peoples, and Tadzhiks (Tajiks; Iranians), as well as small bands of primitive Paleo-Asiatics in northeast Siberia.

It should be noted, however, that as a result of Russian settlement in regions outside historic Russian occupancy the significance of the Soviet minority groups is decreasing. The most dramatic instance of this may be found in Kazakhstan. There, in 1926, Russians constituted 26.5 percent of the total population and Kazakhs, 40.1 percent; by 1959, the ratios were reversed, with Kazakhs representing only 19 percent and Russians 46.2 percent. Since 1959 this trend has continued. Moreover, if the Ukrainians are added to the Russian population, it will be seen that Kazakhstan has become predominantly a Slavic region.

Economic Development

Agriculture

Although the U.S.S.R. has made great strides in its programs of forced industrialization, it remains to an unusual degree an agricultural country. Nearly half of its population is engaged in farming. Yet though the Union has considerably more land in crops and more people involved in agriculture than does the United States, Soviet output on the whole remains below that of the latter country. In fact, in the United States the problem is one of surplus production in most commodities, while in the Soviet Union it is one of deficiency. Indeed, Soviet agriculture has lagged far behind Soviet industrial growth, and for that reason is often called the Achilles' heel of Soviet economic development.

The reasons for the lag in Soviet agriculture are varied. It is true that the U.S.S.R. is a northern land where climate and soil impose serious limitations on crop production. However, the Soviet regime has never invested heavily in agriculture. At the same time, since the beginning of collectivization in 1928 there have persisted shocking deficiencies in agricultural planning and farm management. Moreover, collectivization created hostility and resentment in the peasantry, which remains a factor in peasant or farmer attitudes even today, nearly a half century later. Finally, World War II inflicted heavy damage on the farm structure, especially in the western and southern regions of the European part of the country. Since then, and more particularly after the death of Stalin in 1953, the Soviet regime has attempted through a number of institutional and agronomic changes to boost Soviet output. Though some degree of success has been achieved, the results have not been adequate to meet the needs of the Soviet state, with its growing urban population.

In spite of its huge size, the Soviet Union does not possess enormous agricultural resources. Only 27 percent of the country is suitable for crop production and livestock raising. Only 10.7 percent of the country, or approximately 600 million acres, is considered tillable. This figure is somewhat inflated, however, because it includes farmland that, by American standards, would be considered marginal. At any rate, since 1913 the area sown in crops has been increased by over 200 million acres, an area comparable to two-thirds of the total United States crop area. It would seem at this point that no further expansion of sowing area is possible, since much of this increase has

MAJOR AGRICULTURAL REGIONS

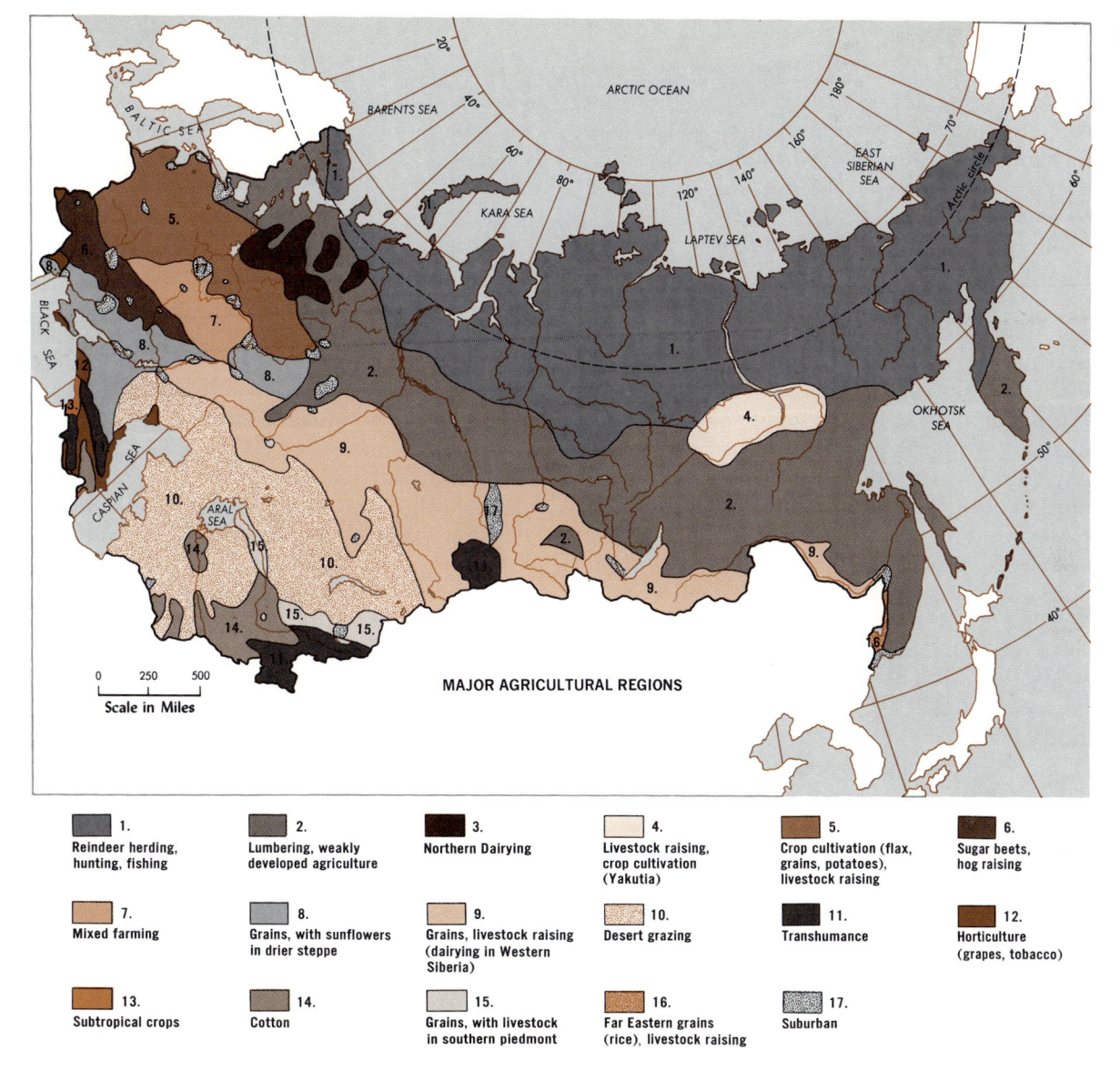

1. Reindeer herding, hunting, fishing
2. Lumbering, weakly developed agriculture
3. Northern Dairying
4. Livestock raising, crop cultivation (Yakutia)
5. Crop cultivation (flax, grains, potatoes), livestock raising
6. Sugar beets, hog raising
7. Mixed farming
8. Grains, with sunflowers in drier steppe
9. Grains, livestock raising (dairying in Western Siberia)
10. Desert grazing
11. Transhumance
12. Horticulture (grapes, tobacco)
13. Subtropical crops
14. Cotton
15. Grains, with livestock in southern piedmont
16. Far Eastern grains (rice), livestock raising
17. Suburban

Fig. 10–11 Owing to latitudinal extent, variation in soils, and differences in climate, the Soviet Union grows a wide variety of staple food crops. Industrial raw-material crops such as cotton, sunflowers, and flax are also grown.

---

been at the expense of fallow land. Thus, in the future, additional increases in output will have to be achieved through higher yields per acre.

Much of the land currently in crops, about 500 million acres, lies within the so-called fertile triangle. The northern boundary of the triangle coincides with a frost-free period of 105 days, representing the thermal limit of wheat culture. On the southern margin, where the annual precipitation drops to less than 12 inches, crop cultivation without irrigation becomes hazardous. Yet, even within the triangle, farming may suffer from drought or frost, or it may involve soils that are low in natural fertility.

Throughout the nonchernozem zone, apart from the acid soils and areas of poor drainage, the major handicap is the short growing season. In the steppe, where the soils are rich, precipitation is generally unreliable. East of the Urals, in western Siberia and northern Kazakhstan, frost and drought may singly or together seriously impair the success of the harvest. Clearly, then, only a relatively small portion of the Soviet farmland enjoys a growing season and precipitation levels sufficient for most temperate crops. Such favorable conditions may be found in the western Ukraine, in parts of the western North Caucasus, and along the Black Sea coast. In the valleys of Soviet Central Asia, a long growing season permits cotton to be cultivated if irrigation is available.

Soviet farming involves, essentially, two types of farm organization; the collective farm (kolkhoz) and the state farm (sovkhoz). The bulk of the peasantry belong to kolkhozes. As a result of mergers and amalgamations, however, the number of collective farms has rapidly declined from 123,700 in 1950 to 36,800 in 1965.

Theoretically, the collective farm is a voluntary association of peasants, working the land in common; their acreage is in effect leased from the state in perpetuity. In return, the collective farm delivers to the state a large portion of its annual harvest. Until 1958, the collectives were not allowed to own and operate heavy farm machinery; instead, the latter equipment belonged to "machine tractor stations" (MTS), which contracted to do the plowing, harvesting, and other heavy work required by the farm. For these services, the collectives paid another portion of their harvest, after deliveries had been made to the state. What was left over, after seed had been set aside for the following year and other obligations were met, was divided among the kolkhoz workers according to their work on the farm. Frequently, the return to the workers was very low. Since 1958, however, collectives have been permitted to purchase and keep their own farm machinery; the MTS's have been converted to Repair Tractor Stations, and their task now is simply to service machines. Moreover, in 1958, the Soviet regime established a system of purchase prices that were higher than the earlier obligatory delivery prices, and the result has been higher farm income. These prices have been raised in subsequent years. Even so, a lack of incentive continues to be a major factor affecting the efficiency and productivity of the Soviet farmworker.

Collective farms vary in size and in number of member households. However, the average collective farm totals over 16,000 acres and contains about 400 farm families, as compared with 165 in 1950. An average of two to five villages are contained within each of the collectives. Individually owned peasant farms are not economically important in the U.S.S.R.,

Fig. 10-12 Collective farm market near Tbilisi (Tiflis), Georgia, U.S.S.R. (Courtesy of Douglas Jackson)

but collective farmers are permitted to work small kitchen gardens or private plots for their own use. These range in size from one-half to two acres, but are adequate to provide for the peasants' needs and permit a significant surplus for sale in the free collective farm market; indeed, the private plots are said to provide up to 30 percent of the gross agricultural produce. Their contribution is greater in certain important commodities (eggs, fruit, potatoes, meat, and milk). Because of serious deficiencies in Soviet transportation and food handling, the government-owned state stores, unlike the farmers' markets, usually have few fresh vegetables. Until 1969, because they were unregulated, farm market prices (i.e., free market) were higher, bringing the collective farmer up to 40 percent of his income. In that year, however, the regime placed a ceiling on the prices the peasants could charge for their produce in the free markets.

The Soviet regime considers the state farm, that is, the sovkhoz, to be the highest form of socialism in the countryside. While there were only 12,783 state farms in 1967, their number has more than doubled in a decade. More specialized than the collective, the state farm is operated like a factory, where the workers are paid entirely in cash. The state, accordingly, is able to take virtually all of the farm's produce. In recent years, many collective farms have been converted into state farms; new state grain farms have also been established, especially in the virgin lands. State farms are gigantic in size, as much as 50,000 acres, and often employ over 350 workers on each.

Because yields per acre have ordinarily remained low during the years of Soviet power, the regime has sought to increase farm output through expansion of the sown area. Much of this expansion has occurred in the eastern steppe region, where there were extensive reserves of long-term fallow or idle land. The most ambitious program of expansion began in 1953 when Khrushchev directed that over 70 million acres of these reserves be plowed for wheat and other grains by 1956. By 1957, the Soviets revealed that, instead of 70 million, almost 90 million acres of new land had been plowed. Since then, additional acreage has been plowed. However, evidence to date indicates that the program has not been an unqualified success. The harvests that Khrushchev anticipated have not been achieved. Fallowing has been drastically cut in an area where, barring other means to conserve soil moisture, it is essential. Moreover, because delivery quotas to the state are high, state and collective farms are compelled to sow wheat year after year in the same field, which rapidly causes soil deterioration. Soil erosion in parts of northern Kazakhstan has already become a serious problem.

In addition to enlarging the area of dry farming, the Soviets have also reclaimed land from deserts and swamps. Since 1913, the irrigated area has been expanded by 13 million acres, to a total of 23 million. More than half the irrigated area is in Soviet Central Asia, in the valleys of the Syr Darya and Amu Darya; other irrigated areas are in Transcaucasia. Because of their long growing season, these irrigated areas produce mainly cotton. In the future, it is to be expected that irrigation agriculture will be more widely practiced in the Soviet Union, particularly in the steppe along the lower Dnepr and in the North Caucasus. Irrigation should do much to raise the yields of corn and other grains here. The Soviet regime has also reclaimed wetland, especially in the Polesye district and in the Colchis Lowland, but in neither case has the effort increased crop acreage significantly. The Colchis Lowland, if properly drained, is important to the Soviet Union because its humid subtropical climate permits a wider choice of crops than elsewhere.

A major feature of Soviet agriculture is the

predominance of grains in the crop patterns. That this is so is mainly owing to the fact that Soviet cropland is suitable, to a large extent, only for grains; consequently, the Soviet diet consists essentially of bread, potatoes, and a few vegetables. Meat and dairy products have, until recent years, been extremely scarce. Nevertheless, under the Soviet regime, there has been a decline in the relative position of grains; whereas in 1913 grains accounted for 90 percent of the total sown area, at present they are only 60 percent of the total. On the other hand, under the Soviet policy of self-sufficiency in basic agricultural commodities, acreages devoted to sugar beets, cotton, and forage crops have been increased substantially.

Of the grains, wheat, a crop of the steppe zone, is the most important, occupying about 54 percent of the acreage. The virgin-lands program has caused an eastward shift in crop acreage, so that almost two-thirds of the wheat is grown in the territory between the Volga and the Altai Mountains. Practically all this wheat is spring-sown; winter wheat is found mainly in the Ukraine and North Caucasus where the winters are milder.

Although the Soviet regime attempted in the 1930s, with little success, to establish a "secondary wheat base" in the mixed-forest zone of European Russia, the soils of that area are more suitable for winter rye. Rye is a hardy, adaptable crop, but in recent decades acreage planted in this grain has been declining. Clearly, consumer preference in the Soviet Union has shifted from the traditional black bread of the peasant to white and wheat bread.

Of the feed grain, oats and barley are the oldest crops; corn has, since 1953, undergone a remarkable increase in acreage. Oats, a crop of the humid central part of Russia, has been declining in importance, especially because the number of horses has decreased. Barley, on the other hand, is a versatile crop grown for human consumption as well as for livestock feed. In

Fig. 10-13 Tea picking on a collective farm in Georgia. (Courtesy of Embassy of the U.S.S.R.)

the north, barley is grown for kasha, a traditional Russian porridge; in the west, barley is used in the manufacture of beer; in the south, in the Ukraine, it serves as high-protein feed. Prior to 1953, grain corn was limited to the humid western Caucasus and the southern Ukraine. In an effort to improve the feed situation, Khrushchev directed that corn be sown widely throughout the southern regions, as well as elsewhere in the country, where, though it may not ripen, it can be cut green for silage.

The basic industrial crops include cotton, flax, sugar beets, and sunflowers. Cotton, the principal fiber crop, is grown entirely on irrigated land in Central Asia and the Caucasus. Moreover, as a result of improved seed, the use of fertilizers, and higher incentives to cotton farmers, Soviet cotton yields have increased substantially in recent decades. Flax, for making linen, is grown primarily in the forest zone of European U.S.S.R. The flax plant requires heavy fertilization; consequently, flax cultivation is found in association with dairying and livestock raising. Sugar beets are the only domestic source of Soviet sugar. Its main region of cultivation is in the northern Ukraine and in the neighboring sections of the Russian S.F.S.R. Acreage has been expanded into other areas, notably in the North Caucasus as well as in southern Kazakhstan, where the crop is irrigated. Sunflowers, a drought-resistant crop, are grown for oil for domestic use in the eastern Ukraine and the lower Volga Basin.

Potatoes and vegetables occupy only a small part of the total sown area. Irrational shipments of potatoes from one region to another have been reported in the Soviet press, but none of this movement involves the transport of early potatoes from southern regions to northern industrial cities. In addition to grapes and other fruits, the U.S.S.R. also grows tea in western Georgia and tobacco along the east coast of the Black Sea.

Collectivization in 1928–1930 resulted in widespread slaughter of livestock by the peasants who opposed joining the kolkhozes. Some improvement in this sector took place during the late 1930s, but the war again inflicted extensive losses. It has only been in recent years, therefore, that the numbers of livestock have surpassed prerevolutionary levels. Nevertheless, the sizable growth in population means that there are few livestock units per capita more than in 1913. The reasons for the slow increase in livestock numbers are to be found primarily in the poor care given collective herds, as opposed to the comparatively few animals fed on the private plots; this neglect is reflected in inadequate housing and above all in insufficient feeding. The expansion in corn acreage, however, is an attempt to provide better livestock nutrition, in the hope of rapidly overtaking the United States in the output of meat and dairy products. Greater output has indeed been achieved, but lack of an adequate distribution system and refrigeration means that the urban dweller has not yet been able to benefit fully from the improvement.

The bulk of the dairy cows are concentrated in the north-central part of European Russia. In the former, almost half the dairy cows were kept on private plots, although under Khrushchev the regime forced large transfers of herds to the collectives. Dairying, a traditional farm activity in the northern European part of the country, has lately tended to shift to the Ukraine. The reasons for this change are not clear, but it may be because of the fact that during the long winters of the north the cows get insufficient exercise, being confined to barns for months. Moreover, the collective farmer in the Ukraine may find it more profitable to put his feed supply into milk rather than into meat. Pigs reach their highest density in the west-central parts of the country, especially where farming is mixed. In the warmer, drier parts of the country, sheep and goats are more numerous than cattle or pigs. Severe losses in sheep and goats were suffered in the 1930s, but some relaxation of control over the formerly nomadic livestock raisers in these areas has contributed to a rapid upswing in numbers. In the case of horses, the general trend toward mechanization of farm operations has largely been responsible for the continuing decline in their numbers.

Mining and Manufacturing

Beginning with the first Five-Year Plan in 1928, the Soviet Union launched a program of rapid industrialization designed to transform the economy of the entire country in the shortest possible time. The objective was the establishment of a base of heavy industry, upon which further industrialization could proceed. The government economic plans placed considerable emphasis, therefore, on mineral extraction and on metallurgical, engineering, and related industries. Until recently, consumer

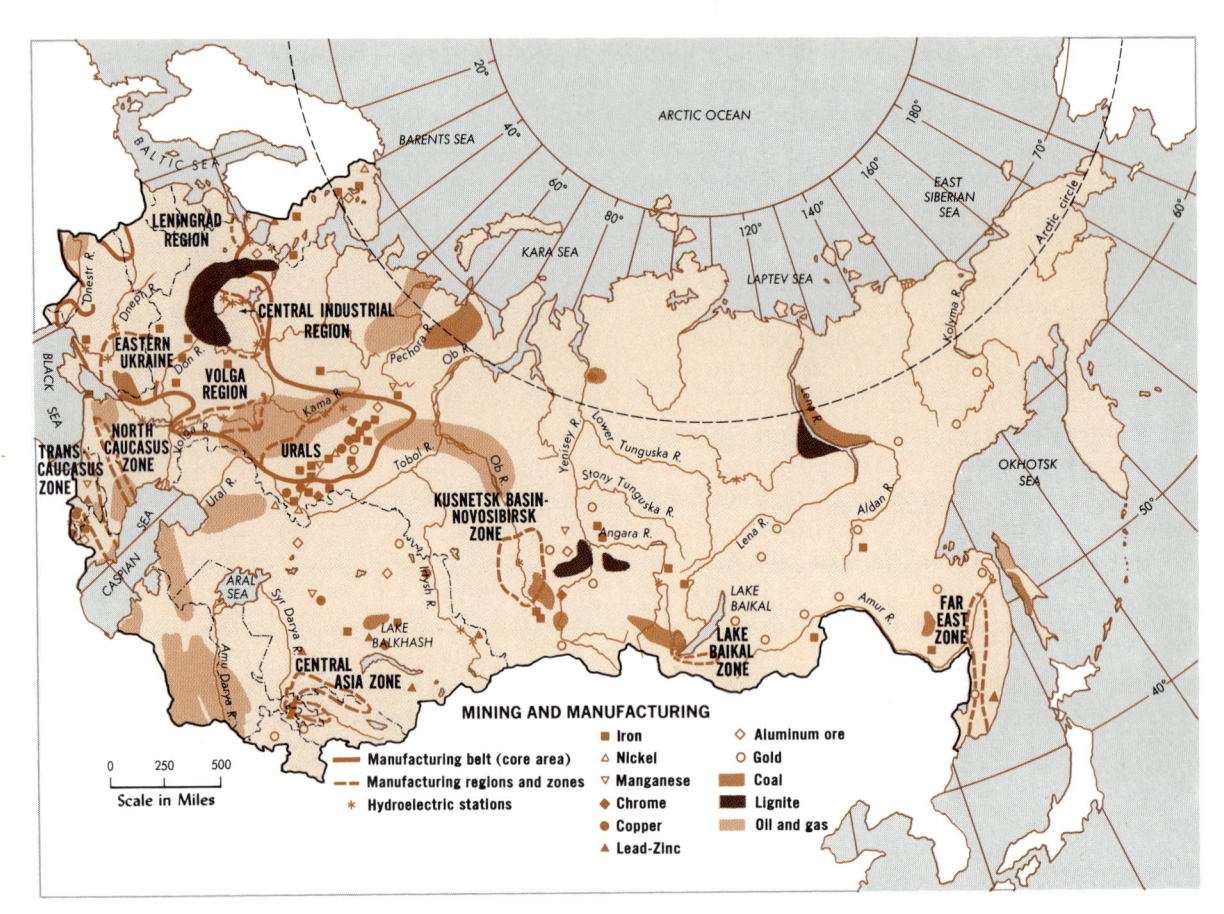

Fig. 10–14 The most highly developed mining and manufacturing regions are in the European sector of the U.S.S.R. Some areas in the Asiatic part, however, are in the process of intensive development.

goods industries have received scant attention and little investment.

The cost of the program has placed a heavy burden on the collectivized peasantry and the urban factory worker, but the U.S.S.R. today ranks second after the United States in output of heavy industrial goods. Moreover, in recent years the Soviet Union has challenged the United States not only to economic competition but also to a race into space and in armaments, with all the attendant military implications. It is quite clear from Soviet pronouncements, therefore, that the U.S.S.R. intends, through its industrial might, science, and technology and its military machine, to dominate the earth. Consequently, it is important that we understand and appreciate the quality and extent of Soviet industrial resources and the nature and patterns of the Soviet industrial base.

In order to build a heavy-industry base, extensive search has been made throughout the U.S.S.R. for new deposits of metal ores and new sources of fuel and power. Owing to the vast size of the country and its varied geological features, extensive investigation has uncovered large reserves of coal, oil, and natural gas, as well as a variety of metals. The hydroelectric power potential, too, has grown to enormous proportions, as surveys of Siberian rivers have revealed the feasibility of this or that major power project.

Fuel and Power Resources

Coal The geological reserves of coal total 8.8 billion tons, only 5 percent of which are classed as proved; yet, at present rates of extraction, the latter are considered sufficient to last for more than 500 years. Coal is used not only for coke in the metallurgical industry, but also in the generation of thermal power. Total yearly coal production now reaches 595 million metric tons, more than double that of 1950.

The geographical distribution of the coal deposits, however, is generally unfavorable relative to population and industry. About two-thirds of the confirmed reserves lie in Asiatic parts of the country. In European Russia, the reserves are not all of high grade, and some of these deposits are located in the extreme north, where conditions are harsh. Since the core area is by far the chief consumer, local coal supplies must be supplemented by large volumes hauled from distant sources. Because it is largely by rail, coal transport tends to be costly. However, to prevent transportation congestion and to minimize the length of coal hauls, more accessible local deposits of soft coal and lignite, though of poor quality, are also worked.

Most of the higher grades of coal and virtually all the coking coal are produced at four centers: namely, the Donets Basin (Donbas) in the southern Ukraine, the Kuznetsk Basin (Kuzbas) of western Siberia, the Karaganda Basin of north-central Kazakhstan, and the Pechora Basin in the European north. Several small scattered coalfields in the Urals have a relatively high total output, but only a small part of this is suitable for coking purposes. There are, in addition, numerous small fields in eastern Siberia, the Soviet Far East, Central Asia, and Transcaucasia.

The Donets Basin, the leading field in the Soviet Union, accounts for about one-third of all coal produced (however, it produced over one-half the total prior to World War II). Moreover, it contributes 60 percent of Soviet metallurgical coking coal. Having been mined intensively since the late nineteenth century, the more easily worked deposits of the Donbas are nearly depleted. Difficulties in extraction have increased, so that costs have risen to levels higher than those associated with the Kuznetsk and Karaganda basins. Nevertheless, because of the high concentration of industry nearby, as well as in the Central Industrial Region to the north, the absolute output of coal in the Donbas continues to expand.

The Kuznetsk Basin ranks second to the Donbas in output, but its importance is increasing steadily, both absolutely and relatively. Much of the coal is shipped to blast furnaces in the Urals. Its present output is equivalent to 17 percent of the Soviet Union's total production. The reserves in the Kuzbas are much larger than those in the Donbas. Moreover, the seams, which are generally thick and more easily worked, permit the lowest extraction cost in the country. The advantages of mining here are offset, however, by the long distance to the metallurgical industries of the Urals.

The Karaganda Basin ranks third, not in total output but in production of good-quality coking coal. Development on a large scale began in the 1930s, when it became a supplier of coal to Ural metallurgical centers. Nearly one-half of its output moves to the Urals; much of the remainder is used at the nearby Temir Tau iron and steel center and as a fuel for metal smelting in Kazakhstan and Central Asia.

Exploitation of the Pechora Basin deposits

was started just prior to World War II. In the early years of the war, the construction of a railroad provided a link with northwestern European Russia. The overwhelming portion of its coal output is moved to various parts of the northwest for use in electric power stations and for rail, river, and sea transport. Only a small percentage of the Pechora output is used as coking coal, the latter type moving a great distance to an iron and steel center at Cherepovets or to coke gas plants in Leningrad.

Some coking coal is obtained from the Kizel field in the Urals; but the need for metallurgical coal in the Urals far surpasses local supplies and accounts for heavy shipments from the Kuzbas and the Karaganda Basin.

Heavy emphasis is placed on the use of coal in the production of thermal electricity, especially in the Central Industrial Region and in the Urals. The lignite fields near Moscow assume importance in this regard because of their proximity to the populated and industrialized core. Peat also serves as a source of thermal power, although it is of low heat value. Large peat deposits in European Russia are favorably situated relative to demand, and this accounts for their use.

Petroleum and Natural Gas In order to lessen dependence on the use of coal for fuel and power in the core region, increasingly attention has been given since the 1950s to oil and natural gas as substitutes. Production of petroleum increased from 38 million metric tons in 1950 to 288 million in 1967. A similar upsurge in natural gas production raised output from 6 billion cubic meters in 1950 to 159 billion in 1967. Coal accounted for 62 percent of the total fuel supply in 1950; oil provided 17 percent, and natural gas a mere 2 percent. But, by 1967, this pattern had changed significantly. In that year, oil met 38 percent of the fuel needs, natural gas 17 percent, with coal falling to 39 percent.

About 80 percent of the oil produced in the U.S.S.R. comes from the Volga-Ural fields, and the recent expansion of Soviet output is largely a result of exploitation of this source. Lesser, relatively ever decreasing amounts come from the Baku district in Azerbaidzhan and from Grozny and Maykop in the North Caucasus. These districts were formerly the chief source of Soviet oil. Some production occurs in the western Ukraine, at Ukhta in the European north, on Sakhalin Island in the Soviet Far East, and in other parts of Soviet Central Asia. New fields along the Ob River as far upstream as Novosibirsk offer considerable potential. There are also reports of oil in central and eastern Siberia, along the Lower Tunguska and Lena rivers.

The Soviet Union has widespread deposits of oil shale, the largest of which are found near the Middle Volga region and in the Estonian S.S.R. About 85 percent of the production comes from the latter region, an output equivalent in total energy value to about 4.5 million tons of coal.

The Volga-Ural oil fields are ideally situated relative to consuming centers. Much of the oil from these fields is moved by pipeline. Pipelines have been constructed to the industrial centers of the economic core region, as well as beyond the western borders of the country to some East European satellites. In addition, crude-oil pipelines have been laid to Irkutsk in eastern Siberia.

The rapid rise in natural gas production in recent years has been accompanied by development of a number of new producing areas. Until recently, production has been concentrated largely in European Russia, with important fields in the Ukraine at Shebelinka and Dashava; in the North Caucasus region at Stavropol; and between the Volga and the Urals. New fields in Soviet Central Asia (Bukhara) and in the northern part of the West Siberian Lowland, however, have begun to alter produc-

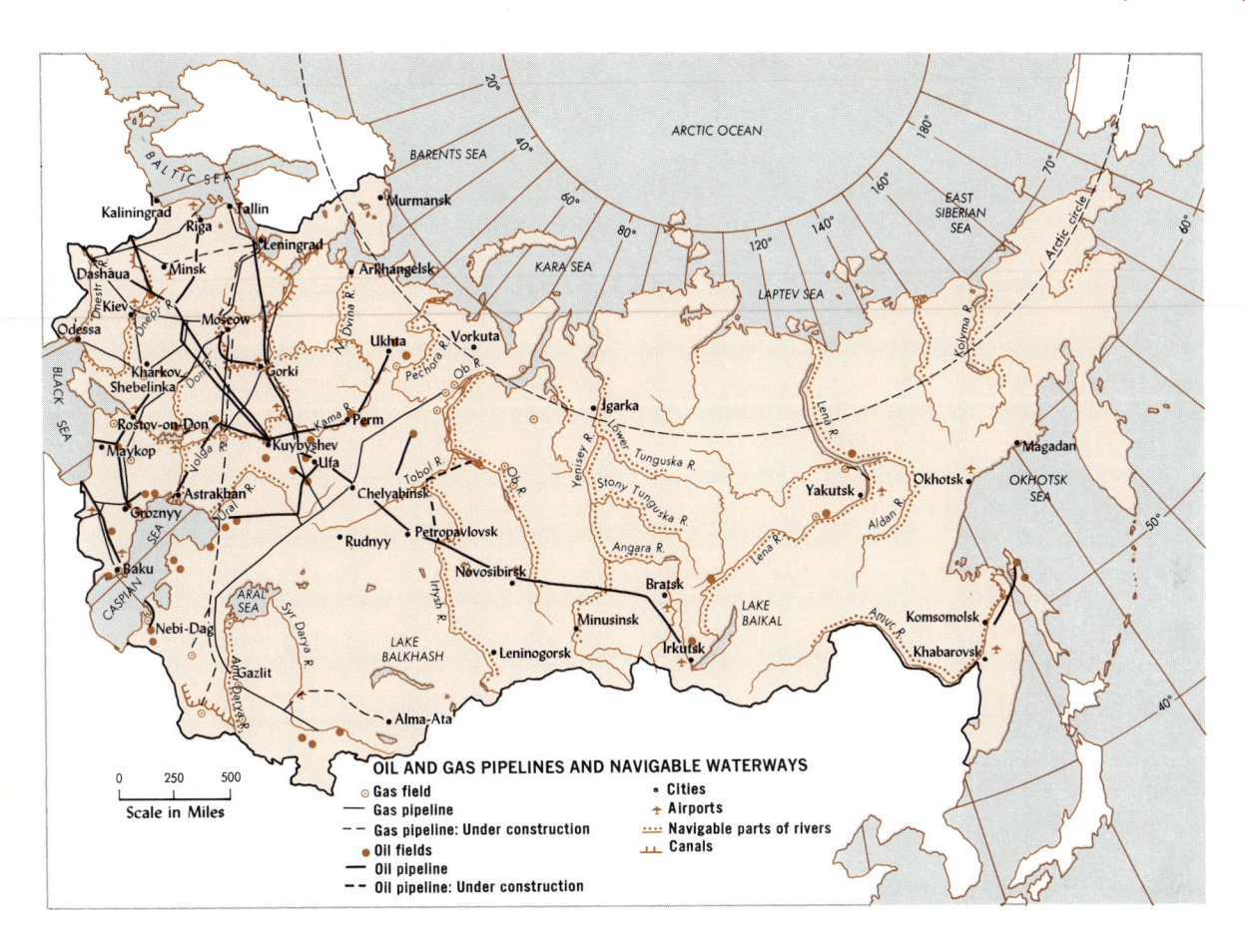

Fig. 10-15 Oil and gas pipeline system of the U.S.S.R.

tion patterns; these new fields are said to contain a major share of the Soviet Union's total reserves. Natural gas is delivered to the major centers in the Central Industrial Region, to Leningrad and the Baltic region, as well as to the Ural Industrial Region. Both domestic and industrial needs are being met increasingly by natural gas, the heat value of which is considerably higher than poorer local fuels such as peat and lignite. Moreover, total costs, including transport, are lower for natural gas than for the latter fuels.

Waterpower The Soviet Union has enormous waterpower potential; but in spite of much publicity given to the construction of hydropower installations, only a small part of the electric power produced at present comes from the U.S.S.R.'s rivers. Indeed, of the 587,700 million kilowatt-hours produced in 1967, hydropower accounted for only 88,500 million. The bulk of the hydroelectric potential lies in the Siberian rivers; but much of the development up to the present has taken place to the west of the Urals.

As early as 1920, well before the beginning of the industrialization drive, a scheme of state electrification, known as GOELRO, was formulated as the basis for industrial and agricultural development. Sizable power facilities were constructed on the Volkhov and Svir rivers, in northwest Russia near Leningrad, while a giant hydroelectric station was completed on the Dnepr River near Zaporozhe in 1933. Swift-flowing streams in the Caucasus Mountains have also been harnessed. In recent years, large dams and power installations have been built on the Volga River, for example, at Kuibyshev and Volgograd. Still others are planned or under construction. When these are completed, the Volga will resemble a long narrow sea, with its waters held in check by a "cascade" of power dams. Work has been proceeding on the development of the Ob, Irtysh, Yenisey, and Angara rivers in Siberia, where giant power stations have been built or are under construction. The Bratsk station on the Angara has been opened and, when it reaches full operation, will have a capacity of 4.5 million kilowatts. The recently completed Krasnoyarsk station on the Yenisey has a planned capacity of 5 million kilowatts, thus ranking as the largest in the world.

The Soviet Union has built a number of small atomic power plants, particularly in those regions such as the Central Industrial Region which require enormous amounts of energy. Others have been planned for elsewhere in the U.S.S.R.

Metalic Minerals

Ferrous Metals and Ferrous Metallurgy Iron and steel are the sinews of modern industry, and the volume of their output is a key to the strength of a country's economic base. Heavy machinery and transportation and construction equipment all need steel, and steel is essential in an ambitious military and space program. Consequently, extensive reserves of quality iron ore are an asset. Soviet reserves are large, but not of especially high quality.

From 1928 to 1967, production of iron ore in the U.S.S.R. increased greatly, from 6.1 to 168.2 million tons. The production of pig iron rose during the same period from 3.3 to 74.8 million tons, and the production of raw steel rose from 4.3 to 102.2 million tons.

Most iron ore, pig iron, and steel has come from the southern Ukraine. Iron and steel had been produced there since the late nineteenth century. The proximity of Krivoi Rog iron ore and Donbas coal continued to favor the expansion of the industry in that region during the Soviet period. While the share of the Ukraine relative to total output has declined as new centers have been opened elsewhere, in 1967 the region still supplied 49 percent of the nation's pig iron, 55 percent of the iron ore, and 42 percent of the steel. Heavy use of Krivoi Rog ores has depleted the higher-grade reserves, but enormous quantities of low-grade ores suitable for concentration assure a continuing high potential. Metallurgical plants in the southern Ukraine are situated near the source of coking coal in the Donbas, as well as, to a lesser degree, in the vicinity of the Krivoi Rog iron ores. A third center at Zhdanov on the Sea of Azov uses low-grade iron ore from Kerch in the Crimea mixed with ore from Krivoi Rog.

With the creation of the Ural-Kuznetsk Combine in the 1930s, there developed a large iron and steel complex in the Urals. Integrated metallurgical plants were constructed at Magnitogorsk, Chelyabinsk, and Nizhni Tagil. These centers form the second most important metallurgical center of the Soviet Union. In 1940, the Urals supplied 18 percent of the nation's pig iron and 21 percent of its steel; but in 1958, the proportions had increased to 33 percent and 35 percent respectively. Having been heavily worked, the high-grade ores of Magnitogorsk have been depleted, so that new sources are being sought. The iron mines at Rudny in northwestern Kazakhstan will contribute to the ore requirements of the Urals.

A smaller metallurgical center is situated in the Kuznetsk Basin. Linked to the Urals, the integrated plant there was originally supplied with Urals ore, but now depends upon locally available or eastern Siberian ores.

As part of a developing "third metallurgical base," extending from Kazakhstan to eastern Siberia, new sources of iron ore are being exploited in Kazakhstan, in the Altai Mountains of western Siberia, and in eastern Siberia, to serve new and expanding plants. The new plant at Temir Tau in Kazakhstan is based on nearby Karaganda Basin coal and on relatively close sources of iron ore.

In central European Russia, Tula and Lipetsk, centers of iron production as early as the eighteenth century, have grown into modern iron and steel centers based both on local ore supplies and ores from the so-called Kursk magnetic anomaly and on coking coal from the Donbas.

The Soviet Union is well supplied with ferroalloys. Manganese, especially important in the production of good-quality steel, is found at Nikopol in the Ukraine and at Chiatura in Transcaucasia. These two deposits currently account for two-thirds of the world's manganese supply.

Nonferrous Metals The U.S.S.R. has uncovered, through intensive geological exploration, adequate supplies of nearly all important nonferrous metals needed in its industries. Since it has earnestly sought to be self-sufficient in this regard, however, the Soviet Union has undoubtedly placed in production mines and processing facilities at costs considered high by Western standards. Such high mining costs may reflect the difficulties of extracting minerals in areas of harsh arctic and desert climates and may be related to problems of transportation both within and from regions of unfavorable geographical conditions.

Large-scale deposits of copper are found in Kazakhstan, the Urals, and Soviet Central Asia. These sources presently permit the U.S.S.R. to be self-sufficient in copper, even though the overall reserves are not considered abundant. Kounradski and Dzhezkazgan in Kazakhstan, the two chief sources of copper ore, both have the disadvantage of remoteness to major markets. Several other deposits are worked in the Urals, and a major center has recently been opened in Almalyk in Uzbekistan.

Bauxite and other mineral resources form the basis of a sizable aluminum industry. Soviet aluminum production dates from the 1930s, starting at Boksitogorsk, near Leningrad, where both the necessary bauxite and cheap hydroelectricity for refining were found. Refining also developed on the basis of Dnepr hydroenergy at Zaporozhe, and other bauxite deposits were worked and processed in the Urals. More recently, bauxite has been mined in northern Kazakhstan at Pavlodar and in Transcaucasia. In the latter region, refineries are found at Yereven (hydropower) and at Sumgait, near Baku (thermal power). Aluminum is produced in the Kola Peninsula, with the industry based on aluminum-bearing nepheline deposits and hydroelectric power available locally. Very recently, aluminum refining has been developed at Krasnoyarsk, Bratsk, and Irkutsk, where the waterpower of the great Siberian rivers has been harnessed.

Total reserves of lead and zinc in the U.S.S.R. are considered modest. The most important deposits, distant from markets, are located near Leninogorsk and Chimkent in Kazakhstan. Other sources of these metals are found in the Urals, the North Caucasus, and the Soviet Far East. Tin deposits appear mainly in remote northeastern Siberia and the Soviet Far East. Eastern Siberia also provides important sources of gold and diamonds.

Industry

Engineering Industries The engineering industries, which include the production of machine tools, agricultural and textile machinery, mining and metallurgical machinery, transportation equipment, and power generators, constitute the largest single group of Soviet manufactures. Together they account for one-third of the U.S.S.R.'s industrial employees. These industries have grown significantly since the 1920s, when the drive for the development of heavy industry began. Initially, the core industrial area accounted for the bulk of machine production, particularly in Leningrad, Moscow, Gorki, Kharkov, Rostov-on-Don, Sverdlovsk, and Chelyabinsk. Since World War II, however, many new plants have been located in other parts of the country.

The production of heavy mining and metallurgical equipment, oriented toward the steel-producing centers, is important at Kramatorsk in the Donbas, Sverdlovsk in the Urals, and Novosibirsk near the Kuznetsk Basin. Kharkov in the northern Ukraine and Nizhni Tagil in the Urals are leading producers of railroad rolling stock and locomotives. Moscow and Gorki were the original Soviet producers of automobiles and lorries, but new automotive works have been built in the Urals, at Ulyanovsk on the Volga and at Kutaisi in Transcaucasia. Machine tools, electrical equipment, and textile machinery are produced largely in the Central Industrial Region, while heavier farm equipment is manufactured nearer the extensive agricultural areas at Kharkov, Rostov-on-Don, Volgograd, and Chelyabinsk.

Chemicals Industry The Soviet chemicals industry has been geared mainly to producing fertilizers, synthetic rubber, and basic chemicals, such as sulfuric acid, caustic soda, and soda ash, from inorganic minerals. Through the expansion of the oil and natural gas industries, however, output of a wider assortment of products, including plastics, synthetic fibers, and detergents, is becoming available.

Chemicals based on salts and sulfur ores are manufactured especially in the Urals and in the Donets Basin. The Solikamsk-Berezniki area of the Urals contains enormous deposits of potassium, sulfur, and common salts and is important for production of mineral fertilizers as well as a variety of other chemicals. The by-products of coal also provide the basis of nitrogen and ammonium sulfate production (fertilizers) in the Urals at Kizel, Nizhni Tagil, and Magnitogorsk, at Gorlovka in the Donbas, at Kemerovo in the Kuzbas, and near Moscow (from lignite). Production of synthetic rubber is concentrated in central European Russia, notably at Voronezh, Tambov, Yaroslavl, and Kazan, where potato alcohol, a chief material in synthetic rubber manufacture, is available. However, new rubber factories, using gas and oil by-products, are being constructed in the Middle Volga region, at Sumgait near Baku, and at Stavropol in the North Caucasus.

Textile Industry The production of textiles formed the most important branch of Russian industry before the Revolution. Factories were found principally in and near Moscow and in St. Petersburg. During the Soviet period, the industry received only a small share of government investment in manufacturing up to the 1950s; and, while textile production has risen since then, gains have not kept up with consumer needs. All branches of the textile industry are heavily concentrated in the Central Industrial Region. The regime has encouraged its spread to other population centers, partly to help broaden and vary the employment base in those areas.

Cotton milling remains concentrated in Ivanovo-Shuya in the Central Industrial Region and in Leningrad, where nearly three-quarters of all cotton cloth is produced. In recent years, however, some processing capacity has been built nearer the cotton fields in Central Asia and Transcaucasia, as well as at consuming centers in the Ukraine, the Middle Volga region, the Urals, and western Siberia.

Linen manufacturing has traditionally been located in the Central Industrial Region, and new factories continue to be established in or near the flax-growing areas of European Russia. The woolen industry is centered in the Moscow and Leningrad areas for the production of finer cloths and in the Ukraine and along the Middle Volga for coarser goods. New mills have been built also in the sheep-raising areas of Central Asia and Transcaucasia. The manufacture of "silk," which includes both natural and artificial fibers, has grown significantly since the 1950s. Again, the Central Industrial Region predominates in this "silk" cloth production, especially that made from artificial fibers. Other important areas of its manufacture include Central Asia and Transcaucasia.

Food-processing Industry The processing of crops and livestock products is found both in the large consuming centers and in the agricultural areas of the U.S.S.R. The major wheat flour-milling and sugar-beet processing plants are located in the black-earth regions of the Ukraine and adjoining areas in the Russian S.F.S.R., along the Volga, and in the North Caucasus. Large meat-packing plants are found mainly in Moscow, Leningrad, Gorki, and Sverdlovsk. Some dispersal of these major processing industries has occurred, however, into Kazakhstan, Soviet Central Asia, and western Siberia. Butter and cheese production is associated with the remote areas of dairy farming, notably in the Vologda district of northern European Russia and in the Omsk-Novosibirsk area of western Siberia. The canning of fruits and vegetables is important in the Ukraine, Moldavia, and the North Caucasus.

Forestry and Fishing Timber is one of the Soviet Union's most abundant resources. Forests cover 32 percent of the U.S.S.R.'s territory, but about 80 percent of the reserves lie in Asiatic Russia. Because of the inaccessibility of most of the Siberian timber resources, however, the bulk of the lumbering remains in the European U.S.S.R. Even so, distances between the timber supply and the areas of consumption are very large. Since most of the timber is floated down rivers, sawmilling is found in the forest zone at the mouths of logging streams, such as at Arkhangelsk, and outside the forest zone at important river-rail crossing sites, such as Leningrad and Volgograd. East of the Urals, the major logging areas are found in central Siberia. Igarka, on the Yenisey River, north of the Arctic Circle, is the chief lumbering center.

The manufacture of pulp and paper, employing coniferous species chiefly, is centered in the European north. Plywood production, using material from the mixed-forest zone, is located in central and western European Russia. The largest centers of the furniture industry are Moscow, Leningrad, and Kiev.

The major Soviet fisheries are located in the Barents Sea, in the Baltic and Caspian seas, and off the Pacific coast. The Caspian Sea, owing both to its shallowness and the huge amounts of organic matter carried to it by the Volga and other rivers, has long been a major source of fish. However, with growing Soviet interest in ocean fisheries, the share of the Caspian fisheries relative to total catch has declined from 66 percent in 1913 to 15 percent in 1956. Also, the annual catch has fallen to one-half that of 1913 because of the declining water level of the Caspian and the diminished inflow of fresh water. The larger dams built

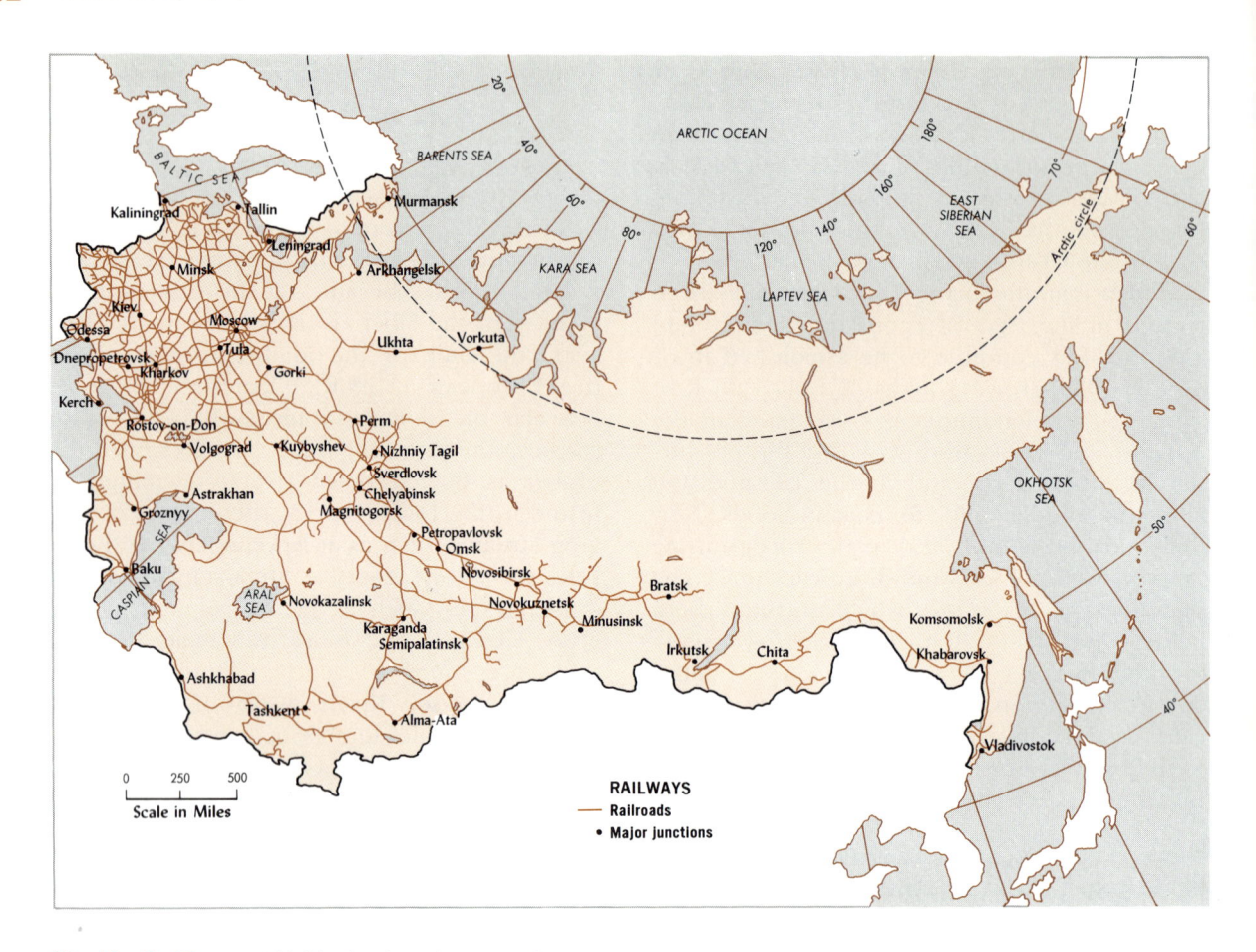

Fig. 10-16 The most highly developed system of land transportation in the U.S.S.R. is by rail, with Moscow as its principal focus. When and where possible, rivers are also used for transportation.

on the Volga have also had a damaging effect on fish spawning. The Pacific fisheries, particularly around the Sea of Okhotsk, and the Barents Sea fisheries are assuming leading roles in the U.S.S.R.'s total production.

Transportation

While Czarist Russian expansion was facilitated by the rivers of the empire, modern economic development in the U.S.S.R. has been based mainly on the network of railroads. This results from the enormous size of the country and the fact that, in many instances and for at least half the year, the frozen rivers offer no alternative. Although river and coastal shipping together account for 21 percent of the Soviet Union's freight total, the railroads carry 68 percent and, in addition, more than 50 percent of the passenger travel. Only 5 percent of the country's goods is handled by truck and 6 percent by pipeline. In recent years the railroads have declined relatively as the use of pipeline and coastal shipping has increased.

The densest rail network prevails in the core region and looks, indeed, like a spiderweb centered on Moscow. However, not all rail lines are of equal importance. Some of the heaviest traffic is found on the trunk line between the Donets Basin and Moscow, between the Donets Basin and the cities of the Dnepr bend, between Moscow and the Volga cities, and between the major industrial sites in the Urals. Within the core, however, waterborne freight also reaches considerable proportions, especially on the Volga system. The Moscow-Volga Canal, linking Moscow with the Volga River, and the later (1952) Volga-Don Canal, between the Lower Volga and the Don rivers, afford water transportation between many of the major industrial cities of the north and south; but the bulk of the river traffic still occurs on the Volga between the mouth of the Kama and Volgograd.

Beyond the core region, the heaviest rail freight movement occurs on the Trans-Siberian Railway between the Kuznetsk Basin and the Urals, in Kazakhstan between Karaganda and the Urals, and in the Caucasus between Baku and Rostov-on-Don. In every instance, these movements involve large quantities of raw materials being shipped to industries in the core, including fuels such as coal and oil.

Since World War II, the Soviet regime has attempted to relieve the heavy burden borne by these long rail lines feeding into the core region. To transport oil and gas, increasing use of pipelines is being made, while extensive additional rail lines have been built or are under construction, for example, in western and central Siberia. Moreover, rail lines are being modernized and electrified, especially the Trans-Siberian and those between Moscow and Leningrad, as well as in the Caucasus.

Both inside and outside the core area, river traffic is hampered by freezing during the long winter period, which in parts of Siberia lasts for seven months. Seasonal irregularity of flow also is often a problem. Consequently, except for the Volga proper and to a lesser extent the Dnepr, overall river freight movements are not great in relation to all freight totals. However, such rivers as the Kama, the Northern Dvina, and the Yenisey are important in the movement of logs to sawmills downstream. In recent years, the freight volume on the rivers of Siberia has undergone a relatively significant increase, owing to substantially heavier movements of lumber and oil.

Although the output of trucks and automobiles in the U.S.S.R. has risen to over 728,000 annually, there are still only about 4.3 million motor vehicles in the entire Soviet Union, compared with over 94 million in the United States. Because of the lack of an extensive network of good paved highways, use of the motor vehicle in the U.S.S.R. tends to be local, principally for hauling from farm to city or from farm to railhead.

The volume of freight carried by the Soviet airlines is less than 1 percent of all freight carried in the U.S.S.R. Nevertheless, the Soviet regime has built up and operates an extensive air service to all parts of the nation, as well as internationally. Passenger service is increasing, too, especially on long-distance flights between Moscow and centers outside the core area. A major air corridor extends from Moscow eastward to the Pacific Coast, paralleling the route of the Trans-Siberian Railway and connecting the major cities of Siberia and the Soviet Far East. There are other major links between Moscow and key centers in the Ukraine, along the Volga, and in the Urals, as well as to Transcaucasia and Soviet Central Asia. The major Black Sea and Caucasus resorts are also served by direct connections with Moscow and other centers of population. The Soviet government airline, Aeroflot, operates international flights to Western Europe and North America, the Middle East, India, and China.

In Perspective

Over the past half century, since the Bolshevik Revolution, the Soviet Union has been transformed into one of the world's major industrial countries. This has not been achieved, however, without a good deal of human hardship and cost. Moreover, because of the strong official emphasis placed on the development of heavy industry, the Soviet economic structure remains out of balance. Agriculture continues to lag behind industry and fails to satisfy domestic requirements. Agricultural yields in the state sector generally remain low, and the output from private plots remains an essential component of the urban food supply. On the other hand, the tempo of Soviet industrial growth has, over most of the past decade, tended to fall below the levels demanded by the regime.

Despite the rapid development of southern Siberia, Kazakhstan, Soviet Central Asia, and Transcaucasia, the economic core of the Soviet Union still lies in the western part of the country, that is, in European Russia from the Urals westward. Still, the economic advances, and particularly the industrial growth, of the Asian part must not be discounted. The rise of new manufacturing and processing centers there has also aggravated the problem of agricultural supply. The growing season throughout southern Siberia and northern Kazakhstan tends to be as mercilessly short as the moisture is deficient. Elsewhere, in the more southerly regions, the longer growing season and possibilities for irrigation encourage the cultivation of non-food crops, notably cotton.

In the foreseeable future, therefore, it must be expected that—the growth of manufacturing notwithstanding—Asiatic Russia will remain essentially a provider, if not entirely a processer, of basic raw materials to be transported or shipped westward to the nation's economic core.

Planned proportional development did not necessarily require that an iron and steel mill be constructed in every region of the country. A basic objective was, of course, to lessen wherever possible the long hauls of fuels (essentially coal) from one region to another by more extensive utilization of such raw materials *in situ*. But with the rapid growth of demand in the industrial core and the depletion of some of its local sources of supply, the average length of haul for fuels (exclusive of pipelines) has increased. The construction of new fuel-consuming centers in Siberia does not lessen the industrial needs of the western U.S.S.R. That is where the bulk of Soviet manufacturing and processing activities are located; the largest cities are situated there, as well as nearly three-fourths of the total population.

How the new oil and gas fields in Siberia and Soviet Central Asia will affect patterns of economic activity in Asiatic Russia remains to be seen. To date, the major direction of these fuels has continued to be toward the core area in European Russia. Moreover, no one can predict what the consequences of prolonged border conflict with Communist China may be: under duress of the Chinese threat, what will be the nature and degree of future economic development in the U.S.S.R.'s eastern regions? This is a question that only time can answer.

REFERENCES

Berg, L. S.: *Natural Regions of the U.S.S.R.*, The Macmillan Company, New York, 1950.

Borisov, A. A.: *Climates of the U.S.S.R.*, Aldine Publishing Company, Chicago, 1965.

Dewdney, John C.: *A Geography of the Soviet Union*, Pergamon Press, New York, 1965.

Gerasimov, I. P. L., Armand, and K. M. Yefron (eds.): *Natural Resources of the Soviet Union: Their Use and Renewal*, W. H. Freeman and Company, San Francisco, 1971.

Harris, Chauncy D.: "U.S.S.R.—Resources for Heavy Industry," *Focus*, vol. 19, pp. 1–6, February, 1969.

———: "U.S.S.R.—Resources for Agriculture," *Focus*, vol. 20, pp. 1–7, December, 1969.

Hooson, D. J. M.: *The Soviet Union: People and Regions*, Wadsworth Publishing Company, Inc., Belmont, Calif., 1966.

Jackson, W. A. Douglas: "The Soviet Collective Farm," *Focus*, vol. 20, pp. 7–12, December, 1969.

Lewis, Robert A., and Richard H. Rowland: "Urbanization in Russia and the U.S.S.R.: 1897–1966," *Annals of the Association of American Geographers*, vol. 59, pp. 776–796, December, 1969.

Karcz, Jerzy (ed.): *Soviet and East European Agriculture*, University of California Press, Berkeley, 1967.

Kingsbury, Robert C., and Robert T. Taaffe: *An Atlas of Soviet Affairs*, Frederick A. Praeger, Inc., New York, 1965.

Krebs, J. S., and R. G. Barry: "The Arctic Front and the Tundra-Taiga Boundary in Eurasia," *Geographical Review*, vol. 60, pp. 548–554, October, 1970.

Mellor, R. E. H.: *Geography of the U.S.S.R.*, St. Martin's Press, Inc., New York, 1964.

Nettl, J. P.: *The Soviet Achievement*, Thames and Hudson, Ltd., London, 1967.

Parker, W. H.: *An Historical Geography of Russia*, Aldine Publishing Company, Chicago, 1968.

11

The Middle East: North Africa and Southwest Asia

The geographical concept of a broader Middle Eastern region spanning Southwest Asia and North Africa (and a bit of Europe) has become widely accepted. Its growing use reflects not only the recognition of bonds of a shared cultural heritage but also, more importantly, recognition of similar environmental, economic, political, and social problems and common, recent historical experiences. Nonetheless, some confusion remains as to what is meant by the "Middle East," where it is, and how it differs from the "Near East" and the "Muslim World."

The term itself follows from European designations for lands that fell beyond its cultural realm. Knowing no indigenous comprehensive name, Europeans began designating by direc-

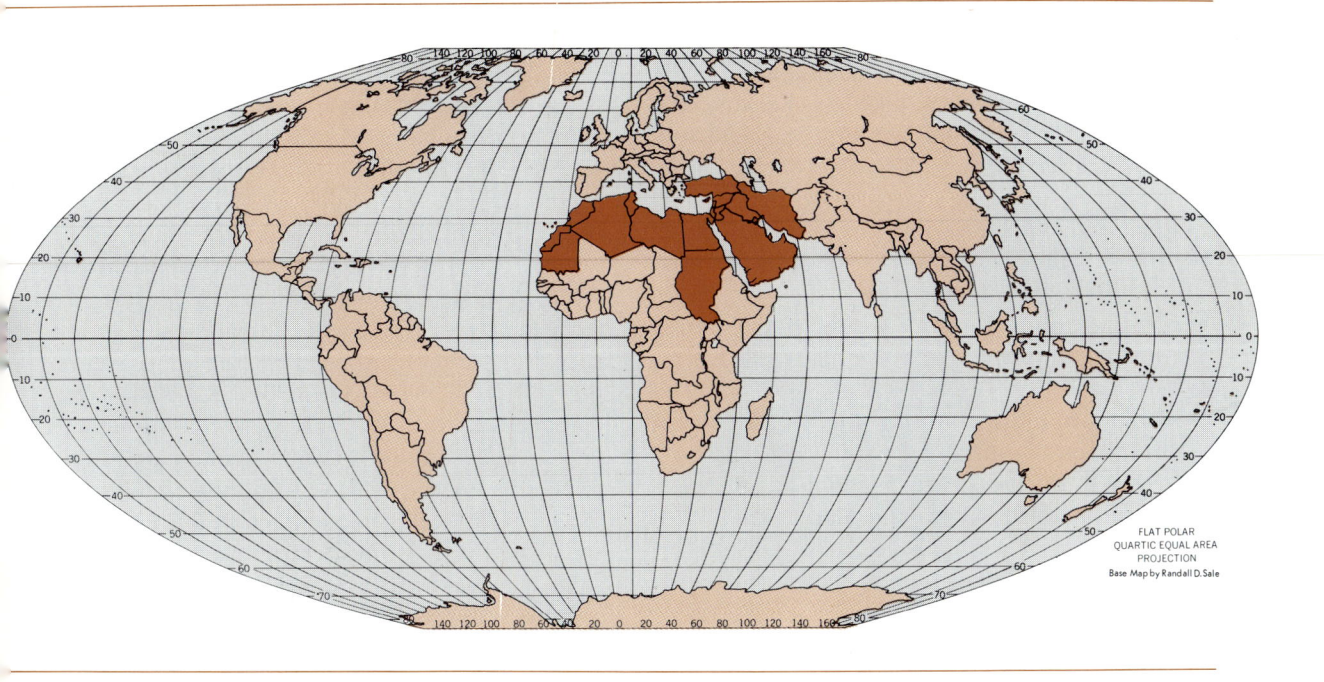

tion those lands which lay to the east, and from Roman times until the age of exploration "the East" sufficed for those dimly known lands. Following the discovery of China, Japan, and the Indies, the realization of a "Far East" provided a twofold, regional identification that lasted late into the nineteenth century. European imperial involvement in the 1890s with the eastern, southern, and western extremities of Asia communicated to Europe the fact that there were several "Easts." So, by the turn of the century the neologism "Near East" was being applied to the Balkans, Ottoman Asia, the Arabian Peninsula, and much of Iran; thereafter, the nameless void stretching from the Persian Gulf to India not illogically became the "Middle East." This tripartite division remained largely intact until World War II when the British combined several separate military commands for the Mediterranean and Southwest Asia into one. With headquarters in Cairo, it was arbitrarily labeled the Middle East Command, and its control extended from Libya to Iran and from Greece to Ethiopia. Having its own military command, India was excluded from this jurisdiction. Later, American military involvement resulted in adoption of the same terminology by the United States. Press reportage so effectively diffused this altered concept of a Middle East that, although not unchallenged, it has remained as a viable regional concept. Subsequent territorial modifications followed postwar political developments and attempts to rationalize the region, thus creating an even greater longitudinal dimension—stretching from Morocco to Iran.

Unfortunately, precise delimitation of the Middle East remains a problem because the region lacks clearly defined physical, cultural, or political boundaries. The region can, nonetheless, be subdivided according to the frequency with which various areas are included within it. The core consists of countries that are always included: Iran, Turkey, Iraq, Syria, Lebanon, Jordan, Israel, the Arabian Peninsula, and the Arab Republic of Egypt. Other countries often incorporated are Cyprus, Sudan, Libya, and the Maghrib (Morocco, Algeria, and Tunisia). Peripheral areas only occasionally included are Afghanistan, Greece, Ethiopia, Spanish Sahara and Mauritania. With the exception of Afghanistan, Greece, and Ethiopia, all these areas will be included in the discussion of the area here.

The Near East and the Muslim World represent alternative regional concepts. Because the former designation still retains its previous territorial meaning, it thereby signifies the core rather than the entirety of the region. The latter term encompasses all lands where Muslims comprise a majority or a significant minority of the population and, hence, includes not only the Middle East but also large areas elsewhere that embrace the majority of Muslims.

Regional Characteristics

Habitat

The widespread stamp of aridity upon the habitat, and consequently upon its human occupancy, is a basic fact of Middle Eastern geography. In almost all areas a strong and persistent water deficiency, caused by losses of moisture through evaporation from the soil surface and through plant transpiration which exceed the precipitation received, lasts from six to twelve months yearly. Few are the regions within this area not requiring biotic and human adaptations to this dryness. Yet, there are important variations in the distribution, amounts, and seasonality of precipitation and in the lengths of the dry season.

The great deserts of the North African and Arabian plateaus, like the smaller, desert plateau hearts of Turkey and Iran, receive only rare and scattered rains, with vast areas averaging under 4 inches annually. Rainfall occurs erratically, with long periods—sometimes years—between downpours. Average daily temperatures in the warmest months may reach 90 to 95°F, with maxima commonly soaring well over 100°F; winter temperatures often drop to near or below freezing. Throughout the year a large water deficiency prevails. Under such arid conditions, exposures of bare ground—usually rocky or gravelly surfaces, occasionally dunes—frequently appear. Characteristically, the adaptive plant cover is poor in species and consists of widely spaced, low woody shrubs, interspersed with patches of low grass and flowering annuals that spring up after the sparse rains. Since cropping requires irrigation, oases of small, compact settlements cluster mostly about wells or springs, although, where rivers run from the mountains or from wet tropical areas, ribbons of agricultural settlements may occur. Here reside the majority of desert dwellers; elsewhere the desert usually supports a sparse population of pastoral nomads. Although comprising the largest part of the region, except for the Nile Valley, deserts contain only a small proportion of the total population.

Between the deserts and wetter coastlands, mountains, and tropical savannas lie the steppes. They have a semiarid climate, averaging some 8 to 20 inches of precipitation per year, with the occasional rainstorms coming in

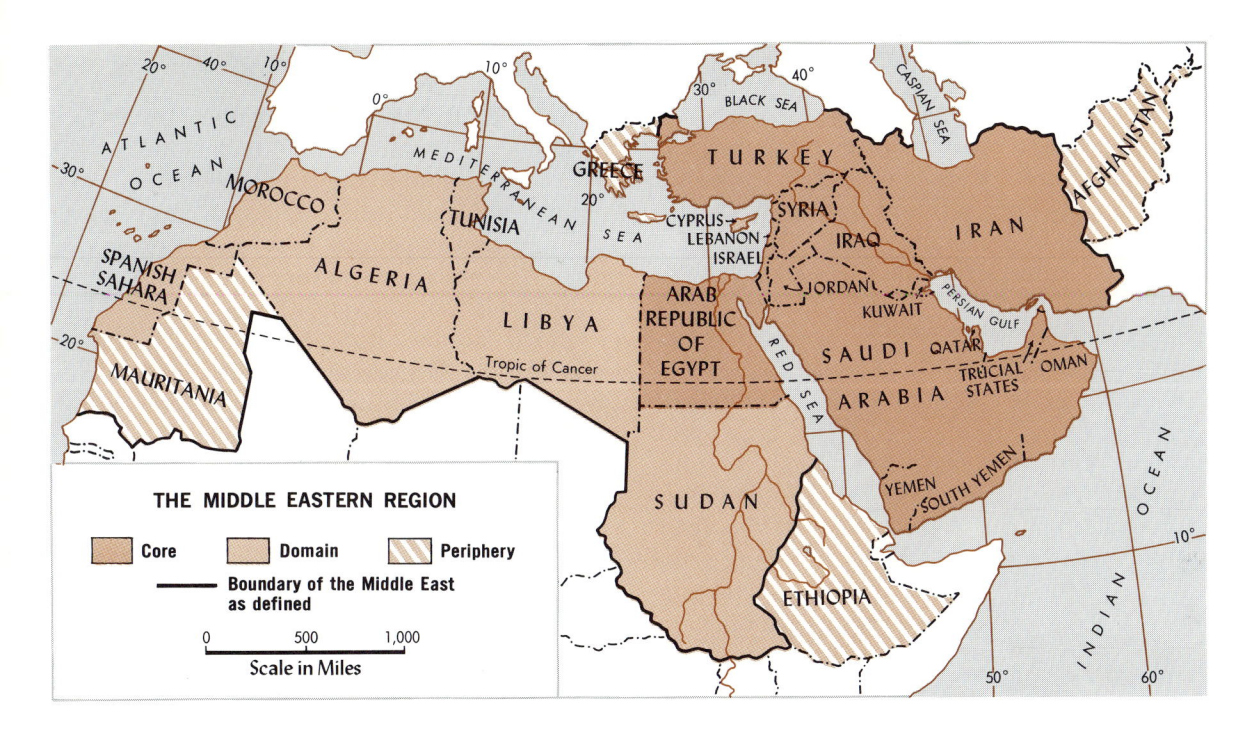

Fig. 11-1 The Middle East extends across North Africa into Southwest Asia. Much of the region is desert, and most of the people are Muslims.

the cool season north of the great deserts and in the hot season south of them. Although temperatures remain high, improved moisture conditions help produce a more extensive wild vegetation, with shrubs and particularly grasses being more plentiful. These grass-dominated plains and plateaus are often good grazing lands and support an expanding dry and irrigated farming as well.

Mediterranean-type climate, with its distinctive regime of winter precipitation and summer drought, occurs chiefly in coastal areas like the Maghrib and the eastern Mediterranean. In winter the coastal plains and mountains beyond intercept moisture-laden, cyclonic storms associated with the Westerlies; annual precipitation normally reaches 20 to 40 inches, though some higher elevations, through orographic intensification, and some higher latitudes receive more. Mountains receiving this precipitation in the form of snow act as reservoirs for many perennial rivers. The associated vegetation commonly runs to degraded variants of Mediterranean scrub forest—a scattered, open woodland of cork oaks, cedars, or bushes, with sparse grass; at higher elevations, something remains of the deciduous woodlands and coniferous forests. Even though irrigation is required for summer field crops and for tree crops, cool-season moisture conditions permit extensive rainfall cultivation. This is a favored habitat for agriculture and settlement.

Similar to the Mediterranean lands are the Black Sea and Caspian Sea littorals, which have a winter maximum of precipitation but also receive considerable summer rainfall. These also are areas of heavier precipitation, with some stations averaging over 75 inches yearly;

in response to such favorable conditions extensive, luxuriant forests have developed. Again, like Mediterranean lands, these littorals are well disposed to agriculture and settlement.

Equatorward of the tropical steppes in Sudan and Yemen is found a savanna climate distinguished by its summer rainfall and winter drought. Annual rainfall averages from 20 to 60 inches, and high temperatures are normally the rule, with maxima in the coolest months exceeding 80°F (in Yemen, higher elevations moderate the temperature conditions). Over most of the savanna, tall grass and drought-resistant thorn trees dominate, but in the extreme south of Sudan are woodlands of broadleafed trees and forests in the river bottoms. In both areas, pastoralism and rainfall-based cropping occur extensively, though cool-season agriculture requires irrigation.

Population

In 1970 the population of the Middle East reached an estimated 190 million, which, if evenly distributed over the region's approximately 5,875,000 square miles, would yield an average density of 32.3 persons per square mile. There are, however, vast areas without appreciable population; whereas other, more confined localities have densities easily surpassing 250 persons per square mile. On a highly generalized level, one can associate certain distributional patterns with particular habitat qualities, especially the availability of rainfall and water supplies: areas of Mediterranean climate and the great river valleys sustain the heaviest population densities; some Mediterranean lands, steppes, and savannas maintain medium densities; and deserts support the light-

Fig. 11-2 Except along river valleys and in areas where there is sufficient water for irrigation, population density is low. As elsewhere, the greatest concentration of population is around the larger cities.

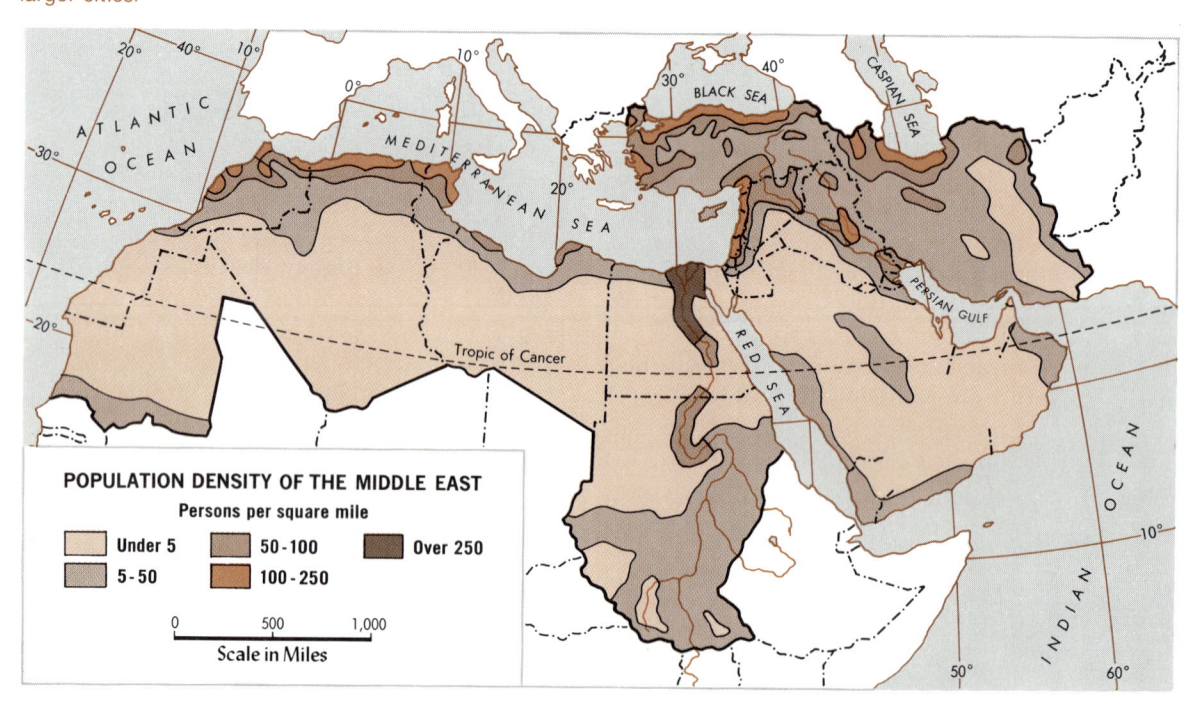

est densities. Although perceived habitat qualities are a prominent factor in the variations in density, numerous anomalies occurring in like environments can only be explained by some combination of economic, historical, and social factors.

Also uneven is the distribution of population changes and associated problems. During the period 1960–1970, population in the Middle East grew at the high annual rate of 2.9 percent, representing an increase of 5.5 million for 1970 alone. Some countries have experienced exceptionally high recent growth rates: for example, Qatar, 7.8 percent; Kuwait, 6.8 percent; and Libya, 3.7 percent. Others instead have grown more slowly: Cyprus, 1.1 percent; Saudi Arabia, 1.7 percent; and Mauritania, 2.0 percent. Certain cities and developing rural localities have grown more rapidly than the national norms, and others have lost population —both thereby furthering regional discrepancies. Factors primarily responsible for increased growth include: suppression of death rate through improved public health and medical care, young populations having high birthrates, more productive agriculture, and intra- and international migration flows. Where such increases outstrip economic growth, greater pressure is placed upon already-exploited resources, leading to overpopulation, in the sense that the presence of additional persons reduces the general welfare. Fortunately, in most countries economic growth has outstripped population growth, so that a rising living standard has ensued.

Human Communities

To identify and describe the characteristic human communities of the Middle East, geographers use religion, language, way of life, and political nationality as major cultural variables. These yield four distinctive kinds of communities, which show congruence in some countries but which manifest discontinuities in others. Accordingly, the population of Saudi Arabia speaks Arabic, practices Islam, includes many nomads, and acquires citizenship at birth; whereas most Lebanese speak Arabic, profess Islam and Christianity in approximately equal numbers, are not nomadic, and deny citizenship to Palestinian refugees and their offspring. Necessarily, one must hesitate in stereotyping Middle Eastern "peoples."

Linguistic Language furnishes an effective if imperfect identification of discrete ethnic communities. Native Arabic speakers, numbering some 99 million, or 52 percent of the total population, predominate from Morocco to Muscat. Modern literary Arabic varies but little regionally, thereby sustaining close cultural contacts among the educated elite of various nations; moreover, classical Arabic, as the language of the Koran and of the Islamic liturgy, historically pushed into Turkish, Persian, and Berber linguistic areas and furthered the influence of the language. On the other hand, significant differences exist among the several regional dialects of spoken Arabic; a major divide between eastern and western dialects occurs along the Egyptian-Libyan border; furthermore, idiomatic pecularities mark the dialects of nomads, villagers, and townsmen. Important ethnic enclaves and intrusions interrupt the continuity of the Arabic realm: some 10 million speakers of Berber tongues in North Africa; 7 million users of Cushitic, Sudanic, and Nilotic languages in Sudan; 2.5 million Hebrew speakers in Israel; and about 1.5 million Kurdish speakers in Iraq and Syria. Turkish (32.7 million) and Persian (18.4 million) language communities dominate Anatolia and Iran, respectively, but are by no means all-pervasive; in the same areas, there are minorities such as Arabs, Kurds (3.7 million), and Turko-Tartars. In Cyprus, where a small Turkish community of about 114,000 dwells, Greeks form the primary linguistic grouping, numbering over 509,000. The widely dispersed community of Armenian speakers totals over 400,000 members.

Religious As with language, religious affiliation represents a basic determinant of communal structure in the Middle East. In size and extent, the Muslim religious community is paramount, numbering some 160 million, or 84 percent of the total population, and incorporating most Arabs, Berbers, Turks, Kurds, and Persians. Being a Muslim means acceptance of the chief tenets of Islam: professing the faith, praying five times daily as well as Friday noon prayers in the mosque, almsgiving, keeping of the month-long fast of Ramadan, and, if possible, making the pilgrimage to Mecca. Islam provides believers not merely with a religion but also with a system of social customs, economic attitudes, legal codes, and governmental rule. The Muslim community, however, is not monolithic, for regional departures from the Islamic ideal exist. First, the Islamic system was grafted upon old cultural traditions —for example, Berber and Egyptian traditions —which, although modified, persist in rural areas as folk variants of Islam. Not only do these folk variants contrast regionally, but also they depart idiosyncratically from the standardized orthodoxy of urban Islam. Second, the religion has experienced fission ever since the first century of Islam (after A.D. 622), when it split into Sunni, the widespread majority faction, and Shi'a, the minority branch found largely in Iraq and Iran. Further fragmentation of both factions into more orthodox or mystical sects of local or regional magnitude tends to localize human activities even more: for instance, Shi'a pilgrimages to shrines at Karbala and An Najaf in Iraq, the austerity of the Wahhabi reform in Arabia, and militant Sanusi resistance to the Italian occupation of Libya.

Although some hundreds of thousands of European settlers lived in Algeria, Morocco, Tunisia, and Libya prior to independence, today only Cyprus, Lebanon, Israel, and Sudan have relatively large numbers of non-Muslims among their populations. Although once much more widespread, comparatively large Christian communities survive now only in Cyprus, with its Greek Orthodox majority, in Lebanon, with its large Maronite and Greek Orthodox communities, and in Egypt with its numerically large Coptic community. In Israel the overwhelming majority of non-Muslims comprises Jewish immigrants from the Americas, Europe, Asia, and Africa and their offspring. Sudan is unique for the Middle East in that a large "pagan" element inhabits its southern provinces. Most countries of the Middle East have several small communities of Christians, Jews, and other special denominations such as the Druze and Zoroastrians.

The Middle East, then, has a plurality of religious communities serving as basic social units. Although some weakening of the religious way of life has followed from modernization, the religious unit still retains much of its social and political value for the community.

Ecological Consideration of ways whereby Middle Eastern populations have adapted to diverse habitats suggests three dominant modes of life, or human ecologies: village agriculture, nomadic pastoralism, and urban commerce. Each displays separate life styles, operates principally within particular habitats, but symbiotically supports the others within the context of the total socioeconomic system.

Villagers comprise approximately 60 percent of the overall population and make up the majority in most countries. Besides ties of ethnicity and religion, villagers are bound by demands of kinship and economy. Ascriptive socioeconomic status, simple technology, and a simple division of labor typify a consciously conservative cultural tradition. The cycle of agricultural activities—tilling, planting, weeding, irrigating, and harvesting—occupies much of village life, and the rhythm of this cycle is intimately tied to the annual biological cycle, which is closely correlated with the seasonal availability of water. Thus, wheat and barley are planted in the winter rainy season and harvested in the dry early summer. Olives, vines, and certain other tree crops (dates, figs) are native to the region and are able to endure the summer drought. With the technical advances of irrigation, other crops such as cotton and citrus can be grown if given the necessary water during the dry season. The agricultural settlements, consisting frequently of squat mud- or stone-walled dwellings, range in size from dozens to thousands of inhabitants. As socially conscious governments institute agricultural development programs, provide paved roads, electricity, and pure drinking water, build schools, and expand medical services, village communities are responding positively.

The dwindling nomadic pastoral community now comprises less than 10 percent of the population. These nomads take their herds on regular seasonal migrations across uncertain habitats to satisfy the pasture and water requirements of the livestock. This necessary mobility results in minimal material possessions; wealth, prestige, and power emanate from the large size of one's herds. In these tribally organized communities, kinship ties remain the key in the sociopolitical realm. There is an economic interdependence between these nomadic and sedentary populations, with the former providing milk, meat, and other animal products, and the latter furnishing grains, dates, sugar, tea, and manufactured goods. Over the past century the nomadic population has declined in numbers and importance; a combination of agricultural expansion and improvements in communications, transportation, and weaponry have contributed to the decline.

By contrast with the declining nomadic population, over 30 percent of the Middle Eastern population now reside in towns and cities, and their numbers are being augmented as the pace of urbanization accelerates. Since World War II, as these urban places have multiplied and expanded, they have diverged economically, politically, and socially even further from the village community and have become increasingly dominant in all spheres of life. Historically, cities have served as local and regional centers of administration, commerce, culture, industry, religion, and transportation. Moreover, this concentration of functions in cities is being intensified as all the institutions newly introduced in the process of Westernization, such as modern factories, hospitals, libraries, and places of amusement, become a part of the cities. Among the responses to these developments have been additions to the main institutional foci of urban life: to the mosque, market (*suq*), government buildings, and residential quarters, still largely segregated along communal lines, have been added the central business district in larger urban places, cafés, and cinemas.

Another kind of response has been the demise of craft industries, especially in those areas which underwent colonial domination by European countries, and accompanied by the difficult, uneven rise of modern industries. Colonial policy normally facilitated the importation of mass-produced goods from the home country, frequently on a duty-free basis, with the result that many traditional industries could not compete. The development of locally owned, modern industries was hindered by colonial policies protective of home industries, lack of investment incentives, poor markets, shortages of skilled labor, inadequate power, and insufficient transportation and communications. With independence, most Middle Eastern governments have consciously fostered the growth of both light and heavy industries particularly oriented to their internal markets. Even more important, (1) as the towns modernize, (2) as the new attitudes, values, and ideologies diffuse into the countryside, and (3) as modern communications and transportation make urban places more accessible, large numbers of migrants are attracted to them.

National-Political Being less than a century old in the Middle East, the national-political community is the most recent to evolve and remains the least-developed. The state alone, as a legal, administrative entity, has never been stable or efficient enough to inspire the devotion and loyalty of its inhabitants. Allegiances have been directed to kin and commune. As in most contemporary states, it is within the urban community that the strongest connections with the nation-state are to be found, the very place where traditional communal ties are most rapidly weakening. In the countryside, particularly among the nomads, where kinship ties are especially intense, allegiance to the state is frequently tenuous.

In most cases, contemporary Middle Eastern states have had their territorial frontiers drawn by foreigners. Many states, such as Libya, have boundaries with little historical or geographical significance. Seldom do the national boundaries circumscribe linguistic/religious communities in such a way as to provide a homogeneous political unit; thus we find the boundaries of Iraq, Lebanon, and Syria holding together several orthodox and heretical, Muslim and non-Muslim, Arab and non-Arab communities. Separatist movements that are positively antithetical to the continuance of certain national-political communities have arisen; in Turkey, Syria, Iraq, and Iran, the Kurds oppose the central governments with their own communal nationalism; in Sudan the southern provinces have demonstrated separatist tendencies in prolonged civil war; and in Israel many Arabs still reject the concept of state nationalism.

Most governments attempt to foster state nationalism, and some—Israel, Turkey, and the A.R.E., representing populations with long, coherent traditions—have achieved considerable success. Efforts to promote a self-conscious nationalism involve education, propaganda, the use of modern media, and governmental actions taken in the name of the state and the people. The development of this nationalism, of this sense of belonging to a political community beyond that of language and religion, is important in that it provides a framework for the economic, political, and social changes underway in the region. Without this new context, the likelihood is that the changes that are occurring so rapidly in the Middle East will lead to communal fission rather than fusion.

The Maghrib

The moderate winter rainfall of the Atlas Mountains of northwest Africa provides a more favorable setting for human occupancy than do the adjoining lands. The Arabs perceived its islandlike qualities, bounded by sea and desert, by calling it *jazirat al maghrib* ("the westernmost island"). Historically, it has been linked as often to other Mediterranean lands as to adjacent Saharan lands; but today, the Maghrib countries of Morocco, Algeria, and Tunisia incorporate within their boundaries various Saharan as well as Atlas territories.

The original Maghrib inhabitants were apparently Berber peoples of ancient and obscure origin. Over the millennia they have mixed with their captives and conquerors—Phoenicians, Romans, Vandals, Byzantines, Arabs, Portuguese, Spanish, Turks, and French—but comparatively pure groups of Berbers are still found in certain humid, mountainous areas, especially in Morocco and in some of the higher parts of Algeria. The Arab conquest of the seventh century had great and lasting effects by introducing a new culture to the Berbers. The Arabs converted the region to Islam and began the Arabization process whereby the majority in each North African country came to be Arab rather than Berber in culture.

Like so many other newly independent countries, the Maghrib states face major problems of economic, social, and political development, complicated by a lack of national unity. It is mainly these problems which we will discuss in the following pages.

The Atlas Backbone

As the dominant landscape feature, the Atlas Mountains trend generally east-west across the Maghrib from the Atlantic to the Gulf of Gabès. The coastal zone, called "the Tell," consists of an interfingering of plains, hills, plateaus, and mountains and comprises the main agricultural region of northwest Africa. In Morocco the Rif Atlas, fronting the Mediterranean, and three overlapping parallel ranges —the northern Middle Atlas, the central High Atlas, and the southern Anti-Atlas—compose

Fig. 11-3 The Maghrib and Saharan nations are located in northwestern Africa. The Atlas Mountains extend across the northern parts of Algeria and Morocco.

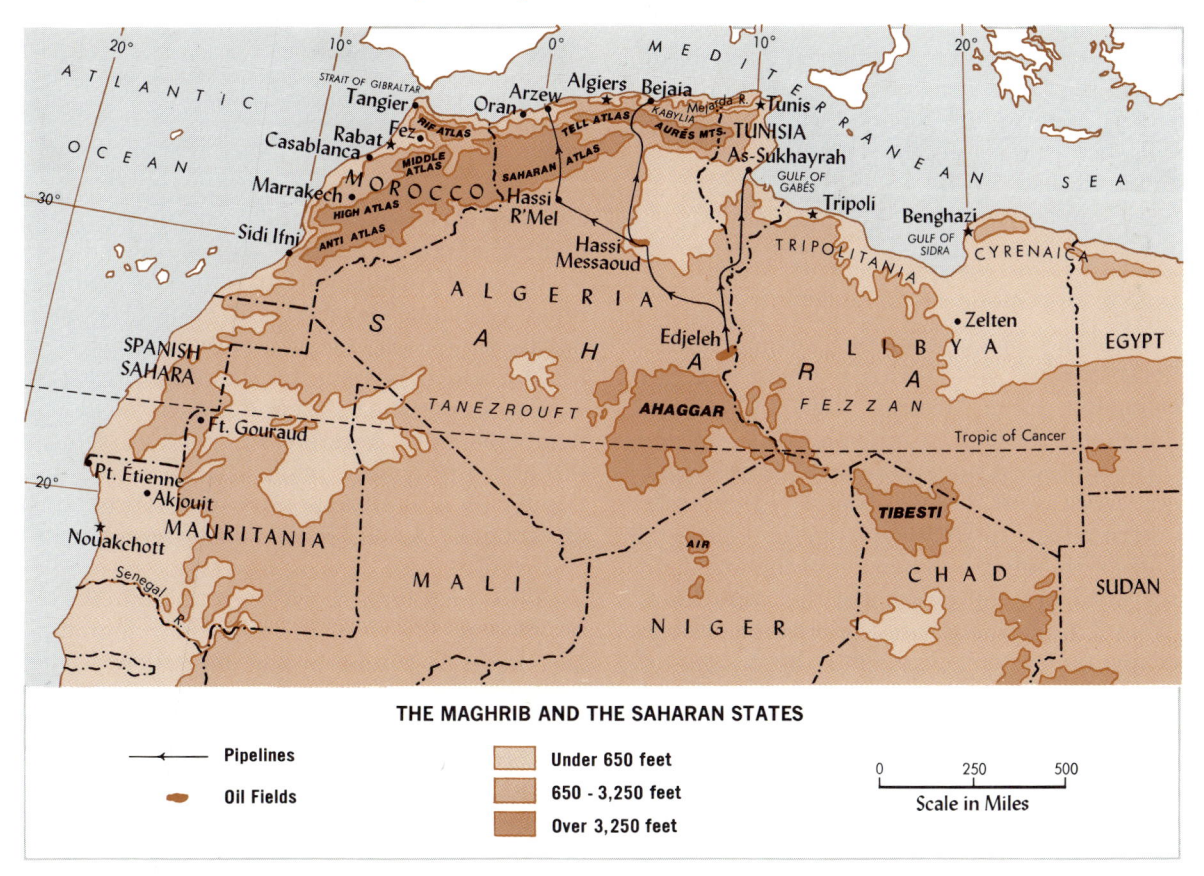

the Atlas system. Here, as elsewhere, a series of wide plateaus, depressions, and deep valleys separate the Atlas ranges from one another. The high, rugged Middle and High Atlas serve as the reservoir of Morocco; their western slopes catch and deflect moist Atlantic winds, causing heavy spring and fall rains. Further east the Tell Atlas, consisting of mountain masses and plateaus detached by plains and river valleys, extends for a thousand miles across Algeria into Tunisia. Inland from the well-watered coastal strip, and separating the Tell from the Saharan Atlas farther south, lie semiarid high plains and plateaus. The two Atlas ranges join in eastern Algeria to form the bold relief of the Aurès block, or massif, and to extend across Tunisia as a single range. The Madjarda River valley splits the Tell Atlas in Tunisia into a lower, wetter north and a higher, drier south. Behind the High Tell and to the south, the relief breaks down into zones of hills and interior plains before reaching the Sahara. To the east is a broad coastal plain.

Population Patterns and Problems: Growth, Migration, and Urbanization

With 15.4 million and 13.7 million, respectively, Morocco and Algeria have much larger populations than Tunisia with 4.9 million. The bulk of the Moroccan population is located on the windward side of the main Atlas ranges, with the heaviest densities found in the northwestern coastal districts. In Algeria the coastal plains, with only 2 to 3 percent of the total land area, hold over 50 percent of the population; and in Tunisia the greatest population clusters occupy the wetter north and the fringes of the east coast.

Each country has a young, rapidly growing population, a fact that is both boon and bane to national development. High birthrates and decreasing death rates result in annual increases of 2.3 to 2.9 percent, or some 400 thousand

Fig. 11-4 Fez, Morocco, is composed of two sectors, the old town of Fez el-Bali and the new town. The city is located just west of the Toza corridor, in one of the most productive agricultural areas of Morocco. (Courtesy of Pan American Airways)

annually in Morocco's case; consequently, the Maghrib's population has tripled since 1900. Such rapid population growth has complicated economic development efforts, because much of the economic gain must go to provide for the added people. Of the Maghrib states, Tunisia in particular has undertaken a national family-planning program designed to reduce fertility. In a more positive view, a youthful population is more responsive to new programs of economic, social, and political development and reform.

Two migration flows have modified the population situation: European emigration and internal rural-to-urban migration. The presence of European colonists—mostly French, Spanish, and Italian—dates from shortly after the French occupation of Algeria in 1830, when the French government fostered colonization. By 1956 the number of European residents totaled 1.8 million, 90 percent of whom were born in North Africa; 1.1 million of these lived in Algeria. It is estimated that about four-fifths of the *colons* resided in towns; most of the rest lived in the fertile, moist plains of the littoral and practiced commercial agriculture. By 1964, more than 1.5 million of the European colonists had emigrated, largely depleting the pool of managerial and technical skills.

Stimulated by war, drought, independence, and economic development, the flow of rural migrants to towns has greatly accelerated over the past twenty years. Even though town life is old in the Maghrib, dating back at least to Phoenician times, the greatest urban growth followed European colonization. In the Tell regions, old towns were rejuvenated and new ones founded, this being so of ports which came to control the import-export trade. Casablanca grew from 20,000 in 1900 to over 1.1 million in 1964; Algiers more than quadrupled its size, from 220,000 in 1936 to 940,000 in 1966; and Tunis grew almost as rapidly during the same period, from 220,000 to 680,000. By 1970, over 20 percent of the Maghrib population lived in cities of 100,000 or more, and probably 30 percent in places of 20,000 or more. This flow arose from a combination of the "pull" of the towns, involving concentration of wealth, jobs in industry, commerce, and services, and more social services, and the "push" from the countryside, with its low living levels, overpopulation, underemployment, and lack of services.

Economic Patterns and Developments

Areas favorable for agriculture are restricted to Mediterranean and steppe climates, with their moderate to light, but unreliable, winter precipitation. Even in areas with favorable rainfall, many places remain unsuited for agriculture, being either too mountainous or too poorly drained, especially in coastal Algeria. Small coastal plains, adjacent hill lands, and sheltered, well-watered inland valleys yield most abundantly. In the Sahara, irrigation is imperative.

In the main agricultural lands three distinctive crop associations are adapted to the seasonal moisture: (1) rain-grown crops of wheat and barley, depending upon winter rainfall, cover the largest cropped area; (2) perennial crops adapted to withstand summer drought include olives, figs, and cork oaks; and (3) irrigated crops such as citrus, deciduous fruits, vines, vegetables, and flowers occupy small land areas but are nonetheless high-value export crops. The limited oasis gardens of the Sahara provide dates, some grains, and a few fruits and vegetables. Additionally, cultivated semiarid steppes support considerable numbers of sheep and goats.

Two disparate agricultural systems emerged in the Maghrib as the outgrowth of European rural colonization. Frequently described as traditional, subsistence, and indigenous, the older, that which was practiced by Arabs and Berbers, concentrated upon such food crops as hard wheat, barley, dates, and olives, with only small surpluses being traded in local markets. Its agricultural techniques involved simple tools, terracing, some irrigation, and the tending of small livestock. The scale of operations was small, with most landholders possessing less than 25 acres. By contrast the modern, commercial European system, which was widely subsidized, was ordinarily a highly organized, capital-intensive, mechanized, more productive, and large-scale operation. Having acquired the better plains and uplands of the Mediterranean climatic zone (often by highly questionable means), European farmers grew such cash crops as soft wheats, vines, citrus, vegetables, and olives.

With independence, rapid decolonization of the countryside ensued, particularly in Algeria in 1962–1963, when European landholdings were nationalized and redistributed as state lands among laborers and former sharecroppers. Tunisia and, especially, Morocco followed this expropriation and land-reform procedure more slowly. Algeria and Tunisia have undertaken rural assistance programs designed to reduce the disruptive effects upon their national economies caused by the resulting decline in agricultural productivity.

The mineral resources of the Maghrib are diverse and, in some cases, substantial. During European rule, mining became a major economic activity and stimulated the construction of railways, mining towns, and ports. Mining still plays a major economic role, even though most production is for export and local processing remains minimal. Aside from petroleum and natural gas, phosphates and iron ore are the main mineral output, with Morocco being the major producer of the former and Algeria of the latter. The mining of lesser minerals such as lead, zinc, and manganese fluctuates with world market prices.

Discovery of petroleum at Edjeleh and Hassi Messaoud in its Sahara regions in 1956 led to a new phase in the economic development of Algeria. Three pipelines carry oil to the ports of Arzeu (Arzew), Béjaia, and As Sukhayrah in Tunisia. Additional discoveries of natural gas fields, particularly at Hassi R'Mel, provide Algeria with yet another major source of fuel. So far, the markets for oil and gas in Algeria are limited but growing, and the bulk of the production is exported to Europe.

As with the agricultural sector of their economy, the northwest African countries possess dual industrial sectors—traditional and modern. Local craft industries producing multifarious consumer goods have existed in Maghrib towns ever since their founding, with some, such as Fès (Fez) and Kairouan, gaining renown for their crafts. Since the late nineteenth century, however, imported, mass-produced goods have severely and permanently damaged the viability of many crafts. Modern industrialization, begun under colonial rule, was minimal until after 1945; yet local industries had great difficulties competing with French ones, whose products entered the Maghrib without duty. Not only did lack of investment incentives, poor market, shortage of skilled labor, and inadequate power hinder North African industrialization, but even more importantly colonial policy was disinterested in developing a balanced economy. What industry there was centered in the ports and concentrated upon exports and the local urban market.

Over the last two decades, and especially since the independence of Maghrib countries, major attention has focused upon developing a broad industrial base, first concentrating on consumer industries such as food processing, metalworking and engineering, and leather goods and textile manufacture. Each country has undertaken some expansion of its heavy industry, but so far only on a modest scale. Further industrialization is deemed necessary to provide employment for urban migrants and for the booming rural populace, as well as to promote a more balanced economy; moreover, it is recognized that investments in industry are often more profitable than any other.

National Unity, Political Modernization, and Territorial Problems

With Algerian independence in 1962, the formal process of decolonization closed; the political mold that was the colonial system had been broken after 132 years in Algeria, 75 years in Tunisia, and 44 years in Morocco. With independence achieved, the exceptional unity that had marked the nationalist struggle at an end, and the new tasks of administration, development, and modernization underway, with their mixed successes and failures, many old forces of disunity have arisen in these nations. Among these, one can recognize conflicts such as Arab versus Berber, nomad versus sedentary, rural versus urban, and the con-

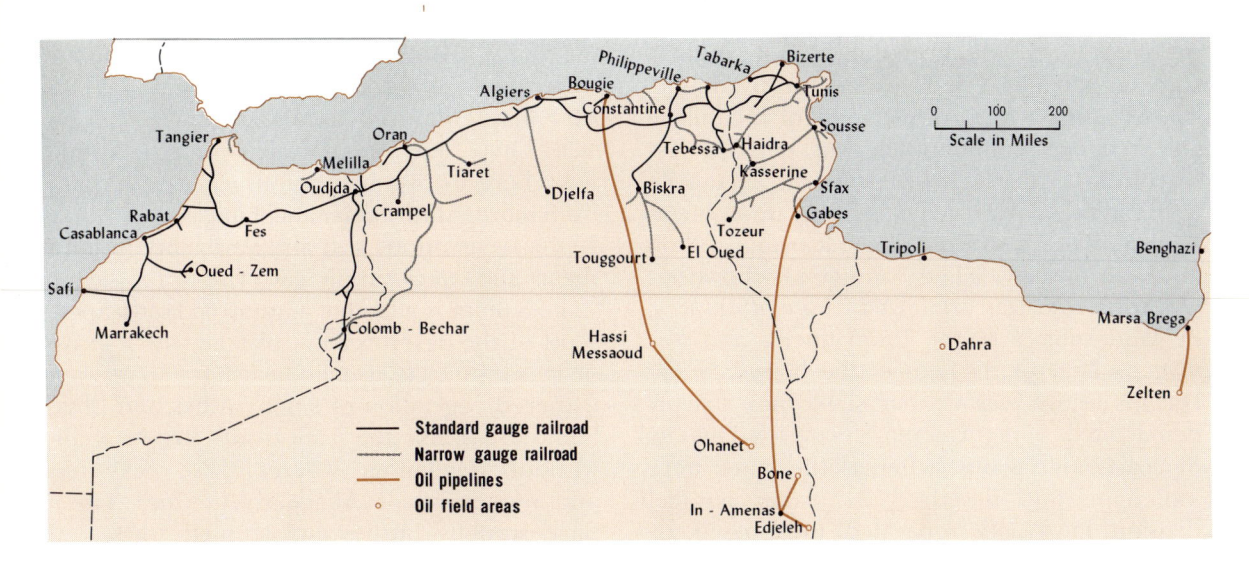

Fig. 11-5 The railways of North Africa serve the productive coastal plain and the mountain valleys. Extension of rail lines southward is limited by the Sahara Desert. Pipelines have been constructed to carry the oil from the desert fields to various cities and ports.

frontation of traditional and modern.

In the case of Algeria, the question remains as to what extent national unity can be maintained in the face of deeply rooted cleavages. No clearly defined Algerian state existed, and no one dynasty dominated over divisive local rulers prior to French colonial rule. Historically, regional and autonomous strong men governed Algeria, while Berbers, living isolated in the mountains of Kabylia and Aurès, retained their complete independence from the Arabs of plains and towns. The French presence after 1830 created a tenuous administrative unity at first, but one which became increasingly effective in the twentieth century until the outbreak of the revolution in 1954. The revolt, however, went a long way to breaking this unity asunder; during the eight years of fighting, relatively independent warlords established local control over various regions. Pressures for decentralization inherited from the distant and the immediate past are great, then, producing conflicting conceptions of how to modernize —authoritarian centralized control over development and political processes or greater decentralization along with popular participation. Significantly, opposition to present governmental centralization has centered in Kabylia.

Although Tunisia shares many of the difficulties of Algeria, the trauma of political transformation has touched it much less. To begin with, France released it from its protectorate status with less damage and disruption to its economy, political institutions, and society. By the time of independence, the Tunisians already had a mature political organization overwhelmingly accepted by the populace. Moreover, ethnic/linguistic rivalries of long standing did not threaten to divide its people, for the process of Arabization of the former Berber populace was largely completed; and as in Algeria, the European settlers departed in large numbers. Additionally, the smaller size of the country, its much more extensive plains area, and its lower and less-isolating Atlas range offered less favorable circumstances for the persistence of particularistic, ethnic communities. Of course, similar problems of political modernization endure, such as nomadic tribalism and the traditional conservatism of certain elements of the village and religious communities, but perhaps less forcefully than elsewhere in the Maghrib.

Traditional Morocco was based upon towns as centers of trade and Islamic learning and on an Arab dynasty. The countryside, however, was tribal and was organized into *bled makhzen*, or land subject to the government, and *bled siba*, or land of dissidence; the boundary between the two was neither stable nor sharply demarcated. The former lands consisted mainly of coasts and plains, whereas the latter were mountains and deserts. A further parallel division in society separated town-based Islamic orthodoxy from tribally-based heterodoxy, especially among Berbers. The establishment of the French protectorate in 1912 led to rapid occupancy of the *bled makhzen* but to a much slower conquest of the *bled siba* and to the first effective unification of Morocco under a centralized administration. (Spain administered the north and the Ifni enclave, and Tangier was internationalized.) Unlike Algeria, Morocco achieved independence in 1956 without devastating the economic and political fabric of the country. Ethnic and tribal conflicts have diminished, so that the principal obstacles to national unity are those cultural gaps alienating modern and traditional segments of society. Morocco, however, faces another set of problems deriving from its irredentist claims to Spanish Sahara, Mauritania, parts of Algeria, and certain Spanish-controlled towns along the Mediterranean coast.

The Saharan States

The Sahara extends from the Atlantic Ocean across the widest part of Africa to the Red Sea, encompassing an area larger than the United States. Politically, only part of the Sahara—that covering Morocco, Algeria, Tunisia, Spanish Sahara, Mauritania, Libya, the Arab Republic of Egypt, and Sudan—falls within the Middle East; here we are mainly concerned with Spanish Sahara, Mauritania, and Libya.

Its tremendous size and variability in landforms and water supply gave rise to the saying, "The Sahara is a land of a hundred landscapes." Most of the desert is a low plateau with a gravel or stony surface, usually bearing a sparse and withered vegetation of small bushes and grass. Some areas, like the Tanezrouft, are level and without vegetation; whereas the rocky hills and mountains of Ahaggar, Air, and Tibesti have scrubby bushes and stunted trees. Although the commonly envisaged sand regions, or *ergs*, are larger in area than Pennsylvania, the sandy surface comprises only about 15 percent of the total of the Sahara. Everywhere the vegetation responds to the rare rainstorms, and the dried-up shrubs and the dormant seeds and roots of grass spring to life to tinge the landscape a pale green. Water supplies vary from the infrequent, brief spates of the shallow valleys, *wadis*, to the springs and wells of oases, to the mighty Nile.

Before the Arab conquest in the seventh century, the oases and mountains of the central and western Sahara were occupied by Negro farmers. The Arabs spread Islam and the use of the camel to all parts of the desert, and the Negroes were driven from some of the northern oases; in other places, they remained as slaves or serfs. Some Berbers, especially the warrior Tuareg, turned nomadic and controlled some Saharan regions. The camel enabled the nomads to participate in a trans-Saharan caravan trade in slaves, gold, ivory, sugar, and cotton cloth; this trade became a basic component of the Saharan economy until its decline in the nineteenth and twentieth centuries with the suppression of the slave trade, decline in gold and ivory markets, and the evolution of modern transport. Today the very sparse population of the desert still consists mostly of oasis dwellers and nomads. Nomads are most numerous on the steppes and the more habitable northern and southern margins of the Sahara; their herds provide milk, wool, mutton, and hides to be traded to oasis peoples for dates, grain, fruits, and vegetables. Most oases depend upon spring or well water to maintain an irrigated, garden agriculture; however, some larger ones are urban in character, with cafés, shops, and marketplaces serving travelers and catering to tourists. Since the discovery of petroleum, some oasis dwellers and nomads have found employment on petroleum projects or in related industries.

Mauritania and Spanish Sahara

Many economic, social, and political questions that confronted Mauritania, formerly part of French West Africa, upon independence in 1960 remain unresolved. Can a country consisting mostly of desert and steppe and populated largely with nomads and seminomads (75 percent) establish economic and political viability? Nine-tenths of all Mauritanians live in the south, especially in the seasonally flooded Senegal Valley, where they grow millet and other food crops and raise sheep, goats, and cattle. Mining of vast reserves of high-grade iron ore at Ft. Gouraud and of copper at Akjoujt, for export through Port Étienne, offers the best economic possibilities for the nation's future. Social and political problems revolve around conflicts between the majority Arab-Berber tribesmen and the Negro cultivators of the Senegal Valley. Saharans dominate the government in the capital, Nouakchott, and their acts, such as raising the Arabic language to official status with French, are sometimes challenged by the southerners. Tangen-

Fig. 11-6 Grazing camels and sheep at the edge of the Sahara in Algeria. (Courtesy of United Nations)

tially, equipped with a minute administrative and technical elite, Mauritania suffers in attempting to impose Western political institutions on a tribal population.

Spanish Sahara extends south of Morocco along an arid Atlantic coast that supports a few poor nomadic pastoralists of mixed Arab-Berber-Negro descent. The prospect of petroleum, iron ore, and phosphate riches has drawn some foreign investment, renewed territorial claims by Morocco and Mauritania, and greater Spanish determination to stay.

Libya

Three developments in this century have fashioned the human geography of Muslim, Arabic-speaking Libya: the Italian occupation beginning in 1911, the achievement of independence in 1951, and the discovery of petroleum in 1958. Before 1911, Libya was an economically backward province of the Ottoman Empire, inhabited by nomadic pastoralists and subsistence cultivators. Although in ancient times, under the Phoenicians, Greeks, and Romans, trading cities had flourished and agriculture had expanded and prospered in the moister coastal areas of Tripolitania and Cyrenaica, under Ottoman rule they had stagnated for over 400 years.

With great difficulty the Italians subdued the Libyans and began to colonize the littoral. In the process they expanded cultivation by digging wells, planting trees, stabilizing dunes, and building roads; olives, vines, tobacco, and barley sustained the agricultural colonies. Elsewhere they built towns and ports and developed small-scale industries in the growing towns of Tripoli and Benghazi. Like the French in the Maghrib, the Italian colonials disturbed the indigenous economy by expropriating some of the best lands belonging to the Libyans.

Reduction of Italian influence, grave economic problems, and serious provincial rivalries accompanied independence. From over 150,000 in the late 1930s, the Italian settler community fell to 35,000 by the late 1960s, when further difficulties and expatriations occurred. The weak agricultural and pastoral base, absence of mineral resources, and few industries did not provide sufficient governmental income, so that Libya had to depend upon foreign aid and military expenditures. Finally, old provincial conflicts reemerged when national leadership fell to the head of the Sanusi religious order based in Cyrenaica. In 1969 this monarchy, in turn, fell to an austere military regime, bent on stringent measures for unification.

Discovery of vast petroleum deposits south of the Gulf of Sidra, particularly at Zelten, has revolutionized the country by suddenly providing enormous wealth and causing great social disruption and transformation. Development of the oil fields required the construction of pipelines to new terminals at the coast, from which tankers carry the oil to European markets. With the enormous oil revenues, much attention is now being devoted to schools, hospitals and medical services, housing, roads, bridges, and communications. The new wealth has prompted a vast drift of rural people to urban places, but unfortunately they lack many of the skills required there. Meanwhile, agriculture suffers from lack of labor and general neglect, despite governmental efforts at development.

The Nile Valley

An irrigation-based civilization began in the northern Nile Valley several thousand years before Christ. Since then, successive Egyptian states, being dependent upon the annual flood for irrigation, have shown an abiding concern for the economic and political affairs of upstream neighbors. In 1821, Egypt began its annexation of the Sudan, seeing it as an economic appendage for supplying slaves, gold, and ivory. Having lost the Sudan to an indigenous revolt (1885–1898) and later recovering it only with British assistance, Egypt acceded to joint control with Britain from 1899 to 1956. The Arab Republic of Egypt has retained its interest in its southern neighbors even after independence, particularly with regard to sharing the Nile waters.

The Nile and Its Development

The Nile has its sources on the equator, some 2,000 miles from its delta. Heavy equatorial rains and the stabilizing effect of Lake Albert, Lake Victoria, and the swampy Sudd region assure a constant flow for its longest tributary, the White Nile. In the flat Sudd, the flow is slow and evaporation rate is high, allowing only half the water to egress; consequently, the flow is steady, and without the other tributaries, there would be no great floods on the Nile. The Blue Nile and the Atbara rise in the Ethiopian highlands, where rainfall is heavy only in summer, and seasonally flow swiftly onto the plains of Sudan, bringing large quantities of silt. Northward from the junction of the Blue and White Niles at Khartoum, the main Nile floods seasonally, with the overflow lasting into the fall. From January to June, while the Ethiopian tributaries abate, the White Nile maintains the critical low-water flow. By planting in the regularly flooded silts of the Nile Delta and the narrow floodplain, the ancient basin system of irrigation yielded one yearly crop. During low-water, a second crop could be grown on small plots immediately adjacent to the river by employing simple water-lifting devices.

Modern Egypt is almost literally the creation of irrigation engineers, who plan and maintain a highly complicated and efficient irrigation system to regulate the flood and furnish con-

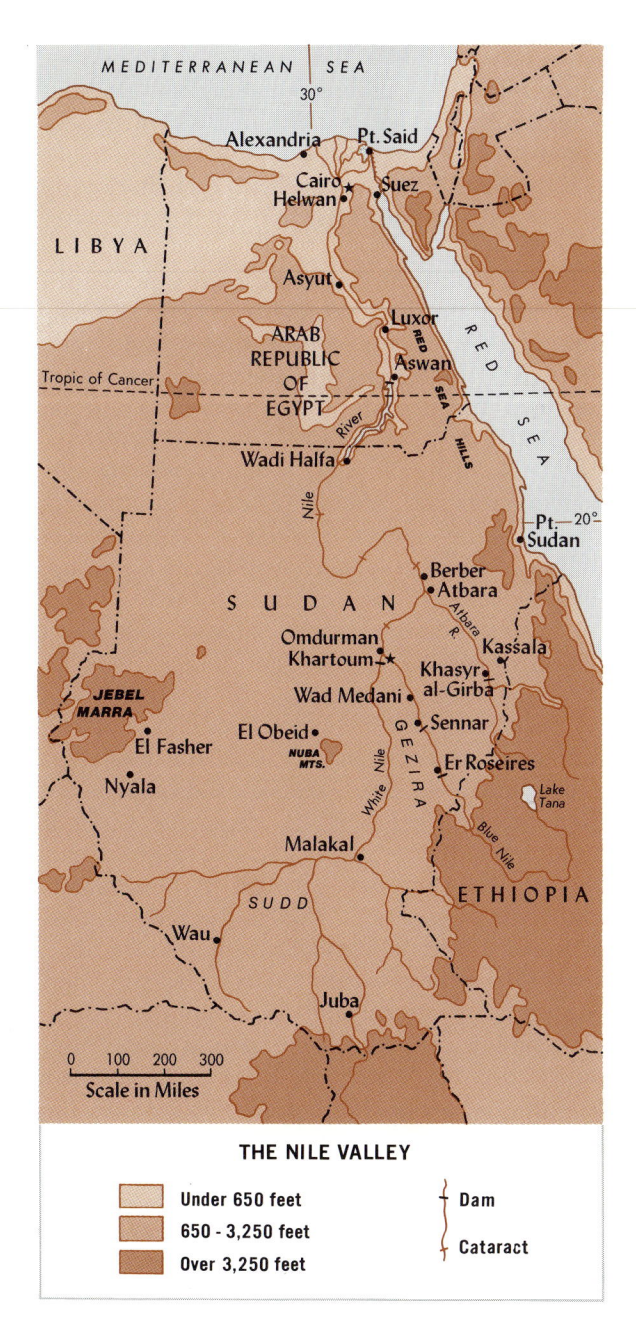

Fig. 11-7 The Nile River is the lifeline of the Arab Republic of Egypt and the Sudan. The added water supply from Lake Nasser, the huge artificial reservoir back of the Aswân High Dam, will enable the Egyptians to irrigate many thousands of additional acres.

trolled amounts of water. This perennial system, involving a complex of storage reservoirs, regulatory barrages, branch and distributary canals, and drainage lines, permits year-round cropping, increases yields, and supports a wider variety of crops. Besides producing food grains—wheat, corn, rice, sorghum—the productive system has made Egypt a major exporter of long-staple varieties of cotton. The completion, in 1968, of the Aswan High Dam, which stores water formerly lost to the sea and eliminates flooding, will eventually add a million acres of new land and will convert another million acres to perennial cultivation.

Although small private and governmental pump schemes line the Nile rivers in Sudan, major efforts at agricultural modernization center upon large irrigation schemes. On the Blue Nile, dams at Sennar and Ar Roseires have made possible large-scale cotton growing in the Gezira and Managil areas between the White and Blue Niles and potential expansion in the Kenana farther south. On the Atbara River, most of the 50,000 Sudanese Nubians displaced by the rising waters of Lake Nasser (Aswan High Dam) have resettled in the Khashm Al Girba irrigation scheme; another 50,000 Egyptian Nubians also were similarly displaced and resettled.

Because Nile water is a scarce resource of paramount concern to both Egypt and Sudan, its use is regulated by international agreements. Egypt has had priority rights, established when both countries were under British domination, under which Egypt was allotted virtually all the low-water flow and Sudan's restricted share was confined to the summer flood. In 1959, a compromise agreement allowed for the creation of Lake Nasser in exchange for a large increase in Sudan's share of the Nile flow.

Rural Patterns Outside the Nile Valley

East and west of the Nile Valley and Delta, the Egyptian landscape stretches arid and desolate. Except for a few favored oasis depressions and some settlements along the Mediterranean, the wide sandy plains and rocky outcrops of the west repel all but a few nomads. To the east, nomads move with their flocks and herds over the more rugged but slightly wetter Red Sea hills.

By contrast the Sudanese landscape is such that rural life does not, except north of Khartoum, focus anywhere near so intensely upon the Nile. In a rough way, the landscape grades through transitional zones from desert to humid tropical forest. Except for the Red Sea coast, rainfall comes in summer and increases steadily in amount and length of onset southward. Consequently, vegetation passes from desert to acacia-scrub, to denser acacia-grass savannas, to woodland-grass savannas, to forests.

Water shortages in the Saharan semidesert margin severely restrict land usage; with its drought-resistant vegetation, it is most suited to raising camels, sheep, and goats. The most numerous herders are Cushitic-speaking Beja tribesmen, who also farm sporadically in the Red Sea hills. Sweeping across the waist of Sudan, the central rain lands have rain-fed agriculture that contrasts sharply with the commercial Nile complex. Village-dwelling Arabs, who constitute the agricultural heart of Sudan, grow drought-resistant millets, sesame, and gum arabic under a shifting, rotational system. Availability of water from springs, wells, and tanks dug to catch surface runoff strongly conditions the distribution of villages. In the southern margins of these acacia-grass savannas, cattle nomads are heavily represented. Arab cattle nomads dominate in the north, and Nilotic tribesmen are predominant in the Sudd. In both instances, nomadic values discourage the sale of animals and thereby inhibit modernization of the industry. In the remote south, Negro villagers practice shifting cultivation; great distances, poor communications, little capital investment, and civil war hinder development there.

Population Problems and Contrasts

More than any other Middle Eastern country, the Arab Republic of Egypt has serious population problems. Its population of over 33.3 million represents a tripling of population since 1900, and its growth rate of 2.5 percent annually adds 830,000 more people each year. At least 96 percent of this population live on 13,600 square miles of the Nile Valley and Delta, which is only 4 percent of the Egyptian land area, thus giving an average population density for this settled area of 2,350 persons per square mile. Because new additions of agricultural land can only be small, the population pressure on agricultural land grows as the area of cultivated land per person declines. Despite the efficiency of the agricultural system and recent improvements, output cannot keep up with population growth at present rates.

Consequent with this growth have been two main streams of migration: from south to north and from rural to urban. The former movement relates to greater rural overpopulation south of Cairo, while the latter pertains to unattractive rural conditions and more attractive urban conditions. The population of many large cities has grown at a rate greatly exceeding that of the country as a whole; the populations of Cairo and Alexandria have approximately tripled since 1937, which represents less than half the time required for the same increase at the national level.

By comparison, the population of Sudan is smaller, widely scattered, less urban, and more heavily nomadic. Its population has just reached 15.6 million, but in the Middle East's largest country this gives a density of only 15 persons per square mile. Over 90 percent of this population is rural, with the greatest concentrations, perhaps totaling 25 percent, living around Khartoum and in the irrigation schemes between the Nile branches. It has not been possible to estimate the size of the nomadic population in southern Sudan accurately, but for the northern two-thirds of the country estimates run as high as 31 percent.

Fig. 11-8 Irrigation agriculture in the Sudan Desert. Tomatoes are being grown under plastic cover. (Courtesy of United Nations)

National Development in Egypt

Since the overthrow of the Egyptian monarchy in 1952, the intensely nationalistic government has attempted to transform the country's political and socioeconomic systems. Recognizing the need to improve the quality of life in Egypt, it has taken as its goals the improvement of living conditions, industrialization, and national economic independence. Steps toward achieving these goals have included land reform, expansion of public services, development of mineral resources, introduction and expansion of industries, and nationalization of foreign assets, including the Suez Canal.

Before the land reform measures of 1952, 2,642 landowners, a mere 6 percent of the total, owned 65 percent of all the cultivated land; about 2 million farmers owned an acre or less; and landless tenants leased about two-thirds of the land at rents that sometimes exceeded the value of the crop. Enactment of the reform measures limited holdings to 100 acres, redistributed almost 1 million acres to former tenants, and regulated tenancies on another 4 million acres. Over the years since, peasant incomes have gradually improved.

In the area of public services a wide range of projects are underway, but with limited available financial resources. Public health measures include minor attempts to provide potable water at the village level, some systematic attacks on diseases, especially malaria, and extension of medical care to the rural populace. The expansion of educational facilities in recent years has been remarkable; but literacy rates, especially in rural areas, remain low.

Mineral production and industrialization have expanded rapidly in Egypt since 1952. Although the mineral-resource base is still poorly known, mining of petroleum, iron ore, and phosphates is thriving, to a point where Egypt has become a net exporter of oil and phosphates. While early industrial efforts were concentrated on typical consumer industries such as food processing, textile making, drug manufacture, and building materials, more recent attention has been focused upon heavy industry. Egyptians point particularly to the iron and steel complex at Helwan; but with the enormous new hydroelectric power generated at Aswan, other recent developments embrace engineering, fertilizers, and machinery assembly plants.

In 1956, Egypt nationalized the foreign-owned Suez Canal, ran it efficiently for a decade, and increased its capacity; its closure since the 1967 war with Israel and the consequent loss of revenue has damaged the economic development efforts of Egypt.

Problems of National Unity in Sudan

The Sudanese populace displays greater heterogeneity than that of any other Middle Eastern country; diversity appears in physical characteristics, language, religion, ways of life, levels of living, and sociopolitical organization. Although four major groupings are often recognized—Nubians along the northern Nile, Beja in the northeast, Arabs in the center, and Negroes in the south—such generalization disguises the nation's cultural complexity. For example, southerners speak a multiplicity of local languages and practice numerous local religions; or, in the midst of the Arabic-speaking, Muslim tribes of central Sudan are found large non-Arab minorities such as the Fur in the west, the Nuba in the south-center, and the Ingessana in the east. Theoretically, the society is culturally plural, since the dominant ruling group, a coalition of Arabs and Nubians, is a minority.

Conflict has long characterized the political relationships between the Arab core and the southern, eastern, and western peripheries. Since before its independence, the most patent expression of such conflict has been civil war in southern Sudan. This war has killed many thousands, exiled perhaps 250,000 refugees, sent an unknown number of villagers fleeing into the bush or into the few towns, and produced wide devastation in the countryside. Elsewhere manifestations of conflict have taken nonviolent political forms, expressed in ethnic parties and minor separatist movements.

At issue are the relationships between the dominant core and the non-Arab peripheries. Is the state to have a centralized or regional form of government? Is cultural uniformity or diversity the goal? Is Arabic to be the national language and Islam the national religion? Is the core to continue to be the focus of the government's developmental effort? As these issues remain unresolved and the fighting in the south continues, the problem may eventually become one of southern secession.

The Levantine Coastlands

The term "Levant" is sometimes used for all the lands around the eastern Mediterranean from Greece to Egypt, but here we use it more narrowly to refer to Syria, Lebanon, Jordan, and

Israel. Throughout the Levant's history, its long-established human communities have experienced great difficulties in uniting themselves politically; in fact, separation of linguistic, religious, ecological, and political communities has stood as a basic structural feature of Levantine society. The political struggles of the last quarter century, and even throughout the mandate period, can be viewed in terms of conflict between various communities. It has frequently been the outside conquerors who have most effectively organized and developed the region as a unit; yet current territorial problems derive from arrangements made by France and Britain, the mandatory powers responsible for the administration of the Levantine territories after the fall of the Ottoman Empire. As well as drawing the political map of the Levant, thereby artificially dividing the Arabs, they also established the conditions leading to the founding of Israel. Since achieving independence in the 1940s, the Levantine nations have encountered grave problems of integration of minorities, of regional cooperation, and, most serious of all, of the nature of the Israeli Jewish political presence in Palestine.

Climate and Landscapes

Climatically, the area's dominant feature is the contrast between a rainy season from November to March and a dry season during the remaining months. Moisture-bearing cyclonic storms bring variable amounts of precipitation annually, thereby having major consequences for nonirrigated agriculture. Rainfall amounts differ considerably with latitude and longitude, diminishing from north to south (Aleppo, 15.5 inches; Amman, 10.9 inches) and from west to east (Beirut, 35.1 inches; Damascus, 8.6 inches), and with exposure to rain-bearing winds, west-facing slopes receiving much heavier totals than east-facing exposures.

Geographers commonly divide the Levant into five general landscape regions paralleling the coast. At the coast itself is an interrupted plain which is alternately narrow or absent in Syria and Lebanon but which widens appreciably in southern Palestine. Favored by mild, humid conditions wherever it occurs in each country, this has become an important commercial agricultural area, especially around Tripoli and Beirut in Lebanon. Backing the coastal plain and sometimes reaching to the sea is a rugged landscape of hills and mountains that receives considerable rain and dominates the region throughout its length. These deeply dissected limestone and sandstone ranges are highest and broadest in Lebanon, where they attain heights of over 10,000 feet; the Ansariya Mountains of Syria and the Israeli highlands have similar features but are lower. Centuries of exploitative farming, overgrazing, and indiscriminate cutting of trees have wantonly denuded these slopes. Next comes an elongated series of narrow, structural valleys, separating the western highlands from an eastern belt. The flat-bottomed, steep-sided Jordan Valley, with its internal drainage ending in the salty Dead Sea, is shared by Israel, Syria, and Jordan. In Lebanon the Al Biqa Valley, about 15 miles wide and 75 miles long, is drained southward by the Litani River and northward by the Orontes, which continues north in a less defined and swampy depression behind the Ansariya. Further east is another line of plateaus, hills, and mountains, including Mt. Hermon and the Anti-Lebanon Mountains. The final landscape consists of the desert plains and hills, except in northeastern Syria, where the plains along and east of the Euphrates, the Al Jezira, are semidesert grasslands.

Population Distribution: Problems and Prospects

The population of the Levantine states has come to occupy their multiform landscapes in a highly uneven fashion. An overview of the distribution reveals a zone of concentration in Lebanese and Israeli districts fronting the Medi-

THE LEVANTINE COAST LINES, MESOPOTAMIA, CYPRUS

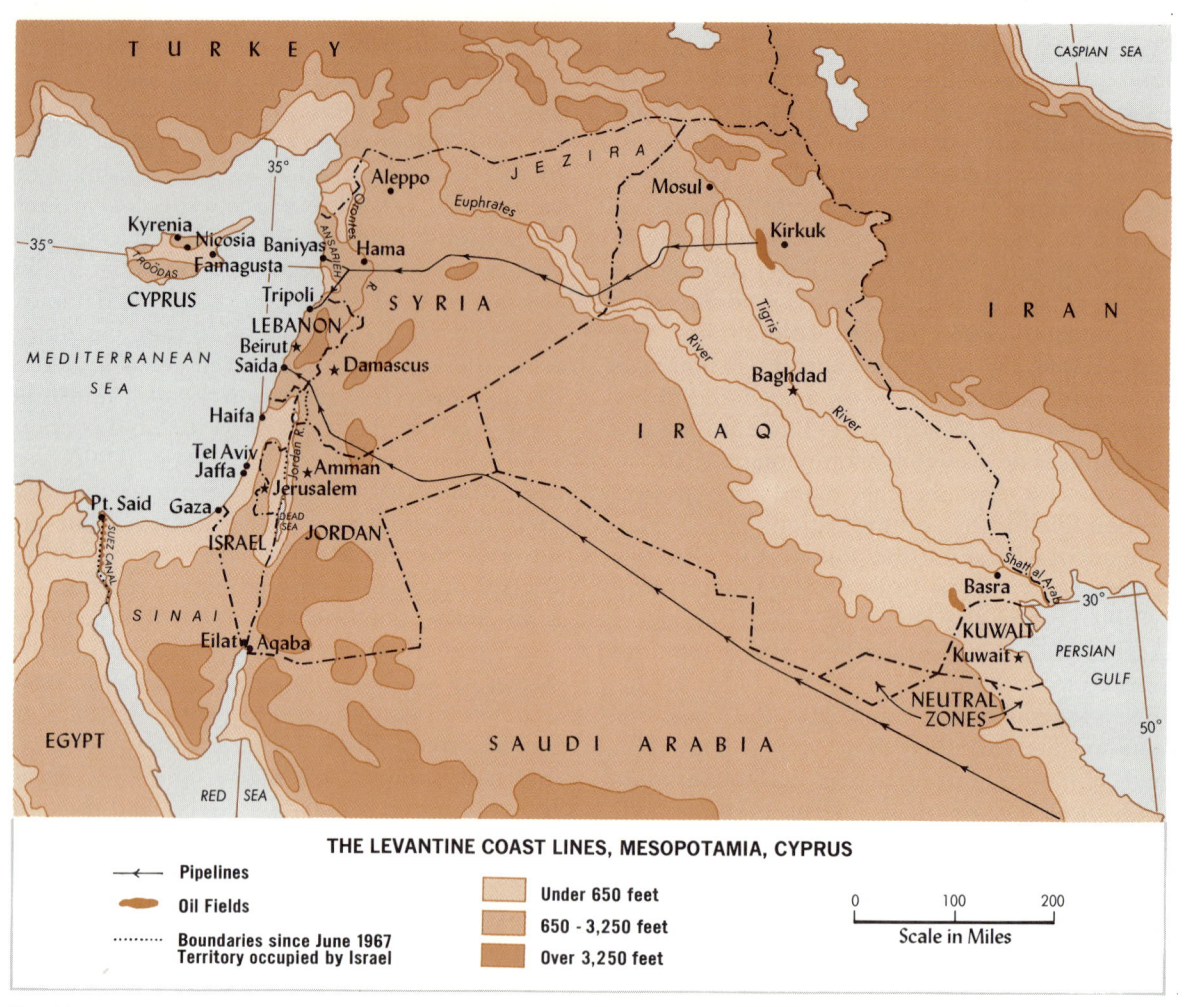

Legend:
- Pipelines
- Oil Fields
- Boundaries since June 1967
- Territory occupied by Israel
- Under 650 feet
- 650 – 3,250 feet
- Over 3,250 feet

Scale in Miles: 0, 100, 200

Fig. 11-9 The Levantine sector of the Middle East is a region of serious political problems. The establishing of Israel in the center of a large Muslim area has raised the difficult problem of two opposing groups claiming the same homeland.

terranean and a second, discontinuous band associated with the interior plateaus and desert margins of Syria and Jordan. In Israel (2.9 million), over three-fourths of the people live in the wetter coastal plains; in Lebanon (2.7 million), most persons inhabit either the littoral or the foreslopes of Mt. Lebanon. Only 10 percent of the Syrians (6.1 million), however, reside in coastal zones, with the majority occupying the Damascus Oasis and the arc of plains stretching from Homs to Aleppo to the Euphrates and beyond. The largest number of Jordanians (2.2 million) dwell in the highlands on either side of the Jordan Valley. Thus, the deserts, the high mountain areas, and portions of the structural valleys are empty or sparsely inhabited.

For the region as a whole, population is increasing at an annual rate of 3 percent and thereby diminishing the effects of economic growth and development. In the Arab countries the natural increase of population remains high, and all countries, but especially Jordan and Israel, have experienced considerable immigration. This has created problems of incorporating foreign refugee elements into the national society. In terms of areal distribution, however, population is not growing uniformly, for internal migration from rural areas to the major centers is proceeding at a quickening pace, particularly in the Arab states. Thus, certain landscapes, such as Mt. Lebanon, are experiencing some depopulation; whereas others, such as Beirut, Tripoli, Saida (Sidon), Zahle, and As Sur (Tyre), are faced with uncontrolled population growth. Already the Levant has a higher proportion of its people living in cities than have other Middle Eastern regions; over one-third of its Arabs live in cities, and Israeli statistics show that over 80 percent of its populace is urban. Current trends seem to point to even further urbanization of the populations of Syria, Lebanon, and Jordan.

Economic Diversification and Development

Considerable disparity marks the economic bases and levels of economic development among Levantine nations, with Israel having the most diversified, developed economy and Jordan and Syria the least. The fundamental fact about Jordan's economy is that the largest share of its economically active population is engaged in crop and animal husbandry in marginal habitats. With only 17 percent of its area cultivable, but 30 percent of this irrigated, and with its large refugee population, Jordan cannot feed itself, even with extensively grown wheat and barley. Its small mineral wealth consists mostly of phosphates, potassium chlorides from the Dead Sea, and other salts; light industries form most of its industrial establishment. Developmental efforts focus upon irrigation and agricultural mechanization projects, expansion of mining, and transportation improvements, including development of 'Aqaba port, Jordan's only direct outlet to the sea.

Like Jordan, Syria has over one-half of its population participating in agricultural activities, but here the endowment of agricultural resources is better, since about one-third of the country is cultivable. Along the humid coastlands, farmers grow olives, soft and hard fruits, cotton, and tobacco; in the dry interior they produce great harvests of wheat and cotton for export. Developmental activities have resulted in large-scale, mechanized farming in the northern and northeastern plains, in private and governmental pumping of water for irrigation, especially from the Euphrates and Khabur rivers, and in swamp drainage and irrigation projects along the Orontes. Completion of the Euphrates Dam, scheduled for 1975, will lead to a vast expansion of the irrigated area. Discoveries of petroleum in the northeast have alleviated some energy problems, but most industrialization still concentrates upon refining agricultural products and on consumer goods.

Since the Lebanese economy is diversified, agriculture plays a lesser part in the national well-being. Landscape variability makes possible much variety in the raising of fruits, vegetables, and grains. Productive cash cropping has surged, responding to investments of capital and skilled labor in terracing, irrigation, fertilization, and improved farming techniques; completion of the Litani project will expand the land under irrigation as well as generate large quantities of electricity. In other sectors, manufacturing, trade, and services continue to expand, especially in Beirut, Saida, and Tripoli. Lacking raw materials for much heavy industry, Lebanon relies heavily on income gained from financial services, remittances from Lebanese émigrés abroad, transit trade, and tourism, divided between summer resorts in the mountains and the entertainments of Beirut.

Israel's economy stands apart from that of the other Levantine states by virtue of being more modern and productive, having had great infusions of Western capital and technology. Its agriculture is intensive, modern, and irrigated and displays the special feature of cooperative farm settlements. As elsewhere, shortage of water is a limiting factor; currently, Israel and its neighbors dispute the disposition of unused Jordan waters. Since emphasis is on specialized cash cropping and animal raising, large quantities of grain must be imported. To provide employment, save foreign exchange, and balance its economy Israel has emphasized industrialization; but because of limited sources of power and industrial raw materials, it must depend greatly upon the skills of its workers. Although over half of Israeli manufacturing is concentrated in and around Tel Aviv, most heavy industry clusters around Haifa Bay: petroleum refining and manufacture of steel, chemicals, cement, and pharmaceuticals. As with Lebanon, over half of the labor force is employed in the tertiary sector of the economy, namely, in services and trade. For its future economic development, Israel certainly will continue to rely on foreign capital assistance.

Communal Patterns and Problems: The Arab States

Although Lebanon, Syria, and Jordan differ in the complexity of their communal structures, they share the problems of disharmony and disunity created by their linguistic, religious, and national-political minorities. These minorities, retaining a high degree of self-consciousness and solidarity, persist as a mosaic of separate communities not fully integrated within the national state. In Lebanon, where most inhabitants are Arabs, no single religious community has a majority. The largest group, the Maronite Christians, constitute less than 30 percent of the total; among the others are Sunnite and Shi'ite Muslims, Druzes, and Greek Orthodox. Areally, Christians dominate Mt. Lebanon; Sunnis, the Tripoli area; and Shi'is, south Lebanon and the Al Biqa region. The fragile political-religious equilibrium between Muslim and Christian communities seems continually threatened by the more rapidly growing Muslim population, part of which desires political affiliation with Syria. Sunnite Arabs comprise nearly 60 percent of the Syrian population, but there is a plethora of linguistic and religious minorities: Kurds, Armenians, Turks, Shi'is, Alawis, Druzes, and assorted Christians. Both the Alawis and the Druzes are rural-dwelling and fairly isolated, with the former inhabiting the coastal plain and the Ansariya Mountains and the latter Jebel Ad Druze, south of Damascus. Jordan differs in that 90 percent of its population are Sunnite Arabs and in having such a sizable community of Palestinian nationals. In 1950, when Jordan annexed what remained of Palestine, it added 400,000 new citizens to the 475,000 refugees from Israel it already accommodated. By and large, these Palestinian refugees remain wedded to the idea of returning to their homeland.

Israel and the Palestine Problem

Jews, Christians, and Muslims all have long historical and religious ties with Palestine, giving each some claim on it. Prior to World War I, Palestine was for centuries Ottoman territory inhabited by long-resident Arab Muslim and Christian communities and a small Jewish minority. In 1920, political control, in the form of a mandate, passed to Great Britain, which had promised Jewish Zionists a "national home" there. Throughout the British occupation the Palestinians unsuccessfully opposed large-scale immigration of Jews into their homeland;

then the 1948 United Nations partition of the country into Arab and Jewish sectors and subsequent wars ultimately left Israel in full control of Palestine.

Mass transfers of population stand out among the consequences of the creation of Israel. Before Israel's independence in 1948, some 430,000 Jews had immigrated to Palestine, 90 percent of whom came from Europe and the Americas; by 1964 another 1.2 million had arrived, with 55 percent of these coming from Africa and Asia. Arab Palestinian refugees numbered 1.4 million before the June 1967 war, which then added another 430,000 displaced persons. Israel has effectively incorporated its Jewish immigrants into its economic, political, and social life; however, the Arab states have not done likewise, and many Arab refugees still live in camps. Long at issue between Israel and its neighbors is the future of these refugees, their possible compensation and repatriation or resettlement. Additionally, there are over 300,000 Arabs in Israel and another 1.3 million in the Israeli-occupied territories; these large groups pose serious problems of administration and future status.

Numerous territorial matters remain in serious dispute between Israel and the Arabs. In the Six Days' War of June 1967, Israel seized the remaining parts of Jordanian-held Palestine, including Arab Jerusalem, Syria's Golan Heights, and the Gaza Strip and the Sinai Peninsula from Egypt. Final disposition of these territories is still an outstanding matter; the Arabs call for their return, and

Fig. 11-10 Old Jerusalem, located within the ancient walls, is a densely populated city with very narrow streets and little open space. Many sites here are sacred to Jews, Christians, and Muslims. (Courtesy of Israel Government Tourist Office)

Israel has refused to do so without a formal peace settlement. For any peaceful settlement of the conflict, the status of Jerusalem will present a particularly thorny problem because of its sacredness to Judaism, Christianity, and Islam alike.

Another problem stemming from the Palestine question concerns the Suez Canal, linking the Mediterranean with the Red Sea. The canal is blocked, opposing armies face each other across its breadth, and no agreements have been reached regarding its reopening. In the past, Egypt has not allowed passage of Israeli vessels or goods through it, and Israel demands this right of free access. Oil tankers from the Persian Gulf now sail to Europe around Africa; consequently, Egypt loses the sizable annual income from tolls and services.

Yet another problem is the apportionment and use of the waters of the Jordan River system by Israel, Lebanon, Syria, and Jordan. Attempts to negotiate an agreement to share and develop the scarce waters have failed, so that the conflicting parties have gone ahead independently: Israel is pumping water out of Lake Tiberias for its national water system, while Syria and Jordan have cooperated to exploit partially their Yarmuk tributary.

Mesopotamia

The name "Mesopotamia" is a regional designation for the land between the Tigris and Euphrates rivers; it served as a cultural hearth for various ancient and medieval civilizations but afterward stagnated and declined during centuries of Turkish rule. When Iraq became independent in 1932, after more than a decade under British mandate, this area once again began to assert itself and has become the core of the country.

The Regional Base

Of the five major subdivisions of the Iraqi landscape, three fall within the Tigris-Euphrates Valley. Al Jazira, a low plateau sloping gently southward, lies between the two great rivers, which rise in the Turkish highlands, south of the foothills and north of Baghdad. Because the twin rivers flow in entrenched channels, most agriculture depends upon rainfall. The central alluvial plains extend from Baghdad to the marshes of southern Iraq. This principal agricultural zone depends heavily upon irrigation from the rivers; summers are hot and dry, and winters are cold and rainy. The marshes of the south extend from north of the Shatt-al-Arab, the effluent of the twin rivers that empties into the Persian Gulf. These marshes consist of several thousand square miles of lagoons, mud flats, reed beds, and small channels. In the northeast, high ground, separated by broad undulating steppes, gives way to mountains ranging from 3,000 to 12,000 feet. This region, much of it part of Kurdistan, receives enough rainfall (15 to 25 inches annually) to support agriculture in the lowlands and in some valleys. More than half of Iraq, the portion lying west of the Euphrates, consists of a wide, stony, desertic plain that supports nomadic life only.

Population and Communal Problems

In contrast to Egypt, Iraq is still sparsely settled and does not suffer from population pressures. The main area of population concentration lies in the center of the country and extends north and south of Baghdad along the Euphrates, Tigris, and tributary rivers. Some heavier densities are found in the southern plains, along the Shatt-al-Arab, and between Mosul and Kirkuk in the north. An important shift of population has come about within the past twenty years: namely, a strong rural-to-urban migration, with cities such as Baghdad and Basra soaring in population. In 1970 the Iraqi population had reached 9.1 million and was growing at 2.5 percent annually. If such a trend continues, and unless natural resources are better used, Iraq may face serious population pressures before long.

Iraq is predominantly an Arab country, whose people are over 90 percent Muslim. This majority suffers from the ancient split between the Sunnis and the Shi'is; the latter are the most numerous and the former the most powerful. The Shi'is are mainly farmers in the southern alluvial plain, where historically they have formed a dissident group. Communal identification is still a strong basis for conflict in Iraq; geographical localization has helped in the formation and survival of certain communities, such as the Kurds in the northeastern highlands. The Kurds, constituting 15 to 20 percent of the Iraqi population, live in isolated villages in valleys along the Turkish and Iranian borders. Kurdish irredentism is strong and remains a sharp issue of conflict, one that sometimes erupts into civil war with the Arab majority.

Land and Water; Petroleum and Industries

About 60 percent of the Iraqi population depends directly upon agriculture for a livelihood. Crop cultivation occurs in two areas, the rain-fed zone of the north and northeast and the irrigated zone of the Tigris-Euphrates plain in central and southern Iraq. Outside the northern zone, the major crops—wheat, barley, rice, and dates—depend on irrigation waters drawn mainly from the Tigris and the Euphrates. Dams and canals have been built to utilize the river system more effectively: to store water for increased, year-round irrigation, to provide proper drainage systems, and to prevent flood damage. Another governmental program involves vast land redistribution; this program has arisen in response to the fact that so many Iraqi farmers were landless sharecroppers working at a bare subsistence level. Capital for such agricultural development has come from petroleum earnings.

Ranking seventh in the world, Iraq is a leading Middle Eastern oil producer, and petroleum receipts furnish the country's principal source of foreign exchange. Most of the petroleum production, as with most other countries, is in foreign hands, and Iraq receives 50 percent of the profits. Petroleum fields are located in the north around Kirkuk and Mosul and in the southern alluvial plains. Pipelines from the northern fields lead to the Mediterranean ports of Tripoli and Baniyas and to Baghdad; petroleum from the southern fields is exported by tanker from Basra. Inevitably, Iraq's most modern industries are mainly related to petroleum: refineries and a by-product gas plant at Kirkuk.

The Arabian Peninsula

As a geographical unit, the Arabian Peninsula is sharply defined on three sides by large bodies of water—the Persian Gulf, the Indian Ocean, and the Red Sea—but on the northern side it merges with the deserts of Jordan and Iraq. Physically isolated and extending over about a million square miles, its territory is divided among several states. Saudi Arabia occupies about nine-tenths of the peninsula's area. The remainder, covering the eastern and southern margins, consists of several smaller territories whose boundaries and very existence are in some instances in dispute. The western shores of the Persian Gulf and the Gulf of Oman are lined from north to south by the state of Kuwait, former neutral territory divided between Kuwait and Saudi Arabia, a stretch of the Saudi coast, the islands of Bahrein, the Qatar Peninsula, seven small Trucial "shaykhdoms", and finally the larger state of Muscat and Oman. The People's Republic of Southern Yemen, composed of the former British Aden territories, occupies the south coast; to the north and facing the Red Sea is the Yemen Arab Republic.

Aridity and Nomadism; Water and Agriculture

The whole of Arabia is a vast, ancient platform that can be divided into seven major regions: a western coastal plain, adjoining mountains, a central plateau, an eastern coastal plain, an extensive southern desert, a southern and far eastern highland rim, and an adjacent coastal plain. Most of these regions are arid, but the north receives a rainfall of 4 to 8 inches yearly; the higher parts of the south and west, especially in Yemen, obtain appreciable falls; and the Omani mountains also have more rainfall than surrounding areas. Accompanying the characteristic aridity and cloudless skies are great extremes of temperature, with summer maxima frequently exceeding 120°F. Except after the sparse rains, when *wadis* run briefly, there is no surface water, and human life is dependent upon springs and wells.

Most of the desert areas are unsuitable for agriculture and remain the preserve of nomadic pastoralists, who move about the country regularly to exploit rain-filled wells and pastures for their camels, sheep, and goats. For most nomads camels are multifunctional, providing meat, milk, clothing, and transportation, as well as serving as the chief source of wealth. However, sheep constitute the main source of meat and clothing. Some parts of the desert—in the north and the sandy Rub Al Khali—are too barren even for camel nomads. It should be noted that, here as elsewhere, the introduction of the truck has ruined the camel market, impoverishing many tribes that formerly had derived both wealth and status from breeding camels.

Probably only 2 percent of Arabia is under cultivation, and much of this is in the south. Yemen has an extensive agriculture, showing a gradation of crops by altitude; cereals, fruits, coffee, and kat (*qat*, a narcotic) are its main products. Other agricultural districts occur in the Hadhramaut region of Southern Yemen, in Oman, and in the large oases of the western mountains. In Saudi Arabia, about 80 percent of the cropped land is irrigated; in the central and eastern oases, agriculture is being diversified and modernized, and this has led to a decline in the traditional role of the date as a staple food.

The People of Arabia

Over 15.1 million people live in the Arabian Peninsula, 80 percent of whom inhabit Saudi Arabia (7.3 million) and Yemen (5.0 million); of the other territories on the peninsula, only Southern Yemen has over a million inhabitants. This population is unevenly distributed over the peninsula, with the greatest regional concentrations occurring in areas of sedentary life: north of Mecca in the central part of Al Hejaz Province, south of Mecca in Asir province, in a band running northwest of the Saudi capital of Ar Riyadh, along the coast of eastern Saudi Arabia, and in western and central Yemen. Apart from the empty areas of the north and the Rub Al Khali, the remainder of Arabia is sparsely settled, mostly by nomads.

The ecological communities for the Arabian Peninsula as a whole are largely rural, even though not more than 20 to 25 percent of the population is nomadic. Because many are seminomadic, it is difficult to assign the rural populace to nomadic or village communities. In any event, among the several countries the

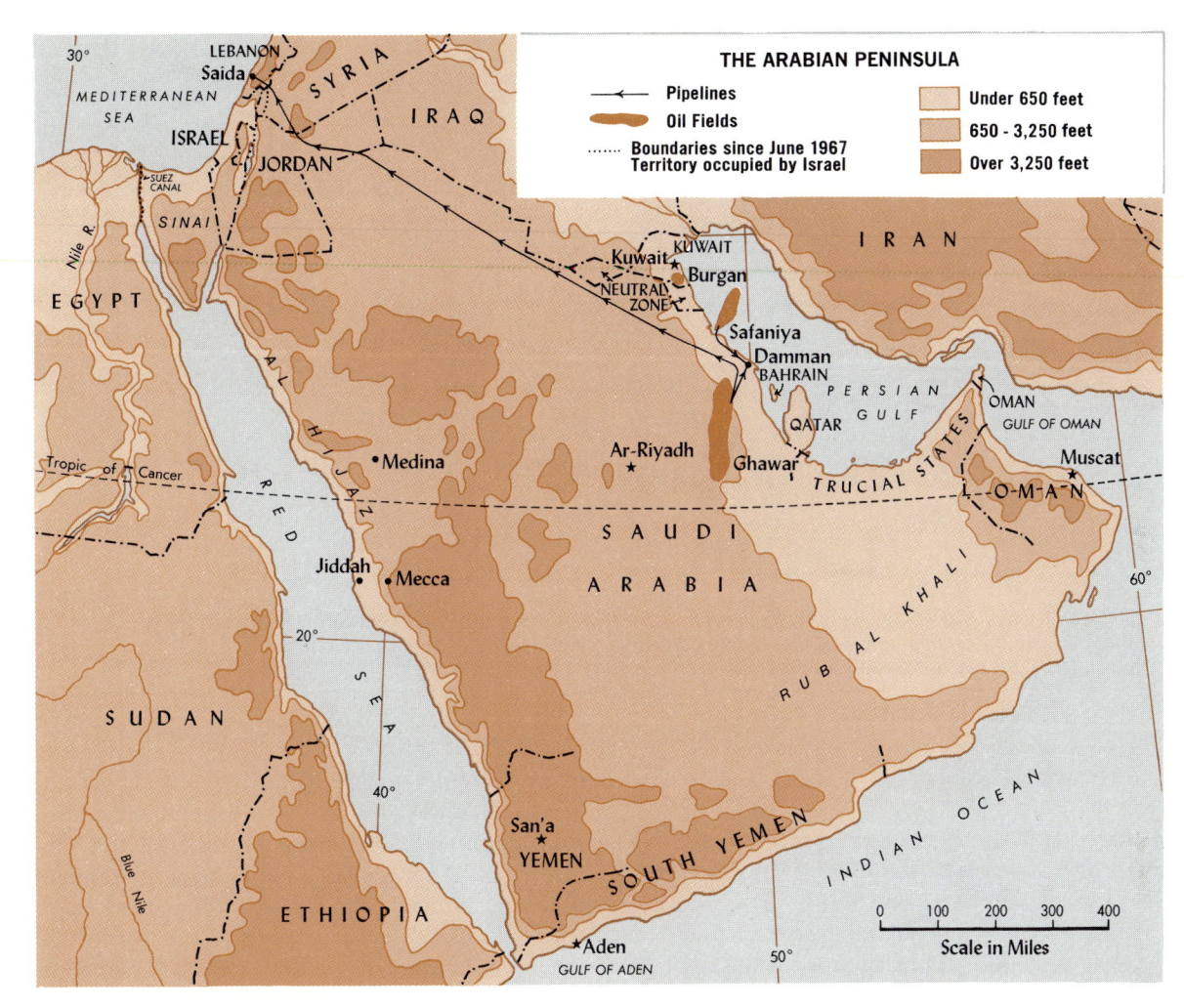

Fig. 11-11 The Arabian Peninsula is an area where one finds extremes of poverty and wealth. Kuwait has an extremely high per capita income; whereas in Yemen, Oman, and some of the other nations average income is extremely low. Almost all of the region is desert, with its main wealth deriving from oil reserves.

proportion of the population in each of the three communities varies widely; over 60 percent of Kuwait's populace reside in Kuwait city, and in Yemen the bulk of the people are village dwellers.

There are no large linguistic minorities, with 96 to 97 percent of the population being Arabic-speaking; an even larger percentage is Muslim, with most of the Persians, Indians, and Pakistanis being Muslims. Sunnite Muslims form a decisive majority in most peninsular countries, but there are several different sects; the Wahhabis predominate in Saudi Arabia and Qatar. Shi'is form the majorities in Oman, Bahrein, and Yemen and are sizable minorities in Kuwait and along the eastern Saudi coast.

Fig. 11-12 Homes of families with limited incomes in Kuwait City (al-Kuwait), Kuwait. This small but oil-rich desert nation has a per capita income in excess of $3,600—among the highest in the world. (Courtesy of Kuwait Embassy)

The Heartland of Islam

In A.D. 570 the prophet Muhammad was born into a leading family of Mecca, which was at that time a large, highly civilized city, a caravan center, and a religious sanctuary of western Arabia. There he spent most of his life, first trading and then meditating, preaching, lawgiving, and organizing; however, the last 10 years of his life were spent in Medina, some 200 miles to the north. In these places he formulated the new creed of Islam and eventually, after considerable difficulties, succeeded in converting, before his sudden death in 632, first the townspeople and then the nomadic tribesmen of western Arabia.

Because of their close association with Muhammad's life and with that of Abraham, these cities have become the holiest places in the Muslim world. The prophet himself made a pilgrimage (*Al Hajj*) to Mecca and required that all Muslim faithful should do so once in their lifetime if they can. A primary focus of this pilgrimage is the Kaaba, Islam's most venerated building, housing a black stone associated with Abraham; it is located in the center of the vast courtyard of the Great Mosque of Mecca. There are other pilgrimage sites in the vicinity of Mecca, and many pilgrims also journey to Medina, where the Prophet is buried. This annual pilgrimage brings the faithful from all over the Middle East and beyond to celebrate the life and faith of Abraham, the first Muslim. Over a million pilgrims come to Mecca each year, and some take years in the process, thereby experiencing and reinforcing their sense of membership in the Islamic community. The impact of *Al Hajj* on Saudi Arabia is manifested in the development of extensive pilgrimage facilities, including the port of Jiddah;

in the revenue derived from the pilgrimages, ranking second only to petroleum receipts; and in placing Saudi Arabia at the center of the Muslim World, rather than on the desert margins of the Middle East.

The Petroleum Industry and Economic Modernization

The Persian Gulf (also called the Arabian Gulf) and Mesopotamia together form a vast sedimentary basin, bounded by the Taurus mountain system of Turkey in the north, the Levantine ranges and the Arabian highlands on the west, and the mountain ranges of Iran and Oman on the east. This area contains the world's largest-known petroleum reserves. Along the eastern margins of the Arabian Peninsula the most important industry has become the production of crude oil and petroleum products. Saudi Arabia and Kuwait rank first and third, respectively, among the petroleum producers in the Middle East, with the neutral zone, Abu Dhabi (a Trucial shaykhdom), Qatar, Bahrein, and Oman also being substantial producers. Most of the main fields are located either close to the coast or offshore; in Saudi Arabia the main fields are Ghawar, which extends well inland east of Dammam, and Safaniya, which lies offshore near the neutral zone; and in Kuwait the extensive Burgan field lies south of Kuwait city. Most of this petroleum is shipped overseas by tanker, either as crude or refined oil; there are six refineries located along the eastern Gulf coast of Arabia. Some Saudi Arabian crude oil is carried by pipeline to the Lebanese port of Saida.

The petroleum industry is a primary factor in the economic transformation occurring in the Arabian Peninsula. In Saudi Arabia the huge earnings from petroleum revenues have financed the construction of roads, ports, schools, hospitals, and new housing, as well as

Fig. 11-13 Oil refinery in northern Saudi Arabia. (Courtesy of United Nations)

the modernization of agriculture; Kuwait has modernized its economy and its settlements and sharply expanded its social services. The other Persian Gulf territories are in various stages of carrying out similar changes.

Cyprus: The Island Republic

Cyprus, roughly the size of Lebanon, lies close to the southern Turkish coast. The island's main landform features are two mountain ranges, the narrow Kyrenia Range in the north and the bulky Troödos Mountains in the southwest, bordered by coastal and inland plains. The flanks of these ranges and the central plain connecting them, called Mesaoria, provide agricultural land dependent upon a modest, irregular winter rainfall of 12 to 20 inches. Once-extensive Mediterranean forests and fertile soils have been depleted by goats and by man's cutting and burning; this situation has been only partly overcome by conservation and legal measures.

The Communal Conflict

After nearly five years of terrorism and guerrilla warfare against British rule, Cyprus became independent in 1959, the product of a compromise between Greek Cypriot demands for union with Greece and Turkey's proposal for partition of the island between the two countries. The internecine dispute at the time of independence has continued until today as just one episode in a long, bloody struggle between Greek Christians and Turkish Muslims in the Middle East, which began in the Middle Ages with the gradual Turkish conquest of the Greek-speaking Byzantine Empire. It was revived and intensified in the last century when the Greeks began winning back their independence. Currently at issue is the political status of the Turkish minority community on Cyprus.

Of the island's 636,000 people, about 80 percent speak Greek and are Greek Orthodox Christians; about 18 percent are Turkish-speaking Muslims. For the most part, the two ethnic elements were not separated regionally, although, as a result of the fighting, there are now clear-cut ethnic enclaves. The largest of these extends from the old Turkish quarter of Nicosia, the island republic's capital, almost to Kyrenia on the north coast. Primarily, both communities dwell in agricultural villages ranging in size from 100 to 5,000 inhabitants. Greeks form a majority of the populace in all towns, but there are significant Turkish minorities in Nicosia, Famagusta, Larnaca, and Limassol.

Economic Problems

The most significant areas in the island's economy are the central plain, which is the center of wheat and barley production, and the mountains, where the mineral deposits, the vineyards and forests, and the principal tourist attractions are located. Agriculture is the most important economic activity, constituting the principal occupation of about 40 percent of the population, but overall productivity is generally low. Almost 85 percent of the cultivated land consists of rain-fed crops and therefore is subject to the vagaries of uncertain precipitation, resulting in serious droughts every few years. One consequence of low and uncertain agricultural incomes has been a steady urban immigration of rural youth.

Development programs are underway not only to improve agricultural output but also to expand forestry, fishing, and industry. In the latter sector particularly, the potentialities for development seem unrealized, for the bulk of Cyprus' imports consist of manufactured goods.

The gradual and long-term development of farmlands, forests, fisheries, and factories is considered especially important because of the Cypriot economy's past dependence upon depleting assets, particularly minerals which are at the mercy of fluctuating world markets and which, in some cases, are nearing apparent exhaustion. Copper, which above all has brought much wealth to the island, is showing signs of such a decline.

Until the intercommunal conflict abates and the two now separate communal economies merge again, economic disruption will continue for Cyprus.

The Northern Highlands

The northern highlands of the Middle East are composed topographically of the Anatolian and Iranian plateaus, with their rimming mountains and marginal plains, and politically of the Republic of Turkey and the kingdom of Iran. By isolating the high plateaus from the valleys and plains of Iraq and Syria, the Taurus-Zagros mountain systems mark off the northern highlands from the rest of the Middle East. Furthermore, the highlands have a basic physical unity in their climatic and topographic conditions.

Although Turkish and Iranian states of the past have controlled vast areas outside the highlands, both nations are now largely confined to their limits. The Ottoman Turkish Empire at its height of power encompassed the Balkans, southern Russia, most of Arab Asia, and a large share of North Africa; but throughout the nineteenth and early twentieth centuries, it lost virtually all its territories outside Anatolia. In 1923, the Turks threw off Ottoman rule and became a republic and, at the same time, began to modernize economically, politically, and socially.

Foundations of the modern Iranian state date from the sixteenth century, but periods of internal dissension and conflict and economic, military, and political weakness led in the late nineteenth and early twentieth centuries to British and Russian intervention. After 1921, Iran became truly independent, with its central government gradually gaining full control over the provinces and beginning the modernization of the country.

Physical Similarities and Contrasts

Anatolia and Iran are roughly comparable in the layout of their landforms, in that each consists essentially of an interior plateau ringed by mountains that mostly fall away to lowlands or to the seashores. Besides being a plateau, Anatolia is also a peninsula embraced by the Mediterranean, Aegean, and Black seas and by the Turkish Straits. The irregular inner plateau varies in elevation from 2,500 to 7,000 feet, with the surrounding mountains reaching 2,000 to 4,000 feet higher on all but the western side. Coastal margins differ considerably: the Aegean coast is long and indented with numerous islands, peninsulas, and rift valleys; the straits are also faulted with successions of cliffs, coves, and landlocked bays; the Black Sea coast is mainly steep and rocky; and the Mediterranean coast consists of isolated plains fronted by the steep Taurus ranges.

Topographically, Iran is the reverse image of Turkey, in that the mountains bordering the inner plateau are high and continuous on all sides except the east. On the north the Elburz chain is high and narrow, with some peaks reaching over 10,000 feet; on the west and south, the Zagros Range is mostly massive, elongated, and regularly arranged. The central plateau contains about one-half of the total area and is occupied by a series of enclosed basins, several containing extensive salt marshes (*kavir*). Iran also contains two important fringing plains areas, one along the Caspian coast and the other a part of the southern Mesopotamian plain.

The surrounding mountains shut off the central plateaus of Turkey and Iran from the effects of the sea, causing arid conditions and producing a so-called "dead heart." Other conditions of marked continentality include high summer temperatures and unusually cold winters. In many respects, the Aegean coastlands of Turkey are climatically the most favored region, with mild, fairly rainy winters and warm summers; in Turkey, rainfall is heaviest along the eastern Black Sea coast, where over 100 inches fall annually. Northwestern Iran also experiences heavy rainfall, and, as in adjacent parts of Turkey, precipitation falls in all seasons; with the exception of higher areas in the Zagros, the poleward slopes of the Elburz and the Caspian coast receive the heaviest totals.

Unlike the heart of Iran, most parts of Anatolia are drained by rivers that reach the sea, but many of these are deeply entrenched and have little utility for agriculture on the plateau itself. Only the northern slopes of the Elburz are adequately drained, by short, torrential flows; apart from the Karun River, which emp-

Fig. 11-14 Turkey is the most powerful nation in the Middle East. Its more northerly latitude and average higher altitude keep it from being a desert land.

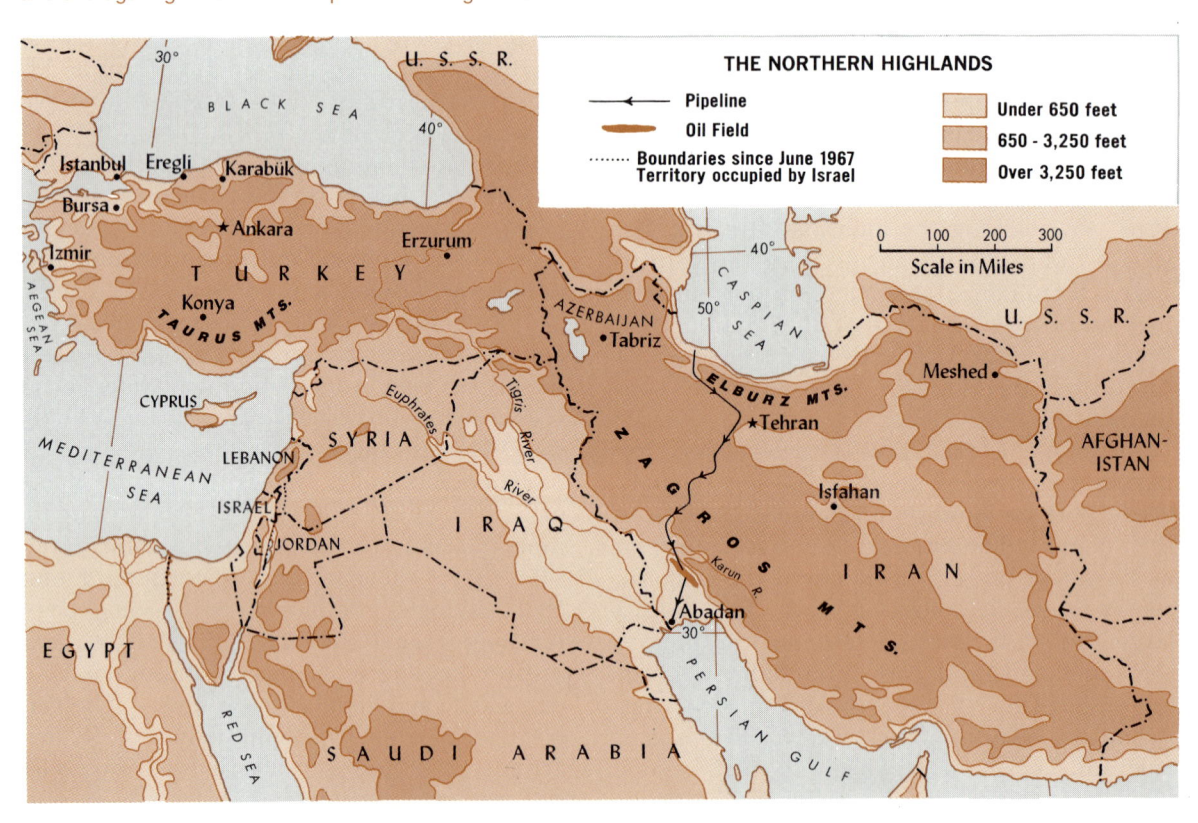

ties into the Shatt-al-Arab, few streams from the Zagros region flow into the Persian Gulf.

Population Contrasts and Changes

Along with the Arab Republic of Egypt, Turkey and Iran, with 35.2 million and 28.6 million people, respectively, have the largest populations of the Middle Eastern countries. Both are growing rapidly, but the problems associated with such growth seem more acute in Iran: greater rural population pressure but fewer agricultural resources; large-scale movements to towns but a smaller industrial base. Each country displays considerable regional variation in the density of population, but in Iran this contrast is especially sharp, with some areas having virtually no inhabitants and other areas with densities of over 100 persons per square mile.

The geographical regions of Turkey having the heaviest population densities are situated in the west, along the Aegean coast and the straits; those with the lowest densities occur in the south, southeast, and east. Most of this population is rural-dwelling, and only since 1950 has the rural-urban migration involved large numbers. The main factors in this change are: a rapid population growth; development of a highway system facilitating population movement; the pull of some big cities such as Istanbul, Ankara, Konya, and Bursa; and migration from Black Sea regions to certain eastern areas such as Erzurum, where employment opportunities are better. It has been estimated that the annual flow into the towns may equal 3 percent of the rural population, which means that the latter has reached a plateau.

The large majority of Iranians live in the north and northwest, with the heaviest concentrations found along the Caspian coast, in Azerbaijan, and around Tehran, all of these being areas with adequate water supplies. Virtually uninhabited areas, comprising the great interior deserts and the most rugged parts of the Zagros and Elburz ranges, equal about 50 percent of the country. Among the natural factors affecting the distribution of population in Iran, water supply is outstanding; a comparison of rainfall and population shows closely similar patterns of distribution. Furthermore, water supply has been the overwhelming factor in determining the size and location of villages. Villages are smallest where water supply is most critical, in and near the central deserts, and largest where it is relatively greater, in the north and northwest. Including the nomadic and seminomadic peoples, the rural population composes well over 60 percent of the inhabitants. Like Turkey, Iran is experiencing an exodus of rural population to the cities; thus, Tehran, with 2.7 million people, Isfahan, Meshed, and Tabriz as well as other regional centers are receiving immigrants faster than they can absorb them, an occurrence that has resulted in the rapid growth of shantytowns.

Communal Groupings and Cleavages

Although Islamicized, neither the Turks nor the Persians ever became Arabized culturally, and each group maintains a strong sense of ethnic identity, pride in its language and history, and intense feelings of cultural nationalism. The ethnological history of Anatolia is of great complexity, for numerous migrating peoples have come here over the millennia; yet most of those who remained became Turkified following the securing of Turkish dominance. In this century the population has become more culturally homogeneous, with the demise and deportation of large numbers of Greeks and Armenians as an outgrowth of events before, during, and after World War I. By the 1960s, at least 92 percent of the country's population was distinguished as Turkish-speaking; among the numerous minorities, the Kurds were the largest, having wide distribution along the Syrian and Iraqi frontiers. The Kurds, like the Armenians and Greeks, have had mutually antagonistic relations with the Turks, and more than once violence has broken out between them. Turkey also has numerous religious communities that have been in conflict at various times.

Iran contains numerous ethnic groups of widely differing origin. Most Persian-speaking (or Farsi) Iranians are situated in the northern and central parts of the country, especially in villages and towns on the Iranian plateau. The mountainous rim is peopled by tribally organized groups—Kurds, Lur, Bakhtiari, and Baluchi—speaking various non-Persian languages. The Arabs occupy primarily the southern coastal plains and the lower Karun River plains. Considerable cultural and political overlapping produces complex patterns; for instance, predominantly Turkish-speaking tribes may include Arabic- and Persian-speaking clans. Further communal groupings are based on religious grounds; Iran has the largest concentration of Shi'ite Muslims of any Middle Eastern country, comprising some 90 percent of that nation's population. Most of the rest of the Iranians are Sunnis, with the greatest proportion of these being Kurds, Baluchi, and Arabs. Among many communities in Iran, loyalties to the communal group are much stronger than to the nation.

Resource Use and Development: Achievements and Problems

The Turkish part of the northern highlands has a more abundant and varied resource base, which in this century has been more effectively utilized and developed than has the Iranian sector. More effective resource use has come after a destructive exploitation of habitat during classical times—a devastating attack on forests, soils, and pastures—followed by a long, neglectful phase during Ottoman times.

The variety of geographical conditions throughout Turkey has contributed to uneven agricultural development—an unevenness intensified by poor communications. Because the coastlines are climatically more favored, traditional forms of cultivation are more productive. The staple grains, wheat, barley, rye, and oats, are grown almost everywhere as rainfall crops, particularly on the plateau. Regional diversification is pronounced: along the Black Sea, hazelnuts and tea are cash crops; in the west, tobacco, figs, and raisins; and in the south, citrus and cotton. Efforts at agricultural improvement and development include large-scale importation of tractors and plows, a more than doubling of land under cultivation, construction of dams and expansion of irrigation, introduction of new and better-yielding varieties of cash crops, and the teaching of modern agricultural techniques.

Turkey has a wide range of mineral resources, which are in varying stages of development: coal, lignite, petroleum, iron ore, copper, and chrome. These minerals are of vital significance to the national economy, particularly to the heavy industry concentrated around Karabük and Ereğli, near the Black Sea. Here are located blast furnaces and steel mills, using local coal and iron ore from eastern Anatolia. By contrast, other industry is widely dispersed, with the exception of some concentration in Istanbul and the vicinity of the Straits.

Serious economic problems still lie ahead, inhibiting further progress and even threatening gains already achieved. Population growth continues to strain food supply, educational facilities, housing, and, in general, the productive sectors of the economy.

Iran's share of the northern highlands suffers greatly from mountainous barriers, isolation, and aridity, making effective utilization of resources a more difficult task. As in Turkey, there has been a long history of despoliation and neglect, which can only be rectified through a costly, extensive effort.

To a greater extent than Turkey, Iran's populace depends upon the land; some 60 percent of the people are village-dwelling farmers or pastoral nomads. At most, 10 percent of the land is cropped, of which about two-fifths is irrigated. The most productive regions lie in the north and northwest, the areas of relatively clement conditions, where cotton, tea, tobacco, and such grains as rice are grown. In most other areas rainfall is less reliable, and consequently irrigation looms more important; horizontal infiltration tunnels (*qanats*), wells, and intermittent streams provide the main sources of irrigation water. Iranian agriculture is much afflicted by a land tenure system wherein whole villages of sharecroppers are owned by absentee landlords. Coupled with water deficiencies and tenure problems is a low technological level, limited use of fertilizers, and lack of capital investment—all factors contributing to a low productivity.

Attempts are being made to remedy these problems. Dams are being built to expand irrigation and to reduce disasters from flooding, and roads are also being built to make markets accessible to more remote villages. An ambitious agrarian reform program, which aims at redistributing land and increasing the villager's share of the harvest, is underway. Other technological innovations are slowly being introduced.

Aside from agriculture, the most important economic activity in Iran is the petroleum industry, upon which the national reconstruction program depends for capital. Iran possesses major deposits of oil and natural gas, and is the second leading petroleum producer in the Middle East. The main producing wells lie in the southwest, not far from the Persian Gulf; oil is piped to Abadan, the site of a huge refinery, or directly to the Gulf, from where tankers carry it to overseas markets. New fields have been located away from the southwest, on the central plateau, and exploitation of these will help to meet the internal needs of the country.

In Perspective

An overview of the Middle East reveals recurring trends, problems, and developments that most, if not all, of the countries of the area share. In order to appreciate the changes shaping the region's geography, one must understand something of the nature and scope of population problems, economic development, and communal conflicts. Emerging from brief episodes of colonial rule and longer periods of economic backwardness, these countries face complex and difficult problems of how to establish a sound economy and raise the living levels of their rapidly growing populations. These economic problems are further complicated by the traditional communal structure of the Middle East, which promotes economic, social, and political cleavages that divide countries and thereby hinders concerted programs of national development and diverts resources needed for the solution of political problems.

Two population trends, rising growth rates and increasing urban immigration, are negative factors in most Middle Eastern countries, exacerbating local, regional, and national problems. While death rates continue to fall as a result of expanded and improved medical facilities, public health measures, and food distribution systems, birthrates mostly continue at their previous high levels. Whether viewed at local, regional, or national levels, the growing populations frequently place inordinate pressures on the land and on the economic system to produce. One consequence is the ever-increasing outflow of rural migrants to the cities, where they flock to burgeoning shantytowns. Unfortunately, these new urban occupants are arriving faster than jobs can be provided for them in commerce, industry, and services, hence resulting in a broad layer of unemployed and unproductive persons.

The governments of Middle Eastern countries and, to an increasing degree, their inhabitants are highly development-conscious. They recognize that their poorly developed resource bases can be made to yield more abundantly and that their people can enjoy substantially improved material existences. Moreover, the governments are becoming aware that, if economic development is to succeed in raising levels of material well-being, it must exceed population growth. Given that most people are cultivators, it is not surprising that considerable effort has gone into the modernization of agriculture—irrigation, mechanization, fertilizers, pesticides, improved and varied crops, and livestock. The need for a balanced economy has led as well to the modernization of mining, industry, and communications. Because these developments are highly uneven in their distribution, they have contributed to rural migration and to communal dissatisfaction in less-favored areas.

The organization of Middle Eastern societies into several kinds of communal structures—linguistic, religious, ecological, and national-political—which have histories of sharp, often bloody conflict, has rendered more difficult the problems of reconstructing life along more modern lines. At worst these conflicts are still extant in the Arab-Israeli wars and in communal civil wars such as found in Sudan, Cyprus, and Iraq. Even when these conflicts are not manifested violently or lie dormant, they may still influence relations among communities. Besides having a negative effect on national unity, these problems also impede developmental efforts in some localities and among some communities.

REFERENCES

Baer, Gabriel: *Population and Society in the Arab East*, Frederick A. Praeger, Inc., New York, 1964.

Barbour, K. M.: *The Republic of the Sudan: A Regional Geography*, University of London Press, London, 1961.

Brice, William C.: *South-West Asia*, University of London Press, London, 1966.

Brown, Leon Carl (ed.): *State and Society in Independent North Africa*, Middle East Institute, Washington, 1966.

Fisher, W. B.: *The Middle East: A Physical, Social, and Regional Geography* (6th ed.), E. P. Dutton & Co., Inc., New York, 1971.

Longrigg, Stephen H.: *The Middle East: A Social Geography*, Aldine Publishing Company, Chicago, 1963.

Orni, Efraim, and Elisha Efrat: *Geography of Israel*, Israel Program for Scientific Translations, Inc., Jerusalem, 1966.

Planhol, Xavierde: *The World of Islam*, Cornell University Press, Ithaca, New York, 1959.

Prothero, R. Mansell (ed.): *A Geography of Africa: Regional Essays on Fundamental Characteristics, Issues and Problems*, Routledge & Kegan Paul, Ltd., London, 1969.

Rivlin, Benjamin, and Joseph S. Szyliowicz (eds.): *The Contemporary Middle East: Tradition and Innovation,* Random House, Inc., New York, 1965.

Sweet, Louise E.: *The Central Middle East: A Handbook of Anthropology,* Human Relations Area Files, New Haven, 1968.

12

Africa South of the Sahara

In 1956, there were only three independent countries in Africa south of the Sahara Desert. By 1971, their number had increased to 36 as most of Africa[1] emerged from nearly a century of European rule to take an active place in the modern world. This great and rather sudden transformation is one of the most significant features in the international power politics of today.

It is at times easy to be optimistic about the future of Africa, especially when one strolls about the gleaming façades of such large new cities as Johannesburg, Nairobi, Kinshasa, or Abidjan. However, these cities do not represent

---

[1] In this chapter the term "Africa" refers to the area south of the Sahara.

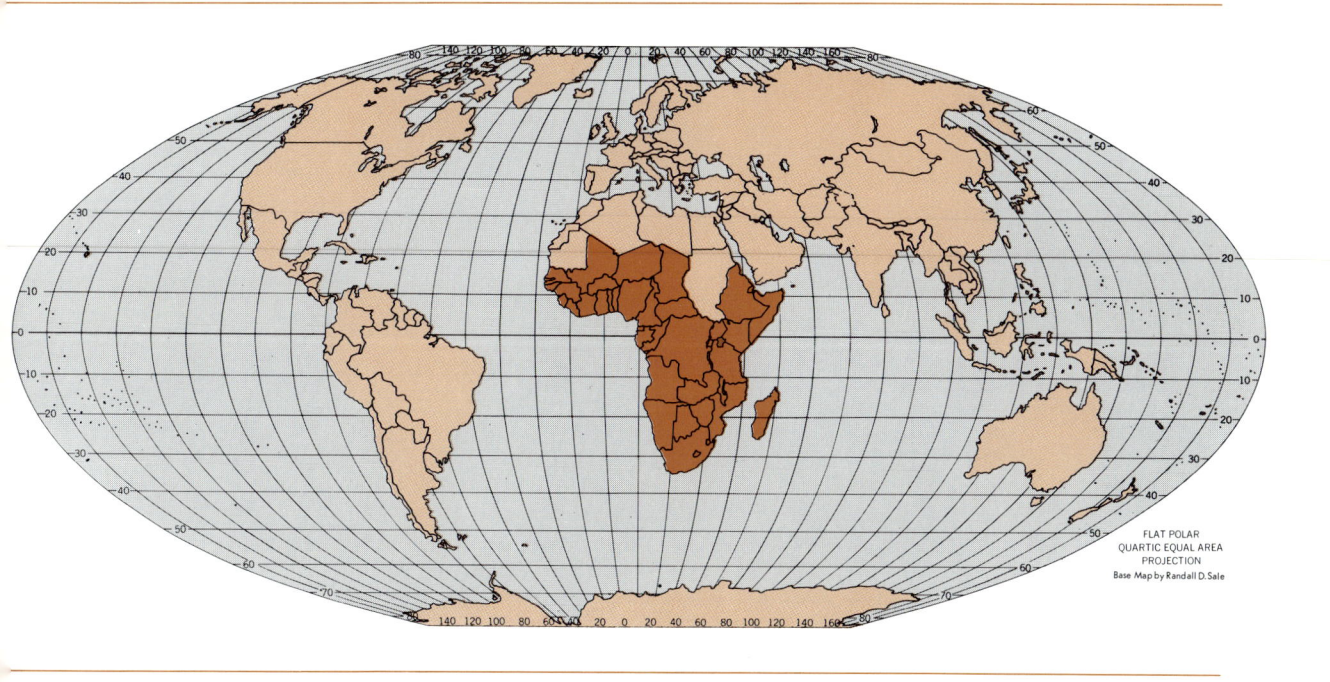

the "real Africa" of the "bush." The fact is that Africa is beset by problems of enormous magnitude. Among these are a harsh and unyielding environment, where good soil and the right amount of precipitation at the right time are at a premium. Disease is widespread among the people, livestock, and crops. The rate of illiteracy is higher than that of any other continent. Population growth is rampant. Africa is the poorest of the major areas of the world, with a per capita income averaging about $100 per year. Its economic growth rate since 1960 has been the slowest of any continent, progressing at an annual rate approximately 1 to 2 percent ahead of population growth. Indeed, with about 8 percent of the world's population, Africa produces less than 2 percent of the world's goods and services. The situation is worsened by the tendency of the industrialized countries to exploit Africa as a source of cheap raw materials.

Many of Africa's problems could be alleviated if more capital were available for development. Unfortunately, each nation has very little surplus in the way of domestic savings, and the small amounts of foreign aid donated by the better-developed countries of the world must be spread too thinly for much benefit. Another source of capital, private investment from abroad, tends to avoid most of the newly independent countries, fearing the seizure of its assets by those which are socialist-oriented. It is chiefly in mineral exploitation, where there are quick and handsome returns to be made,

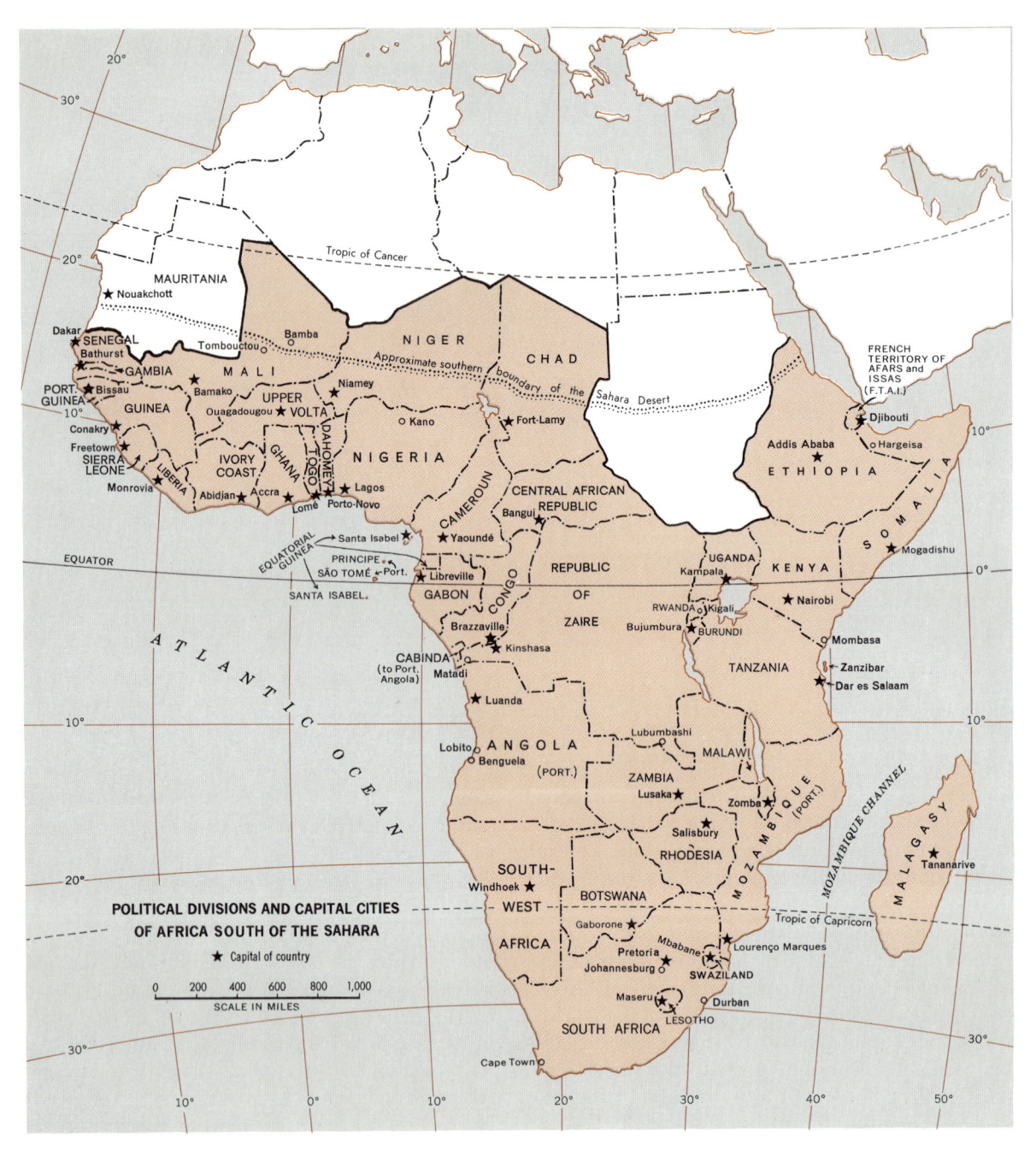

Fig. 12-1 Thirty-three countries south of the Sahara have gained their independence since 1957. Some have had civil wars, and ten are now under military rule.

that private capital has been induced to invest large amounts in Africa.

The problems of Africa would be closer to solution were it not for racial, cultural, and tribal differences. There are over 800 remarkably diverse kinds of indigenous African peoples aside from the "immigrant" races.[2] Within Nigeria alone dwell an estimated 250 cultural groups, and these were recently engaged in a deadly conflict over the question of the right of some of them to secede from Nigeria in order to form the independent Republic of Biafra. In every other African territory, intercultural tensions lie beneath the surface of everyday life.

Africa is fragmented in another way as well. There are over 40 political units south of the Sahara, and intense rivalries have developed among them. Their differences have been accentuated by a wide spectrum of governmental political philosophies, which range from the very conservative to the very radical, though none now consider themselves Marxist or Communist. Moreover, a majority of the sovereign political units are too small in area and population to withstand the political and economic pressures of each other, Europe, and the superpowers. Since fragmentation is tantamount to weakness, Africa is today a "power vacuum" largely unable to determine its ultimate destiny. Consequently, the battle among the superpowers for control of Africa has already begun. At present the Soviet Union would seem to have the upper hand, mainly because of the repercussions of racial discrimination in the United States and of United States support for Portuguese colonialism; whereas Soviet influence continues to expand in adjacent portions of North Africa and the Middle East.

Physical Setting

Relief Features

Africa south of the Sahara is a huge area of 8 million square miles, whose size is well over twice that of the United States. The distance from Dakar, Senegal, on the west coast to the eastern cape of Somalia is some 4,500 miles, or about 1.5 times the east-west expanse of the continental United States. In a north-south direction, Cape Agulhas, South Africa, lies 3,500 miles away from the southern margins of the Sahara. Such great size confers upon Africa a wealth of diversity in peoples, environments, and other assets, but also levies a penalty in the vast distances to be traversed and in the large number of landlocked interior countries.

The surface of Africa south of the Sahara is unique among the continents in not having a great mountain cordillera, or backbone, comparable to the Rockies, Andes, Alps, or Himalayas. Instead, the basic topography is that of a massive plateau which is higher in the east and south, where it averages over 4,000 feet above sea level, than it is in the center and the west. Within this plateau are immense, rather shallow, saucerlike drainage basins. Usually these are separated from each other by broad, low, geologically ancient uplifts or swells that form the watershed divides between basins, including those of the five major rivers: the Niger, Congo, Nile, Zambezi, and Orange.

In the eastern portions of Africa, the gently

---

[2] In order to distinguish racial groups, various terms are commonly used in Africa. The indigenous people are referred to as "Africans." Immigrant peoples in Africa, despite many centuries or years of residence, are usually identified by their place of origin, such as "Europeans," "Asians," "Arabs," "Chinese," "Malays," etc. Persons whose descent is a mixture of African and European are known in English by the term "Coloured."

undulating topography of the plateau is dramatically broken by the reverse of a great mountain cordillera. This is the Great Rift Valley and its various branches, which can be traced almost continuously from Syria to southern Africa. Created by longitudinal parallel faults, these steep-walled troughs split the plateau surface to depths of over 5,000 feet. The valleys range from 20 to 60 miles in width and have floors whose elevations vary from below sea level to 7,000 feet above it. These are occupied in places by long, narrow lakes. One of these, Lake Tanganyika, is the world's second deepest lake (4,710 feet). On both sides of the rift valleys, the plateau has been elevated by faults and then dissected to form pseudo-mountain ranges. In addition, vulcanism has accompanied the rift faulting to produce Africa's highest peak, Mt. Kilimanjaro (19,340 feet), and other high peaks and volcanic lava massifs from Ethiopia to southern Africa.

Along its edges, the African plateau drops precipitously in a variety of escarpments and terraces to relatively narrow coastal plains. Only in Somalia, Mozambique, and Nigeria are there sizable coastal plains that exceed a width of 100 miles. These large plains are backed, as elsewhere in Africa south of the Sahara, by the abrupt features of the higher plateau.

Due to the steep and rugged edge of the plateau, access from the sea to the interior is difficult, especially south of the equator, where an ascent of several thousand feet is generally required. Early travelers could not penetrate Africa by boat to any appreciable distance because of waterfalls and rapids in the gorges cut by rivers through the edge of the plateau. Today the building of roads and railroads up the coastal escarpments is handicapped by the high cost of construction and maintenance. A secondary aspect of Africa's plateau structure is the continent's compact shape and straight shoreline, which results in an absence of the penetrating bays and gulfs typical of other continents and in a paucity of good natural harbors.

According to the Continental Drift Theory, the plate that supports Africa may be unique, for it does not apparently move or drift at the rate of from 1 to 5 inches per year as the other continental plates do. If this is so, an explanation of the anomalies of Africa's structure is possible. Thus, the African plateau, having but little motion, lacks a great mountain cordillera like the Andes because the plate beneath it drifts insufficiently in any one direction to generate the friction and piling-up of material from which a mountain range would be created. Further, the origins of the abrupt coastal escarpments, with their typically narrow coastal plains, are consequently due to the splitting, or fracturing, of the original Gondwanaland plate into smaller pieces that vertically shear the continents above them in the process of drifting away. Finally, the present African rift valleys would seem to be the result of an active, growing fracture in the plate below Africa. This indicates that the portion to the east of the present rift valleys will, like the others, separate from Africa eventually.

Climate

Of all the continents, Africa is the most tropical, with over 75 percent of its land area south of the Sahara situated between the tropics of Cancer and Capricorn. The continent straddles the equator, causing the formation of a roughly symmetrical, or mirrorlike, arrangement of climatic and vegetation patterns between the equator and the northern and southern extremities. The symmetry is not perfect, of course, because of such factors as ocean currents, relief features, prevailing winds, position in relation to other landmasses, distance from the ocean, and elevation. This last factor is of particular significance, for it mod-

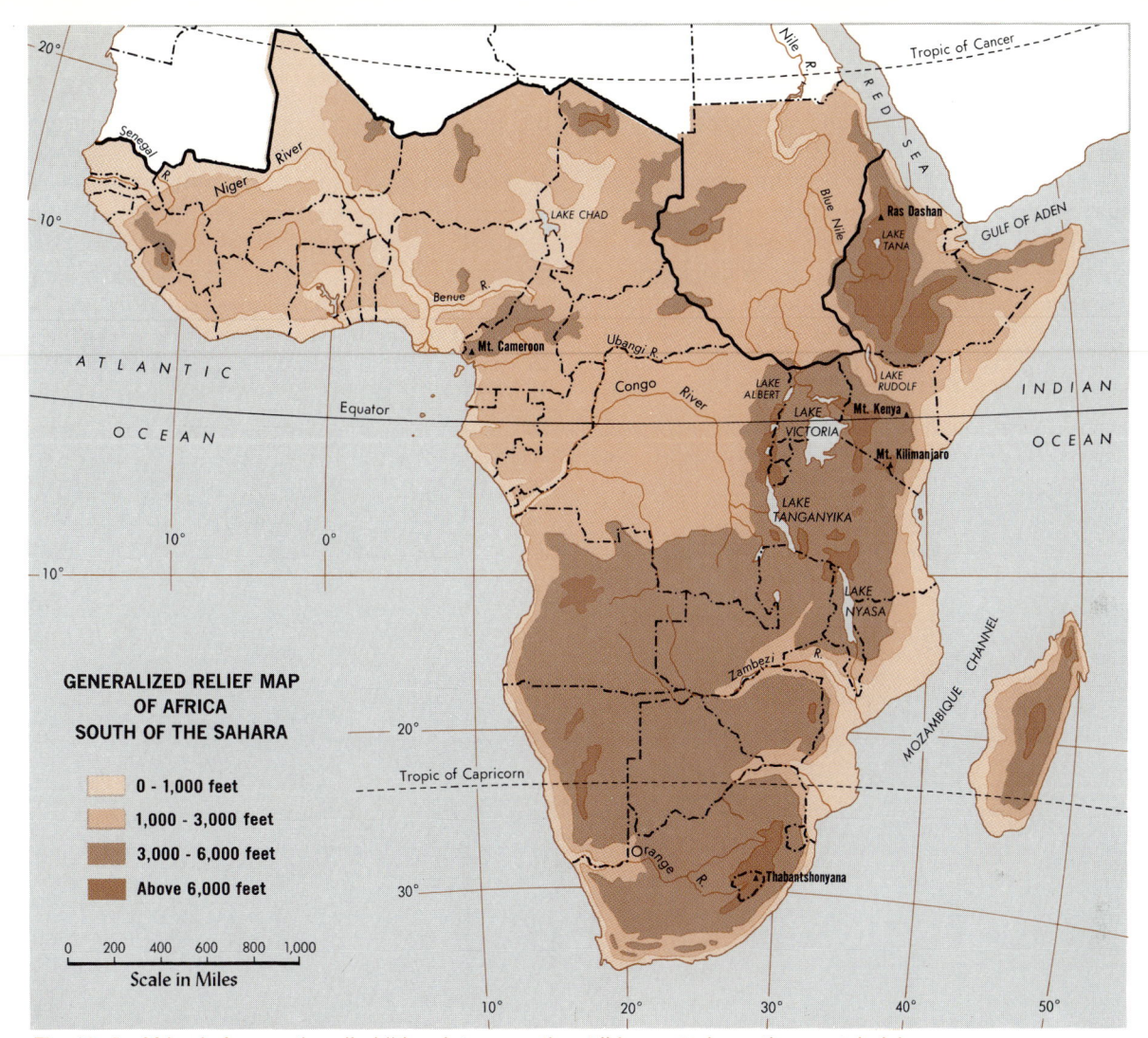

GENERALIZED RELIEF MAP OF AFRICA SOUTH OF THE SAHARA

- 0 - 1,000 feet
- 1,000 - 3,000 feet
- 3,000 - 6,000 feet
- Above 6,000 feet

Fig. 12-2 Africa is frequently called "the plateau continent." In most places the coastal plain is narrow; the rivers flowing from the plateau often have rapids or waterfalls that mark the boundary between the coastal plain and the plateau. The highest peaks on the continent are in East Africa.

erates the tropical heat over large areas of the African plateau, especially in the east and south.

Rainy Tropical This climate is characterized by its steady warmth, abundant precipitation, and its monotony. Monthly temperatures average from about 70 to 82°F throughout the year, rarely exceeding 95°F on the warmest days and seldom falling below 60°F during the coolest nights. There is no true dry season, although some months have less precipitation than others. Average annual rainfall usually ranges from 60 to 150 inches, but one location in Cameroun receives over 400 inches per year. High relative humidity contributes to morning fogs, considerable cloudiness, and the reduced daily temperature ranges in this climate.

For man, the constantly warm temperatures and high humidities have been traditionally protrayed as detrimental to his physical well-being. This thesis cannot be fully accepted because of substantial human development in areas of tropical rainforest climate in Africa, South America, and Asia. For example, some of the most progressive parts of West Africa are found within this climatic zone.

The areal extent of the rainy tropical climate in Africa is much smaller than is generally assumed, comprising perhaps 10 percent of the continent. It is best developed in a wedge centered slightly north of the equator and is bounded on the north and south by a transition to drier savanna climates. On the east, it is

Fig. 12-3 The climatic regions of Africa south of the rainy tropical region are almost like those north of it. Only in the eastern highlands are there great variations.

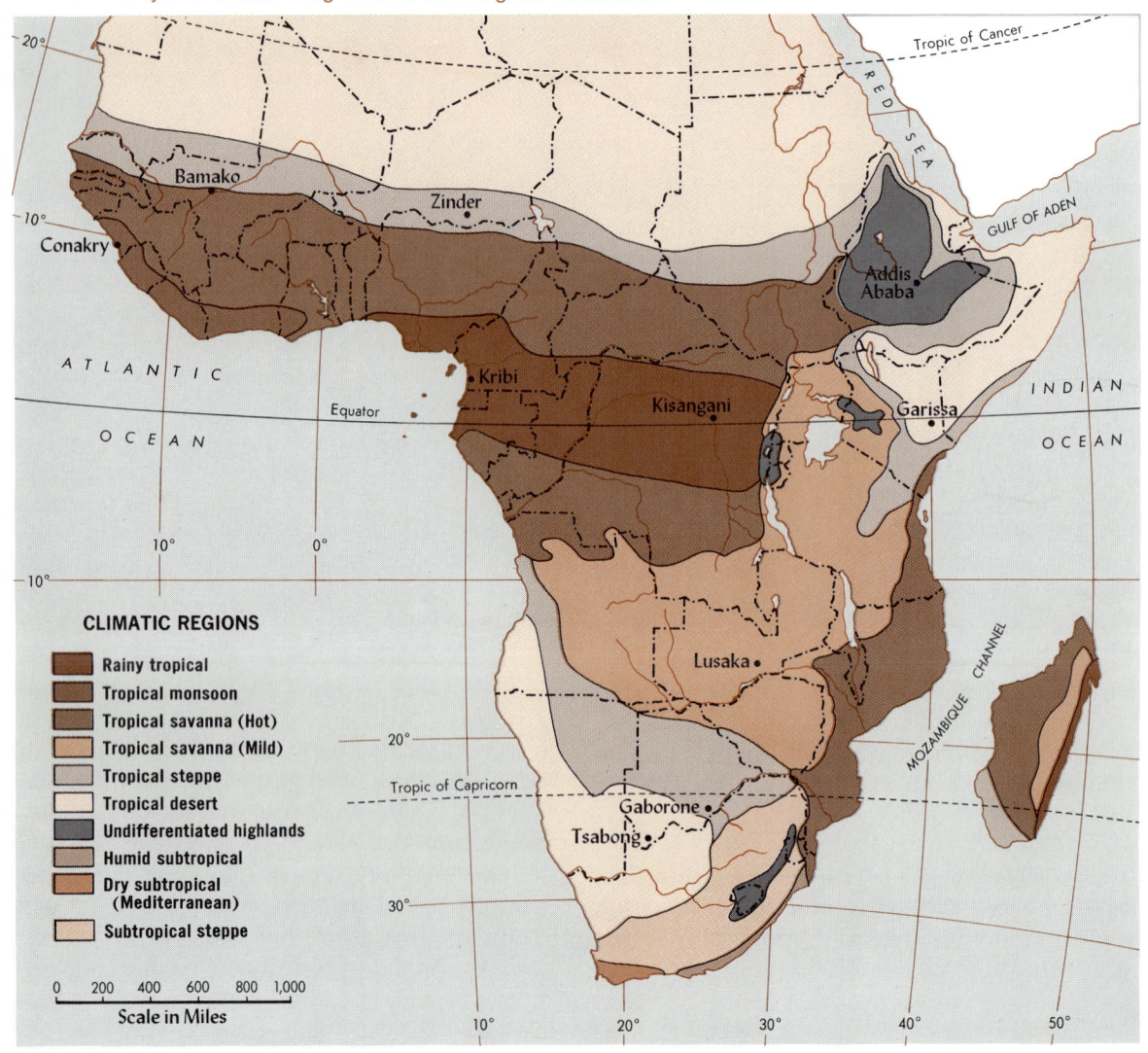

CLIMATIC REGIONS
- Rainy tropical
- Tropical monsoon
- Tropical savanna (Hot)
- Tropical savanna (Mild)
- Tropical steppe
- Tropical desert
- Undifferentiated highlands
- Humid subtropical
- Dry subtropical (Mediterranean)
- Subtropical steppe

cut off by the cooler and generally drier conditions of the East African highlands. Another portion is found along the windward east coast of Malagasy.

Tropical Monsoon Conakry, capital city of Guinea, exemplifies the tropical monsoon climatic regime. Its total average annual rainfall is 171 inches, but four months provide most of this amount. On the average at Conakry, June brings 23, July 53, August 42, and September 28 inches. In contrast, the combined amount of precipitation from December to April is less than 2 inches. As a comparison, Chicago normally receives 33 inches for the entire year. Temperatures at Conakry are essentially the same as those of the rainy tropical climate, but the seasonality of precipitation helps to relieve the usual tropical monotony.

The extent of the monsoon area depends upon statistical definition. It may be found in various parts of equatorial Africa, but its primary area is in western Africa. Here a strong flow of extremely moist air from the southwest is forced over the barrier of high escarpments in the summer months to bring copious rains. To the north and east, away from the steeper escarpments and where the coastline lies less obliquely to the path of the winds, rainfall decreases.

Tropical Savanna The tropical savanna climate is by far the most widespread in Africa south of the Sahara. It includes a variety of temperature and precipitation characteristics, but may be divided into two main subtypes: (1) hot and (2) mild.

The key unifying element everywhere in the savanna is the pronounced alternation of rainy and dry seasons. These seasonal changes are the result of the annual migration of world pressure belts, so that the savanna is subject during the year to both the heavy rains of the rainy tropical climate as well as the dry weather conditions of the desert.

Because temperatures do not vary greatly throughout the year, precipitation is the critical element in the tropical savannas. Rainfall ranges from about 25 to 60 inches annually, with the smaller totals and shorter rainy season generally corresponding to increasing distance from the equator. For example, at Bamako, capital of Mali, average rainfall totals 42 inches per year. There are four very wet months with 36 inches of the average total, followed by eight very dry months. Temperatures frequently exceed 100°F during the daytime for many weeks before and after the rainy season, but decline by about five degrees in the winter (November–February) and again during the months of heavy summer rain (June–September). Thus, April and May, prior to the rains, are the hottest months of the year. Bamako represents the "hot-savanna" subtype found at lower elevations (below 3,000 feet approximately).

South of the equator, the wet and dry seasons are reversed in terms of the calendar. But a more important difference is due to elevation. Lusaka, Zambia's capital, lies at 4,200 feet and typifies the elevated "mild" savannas of vast portions of the southern Congo, Angola, Zambia, Rhodesia, Malawi, Mozambique, Uganda, Kenya, and Tanzania. At Lusaka, average monthly temperatures range from 61°F in July, when cool dry air invades the area, to a high of 76°F in October as the rainy season approaches. Total annual precipitation averages 33 inches, with all but 1 inch of this concentrated in the five months from November through March.

The two very distinct seasons of the savanna climate profoundly affect the rhythm of man's agricultural activities. Man is forced to adjust to the erratic timing of the onset of the rainy season and to the variable amount of rain received each year. The rains usually occur in brief, intense downpours that erode the topsoil and may damage crops. Wind, occasional hail, and lightning accompany severe convective

AFRICAN RAINFALL PATTERN (INCHES)

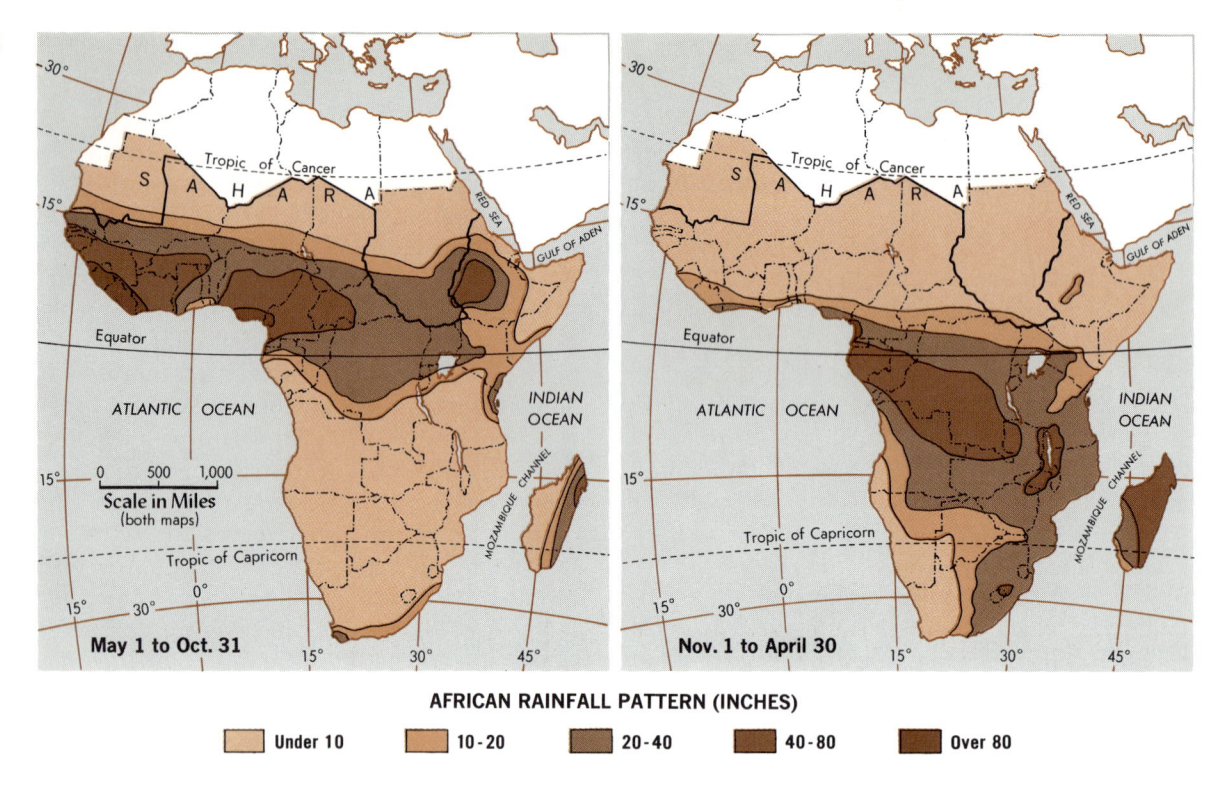

Under 10 | 10-20 | 20-40 | 40-80 | Over 80

Fig. 12-4 These rainfall maps illustrate the variation of rainfall between the high-sun and low-sun seasons.

storms. Somehow, it seems, man and his crops survive to enjoy the best time of the year when the rains end, the crops are harvested, and the landscape appears fresh and green. But as the dry season sets in for anywhere from two to ten months, the earth dries out and cakes, the landscape turns brown, winds cause dust storms, and food supplies diminish. This leads to the worst part of the year in some savanna areas—"the hungry season." Not only is most of the food gone, but both temperatures and tempers soar as the next rainy season approaches.

Tropical Steppe The tropical steppe climate is strongly transitional, separating the savannas from the deserts. The rainy season is short and erratic, usually about three months in duration, so that annual precipitation ranges only from about 12 to 25 inches. Gaborone, capital of Botswana, and Zinder, Niger, are examples of this climate.

Averages mean little in the tropical steppes, owing to wide fluctuations in rainfall from year to year. Crop raising is a perilous venture; consequently, man's chief occupation here is grazing livestock.

North of the equator, the steppe extends in a band from the Atlantic Ocean to the Red Sea. It is also found in eastern Africa, where topographical features, prevailing winds, and ocean currents combine to induce aridity. In southern Africa, the steppe lies too far inland to receive much rain from the easterly winds that give up their moisture as they rise over the plateau escarpments. The Kalahari Desert is actually a steppe in its eastern portions.

Tropical Desert The climatic influence of the Sahara is felt far beyond its actual limits.

The "harmattan," a dry, dust-laden wind that originates in the Sahara, affects the climate of all western Africa in the winter as it sweeps south and west toward the Atlantic Ocean. The Sahara is significant in other ways, for its inhospitable and seemingly endless wastes have long isolated the bulk of Africa from the Mediterranean coastal areas.

Three desert regions exist in Africa south of the Sahara. One lies in parts of Ethiopia, Somalia, and Kenya. The town of Garissa, located close to the equator in Kenya, averages only 10 inches of rain per year. Constant high temperatures at Garissa result in an extremely high evaporation rate; hence, the tropical desert classification.

In southern Africa there are two deserts which merge. The larger is the Kalahari of Botswana, South-West Africa, and South Africa. While part of it is steppe, annual precipitation in its western portions declines to about 5 inches. The Kalahari, with elevations ranging from 3,000 to 5,000 feet, is not as hot as the Sahara or the equatorial African deserts. To the west of the Kalahari, the Namib Desert trends north for approximately 1,000 miles in a narrow strip along the coast into Angola. Virtually rainless, the Namib is caused, in part, by the cold offshore waters of the north-flowing Benguela Current. Namib coastal stations experience cool, foggy conditions throughout the year. Some of the higher escarpment elevations in the Namib of South-West Africa receive sufficient rainfall for limited crop raising and animal husbandry.

Highland The highland type of climate is found in scattered locations in Africa but is expressed over a large area only in the high, deeply dissected plateau of Ethiopia. The wide variety of climates there cannot be classified properly other than as "undifferentiated highland." Addis Ababa, Ethiopia's capital, situated at 8,000 feet above sea level, has a warmest-month average temperature of only 65°F, as compared with July in Boston (72°) or Dallas (83°).

Vegetation and Soils

Man has greatly upset the natural vegetation of Africa. The extent of tropical rainforest is diminishing in the face of man's onslaught by ax, fire, and bulldozer. In West Africa, the forest is thought to be retreating along its northern boundary at the rate of about 1 mile per year. Savanna forms of vegetation are advancing into the forest, but the savanna in turn is being encroached upon by the steppe, and the desert invades the steppe. Man's abuse of the land through overgrazing, poor soil-control measures, and the repeated burning of forest, savanna, and steppe grasses constitutes the chief reason for the continuing deterioration of the vegetative landscape. It is also possible, however, that some desiccation is the result of a gradual change in the climate of Africa.

Africa's soils are notoriously poor. For the most part, they are low in fertility and are difficult to work. This is one of the most serious deficiencies in the African environment and, despite all efforts, will undoubtedly continue to worsen for some time to come as a result of increasing population pressures. Somewhat better soils are found in the higher eastern plateaus, in some river valleys, and in areas of steppe. Large-scale attempts to prevent soil erosion have been successful mainly in areas of European settlement. For most of Africa, however, suggestions for soil conservation make little headway when they run counter to traditional agricultural practices. Also, there is a lack of funds to carry out such projects.

Wild Game

Africa's wild game is much less abundant than is commonly assumed. Over vast areas, the large species of wild game have completely

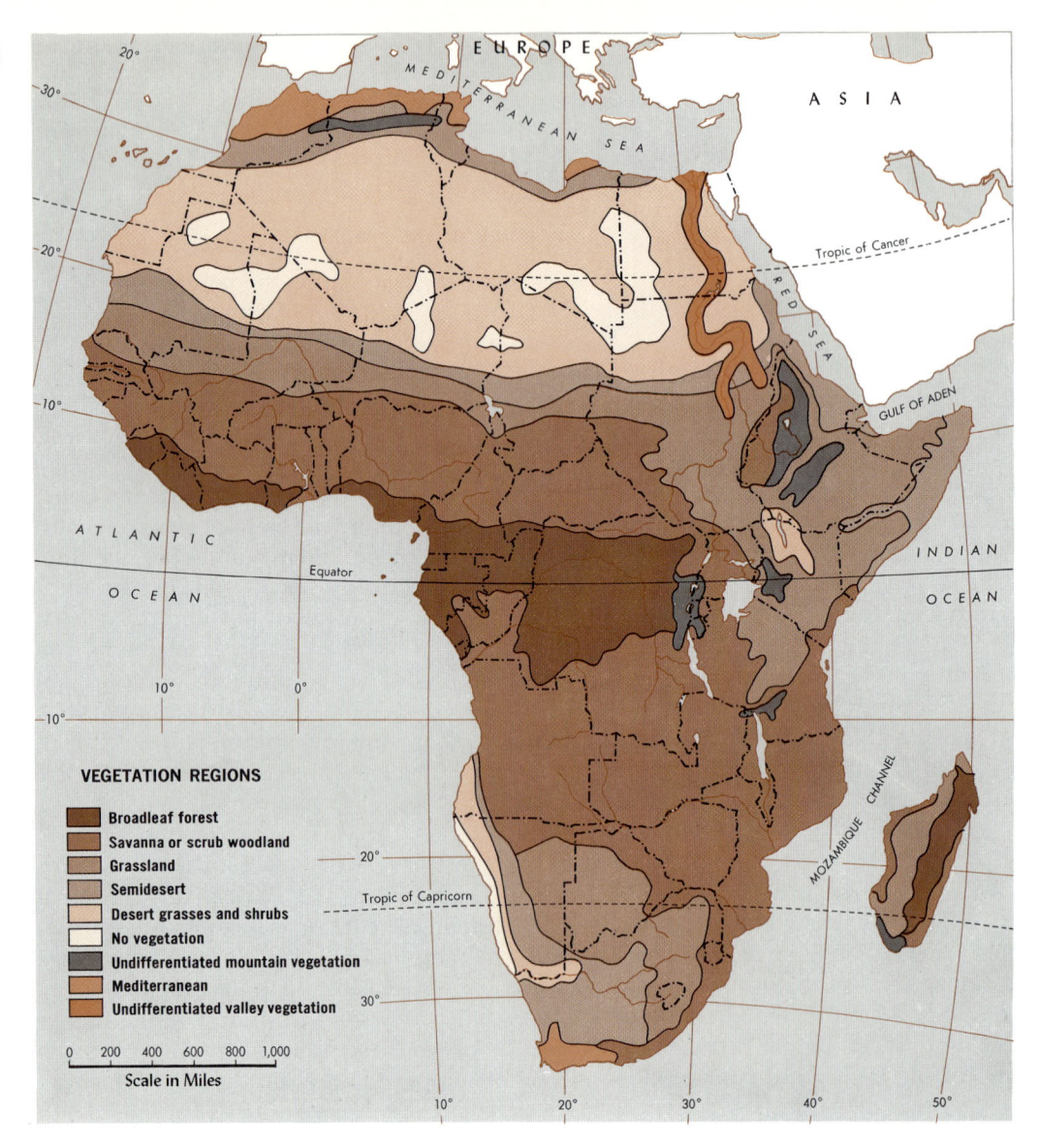

Fig. 12-5 Africa's vegetation regions are directly related to the climatic regions. Note the similarity between the two maps.

disappeared. One estimate for East Africa is that 90 percent of its game population has been eliminated since the arrival of Europeans 100 years ago. Some species, including the mountain gorilla, sable antelope, and white rhinocerous, are not far from possible extinction, while others such as the leopard, lion, and crocodile have been greatly reduced in number by hunters and illegal poachers. Most tourists now hunt with cameras, but poaching continues because of the high prices paid for skins, furs, horns, and ivory.

Africa South of the Sahara **437**

The large herbivorous species—zebra, wildebeest, buffalo, giraffe, rhinoceros, and many kinds of antelope—prefer the savanna and the steppe regions of Africa. Here their constant companions are the carnivores and scavengers. Elephants are equally at home in forest, savanna, and steppe; while crocodiles, hippopotamuses, birds, and fish share the waters.

In western Africa, man's appetite for protein has virtually eliminated big game, leaving a residue of monkeys, baboons, snakes, birds, and small animals. Nor does the Congo Basin contain much big game any longer. Game is relatively more plentiful in the east from Ethiopia to southern Africa, but only in the areas considered less desirable for human settlement. Land pressure is the central problem today in game conservation, for while a few African peoples live in harmony with game, most do not, and their increasing needs for land are in direct conflict with the spatial requirements of the animals.

Fortunately, African governments have shown a keen interest in retaining this part of their heritage and have built a creditable record in maintaining and establishing game reserves. These reserves serve a double pur-

Fig. 12–6 Major vegetation and climatic types: *(top left)* tropical rain forest, Congo (Kinshasa); *(top right)* tropical steppe, Nigeria; *(bottom left)* transitional zone from tropical savanna to Kalahari, Angola; *(bottom right)* tropical desert, South-West Africa.

pose in clarifying the boundaries of the man-game conflict over land and in concentrating game where tourist amenities can also be provided. Reserves are found in nearly all African countries, including western Africa, where it has sometimes been necessary to stock new parks with imported game. A recent problem of some game reserves is an excessive number of animals in relation to the carrying capacity of the land. Because man has upset the ecological balance, it has become necessary at times to kill or remove some of the larger animals, especially the elephant, whose increasing populations trample the land, overgraze the vegetation, and consume the food supply of smaller game varieties.

Cultural Setting

The Peopling of Africa

Over the millennia the ancestors of Africa's four basic racial stocks evolved. Three of these were indigenous: namely, Bushmanoid, Pygmoid, and Negroid. But the fourth, Caucasoid, apparently entered the northern part of the continent from Asia and Europe about ten thousand years ago. At that time, the four stocks were sparsely scattered about the continent, each occupying portions of large areas. The Bushmen ranged over the drier country of the east and south, while the ancestors of the Pygmies hunted and gathered in the tropical rainforests of central and western Africa. To the north and northeast were the Caucasoids. The Negroid group, living in the western savannas, was probably the least numerous of the four.

The Negroid and Caucasoid peoples began to accept the innovation of settled agriculture as early as 5000 B.C. This led to gradually increasing population densities and the absorption and displacement of some of the Bushmen and Pygmies, who rejected innovation and continued as hunters and gatherers. Then about the first century of the Christian Era, a new wave of Negroid and Caucasoid expansion occurred. This seems to have been facilitated by the introduction of ironworking techniques, which led to new tools and weapons, and by the importation of several food plants from Asia, notably the banana and yam. The new crops, suited to wetter climates than those previously used for agriculture, enabled the Negroid stock to expand its numbers rapidly in the rainforests of western Africa, in the Congo Basin, and in parts of eastern Africa. Again the Negroid peoples encountered Bushmen and Pygmies, absorbing some of them and forcing the remnants to retreat into still smaller areas. In eastern Africa, the Negroid stock slowly overcame the competing Caucasoid occupants, but with much admixture of the two races. Next, massive Negroid migrations swept into the moist parts of southern Africa, and these migrations continued into the last century.

By the year 1500, the racial stocks of Africa had met and mingled to produce a great diversity of peoples. Since then, there have been four significant developments in the peopling of the continent. First was the impact of the slave trade, by which African peoples spread by the millions into the Western Hemisphere. A few descendants of these slaves later "returned" to Africa, and their descendants are found principally in Liberia and Sierra Leone. Second, trans-Atlantic trade brought to Africa new crops from the Americas, including corn, cassava, sweet potatoes, peanuts, tomatoes, cocoa, and tobacco. Some of these new foods have now become staples over large areas, permitting the cultivation of additional land and increased population densities. The third important development since 1500 has been the settlement of non-Africans. Though their numbers are small, the Europeans, Arabs, Indonesians, Asians, and others have added to Africa's racial medley, sometimes mixing with each other and with the Africans. Lastly, Africa has begun to experience a population explosion comparable with that of other regions of the world. In 1650 the entire continent of Africa was estimated to have had between 50 and 100 million people; in 1900, the estimate was 120 million. Now it is over 350 million, and forecasts for the year 2000 range from 500 to 800 million persons.

Africa's People Today

Efforts to classify the 800 or more kinds of people south of the Sahara have aroused controversy for years. In view of the bewildering variety of human types and their overlapping physical and cultural traits, the following groupings are suggested as a compendium:

1. Numerically, the largest category of people is the West African Black, totaling over 110 million. The average adult male is a muscular, lively person of medium height, having a very dark skin, woolly black hair, broad nose, thick lips, and other Negroid characteristics. Toward the Sahara, considerable mixing with Caucasoids is evident in lighter features. The West African Black is primarily a cultivator of the soil, but turns to animal husbandry in the drier areas. He is considered the most progressive of the African peoples.

2. The second largest group, classified on the basis of its interrelated languages as Bantu, comprises most of the peoples south of the equator. Physically, the Bantu are quite similar to the West African Blacks, but they seem to possess more Caucasoid, Bushman, and Pygmy characteristics, which result typically in a smaller stature, lighter skin color, narrower nose, and smaller lips. There is some contention, however, that the features of the Bantu are also traceable to processes of biological adaptation to the higher, cooler plateaus of eastern and southern Africa. The Bantu total approximately 100 million and engage in both cultivation and herding.

3. Africa's Caucasoids south of the Sahara number about 35 million in Ethiopia, Somalia, eastern Africa, and along the southern margins of the Sahara in West Africa. Although they may have originally been white, mixing with Negroid stock and the effects of the tropical sun, perhaps, have given them a range of skin color from grayish-white to black. Most are brown in color and have retained the narrow noses, thin lips, and prominent chins of their Caucasoid ancestors. Long, wavy black hair is another feature, and some are quite tall. They tend to prefer herding to crop raising.

4. The Nilotic peoples of Africa comprise still another group. Numbering about 10 million, they live along the reaches of the Upper Nile in the southern Sudan and are scattered among Bantu and Caucasoid tribes in the Lake Victoria area. Nilotes are usually very dark in color, but some tend to have Caucasoid facial features. In the southern Sudan, the men are tall, elegant, and slender, averaging close to six feet in height. They engage in herding, crop cultivation, and fishing.

5. The Pygmoid remnants of Africa survive in scattered, seminomadic bands in the equatorial rainforests. As few as 50,000 of them continue as traditional hunters and gatherers, while perhaps twice that number have rather recently settled down to subsistence agriculture. Adult male Pygmies average 4 feet, 9 inches in height and weigh about 88 pounds. Their skin is a dark yellow-brown, and their large eyes protrude. As might be expected, they have acquired many Negroid physical characteristics; yet they retain certain distinctive features which merit their acceptance as a separate racial classification by most anthropologists. It is not clear why the Pygmies are such small

people. One explanation is based on inherent genetic factors. Another argument maintains that poor diet accounts for their size. A third possibility is that natural selection and evolution in the rainforests have favored small size. This theory is supported by the development in rainforest areas of dwarfed species of cattle, elephant, and hippo, among others.

6. The Bushmanoids are a few inches taller than the Pygmies but are more obviously a separate racial stock. Despite admixture, typical Bushmen have a light yellow-brown skin that wrinkles easily, "peppercorn" hair spirals, and narrow, near-Mongoloid eyes. At one time, they ranged far over Africa, but Bantu and European incursions have forced most of them into the isolated Kalahari region of western Botswana and South-West Africa, where they total about 50,000. Of this number, only a few thousand still maintain their ancient way of life.

Although the remaining true Bushmen are considered the most "primitive" people in Africa, because they do no more than hunt and gather their food, they are actually quite ingenious in sustaining their existence in an area that would soon bring death to unprepared Bantu or Europeans. The Bushmen are superb bow-and-arrow hunters, able to track game for miles on scanty evidence, while the women have an unerring instinct and memory for locating water and edible roots, berries, honey, and other food in what "advanced" peoples would regard as a trackless wilderness. Bushmen can run great distances without fatigue, see four of Jupiter's moons, and hear the rustle of a snake at a distance of 100 yards. Their dwindling numbers have been intensively studied by anthropologists.

7. Another indigenous group in Africa south of the Sahara is the Malgache. There are 5 million of these, who are now a mixed Bantu, Arab, East Indian, and Indonesian stock living in Madagascar. On the high uplands of the island, a purer Indonesian type, with typically Mongoloid features, survives. These are the slightly mixed descendants of migrating seafarers who crossed the Indian Ocean on rafts from the vicinity of Borneo centuries ago.

8. The more recent immigrant groups in Africa total less than 6 million, of whom 4 million are Europeans. Of this number, over 80 percent are now found in the Republic of South Africa. This concentration of Europeans in one country reflects its long settlement by whites, its subtropical climates, and its economic opportunities. Another half-million Europeans have settled in the nearby territories of Rhodesia, Angola, and Mozambique. Elsewhere Europeans are much fewer in number and do not wield political control; nor do most of them consider themselves permanent settlers, for they are assigned to governments, businesses, schools, and missions on a temporary basis.

A second immigrant group is the approximately 1 million people of Indian origin, about half of whom reside in South Africa and the remainder primarily in eastern Africa. The Indians are divided among themselves into Hindu, Muslim, and Christian factions. They arrived in Africa only within the last century, to provide labor for plantations and railroad construction, but have today branched out mainly into business and government service.

Arabs south of the Sahara are engaged mostly in trade and small retail businesses. Few in number, they are found principally at East African seaports from Ethiopia to Mozambique, where, along with Persians, their ancestors have lived for centuries, at times controlling the area. A more recent migration of Syrian and Lebanese Arabs, again small in number, conducts some of the retail trade in West Africa.

9. The final racial group to be noted is the result of African and European mixing, with an additional element of Malay, Asian, Arab, and Bushman. Known as "Coloureds," there

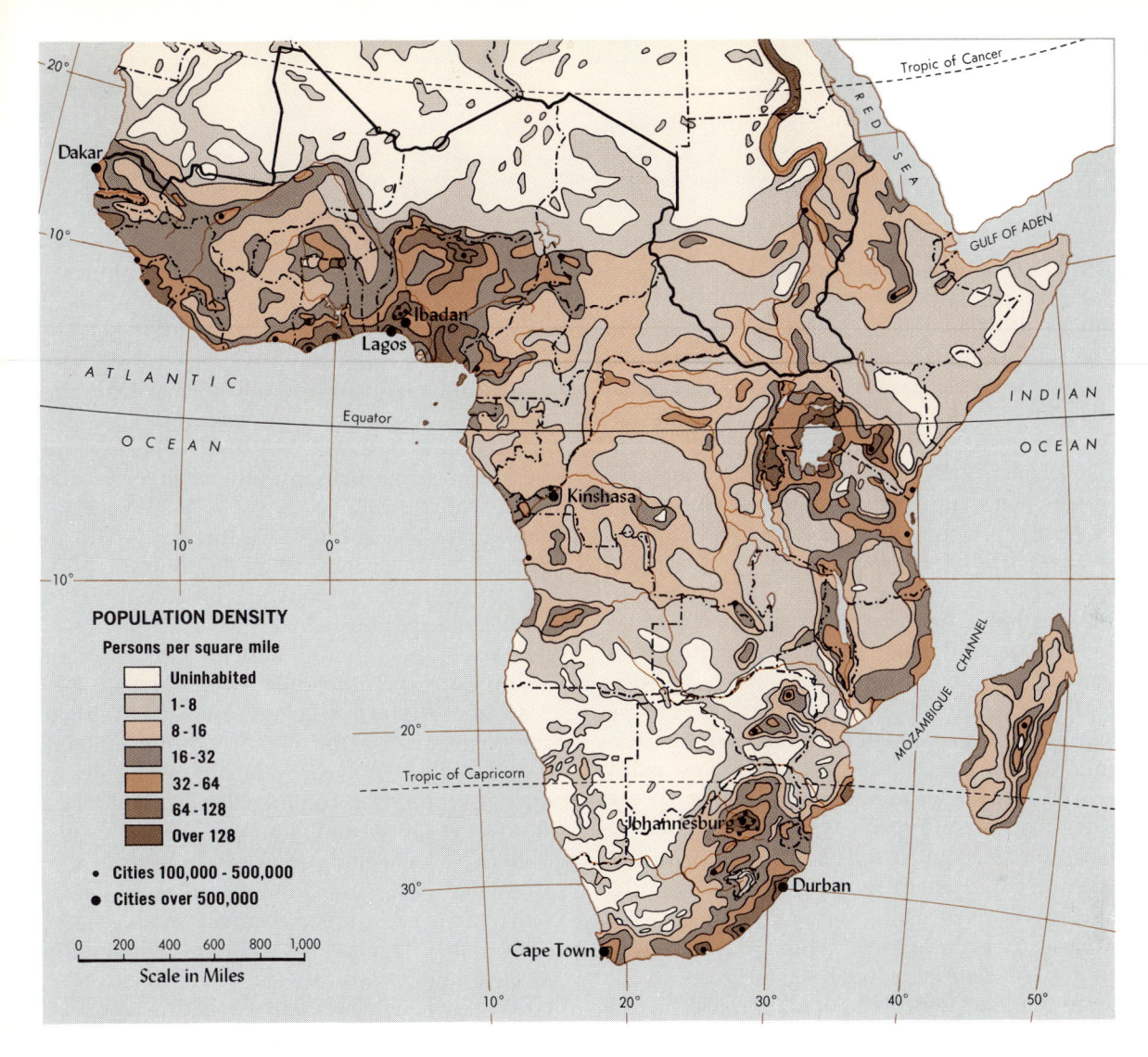

Fig. 12-7 Population varies greatly according to local conditions. In general, the continent's areas of greatest population are found in West Africa, in the lake and highland region of East Africa, and in urbanized South Africa.

are 2 million of these people in South Africa, where mixing began upon the arrival of the first Dutch colonists in 1652. Although South African law now strictly prohibits racial mixing, the process nevertheless continues. Coloureds are relatively few in number elsewhere in Africa.

Distribution of Population

South of the Sahara, Africa had an estimated 270 million people in 1971, amounting to an average density of about 34 per square mile, as compared with 205 million and 56 per square mile in the United States. Rwanda, with about

350 persons per square mile, is the most densely populated country, but the greatest in total population is Nigeria, with some 60 million inhabitants.

Sizable variations in density reflect a combination of physical and cultural influences. Over 40 percent of the people south of the Sahara live in western Africa, particularly in Nigeria, where rural densities of over 500 per square mile occur. The broad central areas of Africa, including the Congo Basin, are rather sparsely populated. Farther south, the Kalahari region averages only about 1 person per square mile. Scattered clusters of moderately heavy densities characterize the eastern regions from Ethiopia to the Natal coast of South Africa. Except for the Johannesburg industrial area, these clusters, such as the one around Lake Victoria, are the result primarily of favorable climatic and soil factors.

In an Asian sense, Africa south of the Sahara is not overpopulated. India, with one-seventh as much area, supports twice as many people. But such comparisons, however, do not recognize Africa's poor soils, its difficult climates for agriculture, the backward agricultural practices of most of its people, and the disease problems of man, crops, and animals. An American geographer, William A. Hance, noting the annual population increase of 2.4 percent and evidences of population pressure in rising malnutrition and soil erosion, concluded that 32 percent of the continent's population in 38 percent of its area is experiencing population pressure. He warns that "population pressure is a much more serious and widespread phenomenon in Africa than has generally been accepted." Significantly, the present rapid rate of growth of Africa's population seems bound to increase, thus intensifying the problems of land, food, education, health, economic development, and social and political unrest.

Religion

Because people practice their religion in various ways, it is possible only to arrive at rough estimates for Africa. South of the Sahara, the population may be divided approximately as follows: 50 percent traditional religion, 30 percent Muslim, and 20 percent Christian.

The modern world is slowly destroying the traditional religions of Africa, and little can be done to save them. Yet there is beauty and order and a rationale in these religions. What might seem "primitive" to the visitor is relevant and significant within the context of African life and in the search for philosophical truths. Most traditional religions are identified with a particular tribe; but whatever forms they take, they have in common a deep and constant meaning for the individual throughout his life.

Of the Western religions, Islam, more than Christianity, is on the march, adding millions of followers each year. Islam propagates a sense of brotherhood among all tribes and races, allows polygamy, and has had a long history in Africa. Throughout western Africa, it is advancing from its savanna base to the Atlantic coast. Senegal, Guinea, Gambia, Mali, Niger, Chad, and Nigeria are now predominantly Muslim, and Islam is spreading in eastern Africa, where it is well entrenched in Somalia, Ethiopia, and along the coast of Tanzania. It is also found among some of the Indians and Malays of South Africa.

Though Christianity has millions of believers, and is growing rapidly in a few places, it is viewed as an European religion. It imposes monogamy and interferes with tribal customs. One large Christian group is indigenous among Africans. In Ethiopia, Coptic Christianity has been practiced for over 1,600 years, and it includes about 40 percent of the population among its adherents.

The variety of religions in Africa is enhanced by several hundred thousand Hindus, perhaps 150,000 Jews, and a few Sikhs, Buddhists, and Bahai. In addition, agnosticism and atheism are growing among Africans who have rejected their tribal religions as well as Islam and Christianity.

Customs and Social Change

Many forces are transforming the people of Africa. The rural village, with its close-knit ties among the members of extended families and clans, its sense of stability, its communal properties and communal festivities, has been disrupted almost everywhere in Africa. Some of the change was forced by colonial administrators for various economic and political reasons. Missionaries have also brought change. Economic opportunities for young men in mines, plantations, and new cities have drawn the youth away, and sometimes they never return to the village. Schools have disrupted the old society, and more recently television, radio, magazines, and newspapers have brought the world to the village. The use of money and the growing of cash crops for world markets have further spelled the ultimate doom of the old ways.

On the other hand, it is surprising to note how many customs remain and how resistant some of the people are to change. Men still usually pay a "bride price." Scarring and tatooing of the body and the filing of teeth, though practiced less, continue among the more isolated groups. Tribal government by a chief and a council of elders is widely maintained, even if half of the men are absent because of work elsewhere.

Africa is a continent of social contradictions. In some ways it appears to be changing rapidly; in others, not at all. It is a land of human contrasts. One may observe tribal dancing through binoculars while atop a 52-story skyscraper in Johannesburg. The traveler stopping at Nairobi, having just finished a delicious Western-style meal, may leave his luxury hotel suite and, within minutes of emerging from the traffic congestion of the city, witness a traditional Masai feast based on cattle milk and blood. How much longer will these customs survive? Will the young people continue them? Indeed, the fact that persons less than eighteen years of age comprise about one-half of the population of Africa almost guarantees widespread change ahead through the unremitting processes of cultural diffusion.

Urbanization

Africa remains the least urbanized among the continents, with only 15 percent of its people in towns or cities, while the rural/urban ratio varies from over 50 percent in South Africa to less than 5 percent in several countries. However, these figures obscure the rapid growth of African cities, particularly since World War II.

Millions of Africans have been attracted to cities and towns by the promise of steady work and the "bright lights." A few have achieved a university education and have become doctors, lawyers, engineers, teachers, businessmen, and politicians. Most, however, came to fill lesser positions, while some "drifted in" and found only sporadic employment opportunities. In the town, the African loosens the bonds that tie him to his tribal village, although he may join a "club" of his tribal kinsmen for social gatherings. Also, in the town, he is subject to new laws, and he works with different kinds of people. After many years, he begins to become "detribalized," perhaps to the extent of marrying a woman of another tribe, speaking a European language, and living in Western fashion insofar as his income permits. His children, in turn, will know only the town and may feel quite out of place in a traditional village setting.

Urbanization in Africa has created several serious problems. In the villages, the departure of able-bodied men to work in the towns or mines has caused distress in maintaining adequate food cultivation. Further, men who have returned to the village are more sophisticated, restless, and less likely to accept the authority of the chief and the elders. In the towns, satisfactory housing for the flood of job seekers cannot be built fast enough, which results in a continuous battle to replace squalid, crowded shantytowns. Another problem is that of the idle and unemployed, especially young men and women who may become discouraged and turn to radical social or political movements. A third kind of problem arises when tribal groups antagonistic to each other in rural areas must live closely together within the confines of an urban area.

Diet and Disease

The amount of suffering endured in Africa as a result of inadequate diets and infectious diseases is incalculable. Many authorities believe that poor health is the continent's most serious problem. Moreover, there are no easy solutions to the problem, for the ratio of medical doctors to population is about 1 to 30,000. Although modern hospitals and clinics are now found in the larger towns and cities, the rural African is often unable to obtain treatment, because he either cannot afford it, is too sick to make a long journey, or prefers to rely on the traditional services of spirit mediums and witch doctors. Ill health, and its

Fig. 12-8 In this village in Uganda, the women are preparing part of a meal. Food storage bins appear in the background. The majority of sub-Saharan peoples still live in similar villages. (Courtesy of Uganda Department of Information)

accompanying waste of human resources, constitutes one of the greatest handicaps to economic development in Africa.

The list of major common diseases is long. Malaria and tuberculosis are the principal causes of death today; but yellow fever, plague, relapsing fever, river blindness, smallpox, typhoid, yaws, cholera, leprosy, venereal diseases, elephantiasis, poliomyelitis, sleeping sickness, meningitis, dysentery, and a large variety of intestinal and parasitic worms (including "bilharzia") also bring death or disablement. To this list should be added the diseases of malnutrition: pellagra, beri-beri, rickets, and "kwashiorkor" (protein malnutrition of young children). These maladies could be avoided by fuller and better-balanced diets, but most Africans must build their diets around foods known as "starchy staples," such as corn, cassava, yams, sweet potatoes, rice, millets, and green bananas (plantains). Hence, their diets are deficient in proteins, vitamins, and minerals. The minds of many young children are permanently stunted by malnutrition occurring up to about the age of seven, for these first seven years are the time when the brain should develop to approximately 90 percent of its total adult size. Malnutrition also causes millions to "drag" through life drained of their vitality and their ability to work or to learn effectively, as well as lowering the body's resistance to infectious diseases.

It is little wonder, then, that Africa has the highest death rate of any continent: approximately 25 per 1,000 people per year. Infant mortality claims one-fourth of all babies before their first birthday, and the average life expectancy after that is only 40 years.

Improved and expanded medical care is obviously one of Africa's crucial needs. Notable progress has already been made, but much remains to be done in educating the African toward better health practices and in bringing medical services closer to the population and at a reasonable cost. It would seem wise, however, to match future reductions in the death rate with comparable reductions in the birthrate, which is nearly triple that of the United States. Otherwise, excessive population growth will nullify or even reverse all attempts at medical, social, and economic betterment.

Historical Development

The earliest African empires outside Egypt originated about 1000 B.C. along the Nile in what is now the northern Sudan and in northern Ethiopia. These were followed centuries later by empires that flourished in many other parts of Africa, notably in the savannas of West Africa, the lower Congo, Uganda, along the eastern coast, and in the south-central portion. It is beyond the scope of this chapter to discuss the glories of such empires or kingdoms as Ghana, Mali, Ashanti, Benin, Kilwa, and Kongo, but today they are increasingly significant for Africans who wish to gain a sense of pride and achievement concerning their past.

The empires of Africa were not able to match the sudden growth of power in Europe that began about 1500. Eventually, Europe's military and cultural momentum engulfed and virtually destroyed the African empires, except for the empire-state of Ethiopia. Among the once-powerful kingdoms and city-states, only a few weak and scattered remnants may still be found at the present time. These maintain some of their ancient traditions and a lingering authority, but have become subservient to national governments.

The year 1498 marked a highly significant turning point in African affairs. It was then that Europeans first sailed around the Cape of Good Hope to pioneer a sea route from Europe to India. This occurred just six years after Columbus had opened up a new world whose future plantations were to demand the labor of millions of African slaves. The combination of new lands, new markets, new trade routes, and the availability of strong bodies introduced an ugly chapter in Africa's history. For well over 300 years, Europeans and Americans purchased slaves along the continent's shores. The shipment of human beings across the Atlantic reached a peak of more than 100,000 per year in the early 1800s before economic problems and a growing clamor from Abolitionists finally brought the trade to a close.

During the centuries of the slave trade, Europeans built numerous forts and castles along the coasts, particularly in the west, from which they dealt with their African suppliers. Only in the area of Cape Town did whites venture very far inland to establish settlements. Thus, the interior of Africa was essentially an unknown, forbidding region, a "Dark Continent" to Europeans as late as 1800. What then occurred in the nineteenth century was an extraordinary period of exploration as Europeans, Arabs, and some Americans rushed in to prowl and penetrate the "Dark Continent." Accounts of their travels made their names famous: Stanley, Livingstone, Park, Burton, Speke, Baker, Caillie, and many others. Although their writings were at times erroneous, exaggerated, and biased, they dispelled some of the fallacies about Africa and solved a number of geographic puzzles, such as the age-old question of the sources of the Nile.

Again in Africa a combination of factors led to a new historic period. The information revealed by the explorers, the end of the slave trade, advances in medicine to protect settlers, and European imperial rivalries combined to bring about very rapid colonization. Between 1850 and 1914, Europe converted nearly all of Africa into a vast patchwork of ill-defined and often rebellious holdings. Each of the colonial powers—Britain, France, Belgium, Portugal, Germany, Italy, and Spain—pursued widely disparate policies after "pacification" was attained. Though they spoke loftily of bringing peace and development to Africa and of the "white man's burden," the stubborn fact that essentially characterized the colonial period was economic benefits for Europe and for European settlers in Africa.

Today, white political control in Africa has decreased to ten possessions, mostly in the south. Since 1957, thirty-two new African-ruled nations have reached independence. In retrospect, the colonial period was relatively short, for effective control lasted approximately seventy-five years. Yet, today it is amazing to find how deeply the colonial influence of Europe has permeated African society in terms of imposed boundaries, political institutions, education, language, religion, art, music, dress, food, and athletics. The visitor to Kenya, for example, will find a considerable layer of British institutions and customs adopted and maintained by Africans since Britain's departure in 1963. In neighboring Somalia, there is a residue of Italian colonial influence. In Tanzania, one encounters a triple cultural overlay resulting from Arab, German, and British rule. In the Ivory Coast, an ambitious young man in government service may emulate the former French colonial way of life to the point of speaking French, eating French-style food, possibly marrying a French girl, and wearing French-style clothes. Only a few miles away but across the border in Ghana, another young man eager to rise in business or government may be trying to be very British. If he lives in coastal Ghana, however, he will never be far from a slave fort to remind him of an earlier part of his heritage; and when he travels to the Ashanti region of Ghana, he may observe the continuing remnants of a great empire in which his ancestors took part centuries ago.

The characterizations given above represent only a small urbanized minority. Most Africans prefer to maintain a semitraditional village existence less disturbed by the colonial impact. Persistent Westernizing currents, however,

TABLE 12-1
GNP AND ANNUAL GROWTH RATE

	Estimated Gross National Product per Capita, 1969	Estimated Average Annual Growth Rates, 1961–1968		
		Gross National Product, %	Population, %	Net Gross National Product per Capita, %
Nigeria	$ 70	2.1	2.4	−0.3
Ethiopia	70	4.6	2.0	2.6
South Africa	650	6.0	2.3	3.7
Congo (Kinshasa)	90	1.8	2.1	−0.3
Tanzania	80	3.7	2.5	1.9
Kenya	130	4.3	2.9	1.4
Ghana	170	2.0	2.7	−0.7
Uganda	110	3.6	2.5	1.1
Mozambique	220	4.9	1.3	3.6
Malagasy	100	2.2	2.4	−0.2
USA	3,980	4.8	1.4	3.4

SOURCE: *World Bank Atlas*, 1970.

seemingly cannot be stopped, even by those in power in some African countries, who issue decrees banning the acceptance of Western fashions in dress, literature, and music. Hence, African society faces the dilemma of harmonizing its ancient heritage and traditions with the onrush of Western materialism and cultural forms. Resolving this problem is not an easy task.

The Economy

Basic Problems

By 1970, it was evident that a decade of independence for most countries south of the Sahara had failed to generate the economic growth so optimistically forecast in 1960. As measured in terms of gross national product (GNP), the overall growth rate was approximately 4 percent per year, a figure barely sufficient to offset the annual population growth rate of about 2.5 percent. This difference of 1.5 percent between the two rates meant that on a per capita basis the average share of GNP in Africa moved from perhaps $90 to $105 during the decade.[3] Since this growth occurred primarily in cities and mining areas, it meant that life for the 85 percent of Africa's people who lived in the rural areas improved very little, in an economic sense, during the 1960s. Table 12-1 indicates estimated annual growth rates between 1961 and 1968 for the ten largest Africa countries, ranked in order of total population, and for the United States. It shows that estimated population growth exceeded economic growth in 4 countries (Congo, Nigeria, Ghana, and Malagasy), resulting in per capita income declines. At the same time, white-ruled South Africa led both in per capita incomes and in the growth rate of per capita incomes.

Africa's economic development, though,

---

[3]Compared with an advance in the United States of approximately $2,800 to $4,500 over the same period. The figures do not take into account inflationary increases and differences among countries in price structures.

cannot be measured solely in terms of visible percentages. The 1960s marked a beginning for many young and inexperienced African-ruled governments, whose primary concerns were political rather than economic. Political problems impeded the flow of private investment capital and the proper use of the relatively small amounts of foreign aid and technical assistance that were offered by the rich nations. Both Nigeria and the Congo (Kinshasa)[4] suffered severely from armed secession movements, and in Ghana large sums of money were misspent for grandiose semipolitical schemes. The new African governments committed many mistakes, both political and economic, during the 1960s; yet it is remarkable how much progress actually was made.

The acceleration of African economic development in the 1970s faces many basic problems. Rampant human population growth is probably the greatest one. Malnutrition, disease, poor housing, and an illiteracy rate of about 75 percent constitute other very serious handicaps. The African traditions involving social customs, systems of land tenure, and a high premium on leisure time are incompatible with modern agricultural and industrial demands.

One of the central problems of African development is the continent's role as a producer of cheap unfinished primary products for the markets of the developed nations. Most African countries must rely on the export of one or two primary agricultural products for their major source of income. Unfortunately, though, prices for these products in relation to manufactured goods have generally declined on world markets since the mid-1950s, owing to oversupply and competition. Thus, Africans must now produce from 25 to 50 percent more coffee, cocoa, palm oil, cotton, and other primary products in order to buy the same quantity of manufactured goods from Europe, Japan, or the United States that they did several years ago. This decline in world market prices has been an important factor in retarding expected economic growth.

The remedy for Africa's present reliance on exports for income seems clear, but its attainment will be difficult. What is apparently needed is a reorientation of the economy toward domestic production for domestic markets while maintaining traditional exports. The problem of creating domestic markets that would buy locally produced agricultural and industrial goods is essentially twofold. First, the mass of the population which is engaged in agriculture must acquire a much greater purchasing power. This will require no less than an agricultural revolution. Second, most African countries have populations of less than 5 million people and, consequently, offer limited internal markets to potential investors, whose costs of output per unit would be excessive in competition with imported goods from the industrialized nations. The apparent solution to the dilemma lies in the organization of regional economic groupings of countries which, by the greater size of their markets, would stimulate investment, whether private or public, in domestic agricultural and industrial production.

Several regional groupings have already appeared, such as the East African Community (Kenya, Uganda, and Tanzania), but their success has been limited by problems inherent in their organizations. Two principal difficulties are: (1) poor internal circulation of goods, money, and people because of inadequate means of transportation and government restrictions, and (2) political differences. If the regional economic blocs are to be truly effective, each member country must yield a part of its sovereignty to the group. It follows, therefore, that economic unity demands considerable political unity. As yet, however, only two small countries in Africa have chosen to sacrifice a large measure of their newly won

---

[4]Renamed the Republic of Zaire in October 1971.

independence for the goal of integration with another country; these were British Somaliland (Somalia) and Zanzibar (Tanzania). No other countries within the present regional economic blocs have given up their political identity. In fact, the various groupings are rife with dissension among the partners and are constantly being threatened with dissolution.

Given this state of affairs, what other major steps might Africa reasonably take to better itself economically? One is to process more of the primary products at their source, thus increasing their value and providing new employment. Another is the training and encouragement of Africans to go into business at all levels. A third is sound government fiscal policy, planning, and reform where necessary. Finally, Africa must continue to seek more investment capital through a combination of domestic savings, foreign aid, low-interest loans, and tourism.

Agricultural Problems

Increasingly, economists have taken the view that if real progress is to be made in most of Africa, it must start "down on the farm," where 85 percent of the people live. Agriculture, however, is more than just an occupation for the African; it is a way of life and it does not yield to rapid change. The problems that beset agricultural progress fall into two general categories: (1) those of the physical environment, and (2) those arising from the nature of African society. With regard to the physical problems, technology and capital investment should ultimately be able to overcome these. Technology already exists by which much can be done to improve the heavy, leached tropical soils. The problem of water supply, so critical in vast areas of the savanna and steppe having severe dry seasons, could be ameliorated by farm ponds, reservoirs, and deep wells. Indeed, a striking feature of Africa south of the Sahara is the widespread lack of the practice of irrigation and the conservation of surplus water from the rainy season.

Another problem of the physical environment, pests and diseases, may not be overcome so easily. Plant diseases destroy ground crops and tree crops; and rodents, birds, monkeys, worms, and rapacious, destructive insects, including locusts, have ruined many a future harvest. Domestic animals are subject to a host of diseases, with cattle being particularly susceptible to "nagana," or sleeping sickness. Sleeping sickness deserves special attention in reference to Africa. It is spread by the tsetse fly, whose bite transmits parasites known as trypanosomes from wild game, which is immune, to people and cattle. Even where such game has been eliminated, the tsetse carries the parasites from the bloodstreams of infected men or cattle to new victims. Unless treated with drugs, the disease eventually brings death through destruction of the nervous system. Over half of Africa south of the Sahara (more than 4 million square miles) is infested by this fly; consequently, vast areas remain only partially occupied by man and useless for cattle raising. A number of methods are available to control the tsetse fly, but these are cumbersome and expensive. Two alternatives having some promise are the widespread use of preventive inoculations for cattle and the development of breeds of cattle resistant to the parasite.

On the human side, agricultural progress is hindered by the conservative economic outlook of rural African society. Some groups refuse to accept any new ideas that will conflict with their traditional customs; whereas a few will readily accept a new method, tool, or a better kind of seed. The majority, however, change only when convinced that they will thereby benefit, provided that their way of life, whose structure is familiar to them and essential for their survival, is not impaired. Much of the rhythm of life is built around the timing of agricultural activities in various rituals and festivals. For most rural Africans, the old ways are relatively secure and comfortable. They do, after all, usually provide enough food; and any needs for nonsubsistence manufactured goods and services can be met by raising a small surplus for sale. Should the average farmer want to improve his techniques radically, he has little money for machinery; besides, his neighbors may ostracize him for his trouble. Not only that, he may be further discouraged from producing surpluses on a large scale because he has no way to take them over a rutted road to a distant market, which generally pays him rather poorly for his efforts.

Africa is also littered with the ruins of government agricultural projects that failed to recognize physical and human limitations. Both the colonial and the new African governments have met defeat in most of their schemes. Those few that succeeded were based on the right combination of advance planning, land-tenure reform, fair crop prices, sufficient capital for equipment, a favorable physical environment, the training of the farmers, adequate transportation facilities, and perhaps most importantly, the psychological readiness of the people involved to cooperate and to prosper.

Agricultural Systems

Rural Africans may be differentiated into five major groups based upon their primary agricultural systems or activities. The first group, which collects its food from the environment, consists of fishermen, hunters, and gatherers. Most African fishermen, whether toiling along the coasts or on inland waters, sell a part of their catch; whereas the principal group of hunters and gatherers, the Bushmen, live a subsistence life built around wild game and wild roots, fruits, nuts, and berries. Altogether, the members of this group comprise less than 1 percent of the rural population.

The second major group, numbering about 5 percent of the rural population, devotes itself to pasturing cattle, sheep, goats, and in the north, camels. Most pastoralists believe that their wealth depends on the numbers of livestock they possess, particularly cattle, and their goal is quantity instead of quality. Actually, pastoralists are quite rational to insist on quantity, for in their nomadic ecosystem of periodic droughts, poor grasses, cattle epidemics, occasional cattle raids, and intertribal feuds over grazing rights and water, quantity is the key to their survival. These pastoralists utilize the drier, tsetse-free parts of Africa immediately south of the Sahara and large areas of eastern and southern Africa.

The modern pastoralist, the ones the tourists find so picturesque, now faces problems arising from the growth of his population. The increasing numbers, both human and livestock, in this sphere require more and more land for grazing; but this is in conflict with other agricultural demands for land and the setting aside of wild-game reserves. Equally serious is the problem of overstocking, which is causing steady deterioration of the carrying capacity of the ranges. Attempts to induce the pastoralists to sell their livestock on a commercial basis have been notably successful in some areas and a dismal failure in others. Certain groups, such as the Masai, have been extremely resistant to change. Yet it would seem that they must surely change if they hope to alleviate their worsening plight in regard to soil erosion and overstocking, and to help feed Africa's expanding protein-hungry population. Fortunately, there are signs that the younger generation of pastoralists will be more amenable in adapting to new and changing circumstances than were their predecessors.

Cultivators of the soil constitute the third major group engaged in a form of agricultural activity south of the Sahara. Occupying the warmer, wetter, tsetse-infested regions, particularly in West Africa and the Congo Basin, they comprise about one-half of Africa's rural population. Cultivators specialize in growing one or two basic food crops, choosing those which best suit their needs and local soil and climatic conditions. These basic food crops include yams, cassava, sorghums, millets, corn, plantains, bananas, and sweet potatoes. Cultivators also raise a variety of less starchy vegetables, fruits, spices, sugarcane, peanuts, and tobacco. Because the tsetse fly precludes the keeping of cattle, goats and chickens are prized for their meat. The lack of cattle also means that cultivators cannot plow their land; consequently, the hoe is their principal work tool, and they are unable to farm much more than an acre per adult at one time.

Some of the cultivators grow just enough food for their own subsistence, but the majority produce a small food surplus for sale at local markets and/or a cash crop destined mainly for overseas export. Although the financial returns are not great, they enable the farmer to pay his children's school fees, enjoy a local trip, or purchase such consumer items as radios, lamps, bicycles, clothing, utensils, or perhaps a galvanized iron roof and kinds of food he cannot grow himself. In a few very favorable areas, cultivators have concentrated their energies on cash crops and display scant interest in growing food for themselves. Some of them have prospered nicely; as testimony of their success, their large homes and automobiles are especially evident in the cocoa region of Ghana and the coffee districts of southern Uganda.

Since cultivators must contend with soils of low natural fertility, they have devised methods of crop rotation that permit the land to rest in order to regain its fertility over a period of fallow years. Where soils are poor and population is sparse, "shifting cultivation" is usually the choice. The essence of the system is periodic relocation of the village and its fields. Typically, village farming exhausts the soil at a location after anywhere from 2 to 8 years of occupancy. Declining crop yields signal the need to move to a new area, where the men clear and burn off most of the natural vegetation. Ashes from the new burn are spread over the soil to enhance its fertility. Crops are then planted, and a new village is built. The destruction of the natural vegetation cover for planting exposes the soil to heavy rains and sunlight, which, along with the demands of the crops on the soil, quickly leaches it of its nutrients. Eventually, rather sooner than later, it is time for the village people to move on and repeat the process.

The other method of agriculture practiced by cultivators is known as "bush fallow," and in Africa today it is more prevalent than shifting cultivation. In this method, the village remains stationary while the surrounding lands are rotated periodically between crops and fallow fields in order to restore soil fertility. Bush fallow has the advantage of permitting the cultivator to build a larger, more permanent home and to raise tree crops, but it requires better soils than those of shifting cultivation, since they are used for crops more frequently. Population densities are greater in bush-fallow areas, which may be advantageous with regard to shorter distances to markets, towns, and schools.

There appears to be, however, an optimum density under the bush-fallow method for, as population increases, the cultivators must plant their land more often. The best (or worst) example of this is found in eastern Nigeria, where rural population pressures of well over 200 people per square mile have forced cultivators into an intensive use of their land that does not allow the fields sufficient time to regain their fertility. This destructive but necessary practice brings about soil erosion, reduced yields, and malnutrition and compels the emigration of some of the region's people.

The fourth major group of rural Africans

to be defined according to agricultural systems centers its activities around a form of "mixed farming" that integrates cattle and cultivation. Usually, a small number of cattle are kept by the farmer or the communal village. These cattle are an important source of protein in the diet and can be utilized for plowing. In addition, their manure is valuable in restoring fertility to the fields. The cattle are grazed on whatever fields lie fallow and in pastures that are generally not far from the village. Otherwise, mixed farming resembles the bush-fallow method of cultivation, although in some mixed-farm societies cattle raising takes precedence over crop raising.

The number of Africans engaged in mixed farming is slightly less than those who are purely cultivators. Primarily, it is the absence of the tsetse fly that permits mixed farming; hence the practice is best developed in areas (1) too dry for the existence of the tsetse fly during most of the year (as in parts of West Africa), (2) too cool for the tsetse because of higher altitudes (parts of East Africa) or higher latitude (parts of southern Africa), or (3) too densely inhabited for the fly to exist.

The fifth major system of human activity in agriculture includes both Africans and non-Africans who work for wages on large, modern plantations in the equatorial regions or on European-owned farms in the southern portions of Africa. The number of such workers is relatively small, probably less than 2 million, but their contribution of cash crops to world markets is of major importance as a source of income for themselves and their countries.

African Export Crops

The production of cash crops for export from Africa is primarily a development of the twentieth century, with a particularly rapid increase noted since World War II. Agricultural exports from south of the Sahara are currently valued at more than 3 billion dollars per year and represent about 40 percent of the area's total exports. For the great majority of African countries, the sale of crops to Europe and the United States constitutes their chief source of foreign income.

Among the export crops, raw coffee is today the most valuable, averaging about 600 million dollars in sales per year. Next are cocoa (or chocolate), with 450 million dollars per year, and peanut products, with 350 million dollars. Several other products whose annual export value is around the level of 200 million dollars are cotton, fruit, palm-oil products, tobacco, wool, and sugar. Further down the list are tea, rubber, sisal, pyrethrum (an insecticide), copra, corn, cloves, wine, and hides and skins. Timber exports, while not agricultural, are also valued at about 200 million dollars per year.

Africa's share of world coffee production rose from 6 percent in 1940 to 30 percent in 1970. In fact, because Brazil consumes a substantial amount of its own coffee production, Africa accounted for about 40 percent of the coffee moving in world trade in 1970. With the United States as the major purchaser, coffee has proved to be a modest financial blessing for millions of African cultivators and the owners of producing plantations. Coffee is the chief export of ten African countries and is very important in the economies of six others. The leading African coffee producer is the Ivory Coast, which usually ranks third in world production. Depending on the particular year's weather, the next most important coffee producers are Angola, Uganda, Ethiopia, Cameroun, Congo (Kinshasa), Kenya, Tanzania, and Malagasy.

For many years, cocoa trees have brought relatively high incomes to the cultivators of the southern forest zone of Ghana. Ghana usually produces 30 to 40 percent of the total world cocoa output, to lead all other nations.

Other African countries combine to raise the total African share to over 70 percent of world production. Next after Ghana are Nigeria, Ivory Coast, and Cameroun. Most of the output in each country is derived from individual cultivators rather than from large estates.

Africa's third most valuable export commodity consists of peanuts, peanut oil, and peanut cake. This is almost exclusively a cultivator's crop, a part of which is processed before overseas shipment. Peanuts are grown in every African country, but Nigeria holds first place in production in Africa and is the world's leader in peanut exports. In the drier savannas of northern Nigeria, peanuts are a staple of the diet, but cultivators also grow huge surpluses as a cash crop. Senegal places second in peanut production in Africa as well as in world exports. For both countries, peanuts are usually the single most valuable agricultural item sent abroad.

African Minerals

The export of minerals from south of the Sahara has a greater value today than does the export of crops, and the gap between the two is widening. Although both have increased in recent years, mineral production, now valued at more than 4 billion dollars annually, has moved ahead to capture about 55 percent of the total export value, as compared with 40 percent for crops. The general prospect for rising mineral output is excellent, owing to world demands, the availability of capital for mining ventures, and the discovery of new deposits in areas being intensively surveyed for the first time.

For a few African countries, such as Zambia, Gabon, and Liberia, mineral exploitation provides a way by which the slower, more laborious process of economic growth through agriculture may be bypassed. Mining ventures usually require the investment of huge sums of capital in equipment, processing plants, railroads, power plants, and port facilities, as well as housing, schools, and hospitals for the workers. These investments have a three-way payoff. First, they provide training and employment for thousands of workers, whose incomes then have a multiplier effect on the country's economy. Second, African governments derive comparatively large revenues from the mining operations in the form of taxes, export duties, and, in some cases, from the profits of full or part ownership. The third kind of payoff results from the mines' infrastructure; that is, the new port facilities, railroads, and power sources built to serve the mines commonly have a side effect in stimulating the production and exchange of other items in the area. Africa is rich in many minerals, and its reserves now appear to be greater than those of any other continent.

Gold Among Africa's minerals, gold has long held the top rank in terms of export value. In 1969, the countries south of the Sahara produced a total of almost 33 million ounces of gold worth 1.1 billion dollars; this amounted to nearly twice the value of the leading agricultural export, coffee. Moreover, Africa dominated world gold output by mining 70 percent of the year's new supply. Much of the gold mined was moved to vaults in the United States and Europe for use as a vital reserve to back national currencies. This fixed use of gold is being challenged, however, by rapidly increasing industrial consumption and by new international monetary systems not primarily dependent upon gold.

Many African countries mine gold. The Republic of South Africa is the world leader, with 95 percent of Africa's output and 67 percent of the world's total output. Ghana occupies second place in Africa; however, its gold production, though respectable, is but a small fraction (1/43) of South Africa's. South Africa's major deposits of gold were created hundreds of millions of years ago when deposition of gold-bearing materials took place on the shores of inland lakes. Forming thin reefs, the gold-bearing formations were compressed, tilted, and buried. Subsequent erosion revealed a part of the quartzitic reefs, and this led to their recognition in 1886 and a wild gold rush to the Witwatersrand district of South Africa. Since then, about 22 billion dollars' worth of gold has been mined in South Africa, in and near the great metropolitan area of Johannesburg. This mining area extends in a general east-west axis along the main reefs for over 100 miles.

South Africa's 47 working gold mines are by far the deepest in the world; deeper, in fact, than many oil wells. A new mine west of Johannesburg has shafts in operation at a depth of 13,500 feet, and another mine is being readied at the incredible depth of 18,000 feet (more than 3 miles). In the former, natural rock temperatures reach 130°F, thereby requiring a massive air-conditioning system for the miners working below. What might seem even more incredible is the actual gold content of the ore: it is so low that from 2 to 5 tons of rock must be hoisted to the surface for processing in order to obtain 1 ounce of gold.

The gold industry of South Africa employs 40,000 Europeans and 368,000 Africans, about 60 percent of whom are imported from nearby countries as contract labor. For their day's work, Africans receive from 2 to 3 dollars, plus food and lodging. By saving their money over a year's contract, they are able to return to their villages laden with new goods and with cash for the purchase of land, cattle, or brides. After a few years, some return to the mines for another contract.

With the U.S. price of gold stationary since 1933 at 35 dollars per ounce, the giant mining corporations of South Africa, because of rising production costs, have been forced to close some mines, increase productivity in others, and diversify their investments. Unless the price of gold is raised, it is expected that South African production will begin to trend downward by about 1975. A by-product of gold mining that has eased the financial problems of the mining corporations is uranium.

Petroleum Production of petroleum from south of the Sahara is experiencing a very rapid increase and may soon be as valuable an export as gold. A recent series of discoveries along the western coast and in adjacent coastal waters confirms the existence of rich oil deposits in an offshore sedimentary belt stretching 2,000 miles from the Ivory Coast to Angola. Offshore drilling rigs are also scattered along the east coast from South Africa to Ethiopia, but as of 1971 there had been no oil strikes in this area.

Nigeria is the leading producer of oil in Sub-Saharan Africa, followed by Angola and Gabon. The Nigerian fields are located mainly in the delta of the Niger River and offshore. If production reaches the goal expected, Nigeria will be among the world's top ten exporters of petroleum, receiving benefits of more than 1/2 billion dollars annually by 1973.

Copper A third large mineral export of Africa south of the Sahara, worth about 900 million dollars per year, is copper. Zambia ranks third in world production (12 percent), after the United States and the Soviet Union, but is first in copper exports. In sixth place in world production is the Congo (Kinshasa). Along with South Africa, South-West Africa, and Uganda, these Sub-Saharan nations account for one-quarter of the world's output of copper.

The principal mining area is the "Copper Belt" of northern Zambia and the adjoining Katanga region of the southeastern Congo. The ores are rich, averaging 7 percent, and the mines are both underground and open-pit. Smelters and refineries process the ores before shipment by rail to distant ports on the east and west coasts of Africa. Very sizable deposits of copper have recently been discovered in Rhodesia, South Africa, and Botswana, but these await large-scale exploitation.

Diamonds Of somewhat lesser value as an export are diamonds (400 million dollars); yet Africa produces approximately 90 percent of the world's supply of both the industrial and gem types. Most industrial diamonds originate in eleven African countries, with the Congo (Kinshasa) mining about 70 percent of the world's total supply by weight. The value of gem-type diamonds is considerably greater than that of industrial-type, and South Africa and South-West Africa produce about one-half of the world's gem-quality output. These are found mainly along the Atlantic coast near the mouth of the Orange River, where dredges comb the sands of beaches and offshore areas. The remaining African producers of gem diamonds in 1971, together accounting for 40 percent of the world total, were Angola, Congo (Kinshasa), Sierra Leone, Liberia, Tanzania, Central African Republic, and Ghana.

Transportation, Power, and Industry

The transportation, power, and industrial sectors of the economy, with one exception, are not well developed in Sub-Saharan Africa. The one exception is the Republic of South Africa. This country, with its mineral and agricultural wealth, steel mills, and thousands of factories producing everything from automobiles to zippers has a booming economy whose magnitude far exceeds any other in Africa. Furthermore, South Africa has the best road and rail system and the largest, best-equipped ports in Africa. South Africans are fond of pointing out that they constitute only 6 percent of the population of the *entire* African continent but produce over 90 percent of the continent's steel, generate 57 percent of its electricity, have 50 percent of the telephones, load 60 percent of the railway freight traffic, and possess 40 percent of the motor vehicles.

These figures, while impressive, should not obscure the significant economic progress that has been made in the other countries south of the Sahara. Despite many obstacles, industry there is growing, with the emphasis being placed on textile mills, oil refineries, chemicals, food processing, construction, and such consumer items as shoes, soft drinks and beer, soap, bicycles, batteries, and small appliances. In general, industrial growth is proceeding at a higher rate than agricultural growth in the Sub-Saharan nations.

Power and transportation are both essential to further development. In regard to power, Africa is estimated to possess 40 percent of the world's hydroelectric potential, with about half of this potential available where the Congo River plunges off the African plateau toward the sea in a gorge west of Kinshasa. Falling about 1,000 feet in a series of rapids, the huge volume of the Congo at this point offers the world's greatest site for the production of hydroelectricity. Many other favorable sites also exist in Africa. Along the Zambezi River, work has begun on the Labora Bassa Dam in Mozambique, a dam whose power output may eventually be twice that of Egypt's Aswan High Dam on the Nile or of the Grand Coulee in the United States. Africa already has a number of small to medium-size hydroelectric dams, and the prospect is for many more in the future. Unlike the rest of the continent, South Africa derives nearly all its electric power from its large deposits of coal.

As for transportation, Africans have taken to flying over the trackless deserts, savannas, and forests with great enthusiasm. Air travel for businessmen, students, tourists, diplomats, and others is vital to African development, for on the ground the building of roads and railroads over vast distances and through difficult terrain has been advancing rather slowly. Although thousands of miles of road are now paved, these are only a beginning, and the traveler on the ground must reconcile himself to many hours of negotiating dusty, bumpy, and dangerous roads that might be washed out at a crucial point in an area far from any assistance.

Most African railroads are known as "penetration lines"; that is, they penetrate inland along the most direct route from a coastal port to a specific mining or agricultural area without contacting another rail line to form an integrated network. In some places, penetration rail lines ascend the coastal escarpments in order to connect with cargo and passenger vessels at ports on inland lakes or on navigable stretches of rivers. Several large railway projects are now underway; the most notable is the 1,000-mile TanZam railway, which will extend between the Copper Belt of Zambia and the port of Dar es Salaam in Tanzania.

The continuing expansion of communications in Africa—air, rail, road, radio, television —has great implications. Even for the most remote village, it introduces unavoidable social, economic, and political change. It also means that Africa will become a more unified, better integrated continent in the future.

Geographical Regions

West Africa: The Four Zones

The term "West Africa" is usually applied to the fourteen countries located between the Atlantic Ocean and the Sahara and from Senegal on the west through Nigeria on the east. Collectively, these countries had a population of about 110 million people in 1970, or 40 percent of the inhabitants of Africa south of the Sahara. West Africa forms a cohesive geographic region, despite its many political divisions. It shares a common history ranging from the centuries of great African kingdoms through the period of slave trading to the European colonial era. The peoples of the region share a broad cultural unity, although they speak many languages, have different customs, and follow a variety of religions. Most of West Africa is oriented in trade and emotional ties toward the Western communities around the Atlantic Ocean; this is unlike East Africa, where, with the Suez Canal closed, the orientation is focused increasingly across the Indian Ocean toward Asia. In landforms and climate, West Africa forms a distinct unit as well. Most of the land is flat or gently rolling and lies at elevations below 2,000 feet. Moreover, the region is dominated by one great stream, the 2,600-mile-long Niger River. In terms of climate, West Africa has a kind of unity, for rainfall here is almost always greatest along the coast and diminishes gradually inland until the Sahara is reached.

Climate is the principal key to the "four zones" of West Africa. Other criteria, especially the human factor and the tsetse fly, have combined with climate to create four distinctive geographical belts, aligned in an east-west arrangement, into which the region may be divided.

Coastal Zone In almost every way, this zone is the modern heart of West Africa. It is a zone characterized by heavy precipitation, constantly warm temperatures, and large areas of rain forest, except in Senegal. It is the home of a dense population of millions of African cultivators, most of whom are engaged in growing starchy staples (yams, rice, cassava) for themselves and cash crops to supplement their incomes. It is also the zone of large ports that serve a hinterland of 2 million square miles,

Africa South of the Sahara 457

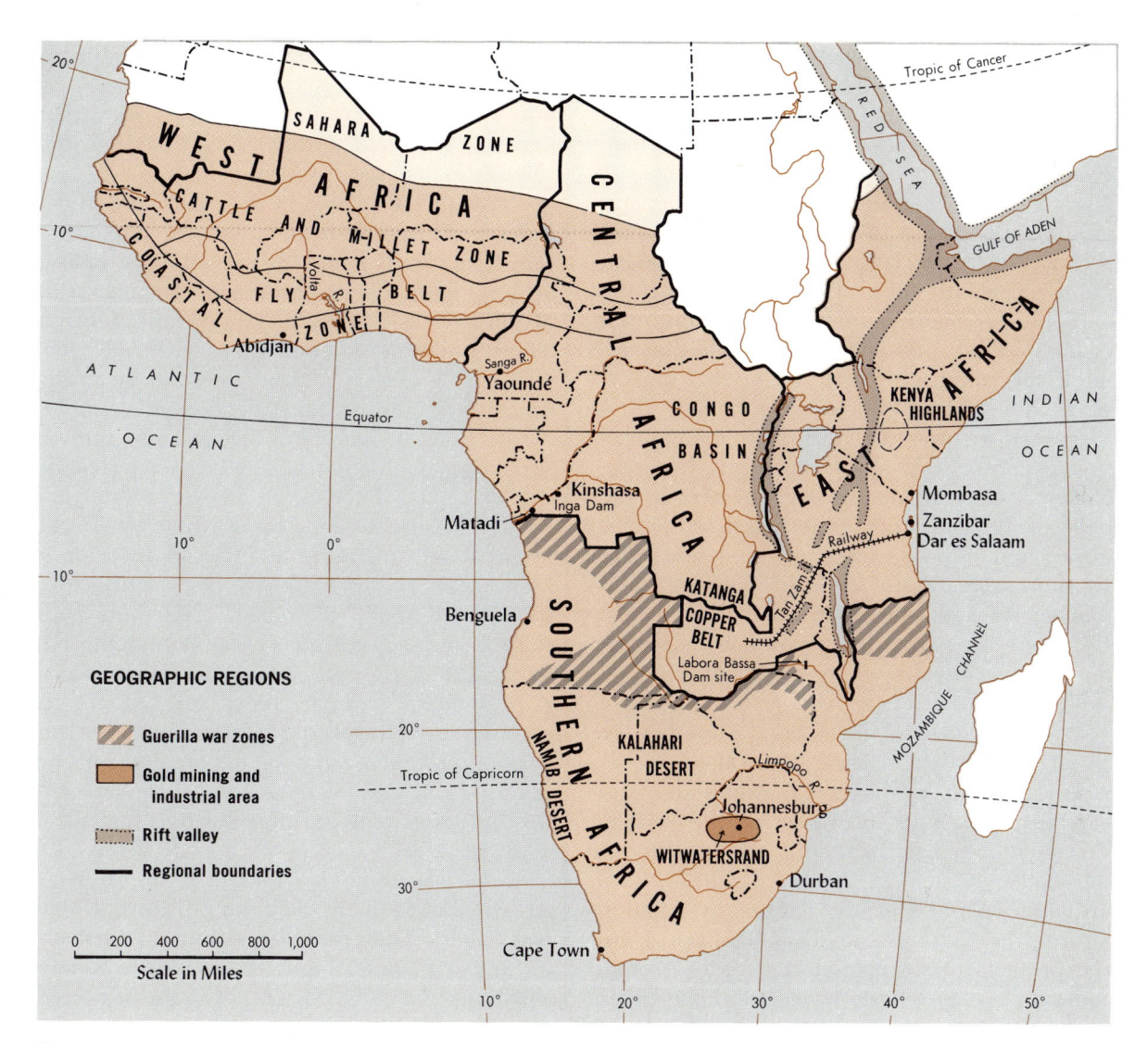

Fig. 12-9 The geographic regions of Africa are based on political, cultural, and physical factors.

and it is an area that contains numerous rapidly growing cities and industries.

The Coastal Zone attains its economic, social, and political peak between Abidjan, Ivory Coast, and the eastern boundary of Nigeria, where it spills over to Yaoundé, Cameroun. Along this 1,000-mile stretch of coast is concentrated the largest collection of modern industrial plants south of the Sahara, with the exception of South Africa. In addition, each country in the zone maintains its political capital at the coast, and there are several universities. The Coastal Zone possesses nearly all of West Africa's presently exploited mineral resources: petroleum, gold, diamonds, iron ore, bauxite, coal, and manganese. Moreover, the bulk of West Africa's export cash crops are grown here: cocoa, coffee, palm oil, and fruit. In short, the Coastal Zone of West Africa is notable for its dynamic human achievements. Its people are pervaded by a spirit of optimism unmatched anywhere else in Africa.

Fly Belt North of the Coastal Zone lies the so-called "Fly Belt," a zone where the tsetse fly is king. Here, population densities decrease rapidly, and rural poverty grips the people. Precipitation is less than along the coast and is unreliable in amount and occurrence. Soils are quite poor, and the savanna vegetation affords the tsetse fly an ideal habitat. Until the problem of sleeping sickness is overcome, the people of this zone have little hope other than to migrate toward the coast.

Cattle-and-Millet Zone North of the Fly Belt, the environment for man improves, and population densities increase again. With less rainfall, the savanna thins gradually to steppe grasslands, and there are far fewer tsetse flies. In this zone, soils are somewhat better, enabling farmers to grow mainly millet and peanuts during the rainy season. Cattle, sheep, goats, and camels are grazed, while the Niger River affords irrigation farming on its floodplain. Three important exports of this zone are peanuts, cotton, and livestock. The livestock raised here are exported mostly to the Coastal Zone, where the demand for meat is high.

Most of the people in the Cattle-and-Millet Zone are Muslims, for whom the pilgrimmage to Mecca in Saudi Arabia is one of life's holiest achievements. They have been in contact with Muslim Arabs across the Sahara for centuries. Their philosophical outlook is generally austere and conservative, and they are generally much less optimistic than the energetic peoples of the Coastal Zone. In Nigeria, where these two large population blocs exist separated by the Fly Belt, the tensions between them are severe. Thus far, they have avoided civil war by extreme tolerance and by their leaders' knowledge that war cannot be a satisfactory solution to their differences. Further, both sides have a vested interest in maintaining Nigeria as a nation-state, inasmuch as the Coastal Zone benefits from northern markets for its manufactured products, while the landlocked Cattle-and-Millet Zone is dependent on the ports of the Coastal Zone for its imports and for the export of its huge peanut surpluses.

Saharan Zone The fourth zone of West Africa consists of the Saharan portions of Mali and Niger. The possibility of mineral discoveries and a planned highway across the Sahara constitute the only prospects for the scattered nomads of this zone, other than improvements in fishing, water transportation, and crop irrigation along the northern arc of the Niger River. A small flow of tourists makes a contribution to the local economy by visiting what remains of the once-great city of Timbuktu in Mali.

Country Surveys of West Africa

Nigeria, with well over half of West Africa's population, is the giant of the region, as indeed it is for all of Africa. Nigeria's population of nearly 60 million far overshadows second-place Ethiopia's 25 million and comprises nearly one-quarter of the total population south of the Sahara. Nigeria is rich not only in agricultural and mineral resources, but in human resources, for there are now tens of thousands of Nigerian college graduates, many of whom hold M.D.s and Ph.D.s, and an industrious labor force. In addition, Nigeria's large internal market for consumer goods has proved very attractive to investors. Although Nigerian annual income per capita is still only about 100 dollars, the future should witness substantial improvements, assuming mainly that the country's 250 tribal groups are able to preserve their present degree of harmony.

Ghana, the world's leading producer of cocoa, also has the richest mineral operations (gold, diamonds, manganese, bauxite) in West Africa, excluding Nigeria's oil reserves. These resources have provided Ghana's 9 million people with the highest annual per capita incomes in the region, now estimated at 200 dollars. Ghanaians in the Coastal Zone are a highly literate people and perhaps the healthiest in Africa. Their economy was boosted in the 1960s by the establishment of West Africa's largest single industrial project in terms of value of output; this is a plant near Accra which converts alumina to aluminum. The success of the operation is based partly on cheap power from West Africa's largest hydroelectric dam, built nearby on the Volta River.

Ivory Coast, west of Ghana, experienced the most rapid annual rate of economic growth (8 percent) of any country south of the Sahara in the 1960s, as exports of coffee, cocoa, and timber increased. Ivory Coast has become Africa's leading source of coffee. Liberia, founded by ex-slaves from the United States in the 1820s, has become the world's tenth largest miner of iron ore, most of which is shipped to the United States. Sierra Leone is a major producer of diamonds and iron ore; its neighbor Guinea exports bauxite and claims to have the world's largest deposits of that mineral. Portuguese Guinea, in 1971, was still embroiled in a long conflict between the Portuguese army and Africans fighting to gain independence. Gambia and Senegal are primarily peanut exporters.

Inland, the countries of Mali, Niger, and Upper Volta are known mostly for their cattle, cotton, fish, salt, and peanuts. These are poor countries, with the per capita income in Upper Volta averaging only 50 dollars per year. On the coast, Togo and Dahomey are so small and economically insignificant that they are barely viable as sovereign nations.

West Africa seems to have real potential for improvement of its human living standards. However, the multiplicity of independent governments acts as a barrier to a more efficient utilization of the region's resources. With or without Nigeria, it would seem that nowhere else in Africa is there a greater need to discard the old boundaries imposed by European colonialism, which now impede economic and political unity and progress.

Central Africa: Sparse Population

The eight countries that compose the Central African geographic region do not form a clearly defined unit like West Africa. Extending 2,600 miles in a north-south direction, they include a great variety of landforms and a wide range of climatic-vegetation types, from the Sahara of northern Chad through steppes, savannas, and dense rain forests. Historically, these countries have less in common with each other than do those of West Africa, and culturally their peoples are more diverse. Two ways in which this region resembles West Africa are: (1) the dominance of a single country in terms of the size of its population and its wealth—that is, the Congo (Kinshasa) and its West African counterpart, Nigeria; and (2) the concept of a master stream—the Congo River and its counterpart, the Niger River.

Perhaps the greatest contrast between the two regions, however, and the only one that provides a major unifying theme for Central Africa as a region, is the sparseness of population. While both regions have approximately equal areas, West Africa in 1970 had a population of about 110 million, as compared with 30 million for Central Africa. On the basis of persons per square mile, West Africa averaged 55 and Central Africa averaged only 15. This difference may be due to the greater extent of poor soils and the tsetse infestation in Central Africa. Another factor could be the much-shorter length of Central Africa's coastal zone, with its opportunities for cash-crop production. In any event, there are relatively few people in Central Africa today; one may fly over the Congo Basin for hours and see scarcely any signs of human settlement. A less important unifying factor is the use of French as the language of government and business in most of Central Africa.

In area, the Congo, with its capital at Kinshasa, is the largest country south of the Sahara (905,000 square miles, or 3.5 times the size of Texas). Its population of nearly 20 million is about 50 percent Christian and 30 percent urbanized, with Kinshasa having grown rapidly since independence, to an estimated 1.5 million people. In 1960, Belgium suddenly withdrew as the Congo's master, thereby plunging the country into seven years of chaos and rebellion. Since the achievement of relative stability in 1967, the Congo has begun to resume its role as one of the richest nations in Africa. From Katanga Province, copper, over half the world's cobalt, and zinc go to world markets. About 70 percent of the world's industrial diamonds, tin, manganese, silver, and gold from other areas add to the country's natural wealth, along with a diversity of agricultural exports: oil-palm products, coffee, tea, cocoa, rubber, and cotton. There is also the Congo River itself, providing 7,000 miles of navigable waterways in the basin as well as the world's greatest known hydroelectric potential, most of it in the rapids below Kinshasa. A large project, the Inga Dam I (first stage), is now under construction on the river above the Congo's ocean port, Matadi. It is expected to generate the world's cheapest electricity and attract new industries.

Across the river from Kinshasa lies Brazzaville, capital of the People's Republic of the Congo. Very sparsely populated (7 per square mile). The major product of Congo (Brazzaville) is timber. The world's largest-known deposits of potash were discovered recently not far from its coast. Gabon deserves special note for having the highest annual per capita income (500 dollars) of any country south of the Sahara, except for South Africa. This is based upon its expanding petroleum output, manganese, uranium, and lastly, timber, which was the only significant export ten years ago. As yet, exploitation of a huge deposit of iron ore, amounting to a billion tons, awaits the building of a railroad and port facilities. Gabon has only half a million people and a population density of 5 per square mile; this helps to account for its high per capita income.

Equatorial Guinea, which was Spanish territory until 1968, consists of the volcanic island of Fernando Po and the former mainland area of Río Muni. The big problem facing the new nation is the building of national unity. This small country exports cacao and coffee.

To the north, Cameroun, Central African Republic, and Chad tend to repeat the four zones of West Africa. Cameroun is the most progressive of the three and exports coffee, cocoa, and bananas. Aluminum is also processed at a plant drawing hydroelectric power from the Sanga River. The exports of Chad, a poor and remote country, are limited to small amounts of cotton. The Central African Republic, also poor, specializes in both industrial and gem diamonds, cotton, and coffee.

The Central African region presents a mixed picture. The Congo (Kinshasa), Gabon, and Cameroun apparently have good economic prospects, whereas the People's Republic of the Congo, the Central African Republic, and Chad face relatively slow growth. As in West Africa, economic and political unity would be beneficial. In Cameroun, a small step toward African unity was taken in 1961 when the Southern Cameroons, which were under British control, voted to join Cameroun, which had been French. Over the last ten years, this merger of the two systems of government, language, and European customs has apparently been successful, and it offers a hopeful model for future unification proposals involving English- and French-speaking parts of Africa.

East Africa: Diversity

East Africa encompasses great volcanoes, high plateaus, rift valleys, great lakes, the headwaters of the Nile, high interior plains, low coastal plains, and the islands of Madagascar, Réunion, and Mauritius. It is a land of breathtaking scenery, numerous game reserves, and fascinating African peoples. In places, it offers a delightful climate. Unfortunately, East Africa is the poorest of the four major geographic regions and apparently will remain so, for its economy is based on agriculture and there are problems of population pressure. Only in Zambia, with its copper wealth, are there any significant mineral deposits that might provide a shortcut to rapid economic growth. Increasingly, the tourist trade is being encouraged as a source of revenue.

The 2 million square miles of the East African region had a population of 85 million in 1970, for an average density of 43 per square mile. But average densities are quite deceptive in East Africa because populations are densely clustered in the cool, fertile, tsetse-free highlands and in a narrow strip along the coast. The region is remarkably diverse also in the kinds of people who inhabit it. There is no typical East African, for the indigenous people range from the tall Watutsi to the Pygmy and from very dark skins to very light. The people are further divided by contrasting languages, customs, and agricultural occupations. More than 90 percent are rural. Compounding

Fig. 12-10 Urban and rural African scenes: older part of a coastal city in West Africa; women gathering around a well to fill water jugs in West Africa.

the cultural mix are several hundred thousand East Indians, Arabs, and Europeans. In Kenya, the government employs a slogan derived from the Swahili word "Harambee," which means "Let's all pull together," in its efforts to unify its many cultural groups and races.

The high, mountainous empire of Ethiopia is the largest East African country, with an estimated 25 million inhabitants. About 40 percent of its people, those who dominate the country, belong to the ancient Coptic Christian church. Another 40 percent are Muslims. Ethiopia's farmers work the richest soils in Africa, soils which develop from the underlying volcanic rocks of the high, deeply dissected plateaus at average elevations of 8,000 feet. Coffee is the chief export of the country, but the full agricultural potential is far from realized because of the very conservative attitude of the landowners toward change, and also because of the difficulties of getting surplus crops and livestock across the mountainous terrain. Ethiopia's future is clouded by the advanced age of its venerable emperor, Haile Selassie. Internally, Ethiopia's society and economy are essentially feudal, and when the Emperor dies or is displaced, conflict involving the church, army, landowners, and progressive-minded youth is expected. Sporadic guerrilla warfare continues in northern Ethiopia, and Somalia claims a large area in the southeast.

Somalia is primarily a grazing land; 75 percent of its people are nomadic. There are no important exports, and its population is perhaps the poorest in the world. A recent uranium discovery may help the economy, coupled with bounteous aid from the U.S.S.R., which recognizes Somalia's strategic position in the Indian Ocean.

Kenya has become Africa's safari center. In 1968, tourist expenditures became the chief source of Kenya's income for the first time, exceeding coffee shipments in value. The fertile Kenya highlands, comprising about 10 percent of the country's area, contain about 75 percent of the people. They form the crowd-

Fig. 12-11 Savanna region, Tanzania. Students are learning new agricultural techniques. (Courtesy of Tanzania Information Service)

ed, modern core of Kenya, leaving the arid remainder to wild game, tsetse flies, and a few hardy pastoralists. Kenya's capital, Nairobi, situated at 5,400 feet above sea level near the equator, has a near-perfect climate and is considered one of Africa's most beautiful cities.

Uganda presents a lush, green landscape through which the Nile threads its way from Lake Victoria to the Sudan. In the southern areas, the farmers have reached relatively high standards of living from the production of coffee. Despite its small size, Uganda has a variety of peoples and numerous wildlife reserves, both of which are a favored attraction for tourists.

Rwanda and Burundi are the most densely populated countries of Africa, with about 350 persons per square mile. Good soils exist in the highlands, and there is abundant rainfall for food crops, cattle pastures, and their chief export, coffee. These, however, are poor, small, and politically unstable countries, with no solution in sight for their high population pressures.

About two-thirds of Tanzania is "ruled" by the tsetse fly. Much of this useless land occupies the dry central plateau, resulting in a pattern of peripheral economic development. Coffee, cotton, and diamonds are the chief exports, although Tanzania leads the world in the production of sisal. Most of the world's cloves are grown on the coastal islands of Pemba and Zanzibar. Tourists make a significant contribution to the Tanzanian economy. Vigorous attempts to modernize African agriculture through the setting up of village cooperatives have been launched. These appear to have been moderately successful in maintaining the traditional African way of life, while at the same time leading to increases in production.

Malawi is a poor agricultural land, but it is rich in scenery, provided by its rift valley features and Lake Malawi. Tobacco and tea are its primary export crops.

Zambia is included among East African countries because of its political orientation; otherwise it is not typical of East Africa, owing to the predominance of its mining industry (98 percent of all exports). The largest exporter of copper in the world, Zambia has few of the financial problems of its neighbors to the east. On the west and south, however, it faces European ruled governments, and this poses a different kind of problem. The TanZam Railway will enable Zambia to export its copper through Tanzania, in order to cut most of its economic links with white-ruled Rhodesia to the south. The TanZam line is under construction with the assistance of the government of the People's Republic of China.

The Malagasy Republic is chiefly agricultural; coffee, sugar, rice, and most of the world's vanilla are its main exports. Its mixed Indonesian population is perhaps its most distinctive feature. Réunion and Mauritius are very crowded islands having a wide assortment of races. Sugar is their dominant crop and source of income. Mauritius is independent, but Réunion is an overseas department of France.

East Africa is spectacular in scenery and peoples but is apparently devoid of possibilities for rapid economic growth, unless more minerals are found or low jumbo-jet airplane fares open up substantial new tourist markets. Politically, East Africa now tends to look toward Asia and the Middle East, largely because the closure of the Suez Canal in 1967 bars the short sea route to Europe. The region is one of undercurrents and rivalries, some of them instigated by the great world powers. The general economic prospect for East Africa is decidedly less optimistic than are the prospects for West and Central Africa.

Southern Africa

This geographic region is of similar size to the others, having about 2 million square miles.

Its population of 42 million gives it the rather low density of 21 persons per square mile. The bulk of the population is found along the eastern coasts and highlands. In the central and western portions, the Kalahari steppe-desert and the Namib coastal desert preclude much in the way of human settlement. Some parts of Botswana and South-West Africa are uninhabited.

The uniqueness of this region in Africa is based upon the determination of its 4 million European citizens to preserve forever their cultural identities, political rule, economic domination, and other privileges. This determination is strongly opposed by African nationalism. In three of the territories, Angola, Mozambique, and Rhodesia, irregular forces representing African nationalism have conducted guerrilla warfare for several years. In three others, Botswana, Lesotho, and Swaziland, peace and independence under African rule has been obtained; but these are relatively unimportant countries. The giant of the region, and by far Africa's richest country, is the Republic of South Africa, which has experienced little overt guerrilla activity as yet. The republic sees itself as a great anti-Communist bastion, guarding the sea-lanes around southern Africa, now more vital than ever with the Suez Canal closed. But underneath this apparently calm, prosperous exterior the African population, especially the large proportion that is urbanized, is seething with hatred for its masters, and the situation is potentially explosive.

The paramount question in southern Africa is the degree to which the white regimes will cooperate in order to survive. This question becomes more acute as the African guerrilla forces slowly continue to expand the areas under their control in Angola, Rhodesia, and Mozambique. South Africa holds the key, for it is the only country with sufficient power to stem the tide in its neighbors' territories. Aside from committing a few token troops, South Africa thus far seems reluctant to become involved in what might well be expensive, bloody, and perhaps endless campaigns on foreign soils. Should its three neighbors fall to African nationalists, as some expect, then South Africa may find the battle for African freedom and equality raging within its own boundaries.

Portuguese Southern Africa

Since their emergence in 1961, Angola's African nationalist forces have steadily enlarged their area of control. Their objective, besides freedom from Portugal's colonial rule, is the considerable natural wealth of Angola in coffee, diamonds, oil, and iron ore. Africans outnumber Portuguese in Angola by a ratio of about 23 to 1.

In Mozambique, the Portuguese face much the same problem as in Angola, with continued guerrilla activities that drain Portugal's manhood and money. South Africa has a much greater interest in Mozambique than it does in Angola because of its proximity to the Johannesburg mining and industrial complex. Moreover, guerrillas could easily infiltrate eastern South Africa under cover of the forests and rugged escarpments of the Drakensberg. To counter this threat, some strategists have advocated a South African advance to the Limpopo River should Portuguese resistance collapse.

Mozambique is primarily an agricultural country, exporting cotton, cashew nuts, and sugar. It also serves as an outlet to the sea for exports from Rhodesia and some parts of South Africa. Another source of income is from the earnings of the estimated 300,000 Mozambique Africans who work in South Africa's mines. The ratio of Africans to Portuguese in Mozambique is about 70 to 1. In order to change this ratio, the Portuguese have begun construction of a huge dam at Labora Bassa on the Zambezi River. The first stage alone will generate more power than any other dam south of the Sahara, and subsequent stages could raise its power output considerably. South Africa would be the major purchaser of the electricity produced here, transported via an 865-mile transmission line. The Portuguese plan a steel mill and also a huge irrigation scheme below the dam that would purportedly attract a million European settlers. The 1/2-billion-dollar Labora Bassa project is a symbol of Portugal's determination to stay in Mozambique.

New Nations

Rhodesia was, until 1965, a British colony; in that year, the 200,000 English-speaking white settlers of Rhodesia issued a declaration of independence, later vowing that they would never permit the African majority to rule the country as Britain had demanded. The Rhodesian whites are outnumbered about 20 to 1. Since 1965, they have diversified and balanced their economy in order to withstand worldwide economic sanctions imposed by the United Nations at Britain's request but defied by their South African and Portuguese neighbors. In normal times, Rhodesia exports tobacco, chrome, asbestos, and gold. As of 1970, the Rhodesian whites had shown remarkable tenacity in maintaining their high standard of living and in challenging the guerrilla forces operating across the Zambezi River from bases in Zambia.

Botswana, Lesotho, and Swaziland, although sovereign nations now ruled by African governments, are very much within the economic orbit of their powerful neighbor, South Africa. They are particularly careful not to offend their neighbor for the reason that South Africa could cripple their economies or conquer them overnight. All three send labor to South Africa's mines, farms, and cities. Among the three, Swaziland has the best economic prospects, with iron ore, asbestos, timber, and citrus fruit production. Lesotho, barren and cold in the high Drakensberg during the winter, is a poor country with small exports of wool, mohair, and diamonds. As for Botswana, its economy is based on the cattle that roam its eroding tropical steppes. A recent copper and nickel discovery may enrich the new nation.

Republic of South Africa

The Republic of South Africa is quite unlike any other country in Africa. It is by far the wealthiest, leading all other countries south of the Sahara in total exports of both mineral and agricultural commodities. It has far more industries than any other nation in Africa, including an annual steel production of 5 million tons. South Africa differs by having almost 4 million Europeans, or about 80 percent of all the whites south of the Sahara, in its population. It is the most urbanized (50 percent) of any African country. Historically, it is different, too, for European settlement and occupation began here in 1652, rather than 100 years ago or less as in nearly all other parts of the continent. In climate, South Africa is more subtropical than tropical. Finally, of course, the country's approach to the problem of racial tensions is unique not only in Africa but in the rest of the world.

South Africa is predominantly an arid land; only 15 percent of its area can be farmed. In the Cape region, a Mediterranean climate is conducive to growing grapes, citrus fruits, and vegetables. Along the east coast, below the Great Escarpment of the Drakensberg, the climate is wet subtropical, and sugarcane is the chief crop. On the high veld, or plateau, of the Transvaal and the Orange Free State, corn is grown during the warm, rainy summers. To the west, the Orange River traverses steppes and deserts suitable mainly for sheep grazing. A large development scheme now underway on the Orange River will generate power and provide more water for irrigation in this dry region. Along the country's coasts, South Africans "harvest the sea," especially the well-known rock lobsters.

South Africa's cities have a contemporary American look about them. Johannesburg, with its 50 and 60-story buildings and freeways, could be Dallas. The oceanfront of Durban bears a resemblance to Miami Beach. Cape Town, where the world's first human heart transplant was performed, has an air of San Francisco with its adjoining mountains and beautiful natural harbor.

South Africa is a land of great variety, scenic beauty, and natural wealth. It would seem to have an unlimited future but for its one great problem: Can the whites, outnumbered 4 to 1 by the nonwhites, indefinitely maintain a system that assures them political, economic, and social domination? Racial discrimination, which began with the early European settlers, has intensified in recent years, partly as a result of South Africa's rapid urbanization and economic development. Known as apartheid, or "apartness," the South African racial system erects barriers to nonwhite advancement in employment and education. Apartheid requires total racial separation in the location of housing, and racial intermarriage is strictly forbidden. Each of the four racial groups must live in its own residential quarters, attend its own schools and theaters, eat in its own restaurants, stay in its own hotels, swim at its own beaches, and ride on buses and trains designated for it. Sports are also rigidly segregated, and this has led to many international complications in rugby, soccer, tennis, boxing, and the Olympic Games. Nonwhites are prohibited from voting in national elections, and none are permitted to sit in the Parliament. One measure of apartheid is evident in the disparity of incomes. Per capita annual incomes for whites average over 3,000 dollars, whereas for nonwhites the figure is about 300 dollars, or 1/10. It must be conceded, of course, that nonwhites in South Africa, although oppressed in many ways, have generally higher standards of diet, medical care, housing, education, and income than Africans elsewhere on the continent.

South Africa's racial dilemma has been parodoxically compounded by its high rate of economic growth, which has been around 6 percent annually in recent years. Boom conditions have attracted millions of nonwhites to urban areas where their labor has become

Fig. 12–12 Johannesburg, South Africa, the largest city in Africa south of the Sahara, has a metropolitan population of over 4 million. Gold-mine dumps appear in the distance. (Courtesy of South African Information Service)

essential to all facets of the economy. The paradox is that many Europeans want the Africans to return to segregated rural areas; but the two groups have become so interdependent that, if this were to happen, the South African economy would collapse.

Nonwhite population growth rates, which are approximately double the rate for whites, pose another problem for continued white rule. Without substantial immigration from Europe, the nonwhite-to-white ratio is projected to advance from 4 to 1 to 5 to 1, or even 6 to 1, by the year 2000. Among the four racial groups, those of mixed ancestry known as Coloureds are increasing at the fastest rate, followed by the Asiatics, who are mainly of East Indian origin. The Africans are also increasing in numbers at a greater rate than the whites.

In recognition of its problems, the South African government has adopted a variation of apartheid, which it refers to as "Separate Development." In order to implement this policy, 13 percent of the country's area has been set aside for tribal homelands or reserves, called "Bantustans." Within these "homelands," Africans have attained a large measure of self-government and are permitted full economic and social rights. The creation of Bantustans in scattered locations, each one designated for a specific tribal group, enables the white government to entrench its power according to the maxim "Divide and conquer." At the same time, it provides the government with a propaganda weapon for those who argue that its Africans are completely oppressed. Moreover, Bantustans serve notice that, in the 87 percent of South Africa held for white rule, Africans are to be considered as no more than imported workers or temporary visitors, subject to rigid segregation and having few rights. The Bantustan scheme is also intended to reduce the flow of Africans into the cities by locating new industry along the boundaries of each homeland. Some Africans also support the concept of "Separate Development" and its expression in Bantu homelands.

For the white Afrikaner, whose roots go back to Dutch or French settlement 300 years ago, South Africa is his only home. Along with the more recent English-speaking settlers, he is determined to remain and to rule. Thus, in the face of the nonwhites' demand for political and other rights, it is quite impossible to predict the future of South Africa. Some prophesy imminent bloody revolution; others foresee the peaceful advent of racial harmony, equality, and cooperation. Most believe that the present system will continue for some time with little modification, regardless of world opinion or economic and diplomatic pressures.

The geographical region of Southern Africa is not without its problems. The possibility of a race war of continental proportions intended to "liberate" South Africa, Rhodesia, and the Portuguese colonies exists. Others visualize an expanded conflict involving several world powers over South Africa's refusal to award independence to the mineral-rich territory of South-West Africa. In 1968, the United Nations voted to rename this territory Namibia and demanded that South Africa relinquish the control it acquired in 1918 under the mandate system of the League of Nations. This issue could lead to a dramatic confrontation between South Africa and the United Nations. The hopeful view for southern Africa is that people might learn to live together in mutual understanding and respect.

In Perspective

One of the most encouraging things about Africa south of the Sahara today is the spirit of optimism among its people. True, there are all kinds of major problems: poverty, disease, illiteracy, rapid social change, overpopulation in places, a harsh physical environment, and tribal antagonisms; not to mention the rather slow rate of economic growth, political instability, narrow nationalism, and European minority rule in the southern areas. Nevertheless, most Africans approach these problems with remarkable enthusiasm, confident that each can be solved eventually. Increasingly, they have realized that development is mainly their own responsibility. In some ways, Africans are beginning to display an attitude reminiscent of the gritty, self-reliant pioneers of the American West, for theirs is psychologically a new land, a new society, a frontier.

If left to themselves, Africans could gradually overcome most of their problems and make substantial progress in the years ahead. But it is fanciful to think that they can be isolated from the rest of the world at a time when the conflict between the superpowers compels each of them constantly and everywhere to seek the upper hand. Africa's great mineral and agricultural wealth and its strategic position are much too tempting for the great world powers to ignore. Moreover, Africa's fragmentation into a multiplicity of small countries is a further invitation to the kind of great-power rivalry that has been so disastrous for the peoples of Germany, Korea, Indochina, and the Middle East.

The challenge of Pan-African unity, with its potential social, economic, and political benefits, is clear. But unity is more than that: it means the strength to preserve recently won freedoms. This is the greatest challenge for Africa in the 1970s, and it will require all the enthusiasm, optimism, and sacrifice that the people of Africa can muster.

REFERENCES

Best, Alan C. C.: "Gaberone: Problems and Prospects of a New Capital," *Geographical Review*, vol. 60, pp. 1–14, January, 1970.

Bohannan, Paul: *Africa and the Africans*, The National History Press, Garden City, N.Y., 1964.

Church, R. J. Harrison, John I. Clarke, P. J. H. Clarke, and H. J. R. Henderson: *Africa and the Islands*, John Wiley & Sons, Inc., New York, 1965.

Gailey, Harry A.: *The History of Africa in Maps*, Denoyer-Geppert Company, Chicago, 1967.

Gleave, M. B., and H. P. White: "The West African Middle Belt: Environmental Fact or Geographers' Fiction," *Geographical Review*, vol. 59, pp. 123–139, January, 1969.

Grove, A. T.: *Africa South of the Sahara*, Oxford University Press, London, 1967.

Hallett, Robin: *Africa to 1875*, The University of Michigan Press, Ann Arbor, 1970.

Hance, William A.: *The Geography of Modern Africa*, Columbia University Press, New York, 1964.

⸻: "The Race between Population and Resources," *Africa Report*, vol. 13, pp. 6–12, January, 1968.

Horvath, Ronald J.: "Von Thunen's Isolated State and the Area around Addis Ababa, Ethiopia," *Annals of the Association of American Geographers*, vol. 59, pp. 308–323, June, 1969.

McEwan, Peter J. M., and Robert B. Sutcliffe (eds.): *Modern Africa*, Thomas Y. Crowell Company, New York, 1965.

Mountjoy, Alan B., and Clifford Embleton: *Africa: A New Geographical Survey*, Frederick A. Praeger, Inc., New York, 1967.

Murdock, George P.: *Africa: Its Peoples and Their Cultural History*, McGraw-Hill Book Company, New York, 1959.

Oliver, Richard and J. D. Fage: *A Short History of Africa*, 2d ed., Penguin Books, Inc., Baltimore, 1968.

13

Central Eastern Asia

Almost one-third of the world's population, or approximately 900 million people, inhabit the large part of the East Asiatic countries which tradition has taught us to regard as the mystery lands of the Far East. Associated with these regions in our minds are such intriguing historical personages as Marco Polo and Kublai Khan, such romantic titles as *The Good Earth*, *Madame Butterfly*, and *The Mikado*, and such world-famous modern-day political personalities as Generalissimo Chiang Kai-shek, Chairman Mao Tse-tung, Chou En-lai, and Emperor Hirohito. Although these lands are no longer remote and mysterious in the old sense, information for much of this area has again become obscure. How true a picture can we acquire of the lands supporting such masses of humanity? What are the geographical and physical settings against which these Oriental peoples live out the drama of their lives? Can we clear away the cobwebs of mystery and romance and the obstructions of official restrictions, in order to discern the true culture, livelihoods, and activities in which the people of these East Asiatic countries engage?

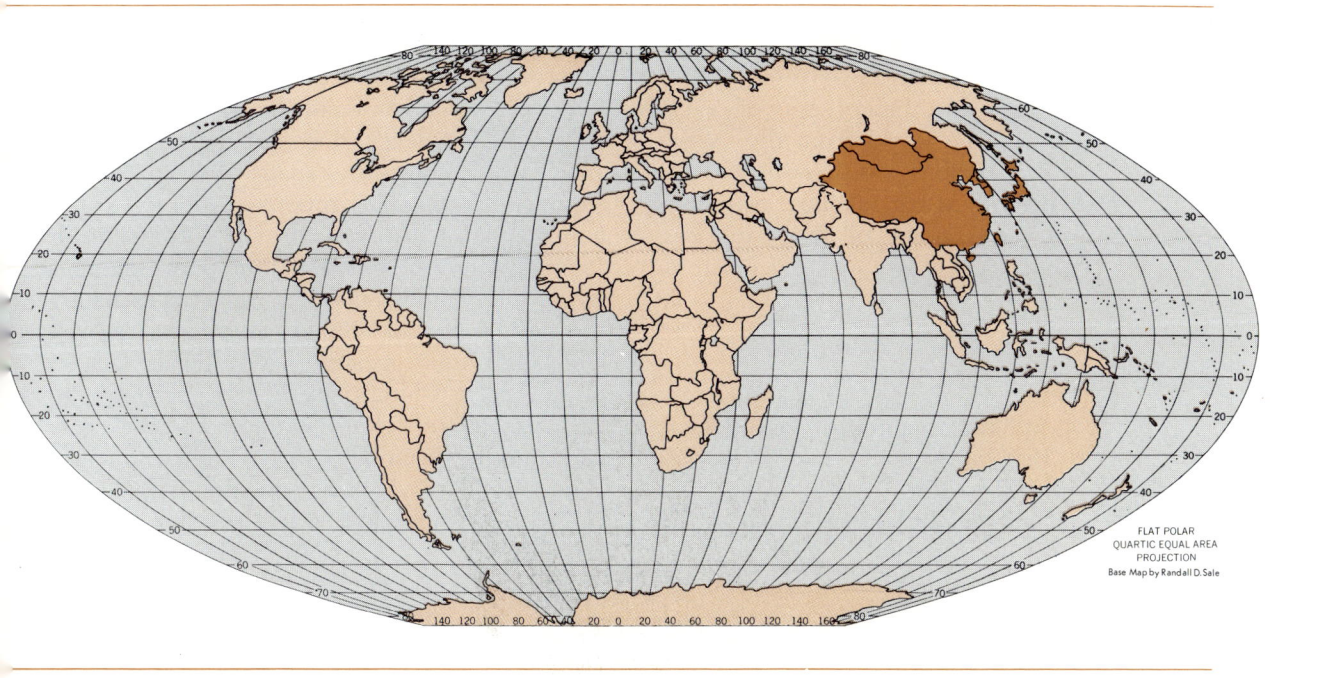

Regional Character

Regions may be regarded as major areas of the earth's surface combining a particular location with particular topographic and/or climatic character, historical development, and cultural orientations to a sufficient degree to impress the observer as an entity significantly differing from adjacent major areas. In the eastern half of Asia one may discern four such regions. These are Northeastern Asia, which is in the Soviet realm; the South Asian and the Southeast Asian regions, which lie south of China; and Central Eastern Asia which includes China, Japan, Korea, and Mongolia. The logical grouping of China, Korea, and Japan as Central Eastern Asia is seen in their similar cultures, embodied in their written language, in their background of Buddhist religion, in Confucian social ethics, and in the careful and intensive methods used in agriculture. The term Far East has been an accepted misnomer for this region, deriving from concepts of geography dating from when Magellan proved the earth was round by sailing westward to the "Far East." Until he did this, the only way of reaching this region of the world from the "Western" or European realms was by sailing to the distant East around Africa. Today, especially from the reference point of the Americas, the term "Far East" no longer makes geographical sense.

Situation

The geographical situation in Central Eastern Asia is that of a continental landmass which merges into Central Asia on its west and which is fringed by several great island arcs of the Pacific to its east. The climates of this extensive region vary from boreal to tropical. The region has appreciable commercial, economic, and strategic advantages. Its eastern parts lie adjacent to one of the major commercial maritime routes of the world; ocean vessels plying between the Pacific ports of the United States and Southeast Asia often follow a great-circle route that runs past the coast of Japan and eastern China. Korea does not have the same advantage in location as do China and Japan, but it does not suffer the landlocked isolation of Mongolia. Strategically, Central Eastern Asia is most significant. Its land surface is one-twelfth that of the entire world, whereas almost a third of the world's total population is crowded together on its more productive lands. At its north and west, the region is contiguous with the powerful Soviet Union; its eastern island chains and coasts command the West Pacific.

Factors of Area and Geographic Shape

The four countries of this region differ in many respects, and these differences also apply to size and shape. The more than 3.7 million square miles of China contrast with the 85,000 square miles of Korea. Japan, with over 147,000 square miles, is less than one-fourth the size of the Mongolian People's Republic, which has some 606,000 square miles. Land area, however, does not necessarily give an idea of a country's capacity for population support. Mongolia supports a mere 1.2 million people, in contrast to over 100 million in Japan, though the former nation is several times larger. The significance of great area lies mainly in the greater likelihood of mineral wealth and often of greater agricultural variety. Strategically, China's great area permits the use of space for defense in depth and for a certain amount of mobility in traditional warfare. Intercontinental ballistic missiles, space surveillance, and nuclear warheads for missiles, however, have greatly reduced this military advantage. Also, large area has its disadvantages in the difficulties of communication and of political cohesion.

Overall shapes are significant in relation to roads and railroads and to other forms of communication, as well as for strategic vulnerability. Long, narrow countries such as Korea and Japan are vulnerable to military segmentation in warfare. With regard to climate, their narrowness, combined with their peninsular and insular situations, allows moderating marine influences to penetrate easily to the interior.

Climates

The 35° latitudinal spread of China brings its northernmost limits to the latitude of southern Labrador; whereas its southernmost territory reaches the latitude of Puerto Rico. Although Japan and Korea do not have such an enormous latitudinal spread, their generally north-south elongation also brings much climatic variety with it. Despite this variety, certain common climatic influences are distinguishable in the region. The huge continental interior of China not only has great continental extremes within itself but also exerts far-reaching climatic effects upon maritime and peninsular neighbors to the east. In these effects, Mongolia shares the position of progenitor with the rest of the Central Asian heartland. The existence of this immense land surface, with its great capacities for heat and radiation and absorption, has brought about the influences creating what is known as the "monsoon climate" of Eastern Asia. It is a monsoon that differs from the Indian monsoon, however, in having dry, cold winter air masses as the dominant element.

Natural Resources

Compared with North America and Northwestern Europe, Central Eastern Asia is faced with a general inadequacy of known resources for large-scale industrial development. There is limited variety in the more important deposits of mineral resources, but a few minerals are concentrated overwhelmingly in a few districts within the region. Most of this mineral wealth lies in China. In terms of accessible forest resources, only Japan has large stands proportionate to her area. China alone has large arable land resources; but because of its large population, that country nevertheless resembles the others in having only small units of cultivable land per capita.

China

China, with an unverified but estimated population of over 750 million in 1970, has less total acreage of land under cultivation than the United States; yet it has a population over four times as great. It has been estimated that by 1978 China's population, if it continues to increase at present rates of growth, will have reached 1 billion. Both natural and social factors have helped China become the most populous country in the world. Favorable natural factors include the large size of the country, the soils, climate, and the native plants in much of her territory. There is a large amount of arable land, including fertile alluvial and loessial soils. Much of China has a long growing season and abundant rainfall, with maximum precipitation occurring in the summer when it aids the crops the most. More than 9,000 plant species, nearly half of them peculiar in origin to that country and many of which are useful, occur in China.

Population-stimulating social factors have included early marriage and the desire for many sons, for both religious and economic reasons. Other influences have included the extensive use of human labor rather than animal power and the predominant use of grains and other vegetable foods in the diet rather than animal products, so that little land is needed to grow fodder. The animals supplying most of the meat consumed are pigs and poultry, which are commonly fed scraps and refuse or are allowed to scavenge for their feed. On otherwise wasted land, ponds are used to raise bulbs and tubers, fish, and ducks for feed. The Chinese are masters of irrigation and of conserving and fertilizing soils to secure maximum yields, even though the pressures of population and a lack of public responsibility have resulted in substantial deforestation and erosion of the land. The Chinese people, willing to toil long hours to support large families, remain cheerful and reasonably healthy under crowded and often unsanitary living conditions that would seem very depressing to Americans and Europeans.

Relief Features

Rivers China occupies the large eastern segment of the Eurasian landmass and shares in the radial system of river drainages extending out from the elevated heart of the continent.

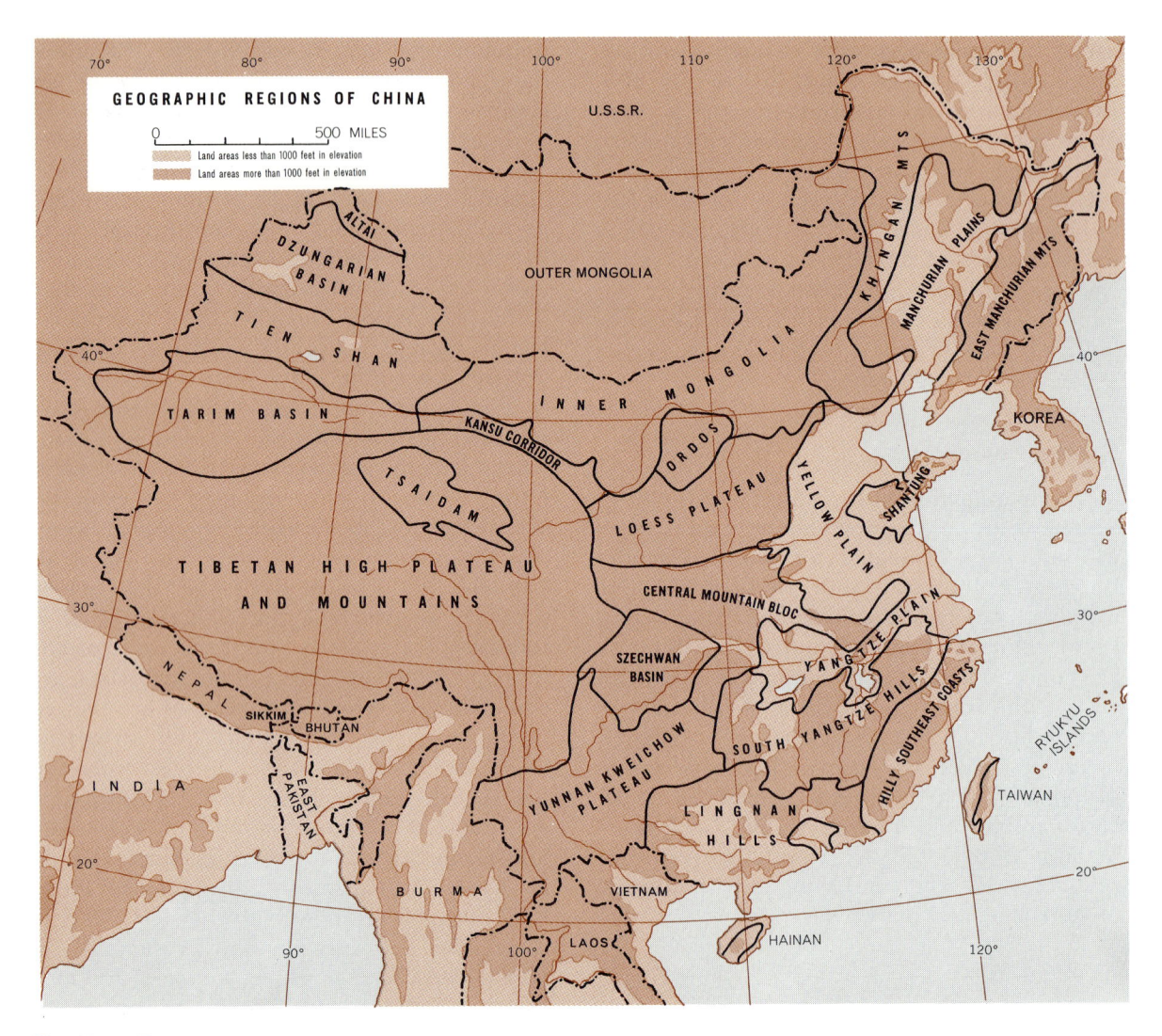

Fig. 13-1 The geographic regions of China are largely the result of physical factors.

Flowing northeastward and out of China into the Soviet Union's Khabarovsk and Maritime Territories is the great Amur River (or Heilung Chiang, which in Chinese means literally "Black Dragon River"). While economically the main stream of the Amur is of greater importance to the Russians than to the Chinese, its tributary the Sungari (Sung-hua) is the most important river of Manchuria; it drains the larger part of the Manchurian Plain and has great hydropower potential, much of which has been developed. In the 1950s the Russians and Chinese Communists jointly developed plans for a vast hydroelectric project on the

Amur. Because of hostile relations that subsequently evolved between the two nations, these plans remain unrealized.

The other large rivers of eastern China flow in great bends and loops generally from west to east, emptying into one of the three seas bordering the China coast. The greatest of these rivers is the Yangtze Kiang (Yang-tzu Chiang); over 3,000 miles in length and navigable by motorized boats for over half of its course, it empties into the East China Sea near Shanghai. Some of its tributaries, such as the Han and the Hsiang (some 900 and 500 miles in length, respectively), also rank as major rivers. Many tributaries are navigable by small boats for hundreds of miles.

Draining the southernmost provinces is the Si-kiang (Hsi Chiang, or "West River"); more than 1,000 miles long, it forms an important navigation net with its tributaries in the mountainous south, where topography makes for difficulties in road construction. The Si system brings much of the trade of the south to the commercial gates of Canton (Kuang-chou) and British Hong Kong, although it is importantly supplemented by the southern railroad system.

In North China, the Hwang Ho (Yellow River) long ages ago laid the silt that filled in the great arm of the sea that now comprises the Yellow Plain. Running out of arid regions and through a dry land of fine-particled, easily eroded loessial soils, the Hwang Ho has created a standing invitation to land-hungry farmers to crowd its plain. At the same time, it has qualified this invitation with the misery inflicted by its frequent rampaging floods. Although a broad, mighty river, the shifting channels and sandbars of the Hwang Ho restrict its use as a navigable artery. Between the Yangtze and the Hwang runs the Hwai (Huai) Ho, which is significant in three important ways. It is a navigable stream now partly controlled by floodgates and dams. It coincides with a zone of diminishing frontal clashes of differing air masses and thus marks the climatic boundary between dry North China and humid South China. It occupies a low-lying trough in the southern part of the Yellow Plain and often receives Hwang Ho floodwaters during major breaks in the south-bank dike along Hwang Ho levees. In spite of strenuous Communist Chinese efforts to control the river, the Hwai Plain continues to suffer occasional disastrous floods. Along the mountainous southeastern coast, rivers a few hundred miles long, but with discharges which may surpass that of the Hwang, form small delta plains of local significance.

The Great Plains Although rivers form important traffic arteries and bring both blessings and calamity to people in their drainage areas, the valley plains they have carved or deposited are of much greater significance for the livelihood of China's large population. The great plains of China are mainly in the north. In northeastern China (Manchuria), the depositional Liao Ho Plain of the south joins with the erosional plain of the Sungari over a hardly noticeable water divide and forms a contiguous surface of some 138,000 square miles. The Liao Ho section is relatively level, and the Sungari section somewhat rolling in topography.

Connected with the Liao Ho Plain by a narrow corridor, which is crossed by the Great Wall at the seaport of Linyu (Shanhaikwan), the Yellow Plain extends southward without interruption around the Shantung hill country to merge with the Hwai and lower Yangtze Kiang plains. With the Hwai as southern boundary, this great plain covers an area of about 125,000 square miles, for the most part level and less than 100 feet in elevation. The Yangtze Plain, although of smaller extent, is more fragmented, more irregular in shape, and extends in a narrow zone much farther inland. Its approximately 75,000 square miles of level surface includes the central plains of the Tungting Hu and Poyang Hu (lakes), as well as the lower Yangtze Plain and the connecting corridors of lowland between them.

The Mountainous South South of the Yangtze, the plain extends in narrow slivers up the river valleys, to be gradually closed in by hills of increasing elevation and ruggedness. Most of South China is slope land, so that level land for farming is restricted to narrow, scattered valleys and basins, which total only a small percentage of the land surface. The largest of these alluvial farmlands occur along the coast.

Separating the Yangtze drainage basin from that of the Si and the small southeast coastal rivers is the great mountain system of the Nan Ling (or Nan Shan; "Southern Range"). Running out of northern Yunnan, it reaches eastward as the Ta Yu Range (or Meiling) until about the 116th meridian east, where the mountain system makes a northeastward bend and becomes the Wu-i Mountains. The general elevation of the Nan Ling system in the eastern sector is from 3,000 to 5,000 feet. From there the land slopes northward to the Yangtze Plain and southward to the Si-kiang Plain. The latter comprises a network of narrow river floodplains until it fans out into its delta south and east of Canton. Here, the so-called Canton Delta spreads over almost 3,000 square miles, comprising the largest and most populous flatland south of the Yangtze Plain.

The Western Plateaus and Basins West of the great plains and the South China hill lands lies the greater part of China's territory. This vast western area generally comprises a number of large plateaus and basin lands of greatly differing character. Occupying an immense area in the southern part of the far western reaches is one of China's colonial realms, the Tibetan High Plateau, which averages over 12,000 feet in elevation and is a cold, forbidding, and sparsely populated upland. North of this, the land drops into the three great desert basins of Tsaidam, Tarim, and Dzungaria: the first is about 9,000 feet in elevation, and the latter two are from 1,000 to 4,000 feet above sea level. Near the eastern end of the Tarim Basin is the down-faulted depression of Turfan, 980 feet below sea level. The Tarim and Dzungaria basins are separated by the 18,000- to 20,000-foot Tien Shan Range, extending a thousand miles eastward into China from the Soviet border. Still farther eastward from the Tien Shan stretch the great desert plains of the Mongolian Gobi (basin), reaching to the Great Wall of China.

South of the Great Wall, east of the Tibetan High Plateau and west of the great plains and the mountainous south, is an intermediate zone formed by four major topographic units. From north to south, these are the Loess Plateau and valleys, the Central Mountain Block, the Szechwan Basin or (Ssu-ch'uan; also called the Red Basin), and the southwest plateaus in Yunnan and Kweichow provinces. Physically speaking, the last of these topographic units continues into northeastern Burma, where it is called the Shan Plateau, and into northwestern Vietnam and Laos. Economically, the Szechwan Basin, supporting an estimated 86 million people in 1970, is most important. In its geographical influence upon the climate of the country as a whole, the Central Mountain Block is of the greatest significance.

Coastline and Islands China's coastline divides roughly into two differing halves. From Shanghai northward, most of the coast is flat and low-lying and often suffers saltwater inundations during the onshore summer monsoon winds. The only important exceptions are the coasts of the hilly Shantung Peninsula and its geological extension northward in the Liaotung Peninsula of southern Manchuria. Ports on the flat coastlands are few and poor, and settlements also are sparse and small along this coast. By contrast, such excellent ports as Tsingtao (Ch'ing-tao) in Shantung and Dairen (Talien) in Liaotung are found along the hilly coasts.

South of Shanghai most of the coast is highly irregular and hilly, and it has numerous bays, headlands, and good harbors such as Amoy (Hsia-men), Hong Kong, and Chan-chiang (formerly Kwangchowan). Off these hilly coasts also are found most of China's numerous small islands, as well as the two great islands of Taiwan (Formosa) and Hainan. Separated by about 170 miles of sea from the mainland, Taiwan forms part of the western Pacific volcanic island arcs. Hainan, on the other hand, is only about 25 miles from Leichou Peninsula and geologically is closely related to the adjoining mainland. About 100 miles southeast of Shanghai is the Choushan Archipelago

Climate and Regional Differentiation

In a general way, China may be divided climatically into a higher, dry western half and a lower, more humid eastern half. Each half in turn subdivides generally into two halves, north and south, with differing climatic characteristics. There are thus four climatic quadrants.

The situation of China is such that most of the moisture obtained by the land is borne in by air masses from the Pacific and, to a lesser extent, from the Indian Ocean. Virtually no moisture is brought in from the Arctic Ocean, and only a very small amount comes from the distant Atlantic to water the western frontiers of China. Mongolia, Siberia, and Central Asia build up great pressures of cold air masses during the winters because of the rapid radiation of heat to outer space. These result in an outward movement of cold, drying winds to much of China. Winter is a season not only of cold but also of drought in the north and of greatly decreased rainfall in the south.

During the summer the intense heating of the interior of the landmass creates a zone of low atmospheric pressure that draws into the interior warm, humid, maritime and tropical air, which furnishes large quantities of rain. Summer in China is the most important wet season. Several obstacles arise, however, to obstruct the penetration of moist air very far inland. The Himalaya Mountains along the southern rim of the Tibetan High Plateau rob the northward-moving air of most of its moisture during the Indian summer monsoon. Although enough moisture blows in along the Brahmaputra River valley and gorges into the Tsangpo (Ya-lu Ts'and-pu) River valley to give a rainfall that at Lhasa has varied from less than 18 to almost 200 inches per year and has averaged 57.5 inches over a ten-year period, most of Tibet has low precipitation and is very dry and barren. Evaporation is high because of the large number of clear days and the low humidity. In the western half of Tibet, where elevations average 16,000 feet, streams do not drain out to the sea but flow instead into salt lakes. The high altitude and southerly location create a climate with great diurnal temperature differences and cold winters. The sunny side and the shady side of a mountain show great temperature differences. The growing season is limited by the short period between killing frosts.

In the northwestern plains and basins north of the Tibetan High Plateau, precipitation is even more limited, reduced to 2 to 5 inches annually except on the higher mountain slopes of the Tien Shan and the Altai, where above 9,000 feet altitude it may increase to as much as 30 inches. The characteristic landscape is desert, except in the line of oases that borders the Tarim and Dzungarian fringes and in reclaimed portions of the Manas Valley in Dzungaria and of the adjoining Ili Valley, situated between two branches of the western Tien Shan. Evaporation far exceeds precipitation in most of this area, and irrigation use takes out more water, so that rivers generally become smaller in volume the farther they flow, to disappear eventually or flow into shallow salt lakes and swamps. Desert gives way to steppe in the moister western part of the Dzungarian Basin and on the northern slopes of the Tien Shan. Even forest growth occurs on slopes from 5,000 to 9,000 feet, above which meadow grasses again furnish feed for grazing livestock.

In the Tarim the climate is more arid, yet warmer because of the natural protection from northern winds and its more southerly location. Growing seasons also are longer in various parts of the Tarim Basin. Here, about two-thirds of the population of Chinese Turkistan of Sinkiang make their living in oasis agriculture based upon irrigation from glacier meltwater.

In the eastern section of the Inner Mongolian region, north of the Great Wall, the outer desert of the Gobi gradually changes to shortgrass steppe as the land comes increasingly within reach of the moist summer monsoon air masses from the Pacific. In the vicinity of the Great Wall runs the 14-inch isohyet. During wet periods this line moves northward into the pastoral nomad land, but during dry years this precipitation mark retreats south of the Great Wall. This forms the transition zone between agricultural and pastoral use of the land. Because of exposure to the full force of the Mongolian winter cold waves, extremes of temperature are great.

In the eastern half of China, moisture decreases generally from the southeast coastal mountain district as one progresses in a northwesterly direction. In this region the summer monsoon operates most strongly, and its dominance brings about uniformly hot temperatures throughout China in summer. Humidity is much lower in the north, however, and summers are less oppressive there.

In eastern China, too, a topographic barrier effects important climatic changes from south to north. Eastward out of the 20,000-foot-high Kunlun Range, stretching across the northern Tibetan High Plateau, runs the lower but still formidable Chin Ling Shan (Tsinling Range). Forming part of the Central Mountain Block, its 12,000-foot peaks separate the Szechwan Basin from the Loess Plateau. The range then forms a great broken, southeastward-trending arc toward the lower Yangtze at decreasing elevations, to end in low hills north of the city of Nanking. The cold, heavy winter air masses of Mongolia cannot surmount the higher portion of this range, so that much of the area to the south is protected during the winter and has a mild climate and long growing season. This is especially true of the Szechwan Basin, where an almost year-round growing season exists. In the lower eastern portions south of the Yellow Plain, cold waves do break through to cool east-central and South China in winter.

In summer, the Chin Ling system operates to rob moisture from the north and northwest, so that these areas have a precarious climate for agriculture. The cleavage between zones is marked. North of the Chin Ling, there is a steppe climate of monsoon type, where precipitation ranges from about 30 inches in the south to 14 inches in the vicinity of the Great Wall. Winters are extremely cold, dry, and dusty. Rainfall typically reaches a peak in July, but its occurrence during the growing season often is irregular, so that crops may suffer early drought with disastrous results. Violent downpours falling on the easily eroded soils create dangerous floods, especially in the lower Yellow Plain north and south of the Shantung hill lands. The growing season varies from 200 to 240 days, depending upon local situations.

Even though northeastern China (Manchuria) does not come under the climatic control of the Chin Ling mountain barrier, its distance from the source of the summer monsoon air masses and the position of the East

CHINA RAINFALL PATTERN
Annual averages

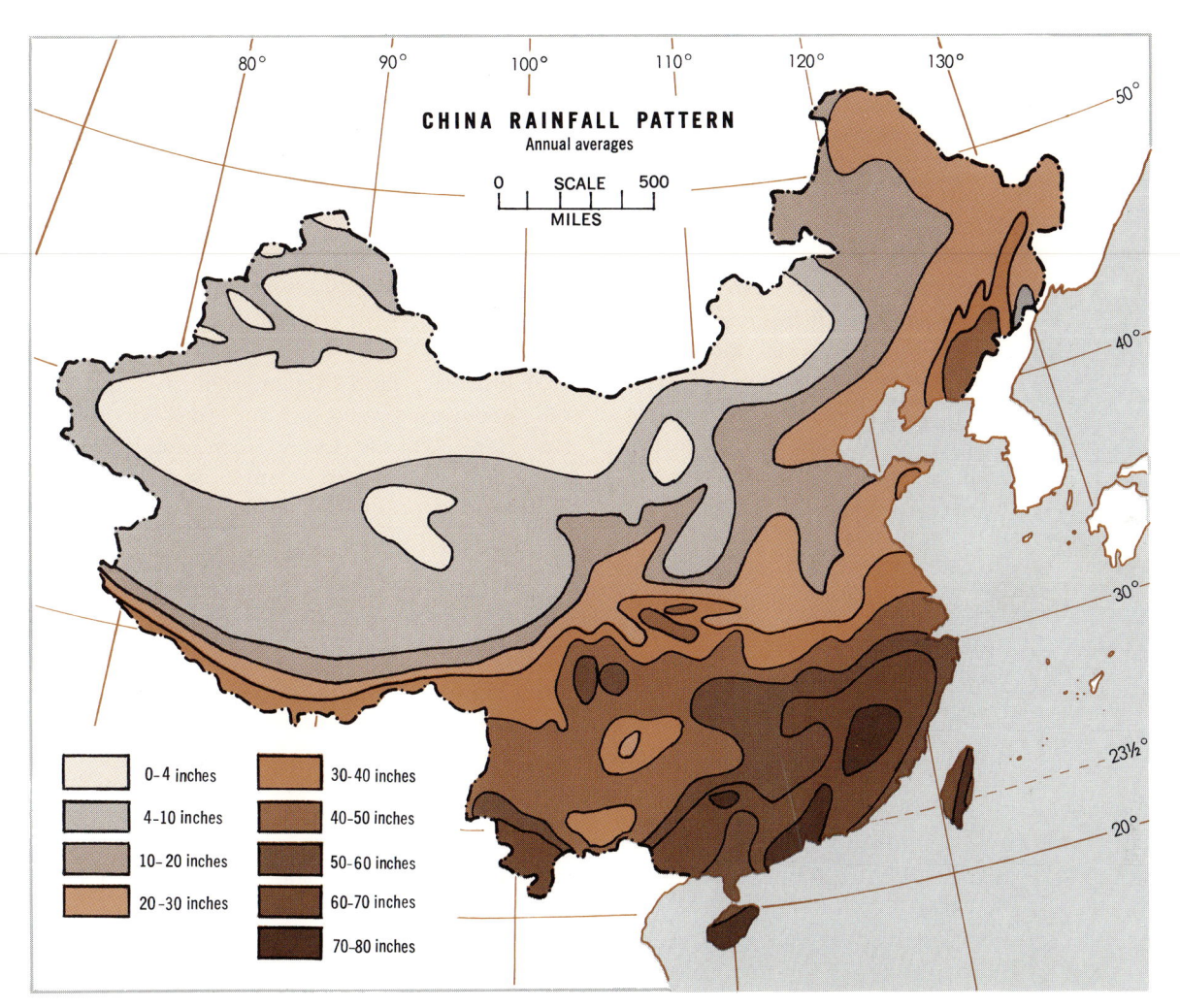

Legend:
- 0-4 inches
- 4-10 inches
- 10-20 inches
- 20-30 inches
- 30-40 inches
- 40-50 inches
- 50-60 inches
- 60-70 inches
- 70-80 inches

Fig. 13-2 The average yearly rainfall of China ranges from over 80 inches along the southeast coast to less than 4 inches in the interior basins and deserts. Much of the rain falls during the summer monsoons.

Manchurian Mountains create a climate similar to that part of North China lying south of the Great Wall, although with lower winter temperatures. Whereas rainfall may be as much as 40 inches on the higher parts of these mountains, on the Manchurian Plain it decreases rapidly from about 30 inches in the lower Liao Valley to 15 in the northwestern part of the Sungari drainage. The growing season decreases over the same area from about 220 days in the south to about 150 days in the northwest.

The more important soils of the north are the calcareous loess of the Loess Plateau, the loess-derived alluvial silts of the Yellow Plain and the valleys of the Loess Plateau, and the light-chestnut and dark prairie earths of

Manchuria. To the north of the Chin Ling line, steppe and prairie grasses form probably the most widespread original vegetation, although in the mountains forests occur in small patches. South of the Chin Ling divide, the earth in uncultivated areas is covered with a blanket of year-round green vegetation. Rainfall is over 40 inches annually in the Yangtze Valley and increases to over 60 inches toward the South China coast, with much higher amounts in the mountains. In the Taiwan mountains, rainfall may reach over 200 inches. Whereas the monsoon brings in most of this moisture, precipitation is caused not only by orography (mountain barriers) and convection but also by eastward-moving cyclonic storms, which are not so large or so frequent as in North America. The winter season is mild and also has some precipitation, although it is definitely the driest season. April and May have excessive humidity and are known as the "moldy period," especially along the Yangtze Valley.

In Ling-nan, south of the Nan Ling Range, there is no severe cold weather, although night temperatures in winter may drop as low as 38°F at Hong Kong and Canton. Here a monsoon type of humid subtropical climate occurs, in which typical tropical plants such as banana, pineapple, and palm grow. On the 6,000-foot Yunnan Plateau to the west and generally in the same latitudes, the climate is among the most pleasant in China. By contrast, the Kweichow Plateau, 2,000 feet lower, often is shrouded with fog, and its valley lands are hot and humid in summer. Off the coast, Hainan Island has a tropical maritime climate, whereas Taiwan farther to the north has a maritime subtropical regime.

As a whole, the soils south of the Chin Ling divide tend to be badly leached. Most of the southern slope lands have sterile red or yellow acid earths. The alluvial river plains have relatively fertile soils, which have been greatly modified through long cultivation, especially in the water-soaked wet rice fields. The Szechwan Basin has an exceptional status because its limestone mixed purplish-red sedimentary earths provide fertile soil for agriculture, making possible the most intensive type of slope cultivation. Terracing for paddy fields is highly developed here.

Historical Development

China is a land of many cultures and ethnic groups, among whom the "orthodox" (or Han) Chinese today form the dominant and most widespread group, over 94 percent of the total population. Some 45 million minority peoples are distributed in the outlying provinces and territories of the south, west, and north. The origin of the Chinese is obscure, but they probably derived from a mixture of ancient immigrant Turanian or Turkic stock from Central Asia with people of the early Tai culture who preceded the Chinese culture bearers in present-day China. These early Chinese established the Shang dynasty, the earliest authenticated Chinese ruling elite, in the region of the lower bend of the Yellow River, east of where the river emerges from the Sanmen Gorge. The earliest reliable Chinese historical records reveal settlement in the Yellow Plain before 1400 B.C. From here, the Chinese expanded eastward into the lower Yellow Plain and later into the rest of modern China.

Historically, in contrast with the steady Chinese population push against the lesser ethnic groups of the south, the Han Chinese have been the object of pressure from the northern and northwestern nomadic pastoral peoples. This pressure may be attributable to the difference in security of livelihood between the climatically favored south and the precarious steppeland conditions of the north. On many occasions, this environmental insecurity im-

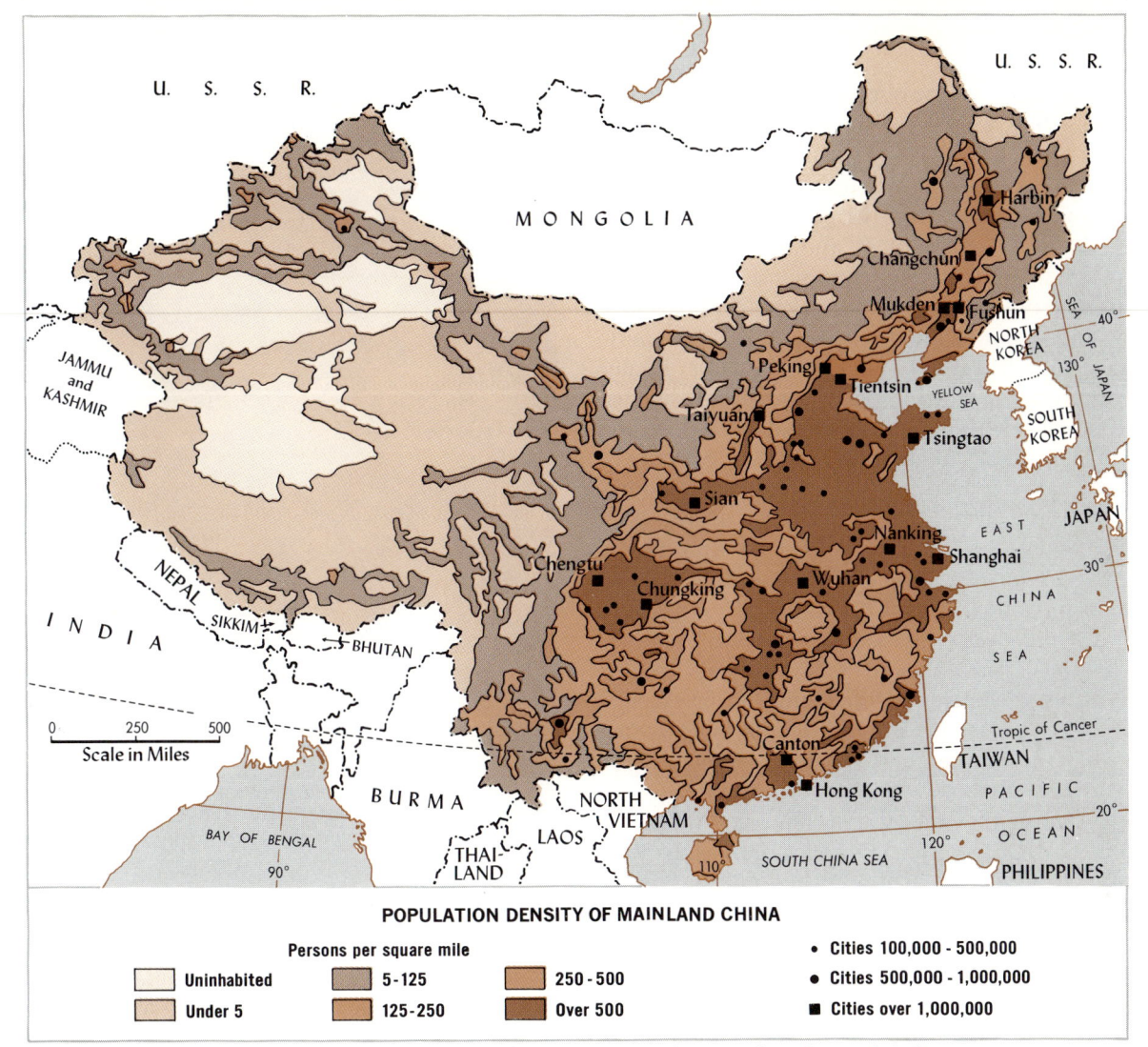

Fig. 13-3 The river valleys and plains of eastern China are among the most densely populated areas of the world. There is a close correlation of landforms, rainfall, and population distributions.

pelled the northern and northwestern nomads to federate themselves and invade the more productive, more densely settled agricultural realm to the south. Thus, for about half of Chinese history, nomadic or seminomadic peoples became the rulers over much or all of China. As a people early accustomed to the saddle, mastery of the strategic value of mobility may have helped them achieve success over their much more numerous southern neighbors. The Mongols and the Manchus were such peoples. The Manchu conquerors, however, retained their dominance over the Chinese only at the reciprocal cost of accepting "conquest" by Chinese culture.

Taiwan has a somewhat special situation,

owing to its relatively recent occupation by the Han Chinese. The original inhabitants were unorganized tribes of primitive Malay-Polynesian peoples. During the sixteenth century, Japanese pirates had their lairs on the coasts of Taiwan. Portuguese and Dutch explorers made settlements on the island, only to be driven out by refugee settlers from China fleeing from Manchu conquest in the mid-seventeenth century. The Manchus finally extended their sovereignty to Taiwan, but no great development occurred aside from gradual migration of mainland Chinese farmers and tradesmen from Fukien and northeastern Kwangtung provinces to Taiwan. More rapid economic development took place after Japan gained control of the island as a prize of the Sino-Japanese War of 1894–1895. However, the population remained largely Han Chinese. On this basis, Taiwan was returned to the existing government of China after World War II; the island is regarded by both Communist and Nationalist Chinese as an integral province of China.

Among the 45 million minority nationals and colonial peoples of China, there are more than fifty officially distinguished ethnic groups with differing languages. Aside from these, the "orthodox" Han Chinese are united by a single written language, but are distinguished and divided by seven major spoken languages, which are more or less mutually unintelligible to their speakers. Most of these different Han tongues are found in the southeastern coastal hill lands, where geographical isolation has been an important factor in their development.

Population and Livelihood

China has had a large population for 2,000 years; even during the Han dynasty in the pre-Christian era, official registers claimed a population of 60 million. That the present population is only some 800 million after 2,000 years of multiplication indicates the limitations and natural checkrein of the existing agricultural technology—good as it was for its time—and the prevalence of other drastic Malthusian checks to population growth. The Confucian ancestral system encouraged procreation of many children, with the emphasis upon sons to help in farm labor and to ensure the family line. Marriage rates have always been high, since there was no place in Chinese society for the unmarried female. Industrialization has been slow in developing, and farm mechanization has not progressed very far. Farm labor demands during the planting and harvesting seasons continue to be high. Birth control

Fig. 13-4 Thousands of Chinese live in "squatter communities" near rivers and at the edges of larger cities, as in this refugee community of Hong Kong. (Courtesy of United Nations)

measures and abortion are not so widespread in China as they are in Japan. Moreover, early Chinese Communist policy vacillated on the desirability of a still larger population for the country. China, thus, has not had the success of Japan in controlling population increase, and China's birthrate remains about twice that of Japan.

Traditional tools, implements, and methods present limitations for the amount of land a farmer can handle, and this handicap forces him to cultivate only the better soils in order to produce enough for his family. The result has been an excessive concentration of farm population on the fertile river plains and irrigable valleys. A very uneven distribution of population, therefore, has occurred.

Since about 80 percent of mainland China's population are farm people, the population distribution corresponds closely to the distribution of cultivated land. Some 290 million people are settled on the great Yellow and Yangtze plains; about 55 million are settled on the plains of Manchuria; and in the Szechwan Basin live another 86 million people. Other smaller, heavily populated plains and lowlands include the Wei and Fen River plains in the Loess Plateau region, and the various coastal delta plains of southeastern China, especially the Canton delta.

About 25 percent of China's population, including some farm people, live in towns and cities; about 10 percent of the population live in large cities. Some of the major metropolitan areas that have developed in China rank with the greatest cities of the world. Shanghai is the largest, with 8.5 million, followed by Peking, with 8.0 million. Five other cities that have passed the 3-million mark are Tientsin (4.5), Luta (4.0), Mukden (or Shenyang; 3.7), Wuhan (4.7), and Chungking (or Pahsien; 3.5). Twenty others have populations in excess of one million.

Intensive use of hand labor, characteristic of all Chinese farming, is especially true of southern paddy agriculture. A smaller unit of arable land can provide a livelihood for a family in the south because of the ample rainfall and long growing season, which permits double- or triple-cropping during one year on the same piece of land. Thus, the greater population density per unit of arable land in the south does not necessarily indicate greater population pressures on the land there. In the dry and unreliable climate of the north, a lower population density per acre actually may correlate to a more marginal standard of living. It may generally be said, however, that China's arable land is already supporting the maximum population that can subsist on its present productivity and maintain tolerable standards of living.

After millennia of trial-and-error learning, farmers in pre-Communist China selected crops suited to the environments of particular localities. Since cultural preferences in food had long centered upon rice, this became the dominant crop wherever water availability, soil type, and growing season have combined to permit it. However, in non-rice lands appreciable variety of crops prevailed because the farmers tried to spread their risk. After the Communist regime took over, there was a considerable reduction in the complexity of crop patterns as the government agricultural planners tried to rationalize the system to fit planned goals.

In the north, where dry cultivation is the rule, there is more variety in important crops. The dominant food crops are wheat, millet, kaoliang (grain sorghum), and soybeans. Most of these are grown in the south as well, and crops of barley, maize, sweet potatoes, and many varieties of beans and other vegetables also are important. In South China and in the Szechwan Basin, where irrigation water can be led to fields, rice is everywhere the summer crop. Rice terraces are common on lower hills and slopes and in mountain

Fig. 13-5 The most highly developed agricultural regions of China are on the eastern plains and in the mountain valleys. Rice is the important food crop grown in the southern part of the nation, wheat in the northern.

ravines. In the winter, wheat, barley, and oil-seed crops may be planted on drained rice fields. South of the Nan Ling Range, two crops of rice, and sometimes three, can be grown on a field in one year.

Regional specialties include the cotton crops in the southeastern part of the Yellow Plain and the northeastern part of the lower Yangtze Plain; tobacco in Shantung; sugarcane in Kwangtung, Taiwan, and Szechwan; sugar beet in Manchuria, Inner Mongolia, and Sinkiang; tung-oil nuts in Szechwan and Hunan; tea in the southeastern coastal provinces; mulberry trees (for silkworm feeding) along the lower Yangtze south of its estuary, as well as on the Canton Plain and in Szechwan; tropical fruits, coffee, and rubber in the southernmost provinces and on the two main islands. In the northwestern desert oases, almost every annual crop in China is represented, and cotton, grapes, tree fruits, nuts, and melons are local specialties.

Since most Chinese live in villages, farmers use much of their time in going to and from the fields on foot. Hand methods of agriculture and the use of simple equipment pulled by an ox or water buffalo remain prevalent, since mechanization has made only limited headway. The maximum yield from land is sought, with disregard for the human labor required. Thus, transplanting tiny rice plants from seedbeds shortens the time the crop occupies the larger fields and permits multiple cropping in areas of short growing seasons. In South China, two crops of rice are grown annually; and between these, quick-maturing vegetables often are grown and harvested. In Central China two grain crops also are grown, but except in limited areas along the Yangtze, only one of these is rice; the other is wheat, another dryland grain, or rapeseed oil plants.

Under the Communist regime, the typical small farms have been consolidated, and work on the resulting larger farms is done by communal groups using more modern techniques and improved equipment when available. Spotty and limited, the mechanization that has been introduced is found mainly in parts of the great plains of the Yellow River and the Yangtze River and in outlying areas such as Inner Mongolia, Manchuria, and Sinkiang. Only 5 to 10 percent of China's agricultural land is worked mainly by machines and other mechanical equipment.

Fig. 13–6   A field of rice on a Chinese commune. Each stock of rice is planted by hand. (Courtesy of Hsin-hua News Agency)

Resources

Because of the demands of a large population, as well as because of destruction for other reasons, most of the forests of China have long since disappeared. As a result, in most mountain regions climatically suitable for forest, no seed-bearing trees remain to reforest the slopes, which most often are covered with grass, bracken, and shrubs. Today, important forests are found only in remote parts of China's southwestern mountains, in the northern and eastern Manchurian mountains, and in mountainous parts of the southeastern coastlands. In South China, bamboo is very widespread and furnishes a most important material for house construction and for a wide variety of other uses.

Fishing is an important occupation along most of the coastal waters and on inland streams and lakes. Because of the limited number of modern fishing vessels, large-scale commercial and deep-sea fishing has had slow development. South of the lower Yangtze, fish pond culture is a sideline on many collective farms. Fish fry from southern Yangtze provinces are sent by air to stock lakes and streams in many parts of north and northwest China.

On the basis of general geological surveys, it is believed that China has a moderately good supply of the basic requirements for modern industrialization; the country has vast quantities of some materials but is deficient in others. Thus, China claims second or third rank in world coal reserves, but known petroleum resources are relatively modest for her needs, and many important oil fields are located far from the main populated consuming regions. Iron ore reserves in 1962 were claimed to be about 20 billion tons, but many of the deposits are small and of low quality. Although China needs most of the minerals it produces, the country has a surplus in certain strategic items such as antimony, tungsten, and tin. In the regional distribution of minerals, the north and northeast have most of the coal and iron; the south and southwest, the major part of the ferroalloys and nonferrous metals. Abundant ores for the magnesium and aluminim industries are found both in the north and in the south.

Petroleum would appear to be found mainly in the northwest desert and semiarid lands and in Szechwan; but a recent oil discovery in central Manchuria, near the village of Taching, is also providing significant oil supplies in a more accessible region. Oil shale is exploited chiefly in southern Manchuria, in deposits overlying the Fushun coalfield, and at Maoming south of Canton. Together with coal, these may form the base for synthetic fuels. In waterpower potential, located mainly in the south, China claims second world rank, but these sources are largely undeveloped. Several

Fig. 13–7 Almost all land that can be put into cultivation on Taiwan (Formosa) has been terraced and is so used. (Courtesy of Chinese Information Service)

excellent sites exist for large-scale power development, the most noted being the Yangtze Gorge, with its reported potential of 13 million kilowatts. The best-developed sites for available hydroelectric power lie along the upper Sungari and lower Yalu rivers in Manchuria; at the Sanmen and Liuchia gorges on the lower and upper Yellow River, respectively; on the Hsingan River of northern Chekiang province; at several sites on the Si-kiang system; and on tributaries of the Yangtze in Szechwan and Yunnan provinces.

In the mineral realm, on the whole, one may say that China has the basis for a considerable degree of industrialization, although its per capita mineral wealth is not high. Retarded development may be attributed to past political instability and internal disunity, the lack of large capital accumulations for investment, and difficulties in communication. Rapid mineral exploitation has occurred since 1960.

Manufacturing

Although modern industries have made much progress in China, it is safe to assume that about three-fourths of China's domestic needs for manufactured and processed materials still are supplied by traditional handicrafts and home industries, which are now organized as communal or collectivized rather than private enterprises. Leading processing industries such as rice and wheat-flour milling are most important in the industrial economy. Rice is milled mostly by traditional methods, often with waterwheels as power sources, but wheat flour is manufactured in the great modern flour mills of such cities as Shanghai, Tientsin, Wuhan, and Tsingtao.

China's masses need cheap cotton cloth, for which household looms formerly supplied almost the entire demand; but the introduction of modern machinery led to a decay in this branch of home industry. Subsequently, foreign as well as domestic interests established large cotton mills in Shanghai, Tientsin, Tsingtao, and other cities to utilize the cheap labor supply there. Communist China has greatly expanded the production of cotton textiles, which comprises China's foremost modern industry, and the three above-mentioned cities, together with Peking, Chinan (T'sinan), Sian, and Urumchi (Tihwa) in the north and northwest, are leading textile producers. Silks are concentrated especially in the southern part of the Yangtze delta and at Canton.

Modern heavy industries such as those connected with chemicals, iron and steel, and machine manufacturing have developed to the greatest extent in southern Manchuria. These industries owe much of their initial growth to the efforts of Japan, which seized the area in 1931 and built up a base for political and military expansion in East Asia.

South of the Great Wall, the region from Peking eastward through Tientsin to Linyu, where the Great Wall comes to the sea, is another important industrial area whose activities are based upon the coal mines located nearby. Cement, chemicals, fertilizer, and glass manufacture and a variety of light industries are found in the larger cities and towns of the area, as well as some iron smelting and steelworking. The great labor concentrations and industrial complexes in Peking and Tientsin are especially important.

Since 1955 the Chinese Communist regime has built an important iron and steel center at Paotow (Paotou), outside the Great Wall in Inner Mongolia. A major oil refining and petrochemical center has developed at Lanchow (Kaolan) in Kansu province.

In central China the Wuhan metropolitan area is an important center for iron and steel manufacture, flour milling, dried-egg processing, vegetable oil pressing, and textile manufacturing. Farther westward in Szechwan, most of the industrial concentrations are around

Chungking, with a less important region located around and south of the provincial capital at Chengtu. Antimony refining and light industries are found at Changsha, south of Tungting Lake (Tung-t'ing Hu). In the triangle formed by Nanking, Shanghai, and Hangchow on the lower Yangtze Plain is an industrial complex specializing in silk and cotton textiles and having a multitude of other light industries. Shanghai, however, is a major steel center, and Nanking is a major chemical manufacturing city.

In the south, the Canton area and British Hong Kong form the nuclei for a variety of similar light industries, ceramics manufacture, shipbuilding, and sugar refining. Sugar refining is even more important on Taiwan. Under Japanese occupation, Taiwan reached a higher stage of industrial development than did the Chinese mainland as a whole. This per capita lead has been maintained since the Nationalist Republic of China established its headquarters on Taiwan in the late 1940s. Nevertheless, although only about 45 percent of the population on Taiwan are engaged in farming, most Taiwan industries are concerned with food processing. This island is one of the largest world exporters of fresh and canned pineapple as well as of sugar. Moderate coal supplies are available on Taiwan, and a large petroleum refinery utilizes imported crude oil. Hydroelectric power also is important on Taiwan, where a long-established plant extracts aluminum from imported ore. Manufacture of cotton textiles is one of the leading industries. The island's one good natural harbor as well as the port for the capital Taipei is Keelung (Chilung), where ships and tankers up to 36,000 tons have been built. Kaohsiung is the major port on the

Fig. 13-8 Peu-chi iron-and-steel mill, Manchuria (Manchow). Much of the heavy industry of China is centered in the northeastern provinces. (Courtesy of Hsin-hua News Agency)

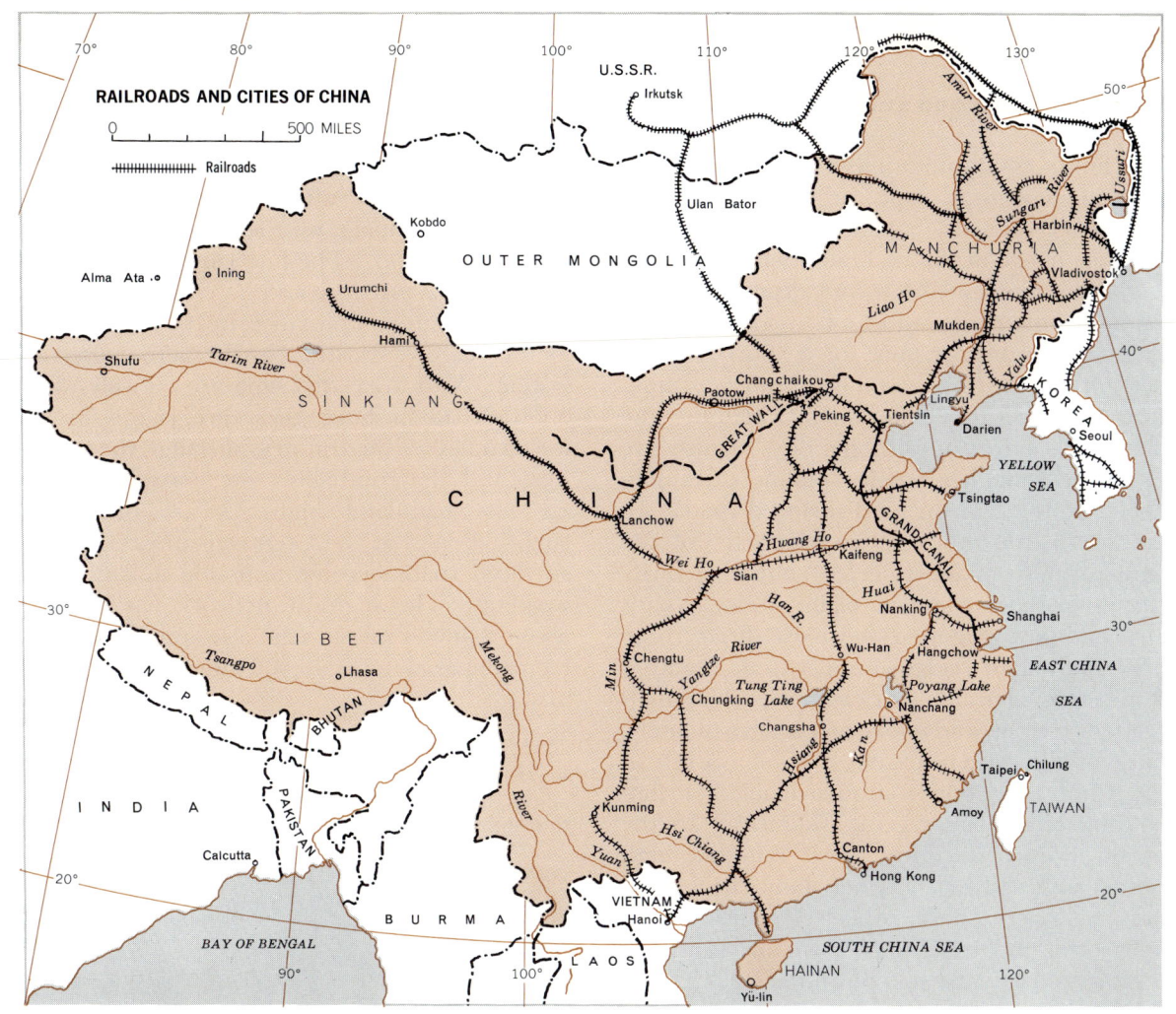

Fig. 13-9 The cities and industrial centers of China, as well as most of the railroads, are located in the eastern provinces. In recent years, however, railway lines have been extended far into the interior areas.

southwest coast. Both on the mainland and on Taiwan some automotive manufacturing has begun, and large oceangoing cargo ships and tankers are constructed at Shanghai and Dairen (Talien).

These examples of modern manufacturing are listed to show that industrialization is making important inroads in Chinese economic life; but in the current Western sense, the country has only begun to industrialize. In certain key commodities such as cotton textiles, village and home industries have been displaced; in other areas, these cottage operations still play a most important role in filling domestic needs. In mainland China, private business enterprise has been supplanted by collective and state control. Overall, the Chinese still are mainly farmers.

Communications and Trade

China's communications system has always been inadequate to cope with its geographical immensity and its notable topographic and climatic irregularities. Traditional means of communications are slow and expensive. The value of bulk commodities often could pay for their own transport only for a short distance beyond the river or coastal boat transportation. Rutted dirt roads in the north, which become impassable mires in rainy periods, require five times the animal power for pulling carts than would be the case on hard-surfaced roads. In the south, the mountainous character of the land formerly made wheeled vehicles a rarity except in the infrequent plains areas. Roads and pathways followed the most direct route over mountain ridges and often required stone steps. Human carriers and pack animals carried freight and passengers. Fortunately, most of the larger streams of China south of the Chin Ling divide are navigable for flat-bottomed craft over great distances. In the north, winter freezing, low rainfall, high silting, and shifting channels limit the usefulness of rivers; here, wheeled vehicles have long been the traditional form of transport.

Railroad construction in China was begun in the late nineteenth century as a result of pressure by Western and Japanese interests that wished to exploit the new materials and markets of China. Because of the strategic and economic importance of Manchuria, historical events have made the railway network here the best in China; about 40 percent of the total railroad mileage of the country is located in this northeastern sector. North China proper has some 23 percent of the mileage, and central China includes 10 to 12 percent; southeastern China, excluding Taiwan, has about 8 percent; Taiwan itself has about 12 percent. The remaining 5 percent comprises the long northwest railroad from Sian to Urumchi and beyond in Chinese Turkistan.

Since 1965, rail connections have given Szechwan its first direct access to an ocean port at Chanchiang, on the Leichou Peninsula south of Canton. The line runs via Kueiyang in Kweichow province and Liuchow in Kwangsi province. Yunnan province, in 1969, had access to the sea only via the North Vietnamese rail system. Vast areas of China are entirely without railroad facilities; these are mainly in the western half of the country, although the southwest and the south also are sparsely served.

Most mainland highways are only dirt-surfaced or, at best, gravel-surfaced; dirt-surfaced roads become unusable during rainy weather. The highway network has reached about 240,000 miles. China is annually manufacturing as many as 40,000 motor vehicles with

Fig. 13-10 Kowloon, a peninsula of the China mainland across the harbor from Hong Kong Island, is a major tourist shopping area. (Courtesy of Trans World Airlines)

three or more wheels; most are trucks and buses, rather than private autos.

As a result of construction by both Japanese and Chinese military forces during World War II, all major Chinese cities and numerous strategic, though lesser, population centers have airfields.

Foreign trade in Communist China is a state monopoly, and postwar imports of the mainland Communist regime have reflected the austerity the Chinese masses have had to face to achieve the planned industrial transformation. Only essential commodities are imported. After the development of Sino-Soviet political differences subsequent to 1959, overall trade with the Soviet-bloc countries dropped to 25 percent of China's total; free-world trade, however, climbed to constitute 70 percent of China's total foreign trade. China's food needs resulted in major purchases of grain from Canada, France, and Australia. At the same time, China has been striving to increase its exports of manufactured goods to the less-developed Asian, African, and Latin American countries.

First among the historically important ports of China is Shanghai, the outlet for the huge Yangtze drainage area; for many years it handled half of the country's foreign trade. Canton has been the major Chinese outlet for southern China, although Hong Kong, with greatly superior harbor facilities, has been at once a rival and a transshipment port for Canton. The recent development of Chanchiang, on the Leichou Peninsula south of Hong Kong, has provided an increasingly important alternate port of entry for South China. Tientsin is the most important gateway for the Yellow Plain, although Tsingtao attracts some of this hinterland trade. In southern Manchuria, Luta (Dairen and Port Arthur) is a commercial port of first rank; urban expansion merged the two formerly separate ports. Dairen, Tsingtao, and Yulin on Hainan Island are important naval and submarine bases.

Mongolian People's Republic

Relief Features

The Mongolian People's Republic is the name given by its Communist rulers to the geographic area traditionally called Outer Mongolia. Its separation from Inner Mongolia, which has no physical basis either in topography or climate, is justified mainly on the rather tenuous grounds of tribal and dialect divisions among the Mongolian peoples concerned. In the west, the Altai Mountains constitute a natural boundary separating the Dzungarian Basin and Soviet Kazakhstan from the Khobdo Basin of Outer Mongolia. The country as a whole is part of the great Mongolian Plateau and ranges in elevation from 4,000 to 6,000 feet, although the central Khangai and Tarbagatai Mountains have elevations up to 9,000 feet and the Altai rises above 10,000 feet. Much of the south and southeast are rather level plains arranged in large shallow basins, or "gobi," often with great patches of barren desert pavement. The western third of Outer Mongolia forms a large basin containing many salt lakes, and the northern frontiers of the country have such mountains as the Sayan, Tannu Ola, and Kentei ranges, in order from west to east.

Since the atmospheric moisture for this area is derived from the distant Atlantic Ocean or from the scanty evaporation from the Arctic, the precipitation decreases the farther south the land extends. Near the northern frontiers, rainfall averages 10 to 12 inches. At Ulan Bator (former Urga), the capital, it reduces to about 8 inches. The southern half receives less than 5 inches; the southwestern Great Gobi and the Khobdo Basin receive only 2 or 3 inches or less.

Only the rivers draining from the northern slopes of the Khangai and the northern mountains find their way to the sea after running into Siberian lakes and rivers. The vegetation pattern corresponds closely to the rainfall pattern. The southwest Gobi region and much of the Khobdo Basin floor have only xerophytic shrubs and grasses. Toward the east of the Gobi, sparse short grass and bunch grass furnish scanty fodder for grazing animals following the early summer rains. Beginning with central Mongolia and proceeding northward, the land gradually provides a more flourishing steppe grass cover, until rather luxuriant grasslands similar to prairies are reached in northern Mongolia. In the higher mountains, even forests thrive where protected from cutting. With the exception of southwestern Mongolia and drier parts of the Khobdo Basin, Outer Mongolia may be described as one vast rangeland, much of it eminently suitable for grazing livestock, in what is a typically continental steppe climate.

The Pastoral Economy

During the first decades of this century, an estimated 13 million domestic animals grazed the Mongolian plains, to furnish a

Fig. 13-11 The Gobi dominates southern Mongolia. The principal rivers and most of the cities are in the northern part of the country.

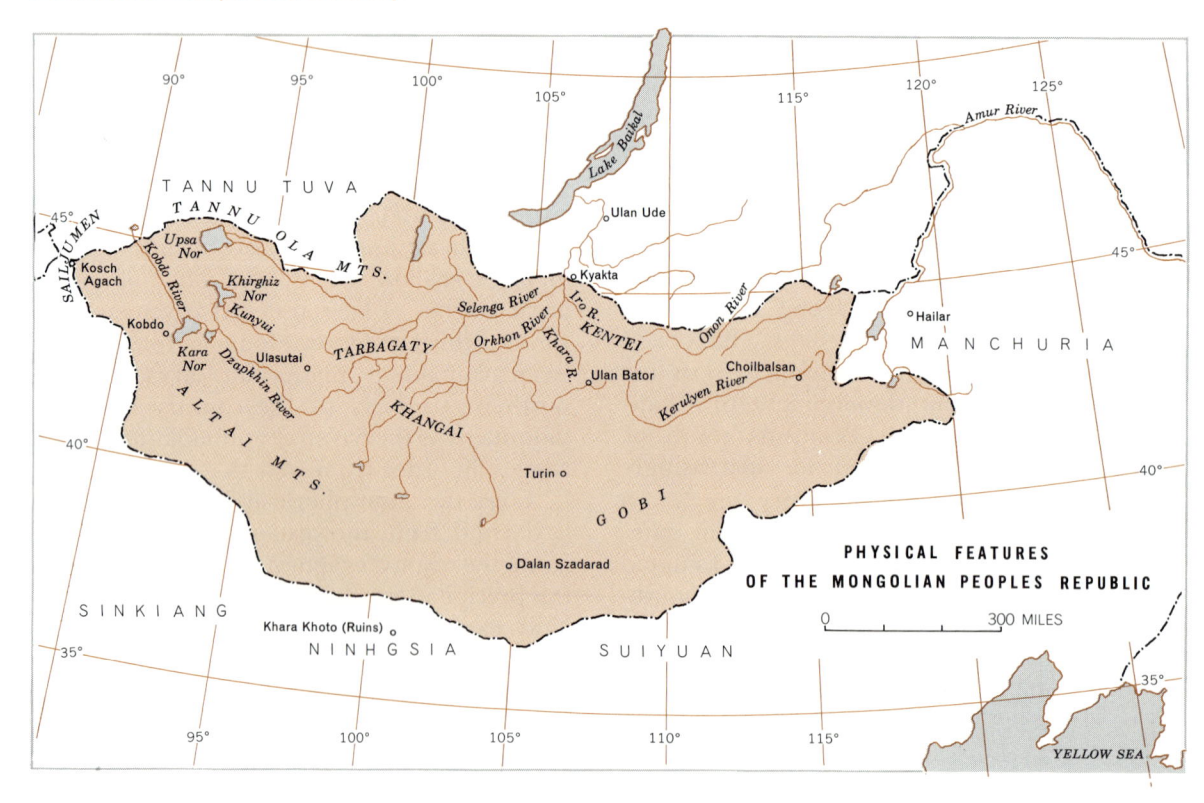

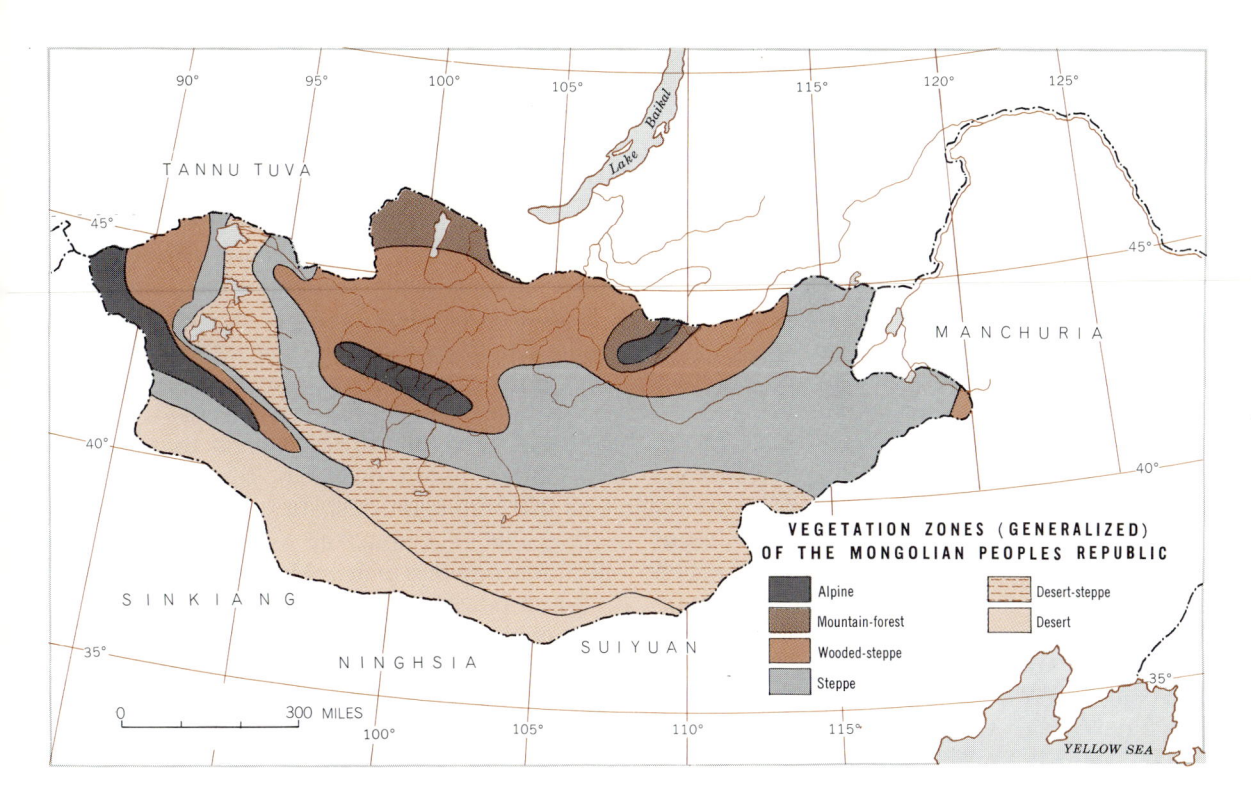

Fig. 13-12 The vegetation zones of Mongolia are transition areas between those of the U.S.S.R. and China; in general, they extend in an east-west direction.

livelihood for about 800,000 Mongols. By 1941, the animal numbers had grown to more than 27 million. However, in 1955 this had dropped to 23 million head, and small success has attended efforts to increase their number. Thus, in 1970 the livestock was reported at 24 million head. More than two-thirds of Mongolia's livestock are sheep and goats. In the 1960s, sheep alone numbered 13.5 million; cattle and horses numbered 1.9 and 2.4 million, respectively; and there were about 864,000 camels. Sheep have a significant position in Mongolian life. The Mongol drinks their milk; eats their meat and cheese made from their milk; uses their wool for clothing, felt shoes, and tents; and burns sheep dung for fuel in cooking and heating his yurt (tent). Because most of Mongolia is climatically unsuitable for agriculture, the Mongol must adapt himself to the needs of his flocks and seek new pastures as old ones become exhausted. Among the characteristic developments of Mongol nomadism is the yurt, a light and collapsible felt tent that furnishes adequate protection from the fierce winter winds and the −43°F temperatures that may occur.

The Mongol meets many problems in his attempt to make a living. He must adapt himself to great continental extremes of heat and cold. Surface water is scarce and must be supplemented by shallow wells, the water of which often may be brackish. Hard-crusted snow and winter blizzards often bring about heavy livestock losses through starvation. Natural enemies such as wolves, animal parasites, and livestock sicknesses such as hoof-and-mouth disease may kill off a large percentage of his flocks and herds. Finally, there is a saturation point in the capacity of the grasslands to support increased numbers of livestock, so that increased human population must find other means of support or starve.

Early Chinese and Russian efforts in Outer Mongolia's agriculture met with limited success. By introducing mechanization and state farms, the Mongols had increased cultivated acreage to 674,000 acres by 1960; in 1969, acreage plowed and sown to grain and fodder crops was over 1.6 million acres. Further expansion has been going on. Wheat comprises the dominant crop, followed by oats and barley. Some 1,660 acres were planted in vegetables, and 3,420 additional acres in potatoes.

Under Soviet Russian influence, the Mongolian nomad families have been forced to adopt a more sedentary mode of life in which social and educational improvements can be brought about more effectively and political control is facilitated. Strenuous efforts in education have resulted in a reported 95 percent literacy rate for the Mongol population. Political theorists and economists, as well as agricultural, mining, and other technicians, from the Soviet Union have aided and directed the development and exploitation of Outer Mongolia's resources. More and more Mongols are becoming trained to carry on such work. Desire for the mineral wealth of the Tannu Tuva sector of Mongolia resulted in its formal annexation by the Soviet Union in 1946, after more than a decade of Soviet occupation. The other parts of northern Mongolia also have significant mineral wealth, including coal mines presently supplying Ulan Bator's thermal electric power generator and its few manufacturing and processing plants. Coal and oil are also found in the southeastern Gobi.

Historical Development

The Mongols became prominent in history at the beginning of the thirteenth century, when their famous leader Genghis Khan united the divided tribesmen of the steppes through skillful political intrigues and military prowess and launched one of the most amazing empires of all time. This empire proved too large and loosely knit to hold together for long, and Mongol power rapidly declined. In the fourteenth and fifteenth centuries, after the Mongols were driven out of China, the Yellow sect of Lama Buddhism took a firm hold among the Mongols. Church power gained at the expense of the ruling nobles, and the tradition developed that every family should have at least one son in the Buddhist priesthood. In contrast to the nomadic existence of the Mongols as a whole, the monasteries became the focus of the only large fixed settlements in the country. Some of these had thousands of monks, who, for the most part, were a nonproductive, parasitic element burdening the country's economy and the rest of the society. Today, most of the lamaseries have been eliminated or greatly reduced, along with the destruction of the nobility.

Because of the region's historical and political-economic ties to China, rivalry between Communist China and the Soviet Union for influence and dominance in the Mongolian People's Republic may well continue, as it has prevailed over the past century. Strategically, Outer Mongolia occupies a dominating position in Central Asia. The northern boundaries are only about 100 miles from the vital Trans-Siberian Railway. From southwestern Mongolia, traditionally, it has been easy to cut off access to Sinkiang from China proper via the Kansu Corridor. In the east, the Great Hsingan

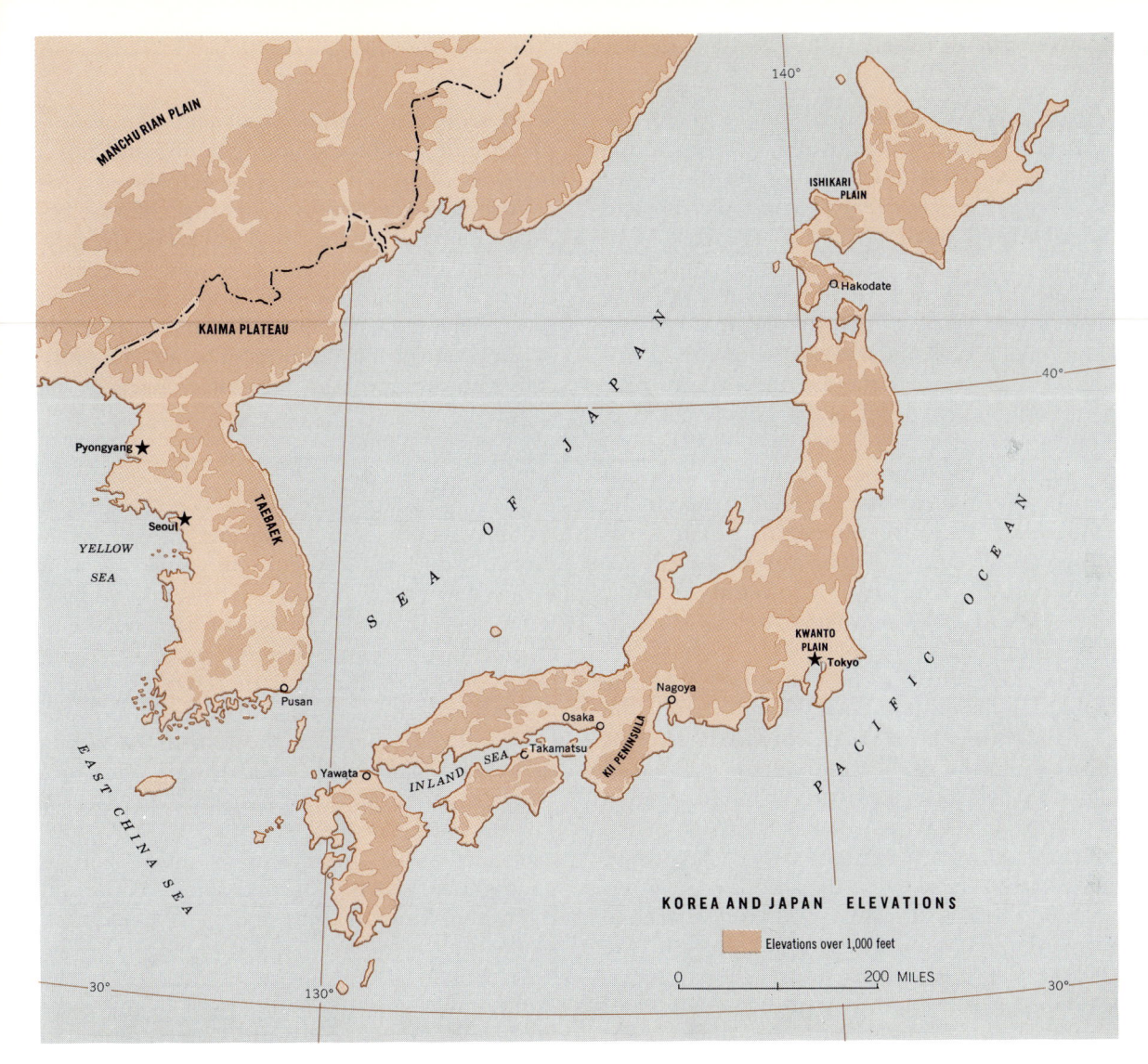

Fig. 13-13 In Korea the highest land is in the north and along the east side of the peninsula. The interior portions of each of the four main islands of Japan have elevations greater than 1,000 feet. In both countries, the highlands present problems to the development of transportation.

(Khingan) Mountains of Manchuria rise a little above the plateau edge and have commanding routes into the Manchurian plains. It is small wonder that Soviet Russia has been and is eager to maintain its dominant role in this strategic country.

Korea

Relief Features

Topographically, Korea is an elongated block tilted from northeast to southwest, with an area of 85,286 square miles. Its greatest length is about 450 miles and its width ranges from 100 to 150 miles. North of the 40th parallel, most of it is a deeply incised plateau and mountainland forming part of the East Manchurian mountain system. In the central part of this north section, at an elevation of some 5,000 to 6,000 feet, lies the Kaima Plateau. A mountain backbone not exceeding this height follows the east coast the length of the Korean Peninsula and is known as the Taebaek Range. The highest elevation is about 9,000 feet, on the dead volcanic peak of Baektu San, which lies on the Manchurian frontier between the Yalu and Tumen river sources. Cutting across the peninsula from Seoul northward to the east coast port of Wonsan is a lowland corridor, or depressed series of valleys, forming a strategic communications route.

The westward tilt and the situation of the highwater divide less than 30 miles from the east coast bring short, steep drops to the streams flowing into the deep Sea of Japan, with small alluvial plains located near their mouths. Most of the drainage flows westward, except in the southern part and where the river valleys trend southward. The slope of the land is more gradual, and the alluvial plains widen out where the rivers deposit their silt-laden waters into the shallow Yellow Sea. Agricultural development, therefore, has been largely in the western and southern parts of the country, where level land is more abundant.

Because of the land connection of northern Korea along a broad front, the continental climatic influence is strongly felt, and the higher elevation of the north also increases the severity of winter temperature extremes there. By contrast, the southern maritime parts of Korea have a mild marine climate, owing to the effect of the warm Japan Current bathing its shores. Along the northeastern coastlands, the cold currents moving southwestward from the Tartar Strait chill the atmosphere and produce damp, gloomy, and often foggy weather that restricts some forms of agriculture. There also is a striking contrast between the tidal rise of 2 to 3 feet along the eastern coast and the differences of over 20 feet between high and low tide along the western coast, where the inrolling tide piles up water in the shallow Yellow Sea. This creates difficulties for shipping in the western parts of the country.

Agriculture

Since here, as in China and Japan, rice is the cultural preference in the staple diet, it is raised wherever possible. The rainfall ranges from less than 20 inches in the northeast coastal region to over 60 in the southern mountain slopes. Paddy rice, therefore, is chiefly a southern crop, although it is found on low-lying alluvial land in the north, including the Yalu Valley. Millet and winter wheat are the most important crops in northern Korea; soybeans are grown everywhere. Barley, which takes up about three-fifths as much land as paddy rice, is the second most important crop; the bulk of this, too, is concentrated in the southwest part of the peninsula. The sweet potato also is an important crop in the south. Wheat is most important in the great peninsula jutting westward between North Korea's Pyongyang and the capital of South Korea at Seoul. From Seoul southward, cotton and tobacco become increasingly significant industrial crops, although some of both is also grown in North Korea. Mulberry and silkworm cultivation is prevalent in the southwest and south-central provinces of the country. The north, with its short growing season, has only one field crop per year; but the south grows two on the same land per year.

Post-World War II development in politically divided Korea resulted in a differing evolution in north and south. North Korea followed the collectivized system of Communist China, and individual ownership has been abolished. In South Korea, land reform brought land ownership and increased prosperity to a majority of farmers. Although moderate progress has occurred in agricultural technology, such as the use of improved seed and plant selection, pest and disease control, chemical fertilization, and mechanization after the Japanese model, food production still has not met domestic needs. South Korean arable land per capita in the early 1960s was only 0.8 acre, and farm size averaged less than 2 acres.

In 1968 the South Korean government ordered a drastic change in farming methods and a rearrangement of ownership to bring a farmer's numerous scattered plots into one piece to increase efficiency. New irrigation systems also are being built to ensure adequate water for larger areas of crops. Recent signs indicate a general economic upswing, leading to increased productivity in both farming and industry and to hopes that the South Korean economy eventually can become self-sustaining.

Fisheries

Off the Korean coasts are important fishing waters, which were largely exploited for Japanese benefit before World War II. The mixing of cold and warm currents in the Sea of Japan and the resulting upwellings have led to exceptionally productive fisheries. Fishing rivalry among Japan, the Soviet Union, and Korea makes for conflicts in the Sea of Japan. In order to avoid seizure, North and South Korean fishing boats must take care not to intrude into waters beyond their own sides of the 38th parallel. Current Korean marine food production has not reached the pre-World War II levels attained when it was under Japanese control.

Historical Development

Korea has served as a cultural link between China and Japan for over 1,500 years. For about 1,000 years of its history, much of Korea formed part of the Chinese Empire. During much of its history it was divided into several kingdoms, so that the present political division is not without precedent. Unification of the country under a separate rule occurred during the tenth century; but, like other countries bordering on China, Korea sent periodic tribute to the Chinese emperor until the nineteenth century.

The struggle between China and Japan for domination of Korea became acute after Japan emerged from its own period of isolation in 1852 and began to emulate its Occidental teachers in the art of empire building. The defeat of China in the Sino-Japanese War of 1894–1895 eliminated China's influence in Korea for 55 years. The struggle over Korea then shifted to a competition between Japan and Russia. As a relative newcomer on the scene, Russia brought her East Asian frontiers down to North Korea in 1858 after acquiring a slice of Chinese-claimed territory on the Pacific shores. Japan's victory in the Russo-Japanese War of 1904–1905 gave her a free hand over a Korea ruled by a corrupt, inept, and backward court. Japan annexed Korea in 1910 and for the next thirty-five years tried, by police-state methods, to stamp out Korean culture and national consciousness. She succeeded only in instilling the deep-seated resentment that resulted in the ousting of all Japanese after Japan's defeat in 1945.

The unfortunate Yalta Agreement among the United States, Great Britain, China, and the Soviet Union, bringing the Soviet Union into the Pacific war against Japan two weeks before the latter collapsed from the pressures of American and United Nations attack, permitted Soviet forces to occupy North Korea and led to splitting the country in half in 1948. This split was demarcated by a boundary, the 38th parallel, meant only as a temporary expediency to maintain civil order by the advancing allied armies from the south and that of the Soviet Union from the north.

The determination of the United Nations to form a united independent Korea met with the equally determined Soviet ambition to control the whole country. In the clash that came in 1950, the failure of the Soviet Union's indoctrinated North Korean armies caused strategic dilemmas for the Soviet Union, which could be solved only by bringing in the forces of its powerful ally Communist China. Thus, newly oriented to an alien ideology, China was restored to its historical position in the struggle for domination over northern Korea. The development of Sino-Soviet hostility after 1962 made for difficult choices for the North Korean Communist leadership, which owed military and political as well as economic debts to both Communist China and the Soviet Union.

Resources and Industry

Korean resources are so modest that there is little hope for a high degree of industrialization in the near future unless Korea emulates the Japanese in exporting labor in the form of manufactures in return for imports of raw materials and fuels. Coal reserves of usable quality have been estimated at only about 34 tons per capita, or about one-seventh that of Japan and a small fraction of that for China. Because of its low quality, much of Korea's coal must be mixed with imported coal to be satisfactory for industry. North Korea, however, is fortunately located to import good-grade coal from Manchuria. For iron ores, Korea is more favored; other important minerals include gold, tungsten, and graphite. The country has significant waterpower potential.

In regional distribution of resources, there is a great disparity between north and south. Korea north of the 38th parallel has 70 percent of the existing forests, 67 percent of the coal reserves, 80 percent of the iron ore, 63 percent of the known gold reserves, and most of the hydroelectric power, as well as better fishing waters. It is not surprising therefore to find that the south, with 75 percent of the 43 million total population in 1970, is mainly agricultural; whereas the north, with the bulk of the industrial potential, naturally has the greater industrial development.

The Korean War resulted in such heavy destruction both in the north and south that mining and manufacturing were almost brought to a halt. Postwar reconstruction with foreign aid—Russian and Communist Chinese aid in the north and American aid in the south—launched both North and South Korea upon strenuous industrializing efforts.

The generally unfavorable contrast for the south in resources and industrial potential is the result not only of the preponderant concentration of minerals, except for tungsten, in North Korea, but also of the location of most of the waterpower potential and the best hydroelectric sites in the higher, rougher topography of North Korea. Although North Korea outshines the south in heavy industry and electric power, the south has a superior position in varied light industries, especially in textiles. The warmer, more moist climate favors cotton growing and silkworm breeding in the south. A larger volume as well as a greater variety of fabrics, including synthetics such as rayon and nylon, are manufactured in the south.

North Korean industry is especially concentrated in and about its chief city and capital Pyongyang, at the port of Chinnampo to the southwest, and at Sinuiju, near the mouth of the Yalu River. Hamhung (Kanko) and Wonsan on the east coast of North Korea are industrial centers supplied with power from hydroelectric dams and plants in the Kaima Plateau gorge region to the northwest.

South Korea's most important and most varied industrial complex lies at and between the capital Seoul and its seaport Inchon. Other important industrial centers are the large southern port city of Pusan and the medium-size cities such as Taegu, farther north in the interior; Kunsan (Gunzan) and Mokpo (Moppo), on the southwest coast; and Samchok, near the east coast. South Korea's large surplus urban population and associated unemployment derive partly from the earlier refugee flight from North Korea as well as from the saturation in farm and rural employment. In spite of earlier gloom over economic progress in South Korea, however, economic observers in the late 1960s expressed optimistic views over the imminence of a South Korean "economic takeoff." South Korea's rise in exports, for instance, has been phenomenal; its 1967 merchandise exports amounted to an estimated 357 million dollars, as compared with only 20 to 30 million dollars at the end of the 1950s. The United States and Japan, South Korea's chief customers, account for 65 percent of its exports.

Communications

Before World War II, Korea's 3,250 miles of railroad provided a fairly adequate network for the economic needs of the country. A 600-mile double-track line ran the length of the country, from the southern port of Pusan to the Yalu railroad bridge into China at Sinuiju. The most strategic and only transpeninsular line then followed the Seoul-to-Wonsan corridor. The division of the country has cut both these important lines near the 38th parallel. Destruction during the Korean War and mechanical obsolescence had reduced the total rail network of Korea to about 2,500 miles by 1958.

Korea's highway system, in contrast to its rail network, has been considerably strengthened, in consonance with the worldwide development of motor vehicle use. Compared with 14,000 miles of dirt highways for all of Korea in 1937, construction by 1964 had brought South Korea's highways alone to almost 17,000 miles; possibly an equal mileage exists in the northern half of the country. Rail and road connections with China provide through routes for trade and transportation. During the Japanese occupation, Pusan was an export port for transit freight from Manchuria, and it is still the leading port, with a 1965 population estimated at 1.4 million.

When Japan took control of Korea in 1910, there were only eleven Korean cities that had over 14,000 population, and urbanization was figured at only 4.4 percent of the total population. In 1964 North Korea reported 40 percent of its population was urban, and South Korea 48 percent. Large cities are few, however, and these are mostly in the south.

Japan

Relief Features

Japan differs from the other countries of Central Eastern Asia in being entirely insular, as part of the volcanic island chains that form scalloped arcs off the mainland of East Asia. Though hundreds of islands are included in Japan, all except four are very small. Among these four, Honshu—with 88,031 square miles, or about 4,000 square miles larger than Minnesota—is by far the greatest and is known as the "mainland" of Japan. Hokkaido, somewhat detached and northernmost, is second in size, with 30,077 square miles, or slightly smaller than the state of Maine. The southern, smaller two islands, Kyushu and Shikoku, together with the southwestern arm of Honshu, form the bounds of Japan's Inland Sea. Kyushu, the southernmost, has 13,768 square miles—about the size of Massachusetts and Connecticut combined; Shikoku, with 6,857 square miles, is larger than Connecticut but smaller than Massachusetts.

Their recent geologic origin has had marked effects on the Japanese islands. They are composed mainly of high and rugged mountains, parts of a gigantic and partly submerged range rising from the seabed. From the elevated parts of the range that form Japan, there extend several island arcs, which reveal the existence of further submarine ranges. From northeastern Hokkaido, the Kurile (Chishima Retto) chain stretches toward the Kamchatka Peninsula of northeastern Siberia. Southward from Tokyo Bay runs the Bonin chain (Ogasawara Gunto). Southwest from Kyushu, the Ryukyu Islands reach almost to Taiwan. (Okinawa, an important island in the middle of the Ryukyu chain, was developed as an American military air base after World War II.) The Korean Mountains extend in a submerged arc toward southwestern Honshu, and from the north the Sakhalin Mountains dip under Soya Strait to emerge again on Hokkaido. Where these mountain chains meet in Hokkaido, in central Honshu west of Tokyo, and in Kyushu, there occur high complex mountain knots, with clusters of volcanoes. The highest mountains are found in central Honshu, with the famed Mt. Fuji (Fujiyama) rising over 12,000 feet. In submarine topography, the 30,000-foot depth of the Tuscarora Deep southeast of Japan contrasts with the shallow Yellow Sea or even the 12,000-foot-deep basin of the Japan Sea. A great part of Japan's coasts is bordered by promontories and sizable indentations, giving the country a coastline of 16,000 miles, or four times that of much-larger China.

Characteristically, Japan's topography comprises a mountainous center enclosing small interior basins and narrow valleys and surrounded by coastal alluvial plains. These plains are surmounted near the mountains by former lowland plains which have been raised by tectonic movements to make high and level, but river-cut, terraces. Rivers are short and have steep profiles, making most of them unnavigable, except for short delta stretches; they have limited backup reservoir capacities. The plains and terraces on which the majority of the population live bring the Japanese people into close contact with the sea. These lowlands are generally small and scattered, with the 5,000-square-mile Kwanto (Kanto) Plain, dominated by Tokyo, being by far the largest. No other plain exceeds a thousand square miles in extent; the next largest is the Ishikari-Yufutsu Plain, having only 800 square miles. Much of the plains are used for agriculture, but urban and industrial land use has increasingly encroached upon agricultural land as cities have expanded. The coastal plains most often are bordered near the sea by dune ridges, which create drainage difficulties but which also serve as settlement sites. Favorable factors affecting occupancy of the coastal lowlands include a long growing season, abundant rainfall, and good nearby fishing grounds.

Climate

Its adjacency to the east coast of the Asiatic landmass brings Japan into the monsoon climatic regime, but with a regime greatly modified by maritime influences. The northward-moving warm Japan Current divides around Kyushu, flows up along the northwest coasts of Japan to Hokkaido, and washes the southeast coast as far as Tokyo Bay. North of this latitude and around Hokkaido, southward-moving cold currents chill the northern coastlands.

The southeastern half of Japan has the familiar midsummer monsoon rainfall peak, with an added peak during the typhoon period in late summer and early fall. Cyclonic storms moving northeastward from the Yangtze Valley also produce the moldy period of the *mai-yu* ("plum rains"), with its oppressive heat and humidity. The winter continental winds bring cool, drier weather to the southeast but heavy precipitation to northwest coastal mountain regions. Much of this precipitation is in the form of deep snow, which thaws rapidly in spring and causes serious floods. Although no part of Japan suffers severe drought as it is understood on the continent, northern Hokkaido and the Inland Sea area have only about 40 inches of precipitation annually, in contrast to an average of 57 inches at Tokyo and 84 inches at Kagoshima in southern Kyushu. Moreover, the mountain-ringed Inland Sea region has mild, pleasant winters. Summers in the southern half of Japan are hot and oppressive. In northern Honshu and Hokkaido, summer fogs are frequent because moist northward-moving tropical air masses override and are chilled by cold polar ocean water.

The long latitudinal stretch of over 1,000 miles between northern Hokkaido and southern Kyushu leads to considerable climatic differences among the islands of Japan. Hokkaido has climatic similarities to Maine in the northeastern United States; Kyushu resembles Florida. In the south especially, altitudinal zonation brings a wide range of vegetation within relatively short distances that equals the change produced by greater latitudinal spans at low altitudes. Vegetation varies from subtropical forests in southern Kyushu, southern Shikoku, and the Kii Peninsula of southeastern Honshu to subboreal coniferous forests in the heart of Hokkaido.

Historical Development

The earliest-known inhabitants of Japan were the Caucasoid Ainu, a short-statured, stockily built race with the hairiness characteristic of Europeans. The Mongoloid Yamato race entered Japan via the southern island of Kyushu. Gradually, with the aid of improved weapons and armor, the Yamato drove back the fiercely resisting Ainu, who were eventually pushed far into northern Honshu and Hokkaido. Today, only a few thousand pureblood Ainu remain on Hokkaido and Sakhalin Island.

Archaeological finds that include Chinese bronze artifacts dating back to about the beginning of the Christian era indicate some Japanese cultural contacts with the mainland as early as that period. Most of the advances of early Japanese civilization were based upon cultural importations from China subsequent to about 2,000 years ago. Korea formed the bridge over which these flowed.

Japan's contact with the West began with the arrival of Portuguese Jesuits in the middle of the sixteenth century. Their numerous converts became embroiled as a group in the political rivalries of the time. Fear of foreign aid to enemies of the reigning *shogun* (military dictator) caused the latter ruler to banish the missionaries and suppress Christianity. For two centuries afterward, Japan withdrew into cultural and commercial seclusion. Serious internal economic, social, and political unrest developed in Japan in the early nineteenth century and came to a head at a time when China's weakness before the aggression of well-organized European naval powers was becoming apparent. Fearful of falling into the same situation, Japan decided to come out of seclusion when, in 1853, Commodore Perry of the United States made a show of force in Tokyo Bay in a move aimed at opening the country to foreign trade and diplomatic relations.

Japan soon exerted itself to develop those aspects of technology that appeared to have made the Western powers strong. Through skillful political maneuvering and military pressure timed to exploit the struggle for power among the Western nations, by 1937 Japan had built up an empire of great magnitude. Between 1876 and 1879, Japan occupied the Ryukyu Islands, then vaguely claimed by China. Formosa (Taiwan) was acquired by the Japanese victors in the Sino-Japanese War of 1894–1895. After a military intervention and threatened war with China, Korea was annexed in 1910. Russia ceded the Kurile Islands in 1875 in exchange for the southern half of Sakhalin Island. When Japan defeated Russia in the war of 1904–1905, she took back southern Sakhalin, which was renamed Karafuto.

In return for entry into World War I on the Allied side, Japan was permitted to oust the Germans from their Pacific island possessions in 1916 and to occupy what became the mandated Pacific Islands under the League of Nations (the present United States Trust Territory of the Marshall, Caroline, and Mariana Islands). Japan already entertained ambitious dreams of a continental empire, which began to take form with the Japanese occupation of Manchuria (renamed Manchukuo) in 1931, when China was too disorganized and weak to resist effectively. These and subsequent Japanese moves in China and Indochina threatened the power balance in the Pacific and the interests of the United States and its allies. Because the United States finally took resolute measures to oppose these advances, at the end of 1941 Japan made the attack on Pearl Harbor and Singapore that led to open warfare in the Pacific. Insatiably ambitious, Japanese militarists overestimated the geographical foundations of their power; as a result, at the end of World War II, Japan was stripped of its empire, thereafter reduced to little more than the long-accepted home islands. Only more than 20 years after the war were the Bonin Islands returned to Japan. The Ryukyu chain, which includes Okinawa, was returned to full sovereignty under the Japanese flag in 1971 (with Okinawa's return negotiated for a later date). With regard to the northern islands of the Kurile chain, not only has the Soviet Union refused to restore the Kuriles proper, but it also has shown no willingness to heed Japanese wishes for the return of Shikotan and Habamoi islands, which are considered part of Hokkaido.

Population and Livelihood

During almost two centuries of self-imposed seclusion, the Japanese population was almost stationary. A number of reasons accounted for this. Owing to feudal restraints on social and geographic mobility, marriage rates were low and infanticide was widely practiced. There was a shortage of arable land because of the large imperial territories withheld from cultivation. Feudal warfare, disease, and malnutrition took their toll. Restraints imposed on trade and commerce and the lack of capital for investment in industries with job opportunities dampened economic development. After 1850, many internal changes related to the opening of Japan to external trade and international relations took place. Feudalism was effectively abolished; imperial lands were made available for settlement; social and geographic mobility loosened; merchants joined with the government to finance industrial development after the Western fashion and thus initiated numerous job opportunities; infanticide was banned. Population, thus, began to rise rapidly.

In the 1930s and 1940s, Japanese militarist policymakers and the industrialists interested in a large manpower supply encouraged the sizable increase of population. By 1970 Japan's population of over 100 million had trebled that of 1850. However, the rate of population increase has been successfully lowered to less than 1 percent per annum. This lowered trend resulted from several factors. World War II brought a hiatus of conception in families whose men had been conscripted and in many instances killed, and interrupted the normal trend of marriages among young people. Economic deprivation following the war and job and food shortages brought a drastic rise in legalized abortion. The subsequent rise of Japan to prosperity concomitant with urbanization and with housing shortages led to the desire for small families and higher standards of living. These new trends, thus, brought Japan's rate of population increase down to about that of the United States.

In 1970 Japan had only 15.8 million acres of cultivated land, compared with 313 million acres of harvested land in the United States—or a ratio of 1 to 20. Japan's 103 million people during the same year compared with 205 million for the United States, or a ratio of about 1 to 2. The Japanese, thus, had only one-tenth the cultivated land resources per capita of the United States; moreover, they did not have the pastoral and mineral resources also available to the United States. The potentially arable land of Japan not already under cultivation, furthermore, is negligible. In 1970 required food imports cost Japan 2,000 million dollars, although there also were some exports of processed foods, mostly of marine origin. Intensive land use and intense exploitation of fisheries, therefore, have been necessary to cope with the country's food needs. Like the British, the Japanese also must manufacture and export to live.

Agriculture

Though the percentage of the Japanese population employed in agriculture has dropped overall from 80 to 33 percent during the past century, the total number increased after World War II. In 1950 there were 6.2 million farm households, with 37.7 million people. However, by 1964 these had again been reduced to only 5.8 million farm households, with 31.1 million people. Continued migration to urban areas from the farms has been substantial. Nevertheless, agriculture still is a leading occupation of the Japanese. Its relative decline, in large part, has come about because virtually no good agricultural land remains to be utilized. Increased productivity per man, as well as per acre of land, is also partly responsible.

More than 55 percent of Japan's cultivated land is planted in rice crops; nevertheless, the area of cultivated land in dry crops far exceeds that used for paddy. This seeming paradox is explained by the fact that winter dry crops are grown on the paddies, whereas rice cannot be grown on most of the land permanently used for dry crops. The long growing season from the Kwanto Plain southward permits such double-cropping, so that the equivalent of 4.48 million acres of cropland is added to the actual cultivated surface. Technical improvements and widespread mechanization since 1950 have brought about a revolution in Japanese farming, and bumper crops have become standard.

Climate is one of the less significant factors limiting agricultural production, although it becomes important in northern Honshu and Hokkaido. About 65 percent of the land is over 15° in slope, and poor drainage and other factors further reduce the land fit for agrarian use. With soils generally well leached owing to high rainfall, production depends heavily upon fertilization. Though maximum use is made of organic fertilizers such as night soil and farm manures, there is increasingly heavy dependence upon chemical fertilizer, most of which is manufactured in Japan, although some is imported.

Twenty-five years ago, 81 percent of the calories consumed by the Japanese came from grains (68 percent from rice alone). Dietary changes since then have involved increasing consumption of wheat and dairy products. In southern Japan, fall-sown wheat and winter barley are associated with summer rice, together with soybeans on field borders and along dikes. A great variety of crop associations occurs on the dry fields. Oats are important only in Hokkaido; barley, naked barley, wheat, millet, and buckwheat are widely grown. Sweet potatoes are conspicuous in the southern parts of Japan, whereas white potatoes are found mostly in Hokkaido. Large quantities of vegetables are grown on the diluvial, upland dry fields.

Some regional crop specialities include tobacco and hemp in the northern part of the Kwanto Plain; tatami reeds for mats in the Okayama and Hiroshima coastal plains; green tea and tangerines in the Shizuoka vicinity; flax, pyrethrum, peppermint, and sugar beets in Hokkaido; rapeseed and sesame seed in Kyushu and between Lake Biwa and Ise (Atsuta) Bay; and mulberry, especially in the west Kwanto upland plain and central mountain basins. In the latter areas, the mulberry has yielded somewhat to the competition in land use from fruits and dairying. Dairying is important in Hokkaido, in central Honshu, and in the hill lands surrounding the Osaka-Kyoto region.

For about the last fifty years, Japan has been one of the foremost fishing nations in the world, an aspect of livelihood partly attributable to its insular advantages, partly to population pressures on limited farmland, and in large measure to the natural productiveness of the seas around Japan. In 1965, the once much larger labor force in fisheries and aquaculture had dropped to 580,000, among whom 439,000 were engaged in coastal fisheries. The recent trend has been toward increased development of large-scale deep-sea operations and a general breakup of the small-scale fishing enterprises. In 1964, for instance, coastal fisheries produced only 40.6 percent of the total value of the marine catch.

The loss of its prewar empire dealt Japan a severe economic blow, and extensive claims to offshore waters by Korea, Communist China, and the Soviet Union have further restricted her fishing waters. Nevertheless, Japanese fishing fleets had, by 1960, surpassed the pre-World War II annual catch of 4 million tons by an additional 2 million tons; the catch in 1965, amounting to 6.9 million tons, was second in the world only to that of Peru. Japan surpassed mainland China's annual catch and caught approximately double the tonnage of fish of each of the next ranking nations, the United States and the Soviet Union.

Industry

Japan cannot live without importing approximately 6 million tons of its food supplies, and it can pay for this food only through exports of other products, mostly manufactures. Furthermore, poor in the minerals required for modern inudstry, the country must import both fuel and raw materials to supplement meager home supplies. Strategically, therefore, Japan is very vulnerable. Virtually all its raw cotton, wool, phosphate rock, crude rubber, nickel, bauxite, magnesite, and tin must be imported. The import requirement for iron ore in 1970 was over 96 percent; heavy coking coal, 51 percent; crude oil, 99 percent; and salt, 80 percent.

Of the important industrial metals, only the zinc deposits are adequate. Except for moderate supplies of chromite, ferroalloys essential for high-grade steels are seriously deficient. Fortunately, sulfuric acid, one of the most important of the heavy chemicals, can be adequately supplied from pyrite and as a by-product in the smelting of copper, zinc, and lead. Much of Japan's needs in copper have been met from domestic mines, but increasing levels of industrial demand have necessitated larger import demand for copper, which at one time was exported in large quantities by Japan.

Nearly half of Japan's coal reserves are located in northern Kyushu. About 40 percent of the supply is found on the northern island of Hokkaido; the remainder is scattered throughout Honshu. Some 55 percent of all coal mined comes from Kyushu, mostly from the Chikuho field in the north; another 25 percent comes from Hokkaido, mostly from the Ishikari field in the valley of that name. Supplies for the Tokyo industrial area come mainly from the Joban field, just north of the Kwanto Plain. Japanese coal is high in waste content when compared with American industrial coals. These and other factors have made coal imported from as far as the United States about 5 dollars per ton cheaper in Japan than are domestic coals. That the Japanese mines have continued large-scale production is in part due to greatly increased efficiency of production, which rose from 5.5 tons per man monthly in 1946 to 20 tons per month in 1968. Coal is imported from India and Australia as well as from the United States and Canada.

Petroleum reserves are meager, and domestic production supplies less than 10 percent of Japan's needs. As early as 1955, Japan was the third largest petroleum-consuming country in the world surpassed only by the United States and the U.S.S.R.

Japan's hydroelectric power potential is about equal to that of Italy and is one-fourth that of China. Over 60 percent of this potential is now being utilized. Most of the present hydroelectric development is along the rivers in the central mountain knot of Honshu. In general, high-cost coal, petroleum, and hydroelectric power constitute an unfavorable factor in Japanese industrial development. Great interest and effort has been focused upon nuclear power generation, which began in 1965 with a 166,000-kilowatt Galder Hall type of reactor. It is estimated that by 1975 Japan's nuclear generation capacity will have increased to 6 million kilowatts.

Iron ore for Japanese industry is largely imported from such suppliers as Malaya, Australia, the United States, Canada, the Philippines, India, Chile, and Peru. Iron ore imports in 1970 totaled 50 million tons. Of the small home production, 60 percent comes from the Kamaishi and Senin mines in northeastern Honshu, north of Sendai; 30 percent comes from the Huchan mines in southern Hokkaido, not far from Muroran. The iron ore goes to the coal regions for smelting, so that about 70 percent of the pig iron is produced in northern Kyushu, the center for heavy industry, where about half of the country's steel is made.

Manufacturing in Japan is concentrated in the belt running along the southern side of Honshu and along both shores of the Inland Sea, from the Kwanto Plain to and including northern Kyushu, where about 75 to 80 percent of the nation's industrial activity is located. Nearly 85 percent of the total value of industrial production comes from this area. The reasons

Fig. 13-14 In Japan the chief industrial centers are along the southeastern coast of Honshu. In Korea the industrial centers are along the western side of the peninsula.

for such a concentration are to be found in the abundant skilled labor, excellent water and land transportation, fine harbors, and the historical focus of the region as the heart of the country, as well as in the agricultural and mineral resources immediately adjacent. Some 60 percent of all industrial activity is divided about equally between the Osaka and Tokyo regions; about

10 percent is found in northern Kyushu, and another 10 percent around Nagoya.

Earlier industrial development was centered mainly on cotton and silk textiles, but the war industry buildup of the late 1930s brought increasing attention to heavy industries and machine making. By 1957, the relative position of heavy and light industries had become the reverse of that in 1935, with heavy industries leading by about 10 percent in value. Textiles, which in 1935 constituted 32 percent of the value of manufactures, comprised only 9.2 percent in 1965. Chemicals comprised 15 percent of the value of manufactures in 1935, but only 10.8 percent in 1965.

Japan's industrial output, which has been growing faster than that of any other country in the world, has made remarkable postwar progress. In production of crude steel, Japan's 47.7-million-ton production in 1966 ranked third among the world's nations, after the United States and the Soviet Union, even outranking the highly industrialized United Kingdom and West Germany. The world's largest shipbuilder since 1956, in 1966 Japan built 47 percent of the world's total ship construction. Its cement production, which in 1966 reached 38.2 million tons, places Japan third in world rank, after the United States and the U.S.S.R. In the last decade the nation also became the world's largest producer of rayon staple, cameras, sewing machines, and transistor radios, and took second rank after the United States in the manufacture of synthetic fibers and television sets. An amazingly rapid increase has taken place in automotive vehicle production. From a postwar beginning of about 15,000 units in 1946, production jumped to over 111,000 in 1957. High rates of growth brought production in 1966 to over 2.2 million four-wheeled vehicles, with almost as large an output of passenger cars as that of the Soviet Union.

Communications and Commerce

As a rugged mountain land with most of its productive areas fringing the sea and with good harbors in abundance, it is natural that Japan should be a seafaring nation. Dependence upon commerce for a livelihood spurred the creation of a large merchant marine, and its strategic ambitions led Japan to build one of the most powerful prewar navies in the world. The merchant fleet reached a pre-World War II peak of 6 million gross tons in 1941, but was reduced to 1.3 million tons after World War II, which also resulted in the sinking of virtually the entire Japanese navy. By 1961, however, Japan had rebuilt

Fig. 13-15 Osaka, Japan, is a large modern city having much industrial development in its immediate vicinity. (Courtesy of Japan Air Lines)

her merchant marine to about 6.7 million gross registered tons, and by 1966 she had doubled this to 13.4 million tons distributed in 1,406 merchant vessels, including the largest tankers afloat. The chief ports for foreign trade are Yokohama-Tokyo, with a combined population of some 13.3 million in 1970; Osaka-Kobe, with some 4.3 million; Nagoya on Ise Bay, with over 2.0 million; and Moji on the Shimonoseki Strait.

In 1965, Japan's rail mileage was 12,886 miles. Japanese railroads have been more important for their passenger traffic than for freight hauls, and passenger transport brings greater gross receipts—a very unusual situation in current-day railroading. In 1964 the Japanese constructed a 320-mile standard-gauge double-track line along the ancient Tokaido postroad between Tokyo and Osaka, with no surface crossings. High-speed trains running at 130 miles per hour operate on this line. In 1969, plans were underway to extend this line westward along the north shore of the Inland Sea, with trains to run at a top speed of 150 miles per hour.

Motorcar and truck transport are increasingly common in Japan. In 1970 there were some 6.5 million registered motor vehicles, overburdening the clogged streets of the urban centers where most are concentrated. Most of Japan's country roads are narrow and poor; several superhighways have been constructed, however, to facilitate commuter traffic in the big cities and travel between Osaka and Tokyo.

The leading exports of Japan are iron and steel products, machinery, cotton fabrics, ships, chemicals and drugs, clothing, fish products, and spun-rayon fabrics. The chief imports are foodstuffs, petroleum, coal, raw cotton, machinery, wool, chemicals and drugs, iron and steel scrap, and iron ore. Japan must import sizable quantities of food and raw materials. To buy these, the Japanese must export labor power and technology in the form of manufactures in large volumes.

Hong Kong

Within this geographical milieu, Hong Kong, located on the southern fringe of Central Eastern Asia, is part of and yet distinct from the other subregions. More like Singapore in being a city-state, it has yet to shed its colonial skin. Long a commercial entrepôt, following World War II it was transformed into an industrial center primarily exporting its own manufactures. Its worldwide banking, finance, and insurance facilities and its vast commercial connections have aided its competitive position. Stable government, law and order, and a general lack of official corruption have attracted investors, many of them being wealthy refugees from mainland China. Low taxation and a measure of Commonwealth trade preference have favored investment.

As a result of these attractions and its large labor market of relatively skilled low-wage workers, Hong Kong's 800-odd factories of the 1950s have grown to over 5,000, employing industrial workers numbering close to half a million. Textiles, which form the largest sector of industry, accounted for over 50 percent of the exports by value. The world's largest ship-salvage operations and scrapworks complement the city's shipbuilding and ship-repair industries. A host of consumer and service industries, as well as thriving tourism, provide a broad base for cramped Hong Kong's economy.

Scarcity of land, housing, and fresh water and problems of pollution, however, plague the crown colony, which is full of economic and social paradoxes. Hong Kong has been described as having the abnormal economic situation of an essentially underdeveloped country living by the export of manufactures. Furthermore, there is the constant specter of the colony's reversion to mainland China's possession at the end of the present century.

In Perspective

Central Eastern Asia's problems are numerous and differ according to regional geographical environment and level of technological development. Generalizations usually require exceptions, but it may fairly be said that in this sphere population increases faster than food production, so that there is an overall nutritional deficiency for the majority of the people, except in Japan and possibly Mongolia. Of the four nations considered in this section, mainland China appears to be suffering the greatest difficulty in providing adequate food and nutrition for its huge population. In spite of initially rapid increases in the first decade of Communist rule, agricultural production there suffered a series of disastrous setbacks resulting from a combination of climatic calamities and governmental bungling, beginning in 1958 (the so-called "Year of the Great Leap Forward"). Inadequate transportation was unable to handle the needs of both agriculture and industry. Mainland China's population continued to increase at a rate of over 16 million per year. By contrast with China's more than 2 percent rate of population increase, Japan has reduced her rate of increase to only about 1 percent. (This still means about a million more people born every year.) This comparatively low rate in Japan means that the age level of the Japanese population is increasing significantly and that the number of young people in the

Fig. 13-16 Victoris, Hong Kong. Overpopulated, teeming with refugees from Communist China, and unsanitary, Hong Kong is nevertheless the center of a highly developed wholesale and tourist trade. (Courtesy of United Nations)

prime work-age bracket will show a marked decrease in the future—a factor that should have important economic consequences for Japanese farming and industry.

Japanese agriculture has made significant production advances; also, because of industrial progress, Japan has the highest standard of living in East Asia. Taiwan with United States military and economic aid, has managed to achieve near self-sufficiency in food production amid an expanding industrial complex. However, the island's population growth rate of 3.6 percent per year arouses grave concern over food deficiencies in the near future.

South Korea's economy was kept from collapse in the postwar era only through American aid. Its imports in 1960 amounted to ten times the value of its exports, and more than two-thirds of these imports were financed by foreign aid. Since 1960 a gradually accelerating economic growth has led to optimistic hopes for future viability. No reliable information is available on the state of the North Korean economy.

In the agricultural sector, there still is strong emphasis on individual crops such as rice, and there is a relatively low state of development in pastoral industries except in Mongolia and in west and northwest China. This is particularly true of dairying, a situation which is partly attributable to an earlier cultural distaste for dairy products but which is changing in Japan. Meat-producing animals, though raised in great aggregate numbers, are few per capita. Except in the pastoral realm, raising animals for food is mainly concentrated on scavenger types such as hogs and fowls, which compete little with man for the available food-growing land. In Japan and Mongolia, however, the pig is not important; Japan has significantly increased the raising of beef as well as dairy cattle.

Progress is being made in industrialization and technical modernization in Central Eastern Asia, but many aspects of society there still follow old traditions. In technology and economic advancement, Communism in mainland China has not demonstrated its superiority over the modified free-enterprise systems operating, for instance, in Japan or on Taiwan, where land reform of less drastic extremes has given farmers land ownership and strong incentives for improvement, and where economic opportunity, given adequate private investment capital, has created conditions for an industrial "takeoff."

REFERENCES

Barnett, A. Doak, and Edwin O. Reischauer (eds.): *The United States and China: The Next Decade*, Frederick A. Praeger, Inc., New York, 1970.

Chen, Cheng-Siang: "The Changing Economy of Taiwan," *Pacific Viewpoint,* vol. 6, pp. 179–190, September, 1965.

Clubb, O. Edmund: *Twentieth-Century China*, Columbia University Press, New York, 1964.

Davenport, John: "Japan's Competitive Cutting Edge," *Fortune,* vol. 78, pp. 90–95, September, 1968.

Fitzgerald, C. P.: *The Chinese View of Their Place in the World*, Oxford University Press, New York, 1964.

Fullard, Harold (ed.): *China in Maps*, Denoyer-Geppert Company, Chicago, 1968.

Ginsburg, Norton (ed.): *An Historical Atlas of China*, Aldine Publishing Company, Chicago, 1968.

Harrington, Richard: "Impressions of Mongolia," *Canadian Geographical Journal*, vol. 74, pp. 64–75, February, 1967.

Hsu, Shin-Yi: "The Cultural Ecology of the Locust Cult in Traditional China," *Annals of the Association of American Geographers*, vol. 59, pp. 731–752, December, 1969.

Karan, Pradyumna P.: "The Sino-Soviet Border Dispute," *Journal of Geography*, vol. 63, pp. 216–222, May, 1964.

Sanders, A. J. K.: "Mongolia," *Focus*, vol. 20, pp. 1–11, January, 1970.

Tregear, T. R.: *A Geography of China*, Aldine Publishing Company, Chicago, 1965.

Trewartha, Glenn T.: *Japan: A Geography*, The University of Wisconsin Press, Madison, 1965.

Wiens, Herold J.: "Change in the Ethnography and Land Use of the Ili Valley and Region, Chinese Turkestan," *Annals of the Association of American Geographers*, vol. 59, pp. 753–775, December, 1969.

14

South Asia

In a total area not much greater than half that of the United States, South Asia contains roughly 790 million inhabitants, or more than a fifth of mankind. This population, more culturally diversified than that of Europe, practices four major and many minor religions; speaks scores of different languages, grouped into at least six sharply differentiated language families; adheres, among the educated, to a wide range of political philosophies; and where traditional values prevail, is divided into hundreds of interdependant, yet often rival, caste and tribal groups. Throughout the region, however, for the mass of the people, there is one significant common denominator—poverty, enduring and profound.

Foremost among the countries of South Asia is the Republic of India. With over 550 million people, India is second only to China in total population and is by far the world's most populous democracy. Its neighbor Pakistan, with less than a fourth the population of India (137 million in 1970), nevertheless ranks as the world's fifth most populous state. Rent from the body of the dying Indian Empire in 1947, at the moment of independence, Pakistan's two widely separated wings were partitioned from the remainder of the Indian subcontinent in order to create an avowedly Islamic state for those regions which contained a Muslim majority. Pakistan today has more people than any other predominantly Muslim nation the world has ever known.[1] Dwarfed by these two giants are the remaining states of South Asia, which include the predominantly Buddhist Dominion of Ceylon (13 million); the nearby independent Republic of the Maldive Islands (100,000); the mountainous, landlocked, overwhelmingly Muslim Kingdom of Afghanistan (est. 17 million); and the three Himalayan kingdoms of Nepal (11 million), Sikkim (150,000), and Bhutan (est. 900,000), where Hinduism and Lamaist Buddhism meet and blend. Independent India, Pakistan, Ceylon, and the Maldive Islands are all fully sovereign members of the British Commonwealth. Afghanistan and Nepal are independent nations. All six are members of the United Nations.

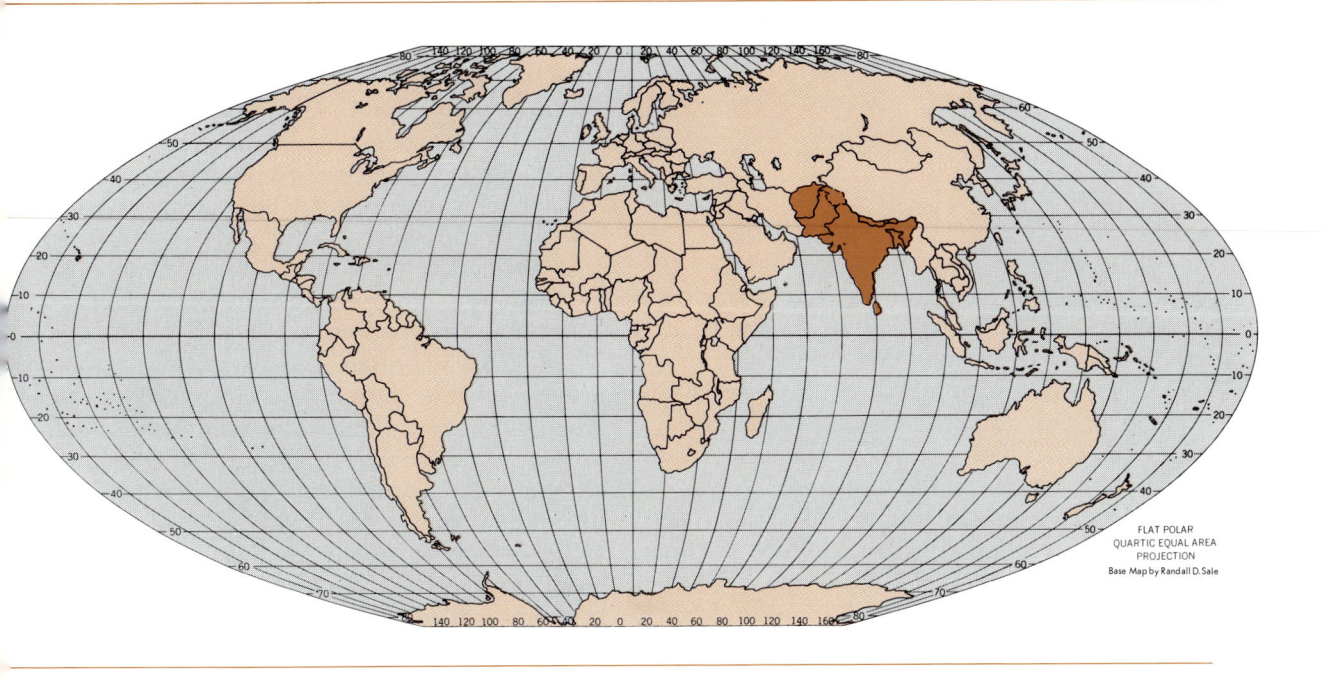

---

[1]*NB.* The text of this chapter and the accompanying maps reflect the situation which existed in South Asia just prior to the outbreak of open war between India and Pakistan in December 1971. Since that date the *de facto* separation of Bangla Desh (East Pakistan) from the Pakistani state has been effected. At the writing of this note an independent state of Bangla Desh, a nation of approximately 75 million souls, has been officially recognized by the Indian government, and there appears to be little prospect of its eventual reunification with West Pakistan.

Sikkim and Bhutan, however, are Indian protectorates.

South Asia as a Region: A Historical Overview

Apart from the objective fact of territorial contiguity, are there valid grounds for treating South Asia as a unit for study? From a purely physical point of view, the answer is not far to seek. Much the greater part of the area, the so-called Indian subcontinent, is sharply cut off from the remainder of Asia by the great mountain wall of the Himalayas and flanking ranges reaching to the sea on both the east and west. Within the subcontinent all physical divisions appear insignificant in comparison. But what of man? We have already noted the tremendous cultural diversity that characterizes the area. And though we have also observed that poverty does provide a unifying factor of sorts, that condition extends to virtually the remainder of Asia as well. Is, then, "South Asia" really a meaningful regional entity? To answer this question about the contemporary geography of the area, to see behind the human diversity, a glimpse into the area's past will be in order.

Early Peoples and Cultures

On the basis of abundant finds of crude stone tools, it is evident that man has been in continuous occupation of South Asia from at least as far back as the beginning of the Middle Pleistocene period of geological history. The findings of archaeology, physical and cultural anthropology, and linguistics all suggest that prehistoric man entered the area from several directions and over a very broad span of time. The racial strains represented include elements of all three of the great racial stocks: Caucasoid, Mongoloid, and Negroid. It is believed that the few scattered survivals of the Negrito racial group, in the Andaman Islands and parts of the peninsular mainland, may represent the descendants of the earliest Homo sapiens to inhabit the subcontinent. That group was followed, presumably, by the proto-Australoids, who were once quite widely distributed and who continue to form an important element in the contemporary racial mix.

Within the present population, Caucasoids are by far the most numerous racial group. Among the several Caucasoid subraces found in the subcontinent, the long-headed Mediterranean type is most prominent. Because it is clearly the dominant type in south India, where languages of the Dravidian family are spoken, the Mediterranean racial strain is sometimes misleadingly called the "Dravidian." It should be borne in mind, however, that Mediterranean peoples, of generally lighter skin color than those found in the south, are also very common in northern India, along with broad-headed Alpine and other groups.

The first-known high culture in South Asia flourished on the Indus Plain and in adjacent areas of modern Pakistan and India from roughly 2500 to 1500 B.C. Though it was positively identified only in the twentieth century, at the site of its presumed northern "capital" of Harappa in Punjab, archaeologists have already discovered scores of additional settlements, including the great southern "capital" of Mohenjo-Daro in southern Sind (West Pakistan) and a number of other clearly urban places. This Indus Valley (or Harappan) culture was remarkable for its relative uniformity over both space and time, leading to the supposition that it had a unified government of a distinctly theocratic and conservative nature. In contrast to its cultural homogeneity, the people of the Indus culture were racially diverse. The predominant strains were Mediterranean

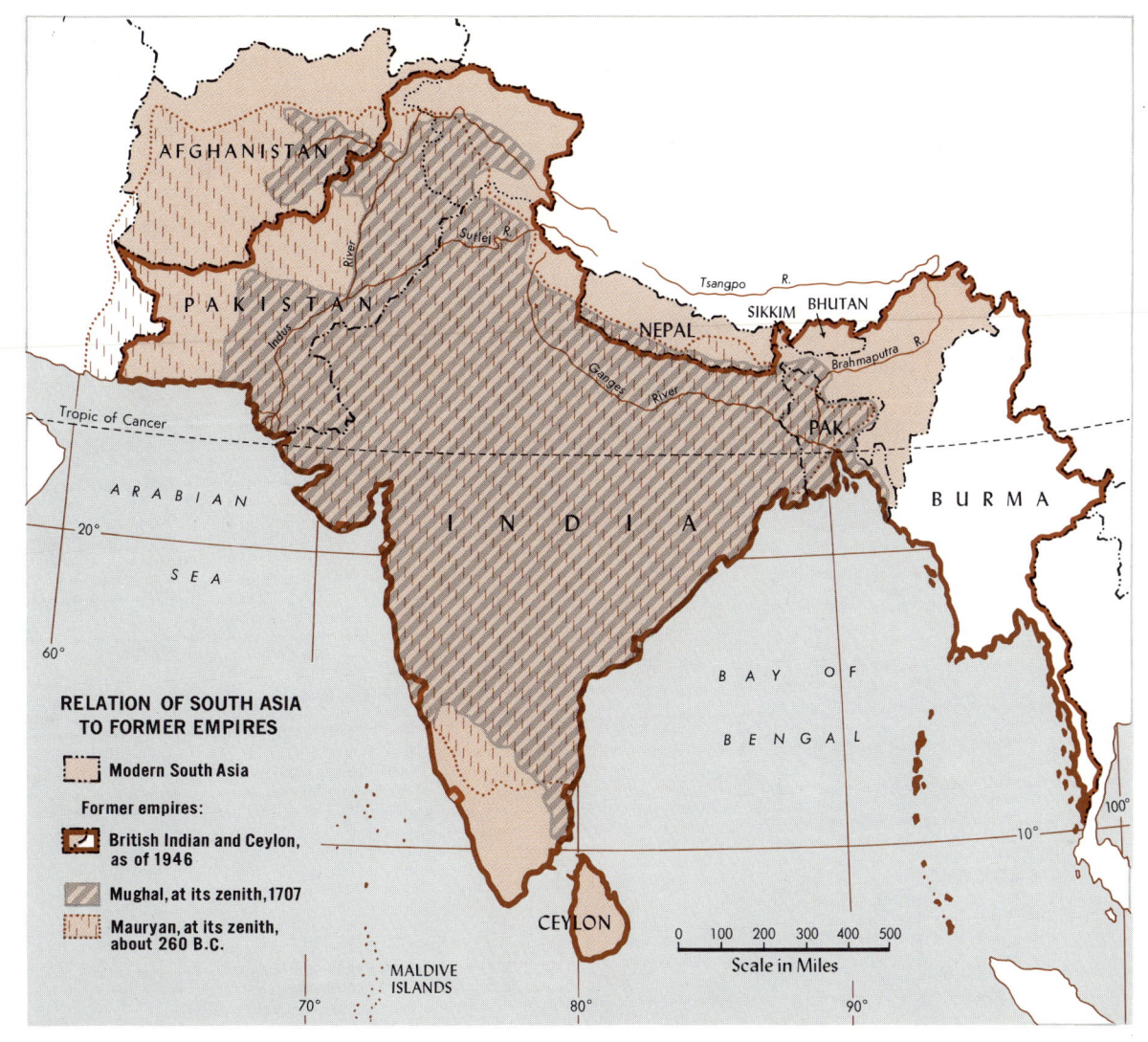

Fig. 14-1 Eight countries make up the region known as South India. India and Pakistan, the two major states of the area, became independent in 1947 on the partition and dissolution of the British Indian Empire.

and proto-Australoid, though others were also present. While there is no doubt that the Indus people maintained maritime commercial ties with the contemporaneous Sumerian civilization of Mesopotamia and borrowed to a minor extent from that area in regard to material culture, there is little reason to suppose that the Indus culture was established as a result of colonization from the Middle East.

The Indus Valley culture was basically agricultural, with a variety of grain staples, including wheat and millet, but not rice; granaries were among the most prominent edifices. It is probable that irrigation was known. Cotton was both grown and traded. Building was of excellently made baked brick, with some use

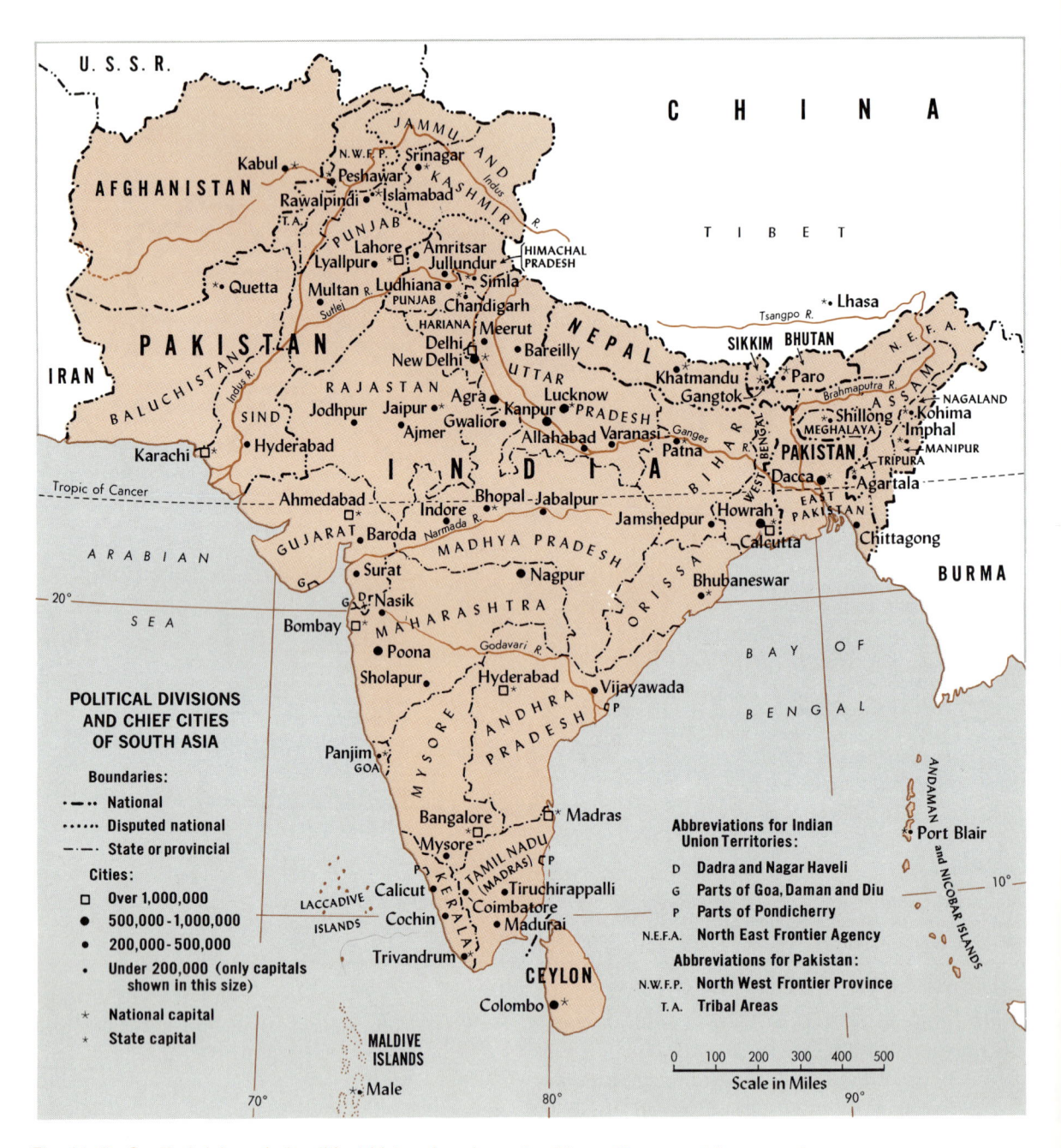

Fig. 14-2 South Asia's varied political history is schematized here. Compare this area today with the Indian empires at the height of Mauryan, Mughal, and British power.

of wood. There was a varied assortment of tools in metal, stone, and wood. Among the distinctive Indus traits were a high degree of centralized planning and control; the building of walled, fortified cities according to a regular grid pattern; the prominence given to bathing and sewage disposal facilities; the general worship of both phallic deities and a Mother Goddess; the more restricted worship of some higher deity; the apparent sacredness of many animals, in particular the bull, and of certain trees; and the existence of a unique script.

That we still know relatively little about the civilization is largely because its script has yet to be deciphered. A team of Finnish scholars, however, claim to have demonstrated that the Indus language is definitely Dravidian in character. Thus, a linguistic link may have been established between modern south India and the northwestern portions of the subcontinent in protohistoric times. One can trace in contemporary Hindu culture, northern as well as southern, many traits with definite Indus analogues; and there is no longer any reason to doubt the continuity of the cultural tradition linking the two.

Various theories purport to explain why the well-ordered, long-established Indus Valley civilization came to an end. Some scholars postulate conquest by militarily superior, chariot-driving Aryan invaders from the northwest; others find evidence of flooding on a vast scale, possibly connected with earth movements, causing a protracted blockage of the established drainage pattern; still others argue that the population gradually exhausted the natural resource base; finally, there are those who believe that desiccation of a once humid environment greatly reduced the carrying capacity of the land. Very likely a combination of causes were at work.

The Aryans did not come in one great wave; rather, they infiltrated into India over a period of centuries. They established their first South Asian hearth in the area between the Sutlej and the Jumna rivers and later expanded it to include all of north-central India, in which area the distinctive outlines of a new Indo-Aryan culture were developed. By about 300 B.C., they had spread the influence of this new culture, if not their direct rule, as far east as Bengal and to the southernmost portion of the Indian subcontinent. The early Aryans were not city builders and left few enduring material remains. What we know of them stems largely from deductions from the rich lore of four great collections of sacred hymns, known as the Vedas. From the Vedas, one can obtain an impression of the continuous wars the Aryans fought with the tribes they encountered in their expansion over northern India, of their gradual conversion from predominantly pastoral to predominantly agricultural tribes, and of their gradual mingling and intermarriage with the prior inhabitants of the conquered territories. One can also surmise a subsequent attempt to maintain "pure" bloodlines by the imposition of a social system with four great social strata, or *varnas*. Three of these, the Brahmans (priests) at the top, the Kshatriyas (warriors) in a second position, and the Vaisyas (commoners), were presumably largely of Aryan stock. The fourth group, the Sudras (serfs), was derived largely from the mass of conquered peoples or those of mixed ethnic derivation.

Hinduism and Buddhism

Unlike other great religions of the world, Hinduism has no founding prophet and no fixed body of dogma. It is, in effect, what its practitioners make it in any given age. Throughout its history, however, its role has been pervasive; that is, it has regulated all aspects of the lives of its followers. In addition to the Vedas, which are still looked to by Hindus for inspiration and guidance,

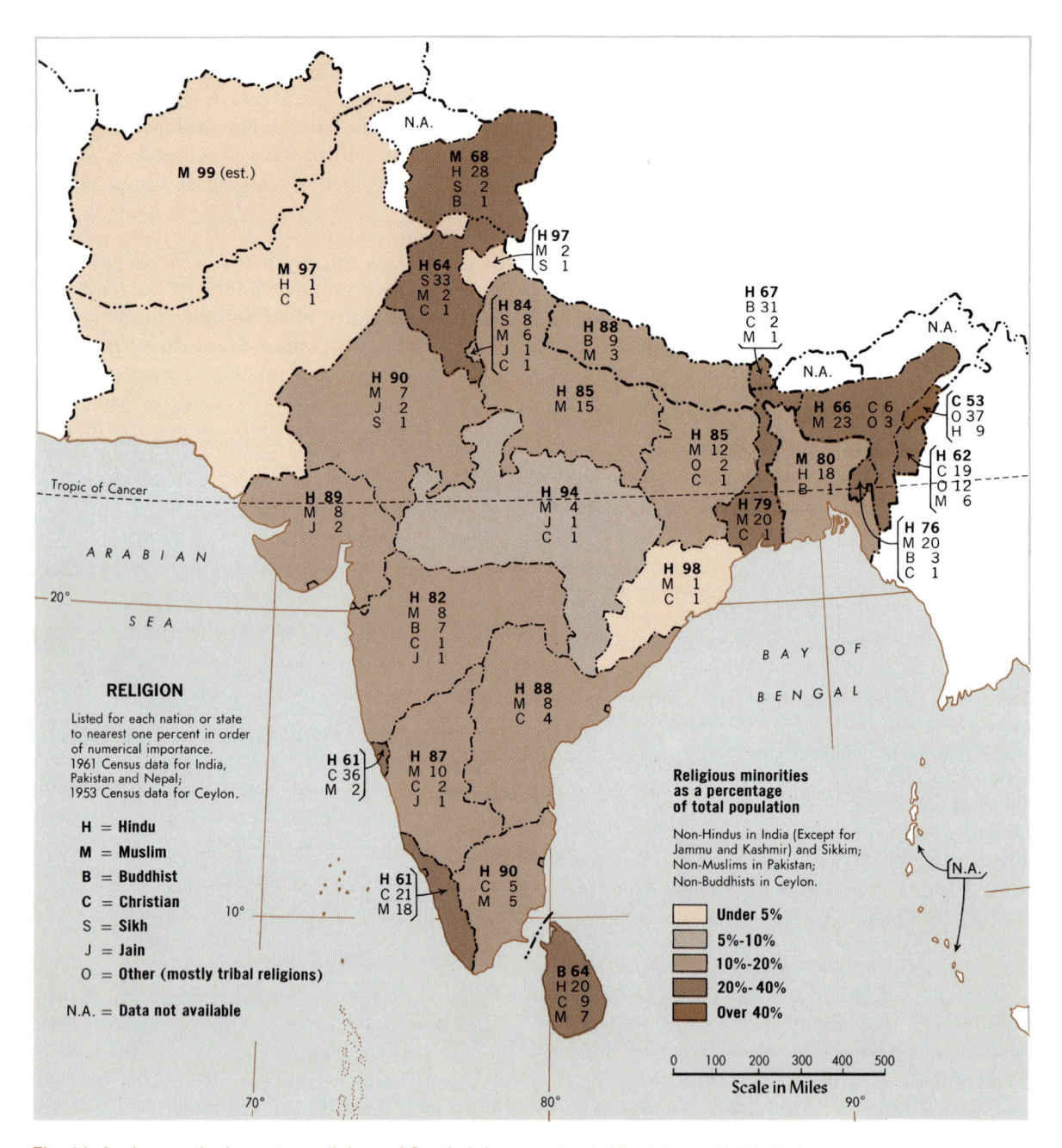

Fig. 14-3 Among the important religions of South Asia are not only Hinduism, which is limited almost entirely to the Indian subcontinent, but also the more widespread faiths of Islam, Buddhism, and Christianity.

there is a vast body of subsequent religious and philosophical texts created within Hinduism. While the cultivation and transmission of the sacred lore remains the special province of Brahmans, there are two great epics, the *Mahabharata* and *Ramayana*, dating perhaps from as early as the tenth century B.C., which more than anything else have formed a continuing basis for Hinduism. These epics are geographically noteworthy not only because of their powerful hold on the common people but also because they provide the first literary indication of a widespread sense of the region now known as India. This is signified by the root "Bharat" in *Mahabharata* ("Great Tale of the Descendants of Bharat"). (Our own word "India" derives from the Greek "Indus," a corruption of "Sindhu," the Sanskrit name for the chief river in the northwest of the subcontinent.) In the *Ramayana* ("Life of Rama"), the god-hero's adventures, like those of Ulysses in the *Odyssey*, take him to the ends of the then known world—in this case Lanka, or Ceylon.

Enduring as Hinduism has been as a unifying factor, it has been repeatedly challenged by other great religions. Earliest among these was Buddhism, which itself arose in northern India in the 6th century B.C. Subsequent to the conversion to the new faith of the Mauryan Emperor Asoka (c. 260 B.C.), the religion spread rapidly throughout his realm. Asoka deserves mention not only as a propagator of Buddhism but also because his was the first of a number of not-quite-successful attempts at political unification of the whole of the subcontinent. Though the Mauryan empire crumbled shortly after Asoka's death, and though Buddhism subsequently declined and ultimately all but disappeared in India proper, that great ruler left a twofold legacy: the dream and ideal of a united Indian nation and the belief in *ahimsa* (nonviolence) as a basis for national, as well as individual, morality. These are enduring aspects of India's cultural heritage.

Indo-Islamic Synthesis

After the coming of the Aryans, many new ethnic streams wended their way into the Indian subcontinent, mainly through the open passes of the arid mountains on its northwest flank. Some were but trickles, and others waves. Just as the early invaders were ultimately absorbed into the mass of people already resident in India, so too were most of their successors. On becoming Hinduized, the invading groups commonly found for themselves a distinct niche in the steadily expanding and diversifying caste system, often obtaining recognition as warrior Kshatriyas, the second-highest of the four great *varnas*.

With the coming of Islam to the subcontinent, however, especially after the establishment of the Delhi Sultanate in 1206, the situation changed. While the invaders probably were not especially numerous, their conquests were more widespread and durable than those of their predecessors. Moreover, as strict monotheists, the Muslims were repelled by what appeared to them to be the idolatry of the Hindus. Assimilation was seemingly impossible and was, in fact, never achieved; yet, each of the two cultures profoundly affected the other. The Muslims brought to or developed in India new forms of architecture, art, music, and literature, which were widely accepted or adapted by the Hindus. They commonly took Hindu wives and, over the centuries, converted millions to their faith. In two areas they became numerically, as well as politically, predominant. One was in the dry northwest, closest to the invaders' homelands, where the pre-Islamic population was not especially large and where the immigrant waves of Muslims were proportionately most numerous. The second area was in the wet, northeastern province of Bengal, where Buddhism maintained its last stronghold. When Buddhism waned there in the face of a resurgent Brahmanical Hinduism, most of the depressed native population preferred conversion to the new egalitarian faith of Islam to enduring the socioeconomic disabilities to which they would have been subject within the Hindu fold.

Just as Islam profoundly altered the established Hindu culture, Hinduism in turn greatly affected Indian Islam. Many mystical movements, known collectively as Sufism, developed within the Muslim faith, inspired by comparable movements that had long been popular among Hindus. The institution of caste, theoretically anathema to the egalitarian faith of Islam, gradually permeated the Indo-Muslim population. (Actually, most Hindus after conversion, like most Indian converts to Christianity at a later date, never were able to shed their former caste affiliation.) At the village level, Muslims adjusted to the established socioeconomic norms by which the Hindus exchanged goods and services. Finally, the language that came to dominate in the court of the Mughal Empire (founded in 1526) and among a very large part of the Muslim population was Urdu, a somewhat Persianized variant of Hindi, the principal language spoken in northern India. Thus, even with much less than a complete synthesis of the cultures, there was a high degree of mutual acculturation and blending, which makes it impossible to draw any simple line indicating where the dominance of one tradition gave way to the other.

The Coming of the West

The last of the great cultural waves to sweep across the subcontinent, and the only one which arrived by sea, was that from the West. Decolonization notwithstanding, the impact of this wave today is probably greater than at any previous time; yet it was slow to gather momentum. For several centuries following Vasco da Gama's initial landing in India in 1498, the Europeans who came to India—Portuguese, Dutch, French, and British—were confined largely to coastal trading stations, called "factories," with their privileges narrowly defined by grants from the dominant Moguls and other local sovereigns. With the conquest of Bengal in 1756, however, England emerged as the leading foreign power in India. Moreover, the British then found the way open, via the yawning Ganges Plain, for conquest of the subcontinent. That stupendous task was effectively carried out in the following century, and subsequent expansion was confined to annexations along the mountainous periphery to protect the gains within.

While certain parts of South Asia, most notably Afghanistan and Nepal, were never formally annexed by Britain, these were long regarded as within her sphere of influence, were treated for certain periods as virtual protectorates, and have, though not to the same extent as India and Ceylon, been significantly affected by the spread of Western culture.

The "Pax Britannica" imposed throughout South Asia endured till 1947. Its results were manifold. First, English still provides a universally used language of government, commerce, and higher education in the area. Second, in India, Pakistan, and Ceylon, parliamentary government, the civil and administrative services, and the judicial and educational systems are patterned in theory, if not always in practice along British lines. Most important is the impact of new ideas and new secularity, an impact heightened by the modern development and spread of the mass media, especially motion pictures and radio. In politics, proponents of democracy and the various Western "isms" have generally won out in power struggles with advocates of traditional religiously oriented states. Keynes and Marx vie successfully with

Gandhi, Muhammad, or Buddha in the economic realm. Science and technology, which have already profoundly altered large parts of the countryside, hold out the hope of one day providing a materially better life for the masses who dwell therein.

The Physical Stage

Form of the Land

South Asia is composed of three great structural divisions: the northern mountain wall, which we have already noted (to which we may add the piedmont region of northern Afghanistan); the Indo-Gangetic Plain, with an eastern extension in the valley of the Brahmaputra; and the southern Peninsular Massif, including narrow coastal plains to the east and west, and the physiographically similar island of Ceylon.

The central and principal component of the northern mountain wall is the Himalayan system, running in an unbroken arc roughly 1,500 miles in length from Kashmir in the west to the North-East Frontier Agency, where India, China, and Burma meet. This mountain system comprises several parallel ranges, in a belt averaging over 100 miles in width. The Great Himalayas, the highest and most continuous of these ranges, include not only the world's highest peak, Mt. Everest (29,028 feet), but dozens of others over 25,000 feet in elevation, as well as scores of glaciers. The landscape is typically Alpine; yet its scale dwarfs that of the Alps of Switzerland. Passes through the Great Himalayas commonly exceed 12,000 feet in elevation, though several rivers cut gorges through the mountain barrier at considerably lower levels.

To the northwest the Himalayan arc terminates in the complex terrain of the Pamir Knot, commonly called the "Roof of the World."

From that mountain node extend not only the Himalayas but also the almost equally high Karakoram range of north Kashmir; the Kunlun Mountains, between Kashmir and Sinkiang; the Hindu Kush, thrusting westward to form the spine of Afghanistan; and a series of lesser ranges fanning out between the Hindu Kush and the Afghan-Pakistani border area and extending southward through Pakistan, all the way to the Arabian Sea.

At their eastern extremity, just beyond the great gorge of the Brahmaputra, the Himalayas give way abruptly to a series of generally north-south trending mountain ranges that run out of southwest China into the Indo-Chinese Peninsula. There they diverge and mark off the great river basins of that part of Asia. The westernmost of these ranges form an arc along the Indo-Burmese border and continue, as the Arakan range of Burma, along the coast

Fig. 14-4 The Khyber Pass, on the Afghanistan-Pakistan border, is about 33 miles long and lies in the Safed Koh Range; it has been traversed for centuries by peoples and armies. (Courtesy of Pakistan Air Lines)

of Bengal for several hundred miles. Though not especially high or rugged, and designated in many localities as hills rather than mountains, the ranges on the eastern flank of the subcontinent have proved, because of their dense cover of forests, tangled brush, and bamboo thickets, to be an exceedingly difficult zone of passage. It was here that the Japanese tide of conquest ground to a halt in World War II.

It should not be assumed that the environment of the northern mountain wall is uniformly hostile to human occupancy. Within it are many valleys quite favorable for settlement and a number of large structural depressions in which there have arisen highly distinctive cultures. Often of great scenic beauty, these depressions include the renowned Vale of Kashmir, the Katmandu Basin in Nepal, the Vale of Manipur in northeastern India, and the Kabul Basin, forming the core area of Afghanistan.

The Indo-Gangetic Plain, possibly the world's largest continuous expanse of alluvium, is the lowland formed by the plains of the Indus, Ganges, and the Brahmaputra rivers. It extends from the Indus Delta on the Arabian Sea coast of Pakistan northwestward to an imperceptible divide not far west of Delhi, and thence for over 1,500 miles along the southern flank of the Himalayas to the eastern end of that mountain system. The huge fertile Ganges-Brahmaputra Delta forms a southern projection of the plain to the head of the Bay of Bengal. Sharply set off from the mountains to the north, the lowland is much less clearly bounded to the south. The Thar (or Great Indian) Desert in the northwest of the subcontinent, while no longer an area of alluvial deposition, can be considered an extension of the adjacent plain because of its generally low relief.

Though the gradients of the major streams of the plain all average less than a foot per mile, the plain is not so featureless as might be supposed. One can distinguish between narrow ribbons of fertile floodplain, still subject to occasional deposition, and more extensive, less fertile terraces rising well above the river level. Canal irrigation, exceedingly important in the western portions of the plain, is confined mainly to the terraces. The volume of the water removed for this purpose is so great that normal inundation and rejuvenation of the floodplains has been markedly curtailed. In some places, especially south of the Ganges, and along some of the upper Indus tributaries, the break between the two levels is fairly sharp, and locally it may be intricately eroded into a badlands topography. In the lower river courses, the terraced uplands become less and less prominent, while features associated with flooding are increasingly important. These include numerous swampy depressions marking abandoned watercourses, broad swaths of infertile sandy soils tracing the courses of once uncontrollable floods, and, particularly in the deltas, a maze of distributaries.

Each year, in the high-water season of the summer monsoon, much of the Ganges-Brahmaputra Delta, especially in East Pakistan, lies below a continuous sheet of muddy water several feet deep, above which there barely protrude hundreds of thousands of islands of human settlement built on low earth platforms. At its seaward margin the delta landscape abruptly changes from one of the most thickly populated on earth to that of a tidal marsh, the Sundarbans, supporting a dense forest of salt-tolerant mangroves but virtually devoid of human habitation.

In striking contrast to the youthful topography of the Himalayas are the landforms of the Peninsular Massif, a very old and stable landmass. A region of very diversified terrain, the massif comprises a large number of fairly open, flat to gently rolling areas separated by low, variously oriented hill and mountain ranges and tabular uplands. The highlands include the Aravalli Range, along the northwest margin of the massif; the hill and plateau area of the Chota Nagpur in the northeast; the east-west trending Vindhya and Satpura ranges toward

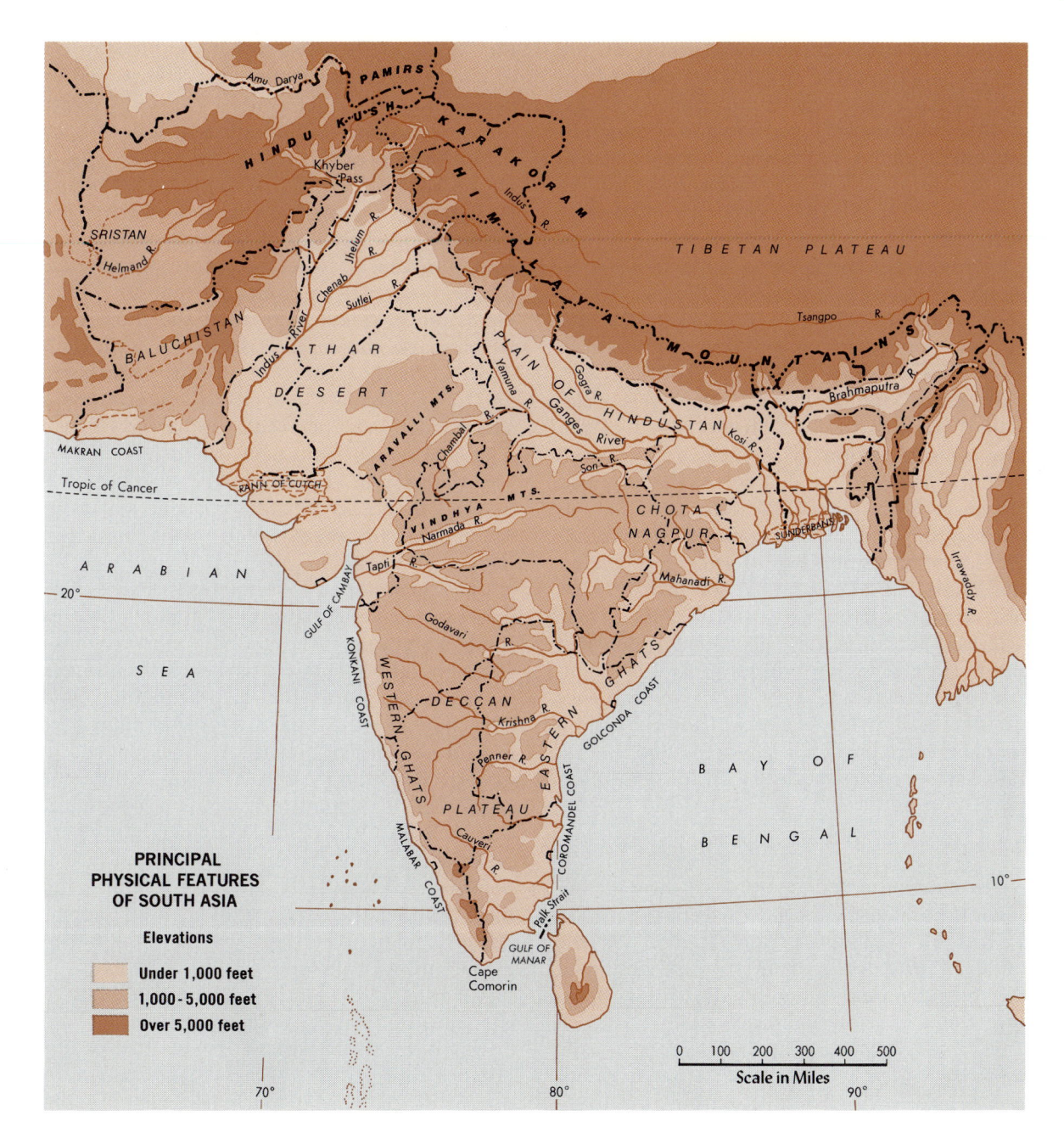

Fig. 14–5 The physical geography of South Asia ranges from tropical rain forests to mountain ranges.

the center; the discontinuous group of hill ranges paralleling the shore of the Bay of Bengal, known collectively as the Eastern Ghats; and the much more impressive and continuous, intricately eroded west-facing escarpment called the Western Ghats, clearly visible from and occasionally reaching the shores of the Arabian Sea.

Crest lines in the Peninsular Massif are commonly no more than several thousand feet above sea level. Only a few peaks, in the Nilgiri and Amaimalai ranges, near the southern end of the Western Ghats, and in the geologically similar highlands of Ceylon, reach elevations as high as 8,000 feet. Much of the west-central portion of the massif is covered by thick layers of basaltic rocks, the so-called Deccan lavas, which erode into soils that, in contrast to those of most parts of the massif, are often of rather high fertility. Flanking the Peninsular Massif are coastal plains. That on the west is particularly narrow and is characterized by numerous small embayments in the north and by sand dunes backed by palm-fringed lagoons and very fertile marine soils in the south. The eastern plain contains a number of important delta areas and has an unusually regular coastline with few good harbors.

Climate

Despite an enormous range in average annual precipitation, there is over most of the Indian subcontinent a basic similarity in the annual climatic cycle, marked by the dominant role of the summer monsoon. Generally speaking, there are three distinct seasons: "hot," "wet," and "cool." The hot season, usually from early March to late June, is one of desultory winds, occasional dust storms, gradually increasing humidity, and very high temperatures, relieved only occasionally by convectional showers and thunderstorms. The wet season typically begins with what is known as the "burst of the monsoon," a sudden inrush of cooler maritime air, usually accompanied by very heavy rains. This arrives as early as late May in southern India and as late as August in northwestern Punjab. Early in this season, almost daily rains may be expected. Though these taper off as the season advances, anywhere from 80 to 90 percent of the total annual precipitation can be expected before the reversal of the wind pattern from maritime to continental, the so-called "retreat of the monsoon," which marks the onset of the cool season. This occurs as early as late August in the northwest and by mid-October over all of the subcontinent save for southern peninsular India and Ceylon. The cool season is normally the driest and the pleasantest. In the far north, however, it is marked by occasional weak cyclonic storms of Mediterranean origin, which may bring a few inches of rainfall.

Two major areas vary notably from the characteristic seasonal pattern of India. One of these comprises Afghanistan, northernmost Pakistan, and Kashmir. In those areas, winter is the season of maximum precipitation, and snow and severe frost are not uncommon. The second exceptional climatic area includes southeastern India and eastern Ceylon, where the season of maximum precipitation commonly occurs from October to December and where two distinct rainy seasons may be experienced. Southern coastal India and Ceylon are also distinctive for their very slight average temperature range between the coldest and the warmest month, typically under 10°F.

Over most of South Asia the variability in the time of arrival, intensity, continuity, and duration of the monsoon is a matter of the greatest concern. A late or weak monsoon, or one in which there is too long a gap between rains, can spell widespread crop failure and hardship to the point of mass starvation. A favorable monsoon, on the other hand, may lead to a bumper harvest. As a rule, the vari-

South Asia **525**

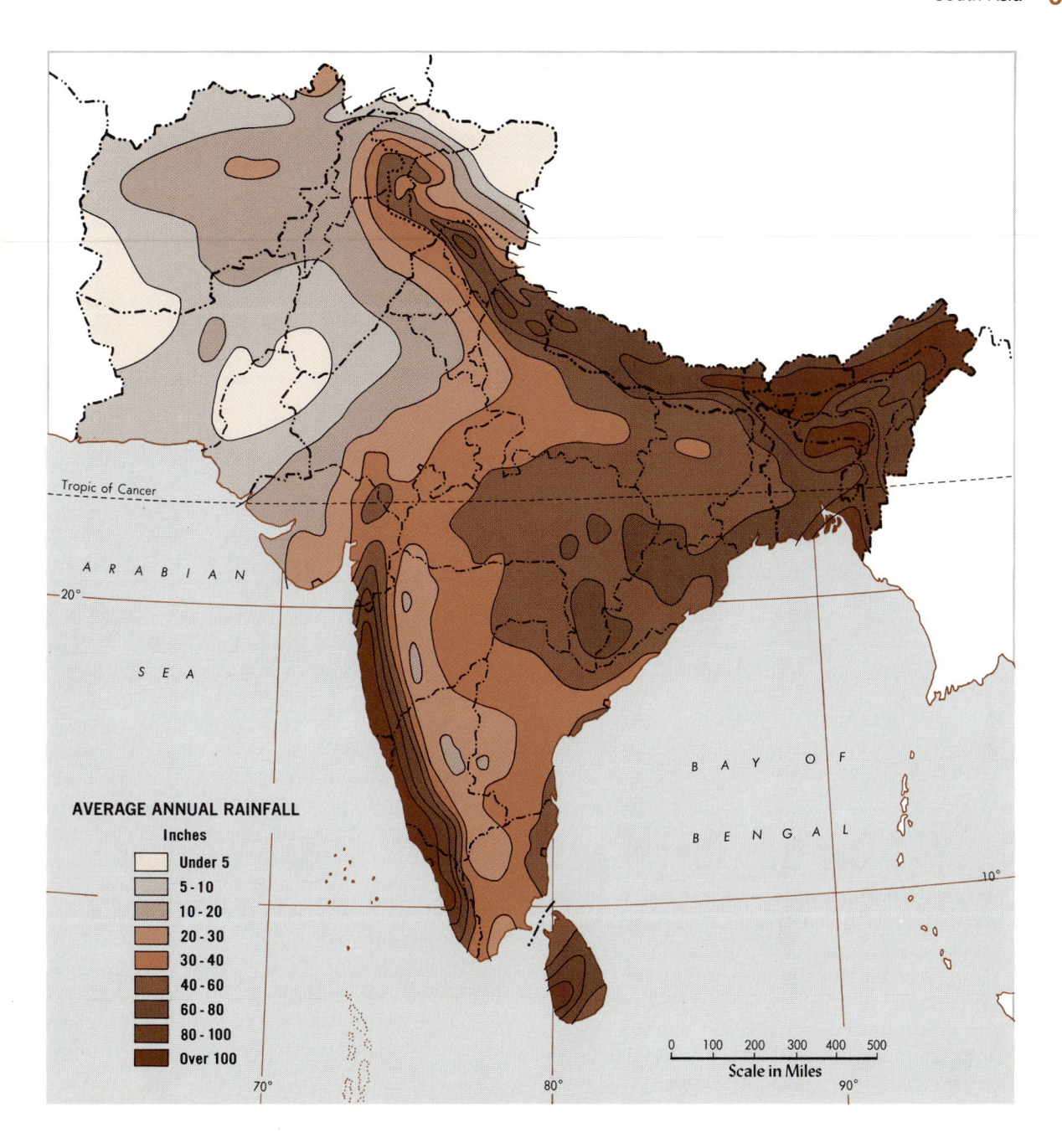

AVERAGE ANNUAL RAINFALL
Inches
- Under 5
- 5 - 10
- 10 - 20
- 20 - 30
- 30 - 40
- 40 - 60
- 60 - 80
- 80 - 100
- Over 100

Fig. 14-6 South Asia, to a large extent, is under the influence of the monsoon. The summer monsoon winds, coming from the southwest, bring much moisture to the Malabar Coast and in areas of northeastern India and East Pakistan where they are forced upward by the topography.

ability of rainfall decreases as the total amount increases. Yet, the only parts of South Asia where there is virtually no danger of insufficient rains are in East Bengal, Assam, the Malabar Coast, and southwestern Ceylon.

The average annual precipitation in South Asia varies from a low of about 2 or 3 inches in the Thar Desert or in the Seistan Desert of southwest Afghanistan to a high of 426 inches at Cherrapunji in Assam. Areas receiving over 80 inches, enough to keep the soil moist throughout the year notwithstanding the seasonality of precipitation, can usually grow two grain crops a year. Areas averaging over 40 inches normally have enough rain to grow one successful crop of rice; and 20 to 40 inches suffice to grow one crop of grain other than rice. Finally, in areas receiving less than 20 inches, desert conditions prevail and cultivation may be practiced only with the aid of irrigation, or at great risk otherwise, even with ample fallowing. In such regions of scant rainfall, livestock herding characteristically assumes an important role in the rural economy.

The Human Mosaic

Given the physical diversity of South Asia and the historical influences which have been at work there, it is to be expected that the patterns of culture and human occupancy will vary greatly from one part of the area to another. The principal vehicles through which these differences are manifested are the social triad of religion, language, and caste or tribal affiliation. Compounding the social diversity is a great regional variety in patterns of settlement.

Religion

The social and political implications of the distribution of religions in South Asia are profound. While it is clear that in each of the countries of South Asia a single religion occupies a commanding majority position, there are some very significant minorities. Not only are these minorities widespread, but in certain areas, of India and Ceylon in particular, they constitute local majorities.

The problem of effectively integrating their minorities into the dominant social stream is among the major internal difficulties that the nations of South Asia face today. It was the failure to solve this problem in regard to Islam which led to the bitter communal discord in India prior to independence, to the creation of Pakistan, and to the subsequent tragic two-way exodus of over 13 million refugees across the newly created Indo-Pakistani frontiers. While a considerable measure of religious homogeneity was thereby attained in West Pakistan within a year of partition, East Pakistan long retained a substantial Hindu minority. This minority, however, has been greatly reduced by renewed emigration in the late 1960s and by a mass exodus, largely of Hindus, in the wake of interwing strife commencing in 1971.

Today Pakistan as a whole encompasses only slightly more than two-thirds of the Muslims of the subcontinent. Most of the hotly disputed, predominantly Muslim state of Kashmir remains with the Republic of India; and save for East Punjab, where the Muslim exodus was nearly total, adherents of Islam constitute a significant minority group all across northern India and in large parts of the south.

Just as Muslims still constitute the principal minority group of India, so do Hindus remain the principal minority group of Pakistan. They are, however, overwhelmingly concentrated in the eastern sector of the country. Millions fled from West Pakistan just after partition, while a mere 500,000, mostly in the province of Sind, chose to stay. Of the Hindus now remaining in Pakistan, both East and West, a majority are of low caste; politically and economically, they are of very little importance. Hindus also constitute an important minority in Ceylon, where they comprise nearly a fourth of the total population. In the north of that country they are an overwhelming majority, and along the east coast comprise a plurality. On the tea plantations of the island's central highlands, the descendants of immigrant laborers, for the most part disenfranchised, also form local majorities or large minorities.

Of the other minority religious groups of South Asia, the Sikhs warrant particular attention. An eclectic, monotheistic faith, Sikhism began in the fifteenth century as a reform movement within Hinduism, but borrowed. heavily from the tenets of Islam. Its center has always been in Punjab, the area in which the two major faiths of India contended most strongly for supremacy. By their manner of worship, style of dress, and unshaved beards, the Sikhs have intentionally set themselves apart from other Indians. They form a highly cohesive and enterprising people, proud of their martial traditions and of the fact that they were the last significant power to oppose the British conquest of the subcontinent. Prior to partition they had spread throughout Punjab, being foremost among the pioneer settlers of the newly irrigated "canal colonies" of what is today West Pakistan. After the post-1947 exodus, they were again concentrated in East Punjab, where after much agitation they succeeded in having that state partitioned in 1965 into a new Hindi-speaking, Hindu-majority state of Haryana and a truncated Punjabi-speaking, Sikh-majority state of Punjab. Some observers fear that it is only a matter of time before the latter strategically located border area, astride the principal corridor between the core regions of India and Pakistan, will demand and possibly obtain complete independence.

Christianity in South Asia has both ancient and modern roots and assumes many forms. On the Malabar Coast in the southwest are several million Christians of the Eastern rite, allegedly descended from converts of the Apostle Thomas, who, it is said, was martyred near Madras. In that same area, in Goa and in other coastal districts of both India and Ceylon, are additional millions who adhere to Roman Catholicism, their ancestors having been brought to that faith mainly by Portuguese missionaries. Protestants of numerous sects are scattered over many parts of India. Primarily of tribal or untouchable origin, they represent the results of intensive missionary efforts in the nineteenth and twentieth centuries. In northeast India, in the state of Nagaland and in adjacent areas of Assam, an armed struggle has been in progress for many years in an attempt by the largely Christian tribal peoples to achieve independence from India or, at least, to gain a much larger measure of local autonomy.

Finally, we must note two small but highly influential religious groups—the Jains, a commercially oriented community of 2 million persons, and the Parsis, who, though numbering barely 100,000, have provided outstanding leaders in many fields.

Language

Throughout South Asia, language has recently proved to be a powerful divisive force. Within less than a decade of attaining freedom, both India and Pakistan had completely redrawn their political maps with linguistic considerations uppermost in mind.

In West Pakistan, to discourage the forces of regionalism, the four linguistically distinctive provinces of Punjab, Sind, the Northwest Frontier, and Baluchistan were amalgamated into one, and Urdu was proclaimed the official language throughout the country. While very widely understood in West Paksitan and quite acceptable to the educated elite, largely emigrants from India, it is the mother tongue of only a small fraction of the population. In response to parochial pressures, the four western provinces were reconstituted in 1970.

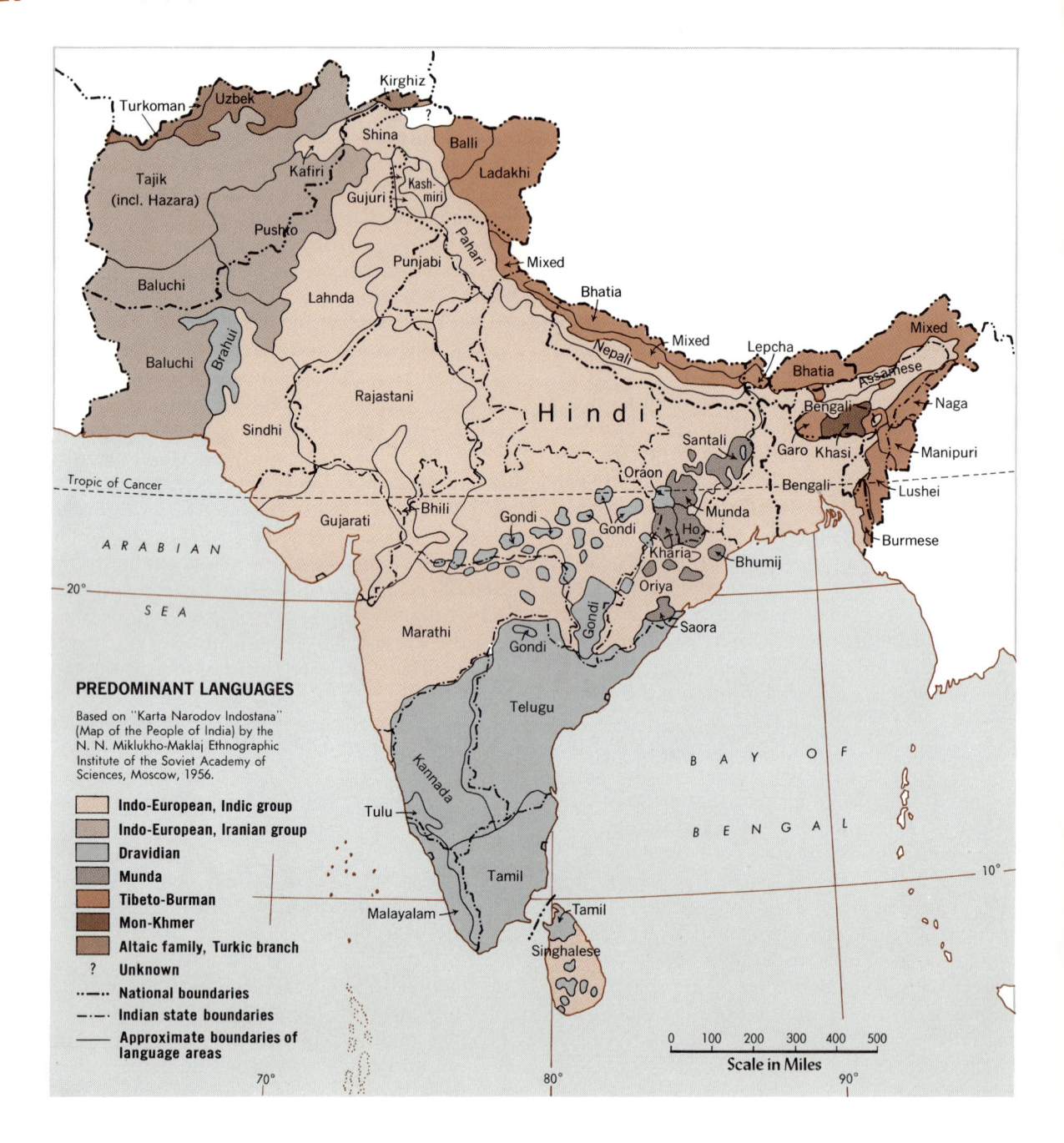

Fig. 14-7 Languages in South Asia, like religions, are numerous. Many tribal dialects are still spoken, and the nations of this area are having difficulty in their efforts to establish or standardize national languages.

In East Pakistan the universally spoken language is Bengali. The literature in that language rivals in richness that in Urdu, a language which few Bengalis understand and which even fewer are willing to accept. In order not to alienate the eastern sector of the new nation, the government ultimately recognized the necessity of making Bengali a second national language, on a par with Urdu. This belated administrative gesture alone, however, could do little to bridge the cultural gap between the two wings of the country as the recent outbreak of hostilities over the issues of independence for "Bangla Desh" (the Country of Bengal) so tragically demonstrates.

India's handling of its language problem following the partition was quite different from Pakistans's. At the outset its constitution specified fourteen official languages (not including English), subsequently expanded to sixteen. It was stipulated, however, that after a fifteen-year period, Hindi, the chief language of the north, was to replace English as the sole administrative language of the central government. But Hindi is spoken by only about 40 percent of India's total population (even if one includes Urdu and other closely related languages and dialects). The decision to adopt Hindi has been vigorously, and often violently, opposed, particularly in West Bengal and throughout the South. While Bengali is no more different from Hindi than English is from German, let us say, the languages of south India belong to a totally different family. English, in fact, an Indo-European tongue, is more closely akin to Hindi than are any of the Dravidian languages of Southern India. Largely on linguistic grounds, there is a very genuine movement in favor of Southern secession in order to create a new state of Dravidistan. Given the gravity of that threat, the central government has very wisely put off the imposition of Hindi. Even more significantly, the government early acceded to the widespread desire to reorganize the country into a federal union of essentially linguistic states. The major changes came in the year 1956; but pressures for new state partitions have resulted in several subsequent alterations.

India is not the only country of South Asia handicapped by the fact that the majority language is of a completely different family from some of the major minority tongues. In Ceylon great tension exists between the Indo-European, Sinhalese-speaking, mainly Buddhist majority and the large, Dravidian, Tamil-speaking, mainly Hindu minority. Roughly half the latter group, the so-called "Indian Tamils" are recent immigrants, or are descended from immigrants, who came to Ceylon as contract plantation laborers and who have not been granted Ceylonese citizenship. The others, the "Ceylon Tamils," are a long-settled group who dominate the north of the country. In Nepal the Indo-European, Nepali-speaking majority must cope with a large number of mountain peoples speaking various languages of the Tibeto-Burman family. And in Afghanistan the dominant Indo-European, Pushtu-speaking group is faced with the existence in the north of the country of several million speakers of Turkic languages.

Caste and Tribe

To the majority of Westerners, the caste system is exceedingly difficult to comprehend. If, however, one examines certain fundamental aspects of the Indian world view, one can at least begin to appreciate why a caste-based social order has evolved and persisted in South Asia. To begin with, Hindus, Buddhists, Sikhs, and Jains all accept reincarnation as axiomatic. Simply stated, reincarnation entails the transmigration of the soul from one body after death to another at birth. The body into which the soul moves may be of high estate or low; it may even be that of an animal. If one has lived his life virtuously, the estate for his soul in the next life will be commensurately higher than in the present; the opposite also holds true. One's status, thus determined, is called his *jati*, generally translated as "caste" but literally meaning "birth." Briefly, a *jati* is an endogamous group of individuals who live by a given set of rules and who occupy a certain position in the social hierarchy. The good and the bad which befall a person as he goes through life, as well as his *jati* itself, are his *karma* (roughly, "fate"), conceived as reward and punishment for deeds done in previous existences. Logically, *karma* also relates to one's net accumulated merit in his present life, to what will be passed on with the soul at the moment of rebirth.

If a person's lot in life is hard and his *jati* is low, he has no justifiable cause for complaint, for he is merely reaping retribution for the bad *karma* bequeathed him. Rather than bemoan his fate or struggle to escape it, what one ought to do, in the Hindu view, is to follow the path of *dharma* (roughly, "duty"), by obeying punctiliously the rules of his religion in general and of his caste in particular. If, therefore, one happens to be a member of the untouchable caste of Bhangis (sweepers), one strives, ideally, to live as a good Bhangi should. For a Bhangi to try to emulate the behavior of a Brahman would be misconduct of a high order. For those who cling steadfastly to their *dharma*, rewards will surely come: a modicum of respect in this life and a better birth in the life to come.

Over the last two or more millennia, there has been a continuous proliferation in the caste system. Creation of new *jatis*, often by fission among the old, is common; on the other hand, fusion of castes is rare. While no meaningful figure can be given as to the number of *jatis* today, it is surely in the thousands. The system also expanded spatially throughout the subcontinent (including the more important parts of what is today Pakistan), to most of Nepal, and to all of Ceylon. In South Asia it has been adopted, paradoxically, by Muslims, Christians, Sikhs, and Buddhists, to all of whom it should be anathema. (Though Sikhs and Buddhists do believe in the concept of reincarnation, the founders of their faiths explicitly proscribed a caste-ordered society.) Today, however, the hold of caste is undoubtedly waning rapidly among Muslims not only in Pakistan society but also in India itself. In Buddhist Ceylon, caste persists, though its importance has been greatly reduced in the modern era.

A few castes, generally of cultivators, are sufficiently numerous over a broad region to play a role in shaping the local cultural style and to act as arbiters of the local social order, notwithstanding the authority of the Brahmans everywhere in matters of religion. These groups are described by anthropologists as "dominant castes," and to knowledgeable persons the dominant caste group comes automatically to mind when one discusses the social system of those regions where such groups exist. Thus, if one thinks of Punjabi society, one thinks of Jats: Muslim Jats in the area that fell to Pakistan; Sikh Jats in the present Indian state of Punjab; and Hindu Jats in the state of Haryana, which as noted, was partitioned off from Punjab in 1966. Similarly, in Maharashtra one thinks of Maratha Kunbis in a great part of Andhra Pradesh of the Reddis, and so forth. Over considerable areas, however, especially in Arya Varsha, as the Aryan heartland of north-central India is called, there is no clearly dominant caste group.

In most peripheral areas of South Asia, as well as in isolated interior regions, the dominant group is not a caste but a tribe, or a group of allied tribes. In such areas the society is, on the whole, simpler and more homogeneous than in those in which the caste order prevails. Some tribal groups are virtually incipient nations, such as the Pushtu-speaking, Indo-Iranian Pathans of the North-West Frontier area of Pakistan or the Nagas, who live astride the Indo-Burmese frontier.

Most castes traditionally follow a certain

occupation; and while virtually all may and do, to some extent, engage in agriculture, such nonagricultural occupations as are pursued are generally consistent with traditional callings (e.g., mechanics are often of the blacksmith caste). Castes generally occupy a fairly clear niche in the social hierarchy of a particular region. In accordance with their status, they show deference to higher groups and expect deference from lower ones and perform prescribed social functions, especially at festivals and on other ceremonial occasions. Each *jati* has its own myths, its own folk heroes, its own polity (decided by a caste council, some of which are now nationally organized), and, commonly, its own gods in addition to those of the general Hindu pantheon.

Not only do castes not intermarry, but they also are supposed to observe an elaborate set of prohibitions with regard to interdining and other forms of personal contact. The basic point of these taboos is to avoid the varying degrees of pollution which lower castes are presumably capable of transmitting to those higher than themselves. While such rules are generally sanctioned by tradition, many specific prohibitions have been ruled illegal since Indian independence, in particular the disabilities of untouchability, such as exclusion from temples and from the use of public wells. In fact, the government has adopted a compensatory policy to make up for past abuses, reserving at least a fair proportion of seats in the central and state parliaments, a certain number of jobs, and a large number of scholarships for members of what are officially designated as the "scheduled castes" and "scheduled tribes."

In some parts of India and in larger cities throughout the country, secularism, backed by legal authority, has gone far to erode the traditional order; elsewhere change is as yet negligible. The situation resembles, in many ways, that in the United States with regard to desegregation and the struggle of nonwhite Americans for full civil rights; but there are also important differences. Whereas the current trend of minorities in the United States is to develop and affirm a cultural individuality distinct from that of the white middle-class mass of the population, in India the low-status groups, through the corporate decisions of their caste organizations (rather than through individual action), generally seek to emulate the life-style of the higher *varnas*. This process, known as "Sanskritization," involves abandonment of the more polluting occupations (e.g., sweeping, laundering, and leatherworking), adoption of vegetarianism, and spending more money on dowries and ceremonies. To Western-oriented persons, including many Indians, the results do not appear altogether beneficial. Sanskritization accepts, rather than seeks to change, a conservative social order, more or less based on the Brahmanical ideal. Paradoxically, its thrust runs counter to the process of "Westernization," which is simultaneously altering, and often radicalizing, the life-style of much of the population already in the higher-caste, better-educated, and more urbanized segments of society.

Patterns of Settlement

From the Himalayas to Ceylon, South Asia is overwhelmingly a region of village-based peasant agriculture. From close to a million villages, farmers go forth to till their fields in the open, unfenced surrounding countryside. Isolated farmsteads, such as one finds in most parts of North America, are exceedingly rare. Only in a few areas of unusually high density and very abundant rainfall, most notably in southwestern Ceylon, on the Malabar Coast of southwest India, and in the Ganges Delta, is it common to find agrarian households living on their individual farm holdings. But in such areas the average distance between farmhouses is so slight that it would be misleading to speak of them as regions of isolated rural settlement.

The look of a village varies considerably

from one area to another. Generally speaking, villages are largest, most widely spaced and most compact in the northwest of the subcontinent, where they strongly resemble the settlement of the adjoining culture realm of Southwest Asia and have a distinctly urban appearance. Their spacing and individual siting reflect the general scarcity of water in the desert or steppe environment and the difficulty of digging sufficiently deep wells to supply the local human and animal population. Within the villages, lanes are narrow and tortuous and houses face inward on high-walled compounds where the seclusion of the womenfolk is assured. The practice of *purdah*, or female seclusion both in public and within the home, has been widely adopted among the Hindus of northwest India in imitation of their former Muslim masters. Shortage of building space often requires that expansion be upward, by adding stories onto the existing, flat-roofed structures. As one moves away from the northwest, the size and spacing of settlements tend to decrease, and their plan becomes more and more open. With increasing rainfall, the pitch of roofs gradually becomes steeper, and multistoried dwellings become increasingly rare. In northeast India and in parts of the south, villages are often comprised of a number of hamlets. While in the south the cluster of dwellings comprising the village core is usually quite distinct, in the northeast it is not always apparent.

The holdings of agrarian households are usually quite small and are distributed in several distinct plots of land. There is a continuous and wasteful fragmentation of properties, resulting from the inheritance practice of dividing each of a cultivator's parcels among all his sons. In some areas, especially in Punjab (both Indian and Pakistani), widespread reconsolidation of holdings has been legally effected. While average farm size varies considerably with the carrying capacity of the land, it is not often greatly in excess of what will supply a bare subsistence livelihood in a normal season. The all-India figure is about 2 1/2 acres. In addition to privately owned lands, many cultivators are able to rent additional holdings on a share basis from large landholders. The willingness of the latter group to rent their land, rather than to supervise its cultivation by hired laborers, has recently been greatly diminished as a result of land-reform legislation granting certain tenants permanent occupancy rights. Some lands, neither owned nor rented, are utilized as a matter of traditional right in return for services rendered to the village in general or to a particular landholder. Still other lands, village groves, unimproved pastures, and water tanks are held in common and may be used by all who need them.

Roughly two-thirds of the villages in South Asia contain fewer than 500 inhabitants, and nearly half the total rural population will be found in villages with populations of under 1,000. Yet, small as it is, within the typical village one finds, in addition to cultivators, a

Fig. 14–8 In Calcutta and other Indian cities, bamboo poles tied together form the scaffolding for the construction of most large buildings.

number of artisans and servants of various sorts and, not uncommonly, one or more priests and shopkeepers. Few villages, however, can be considered self-sufficient, notwithstanding the widely held myth to the contrary.

In the villages of India and in much of Pakistan, a very large proportion of the basic non-agricultural goods and services is provided by members of occupationally specialized castes. But, even in India, there is not enough demand to provide adequate employment in the traditional occupations for the millions of persons who by caste are potters, blacksmiths, carpenters, leather-workers, sweepers, laundrymen, barbers, and entertainers. (The list seems endless.) As a consequence, most such persons must depend, at least in part, on cultivation, if they are lucky enough to own or rent land, or on agricultural labor, if they are not. In a typical Indian village of, say, about 750 persons, one might expect to find, in addition to the dominant landholding caste, a dozen or so others, each with its own specialized calling. Customarily, these groups occupy distinct sections or wards of the village. The major landowning group and the higher castes normally live toward the center, while untouchables are found on the outskirts, if not in a separate hamlet or hamlets some distance from the village proper. In villages situated along watercourses, untouchable hamlets would typically be situated downstream from the village core to avoid polluting the waters used by higher castes.

Rural population densities in areas of sedentary agriculture range from less than 100 per square mile at the arid fringes of agriculture in the northwest to well over 1,000 per square mile in such naturally favored areas as the Ganges Delta and the Malabar Coast. In general, there is a marked correlation between average annual precipitation and rural densities, though the population figures one finds for any given rainfall range are vastly higher than would be encountered in North America. The principal areas where this correlation does not hold are those of perennial canal irrigation, especially in northwest India and West Pakistan. There population densities may be several, if not many, times greater than would be possible if man depended on rainfall alone.

Not all of South Asia's rural population is sedentary. Where aridity makes cultivation impossible, as in parts of the Thar Desert and Kashmir, in most of Baluchistan, and much of Afghanistan, nomadic herding, sometimes in combination with meager seasonal cultivation, is practiced. In strikingly different environments, mainly in wooded, mountainous areas of abundant precipitation, many among the tens of millions of South Asia's tribal population still carry on shifting, slash-and-burn cultivation. Rural densities in both types of areas are commonly quite low.

While no more than a sixth of the population of South Asia is classified as urban (that is, as living in towns and cities of over 5,000 population), this still accounts for nearly 125 million souls. As of 1961, some 45 million of these were found in the 128 cities of over 100,000 inhabitants, a figure about equal to the population of France at that time, and as many as 20 million were found in the nine cities or conurbations with populations exceeding 1 million.

Apart from their size, many of the towns of South Asia are, in many ways, barely distinguishable from villages. For the most part, they are strongholds of cultural orthodoxy. By contrast, the larger cities are centers of social and economic ferment in which there is typically a complex amalgam of Western and indigenous ways of life. There are usually sharply differentiated old and new urban cores. The old sections of the city are incredibly congested. So acute is the shortage of housing that, in the largest cities, many tens of thousands of people must literally sleep in the street. Wholesale, retail, and small-scale manufacturing activities

POPULATION DENSITY

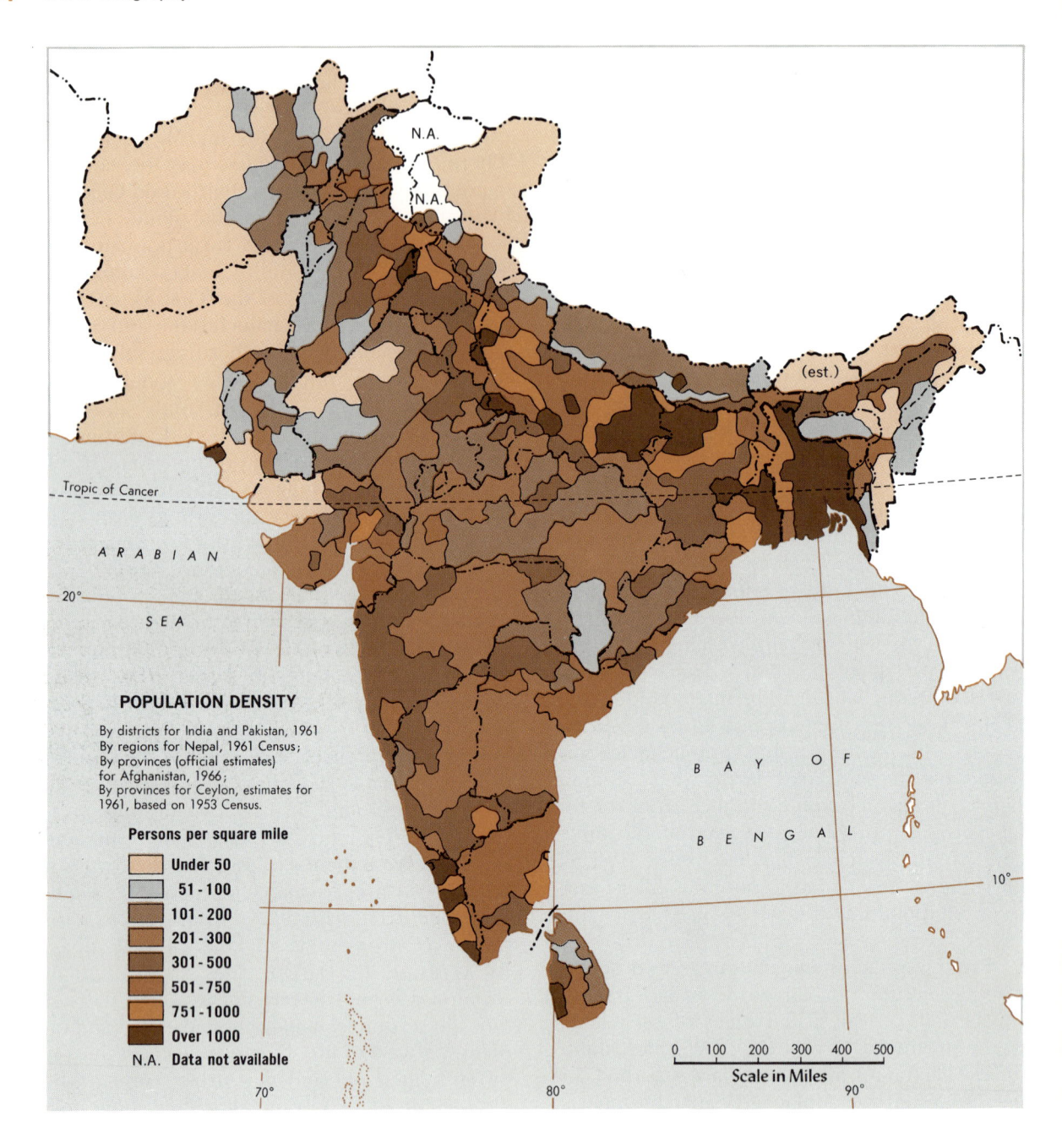

By districts for India and Pakistan, 1961
By regions for Nepal, 1961 Census;
By provinces (official estimates)
for Afghanistan, 1966;
By provinces for Ceylon, estimates for
1961, based on 1953 Census.

Persons per square mile

- Under 50
- 51 - 100
- 101 - 200
- 201 - 300
- 301 - 500
- 501 - 750
- 751 - 1000
- Over 1000
- N.A. Data not available

Fig. 14-9 India ranks second and Pakistan fifth among the nations of the world in total population. The most densely populated areas in these nations are those associated with alluvial and coastal plains.

are not clearly separated from residential neighborhoods. Those neighborhoods, however, are to a large degree themselves segregated, as in the villages, along lines of religion and caste, rather than according to economic class. Hence, rich households often occupy quarters in what outwardly appear to be slums.

Excluding suburbs, the newer sections of many cities were established to accommodate the small population of resident Europeans and their retinues. Today these are the residential quarters of newly created indigenous elites and contain most of the more fashionable business establishments and recreational facilities. Spaciously laid out and built in a distinctive colonial style of architecture, they are commonly separated from the old cities by considerable open spaces. Since independence, especially around the great metropolises, there have sprung up vast suburban developments. Much like those of Eastern Europe, these new dormitory settlements consist mainly of large apartment blocks and depend on a woefully inadequate mass transit system to link workers to their jobs in the urban core.

Economic Development

Economic development in South Asia is far from evenly distributed. Fig. 14-11 indicates the characteristic aspects of the economy in the various parts of the region. The types presented provide the basis for the discussion that follows.

Tribal Economies

In the whole of South Asia, an estimated 40 to 50 million persons may be considered as living within a tribal economic system, in small, relatively self-sufficient communities. These communities are largely isolated from the economies of the countries within which they lie and are little affected by the fluctuations experienced therein. Some tribal groups remain so isolated that they are barely aware of the nature of the ruling power of the state of which they are citizens. A few, in fact, such as the Onge of the Andaman Islands, can scarcely be said to have left the stone age. For millions of others, nominally described as "tribals," the tribal way of life is a thing of the past. Because of mounting pressure on the land, these aborigines or their ancestors have drifted out of their native hills, mountains, and deserts into settled peasant villages and even into towns and cities. Like so many detribalized North and Latin American Indians, they have found work in their new environment mainly as agricultural laborers, as unskilled construction workers, or in a variety of other menial occupations.

The majority of the so-called tribal peoples fall between the two extremes. The chief areas in which they continue to be numerically, if not politically, predominant are in the largely forested hill country of the central and northeastern portions of the Indian Peninsular Massif and in the mountain belt extending along the entire interior margin of the subcontinent. Much smaller groups are desert nomads or occupy remote hilly localities in southern India and the Western Ghats.

The tribes of the Indian Massif include several groups—the Santals, Bhils, Gonds, and Oraons—of over a million persons each and numerous others numbering in the hundreds of thousands. These tribes have had, on the whole, much greater contact with the plains societies than those of the northeast. Most have been Hinduized to varying degrees, though many have also converted, as noted, to Christianity. Most are under sufficient government control as to be required to pay land taxes and to be prohibited, in the interest of conservation, from practicing slash-and-burn agriculture. While thus forced largely to rely on sedentary farming, they may supplement their livelihood by seasonal hunting, fishing, and gathering and by the collection for sale of such forest products as wild silk, lac, and firewood. Many young men and women leave the hills seasonally to do agricultural labor in the eastern plains of Bengal and Orissa, where the monsoon arrives before coming to the hills. Others are employed in brickworks or on railroad, road, and dam construction.

In the hills of northeastern India and East Pakistan, the tribal economies and societies are less disturbed than in the Peninsular Massif. Racially and linguistically, the population is more distinct from the people of the plains than is generally the case among tribes in India. The affinities to Southeast Asia are pronounced. Most groups are clearly Mongoloid and speak languages of the Tibeto-Burmese family. While Hinduism, Buddhism, and Christianity have all made significant inroads in certain areas, the majority of the population continue to practice a wide variety of animistic religions. Government controls are at once lighter and yet more resented than elsewhere in India. Slash-and-burn agriculture continues to be the prevalent type; but certain groups have adopted the plough and taken, at least in part, to sedentary cultivation. Among a few tribes, elaborate and highly productive terrace agriculture has long been practiced.

From Bhutan west to Kashmir, the tribal economy within the Himalayan area combines cultivation with subsidiary pastoralism. In the higher mountains a part of the population moves seasonally with herds of yaks and other livestock, while others tend to the raising of such crops as barley and potatoes. In the generally arid northwest, especially in Afghanistan, but also in West Pakistan and the Thar Desert of India, the tribal portion of the population practices either pastoral nomadism or transhumance, with or without supplemental cultivation. Sheep and goats are the principal livestock.

Subsistence Peasant Agriculture

If we characterize subsistence agriculture as an operation in which the primary aim of production is to feed the farmer, his household (both kin and servants), and employed farm labor and in which the major part of the value of output normally serves that end, then it is probable that a substantial majority of South Asian peasants would be classed as subsistence cultivators. Even in certain of those areas where the greater part of aggregate production is sold off the farm, the inequalities in landholding are such that most individual cultivators are basically operators of petty subsistence farms.

The typical subsistence cultivator devotes the greatest portion of his land and his energies to growing staple grains. Where there is sufficient moisture, rice is almost always the preferred crop. Though, apart from carbohydrates, its food value is low, rice provides the highest yield in terms of calories per acre of any of the grains and is hence capable of satisfying the hunger, if not the other nutritional needs, of the largest number of persons per unit acre of cultivation. In regions too dry to grow rice, wheat, jowar (sorghum), bajra (millet), and maize are the most common grains. Pulses (various types of legumes, the main source of protein for most South Asians), oilseeds, and a few vegetables may also be grown, quite commonly interspersed with grain, and possibly a bit of sugarcane. In season, mangoes and other fruits grown in the village groves supplement the diet.

Animals raised for meat are of little importance, not only because of widespread religious taboos on eating flesh, but also, save for the arid northwest, because of the scarcity of grazing land. Cattle, including water buffaloes, are kept as draft animals and for milking but are usually ill-nourished and not highly productive. Their manure, incidentally, provides the principal household fuel, and little of it is left for fertilizer.

In areas of subsistence cultivation, there is not normally sufficient demand to keep all households of non agriculture castes fully employed in their traditional specialization. Many supplement their incomes by or live entirely from agricultural pursuits, often as agricultural laborers. Those who do follow their traditional calling, however, and many agricultural laborers as well, are normally bound to households of cultivators in a complicated socioeconomic arrangement known as the *jajmani* system. According to this system there are hereditary patron-client ties between families of *jajmans*, who receive certain types of services at specified intervals, and families of *purjans*, who render those services in return for customary payments of grain and gifts at harvest time and on particular ritual and ceremonial occasions (e.g., religious festivals, births, marriages). Ideally the system has no need for money; but, in fact, the gifts proffered often do include token cash payments, as well as cloth and other commodities.

Commercialized Peasant Agriculture

Commercialized agriculture in South Asia is still, by and large, a far cry from what one thinks of when he uses that term in a modern Western context. In much of the area so classified, a majority of farmers are, as we have noted, actually subsistence cultivators, with little salable surplus in normal crop years. The larger operations typically have enough of a surplus so that more of the aggregate yield is sold outside the village than is consumed within it. Over most of the areas of commercialized agriculture, as in those of subsistence cultivation, the principal crop is grain; the variety of other food crops grown however, and their proportion to the total food output will normally be somewhat greater. Animals raised for meat are scarce, for the same reasons as cited earlier; but draft animals are typically stronger and milk cattle slightly higher-yielding. The chief crops grown predominantly for cash are cotton, especially in the Deccan lava country and in Punjab; jute, almost entirely in Bengal and Assam; a great variety of oilseeds, sugarcane, tobacco, chilies; and certain fruits and vegetables.

Of considerable interest is the high degree of correspondence of areas of commercialized agriculture with those in which the leading grain is something other than rice and, conversely, of subsistence agriculture with areas where rice cultivation predominates. A number of reasons may be advanced for this apparent dichotomy. In the rice-growing areas, one might argue, the hot, humid climates induce in the population the characteristic torpidness of the tropics and a consequent failure to produce more than the bare minimum needed to survive. Whether or not such simple, direct physiological effects are real, one cannot deny that the climate does have an important indirect role in debilitating the population insofar as it is conducive, in combination with stagnant water and muddy soils, to high levels of endemism of such diseases as cholera, typhoid, dysentary, and, until recently, malaria. Also of note is the fact that the low levels of nutrition in a rice-based diet do not make for high levels of individual productivity.

In contrast to the rice-growing areas, those where other grains predominate experience greater average variability of rainfall. Thus, there is, on the one hand, higher probability of scarcities, sometimes to the point of famine, and of considerable surpluses, on the other. Given the long-range tendency formerly existing within traditional societies for populations to rise to levels close to the carrying capacity of the land (with the then-current state of technology), severe famines not uncommonly accounted for losses of life in the millions. For a number of years following such losses, however, the reduced pressure of population on the land would mean that what might earlier have been considered only an average harvest would now yield considerable surpluses. In the long run then, there would be more seasons with a net surplus than without. Under such circumstances, societies naturally evolve marketing mechanisms sufficient for gathering up available surpluses and redistributing them either to urban populations or to other areas where deficiencies existed. This, in fact, has been for millennia the case with the bazaars and caravans of the Middle East; and it would appear that the trading patterns of that major cultural realm have over a number of centuries established increasingly widespread and firm roots in Indian soil.

It would, of course, be simplistic to suggest that climate in itself could anywhere provide an adequate explanation for the development of a commercialized agricultural economy. Many other factors may play a role either in initially establishing commercialization or in reinforcing it where it already exists: improved agricultural technology, especially irrigation facilities and, more recently, improved seed varieties and fertilizers; improved transport facilities enabling certain crops from remote areas to reach markets at competitive prices; the implantation in the countryside of sugar mills, cotton ginneries, oil-crushing plants, and other establishments for inexpensive mass-processing of agricultural produce; involvement in expanding world markets, for such products as cotton, jute, and oilseeds; and, finally, the expansion of domestic demand occasioned by the related phenomena of industrialization and urbanization.

One might suppose that the increasing involvement of villages in an extended market economy would automatically entail turning away from the traditional *jajmani* system for the exchange of goods and services within the village itself. This is not always the case, however, particularly—and paradoxically—in some of the most highly commercialized and prosperous regions of India (e.g., Gujarat and western Uttar Pradesh). Irrational as it may appear to outsiders, where villagers in authority can afford the luxury of maintaining the system, wasteful as they realize it is, they commonly choose to do so because it preserves the sort of socioeconomic relationships they deem proper and, in the light of their culture, personally advantageous.

On the other hand, there are many areas where commercialization and widespread improvement in levels of rural prosperity do not go hand in hand. Commonly, in such regions, moneylending is rife. Land is used as security, and in bad years, when the exorbitant interest on debts cannot be paid, foreclosures result in the concentration of land in fewer and fewer hands. This leads to an increase in tenancy and agricultural wage labor, an increasing proportion of total acreage in cash crops (often non-edible), an inability of *jajmans* to set aside grain and gifts for payment to *purjans*, and a virtual disintegration of the traditional socioeconomic order.

Summarizing the matter in the form of a hypothesis: the degree to which the traditional *jajmani* system will be maintained as Indian peasant agriculture becomes commercialized is, normally, a direct function of the effect that commercialization has on the level of rural prosperity.

Plantation Agriculture and Mining

While the commodities they produce could scarcely be more different, there are many reasons for considering plantation agriculture and mining together in the South Asian context. Both contribute to the economy out of all proportion to the number of workers employed and, even more so, to the area on which they are carried out; both are, for the most part, highly capitalized, corporately managed industries;

both depend on masses of wage labor, often recruited in gangs from distant areas; both industries are now highly unionized and closely regulated by government labor legislation; and both were originally mainly foreign-owned and are now increasingly under local control.

Many large plantations and certain mines are discrete social microcosms. All employees are provided company housing, nicely graded to conform to salary scales, with the manager's bungalow at one extreme and the drab "coolie lines" at the other. Shops, schools, dispensaries, and clubhouses are all on the premises. In place of the gross exploitation that characterized these industries not many generations ago, there are now reasonably good wages by local standards and a measure of social security. The price which most workers pay for this security is alienation from the larger society.

There are two principal plantation regions in South Asia, both largely hilly tracts with moderately high to very high precipitation. One, lying on both flanks of the Brahmaputra lowland in Assam and in northern West Bengal, with an outlier in the northeast corner of East Pakistan, is given over almost exclusively to the production of tea. It is from this area that the famous Darjeeling blends originate. The second area includes the Malabar Coast, both slopes of the southern portion of the Western Ghats, the associated Nilgiri and Anaimalai Ranges, and comparable terrain in Ceylon. Its products reflect the physical diversity of the several subregions. Tea, grown in highland Ceylon and on the high western slopes of Kerala, ranks first. Coffee, grown on lower leeward slopes, in the state of Mysore, is also important. So is rubber, the production of which is confined to low, hilly zones in Ceylon and Kerala. Between these zones and the coast, coconut palms are grown both on plantations and on small peasant holdings. These provide a diversified array of products: fresh coconut meat and milk, dried copra, oil, coir (coconut fiber), roof thatching, fuel, and local building wood. Spices such as pepper, cardamom, and nutmeg grow in a variety of locales. They have been significant exports for centuries, and provided much of the initial impetus for South Asian trade to both European and Arab commercial interests.

Fig. 14–10 Loading sugarcane for Darsana sugar mill in West Pakistan. (Courtesy of United Nations)

The areas of South Asia in which mining is carried on are quite widespread; only in India, however, are they of considerable importance. Enormous reserves of iron ore of very high quality and a wide variety of ferroalloys, especially manganese, are found in many peninsular areas of ancient crystalline rock. Reasonably good coal supplies are concentrated in the Jharia and Raniganj fields of the Chota Nagpur, in the states of Bihar and West Bengal. Unfortunately, Indian reserves of coal of coking quality are not great. This disadvantage is offset by the fact that the best and most abundant coal for metallurgical industries is found in very close proximity to excellent deposits of iron ore, abundant limestone, and ample fresh water. The result is that India produces some of the cheapest steel in the world.

Apart from the minerals just mentioned, India is the world's chief supplier of mica and is self-sufficient in lead, zinc, gold, gypsum, most industrial salts, and a host of other minerals. The country has building stone of excellent quality and wide variety. The principal shortage is in petroleum, which is in short supply in the rest of South Asia as well. Production of oil is limited to small fields in Assam and Gujarat, while West Pakistan exploits a major natural gas field in the lower Indus Plain. To date, less than half of domestic petroleum needs are being supplied from local resources. Offshore exploration near both the Ganges and Indus deltas and in the Gulf of Cambay is being vigorously furthered, however, and may hold the key to alleviating South Asia's most critical mineral deficiency.

Manufacturing

Though in absolute terms India ranks among the top ten industrial nations of the world, only about 2 percent of the 200 million persons in its labor force are employed in factories. Even lower percentages characterize the other countries of South Asia. Of roughly 25 million workers in that region who are engaged in some form of processing or manufacturing activity, over half are in village-based "household industries," while most of the remainder are in urban-based but small-scale, nonfactory industrial occupations.

The household and other small industries command our attention not only because they account for such a disproportionately large share of all workers in the secondary sector of the economy, but also because of their special place in the Indian socioeconomic order, a place we have already examined in the context of the *jajmani* system. Many of these workers who are in occupations such as blacksmithing might better be classified as providing services rather than as engaged in the regular manufacturing sphere. Millions of others do, in fact, still produce a great variety of goods for local, urban, and even foreign markets.

Small-scale industries, such as cotton spinning and handloom weaving, were long championed by Gandhi and his followers, in part because of their dedication to a simple, village-based social order and in part on ideological grounds as a means of opposing the British and their protected factory industries, which flooded the Indian market to the detriment of local producers. Patriotic individuals changed over from mill-made to *khadi* (homespun) garments as a part of the independence struggle, and many who can afford to do so still cling to such coarser, and more expensive, attire. Cottage industries of all sorts were furthered and subsidized by the government in the first decade or so after independence. Today India pursues a more selective policy, favoring products with large inputs of relatively skilled labor. Fortunately there exists, overseas as well as in India, a substantial market for the better wares of the cottage industries. Some of these, such as Kashmiri shawls and woodwork, Rajasthani metalware, and Benares saris, are truly outstanding in their workmanship. It is increasingly apparent, however, in the light of the demand for rapid economic development, that most small-scale enterprises will ultimately have to give way before the inexorable trend toward more efficient, low-cost factory production.

While factory manufacturing in India had its beginnings more than a century ago with the establishment, by both Indian and English entrepreneurs, of the first cotton textile mills in Bombay and Ahmadabad, until about World War I manufactures were confined almost entirely to such basic raw materials as cotton, jute, wool, and leather. In the areas that were to become Pakistan, especially so in the Eastern wing, even these simple industrial lines were little developed until after the 1947 partition.

Though textile manufacturing still accounts for roughly half of Indian factory employment, there are now few things which that nation can no longer manufacture for herself. The first steel mill was constructed at Jamshedpur, in

South Asia **541**

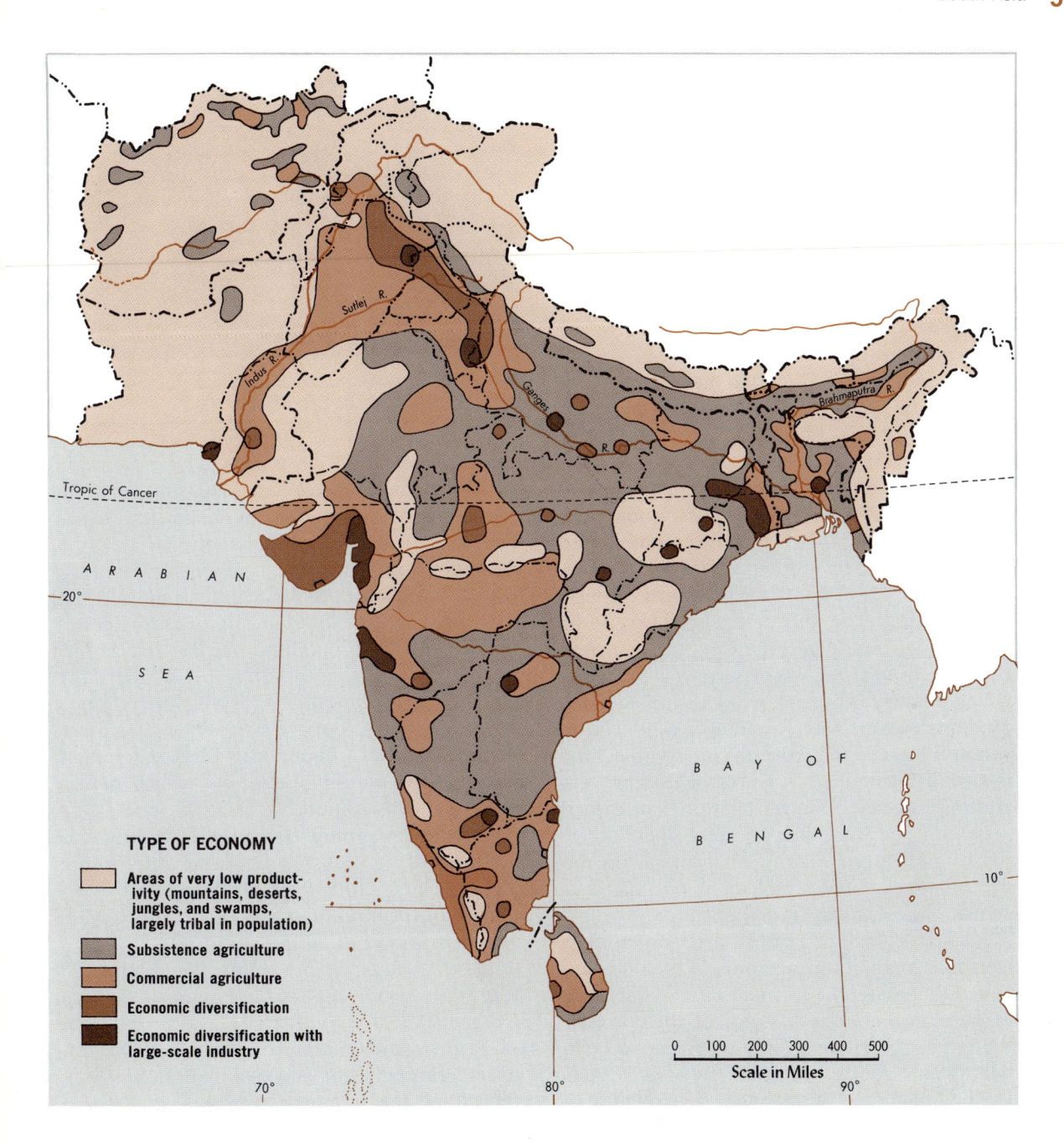

TYPE OF ECONOMY

- Areas of very low productivity (mountains, deserts, jungles, and swamps, largely tribal in population)
- Subsistence agriculture
- Commercial agriculture
- Economic diversification
- Economic diversification with large-scale industry

Fig. 14-11 Agriculture and grazing form the basis for most economic activities on the Indian subcontinent, many of which are at the subsistence level. India is the most advanced country of South Asia in industry and manufacturing.

a new city carved out of the Chota Nagpur jungles, early in the present century. Today there are six steel mills, three of them government-owned and -operated, with a total production capacity of over 6 million tons per year. A seventh mill, planned to have a 4-million-ton capacity, is being constructed with Soviet aid. The engineering, metallurgical, and chemical industries have steadily expanded. Cement production has soared and thereby reduced dependence on steel for building. Fertilizer production is belatedly being given special governmental emphasis; while nitrogenous and potassium fertilizers can be produced in abundance, phosphates unfortunately have to be imported.

For the time being, a number of Indian industries, such as the production of automobiles, aircraft, and electrical goods, must be protected by heavy tariffs and other forms of government aid in order to survive against foreign competition. More and more, however, India is changing from total importer, to inefficient producer, to competitive exporter. The range of her major industrial exports has steadily broadened, from such commodities as cotton cloth and bicycles to textile machinery, machine tools, and locomotives. Nevertheless, her expanding economy requires quantities of capital goods far beyond her present productive capacity, and despite stringent controls on imports of nonessential goods, the nation has consistently run a serious trade deficit.

Pakistan's manufactures are much less diversified than those of India. Jute and cotton still dominate, the former in the eastern and the latter in the western sector of the country. While the government has sponsored certain small capital-goods industries such as a steel rerolling plant at Chittagong in East Pakistan and petrochemical industries based on natural gas in West Pakistan, it has intervened less than India in the economy and permitted a relatively greater emphasis on the production of con-

Fig. 14-12 Blast furnaces of the Bhilai steel plant, India. (Courtesy of Consulate General of India)

sumer goods. Television sets, for example, have been manufactured in Pakistan since the mid-1960s, whereas that industry has barely begun in India. Unfortunately for Pakistan, its mineral raw-material supply is not favorable for the creation of a heavy industrial base, and it is not likely to develop one of consequence, despite the large potential market that its own population represents. In view of this shortcoming, the nation is turning increasingly toward Iran and Turkey, its partners in a three-nation Regional Cooperation for Development pact, to work out schemes of mutually beneficial industrial cooperation.

In the remaining countries of South Asia, manufacturing is still of little importance. Tea packing and the processing of other plantation commodities are of note in Ceylon, and a variety of other industries to serve domestic needs have recently been initiated. Rugs of good quality are exported from Afghanistan.

Prior to independence, Indian industry was overwhelmingly concentrated in just a few cities: the three major ports of Bombay, Calcutta, and Madras and a handful of inland cities such as Ahmadabad, Bangalore, and Kanpur, the last of which owes its industrial origins to the need to supply the British Army with woolen blankets, uniforms, boots, saddles, and other leather goods. The commercial legacy of overconcentration in a few centers persists, particularly with respect to light consumer-goods industries. On the other hand, primary processing plants for agricultural goods and industrial raw materials are now widely dispersed; and the governments of both India and Pakistan have generally pursued a policy of industrial decentralization. In part, this is based on sound economic considerations; and in part, it reflects an unavoidable political necessity, that of placating the demands of all the states of the Indian federation and of the two sectors of Pakistan. It is no accident that India's steel mills are scattered through West Bengal, Bihar, Orissa, Madhya Pradesh, and Mysore; and that at least three other states are actively vying to have the next mill erected in their own territory. Economic policy and politics notwithstanding, one major new heavy industrial region is emerging, the advantages of which, in terms of resource availability and market accessibility, are so great as to be beyond challenge. This area, stretching from Jamshedpur and the Jharia coalfield on the west to the Calcutta-Hooghly conurbation on the east, not only is assuming the role of dynamo in India's drive to industrial maturity but also seems destined someday to become one of the great industrial regions of the world.

In Pakistan, despite officially stated policy to the contrary, most new industrial development, both politically and privately financed, has taken place in the more prosperous but less populous Western wing. This is a major source of resentment in East Pakistan and is one basis of that area's demand for greater autonomy especially in economic matters.

While South Asia's expanding industrialization holds out hope for considerably improved living standards at some future date, for the mass of factory workers the reality of the present is not a happy one. A very large proportion of the factory labor force, especially those in the larger cities and in the older industries such as cotton and jute milling, are lone male immigrants from the countryside whose wives and children are left in the village. Such workers usually manage to share lodgings with kin or caste fellows, in incredibly crowded tenement houses. Though protected to a degree by minimum-wage and other social legislation, factory workers are typically victimized by land-

Fig. 14-13 Beri Bunder (Bombay), India. Bombay is the largest city and principal industrial center of western India. (Courtesy of Government of India)

lords and have somehow to cope with the high cost of urban living. Apart from the ubiquitous cinema, opportunities for wholesome recreation are meager, while purveyors of vice are never far to seek. Hard though their lives may be, most factory workers, and other urban immigrants as well, do manage to send regular remittances to their native villages. It is not uncommon to encounter settlements in which a fourth or more of all households are primarily dependent for their livelihood on the meager sums sent home by family members working in distant cities or possibly on plantations, in mines, or in the armed forces.

Economic Infrastructure

A part of the legacy of India, Pakistan, and Ceylon from their preindependence days was a remarkably good economic infrastructure. India's railroad system was and remains the world's fourth largest (after those of the United States, the U.S.S.R., and Canada), though unfortunately hampered by the existence of three gauges. The road system was fairly good and, since independence, has been tremendously expanded. The well-developed banking and financial system included a number of stock and commodity exchanges. Labor unions and the cooperative movement both had made good beginnings. A communications network, however inefficient it may have been, linked all parts of the area. If Britain's erstwhile colonies had to provide all these facilities for themselves, their economic progress would undoubtedly have been greatly impeded.

Among the greatest assets of India, Pakistan, and Ceylon are their well-developed, well-ordered, and generally dedicated civil services, patterned on the British model. All three countries, and recently Nepal as well, have created networks of rural community development blocks, normally consisting of about a hundred villages each. Through these and the various organs of urban government, the several administrations have sought to reach out to the people, ascertaining their needs, and, within their limited means, responding to them. Noteworthy success, however, appears to have been attained only by India, which perhaps largely explains why, through more than two difficult decades since independence, that country has been able, unlike most developing nations, to adhere to the democratic path of economic, social, and political development its constitution prescribes.

Political Geography of the South Asian Borderlands

Progress in the absence of peace is exceedingly difficult; yet, the nations of South Asia have at no time since Indian independence in 1947 been entirely free from the specter of catastrophic war. There has been unremitting tension and skirmishing along the Indo-Pakistan frontier ever since the partition and along the Sino-Indian frontier since 1962. Serious, though limited, wars were fought along the borders in 1947–1948, 1962, and 1965. In East Pakistan large-scale violence erupted in March 1971 over the political future of that province. What began as a move for increased autonomy has, in the wake of its suppression by military force, burgeoned into a movement for secession. At this writing the situation is still too fluid and clouded by propaganda and censorship to permit an objective analysis; but the flight of millions of refugees into India has once again greatly heightened the tension between that country and its neighbor. Should the major disputes in South Asia one day lead to full-scale conflict, it would undoubtedly bring tremendous economic losses, if not ruin, to the participants and simultaneously endanger the precarious peace of the world as a whole.

The Problem of Kashmir

In 1849, the then newly united state of Jammu and Kashmir entered into a "subsidiary alliance" with the British. In doing so, it became the second largest of the princely states within the Indian Empire. Though ruled by a Hindu dynasty from its southern province of Jammu, a substantial majority of the state's population were Muslims. In 1941, when the last preindependence census was taken, Muslims constituted 77 percent of the population of 4 million and about 93 percent within the fertile Vale of Kashmir, the historic core region of the state. Hindus totaled about 20 percent and formed a slight majority in Jammu. Buddhists, though fewer than 2 percent of the total population, constituted the predominant religious group in the very thinly settled eastern third of the state, known as Ladakh or "Little Tibet." Linguistically, the division was similar.

When the British quit India in 1947, each of the more than 600 princely states (collectively comprising roughly 45 percent of the Empire's area and 24 percent of its population) reverted, in theory, to independence. Almost all, however, had agreed to join either India or Pakistan, depending on the faith of the majority of the people, which was normally the same as that of the ruler. Among the few holdouts, understandably, was Jammu and Kashmir; but it could not hold out in peace. The political turmoil and violence that came in the wake of partition in the adjoining province of Punjab soon spilled over its borders. Unable to maintain order within his domains, the maharaja appealed to the Indian government for help. Such help was granted, however, only after he agreed to accede to the Indian Union. Till then, Pakistan had been counting on the maharaja's acquiescing to the presumed desire of the Muslim majority among his people for union with the new Muslim state. Thus, when Indian troops were sent into the state, Pakistan felt compelled to send in its own troops, and a small-scale war ensued. Only after much effort was the United Nations able to halt hostilities. The ceasefire line established as of January 1, 1949, has remained the *de facto* boundary between the areas of Indian and Pakistani control ever since. It leaves India in control of the Vale of Kashmir and nearly three-fourths of that state's population.

Numerous efforts have been made by the United Nations and other intermediaries to bring about a withdrawal of Indian and Pakistani troops and to prepare the way for a plebiscite to determine the fate of the area. All such initiatives were futile; so too was a renewed attempt to settle the matter by force in the Indo-Pakistani conflict occurring in 1965. Meanwhile India gradually integrated her portion of the state, including the Vale, within the Republic. In 1956, India sponsored her own referendum, which resulted in a vote in favor of complete integration with India; and the country's constitution was amended to seal the fact. Since then, the Indian position has been that the entirety of the state is part of the Indian Union and that the question is no longer open to solution by plebiscite. The Pakistani-held portions of the state, north and west of the Vale, since 1947 have been governed as the theoretically independent Republic of Azad (Free) Kashmir. In fact, rule over this area has always been exercised by Pakistan.

Compounding the difficulties in the area, the Chinese lay claim to and occupy considerable portions of the desolate region of Ladakh, through which they have already built two roads linking Tibet with their far-western province of Sinkiang on the Sino-Soviet frontier. The Indian government, after repeated diplomatic efforts to effect a withdrawal, attempted in vain in 1962 to drive the Chinese troops from Ladakh by force. Subsequently, substantial forces of each nation have dug in to keep the other in its place, and China has made common cause with Pakistan, despite that nation's membership in both the CENTO and SEATO pacts. China not only has agreed on the delineation of the border along the north of the Pakistani-held portion of the state but also has helped Pakistan build a new road link across it to Sinkiang. Along that road, caravan trade has already begun to flow.

Other Areas of Indo-Pakistani Contention

While Kashmir was long the chief bone of contention between India and Pakistan, there have emerged a number of other disputes between those two states. Among the most vital of these was the problem of allocation of the waters of the Indus River system. It so happens that the Indus and four of its five major tributaries flow into Pakistan from the Indian-held portion of Kashmir; the fifth flows out of India proper. Prior to partition, by far the greater part of the irrigated area along the Indus system lay in what was to become Pakistan, while many of the headworks for the canals running through that area were constructed at sites that were later to fall to India. Extensive though the irrigated area already was, the demand for additional water was keen throughout all parts of the system. Naturally, Pakistan feared that India, the upstream riparian power, would divert the water through its own territory (including Kashmir) to farmers in East Punjab and Rajasthan. It was, in fact, feasible for India to deny the downstream users in Pakistan not only the waters they desired for agricultural expansion but even a large part of the supplies on which they were already dependent.

Fortunately, Pakistan's worst fears were never realized. Following a series of temporary water-sharing arrangements, negotiations began in 1952 for an equitable solution to the problem. After eight years of bargaining, an agreement was reached. According to the Indus Waters Treaty of 1960, the waters of the three easternmost tributaries, the Ravi, Beas, and Sutlej, were allocated to India to whatever extent she is capable of tapping them; the remainder of their flow (if any) and that of the Indus, Jhelum, and Chenab, a much larger aggregate volume, were reserved for Pakistan. That division, however, was in itself not an adequate solution to the problem; for it was precisely along the three eastern tributaries that most of Pakistan's irrigated area lay. To keep those streams and the adjoining area supplied with water, it was necessary to construct costly link canals from the more westerly tributaries and from the Indus itself. The costs were to be borne in part by India and were to be financed in even greater measure by loans from the World Bank.

The development of irrigation along the Indus system by both India and Pakistan since partition has been truly stupendous and is still expanding. The total irrigated area exceeds 30 million acres, and considerable expansion is still planned. Though the expenditures have been high and though greater international cooperation might have resulted in considerable economies, it is remarkable that the two nations were not only able to reach an agreement on dividing the waters, but have subsequently scrupulously adhered to the terms of the agreement, even during their armed conflict of 1965.

Among the multitude of points at issue between India and Pakistan have been a number of relatively minor disagreements as to the precise delimitation of their lengthy eastern and western borders, a weightier dispute relating to the border in the area of the Rann of Cutch, and a controversy over India's proposed damming of the Ganges just above the point where it flows into East Pakistan. Happily, all these disputes, save for the last, have already been settled in a reasonably amicable and equitable manner.

The Question of Pakhtunistan

Straddling the boundary between Pakistan and Afghanistan is the home area of a group of closely related tribes collectively known as Pathans, Pushtuns, or Pakhtuns and numbering more than 10 million persons in the aggregate. About a third of this total are fully or partially nomadic herdsmen, of whom a significant number migrate each year between summer pasture in Afghanistan and winter pasture in Pakistan. Approximately 5 million Pathans normally dwell in Afghanistan, where they form the dominant political group. The remainder, in Pakistan, are overwhelmingly concentrated in what was formerly called the North-West Frontier Province. The tribes in the latter area, controlling the Khyber Pass and the other main passes into the Indian subcontinent, are a particularly fierce and independent lot. Under British rule they were pacified only by a system of institutionalized bribery; under Pakistan, over a portion of their area, they are granted special legal privileges, tax exemptions, and other forms of deferential treatment.

During the early nineteenth century all the Pathans were, nominally at least, subject to Afghan rule, and the government of Afghanistan never ceased to concern itself with Pathan affairs. When Pakistan became independent, Afghanistan became the advocate of a separate state of Pakhtunistan, ostensibly as a homeland for the Pathans of that area. The territory claimed for Pakhtunistan, however, comprised more than half the total of West Pakistan, including not merely the North-West Frontier Province and adjacent lowlands up to the Indus but also the whole of Baluchistan and a number of princely states lying between the Frontier Province and Kashmir. Throughout most of the latter areas the Pathans were a fairly small minority of the population. Seemingly, Afghanistan's interests in sponsoring an independent Pakhtunistan go beyond considerations of ethnic self-determination. As a landlocked nation, Afghanistan is concerned with obtaining a guaranteed outlet to the sea. That aim, presumably, might be attained through the creation of the new puppet state she envisages. A friendly buffer between herself and Pakistan appears to be desirable also from the point of view of defense.

It is difficult to gauge the degree of support for Pakhtunistan among Pakistan's Pathan population. As long as a separate North-West Frontier Province existed in which they were the dominant group, most Pathans were undoubtedly reasonably satisfied with their lot. But when, in 1955, West Pakistan was consolidated into a single province, largely to bring about a greater measure of parity between the Eastern and Western sectors of the country, many Pathans sensed that their own regional interests would thenceforth be submerged, and the Pakhtunistan movement gained in appeal. Further strengthening the movement was the overt support it received at various times from India and the Soviet Union for purposes of political expediency. At its height the movement was the source of sporadic violence along and across the Afghan frontier and gave rise to a rupture of diplomatic relations, an embargo on trade, and a ban on the transit of goods across Pakistan to and from Afghanistan. The result was a reorientation by Afghanistan. toward the Soviet economic orbit. In 1963 more normal relationships between Pakistan and its northern neighbor were again established. The future of the proposed independent state of Pakhtunistan remains in limbo. While a majority of Pathans are presumably now willing to live under Pakistani rule, their continued satisfaction with doing so depends in part on the external pressures brought to bear on the region and in part on Pakistan's ability to create the kind of state and society in which they will feel spiritually at home and economically better off than in a separate Pathan nation.

The Sino-Indian Confrontation in the Northeast

No section of the Himalayan frontier between China and India has ever been demarcated, and differences of opinion about where the boundary should be located exist along virtually its entire length. In the sector between Kashmir and Nepal, these differences are of negligible importance. In Kashmir, however, and in the area of India's North-East Frontier Agency (NEFA), they are causes of the utmost concern. Since the former of these two areas has already been considered, the following remarks are focused on the Sino-Indian confrontation in the NEFA area.

Prior to the nineteenth century, NEFA had been almost totally neglected by outside powers. Each of the numerous Mongoloid tribes inhabiting the region was, in effect, an independent political unit, with its own distinctive tribal language or dialect of the Tibeto-Burmese language group. A few of the tribes, especially in the west, were influenced by Tibetan culture and adopted Lamaistic Buddhism; but most remained animists. Until 1914 even British contacts with the area were tenuous. In that year, British, Tibetan, and Chinese diplomats, meeting in Simla primarily to decide on the proper boundaries of Tibet, agreed to the so-called McMahon line, which placed all of what is now in NEFA within the Indian Empire. Although the then new Nationalist Chinese parliament never ratified the Simla convention, that accord remains one of the chief points in the Indian case for control of the area. More germane, perhaps, is the fact that from 1914 onward Britain and then India imposed, to an ever-increasing degree, the only effective state jurisdiction the area has ever known.

Both Nationalist and Communist Chinese maps show the area of NEFA, down to the Bramaputra lowland as Chinese territory. Though the weak Nationalist government took no measures to press its claim, the Communist regime began infiltrating troops into the area in the early 1960s. An Indian attempt to expel these detachments was followed by the unanticipated and massive Chinese drive of late 1962, which shattered forever the myth of the impenetrability of the Himalayas as a defensive frontier. After reaching the plains of Assam, the Chinese withdrew their forces as unexpectedly as they had launched their attack. The usual explanation given is that, having taught the Indians a lesson, they had no further mission to perform. It must be remembered that, had the Chinese chosen to remain in India, their advance troops would have been totally cut off from supplies, except for air drops, once the high Himalayan passes into NEFA were closed by snow. Under these circumstances, withdrawal was expedient. Since November 1962, a precarious and often violated truce has been in force. Both sides are entrenched and seem determined to hold their ground. Meanwhile no progress toward negotiating a territorial settlement has been made.

The Himalayan Kingdoms

Sandwiched between India and Chinese-occupied Tibet, the Himalayan kingdoms of Nepal, Sikkim, and Bhutan are potentially strategic pawns in the cold war between their giant neighbors.

Nepal Though Nepal's history is an ancient one, the present kingdom owes its existence to the conquest and consolidation in the eighteenth century of numerous tiny mountain principalities by the martial Gurkhas, an Indianized, Hindu people from the western part of the country. In 1769 they annexed Khatmandu, the traditional seat of power, and established it as their own capital. They continued to expand their rule both eastward and westward, over a diversity of weak Hindu and Buddhist tribes, until being decisively beaten in a war with Great Britain in 1814–1816. By the treaty of peace, Nepal's boundaries were reduced to approximately their present limits, encompassing not only the central section of the Himalayas but also a 500-mile strip of lowland Tarai immediately to the south. Apart from their discontinuous forest cover, there is, significantly, no barrier between these lowlands and those of India's Ganges Plain.

Somehow Nepal managed to escape absorption within the Indian Empire and to maintain cordial relations with its rulers. Gurkha troops formed important contingents of both the British and the Indian armies and have fought with unmatched distinction in numerous wars. Yet, despite the knowledge of distant lands brought back to the country by returning veterans, government policy until 1951 was to keep the country closed to virtually all foreigners. Only in 1956 was the first motorable road linking Khatmandu to India constructed.

Since making its decision to open its windows to the outside world, Nepal has witnessed remarkable changes. The government has been modernized, a community development system developed on the Indian model, and a national university established. A vigorous road-building program is now underway, and a national airline was started in 1958. Foreign aid has poured in from many quarters, most notably from the United States, the Soviet Union, India, and China. Basically, however, the country's economy is still one of subsistence agriculture, and literacy levels have yet to reach 10 percent.

Currently no nation has any territorial claims on Nepal. Her borders with India were fixed at an early date, and a favorable boundary treaty with China was concluded in 1961. Since then, the entire northern frontier has been demarcated. Nepal has tried scrupulously to avoid leaning too far toward either China or India, in order not to antagonize the other, and this balancing act has produced interesting results. Because India was allowed to build a road from her borders to Khatmandu, the Nepalese government could hardly refuse a similar offer from China. Accordingly, an excellent low-level road was constructed through the Himalayas via the Kosi River Valley. This highway now links Khatmandu with the main east-west road from China proper through Tibet to Sinkiang. Their Nepal highway offers the Chinese a prime artery for southward expansion, should they ever choose to use it for that purpose. The governments in both Khatmandu and New Delhi are, of course, poignantly aware of the possibility.

Sikkim The state of Sikkim was at various times under Tibetan or Nepalese suzerainty. In 1890, however, the Manchu Empire recognized a British protectorate over the area; that protectorate lapsed in 1947, but was reassumed by the Indian government in a treaty negotiated in 1950. The treaty retains Sikkim's internal autonomy, while entrusting its defense and foreign affairs to the protecting power.

Sikkim's ruler and its chief indigenous group, the Lepchas, are Buddhist and speak a language closely related to Tibetan, though a majority of the population are Hindus of Nepalese origin. The economic importance of

Fig. 14–14 Gateway to Hanuman Dhoka Square, an important shopping center of Khatmandu, Nepal. (Courtesy of United Nations)

Sikkim is minor, but its strategic importance is great. Through it, via the Jelep La and Natu La passes, both about 14,000 feet above sea level, run the principal routes connecting eastern India and Tibet, routes now capable of carrying heavy vehicular traffic. Though there was no fighting along the Sino-Sikkimese border during the hostilities of 1962, there have been numerous skirmishes and raids across it in later years. If major hostilities were to begin, this would undoubtedly be a key combat zone.

Bhutan In the 1500s Bhutan was conquered by Tibet and became a Buddhist theocracy recognizing the religious authority of the Dalai Lama. Though nominally a vassal of Tibet, and later indirectly of China as well, Bhutan functioned, for all practical purposes, as an independent federation of local chiefdoms until 1907. In that year one chief, with British support, proclaimed himself hereditary maharaja and in 1910, despite Chinese opposition, negotiated a treaty with Great Britain giving that country control over Bhutan's foreign affairs. This protectorate, like that over Sikkim, lapsed in 1947 and was reassumed by India under a new treaty signed in 1949. Under this treaty Bhutan agrees to accept Indian guidance in foreign affairs; otherwise it is fully sovereign. In 1971 Bhutan, with Indian sponsorship, was admitted to the United Nations.

The dominant ethnic group of Bhutan is Tibetan. Among its minorities are a large Hindu Nepali element and a number of indigenous tribes speaking languages distinct from Tibetan. The latter have assimilated Buddhism and a considerable measure of Tibetan culture.

As yet, the economic importance of Bhutan is negligible. Apart from Indian civil and military aid teams, who are building the country's first motorable roads, staffing a number of schools and dispensaries, and rendering other forms of assistance, the country remains almost totally closed to foreigners. Chinese maps show eastern Bhutan as Chinese territory, and it is alleged that Chinese troops crossed that area in their advance into India in 1962. Unlike Sikkim, Bhutan guards its borders with its own militia, though India is pledged to defend Bhutan in the event of external aggression.

In Perspective

Do the nations of South Asia have the physical intellectual, and spiritual resources to continue to cope with and perhaps eventually solve the problems confronting them? Or will those problems prove to be insurmountable? While no attempt to provide any firm answers will be made here, one can summarize what the problems are, what the means of confronting them are, and what the likely consequences of success or failure will be.

Production

Per capita incomes in South Asia are among the lowest in the world. In the late 1960s they were about 90 dollars per year in India and Pakistan, 70 dollars in Afghanistan and Nepal, and 150 dollars in Ceylon. For the mass of the population, the standard of living is so low as to be inconceivable to most Americans. Diets are sadly deficient in quantity and even more so in quality. Shelter is squalid, crowded, and almost totally lacking in amenities. Though most of the "good things of life," from a material standpoint, can now be produced in South Asia (in India at any rate), the volume of production is low and the unit costs high. Consequently, such new wealth as the economy is generating is far from evenly distributed. Such commonplace commodities as radios, for example, can be afforded only by a very small fraction of the population, and private automobiles are available to fewer than one family in a thousand.

The most critical shortage inhibiting increased production is capital. Earnings from exports, except for Ceylon, are meager and are more than counterbalanced, in most years, by outlays for imports. Foreign aid, mainly from the United States, has over more than two decades averaged little more than 1 dollar per capita per year, part of which is in repayable loans. Thus, capital has had to be obtained mainly through domestic savings and at the expense of current consumption. Yet, with standards of living as low as they are, such savings can be obtained only at a sacrifice that it is difficult to ask the people to accept. Despite this painful fact, India and Pakistan have managed to achieve very respectable average rates of savings and investment and have thereby steadily increased their industrial capability. India in particular has built up her capital-goods industries to include virtually all basic types of production—metallurgical, chemical, and engineering—and has become increasingly independent of foreign sources for capital goods. It is not likely, however, that any other country in South Asia will achieve a comparable position in the foreseeable future.

Given the facts that about 70 percent of the population of South Asia still depend directly on agriculture for their livelihood, that the area cannot afford large imports of food, and that the average farm holding is under 3 acres, one might argue that the shortage of arable land is no less crucial than the insufficiency of capital in retarding economic growth. While it is undoubtedly true that, with more land available for cultivation, production could be significantly raised, that observation is largely academic; for in India and Pakistan the margin for expansion through various forms of reclamation is meager. Happily, Ceylon, Afghanistan, and the Himalayan kingdoms are somewhat more favored in this respect. What is really most needed is not so much an extension of cultivated acreage as a great intensification of effort and technology applied to the land already under cultivation. This means more fertilizer, more tube wells (and more engines, pumps, pipes, and cement with which to construct them), more electric power, improved seeds, better implements, and so forth. Providing most of these need inputs without resort to imports presupposes an increase in capital-goods industries.

The ability to create capital and apply improvements in technology to production depends in large measure on the quality of the nation's human resources. Sheer manpower may be a useful asset for some types of development efforts, such as the building of roads and dams; but normally the availability of skilled workers is of greater significance. While, as we have seen, India is particularly well endowed with high-level skills, there is throughout South Asia a shortage of middle-level technicians. As for the ability of the masses to understand new technology and apply it to agriculture, the rural literacy levels of 19 percent in India and 12 percent in Pakistan offer a not very encouraging guide.

Politics

For better or worse, the independent states of South Asia have accepted as a matter of principle the need for a large measure of governmental guidance of and intervention in the national economy. The course of Indian development has been largely charted since 1950 by a series of Five-Year Plans; comparable planning for Pakistan commenced in 1955. In framing their plans, the nations of South Asia must take into consideration not only the economic realities that have been briefly outlined but a number of basic political realities as well. Unlike private entrepreneurs, they cannot simply allocate investment to those activities and at those sites where the highest financial returns will be realized; rather, they must continually weigh the relative benefits to be derived from investments in social welfare and the improvement of human resources against direct capital investments. Even when deciding for the latter, they must be responsive to the demands of the various regions of the country to have a share in the increase in national wealth. This has already been forcefully and tragically demonstrated with respect to East Pakistan. It is equally true for India, where the Dravidian south, already largely disenchanted with what it perceives as northern domination, especially in regard to the language issue, cannot be allowed to add the fuel of alleged economic descrimination to the fires of separation already smoldering there.

The needs for policies of domestic and international conciliation in regard to the political disputes that currently threaten the peace and internal stability of South Asia are obvious. It is also obvious that the emotional involvement of the parties to certain disputes, such as those of Kashmir, NEFA, and, most recently, "Bangla Desh," is so intense as to render difficult even an approach to peaceful solution. There is no way to judge in advance whether passion or reason will ultimately prevail.

Population

In the context of production, there are definite economic advantages accruing to a nation with a large population. Other things being equal, the larger the population, the larger and more varied the pool of labor and skills on which its economy may draw, and the greater the likelihood that a critical economic threshold of demand for particular goods or services will be reached. Hence, in very populous countries, poor though they may be, entrepreneurs may find it profitable to produce certain goods that would not be profitable, without assured foreign markets, in wealthier countries with small populations. To the extent that this is true, India and Pakistan do derive some benefit from their large populations.

There is, however, another side to the coin. People are consumers from the moment of their birth and producers for only a limited portion of their life-span. Given the conditions of public health, private hygiene, and nutrition in South Asia, much of the population never attains the age when they become efficient producers. Moreover, as the economies of South Asia are organized, even in their working years the productivity of many persons is far less than their potential. One obvious reason for this is the low level of skill and training of most of the labor force. An additional basic cause is the inordinately high ratio of labor to both land and capital. Though levels of outright unemployment are not especially high, there is an incredible degree of underemployment and inefficient use of labor. On a typical small farm, for example, five laborers may be available to do the work that two could easily perform alone. Or, in factories, the abundance and cheapness of labor may induce managers to have many operations performed by hand that in other parts of the world would be fully automated.

The situation at present is bad enough; prospects for the future are frightening. Since about 1921, the death rate in South Asia has been steadily declining, from roughly 45 per thousand per year (about equal to the birthrate) to current rates of about 8 in Ceylon (as against 10 in the United States), 17 in India, and 18 in Pakistan. Over the same period, however, birthrates have fallen very slowly. Current estimates of birthrates in India, Pakistan, and Ceylon are 42, 50, and a very encouraging 32, respectively, making for annual rates of natural increase of 2.6, 3.3, and 2.4 percent. The rates for Pakistan are among the highest in the world. At current rates of growth, the populations of the three countries would double in only 28, 21, and 29 years.

It is painfully obvious that South Asia's population explosion has greatly diminished the impact of increases in productive capacity in the postindependence era and threatens to nullify whatever gains might be expected in the future. It is particularly problematic whether the growth of population will outstrip the capacity of the economy to feed its own population, in view of the limited acreage available for cultivation. If substantial controls are not soon applied to the growth of population, one or more of the Malthusian checks of famine, pestilence, and war may well be unavoidable. Within less than six months following the devastating cyclone of November 1970, East Pakistan was subjected to all of these in frightful measure.

Fortunately, the governments of India, Pakistan, and Ceylon are cognizant of the demographic problem confronting them, and have taken at least the initial steps to deal with it. After a very modest beginning with the first Five-Year-Plan, India has by now enlarged the scope of governmental activity to include the creation of a nationwide network of family-planning clinics, distribution of contraceptive devices at little or no cost, free insertion of intrauterine loops, and a program of voluntary male and female sterilization, with modest bonuses paid to participating males. The Pakistani family-planning program commenced later than that of India and is still generally less comprehensive in its approach to the problem.

It is not yet apparent when the rate of growth of the population of the various countries of South Asia will begin to decelerate, though Ceylon's has done so already. Whatever the time may be, it appears probable that the total population of India alone, 547 million as of the 1971 census, will pass 1 billion well before the year 2000.

REFERENCES

Basham, A. L.: *The Wonder That Was India*, Grove Press, Inc., New York, 1967.

Brown, W. Norman: *The United States and India and Pakistan*, Harvard University Press, Cambridge, Mass., 1963.

Cook, Elsie K.: *Ceylon, Its Geography, Its Resources and Its People*, Macmillan and Company, London, 1951. (A new edition of *A Geography of Ceylon*, revised and brought up-to-date by K. Kularatnam).

Ginsburg, Norton: *The Pattern of Asia*, Prentice-Hall, Inc., Englewood Cliffs, N.J., 1958.

Hagen, Toni: *Nepal: The Kingdom in the Himalayas*, Kummerly & Frey, Geographical Publishers, Berne, 1961.

Lamb, Alastair: *Asian Frontiers, Studies in a Continuing Problem*, Frederick A. Praeger, New York, 1968.

Neale, Walter C.: *India: The Search for Unity, Democracy, and Progress*, D. Van Nostrand Company, Princeton, N.J., 1965.

Spate, O. H. K., and A. T. A. Learmouth: *India and Pakistan, a General and Regional Geography*, E. P. Dutton and Co., Inc., New York, 1967.

Wilbur, Donald N. (ed.): *Afghanistan*, Human Relations Area Files, New Haven, Conn., 1956.

15

Southeast Asia

Since 1950, the peninsular and insular lands south of the two Chinas and east of India are frequently referred to as the "Southeast Asia realm"—with the East Asia realm to the north and South Asia to the west. There are ten countries in this realm. Six are on the mainland: Burma, Thailand, Laos, Cambodia, North Vietnam, and South Vietnam. Three are insular: namely, Singapore, the Philippines, and Indonesia. The tenth, Malaysia, sits astride the waters between the Malay Peninsula and the island of Borneo. Extending southward, the Malay Peninsula forms a land bridge to Sumatra in Indonesia, and from there the islands in the Indonesian Archipelago comprise stepping stones leading to both Australia and the Phillipines. Historically, many of these lands and their cultures are centuries old. Politically, during the nineteenth and twentieth centuries, all these countries except Thailand were colonies up to the post-World War II period. Since they became independent, they have attempted to gain stability, diplomatic respectability, prosperity, and a more affluent society based on developing resources and on using twentieth-century technology. Newest of these states is Singapore, an island-city nation, established in 1965 after its separation from Malaysia.

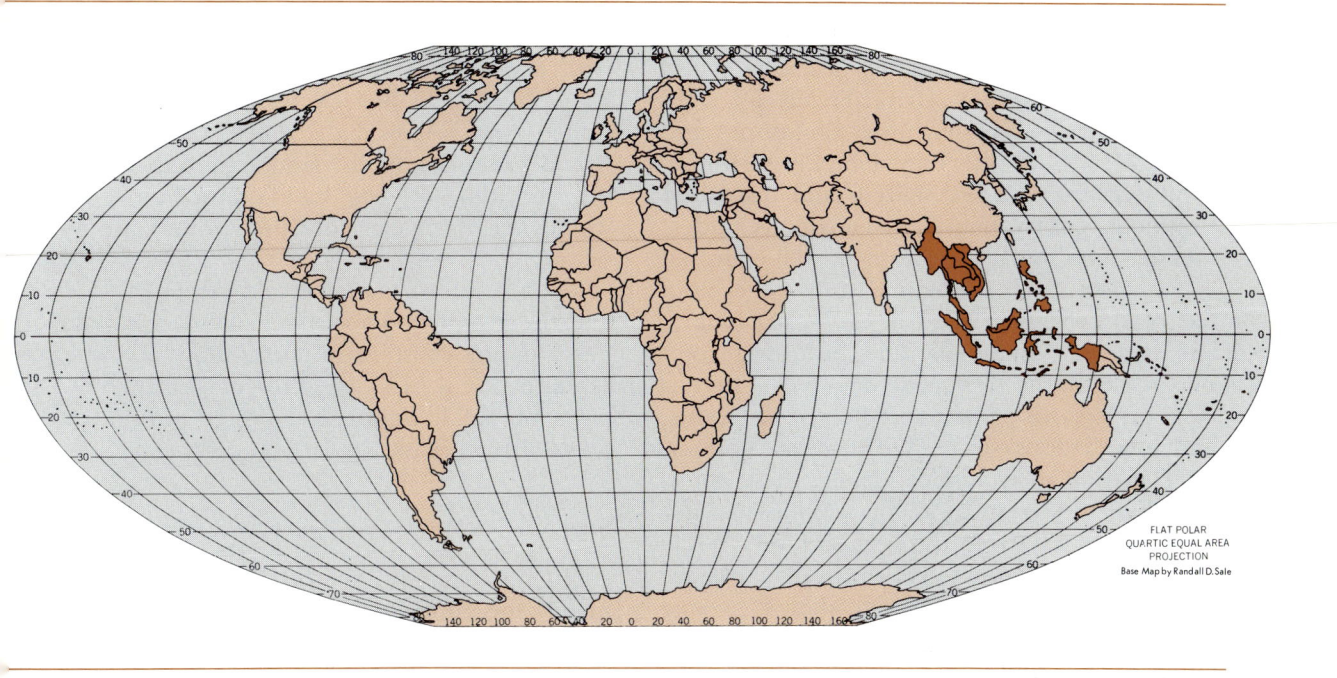

These relatively young and primarily small countries have had one experience in common: they were real estate pawns in a worldwide power struggle in the pre-World War II colonial period. World War II and the Japanese occupation resulted in enormous economic damage and political and social disruption. Since all forms of production were greatly reduced or temporarily stopped, not a single major facet of economic life remained unchanged.

Moreover, the pre-World War II political hierarchy had crumbled, and returning colonial governments were met with demands for immediate independence. The people in these countries preferred to be governed by their own Asian leaders, regardless of their administrative inexperience and lack of know-how. Within two decades following the end of World War II, this realm had ten independent countries with varying degrees of political stability and efficiency, economic development and

555

prosperity, technological advancement, and urbanization. To change a predominantly agricultural, preindustrial, poorly educated, and rural-dwelling society characterized by eighteenth- and nineteenth-century philosophy and a simple way of life into a highly technical, commercialized, industrialized, and urbanized twentieth-century one is an enormous undertaking. The shortage of capital, well-educated and experienced personnel, universities and technical colleges, skilled laborers, machinery, adequate raw materials, and industrial complexes makes the task more difficult. There is an abundance of people, and their needs, wants, and expectations soar. With few exceptions, after getting their independence, these countries had to start with chaotic conditions and to build literally from the ground level. While these people struggle to get the bare necessities of food, water, clothing, and shelter to stay alive, they must also defend themselves against opportunists, infiltrators, and saboteurs and learn to live harmoniously with their neighbors.

Physical Setting

Not only does Southeast Asia have archipelagos larger than any other realm in the world, but it also has over one-half of its land area broken up into thousands of islands. Located here are the world's second and third largest islands, New Guinea and Borneo, as well as thousands of smaller ones and thousands of islets of less than 1 square mile each. Surrounding these islands and on three sides of mainland Southeast Asia are vast expanses of water broken into large and small seas, straits, bays, and inlets. These waters are so shallow that, if the Sundra and Sahul continental shelves were lifted only 500 feet, mainland Southeast Asia—sometimes called the Indo-Pacific Peninsula—would extend southward to include what are now the islands of Singapore, Sumatra, Borneo, and Java. Moreover, if the ocean floor were raised 500 feet, New Guinea would be united with Australia. Such an elevation would about double the land area of the realm, but the Sulu, Celebes, Banda, and Flores seas would not disappear since their waters are too deep.

As a result of both the vast extent of water and the irregular shapes of the land, this realm has a longer coastline areawise than any other realm in the world. Here the most productive parts of the lowlands are located within 150 miles of the ocean.

Size

In square miles, Southeast Asia is approximately the same size as South Asia and Southwest Asia and only about one-third the size of East Asia. All the area of Southeast Asia, including the land and adjacent seas, is about equal to the land area of the United States—approximately 3.6 million square miles. This realm's land area amounts to only about 1.74 million square miles, or a little less than half that of the United States.

Southeast Asia covers a large longitudinal and latitudinal spread. It extends from 92° to approximately 141° east, a longitudinal distance of over 3,000 miles, and it stretches from approximately 11° south in southern Indonesia to about 28°30′ north latitude in northern Burma, a latitudinal distance of about 2,700 miles. Indonesia is the area's largest country, and its longitudinal extent is over 3,000 miles. If an equal-area transparent map of this country were superimposed on a map of North America, part of Irian Barat in western New Guinea would lie over the Atlantic Ocean east of the United States and part of Sumatra would extend west of California. Thus, longitudinally Indonesia is longer than the contiguous states of

Fig. 15-1 There are ten independent countries as well as possessions of the United Kingdom and Portugal in the Southeast Asia area. Indonesia is the largest country (735,865 square miles); Singapore, the smallest (225 square miles).

the United States. In contrast, the Phillipines has a north-south extent of 1,152 miles. A transparency map of mainland Southeast Asia superimposed on a map of the Mississippi River watershed would put northern Burma north of Lake Superior; part of North Vietnam would be north of lakes Erie and Ontario; and the Malay Peninsula would terminate at a more southerly location than Florida.

By world standards, most of the countries in Southeast Asia are small; but if compared with states in the United States, they are relatively large. Singapore, sometimes called an island-city state, contains only 225 square miles; all the other countries are larger in size than Illinois. Since it acquired Irian Barat, Indonesia has 736,512 square miles of territory, and this gives it control of over two-fifths (approximately 42 percent) of all the land in this realm. It is now about the size of Alaska and Texas combined.

Landforms

The land with elevations of more than 1,500 feet is less extensive than that of lower relief. The topography above 1,500 feet is very conspicuous in the landscape and plays an important role in explaining the uneven distribution of both temperatures and rainfall. Backbones of hills and mountains extend nearly the full length of the large islands. These sometimes rugged but comparatively low ranges, with only their upper slopes above 1,500 feet, are often capped with volcanic peaks rising majestically to over 10,000 feet above sea level. The highest and most extensive mountains in the archipelagos are on the islands of New Guinea, Borneo, and Sumatra.

A tripartite system of mountains separates mainland Southeast Asia, or the Indo-Pacific Peninsula, from the South Asian and the East Asian realms. The longest north-south and the central component of this tripartite framework of mountains extends from the river-frayed southeastern margin of the plateau of Tibet southward into the tropics and to almost the full length of the Malay Peninsula. The distance from the northern tip of Burma, which lies north of the Tropic of Cancer, to the tip of the Malay Peninsula, south of 2° north latitude, is approximately 2,000 miles. The western component of this major tripartite mountain pattern includes the Arakan Yoma to the south and the Naga Hills to the north and associated ranges with elevations up to 8,000 feet along the Indian border in western Burma. These ranges separate the basins of the Irrawaddy and Chindwin rivers from India and Pakistan to the west. Along the Burmese-Chinese border, there are elevation up to 18,000 feet. On the opposite side of the Indo-Pacific Peninsula, the eastern primary component, the Annamese Cordillera, forms a claw-shaped pattern extending first southeasterly and then bending southwesterly until the mountains terminate northeast of Saigon. Here the high relief gives way to the alluvial lowlands formed by the Mekong River and its distributaries. This range extends so close to the coast near the 18th (north) parallel that it almost separates North Vietnam from Southeast Asia.

The horsehoe-shaped mountains, including hills and tablelands of the Annamese Cordillera and the central mountainous component forming the boundary between Burma and Thailand, provide a high relief boundary that encircles on three sides the largest contiguous plain area in Southeast Asia. On this plain is located most of Cambodia and Thailand, approximately half of South Vietnam, and the Mekong River valley, where most of the population and cultivated land of Laos is concentrated.

The long rivers with large seasonal volumes such as the Chindwin-Irrawaddy, Salween, Chao Phraya, Mekong, Black (Song Bo), and Red (Song Koi) influence the realm's economic and social life. Rivers provide domestic, irrigation, and industrial water as well as serve as transportation corridors and sewers. In their upper reaches, these rivers have cut deep gorges with steep precipitous banks, while in their lower courses they have wide floodplains and large deltas. Irrigated agriculture, national populations, towns, and cities are concentrated on the alluvial soils of these coalescing floodplains, deltas, and coastal plains.

Transportation on rivers and oceans tends to unite various parts of each country and enables each country to trade with every other country, with the exception of Laos. Rivers, often paralleled by railroads and/or highways, facilitate the flow of raw materials from forest, mine, and farm downstream and oceanward to the cities and river-ocean ports for consumption, export, or manufacture. Manufactured goods move in the opposite direction. Coastwise shipping also provides linkages between smaller cities and the primate one in each country. Today, ports accommodating oceangoing ships exist in all these countries but Laos.

Climate

Vital keys to understanding Southeast Asia's diverse weather and climatic types are: (1) its position astride the equator; (2) its general

Fig. 15-2 The areas of heaviest rainfall are adjacent to the higher mountains, where the orographic effect is dominant.

location between the world's largest landmass, Eurasia, and two of the world's three largest ocean masses, the Indian and Pacific; (3) so much of its land is arranged in peninsulas and islands; and (4) relief ranging often within short distances from sea level to thousands of feet in elevation. As for its temperature, in no other part of Asia and in few other areas of the earth's surface can such uniform year-round high temperatures be found as exist in the lowlands of this realm. In attempting to generalize about climatic conditions, climatologists have categorized most of the land here in two types of tropical climates—tropical rain forest and tropical savanna.

In a tropical rain forest climate, the coolest month has an average temperature above 64.4°F and each month is moist. Since climatic characteristics are based on averages, there can be and are exceptions. Tropical rain forests thrive in areas with 80 or more inches of rainfall which is well distributed throughout the year or occurs where soil and/or groundwater provide enough year-round moisture for abundant vegetation. There are variations or subdivisions of the tropical rain forest zone. The monsoon tropical rain forest is concentrated in the Philippines and in mainland Southeast Asia, where it is farther from the equator than the tropical rain forest type. This monsoon type has a distinct dry season, but it also receives enough rainfall to sustain tropical rain forest vegetation if enough water is stored in the ground and the vegetation is not removed by man.

The tropical savanna is found in eastern Java and islands to the east of it and extends from southern South Vietnam across most of Cambodia, southern Laos, and Thailand to Burma, where it covers most of that country. The tropical savanna has at least two rainfall seasons—a wet season and a distinct dry one, with which is associated cooler temperatures. Because the tropical savanna's dry season is longer than that of the tropical rain forest climates and because it also receives a smaller amount of annual rainfall, deciduous forests and grasslands called savannas are associated with this climatic type. Irrigation is required for good rice yields.

The two minor climates in this realm are the humid subtropical, with warm summers and dry winters, and the undifferentiated highlands. The former is the primary climate in North Vietnam, with the exception of its southern panhandle, in approximately the northern three-fourths of Laos, and in northern Burma. The latter type, found in the mountainous core of Borneo and Irian Barat and in a very minor part of mountainous northern Burma, has such a diversity of climatic conditions that, in accordance with a world classification, it is referred to as "undifferentiated highlands."

Soils and Vegetation

Extensive areas of Southeast Asia are occupied by tablelands and mountains with elevations above 500 feet. The soils generally associated with these landforms are too rocky and sandy, porous, highly leached, and deficient in water and minerals to be agriculturally productive. Generally these landforms are covered with evergreen tropical rain forests and jungles where rainfall is abundant and well distributed throughout the year. Where the rainfall is less, open deciduous forests exist; and where there is an excess of evapotranspiration over precipitation, there are tropical savanna grasslands. In cleared patches on slopes some tea, coffee, cinchona, bananas, rubber, upland rice, and other tropical crops such as opium poppies are grown.

The rich alluvial soils are estimated to cover only about one-tenth of the land surface. Most of the people in Southeast Asia are engaged in subsistence or commercial agriculture, and most of the farm and plantation production is located on the alluvial soils of floodplains, deltas, interior basins, and coastal plains. Crops, especially rice, sugarcane, coconut and date palms, cotton, and other food and fiber crops, can be grown with irrigation or watered by natural subsurface flowing waters. Vegetables and other crops are also grown on the colluvial and alluvial soils of alluvial fans and on terraces built on slopes. There are limited areas of good volcanic soils especially in Indonesia.

The Realm's Subdivisions

There is no universal agreement about what countries and land areas make up Southeast Asia. Nor is there one on where the boundaries should be drawn between East Asia, Southeast Asia, South Asia, and Australia. There is, however, an understanding that the mainland core of the Southeast Asia realm consists of the countries of Thailand, Cambodia, South Vietnam, and Laos in the Great Lowland with its mountainous fringe. These countries are also commonly included in the area referred to as the Indo-Pacific Peninsula.

Whether Burma, North Vietnam, the Phillipines, and all of the island of New Guinea should be included in this realm has been questioned. Because Burma is (1) located on the northwest periphery, (2) situated primarily north of 20°N, (3) the only country with much of its territory north of the Tropic of Cancer, and (4) so nearly cut off physically by mountains and economically and politically by a policy of neutrality and relative isolation, some geographers would like to exclude this second largest of the ten Southeast Asian countries. Because North Vietnam (1) is physiographically almost closed off from the rest of Southeast Asia by the Annamese Cordillera, (2) has a much wider coastal plain connection with the Kwangtung province of China than with South Vietnam south of Phu Loc, (3) watersheds, economic activities, and culture are all oriented eastward toward the Gulf of Tonkin, (4) is the most Sinicized of the countries here, and (5) is the only one of the ten primarily under the influence of a mid-latitude type of climate, some geographers would prefer to place North Vietnam in the East Asian realm. However, if in the future North Vietnam is able to gain political control of the territory in former French Indochina, or the entire area is in some way united, this presently peripheral country would definitely lie in the Southeast Asian realm. Others, who have raised questions about whether the Philippines should be placed in Southeast Asia, stress that the Philippines occupy a marginal position because of its geological structural linkages with the offshore island chains to the north and historical ties across the Pacific. But others maintain that the Filipino peoples and their cultures have numerous linkages with this realm. For example, the government of the Philippines claims that down through the centuries, before the European colonial period of interference, the Sulu Sea was a Filipino cultural unit as well as a Filipino sea and that the northern part of the island of Borneo rightfully belongs to the Philippines. A Pandora's box of heated discussion can be released by asking why the western half of the island of New Guinea, Irian Barat, is in the Southeast Asian realm but the eastern part is not, when only a straight north-south non-Asia political boundary line separates the two.

For detailed treatment, the ten countries and two minor territories of this realm are grouped into three major subdivisions and one minor one: (1) the Great Lowland and its mountainous fringe (Thailand, Cambodia, South Vietnam, and Laos), which faces southward to the China Sea; (2) two peripheral countries (North Vietnam and Burma); (3) the bridge and insular countries (Malaysia, Singapore,

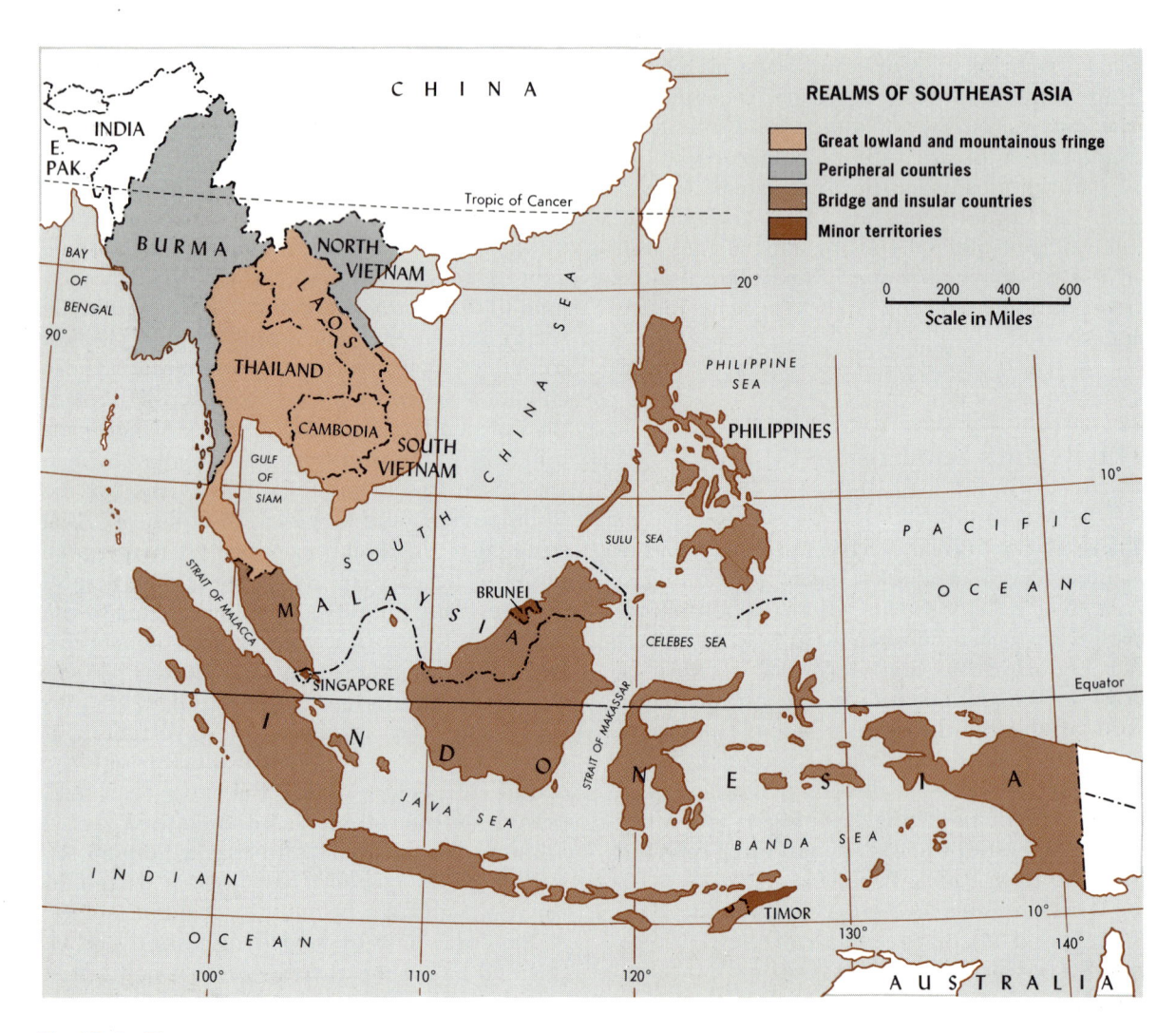

Fig. 15-3 The countries of Southeast Asia can be divided into four large geographic realms.

Indonesia, and the Philippines); and (4) scattered minor territories (Brunei and Portuguese Timor).

The Great Lowland and its Mountainous Fringe

This subdivision contains about one-fourth of the land area and over one-fifth of the 1967 population of the realm. Within this subregion, Thailand holds approximately 46 percent of the territory and about 56 percent of the population. At the start of the nineteenth century, Thailand had most of this lowland under its control. During the forty-year period between 1867 to 1907, according to Thailand's account, it lost approximately 180,455 square miles of its territory to the French.

THAILAND

With an area of approximately 198,000 square miles, the Kingdom of Thailand is the second largest country in mainland Southeast Asia and the third largest in the entire realm. With nearly 35 million people, Thailand ranks first in population among the Indo-Pacific Peninsula countries and third in Southeast Asia—exceeded only by the populations of Indonesia and the Philippines. About the size of France, Thailand has 2,614 miles of shoreline, comprising a 740-mile frontage on the Andaman Sea in the Indian Ocean and a 1,875-mile frontage on the China Sea in the Pacific. The country's shape has stimulated a number of descriptions, such as the comparison with the head of the symbolic white elephant, with the narrow peninsula resembling the animal's trunk.

The Thai stock accounts for over four-fifths of Thailand's population. The principal minority groups are: (1) 4 million ethnic Chinese, who live primarily in Bangkok (Krung Thep) and other cities and in peninsular Thailand; (2) about 1 million Malay-speaking Muslims in the peninsular provinces adjacent to or near the Thailand-Malaysian border; (3) about a third of a million hill tribes living in the mountains in the western and northern parts of the country; and (4) about 40,000 Vietnamese living in several northeast provinces adjacent to the Mekong River.

Thailand can be divided into five distinct regions: the Central Plain, northeast, north, south, and southeast. Southeast Thailand, south of the Dang Raek Mountains and east of the Bight of Bangkok, is unlike the Central Plain in topography, climate, and natural vegetation, but it is such a small area that for statistical purposes Thailand places it in the Central Plain district.

The Central Plain, the heartland of Thailand, is about the size of Virginia and contains twenty-six of the seventy-one provinces, or about one-fifth of the country's territory and two-fifths of its total population. The average population density in Thailand is 132 persons per square mile, but in the Central Plain the density is the highest of any region, 152 per square mile. Despite receiving a relatively low rainfall of somewhat over 50 inches a year, the Central Plain has a rather rich alluvial soil, undergoes annual flooding with waters from north Thailand, and has extensive use of irrigation, making this region the country's rice basket and Thailand one of the few rice-exporting countries in the world. For irrigation purposes, the clay soil of this floodplain-deltaic region is the best in the country. Wet plowing at the same depth year after year helps develop a clay pan, reducing to a minimum water seepage from the paddy fields. Here one finds over two-fifths of Thailand's acreage devoted to rice, producing over half of the country's crop. Here also is concentrated about four-fifths of the cassava production, two-fifths of the maize, and over half of the castor beans.

The major part of the commercial, industrial, and service activities are found on the Central Plain, with the country's hierarchy of activities peaking in the Bangkok metropolitan area. Thailand's heaviest population densities are found in the provinces adjacent to and occupied by the Bangkok metropolitan area. Bangkok and Thon Buri have over 2.1 million people, which accounts for over half of the country's population living in municipalities. With the exception of mining, all economic, political, social, religious, educational, and other phases of Thai life centers in this metropolitan area and the Central Plain region.

Northeast Thailand, with an area of 65,722 square miles, is larger than North Vietnam, is about 3,000 square miles smaller than South Vietnam, and accounts for about 33 percent of Thailand's territory. Whether overpopulated or underdeveloped or both, this region is the poorest; apparently for lack of better work opportunities it has the largest backlog of workers, who are only unproductively employed in farming. The northeast has a population density of 137 per square mile, or about one-third of the country's people who would benefit most from development of the slow-moving Mekong River project.

In spite of a regional average of 57 inches of rain per year, most of the land of northeast Thailand is not well suited for rice or other intensively cultivated crops, partly because of high evapotranspiration, erratic rainfall, and soils that are highly leached and too sandy to retain the precipitation that does fall. The only great quantity of water coming to this area is in the Mekong River, which flows past the northern and eastern rims of the region, but there is no dam across this river to store water for potential irrigation and other uses. Rice is the primary crop, despite the lack of water for good yields of long-maturing varieties. Yields are low, and not enough rice is grown here to adequately feed the people, especially in drought years. Although between 40 and 45 percent of the country's rice acreage is in this region, it produces only about 30 percent of the annual crop. In the production of fiber used for export and manufacture, primarily gunnysacks and rope, the northeast surpasses other regions. Here is grown over 95 percent of the jute, 92 percent of the kenaf, nearly 60 percent of the ramie, and some cotton and kapok.

Although north and northeast Thailand are about the same size, receive nearly the same amount of rainfall (51 inches annually in the north), are located in higher latitudes than the Central Plain, and both grow lac, sugarcane, and cotton, these two northern regions are different in many physical and economic ways. The physical differences are most conspicuous in the landscape. In contrast with the northeast, which is primarily an elevated and tilled plain with local relief of less than several hundred feet and with its rivers flowing southeast into the Mekong River, north Thailand is a mountainous region with a few intermountain basins and narrow river valleys and floodplains cut and built by the four large rivers that join to form the Chao Phraya. These rivers are vital to both the north and the Central Plain because they provide large volumes of water for irrigation, domestic and industrial use, navigation, hydroelectric power, and replenishing the fertility of fields with alluvium. Whereas most of the land in the northeast has been cleared for cultivation, most of the north is covered with forests. One of the best soils in Thailand, the Chiang Mai soil, is more abundant in the north than in the northeast.

The north has a population density of only about 92 persons per square mile, which is less

Fig. 15-4 The floating market of Bangkok, Thailand, is an assembly point and shopping center for all types of produce. (Courtesy of Government of Thailand)

than the average for the country. The per capita amount of cultivated land is less than that either in the Central Plain or in the northeast. Both population and cropland are concentrated in the few large valley floors and intermontane basins, such as the one in which the largest city in the north, Chiang Mai, is located. Although the north has only about 7 percent of the total rice cropland, it accounts for about 12 percent of the annual production. Yields are higher here than in the northeast, partly because evapotranspiration is lower, the soils retain water better, and double-cropping is more prevalent. This region produces nearly two-thirds of the nation's groundnuts, nearly half of the maize, and about two-fifths of the cotton. During the past decade, as teak has decreased in importance as a national export, maize has increased in both the region and the nation. Besides supplying large volumes of water to the Central Plain, three of the greatest contributions the north makes to the national economy are to supply most of the country's hydroelectric power from that produced by the Bhumibhol Dam generators at the highest dam in Thailand (over 500 feet), supply recreational opportunities, and provide most of the middle-latitude fruit and vegetables consumed in Thailand.

South Thailand, both population and area-wise, is the smallest region, with only 14 percent of Thailand's area and about 12 percent of its population. With a tropical rain forest climate on the leeward east side and a tropical monsoon climate on the windward west side and in the mountains, the south receives more rainfall than any other part of the country; in fact, the lack of a distinct dry season and a paucity of good soils are handicaps to rice production. In this region a higher percentage of the people is engaged in fishing, coastwise shipping, extracting forest products, and mining. Most of Thailand's tin comes from the south, and a modern smelter is located in Phuket. This region produces over nine-tenths of the nation's natural rubber. Hat Yai, a railroad junction, the region's largest city, and the fifth largest municipality in Thailand, is noted for processing of raw rubber sheets and manufacturing of rubber products. A recently built thermoelectric power plant at Krabi, operated by the Lignite Authority, helps provide electricity. All-weather, hard-surfaced highways have been built connecting the region and the cities on both coasts with the Central Plain and the Bangkok metropolitan area.

The Thai economy is diversified, fast-growing, and stable. As late as the early 1950s agriculture accounted for 50 percent of the gross national product. In 1963, however, agricultural income (including the income from fishing and forestry) amounted to only about one-third of the country's GNP. In addition to agriculture, the four largest contributors to the GNP, in descending order, are: (1) retailing, wholesaling, and banking; (2) manufacturing; (3) utilities, communication, and transportation; and (4) services.

Although agriculture (including forestry and fisheries) generates only about one-third of the GNP, it provides basic raw materials for most of the industries and direct employment for about four-fifths of the working population. While only about one-fifth of the total land area was farmed during the past decade, agricultural production increased by about 62 percent, and there was diversification in the crops grown and exported. In recent years rice production nearly doubled: it increased from 5.57 million metric tons to 11.84 million tons. This increase was partially owing to: (1) reclamation, by clearing the land and building irrigation projects the area under rice cultivation has doubled; (2) utilization of two-thirds of the cropland for rice; and (3) adoption of improved higher-yielding varieties of rice. Secondary commercial farm crops may be grouped roughly into two categories: sugarcane, cotton, jute, and tobacco for the expanding domestic processing industries, and rubber, maize, cassava, groundnuts, and kenaf for export. Production increases in some of these crops have been large in recent years. For example, during the last fifteen years, maize production rose from 115,000 to 665,000 metric tons; kenaf, used in making rope, from 17,000 to 134,000 metric tons; and pararubber from 136,000 to 198,300 metric tons.

Despite the good record of economic growth during the past decade, per capita income is only about 157 dollars per year. This situation is true in part because of the rapid population growth, resulting in the economically active population increasing from 8.99 million in 1947 to 12.7 million in 1961. Owing to both inflation and an increase in the quantity of some exports, the value of Thailand's exports has tripled—increasing from 223 million U.S. dollars in 1948 to 694 million in 1966. Although the ranking of money earners varies from year to year, the primary exports in descending order of value in 1965 were rice, rubber, tin, teak, cassava products, fiber, (jute, kenaf, kapok), and maize. Thailand's chief trading partner is Japan, which in 1966 took 21 percent of its exports and provided 26 percent of the imports. The United States ranked second as a trading partner, providing 37 percent of the imports but taking only 7 percent of the exports. Thailand has an unfavorable balance of trade, with the value of imports soaring from 144 million U.S. dollars in 1948 to 1,150 million in 1968. However, the balance of payments is favorable primarily because of United States military expenditures, economic aid, private capital inflow, and tourism. The country's foreign exchange reserves are high.

LAOS

The landlocked Kingdom of Laos is the most isolated and least economically developed country in Southeast Asia. It is cut off physiographically in an easterly direction from the South China Sea by the Annamese Cordillera and separated by Cambodia and South Vietnam from access to the same sea by way of the Mekong River. The country does not have a good network of any single type of modern transportation. When all the types of air travel, highway, and rail are combined, only then does its largest city, Vientiane, and its capital, Luang Prabang, have ready access to the outside world. An all-weather highway extends from Vientiane to the Mekong River ferry that crosses at Nong Khai, Thailand, to connect with the longest railroad network in mainland Southeast Asia. Freight moving from Laos can reach the free-port facilities in Bangkok in a few days. From Nong Khai, passenger service is available to Phnom Penh, with a change of trains at Poipet, or to rail-located cities in Thailand, West Malaysia, and Singapore without having to change cars after leaving Bangkok. Although for the Laotians rail service is available from Nong Khai, most intercountry passenger service is by plane. Most of the Laotian exports and imports, which are not too many, move by rail the 311 miles from Vientiane to Bangkok in less than a week. By way of the Mekong River, the distance from Laos' largest city to Saigon is over 3.5 times the rail distance, or about 1,100 miles. Moreover, the river route is subject to interruption due to erratic flows; transshipments due to rapids and falls upstream from Kratie, Cambodia; and banditry and guerrilla raids in Laos, Cambodia, and South Vietnam. In fact, in the 1960s the flow of trade by water from Laos to Saigon virtually dried up.

Since no accurate census has ever been taken, the statistics about this country are estimates. The population was estimated to be about 2.9 million in 1969. The amount of arable land in 2.5 million acres, and the percentage of productive land is 6.3 percent of the country's total area. Areawise, this country is larger than Minnesota or Great Britain,

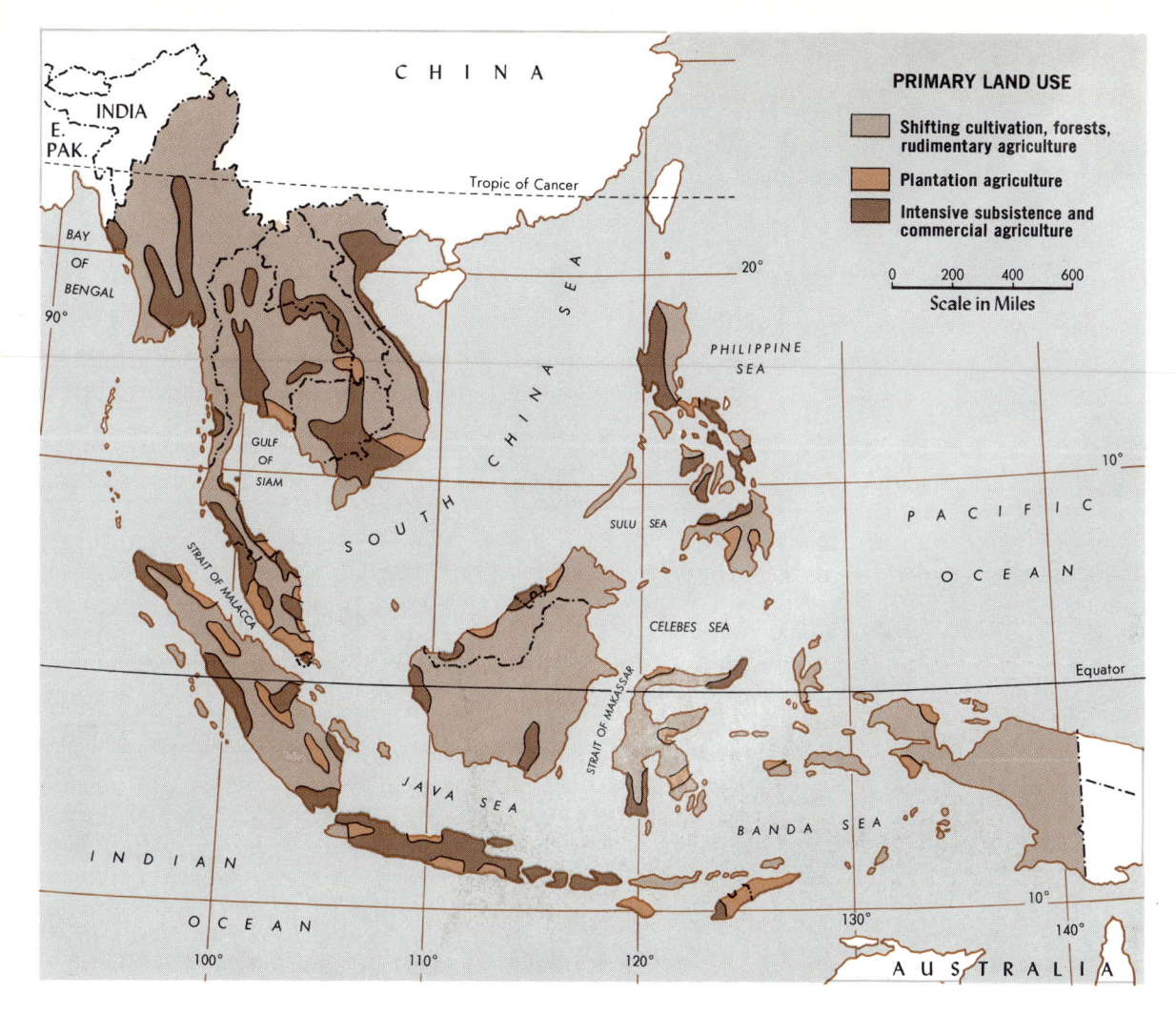

Fig. 15–5 The dominant land uses in the region of Southeast Asia are shifting cultivation and subsistence cropping and grazing. Most manufacturing activities are of local importance only.

and over nine-tenths of it is covered with forest and savanna-cloaked highlands, which reach elevations of over 9,000 feet in the north. Except on the high slopes of the mountains, the climate is monsoon, with three seasons: the five-month (May to October), heavy rainfall warm season; about a four-month cool season (October through January); and a hot, dry season (January to May). Population densities are low on the mountains and tablelands. The inhabitants in the southeastern part of this rugged topography are ethnically non-Lao hill tribes. Unlike the deltaic floodplain core areas such as those of Thailand, Cambodia, North Vietnam, and South Vietnam, Laos is without a socioeconomic-political core area. Instead, most of the population, farms, agricultural production, villages, and cities are concentrated in a disconnected strip about 40 miles wide, or narrower, paralleling the Mekong

River most of the distance from Vientiane in the north to Paksé in the south. Within this narrow band are the irrigated garden spots and cities associated with, or near, the junctions of tributaries and the Mekong River and the Lao people (ethnically Thai).

Laos is the most agrarian and least-developed country in Southeast Asia. In fact, there is some justification in the saying "Laos is a country in territory and name only and, were it not for the United States domestic and military aid and help, would cease to exist." Laos does not have a large-enough GNP to support a police force large enough to regulate its own domain. Neither can it support armed forces of sufficient size to guard its 1,949-mile-long common boundary with Communist China, North Vietnam, and Cambodia and prevent infiltration of Communist-inspired subversives and guerrilla bands. The Laotian army and paramilitary forces divert manpower desperately needed in agriculture and other industries.

Most of the people live a subsistence type of life. Their diet consists of rice, fish, poultry, and fruits. Villages are almost self-sufficient. A few small industries making clothing and household furnishings are located in Vientiane and Luangprabang.

CAMBODIA (KHMER REPUBLIC)

Cambodia, about the size of Missouri and larger than South Vietnam, is an alluvium-covered topographical basin surrounded on the west, north, and east by a horseshoe-shaped area of hills and mountain peaks. The highest relief is in the Cardomo Mountains, which form the boundary between Cambodia and southeast Thailand. The alluvium that covers the basin floor was and is being deposited by the Mekong River tributaries and distributaries and by the annually pulsating lake called Tonle Sap. On the southwest, Cambodia has a short coastline on the Gulf of Siam; this shore, irregularly fronted by numerous small islands, gives the country an opening to the sea through the port of Kompong Som (Sihanoukville). This port was built with French aid during the post–World War II period and linked to the capital Phnom Penh, the country's largest city and Mekong River port, by an all-weather highway constructed by an American company with United States aid money. The country has a puffball shape, with the stem extending southeastward toward Saigon.

Cambodia has a tropical savanna type of climate, with distinct wet and dry seasons. Most of the country receives between 40 and 80 inches of rainfall, heavily concentrated in the hot season. There is a cooler dry season from December to May. During January, most of the country receives less than 1 inch of rainfall. Fortunately, the areas with the lowest annual rainfall flanks both sides of the Tonle Sap and the Mekong River, which provide abundant water for irrigation. At Phnom Penh, an annual average of approximately 58 inches of rain falls. The heaviest rainfall, over 100 inches, falls along the coast and on the seaward side of the mountains. Average daily temperatures vary from about 68°F in the cool season to 97°F in the hot season. Diplomats and tourists find December and January the most comfortable months.

Forests cover nearly half of Cambodia. In acreage, the tropical rain forests rank first, and the tropical deciduous forests and wooded savannas rank second. It is estimated that nearly half of the country could be cultivated but that less than one-tenth is. The cultivated land is concentrated on alluvial floodplains in a zone about 30 miles wide on each side of the Tonle Sap and 60 miles wide on both sides of the Mekong River.

With an estimated population of 6.7 million and a reported arable area of 6.1 million acres, Cambodia ranks second only to Laos in arable land per capita, and it ranks third in the lightest population density. Cambodia has a population density of 90 per square mile; Malaysia, 79; and Laos, only 31. In population growth, Burma and Cambodia are tied for last place—2.2 percent a year. And Cambodia is tied with the Philippines in having the second largest percentage of its population under fifteen years of age—44 percent, as compared with 47 percent in the Philippines.

About 85 percent of the Cambodians are ethnically Khmers. Chinese and Vietnamese number about half a million each. With less than 1 percent of the total population, other small minority groups are the Cham-Malays (Muslims who descended from the inhabitants of the ancient kingdom of Champa), Thai, Laotians, and French. Many of the Chinese and Vietnamese either migrated to the country during French colonial rule or are descendents of immigrants. Cambodia's social cohesiveness and relative political stability since World War II are partially due to a largely homogenous ethnic population, over 90 percent of whom are estimated to be followers of Theravada Buddhism. About 85 percent of the population are rural dwellers engaged in farming and living in villages. By Southeast Asian standards, practically all the people are well fed and adequately clothed and housed.

The Cambodian economy is based primarily on farming, stock raising, fishing, and lumbering. Many of the rural dwellers are engaged in subsistence agriculture, with only limited participation in the commercial life. Although agrarian methods are primitive, farms small, and yields low, in 1968 a record-breaking crop of 3.2 million metric tons of rice was grown, and 250,000 tons of this were exported. The second most important commercial export crop is rubber; 50,000 tons were exported in 1968. In order of value to the country, other products are maize, livestock, timber, pepper, haricot beans, soybeans, and fish.

The commercial fishing industry is concentrated on and adjacent to the Tonle Sap, located in the northwest quadrant of the country. At low water in the cool season, the lake covers some 1,000 square miles. In the wet season, when the Mekong River is in flood, the water flows into the Tonle Sap and rises until the lake covers about 3,500 square miles. Then in the dry season, the lake again drains into the shallow Mekong River. The marketing of fresh, dried, and salted fish, as well as fish products of paste and oil, amounts to millions of dollars yearly.

Phonm Penh is approximately ten times larger than the country's second largest city, Battambang. The political, economic, social, and cultural core of Cambodia is within a hundred miles of Phnom Penh, where all activities peak. Cottage industries are found in the heavily populated villages lining both sides of the Mekong River from Kratie to the Cambodian–South Vietnamese border and on both sides of the outlet of the Tonle Sap, from Kompong Chhnang to Phnom Penh.

SOUTH VIETNAM

With Cambodia and Laos on the concave side and the China Sea on the convex side, South Vietnam is a relatively narrow, crescent-shaped area varying in width from just over 25 miles at its northern boundary to over 120 miles in a few places at its waist. About two-fifths of the southern part is a huge delta floodplain formed by the Mekong River tributaries and distributaries and by smaller rivers such as the Vaico Occidental, Vaico Oriental, Song Be, and Dong Nai. This lowland, with elevations ranging from sea level to 200 feet, has a high water table, comparatively rich alluvial soil, frequent flooding, swampland, and rice paddy fields. It has dense populations living in villages located on the natural levees adjacent to the river channels and on artificially created mounds paralleling canals. This area was formerly the territory of Cochin China, one of the five divisions of French Indochina.

About half of the country consists of highlands, sometimes called the Southern Mountain Plateau, which stretches northward from the southern lowlands to the northern border of the country at 17° north latitude. The base of this highland area is fringed with land which rises to heights of from 200 to 500 feet above sea level. Above these fringe slopelands, the plateau with associated mountains is partially divided into unequal northern and southern halves by the narrow valley of the Song Ba.

The remainder of the country is a very narrow discontinuous coastal plain, extending from approximately 11° to 17° north latitude. Where short east-flowing rivers have cut lowlands, the combined coastal plain and lowlands extend interiorward, for a score or more of miles, provide a sufficiently large habitat for farming, and a hinterland of sufficient size to help support small cities.

The large alluvial plain in the southern part has a year-round tropical climate, with maximum rainfall coming in the warm season. Not so much rain falls here as it does along the narrow coastal plain to the north, where the maximum comes a few months later; but there is sufficient water for rice crops to be grown under irrigation. From about October to March the highlands are cool, with temperatures dropping sometimes to 50 to 55°F.

Although South Vietnam is the smallest country in the Great Lowland group, about twice as many people live here as live in Cambodia and Laos combined. Despite South Vietnam's sparse population density in the highlands, the country averages 256 people per square mile and is exceeded only by the 317 persons per square mile in North Vietnam and the staggering 8,693 in Singapore. Because of the large population and the limited amount of arable land, South Vietnam and North Vietnam tie with Indonesia for having the lowest number of arable acres per capita in Southeast Asia—only 0.4 arable acre per capita, which is less than that of China and India. Reports show that the rate of population increase is 2.8 percent annually.

One of the stabilizing factors in South Vietnam is the fact that about 90 percent of the population is ethnically Annamites. Three of the minority groups are the Chinese, Montagnards, and the descendants of Cambodians. More than a million Chinese, the majority with Vietnamese citizenship, live in the cities, with the greatest concentration found in Cholon. About 600,000 Montagnards, composed primarily of two ethnic groups, Malayp-Polynesian and Mon-Khmer, live primitive, seminomadic lives in the highlands. Descendants of Cambodian origin are farmers in the lowland provinces adjacent to the Cambodian national boundary. Tertiary ethnic peoples are a few thousand Chams (the only indigenous Muslim population), Indians, Malays, and French.

Since this country was established in 1955, the primary goal of the Viet Cong and their allies the North Vietnamese has been to sabotage and, if possible, to undermine, weaken, and destroy its economy by interfering with its agricultural, industrial, commercial, and transportational activities and other facets of its economic and social development. Despite the heavy toll that sabotage and disruptive war activities have taken, with free-world assistance the South Vietnamese government appears to be increasing its economic strength.

With an estimated 70 percent of the total population living in rural areas, the country's economy is primarily agricultural. Before World War II, three factors—(1) the fertile land, (2) land improvement projects of irrigation, drainage, and navigation, and (3) its industrious farmers—made it possible for the lowland area of South Vietnam to be one of the three largest rice-exporting areas in the world. There was a marked decline in rice production during World War II; but with the increase in rice production afterward, exports in 1963 amounted to 340,000 metric tons. Then, as the intensity of the civil war increased during the 1960s, rice production was again interrupted. By 1966, more than 400,000 metric tons of rice had to be imported. Leading secondary agricultural products are kenaf (used in similar ways as jute), sugarcane, maize, tea, coffee, tobacco, and rubber. Fishing is the second most important industry, and the catch continues to grow. Production increased from 165,000 tons in 1959 to more than 400,000 in 1966. Although fish is the primary protein supplement to the local rice diet, exports increase annually.

A diversified industrial development is of recent origin in South Vietnam. Under French rule, the territory now in South and North Vietnam was governed as three major subdivisions: Tonkin, the lowland in the north, extended southward approximately to the Ma River; Cochin China, the lowland in the south, stretched from the Gulf of Siam to the highlands on the north; and Annam, the central division, included most of the highlands north to the Ma River and the bordering narrow coastal plain. Hanoi and Haiphong were the regional urban centers in Tonkin; Hue and Da Nang in Annam; and Saigon and Cholon in Cochin China. Under French management, Cochin China produced a surplus of food and fiber crops and exchanged part of these for industrial products from Tonkin. The principal industries in Cochin China were agricultural processing plants. When South Vietnam and North Vietnam became independent countries, the former was almost without industries. By the end of 1967, South Vietnam had more than 800 new rehabilitated factories in operation, employing 75,000 workers and accounting for about 25 percent of the GNP. Some of these industries are located in a recently established industrial park near Da Nang. Then the Communist Tet offensive in January and February of 1968 damaged 34 plants, 19 of which were textile factories, and left nearly 11,000 workers unemployed. In spite of these handicaps, the *World Bank Atlas* reported that South Vietnam's GNP per capita in 1968 was 150 U.S. dollars.

Two Peripheral Countries

NORTH VIETNAM

North Vietnam is the only Communist country in this geographical realm, and since its origin

Fig. 15-6 The water buffalo is the common work animal of Southeast Asia. (Courtesy of Philippine Embassy)

it has released few statistics. With an area of approximately 63,344 square miles, it is somewhat smaller than the state of Washington and ranks ninth in size in this realm, exceeding only Singapore. North Vietnam can be divided into three physical regions. The first region, the Tonkin hills, borders on mainland China to the north and Laos to the northwest and west and forms a horseshoe-shaped zone almost around the second physical region, the Red River delta. This Red River delta, though smaller, is the present core of the country and the cradle of Vietnamese culture. Chinese chronicles written in the pre-Christian era refer to the delta as Nam-Viet ("Southern Land"), and the dominant cultural group was then the Han Mongoloids. Starting with the Ma River watershed plain and extending southward to the South Vietnam boundary, the third physical region is a narrow coastal plain generally referred to as the Annam coastal plain, since it was formerly in Annam. This region forms the handle of the open-fan-shaped territory of North Vietnam.

North Vietnam ranks fifth in total population in this realm, having 20.1 million in 1967. Averaging 317 persons per square mile, it ranks second in population density. The country compares with Thailand and Malaysia in present growth rate, about 3.1 percent per year, and this could double the population in 23 years. Most North Vietnamese live on the Red River delta and the Annam plain, where the population density per square mile is over 400. In highlands southwest of the Red River valley, population density averages less than 25 people per square mile.

Numerically, the Vietnamese peoples are the predominant ethnic group. In the highlands, however, Thais and other tribal peoples predominate. Another minority group, the Chinese, lives in cities, towns, and mining regions and is also found concentrated in the Red River delta and in the northeastern part of the country east of the Cau River. The Annamese peoples are concentrated in North Vietnam's panhandle.

Agriculture is the primary facet of the economy, with fishing, mining, and manufacturing being secondary. Agricultural activities may be grouped in three types and located in regions: the Tonkin delta (coalescing deltas of the Red, Ma, Chu, and Ca rivers); the Annam coastal plain; and the highlands, with upland shifting agriculture. Types of farming are similar in the Tonkin delta and on the Annam plain, but there are also differences. Land use is more intensive on the Tonkin delta, where market gardening helps feed the large city populations of Hanoi and Haiphong. In some places on the Tonkin delta, an acre of arable land is divided into a dozen parcels or more. Landholdings are not so divided up on the Annam coastal plain; southward, and especially in South Vietnam, land fragmentation decreases. By following Chinese practices of diking fields, applying animal manures and night soil (human) in liquid form to individual plants and using time-consuming manual labor, the North Vietnamese have the largest crop yields in mainland Southeast Asia. The combined process of raising hogs, fish, and vegetables reaches a peak here. Most North Vietnamese are not Buddhist, and they have developed a system of raising meat and vegetables which involves butchering animals. Hogs are raised on slightly inclined cement platforms with low walls; these pens contain a half dozen hogs or less. The hogs are fed on discarded plants selected from intercultured rows of vegetables. Buckets of water splashed on the hogs and into the pens washes the manure and unconsumed matter into a hole in one corner, where it drains into a sunken barrel. The contents are stirred before carried to the gardens and ladeled onto the base of the plants. In many places the houses are surrounded by these vegetable patches rather than by lawns. In raising hogs, a more complex process is to have them on a wooden platform above a fishpond. The manure and leftover vegetable matter, sometimes garbage, fed to the hogs falls into the pond to feed the fish. Then the sediment dredged from the bottom is used as fertilizer.

Since the beginning of World War II, some North Vietnamese have migrated through the Annamese Cordillera, across Laos and the Mekong River, into northeast Thailand. Their villages are models of prosperity, while those of the Thais, found in similar conditions of soil, terrain, and climate but based on a monocrop economy of rice, are not nearly so prosperous. The Buddhist Thai do not kill animals, and if hogs or other animals are raised, the buyer pays a very low price. On the Tonkin delta and the Annam plain rice, grown under irrigation, is the dominant crop, and the major share of the cultivated land is devoted to this crop. In both regions maize is a secondary crop, although the greatest amount grown is on the Tonkin delta, where hog and poultry production is also concentrated. Some groundnuts, castor beans, haricot beans, sugarcane, coffee, cotton, and tuberous root crops such as yams, taro, and cassava are grown in both regions.

In the highlands the typical slash-and-burn type of agriculture practiced throughout the highlands of Southeast Asia predominates. Upland rice, grown without irrigation, vegetables, and a few domesticated animals, primarily hogs or poultry, supplemented with some game, nuts, and wild fruits and roots provide the food for a subsistence economy. The forest-covered highlands have been stripped of their good timber and no longer supply the lowlands with adequate lumber. Some forest products here as well as throughout Southeast Asia are bamboo (used in innumerable ways, from construction to firewood and charcoal), lac, gums, turpentine, resins, and a variety of other products.

Although manganese, lead, zinc, tin, bauxite, iron ore, tungsten, chromium, graphite, phosphate, and coal have been found primarily in the highlands and several of these minerals have been mined in small amounts north of the Tonkin delta near the Chinese border by the Chinese for centuries, only anthracite coal is mined in relatively large quantities in North Vietnam. Before World War II, more coal was mined in Tonkin than in all the other countries of Southeast Asia. If mining reports are correct, Tonkin has coal reserves once estimated at 20 billion tons. If these data are true, North Vietnam has coal reserves on a scale with those of Japan.

BURMA

The northern part of Burma is located between mainland China on the east and India and East Pakistan on the west; consequently, it shares boundaries with the three largest countries in both East and South Asia. The British and French, who drew most of the present boundaries of countries in mainland Southeast Asia, designed them so that it gave Burma and Laos a short mutual boundary and Thailand lost its precolonial boundary with China.

With an area of about 262,000 square miles and a population of 25.8 million in 1967, Burma is the largest country of mainland Southeast Asia and ranks second only to Thailand in population. Ethnically, Burma has perhaps the most diverse population in Southeast Asia. Over 70 percent of the population are Burmese, who live primarily in the lowlands. Approximately 3 million Karens are distributed throughout eastern and southern Burma, primarily in the hills and mountains. About 1.5 million Shans, ethnically related to the Thai, are concentrated in the eastern tablelands north of Thailand. Other major ethnic minorities are the Kachins in the north and the Chins in the northwest, with the combined groups totaling about 1 million. Altogether, there may be a million Chinese, Indians, and Pakistanis. The rest of the Burmese population is divided among a large number of ethnic groups or subgroups. These diverse ethnic groups speak many languages, of which there are many subsidiary dialects. Up to now, differences in ethnic origins, language, and culture have helped prevent strong political unification of the country.

There are two general cultural characteristics. The official language and about three-fourths of the people are Burmese. About 85 percent of the people adhere to Theravada Buddhism, an older, simpler form of Buddhism. Secondary religions are Islam, primitive animism, and Christianity. The well-educated people of Burma can use English as a second language. Population density is about 102 per square mile; the birthrate is 50 per 1,000; the death rate is between 25 and 30 per 1,000, and there is an estimated growth rate of 2.2 percent annually, which if it continues, will double the population in 32 years. About 40 percent of the population is under fifteen years of age.

Lower Burma is about 275 miles wide where the deltas of the Irrawaddy, Sittang, and Salween rivers coalesce, and it narrows and terminates about 200 miles to the north near Prome in the Irrawaddy Valley and north of Lewe in the Sittang Valley. The region is divided into separate lowlands in the northern part by the Pegu Yoma, peaking at elevations of nearly 5,000 feet. Lower Burma is a region of (1) alluvial soils on floodplains, deltas, and coastal plains; (2) colluvial soils on the slopes of low hills; and (3) heavy rains. Usually over 80 inches of rain falls directly into the rice paddy fields during the months of May to September. For natural irrigation the annual floods of the three major rivers, but especially the Irrawaddy and its largest tributary the Chindwin, provide additional water. Most of Burma's cropland, rice production, urban population, commerce, services, and manufacturing, cultural, and educational activities and its transportation network are concentrated in this region. Here, in the core area of the country, is located the capital Rangoon and also the third largest city, Moulmein.

In the 500-mile-long Burmese panhandle, the narrow discontinuous coastal plain at the base of the Tenasserim Mountains, there is some valley cultivation, with the same type of agriculture as found in Lower Burma but on a smaller scale. Here lumbering and fishing are relatively more important to the economy than in Lower Burma.

Middle (or Dry Belt) Burma is an alluvial lowland basin formed by the Irrawaddy and Sittang rivers and their tributaries. This basin is cut off physiographically from Lower Burma to the south by the Pegu Yoma, through which the Irrawaddy and Sittang rivers have cut gorges. This region roughly coincides with the areas receiving less than 30 inches of annual rainfall and a surrounding band receiving from 30 to 50 inches. Despite the lower rainfall, this region ranks second only to Lower Burma in the production of lowland rice and sugar and exceeds the core region to the south in acreage devoted to cotton, millet, maize, groundnuts, sesame seed, and pulses. In Middle Burma are located the nation's oil fields. This area also has the second-heaviest concentration of population, transportation networks, industry, cultural activities, wealth, and ethnic Burmese in the country. The hierarchy of economic and cultural activities peaks in Mandalay, a city of more than a third of a million inhabitants and the country's second largest city.

A wide rim of hills, mountains, and tablelands surround Lower and Middle Burma and effectively isolate this country from South Asia on the west, East Asia on the north, and Thailand on the east. No railway or well-developed all-weather hard-surfaced highways connect Burma to the rest of Asia; and its economic, social, and political linkages are few and weak. The highlands cover most of Burma's territory, and here are the homes of most of the minority ethnic groups as well as the source of (1) lumber products, especially teak; (2) a variety of minerals, such as lead, zinc, and tungsten, of national significance but unimportant internationally; and (3) runoff water for the lowlands, especially for irrigation and domestic and industrial use.

Burma is an agricultural country, with approximately nine-tenths of its people dependent on agricultural crops and their processing for a livelihood. Rice not only is the primary food crop but also utilizes the greatest amount of cultivated land, about 12 million acres annually, and it is the most valuable export, accounting for more than half the value of total exports in 1968. In that year 7.6 metric tons were grown. This small amount contrasts sharply with the over 3 million tons exported in 1939. Some of the factors involved in this drastic decline are an increasing and perhaps a better-fed population, internal political instability, failures to increase yields and planted areas, and the lack of monetary incentives in a socialist state. Other important exports are forestry products (primarily teak), cotton, rubber, metals, and ores.

Bridge and Insular Countries

In this geographical subdivision, the four bridge and insular countries of Malaysia, Singapore, Indonesia, and the Philippines contain somewhat less than two-thirds of the land area of Southeast Asia and about three-fifths of the total population. Two of these, Singapore and Malaysia, are the youngest nations in this region; the only two political remnants of European colonialism, Brunei and Portuguese Timor, are also found here. In part, because such a high percentage of the fertile soils are used for plantation agriculture, the countries in this subdivision have to import food products.

MALAYSIA

Malaysia, the second-youngest nation, is the only country whose two territorial parts are separated by nearly 400 miles of ocean—the southern part of the South China Sea. The former Federation of Malaya, consisting of eleven states located on the lower part of the Malay Peninsula, is now referred to as West Malaysia. Sarawak and Sabah (formerly North Borneo), on the island of Borneo, are known as East Malaysia. With an area of 128,553 square miles, Malaysia is somewhat larger than the British Isles, which formerly ruled it. About three-fifths of Malaysia's land area is in East Malaysia, on the island of Borneo. In 1967 less than 15 percent of the 10 million Malaysians lived there, and the population density was less than 20 per square mile. During the same period, the population density in West Malaysia was 79 per square mile, and the average was one person per arable acre. West Malaysia's population increased from 4.9 million in 1947 to 8.5 million in 1967.

Malaysia is the only country in Southeast Asia that does not have a predominant ethnic group amounting to over 50 percent of its total population. In no other country are the foreign and native ethnic groups so evenly balanced. It is estimated that 48 percent of the population are Malayans; 45 percent, Chinese, Indians, and Pakistanis; and 7 percent, non-Malay indigenous peoples. The Malays are Muslims and speak the Malay language. About one-third of the Malays in West Malaysia are first- or second-generation immigrants from Indonesia, primarily from Sumatra. The Malays, Chinese, and Indians consider themselves Malaysians, but each group tends to maintain its own cultural identity.

When the Federation of Malaysia was formed on September 6, 1963, East Malaysia was included so that its native ethnic population would help balance the overwhelming percentage of Chinese in Singapore. Had the Singapore population been added to that of the Federation of Malaysia, and had East Malaysia been excluded from the new nation, the Chinese would have had an approximately five-to-three majority over the Malays in Malaysia. In 1965, by common agreement, Singapore separated from the Federation of Malaysia, but East Malaysia remained in the Federation.

In the southern part of the Malay Peninsula, West Malaysia has a mountainous core oriented through the center of the northern two-thirds of the country roughly in a north-south direction. This highland area is split into three elongated north-south parts by river valleys produced by both north- and south-flowing rivers. A U-shaped area of coastal plain and lowlands surrounds the highland core on the east, south, and west and is crossed by short rivers originating in the highlands. West Malaysia has a coastline of over 1,200 miles, and with minor exceptions is lined with mangrove swamps and shallow-covered continental shelves extending far out from shore. Boats of shallow draft can penetrate the mangrove swamps and form a link between the shallow ocean waters and the rivers. The swamps and shallow seas, plus heavy silting, prevent large vessels from nearing the coast and entering the interior. Extensive swamps and silting hold up construction or make the maintenance of artificial harbors very expensive. Sandy shorelines are common on the east coasts, and mud flats are more universal on the west; extensive swamps are located behind all three shorelines.

Most of West Malaysia is covered with tropical rain forests and swampy vegetation. When the native vegetation is removed and the land drained, three general types of soil are exposed: the alluvium and colluvium soils in the river valleys and at the foot of slopes, which are the least extensive but the most agriculturally productive; the swampy and peat soils, which have little value; and the forest soils, which are easily destroyed by fire during the clearing process and by unchecked soil erosion. Since West Malaysia is located between 2° and 6° north latitude and most of the land is less than 100 miles from the sea, the area is characterized by uniformly high temperatures, usually in the eighties, and by 80 or more inches of rainfall during the year.

With the primary facts of the physical setting of West Malaysia in mind, it is not surprising to learn that only about one-sixth of the country is under cultivation. About four-fifths of the area is covered by dense tropical rain forest and jungle, mountains, and swamps. Over nine-tenths of the cultivated land of West Malaysia is devoted to three crops: about 12 percent of the country to growing rubber, about 3 percent to rice, and another 2 percent to oil and coconut palms.

Because of the agricultural, commercial, and industrial development of West Malaysia, this country—if judged on per capita income, rate of investment, and growth of GNP—is one of the most developed and the one having the second most prosperous economy in the entire realm. In 1968, per capita income was reported to be 326 U.S. dollars, or more than twice that of the Philippines or Thailand and exceeded only by Singapore and Brunei.

The economy of Malaysia is still based primarily on the output of field, mine, and forest, but West Malaysians are adopting diversification methods in the development of agriculture, manufacturing, and services. Rubber and tin are still the nation's largest economic generators. Grown primarily within 50 miles of the west coast, rubber accounts for 18 percent of the GNP, 38 percent of the total value of exports, and about 20 percent of those employed. About 22 percent of Malaysia's export earnings come from tin. Malaysia, the world's largest

Fig. 15-7 Near the capital city of Malaysia, Kuala Lumpur, are several giant floating tin dredges which work their way slowly across the countryside. (Courtesy of the *Straits Times*)

source of tin, accounts for one-third of the world's ore production and two-fifths of the refined metal. Palm oil, pepper, and coconuts are also exported.

In this rice-deficient country, emphasis has been placed in recent years on growing more rice by using higher-yielding varieties, more irrigation, double-cropping, and better cultivation and administrative techniques. During the past decade, yields were raised from 1,000 pounds per acre to 1,500. During the same decade, the amount of rice milled increased from 350,000 to 570,000 tons. Still, Malaysia must import about 40 percent of its rice. Its minor food crops are sweet potatoes, sago, sugarcane, and cassava. About one million tons of timber are cut each year from West Malaysia's forests, and about two-thirds of this is exported.

The government of Malaysia is attempting to encourage manufacturing. For example, in August 1967, it completed a 26-million-dollar fully integrated steel mill at Prai, on the west coast of the country. West Malaysia produced 5.4 million tons of iron and steel products in 1967. The ore mined contains over 60 percent iron. West Malaysia also has oil refineries, tin smelters, chemical plants, and tire factories.

Containing 77,638 square miles, Sarawak and Sabah, known as East Malaysia, extend along practically the full length of the northwest coast of Borneo. Sabah occupies the complete northern end of the island, which borders on the Sulu Sea. Sabah alone has a 900-mile-shoreline, and its highland core, which includes all the land above 660 feet, capped with mountaintops above 10,000 feet, is the northern part of the highlands forming the eastern and southern boundaries of Sarawak. The watershed of these highlands forms a natural boundary between East Malaysia and the rest of Borneo, called Kalimantan by the Indonesians. Between these interior highlands and the ocean is a relatively narrow coastal plain, varying primarily from 10 to 100 miles in width, often being wider in Sarawak and narrower in Sabah. Tropical rain forests and swamp vegetation cover most of the area. Cultivation is concentrated on alluvial soils located primarily on the coastal plains. Built somewhat parallel to the coast, the only railroad in East Malaysia is on the coastal plain extending from Kota Kinabalu (Jesselton) southwest to Weston.

In general, the standards of living, social services, and public utilities are lower in East Malaysia than in West Malaysia. The former's economy is linked more to the production of raw materials from farm and forest than is that of the latter. There is little manufacturing in East Malaysia, except for cottage industries, and trade per capita is low. The indigenous peoples, such as the Iban, Kedazan, and others, live a subsistence type of life. Commercial activities are in the hands of the Chinese and Malays, who live primarily in a belt less than a score of miles from the coastline or in concentrations in interior urban areas.

Timber is the major export, and forest growth is guaranteed by commercial cutting on an 80-year rotation plan. The four major agricultural products are rice, rubber, pepper, and coconuts. Despite the country's devoting large cultivated acreage to rice, about two-fifths of that consumed must be imported. Only about one-fourth of this crop is grown under irrigation, and primitive methods of cultivation are still common. The second largest acreage of cultivated land is utilized for rubber production, but in contrast with West Malaysia, where rubber is grown primarily on plantations, in Sarawak 95 percent of the rubber is grown on holdings of less than 5 acres. In recent years Sarawak exported 45,000 tons of rubber; Sabah, 23,000 tons. Oil is a major export, not because it is produced in Sarawak but because crude oil from Brunei is refined at Lutong in Sarawak.

SINGAPORE

This country, the youngest and smallest in the realm, became an independent republic on August 9, 1965. It consists of one small island and forty adjacent islets and is jokingly reported by Singaporeans to have about 225 square miles at high tide and several more square miles when the tide is low. Singapore, a roughly diamond-shaped island, extends 27 miles from east to west and 14 miles north to south at its maximum breadth. Lying just south of the southern tip of West Malaysia, this island is separated from

Fig. 15-8 Many low-rent apartment houses have been built by the Singapore government.

the mainland by the Strait of Johore. A three-quarter mile-long causeway supporting a railway, a road, and a huge waterpipe to supply fresh water to the island connects it (the island) with the mainland.

Most of Singapore island has both a low local relief and low elevation. Its highest point, Bukit Timah, rises only 581 feet above the sea. Mangrove coastlines are on the west and north, and the east coast is primarily a low cliff. At the beginning of the nineteenth century, most of the island was largely a swamp and jungle. Much of the vegetation that covered the island when the British bought it in 1819 has been removed or drastically altered. A central upland of about 13 square miles has been reserved as a water catchment area, with three reservoirs and a nature preserve. The city of Singapore, occupying an area of nearly 40 square miles located on land reclaimed from swamp, gulf, and sea, is the only large city and the capital.

Located between 1° and 2° north latitude, Singapore has an average maximum temperature of 87°F, and an average minimum temperature of 78°F. The average annual rainfall of 96 inches is well distributed throughout the year. Both the large amount and the regular monthly distribution are favorable factors in water supply and storage.

Of Singapore's nearly 2 million people, approximately 75 percent are Chinese, 14 percent Malays, 8 percent Pakistanis and Indian, and the remaining 3 percent are of numerous ethnic groups. Because the government emphasizes a family-planning program, and perhaps in part because of a rising standard of living, the annual rate of population growth has dropped to 2.4 percent. Singapore has the highest life expectancy in this realm, 62 years. As a whole, the people are better-fed and -housed. Approximately half of them live in publicly-built and operated housing, and fewer live in slums. Health standards are high.

The annual GNP in 1968 was 700 U.S. dollars per capita. Despite old-age expectancy, half the population is younger than twenty-one years of age.

Singapore's economy is based on business activities, with commercial activities predominating. Service activities have increased rapidly since 1950. As in the preindependence days, the major part of Singapore's economy is still based on entrepôt trade and the utilization of its harbor, one of the five best natural inlets in the world. Both the superior harbor and its strategic location on the Strait of Malacca, the chief passageway between the Indian and Pacific oceans, have enabled Singapore to be the most important port in Southeast Asia for over a century. Trade activities may be grouped into three categories. First is the collection, processing or semiprocessing, packing, and transshipment of the primary products from the realm's plantations, farms, mines, and forests, such products as rubber, copra, tin, timber, petroleum and its by-products, coffee, tea, spices, and palm oil. Second is the distribution in smaller quantities, within the realm, of manufactured products bought in large quantities from industrialized countries. Third are such ancillary activities as shipbuilding and repair, storage, banking, and insurance.

In 1968, total exports reached 1.27 billion U.S. dollars, and imports were 1.66 million dollars. The entrepôt trade generates approximately 15 percent of the GNP, but experts believe this type of trade can only decline relatively, although absolutely it may increase for some years. It is estimated that the British military bases account for some 20 percent of the GNP. Singapore's most important trading partner is West Malaysia, which takes about one-fourth of Singapore's imports and provides somewhat over one-fourth of its exports. Other important trading nations are Japan, the United Kingdom, and South Vietnam.

The principal exports are rubber, petroleum and petroleum products, and ship and aircraft stores, which generate about half the total value of exports.

Agriculture consists of intensive market gardening and the production of some pineapples, coconuts, tobacco, and rubber grown on plantations. Hogs, fish, and poultry are sources of meat, with some hogs being raised on platforms over fishponds.

In an attempt to attract modern industrial complexes, the government has established the Jurong Industrial Estate on the west end of the island. This is a 17,000-acre industrial park which incorporates new deep-water wharves, adequate protected factory sites, and modern housing for workers. These new wharves also provide new facilities for a fishing fleet. Singapore planners expect and are making provision for the population to increase to 4 million in future decades. Singapore's development both to the east and west along the southern coast is being aided by mass-transit facilities that will eventually extend the full length of the island.

INDONESIA

If considered from a worldwide point of view, the Republic of Indonesia, with an area of 736,500 square miles and an estimated population of 110.1 million in 1967, is (1) the largest insular country, (2) the most important equatorial country, (3) the tenth largest world political unit, (4) the fifth or sixth most populous country. This country accounts for over two-fifths of both the land area and the total population in the context of the Southeast Asian realm.

Indonesia is composed of some 13,667 islands in a wide range of sizes, shapes, and population densities. However, all the Indonesian Archipelago is not under the jurisdiction of Indonesia. For example, only the western half of the island of New Guinea, the second largest island in the world, is part of Indonesia. Irian Barat, formerly Dutch New Guinea, containing 161,000 square miles, is the largest single territory in Indonesia; but the eastern half of New Guinea is governed by Australia. Also, only about two-thirds of the world's third largest island, Borneo (208,298 square miles), belongs to Indonesia; this part is called Kalimantan. The two largest islands in the Indonesian Archipelago belonging entirely to Indonesia are Sumatra and Celebes (Sulawesi). Sumatra, with an area of 164,129 square miles, is larger than Malaysia or Japan. With an area of 72,890 square miles, Celebes is almost the size of Great Britain. At the other extreme of size are hundreds of islets less than a square mile in area, some without names.

One way of grouping the Indonesian islands is a threefold physiographical division based on depths of water and age of rocks and erosion. In the eastern and western parts are platforms covered by shallow seas averaging less than 150 feet deep, with a large area of deeper water between. The eastern shallow-sea area is called the Sahul Platform, and on this is located New Guinea and associated nearby islands. The western is called the Sunda Platform, and on this are located the islands of Sumatra, Borneo, and Java (Djawa). The Celebes, Moluccas (Maluku), Lesser Sundas (Nusa Tenggara), and associated islands are surrounded by waters over 150 feet deep and with associated "deep seas" with depths of over 15,000 feet. This is a much younger geologic area created by recent tectonic processes of vulcanism and faulting. Most of the islands have highland and mountainous interiors with narrow coastal plains, across which flow numerous short, steep-gradient rivers.

All of Indonesia has a tropical rain forest climate, which is constantly moist throughout the year except (1) in the mountains of central Borneo and Irian Barat, which are classified

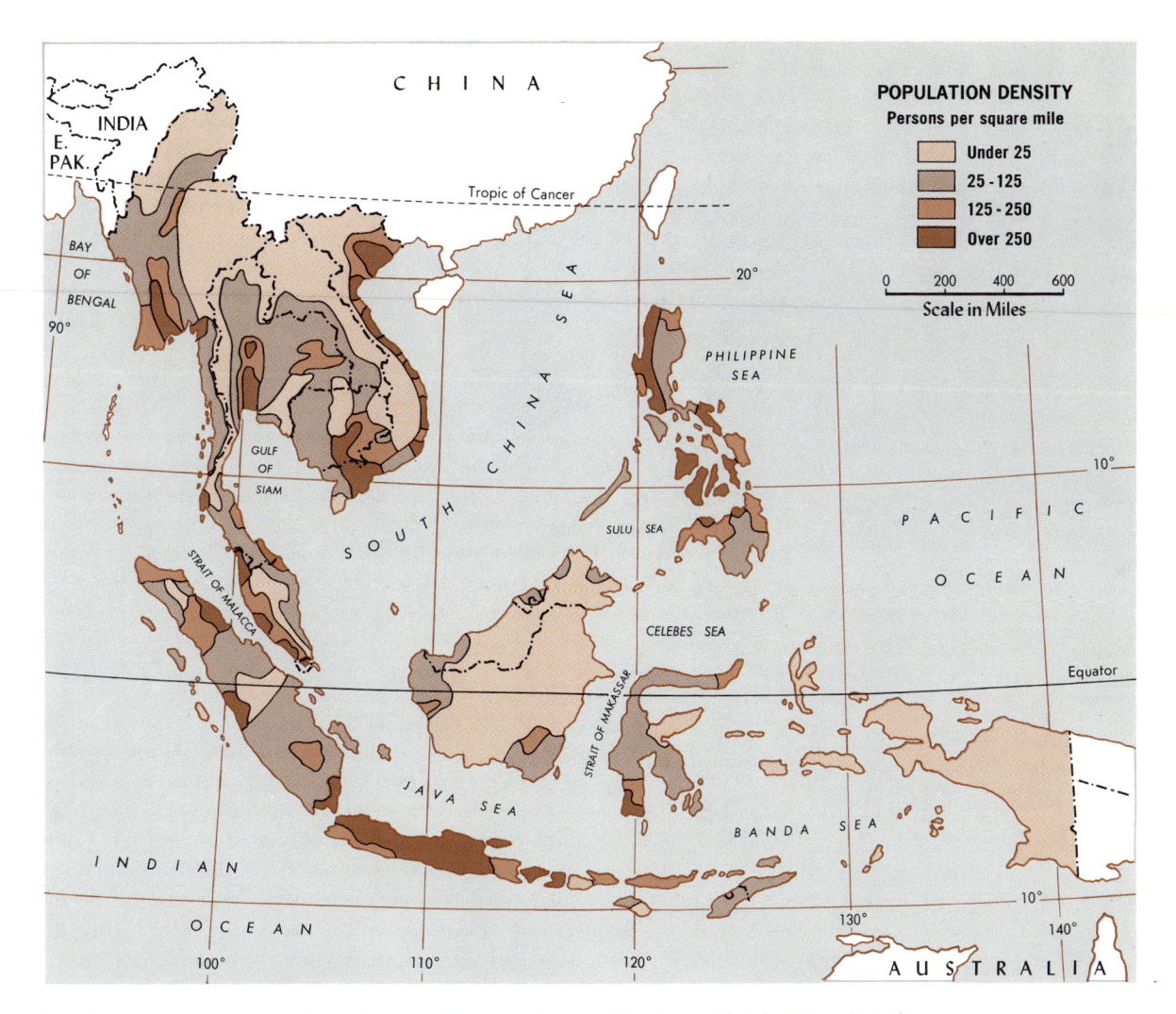

Fig. 15-9 The most densely populated area of Southeast Asia is the island of Java in Indonesia. On the continent proper, the delta areas and coastal plains are densely populated, since these offer more opportunity for growing rice and other food crops.

as undifferentiated highlands; and (2) in the eastern one-third of Java, Madura, and the Lesser Sundas, which have a tropical rain forest climate with a dry season during the cool period. Because of high mountains, warm ocean waters, and prevailing wind directions, topographical differences have greater influence on variations in both temperature and rainfall than do differences in latitude. Tropical evergreen forests, savannas, and swamps cover from 60 to over 90 percent of the surface of Sumatra, Borneo, Celebes, the Moluccas, and Irian Barat.

Over 95 percent of the Indonesian population is of Malay stock. Although there may be 2.5 million Chinese and 0.7 million Papuans in

Fig. 15-10 During the wet season, the terraced hillsides on Bali are flooded for rice crops; in the dry season, they are planted in nonirrigated crops such as maize or beans. (Courtesy of Government of Indonesia)

Indonesia, these numbers account for a very small percentage in a population estimated at 110.1 million. No other country in Southeast Asia has as high a percentage of its population in the predominant ethnic group as does Indonesia. There is great uniformity in ethnology, but there are extremes in population density from island to island. Java averaged over 1,100 per square mile at the last census, while at the other extreme Irian Barat had only 6 persons per square mile. With an area about the size of New York, Java has a population of about 75 million. It has been estimated that seven-tenths of Indonesia's people live on less than 10 percent of its land surface.

Under President Sukarno the stable, relatively prosperous economy the Indonesians inherited from the Dutch colonists was almost destroyed by governmental mismanagement, an unrealistic order of priorities, the pursuit of antidevelopmental policies, waste and corruption, confiscation of foreign capital and property, dismissal of foreign personnel, and, not the least, inflation. In 1966 alone prices rose by 635 percent. After Suharto and his associates took over from Sukarno, they had the enormous task of trying to bring order out of economic chaos. During 1967 and 1968, the new government made encouraging economic gains; inflation was drastically reduced, and the budget was roughly balanced. Moreover, the decline in exports in 1967 and previous years was reversed; In 1968 exports were estimated at 840 million U.S. dollars. In contrast with 1959, when petroleum and petroleum products generated 27 percent of the total value of exports, by 1968 the percentage had increased to approximately 35 percent, or 290 million U.S. dollars. In the same year the value of rubber exports fell 11 percent, primarily because of a decline in world market price rather than in quantity of production. Rubber no longer is the largest generator of export value as it was in 1959, when it accounted for 48 percent. Yet, keep in mind that until 1969 rubber generated larger net exports than petroleum, since its production involved a large import component. (Moreover, there is authoritative opinion that underdeclaration and smuggling made rubber statistics appear much lower than was really the case.) In 1959, exported tin and tin ore amounted to about 4 percent of the total value but in 1963 Thailand replaced Indonesia as the second largest tin producer in Southeast Asia. Also in 1959, the export value of copra and copra cake amounted to 4 percent of the total, but the value declined under Sukarno in the 1960s; yet the value of this export increased 125 percent in 1968 over 1967. The export value of tea also rose 100 percent in 1968, and timber went up 80 percent over the total of the previous year.

Most Indonesians have little involvement in the commercial activities of the country. In 1962, the last time the *United Nations Demographic Yearbook* reported on rural and urban dwellers, over 85 percent of Indonesia's people were rural dwellers. The majority of Indonesians live a subsistence or semisubsistence type of life. They are almost self-sufficient in food and shelter, and the exchange of products and work without the exchange of money is still widely practiced in many communities and islands. This type of self-sufficiency is declining, however, as people move to the cities, where urbanization plus industrialization are needed to support a population that is projected to double within 29 years.

In spite of intensive cultivation techniques of irrigation, slope terracing, and improved varieties in Java, Madura, and Bali, agriculture lags behind population growth. During the twentieth century, the Netherlands Indies had a food deficit, and this situation has continued since Indonesia gained its independence. In a good year under Dutch rule, the islands could just about be self-sufficient in rice, but in many years rice imports were heavy; in recent years, Indonesia has imported approximately 500,000 metric tons of rice annually. Annual yields of other food crops such as maize, root crops of yams and cassava, peanuts, and soybeans have increased, but per capita yields have declined. Plantation yields of rubber, coffee, tea, sugar, palm oil, and cinchona have declined. Higher yields of rubber by small landholders have enabled Indonesia to remain the second largest producer of natural rubber; but most of its trees were planted in the prewar years, and replanting of new high-yielding strains has been neglected. In the future it may be difficult, if not impossible, for Indonesia to maintain its present annual production. In contrast, Malaysia, ranking first as a natural rubber producer, has been rapidly planting new higher-yielding varieties; consequently, in future years the production gap between the two countries may widen as Malaysia moves into a better competitive position. In the mid-1960s, it was estimated that there were over 30 million head of livestock, including water buffalo, cattle, pigs, goats, sheep, and horses, on the three relatively small but heavily populated islands of Java, Madura, and Bali. All these animals provide food, except the horses, which are used only as work animals.

Just as elsewhere in Southeast Asia, fish and poultry are almost as common items in the Indonesian diet as rice. About 500,000 metric tons of deep-sea fish are caught annually, and about the same amount is taken from the inland waters of the rivers, lakes, swamps, artificially constructed farm ponds, canals, and irrigated rice paddy fields.

Since under Dutch management Indonesia was chiefly the source of raw materials from land, mine, and forest and was also a potential market for some manufactured products, twentieth-century industrial complexes were not built. After independence, a few small factories were constructed to make consumer goods such as textiles, fertilizers, and building materials. When Suharto took over in 1966, it was estimated that the factories were operating at less than 20 percent capacity. Perhaps one of the most revealing economic facts is that Indonesia's annual per capita income is currently one of the lowest in Southeast Asia and in the world—only about 96 U.S. dollars in 1968.

THE PHILIPPINES

The Philippines is the fifth largest Southeast Asian country and the second most populous; yet it is relatively small when compared with Indonesia. The Philippine Archipelago contains more than 7,000 islands, has over 115.7 thousand square miles of territory, and in extending for 1,152 miles in a north-south direction helps form the island rim east of Asia. Thousands of the small islets are just pieces of rock jutting above the ocean and are not named. Some 463 islands have an area of 1 square mile or more. The largest eleven islands account for about 95 percent of the area on which most of the population live. Luzon, the largest island—about the size of Virginia and covering 40,400 square miles—is located at the northern end of the archipelago. With 36,500 square miles, Mindanao is second in size and is the largest island at the southern end of the archipelago. Over two-thirds of the country's land area is on these two large islands.

As is true on most of the large islands in the Southeast Asian realm, nearly all the larger islands of the Philippines have interior highlands surrounded by coastal plains. The coastal plains are not always continuous, and even on the largest islands the coastal lowlands seldom exceed 10 miles in width. The highland cores often resemble or are the tops of mountain ranges, since the lower slopes of former mountains are now below sea level. On the larger islands, extensive uplands rise to over 4,000 feet, and in a few places to over 9,000 feet.

Most of the good agricultural land is concentrated on interior lowlands, but these are few in number and small in size. The largest and most productive of the lowlands are the central plain, Cagayan Valley, and the Bicol plain of Luzon; the Cotabato and Agusan lowlands of Mindanao; the western part of Negros; and the southeastern part of Panay. Unfortunately, the two large lowlands of Mindanao have extensive swamplands.

The insular and fragmented character of the Philippines land territory results in a long coastline. With a shoreline about as long as that of the United States and for the most part unhampered by mangrove swamps, the Filipinos have innumerable access points for fishing and shipping. The interiors of the islands are never far from the sea. With the exception of the two largest islands, where all areas are less than 75 miles from the coast, the maximum distances are generally less than 30 miles.

The Philippines has a tropical monsoon rain forest climate, but the amount and distribution of rainfall and the temperatures vary greatly, depending in part on topography and prevailing wind direction. The lowland areas are warm, with average monthly temperatures above 64.4°F for the warmest month, and humid, with rainfall generally above 80 inches. In the interior uplands, temperatures become generally cooler with elevation. Countrywide rainfall varies partly because of rainfall shadows on the leeward side of mountains. The eastern coasts of the islands generally receive more than the western sides. In some places the annual rainfall drops to less than 60 inches, and in other places the amount is four times larger. On the east side of the islands, the rainfall is usually well distributed throughout the year; but on the western sides of the larger islands with higher mountains, the rainfall may be concentrated in the summer and fall, and there may be a long dry season in the winter.

Fig. 15-11 Experimental farm under United Nations supervision in the Philippines. (Courtesy of United Nations)

Where the native vegetation has not been removed and the land brought under cultivation, there are extensive forestlands. The forests are of three types: tropical evergreen with commercial value, tropical evergreen without commercial value, and pine forests.

The dominant ethnic group in the Philippines is Malay. In descending order of numbers, the minority ethnic groups are Chinese, Americans, and Spaniards. Perhaps there have been more intermarriages here during recent centuries than in any other country in this realm.

Over half the Filipinos do not have a common language whereby they can speak, read, and communicate with each other. Approximately two-fifths of the people understand English. Only about half a million, primarily the social elite, still speak and understand Spanish. The three most widely used indigenous languages are Tagalog, spoken around Manila; Cebuano, used in the central smaller islands; and Ilocano, used in northern Luzon. The national government has supported the creation of a national language called Filipino and requires that it be taught in the schools.

The Philippines has one of the four highest birthrates in Asia and the highest in Southeast Asia. At the present rate, the population will double every twenty years. There exists a sharp controversy over whether this rate should and could be drastically curtailed. Formerly the chief concern centered around the amount of potential arable acres of land; according to some recent estimates, over 27 percent of the land is cultivated, and 54 percent could be. If this high percentage of land were intensely cultivated, then the islands could support 70 to 80 million people. Some believe that such estimates are overly optimistic; they stress that the population has been growing faster than arable acreage was being made available, that the ratio of one person per 0.9 arable acre in 1960 no longer exists, and that large quantities of foodstuffs have been imported annually since World War II. Then, in the last part of the 1960s the Green Revolution, with greatly increased yields per acre, was so successful that now it seems the Philippines may not only become self-sufficient in food but that the country may become a food exporter. The problem now revolves around the quantity and variety of the diet and the amount of potential arable land. The Philippines have an average population density of about 240 per square mile, exceeded only by Singapore, North Vietnam, and South Vietnam. The average density statistics, however, obscure the fact that the population density on the island of Cebu and also on the central plains of Luzon, even when the population of Manila and its suburbs are excluded, is about 800 per square mile.

Agriculture's primary position in the economy of the country is supported by the following facts: (1) over one-fourth of the country is under cultivation, (2) nearly two-thirds of the population lives in the countryside, (3) approximately three-fifths of the people are employed in producing and processing agricultural products, (4) agricultural production accounts for over one-third of the net domestic product, (5) agricultural products provided about three-fourths of the value of exports in 1967, and (6) nine of the country's leading export products (copra, sugar, coconut oil, abaca, beverages, tobacco, dessicated coconut, molasses, and canned pineapple) originate on the farm. With nearly two-fifths of the country covered with commercial forest, it is not surprising to find that two of the leading exports are logs and lumber and plywood.

One of the major economic goals is to diversify and expand industrial production, but development is slow. Although the Philippines is the world's greatest source of refractory chromite, copper and iron ore are greater dollar earners in the export trade.

During the five-year period 1963-1967 inclusive, the GNP increased at an annual estimated growth of 5.6 percent. In the last year of this period, the GNP was at 6,228 million U.S. dollars. But a per capita GNP of only 203 dollars in 1968 is low by world standards and represents less than a third of the per capita income in Singapore (700 U.S. dollars).

Scattered Minor Territories

Brunei (2,226 square miles) and Portuguese Timor (5,763 square miles) are two small remnants of the vast pre-World War II colonial period. Located on the northwest coast of Borneo, Brunei is split into two enclaves and, except on its seaward sides, is almost surrounded by Sarawak. When Malaysia was formed, Brunei was the only Malay-inhabited British dependency which refused to become a part of the new country and which remains today a protected state. Here in mid-1968 lived 112,000 people, with about 30,000 concentrated in the capital Bandar Seri Begawan (Brunei Town). In Brunei is found one of the highest standards of living in Southeast Asia because of the wealth secured from petroleum. In 1969, Brunei produced 6 million tons of oil, much of it coming from offshore wells. In descending order of numbers, the ethnic groups of Brunei are Malays, Chinese, and indigenous groups. In 1968 exports amounted to 89 million U.S. dollars and imports to 68 million dollars.

Portuguese Timor comprises the eastern end of the island of Timor and two adjacent islets. These territories have been Portuguese since 1586. Timor is the easternmost island of the Sunda chain in the Indonesian Archipelago. Portuguese Timor's population are primarily rural dwellers engaged in subsistence agriculture, but there is some commercial farming. Small quantities of farm and forest products such as coffee, copra, rubber, cassava, and maize are exported. The imports are primarily manufactured products.

The Realm

More meaningful generalizations can be made about Southeast Asia if Singapore is excluded from consideration. Because of its small territorial size, small amount of arable land, high percentage of people living in one city, and economic dependency on commercialism, industrialism, and world trade, geographic characteristics concerning Singapore are often the exception and not the rule. Consequently, the following generalizations apply primarily to the other nine countries of the Southeast Asian realm.

Agricultural Economy

Today, as has been true for centuries, life and its economic activities depend upon agriculture and forestry. Most of the people are employed in farming and lumbering or in processing the products from farm and forest. The ubiquitous composite landscape is one of rice grown under irrigation in small diked fields, where the ground is plowed and harrowed by water buffalo or oxen. The young rice stalks are planted in neat rows, and later the ripening grain is cut by hand; various hand methods are used in threshing and winnowing the grain. Another extensively used technique is that of driving buffalo or oxen in circles over stalks of grain scattered on an earthern threshing floor.

Most of the farmhouses are made of relatively light materials, with their sides having high exposure to the atmosphere. They are not heated and do not have water, sewage, artificial lighting, or cooling. Glass windows, window screens, and permanent walls are the exception. The framework of the house is bamboo, the floor may be of bamboo, and the walls and floors may be covered with split-bamboo panels and mats. Palms and other leaves are the most common roofing materials. Many houses are built on stilts; the preparing of food, cooking, spinning and weaving, washing, and other household activities often are carried on below the house or in the yard nearby.

During the era of European colonialism, plantation agriculture was introduced into nearly all the present countries, but it thrived best in Indonesia, the Philippines, Malaysia, and South Vietnam. Plantation crops are rubber, coconut and oil palms, sugarcane, pineapples, tea, coffee, cinchona, and abaca. Commercial agriculture, whether of the small farm or plantation type, is (1) concentrated on the lowlands—the flood, deltaic, and coastal plains—and (2) adjacent to or surrounding the largest city or capital located in the core area of the country.

Except for small areas of intermontane basins and tablelands, the highlands covering over half of the surface are covered with forests and grasslands. Here live most of the tribal peoples or displaced lowland dwellers who retreated into the highlands for safety. These people have a shifting or subsistence type of economy, based on temporarily clearing the land by the slash-and-burn technique. A diet based on crops of dry (nonirrigated) rice, maize, yams, bananas, and millet and meat from a few head of livestock, consisting of pigs, poultry, goats, dogs, and cattle is supplemented by gathering fruit, nuts, roots, leaves, and stems from undomesticated plants. In these highlands, in contrast with the diet of those living in the lowlands, wild game replaces fish, and honey replaces cane sugar.

Mining

Despite the fact that this realm supplies over one-half of the world's refined tin, its mineral production is quite low. Malaysia, Singapore, and Thailand mine and refine most of the tin. In 1966, the Philippines ranked third among the world's chromite-mining countries, but to the Philippines' GNP this metal contributes less than does the mining of copper. In this same year, Malaysia and Indonesia accounted for about 4 percent of the world's bauxite and about 2 percent of the world's petroleum. Small amounts, less than 2 percent of the world's supply, of other minerals are also mined, but production fluctuates greatly from year to year. Since the end of World War II, some Southeast Asian countries have built refineries to use imported crude petroleum and to make petroleum products for their home markets.

Handicrafts and Manufacturing

Manufacturing can be divided into two general types, cottage and factory. Cottage industries are common in villages, small cities, and large metropolitan centers. The tradition of subsistence farming and cottage industry go hand in hand. An almost innumerable number of products for house construction and furnishings are made from bamboo, plaited palm leaves, wood, and salvaged materials such as discarded tires and tin containers. Except for the iron tip on his plow, knives made in city blacksmith shops, and cloth goods, the members of the farm family provide their own farm equipment and household furnishings—from shoes to plows, harrows, and boats. Both rural and city families also make in their homes articles or parts of articles in quantity either from raw materials they have or which are supplied to them. A family may collect a truckload of nipa palm leaves and make roofing shingles, or someone may deliver the leaves and pick up the finished product. Specialized handicrafts are sometimes made in a large room of a non-residential building. For example, a dozen or several dozen persons may work together to make articles such as batik in Indonesia, nielloware and lacquerware in Thailand, and art objects of gold and jewels in Burma.

Electricity and steam-operated factories in large buildings especially designed and constructed for mass-production methods and producing items for nationwide distribution are mainly of post-1950 origin in Southeast Asia. In nearly every country, there is a variety of food and drink processing plants such as rice mills, sugar refineries, fish and fruit processing, distilleries, soft drinks, and often pineapple canneries. Each country makes its own household items and building materials; these include lumber and wood products, cement, brick, tile, and other ceramic products, electric wiring and fixtures, paints, varnishes and lacquers, plastic and rubber products, paper and paper products, toilet articles, and furniture. Each country also makes most of its clothing, from footwear to hats, although large qunatities of yard goods still need to be imported. Except for railroad equipment, cars, trucks, and airplanes, transportation vehicles ordinarily used by the masses, such as boats, animal-drawn carts, and bicycles, are made in the individual countries. Especially since the end of World War II, bicycle, motor-cycle, and small-car assembly and tire plants have been put into operation in the larger cities. A long list of miscellaneous manufactured goods, from tobacco and rope products to jute bags and pharmaceutical products, could be added. About nine-tenths or more of the factories hire less than 100 people—large industrial complexes typical of the European and Anglo-American countries in the last part of the twentieth century do not exist. Some of the limitations to major domestic industrial development are: (1) a lack of trained and experienced native managers and industrial administrators; (2) too few laborers having the necessary industrial skills or desire to work in factories; (3) a shortage of investable surplus capital; (4) despite low hourly wages, new industries generally find it difficult, if not impossible, to produce at costs enabling their products to compete in world markets; and (5) either the domestic demand is too small or the countries are too small and/or too poor to provide adequate markets for mass-produced goods made in large industrial complexes.

Laos, East Malaysia, and Cambodia are the least-industrialized areas. Singapore, West Malaysia, Thailand, and North Vietnam have much greater industrial maturity. Especially in Singapore and Malaysia, a large proportion of industrial employees work in a relatively small number of large concerns. According to the United Nations' *Economic Survey of Asia and the Far East*, published in 1967, manufacturing amounted to: 19 percent of the GNP in the Philippines in 1966; 15 percent in Burma in 1964; 14 percent in Thailand in 1966; 12 percent in Indonesia in 1964; and 10 percent in West Malaysia in 1965.

Cities

Except for the processing of mineral products and the production of hydro- and thermo-electric power, factories are concentrated in cities with populations of over 100,000, especially in the millionaire political cities. The seven Southeast Asian cities with over a million inhabitants are: Djakarta (2.9 million), Singapore (1.9 million), Bangkok (1.6 million), Rangoon (1.5 million), Cholon-Saigon (1.4 million), Manila (1.4 million), and Surabaja (1.0 million). If one defines a metropolis as a city with at least 500,000 people, then in addition to the millionaire cities there are four metropolises—Bandung, Thon Buri, Semarang, and Quezon City. Four of these eleven cities are on the relatively small island of Java, and two are in the Philippines. Three countries—Cambodia, Laos, amd Malaysia—are without a city of half a million. Urbanization varies greatly from country to country, with Singapore and West Malaysia being the most urbanized and Laos, North Vietnam, Cambodia, and South Vietnam being the least. Both Cambodia and Laos have only one large city (their capitals) with more than 50,000 inhabitants. In 1967, there were only 43 cities in Southeast Asia with populations between 100,000 and 1,000,000.

Except for Malaysia, the other nine countries have a primate political city. A primate city is one that is at least twice the size of that of the second largest city in the country, and it is often several times as large. It is usually multifunctional in character, being the economic, social, religious, financial, educational, political, and transport center of the country. There are twenty times as many people in Vientiane as there are in Luangprabang; ten times as many in Phnom Penh as in Battambang; six times as many in Saigon as in Da Nang; and five times as many in Rangoon as in Mandalay. The primate city often has a high percentage of the country's urban population. The urban dwellers in the country of Singapore live in the city of Singapore. Two-thirds of all the urbanites in Cambodia live in Phnom Penh. Over half of the Thais who live in municipalities live in the Thon Buri–Bangkok metropolitan area.

The statistics used in the preceding two paragraphs are taken from United Nations publications or are computed from data in their publications. The nations in this realm do not have a uniform time for collecting ten-year periodic census data. Some cities in Southeast Asia are growing rapidly, and these undoubtedly have much-larger populations than data collected five or more years ago would indicate. One Southeast Asian authority believes Djakarta already has a population of over 4 million. Moreover, the statistics are for political cities. Since 1962, however, the United Nations has not published in its *Demographic Yearbook* the populations of agglomerated cities, which include the population of the largest city and adjacent areas. In some countries the primate city and the second largest city are really twin political cities, located side by side. Bangkok and Thon Buri in Thailand are separated only by a bridged river. The combined population of Bangkok and Thon Buri in 1964 was over 2.1 million. If this agglomerated population were designated as the primate city, it would be twenty-eight times as large as Chiang Mai, the third largest political city. Bangkok has spread out in different directions and coalesced with other smaller cities. A uniform set of criteria for delimiting the boundaries of an agglomerated city in Southeast Asia is not available. At the present rate of city growth, it appears that Laos, Brunei, and Portuguese Timor will be the only political areas without a city at the half-million level before the 1970s end.

The growth of the eleven metropolises and of many smaller cities is based in part on river, coastal, and ocean transportation and trade. Five of the metropolises have coastal sites, and the other six are located on rivers less than 50 miles inland from the sea. Because of the insular and peninsular nature of the countries, the failure of the mainland countries to build connecting first-class transportation links between countries, and the fact that the core areas of countries are located on rivers and coastal plains, water transportation for commerce and air transportation for people provide the best linkages between countries and between their large cities.

Trade

In 1966, the value of trade in this area was only slightly under 5 percent of the total world trade. By world standards, this amount is low but is nonetheless vital to the needs and the economy of the realm. The relative importance of exports is shown in their contribution to the GNP of five countries: Malaysia, 42 percent; the Philippines, 21 percent; Thailand, 20 percent; Cambodia, 17 percent; and Burma, 16 percent. In 1968, the realm's total trade amounted to 12,050 million U.S. dollars consisting of 5,262 million dollars in exports and 6,788 million dollars in imports. Singapore and Malaysia provided 50 percent of the exports, and the Philippines, Indonesia, and Thailand ranked third, fourth and fifth, respectively. About half of the total tonnage of regional exports goes to Europe; one-third to Asian destinations, especially Hong Kong and Japan; and nearly all the rest to the United States. The amount of trade between Southeast Asia and Australia is minor. The trade gap between exports and imports widened between 1960 and 1968, with all countries having an unfavorable trade balance.

Population

During the twenty-year period 1947–1967, each of the ten Southeast Asian countries had a population growth of 44 percent or greater. Statistics for this twenty-year period are adjusted to present national boundaries. Population growth percentages vary from 44 percent in Indonesia to 132 percent in Laos. During the same two decades, Singapore's population more than doubled (108 percent) and Cambodia's nearly doubled (99 percent). The growth of population in Thailand, Burma, and Vietnam during 1947–1967 was 88, 64, and 52 percent, respectively. Although Indonesia's population increase during the two decades between 1947 and 1967 was only 44 percent, that for the realm's ten countries—increasing from 164 million in 1947 to 262 million in 1967—recorded an overall 63 percent increase.

In Perspective

Prior to 1940 the British Empire, France, the Netherlands, and the United States, with most of the land and sea area under their control, maintained a relatively stable political and economic situation in Southeast Asia. Only Thailand was not a colony, but having lost large land areas of its nineteenth-century empire to the British and French, it was relatively weak. Within two decades after World War II, nine new countries evolved. Looking back from the 1970s, it becomes clear that Southeast Asia had few unifying characteristics. Even today it is uncertain what land and water areas should be included in the realm. There are ten countries, but all of these are in different stages of political, economic (especially agricultural and industrial), social, educational, and urban development. Moreover, the economies and cultures of these are going through such rapid transitions that social convulsions predominate and are almost continuous. These countries have so recently gained their independence that, with the exception of Singapore, they have not as yet unified the various parts of their territory nor welded the geographic areas and ethnic groups into politically viable nations. Some authorities on Southeast Asia feel that only the Philippines, Malaysia, and Singapore have democratic forms of government, but others do not agree.

Some of the leaders inside these countries and some outside are attempting to set up inter-country political structures in this realm before the political infrastructures of these countries operate effectively. If the ten countries in this region would cooperate and present a united front to the rest of the world through one or several realm organizations, advantageous results such as larger markets, more trade and industrial development, greater efficiency, political strength, joint endeavors in health and education, and others could be realized. Such a situation, however, seems remote. Even organizations established with less than half of the countries in the realm, such as the Association of Southeast Asian Nations (ASEAN), formed in Bangkok in August, 1967, by representatives from Malaysia, the Philippines, Singapore, and Thailand, have been unable to accomplish much more than to generate hope.

Turmoil, war, and isolation are the "order of the day," rather than realm cooperation. North Vietnam repeatedly and loudly avows it will rule all the territory in former French Indochina. Consequently, with war, all parts of Indochina are apparently now in a stage of stagnation or economic decline. War in Laos and South Vietnam and the unwillingness of Prince Nordom Sihanouk, while in power, to have Cambodia give full cooperation, stalemated development along the Mekong River's main stream and almost stopped the Mekong River Project. Burma isolates itself from the outside world as it attempts to unify the various ethnic groups and geographic regions in its territory and establish a socialist state. As Sukarno had tried to take Indonesia back to the "good old days" that existed before European colonialism and still retain the benefits of the twentieth century, economic development ground to a halt and inflation nearly ruined the country. Now Suharto must try to bring order out of economic chaos. In Cambodia, with the war spreading throughout it in the spring of 1970, there were international reports of the balkanizing of this small country.

In contrast with the isolation and economic stagnation of Burma, Cambodia, Indonesia, Laos, North Vietnam, and South Vietnam, in Singapore, West Malaysia, Thailand, and the Philippines important economic, social, and educational growth has developed. According to Hla Myint, an Asian specialist, these last four countries are "outward-looking" and recognize that the "good old days" of precolonialism cannot be restored; they are pushing their countries into the last half of the twentieth century by stressing economic growth and welfare development. Since the processes of commercialization, industrialization, and urbanization should and will be accelerated, guidelines for these developments and accompanying problems have and are being established.

Problems abound, problems associated with the need for and/or the accelerated growth of agrarian reform, industrialization, urbanization, and population control. To adequately describe these problems and suggest alternate solutions for each would require a chapter as long as this one, because conditions vary so greatly from country to country. For example, drastic land reform was implemented early in the Communist regime of North Vietnam; in contrast, in Thailand, the peasants have owned most of the cultivated land for decades. In all countries, rapid population growth and the movement of people to the cities are creating problems of housing, inflation, unemployment, high rents, ethnic groupings and pluralistic communities, slums, traffic, pollution, police, fire, and problems associated with other city services. There are growing expectations for better education, health conditions, and annual incomes. All these problems and others are more difficult to solve in relatively poor, less developed nations lacking economic and social stability than in the Western countries that underwent drastic agricultural changes, industrialization, and urbanization long before World War II. Whether these problems can be considered adequately and resolved without (1) drastic political and social changes and (2) the development of secondary and tertiary industries and services remains to be seen. Changes are coming both peacefully and with violence and will continue to do so.

REFERENCES

Barton, Thomas F., Robert C. Kingsbury, and Gerald R. Showalter: *Southeast Asia in Maps,* Denoyer-Geppert Company, Chicago, 1970.

Boyce, P. J.: "Singapore as a Sovereign State," *Australian Outlook,* vol. 19, pp. 259–271, December, 1965.

Buchanan, Keith: *The Southeast Asia World: An Introductory Essay,* Taplinger Publishing Company, Inc., New York, 1967.

Cutshall, Alden: *The Philippines: Nation of Islands,* D. Van Nostrand Company, Inc., Princeton, N.J., 1964.

Fisher, Charles A.: *Southeast Asia: A Social, Economic, and Political Geography,* E. P. Dutton & Co., Inc., New York, 1964.

Fryer, Donald W.: *Emerging Southeast Asia,* McGraw-Hill Book Company, New York, 1970.

Kalab, Milada: "Study of a Cambodian Village," *Geographical Journal,* vol. 134, pp. 521–536, December, 1968.

Ng, Ronald C. Y.: "Recent Internal Population Movement in Thailand," *Annals of the Association of American Geographers*, vol. 59, pp. 710–730, December, 1969.

Pauker, Guy J.: "Toward a New Order in Indonesia," *Foreign Affairs*, vol. 45, pp. 503–519, April, 1967.

Pelzer, Karl J.: "Man's Role in Changing the Landscape of Southeast Asia," *Journal of Asian Studies*, vol. 27, pp. 269–279, February, 1968.

Ward, Marion W.: "A Review of Problems and Achievements in the Economic Development of Independent Malaya," *Economic Geography*, vol. 44, pp. 326–342, October, 1968.

Weatherbee, Donald E.: "Portuguese Timor: An Indonesian Dilemma," *Asian Survey*, vol. 6, pp. 683–695, December, 1966.

16

Australasia, Oceania, and Antarctica

The Pacific Ocean is the largest single geographical feature on the earth. Covering two-fifths of the surface of the globe and occupying an area larger than all the landmasses combined, it lies between the Old World continents to the west and the Americas to the east. Although so vast, the Pacific was long unknown to Europeans and was among the last parts of the world they explored. The first travelers' accounts established myths of Edenlike islands of great beauty, ease, and plenty which still persist. Although the United States for decades has governed several islands, notably Samoa and Guam, few Americans were aware of the Pacific islands until World War II.

The regional unity of the Pacific arises from its vast extent of water. Its land areas are very small in comparison; even the island continent of Australia is smaller in area than the entire conterminous United States. Extending southeast from Asia is a series of submerged mountain folds, the protruding tops of which form the elongate islands of Indonesia, New Guinea, the Bismarck Archipelago, and the Solomon group. Similar mountains, the visible folds of a submerged Australasian continental shelf, appear as the islands of New Caledonia, the New Hebrides, Fiji, and New Zealand. Further east and north, the islands are of volcanic basaltic origin and appear as mountaintops above the sea or as coral islands perched on volcanic bases.

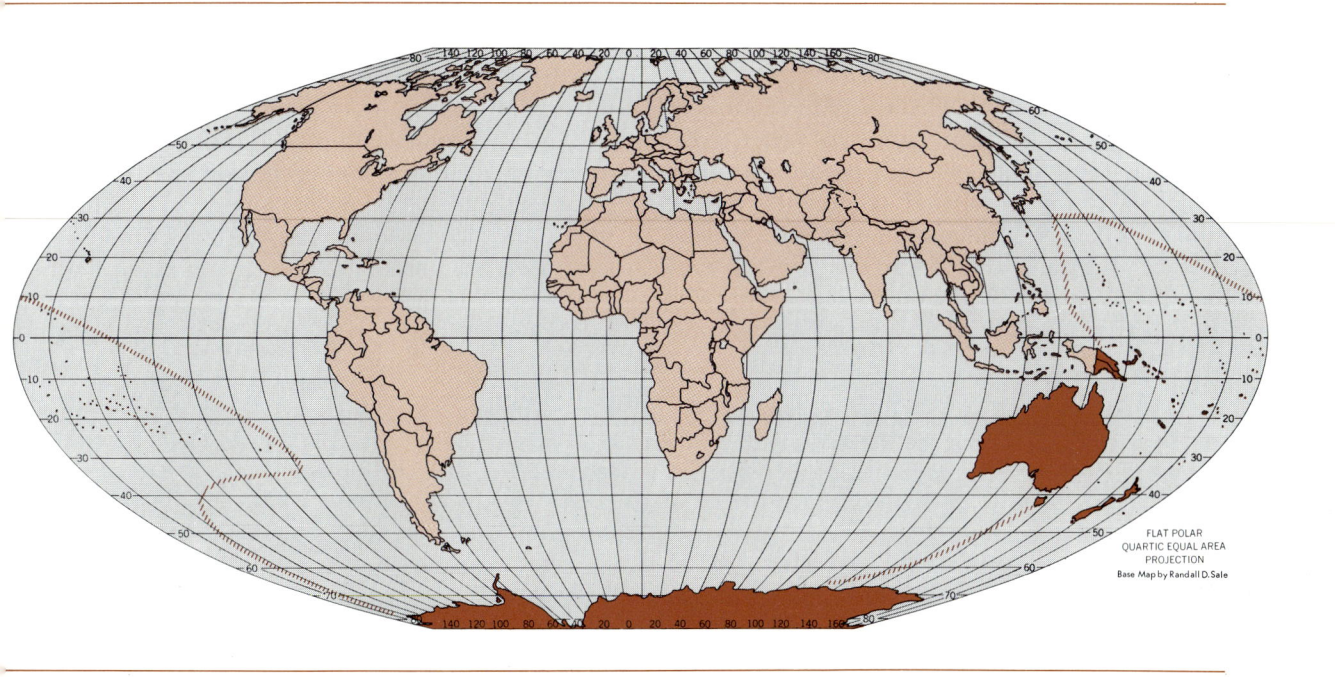

The volcanoes generally form high islands; those composed of coral are low-lying and, except for a limited number of raised coralline platforms, generally have a height of only a few feet above sea level. Characteristically, on a map the oceanic islands appear to be arranged in arclike festoons because they rise from chains of volcanoes, marking lines of weakness in the earth's crust, submerged in deep waters. Compared with the eastern Pacific, in which islands are rare, the western Pacific has numerous islands. The North Pacific, the portion lying between the Aleutian chain and the tropics, has practically no islands except offshore groups near the continents. In the vast southern ocean between Australia, New Zealand, and Antarctica, only a few isolated volcanoes rise above the apparently boundless sea.

Australia

Australia may well be called the island continent because of its isolation from the other great landmasses of the globe and because it is the smallest of the populated continents, having an area of about 3 million square miles. "Downunder" Australia and Antarctica are the only continents wholly south of the equator. The similarity of conditions and problems in some parts of Australia to those of the United States and Canada and their common language make North Americans feel at home in the Australian milieu and among Australians.

The Commonwealth of Australia was established in 1901 by the union of six British colonies. The former colonies, now called states, are New South Wales, Victoria, Queensland, South Australia, Western Australia, and Tasmania. In addition, there are two significant internal territories: the vast Northern

Fig. 16-1 Australia, the smallest of the continents—islands ranging in size from New Guinea, the second largest island in the world, to others with less than an acre—and the island continent of Antarctica, with no permanent population; all these areas make this a region of strongly contrasting environments.

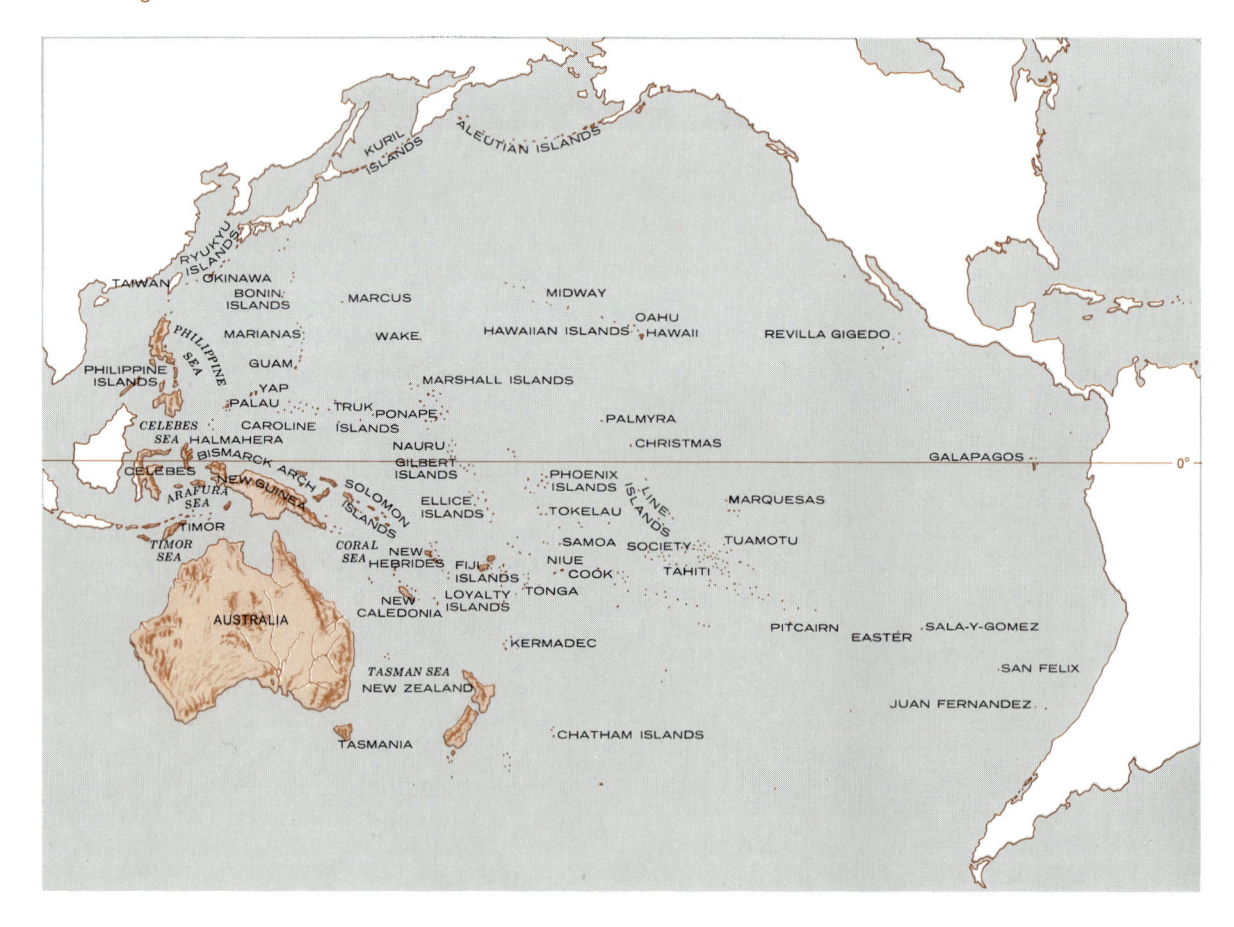

Territory, which is twice the size of Texas, and the Australian Capital Territory, which includes the federal capital of Canberra. Australia governs several external territories, by far the most significant of which is the Territory of Papua and New Guinea, encompassing the eastern half of the island of New Guinea, the Bismarck Archipelago, and the northernmost islands of the Solomon group.

Physical Setting

The physical environment of Australia is in many ways unusual, differing significantly from that of any other continent. Much of the landmass is very ancient, with old crystalline rocks predominating, and it has the lowest average elevation, hardly 1,000 feet, of any continent; only 6 percent of the country has an elevation of over 2,000 feet. Low elevation, combined with the significant influence of high atmospheric pressure conditions and the "blocking" effect of topography on easterly winds from the Pacific, precludes much rainfall; hence one-half of the continent is arid and one-fourth semiarid. No drainage reaches the ocean from this dry interior, and intermittent streams and lake basins contain water only after occasional rainstorms. Barely one-fourth of Australia is characterized by a humid or subhumid climate. Most of the streams in the humid regions are short; but some, notably the Snowy River, have been developed for power. In the southeast interior, the Murray-Darling-Murrumbidgee drainage area supplies much water for irrigation. In tropical northern Australia the monsoon effect is strong, and rivers vary in volume from flood stages during the summer rainy season to intermittent flow during the dry "winter." Australia extends almost 2,600 miles from east to west and 2,000 miles from north to south. It is a compact landmass, and the infrequency of indentations makes Australia's coastline the shortest, relative to land area, of any of the continents. There are few good harbors, no active volcanoes, and the continent's highest peak, Mt. Kosciusko in New South Wales, is only 7,316 feet above the sea.

Regions Australia may be divided into three major physical regions: the Western Plateau, the Central Lowlands, and the Eastern Highlands. The Western Plateau occupies three-fifths of the continent; most of it consists of very old rocks that have been worn down by erosion to relative flatness. Only low knobs and a few ranges rise above the monotonous surface, most of which has an elevation of about 1,200 feet. Around Spencer Gulf, in South Australia, earth movements have resulted in elongated ridges and depressed blocks called "rift valleys." Only in the north and the southwest is the rainfall adequate for trees and good pasturage for livestock. Wheat and other crops are also grown in the southwest.

The Central Lowlands include the flattish plains extending from the Gulf of Carpentaria on the north to the Murray River Basin and the Great Australian Bight on the south. From north to south, the Carpentaria, Lake Eyre, and Murray basins together include 800,000 square miles, or over one-fourth of the country's area. These basins are underlain by dipping layers of sedimentary rocks, chiefly shale and sandstone. The Carpentaria Basin has generally poor soils and is covered with grass and scrubby trees. Beef cattle are grazed extensively in this area.

The Lake Eyre Basin encompasses a vast lowland of interior drainage, with long intermittent surface streams "flowing" southwestward to Lake Eyre, whose usually dry bed is 39 feet below sea level. Despite a lack of surface water, most of the area is underlain by the largest artesian water supply (deep subsurface water under strong natural pressure) in the world. When these artesian reservoirs are punctured by bores (drilled wells), the pressure forces the water to the surface in a sustained flow. Unfortunately, most of the artesian water is too hot and too salty for irrigation or domestic consumption, but it supports extensive grazing of sheep and cattle. Some crop growing, mostly of grains, is possible along the wetter eastern margin of the basin.

The southern portion of the Central Lowlands is generally referred to as the Murray Basin, although much of it is actually drained by the Murray's principal right-bank tributaries, the Murrumbidgee and the Darling. Relatively fertile soils, several long permanently flowing rivers, and higher (though still moderate and often erratic) rainfall totals enable agriculture and pastoralism to be more significant in this area. The raising of cattle and sheep is widespread, and grain growing (mostly wheat and barley) is common along the eastern and southern margins. In addition, there is considerable irrigated farming of citrus fruits, grapes, hay, and other crops along the Murray and Murrumbidgee rivers.

The Eastern Highlands cover about one-sixth of the continent. The region consists of uplifted plateaus, tilted crustal blocks, and folded zones of various rock types, along with small disconnected areas of lowland and coastal plain. Although much of the highlands is thinly populated, the region contains two-thirds

Fig. 16-2 Although Australia is a continent, it is also an independent nation. Deserts and semi-arid areas cover much of the country. The most densely populated parts are along the eastern coast, around the larger state capitals, and in the Perth area of Western Australia.

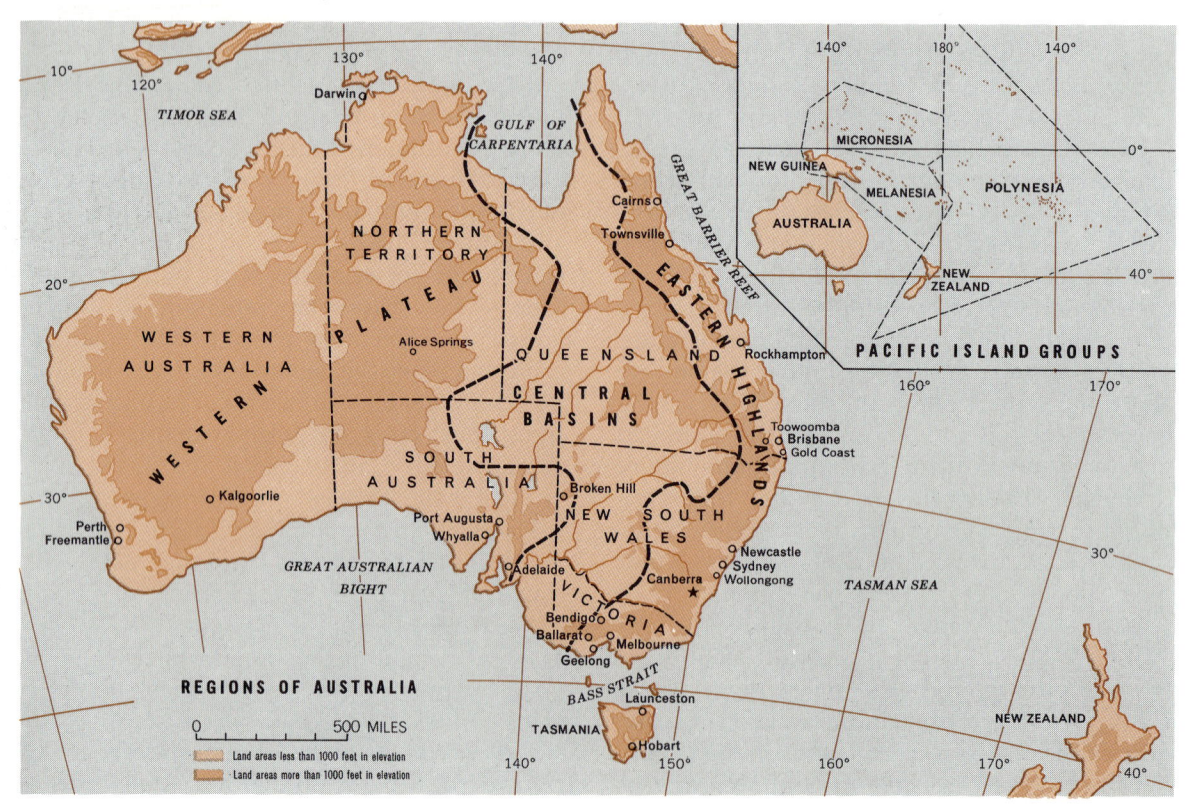

Fig. 16-3 Cooper's Creek (Barcoo) entering Lake Eyre, South Australia. (Courtesy of Australian News and Information Bureau)

of Australia's inhabitants, in part because it is the best-watered area on the continent. Elevations are moderate and vary from under 1,000 to over 3,000 feet, with summits above 7,000 feet in the southeast. The highlands continue across Bass Strait into Tasmania, which is a rugged island composed of uplands penetrated by flattish river valleys. In southern Victoria fertile lowlands, trending east-west, lie south of the highlands; here lives the large majority of the state's population. A remarkable feature off the Queensland coast is the Great Barrier Reef, which parallels the mainland for over 1,000 miles.

The highlands have a rainy climate throughout their length. The temperature varies, however, becoming warmer from south to north. Forests cover most of the region, but there are also well-watered grasslands for grazing. Dairying and other livestock industries are important, and some specialty crops are grown, such as sugarcane and pineapples in coastal Queensland and bananas in north-coastal New South Wales. The drier inner slopes of the ranges are devoted to grains and livestock. Most of the population concentrations of Australia, and all but two of its significant urban areas, are in this region.

Climate The climate of Australia is affected by its location between 10° and 40° south latitude, the generally low elevation of the land, and the position of the Eastern Highlands, lying across the path of winds from the Pacific. The dry interior and western regions are bordered by wetter areas to the north, east, and south; the seasonal pattern of the rains, however, differs from one area to another.

In summer, heavy rains brought by persistent northerly winds from moist seas drench Australia. Winter is the dry season, when the tropical rainy belt shifts equatorward. Throughout the year the temperatures remain generally high. Northern Australia, then, has a wet-and-dry tropical or savanna climate, except for the east coast of Cape York in the northeast, where rain falls throughout the year and the climate becomes rainy tropical. Rainfall diminishes toward the interior. From central Australia to the west coast is a warm trade-wind desert, dry in all seasons. Not only is the rainfall scant, but it is also extremely unreliable. The desert touches the south coast in the Nullarbor Plain, but north and east of the arid interior is a transitional zone of semiarid country.

Southwestern Australia and much of the coast of South Australia have a mediterranean type of climate, with mild, rainy winters and hot, dry summers. In winter the prevailing westerly winds shift north, and the accompanying cyclonic storms furnish the rains. Cyclones "down under" whirl opposite to those in the Northern Hemisphere, with the winds rotating about a low-pressure area clockwise instead of counterclockwise. Inland, the steppe climate begins at about the 10-inch annual rainfall line, in contrast to northern Australia, where the savanna is replaced by steppe at about the 20-inch rainfall line. The difference is accounted for by the higher rate of evaporation toward the equator.

Southeastern Australia has a humid subtropical climate, with rainfall well distributed throughout the year, mild winters, and warm summers. The winter rains are cyclonic; the summer rains are associated with easterly winds from the Pacific or with convective thunderstorms. Tasmania and southern Victoria lie in the westerly wind belt and have a marine west-coast climate resembling that of western Oregon. Winters are cool and wet; summers are mild and somewhat less humid. Snow may occur in highlands during the winter months.

Natural Vegetation Because of Australia's long isolation from other continents, most of the trees are peculiar to the continent and dominantly of two types: acacias and eucalypts. There are hundreds of species of each, which have adapted themselves to the various Australian climates, as have the many unique flowering plants and other vegetation. Scattered patches of rain forest occur at wetter sites along the east coast, particularly in Queensland. More widespread is an open growth of eucalyptus woodland that is the dominant natural association of the Eastern Highlands and much of the north-coastal country. The finest of the eucalyptus forests are found in the far southwest of the continent, where they constitute a major resource.

Most of Australia, however, is treeless. The desert heart of the island continent is largely barren, but there are sizable expanses of grass and shrub around its margins. Notable features of the vegetation of inland Australia include spinifex, a tall porcupine grass that grows in circular clumps; saltbush and bluebush, vigorous shrubs that provide excellent livestock forage; mulga, a low-growing variety of acacia that is widespread in semiarid areas; and mallee, a low tree which looks much like mulga but which is a member of the eucalyptus family.

Animal Life Most of the native animals of Australia are unique in that they lack close relatives on other continents. This condition is the result of the long separation of Australia from other lands; the higher animals never reached the island continent. Two-thirds of the species in Australia are marsupials, primitive creatures whose young are carried in an external pouch. Examples include many species of kangaroo, various tree-climbing possums, carnivorous marsupials such as the "native cat" and the Tasmanian devil, the badgerlike wombat, the omnivorous bandicoot, the ant-eating numbat, and a variety of marsupial "rats," "mice," and "moles." Australia has the only egg-laying mammals in the world, including the platypus, which has the bill of a duck, the tail of a beaver, webbed feet, and a poison spur and which is covered with fur, lays eggs, and nurses its young. The only large nonmarsupial land mammal is a wild dog, the dingo, which was brought into the country by the early Aborigines. Reptiles, especially snakes and lizards, are numerous; but amphibians are rare, as might be expected on an arid continent. Birdlife, remarkably varied, includes such large flightless species as the emu and cassowary, the world's greatest collection of psittacines (parrotlike birds), and the famous kookaburra, or "laughing jackass."

A large number of exotic (i.e., non-native) animals have been introduced into Australia, usually with unfortunate results. Most disastrous was the release of European rabbits, which spread over the southern two-thirds of the continent. like a plague of locusts for almost a century, eating forage in direct competition with sheep and cattle. Despite concerted efforts at control (involving guns, traps, poison, and exclusion fences), it was not until the development of a virulent and highly contagious disease, myxomatosis, which occurred in the 1950s, that the rabbit menace was finally eased. European foxes have also been introduced, much to the detriment of the small, vulnerable native marsupials. Another unusual element of the Australian fauna consists of feral livestock (domesticated livestock that have reverted to the wild), the most numerous of which are pigs, water buffalo, donkeys, and camels.

Economic Development

The Australian economy has all the characteristics of a modern, industrialized nation. Primary activities, mostly agriculture and pastoralism, employ only 12 percent of the labor force; whereas manufacturing employs 26 percent, and tertiary activities (mostly trade and services) employ 62 percent. Nevertheless, it is a great paradox of this sophisticated economy that products of its rural industries provide most of the export earnings. Wool, wheat, and meat have long been the mainstays of Australia's export trade; thus it is that a highly industrialized nation is heavily dependent upon its rural industries to obtain needed foreign exchange. Within the last decade, however, the export of ores and manufactured goods has assumed increasing importance.

Pastoral Industries Australia is world-famous for its sheep and important for cattle. Since the earliest years of European settlement, pastoralism has been the dominant rural activity, and the Australian economy has been said to "ride the sheep's back." Australia leads all other countries in the production and export of wool. Sheep number over 170 million, a ratio of fourteen sheep to each person in the country; but the number of sheep may vary by millions from year to year, depending upon whether there is drought or adequate rainfall. Sheep are raised in three types of operations: (1) In arid and semiarid portions of central and southern Australia are found extensive ranches (called "stations"), which cover up to 5,000 square miles and may support tens of thousands of sheep on natural pastureland. These are strictly wool-growing operations, and the sheep are all Merinos, which have proved to be the best producers under Australian conditions. (2) In a crescent of subhumid lands extending from northern New South Wales through Victoria into South Australia and occupying a similar climatic environment in the southwest sector of Western Australia is the sheep-wheat belt, where most properties consist of large mixed farms that produce both wool and grain (usually wheat, but also barley and oats). The wool clip here is even greater than in the extensive pastoral zone, and the sheep are mostly Merinos. (3) In more humid coastal areas of southeastern Australia, especially in Victoria, the emphasis is on meat production, primarily fat lambs. The farm properties are smaller and much more intensively stocked, and they depend upon improved

Fig. 16–4 Hamilton, Victoria. Sheep and cattle near a waterhole. Western Victoria is one of the wealthiest grazing and agricultural areas in Australia. (Courtesy of Australian News and Information Bureau)

varieties of pasture grasses that have been introduced. Wool is a secondary product, since the sheep are raised primarily for meat; hence, varieties other than Merino are dominant.

In all three of these pastoral zones the sheep are sheared once a year by teams of itinerant shearers. The wool is packed into huge bales and shipped to one of a dozen gigantic auction centers scattered around the country (mostly in the large cities), where it is sold to buyers from all over the world. Australian output amounts to more than one-fourth of the world's total wool clip. Furthermore, sales of lamb and mutton are often greater than those of any other country, although in some years New Zealand ranks first.

Australia has approximately the same number of beef cattle as it does people, and beef production is its second significant pastoral activity. About half the beef cattle are kept on extensive stations in the north and in the interior, where the animals shift for themselves on natural herbage. The other half are raised on smaller properties in the higher-rainfall zones of the east coast, where improved pastures are grazed. Queensland is the principal cattle state, pasturing nearly half of Australia's beef animals. There is considerable export of processed beef from Australia, particularly to the United States and Great Britain.

Dairying is carried on chiefly in the moist coastal lowlands and lower Eastern Highlands in southeastern Queensland, New South Wales, Victoria, and Tasmania. The industry is favored by mild weather and good grazing. There are some 5 million dairy cattle in the country; more than one-third of which are found in Victoria. Butter is an important Australian export, but its inefficient production on too-small holdings is maintained only by considerable government subsidies.

Agriculture Wheat is the most important crop in Australia and occupies well over one-half of the tilled acreage. It is grown primarily in the sheep-wheat belt, a crescent-shaped area extending from the Darling Downs region of southeastern Queensland through New South Wales and Victoria to the Eyre Peninsula of Australia, with a second important area in southwestern Western Australia. This is a subhumid to semiarid zone where the annual rainfall is 12 to 25 inches; however, most of the rain comes in winter, which is the growing season for Australian wheat. Wheat farming is highly mechanized, and yields are moderate (15 to 20 bushels per acre); but total output is great, and Australia is one of the top five wheat exporters of the world. Other grains, including barley, oats, grain sorghums, rice, and corn, are grown, but on a much smaller scale than wheat.

Sugarcane, the country's leading specialty crop, is grown intensively on the discontinuous patches of fertile lowland along the coast of Queensland and northernmost New South Wales. There are some 9,000 small sugar farms. Since Cuba's decline as a sugar exporter, beginning in the early 1960s, sugar production has increased markedly in Australia, and the country is now the world's second largest exporter.

Fruits that need a tropical climate, such as bananas, pineapples, and papayas, are grown in coastal Queensland and northeastern New South Wales. The chief citrus region is along the lower Murray River in South Australia, New South Wales, and northern Victoria, including sections of the Murrumbidgee Valley. Fruit is grown with the aid of irrigation in all these inland locations. In Victoria, Mildura is a chief center for citrus fruits, currants, and raisins. Vineyards also occur in these valleys, particularly in the Barossa Valley of South Australia.

Apples, pears, cherries, and berries are grown in Tasmania, in the cool southern lands of Victoria, in Western Australia, and in the uplands of New South Wales. Fresh apples are exported in quantity to Europe. Fruits are often canned or dried, and much of the output is exported.

Two nonedible money crops are being cultivated increasingly in Australia, and both of these enjoy the support of protective governmental tariffs. Cotton is grown with irrigation methods in several places, but most significantly along the Namoi River in northern New South Wales. Tobacco is grown in widely scattered localities, from northeastern Queensland to southwestern Western Australia.

Minerals The extraction and processing of minerals has been a major factor in the economic development of Australia, almost from the earliest days of European settlement. Probably only two other countries in the world, the United States and the Soviet Union, have a greater variety and quantity of ores and other economically significant mineral deposits. Coal, the first mineral to be mined and the first to be exported, has long been the keystone of the continent's mineral economy, and coal mining still employs about one-third of the nation's miners. Good bituminous coal is mined in a number of localities, but especially in deposits to the north, west, and south of Sydney. Australia is easily self-sufficient in bituminous coal and exports an increasing amount of its production, especially from deposits in Queensland. Lower grades of coal (subbituminous and lignite) are also present in large quantities and are used substantially for electricity generation, especially in the Latrobe Valley of Victoria and at Port Augusta in South Australia. Gold has also been a significant mineral resource of Australia exploited for a long time. Early gold rushes in Victoria and New South Wales and later ones in Western Australia had pervasive influences on the economy and settlement patterns of those three colonies (now states). Australia is still the fifth-ranking gold producer of the world, and gold mining occupies nearly 10 percent of the nation's miners.

Australia is also a major producer of several

nonferrous metals, ranking among the half-dozen leading producers of lead, zinc, and silver and possessing some of the world's largest deposits of bauxite, manganese, and nickel. Broken Hill in western New South Wales and Mt. Isa in western Queensland, both yielding a variety of nonferrous ores, are the largest mining towns in the country. Although Australia has been self-sufficient in the output of iron ore for several decades, discoveries of tremendous iron reserves in the northern sector of Western Australia since 1965 have made the nation one of the world's leading exporters of iron ore, especially to Japan. Major lacks among Australian mineral resources in the past have been petroleum and natural gas. Discoveries of commercial fields at Moonie (Queensland), Barrow Island (Western Australia), and Bass Strait (offshore Victoria), however, promise to make Australia self-sufficient in these fuel sources in the forseeable future.

Manufacturing Secondary industry (manufacturing) developed in Australia under several handicaps, the most important of which was isolation. During much of its formative history the nation had to depend upon overseas factories, primarily in Great Britain, to supply the great majority of manufactured goods. This costly dependence has now been largely overcome, but the process has been a slow and tortuous one, possible only by virtue of government protection in the form of restrictive tariffs.

The contemporary Australian industrial structure is well diversified, reflecting the fact that domestic manufacturers supply most of the wide range of demands of the consumer market. The manufacture of transportation equipment, food processing, and the production of machinery are the leading classes of secondary industry. Automobile manufacture is the principal component of the transportation equipment industry, with major production facilities being located in the large cities, particularly Melbourne, Sydney, and Adelaide. There are a great many kinds of food processing plants, widely scattered over the settled portions of the continent. The machinery industry is concentrated for the most part in Australia's two metropolises, Sydney and Melbourne, although there is increasing production from the steelmaking centers of Newcastle and Wollongong.

Overall, the distribution of factories in Australia is strongly urban-oriented. Most of the factories, like most of the people, are agglomerated in the southeastern corner of the country, particularly in the larger cities. Sydney and Melbourne combined contain nearly 60 percent of the nation's industrial output; other significant industrial centers are Adelaide, Brisbane, Newcastle, Perth, Wollongong, Geelong, and Hobart.

Transportation The concentration of population and industries in southeastern and southwestern Australia and the great expanse of unpeopled desert elsewhere markedly affect the location and construction of railroads and highways. Furthermore, until 1901 each state developed its transportation routes independently, focusing on the capitals, which were also the chief seaports. This individuality is shown by the prevalent use of three railroad gauges—broad, standard, and narrow—though some standardization of rail gauge has been accomplished. Only one railroad, built well to the south, crosses the entire continent from east to west. A narrow-gauge railroad to Alice Springs, in central Australia, and a highway from there to Darwin, on the north coast, form the only well-traveled route across the Outback of the interior from south to north. The government owns and operates about 90 percent of the 28,000 miles of railroads.

Many of the highways in Australia are unimproved, but 20,000 miles are called improved,

Fig. 16-5 Stuart Highway between Alice Springs and Darwin, Northern Territory. The road trains, about 42 feet long, pull up to four trailers; the trailers are known as "dogs." (Courtesy of Australian News and Information Service)

and most of the major roads are paved. Over 4 million motor vehicles are licensed, and the Commonwealth ranks third among the countries of the world in per capita ownership of motor vehicles. All the principal cities except Canberra are situated on the coast, and freight exchanged among them is frequently carried by coastwise ships rather than by rail or truck. Sydney, Melbourne, Adelaide, Brisbane, Fremantle (the port of Perth), and Hobart have freight and passenger service by sea to Europe, Asia, the Americas, and other places.

Airlines connect the principal cities, and air travel in this thinly populated country of large size is greater (per capita per mile) than in any other nation. In general, Australian civil aviation service is efficient, relatively inexpensive, heavily trafficked, and characterized by a splendid safety record. Furthermore, there is much international air travel to and from Australia, with the major routes extending across southern Asia to Europe, northward through the Philippines or Singapore to Hong Kong and Japan, and across the Pacific to North America. The use of aircraft in the Outback is also significant and appears most dramatically in the Flying Doctor Service, which provides speedy medical assistance for any remote station in the land.

Population

Of Australia's nearly 12.5 million people, more than 80 percent are native-born, and the large majority are of British ancestry. Since 1945, over 2 million immigrants have moved to Australia; about one-half of them have been British, but this figure also includes sizable contingents of Italians, Dutch, Poles, Greeks, Germans, Yugoslavs, and several other European groups. Chinese, altogether about 5,000 immigrants since 1945, comprise the only significant non-Caucasian arrivals. Some 150,000 Aborigines and half-castes live mostly in the interior dry lands and in the tropical north, where their numbers are again increasing.

For a country only partially developed, Australia has a surprisingly large urban population. The six state capitals—Adelaide, Brisbane, Hobart, Melbourne, Perth, and Sydney—contain more than one-half of the population of the Commonwealth; Sydney and Melbourne together have two-fifths of the continent's people. Except for Port Adelaide, outside the city limits of Adelaide, and Fremantle, a few miles from Perth, the state capitals are also the chief ports, handle most of the exports and imports of their respective states, are the principal railroad centers, lead in the manufacture and distribution of goods, and are paramount in business management, education, government, and cultural life. Important for central governmental activities is Canberra, the capital of the Commonwealth. In addition to the six state capitals, there are about 400 smaller cities. Less than 2 million people, hardly 16 percent of the population, are classified as rural dwellers, and even some of these live in villages rather than on farms or back-country stations. In each state, the capital city has the major population concentration and, in most, is decidedly the "primate" city. For example, two-thirds of the population of South Australia and Victoria live in Adelaide and Melbourne, respectively. Similar proportions are found in the other states: Perth, 60 percent of Western Australians; Sydney, 58 percent of New South Welchmen; Brisbane, 43 percent of Queenslanders; and Hobart, 32 percent of Tasmanians.

The dominance of Australian capital cities, which has been pronounced since the earliest days of settlement, persisted through colonial times to the formation of the six-state federation. In each colony an early coastal settlement, designated as the seat of government, had the dual advantage of being both administrative center and principal port in the formative days. This led to their development as major commercial centers for the respective colonies, and land transportation routes developed in a more or less radial pattern outward from them. With all these initial advantages, the capital cities have continued to attract the major share of the population, secondary industry, and the full range of urban economic activities.

Sydney and Melbourne, which are clearly the dominant metropolises on the Australian scene, rank among the fifty largest cities in the world, having more than 2.7 million and 2.4 million inhabitants, respectively. Each is a modern, sprawling metropolis with substantial public buildings, high-rise office and other business structures, big warehouses on the waterfront, extensive industrial areas, expansive residential districts, and attractive parks. Sydney ranks with San Francisco for its fine well-protected, landlocked harbor, which, like its American counterpart, is crossed by a huge bridge at the narrows. In truth, then, most Australians are urbanites. They have a high standard of living, a highly industrialized economy, an affluent and sophisticated way of life, and an urban orientation in most of their activities.

Fig. 16-6 Sydney, New South Wales, has one of the best natural harbors in the world. The downtown skyline is constantly changing; the city's famous bridge and operahouse are seen in the foreground. (Courtesy of Quantas Airways)

New Zealand

New Zealand is another land where Americans feel at home, for the climate and landscape resemble parts of Western Europe and the United States, and English is the major language. The industries of the country are familiar, and many of its economic problems are similar to those in certain parts of the United States. Situated about 1,200 miles southeast of Australia, New Zealand consists of two large islands and several smaller ones, having an area of about 104,000 square miles and a population of about 3 million.

The natural environment of the islands varies greatly. With a latitudinal extent approaching 1,000 miles, the country has a generally rainy marine climate, with temperatures modified by the surrounding ocean but getting progressively cooler toward the south. North of Auckland, snow is unknown and the climate is subtropical. Snow and frost are rare on the lowlands of the North Island, but the central highlands and the southern end of the South Island have cold winters. The west coasts are cloudy, with high precipitation, and in the south snow piles up in the high mountains to feed numerous glaciers. The eastern side of the South Island, leeward of the high mountains, has less rainfall; most of it has a subhumid grassland climate. New Zealand lies in the belt of prevailing westerlies, and for this reason extratropical cyclones provide much of the rainfall. Variations in elevation and exposure cause significant local differences in both rainfall and temperature.

The backbone of the South Island is a high and rugged mountain chain called the Southern Alps, which culminates in Mt. Cook, 12,349 feet above sea level. Some fifteen peaks reach elevations of 10,000 feet or more. The southwest coast, called Fjordland, is indented by a score of fjords gouged out by prehistoric glaciers. Beautiful lakes, snowfields for skiing, glaciers, serrated peaks, and fjords are scenic tourist attractions on the South Island. Both islands offer the tourist excellent fishing, sandy beaches, and protected harbors. The North Island mountains are not so high as those on the

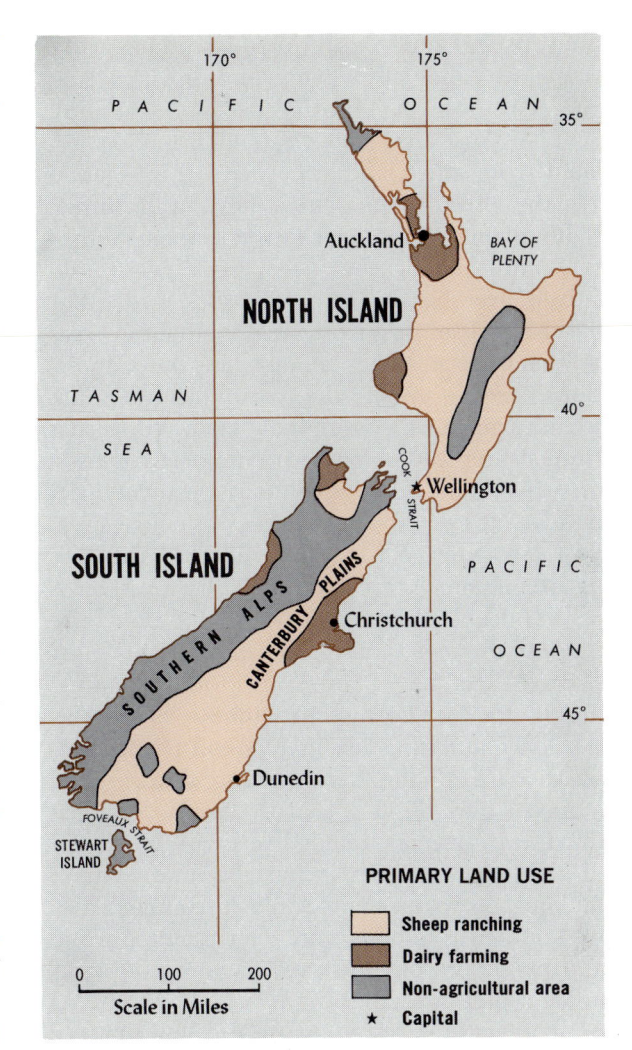

Fig. 16-7 Much of the land in New Zealand is used for pasture, and sheep and cattle contribute greatly to the economy of the country.

South Island, but central North Island has active volcanoes and also many hot springs and geysers; those of Rotorua are world-famous.

Like Australia, New Zealand was long isolated from other landmasses, and this isolation greatly affected the native flora and fauna. Many plants are peculiar to the islands. Southern pines and beeches, tree ferns, ferns, and laurels are characteristic. Tussock and other natural grasses occur on the rain-shadow areas of the eastern slopes, notably in the Canterbury Plains. On both the North and the South islands, much of the original "bush" forest has been cleared and planted with imported grasses, on which New Zealand's intensive farming depends.

Except for marine animals and the bat, native mammals are lacking. Flightless birds were once common; the most famous were those of the moa family, some varieties of which stood 12 feet tall. They were hunted to extinction by the first Polynesian settlers. The flightless kiwi and the kea, a type of predaceous parrot, are unique New Zealand fauna. The tuatara, one of the lizards, has a vestigial third eye, and its family tree can be traced directly to dinosaurs that became extinct millions of years ago. Some introduced animals, such as rabbits and deer from Europe and possums from Australia, have become significant pests.

The natural resources of the country are more limited and less varied than those of Australia. Coal is easily most significant of the economic minerals. There are relatively large bituminous fields on the west coast of the South Island and subbituminous reserves in the Waikato district of the North Island; but most of the supplies occur in a fragmented pattern, and both mining and transportation costs are high. Coal mining is now in decline. Gold is the only other mineral that has had much economic significance. The iron sands of the west coasts of both islands are being utilized as a base for a steelmaking industry, as yet still in its formative stages. The forest resources that originally covered much of the country have been depleted by logging, fire, and careless grazing; plantings of exotic coniferous trees (mostly *Pinus radiata* from California) have been established in the central part of North Island, providing the basis for an expanding and important lumber and pulp industry.

New Zealand's rivers are short, but most are swift and carry a large volume of flow, thus providing considerable potential for water-power development. Hydroelectric power generation is the major source of energy, although recent discoveries of natural gas in the Taranaki district of North Island and offshore petroleum deposits promise to diversify New Zealand's power sources. An interconnected power grid blankets both islands, with the result that per capita consumption of electric energy in New Zealand is one of the highest in the world. An interesting aspect of power generation is the harnessing of geothermal steam power in the geyser and hot-springs area of the North Island.

Animal industries form the chief economic base. The generally mild, rainy climate favors both native grasses and the introduced species that give the grasslands a high sustaining capacity. Farmland is mostly devoted to grazing. Even most crops are feed for livestock; of 45 million acres of occupied farmland, only 1.5 million acres are in field crops. Animal products such as wool, meat, dairy products, and hides constitute over 90 percent of the exports from New Zealand.

Dairying is the prime activity for a majority of the farmers of the North Island, especially in the low-lying areas. Since the climate is mild, no shelter is required and little supplemental feeding is needed. Purebred animals predominate among the nearly 4 million dairy cows. Methods are sanitary and efficient, and New

Zealand dairy products have a good international reputation. Cheese, butter, and dried and canned milk are exported, in particular to the United Kingdom.

Sheep are raised extensively on both islands, with the North Island having 60 percent and the South Island 40 percent of the country's 60 million sheep; they outnumber the human population by a ratio of 20 to 1. New Zealand follows Australia in wool exports but usually leads all countries in exports of mutton.

Most of New Zealand's wheat is grown on the Canterbury Plains, which have a subhumid climate. Production is adequate to supply local demand. Oats, potatoes, and apples are secondary crops, but only apples are exported in any quantity. New Zealand has had a recurring problem in recent years in disposing of or obtaining an economic market price for some of her agricultural products, notably wool and butter, both of which are under pressure from competitive producers and substitute commodities.

Although more than 200,000 New Zealanders are employed in manufacturing, there is little that is distinctive about this segment of the economy. About one-fourth of the entire manufacturing workforce is employed in processing pastoral and agricultural products. Other types of manufacturing are oriented almost entirely toward supplying the domestic market. Whereas the manufacturing sector of the economy is becoming increasingly important, its position continues to be insulated by government paternalism in the form of protective tariffs, customs duties, and import quota restrictions.

The population of New Zealand is increasing at a rate of 60,000 to 70,000 per year. The majority of the people, about 70 percent of the total, live on the North Island, following a trend initiated about a century ago. As in Australia, the population is predominantly urban; about 75 percent of the people live in cities and towns. There are four principal urban areas in New Zealand, which together contain nearly half of the nation's population. Auckland, with a population exceeding 600,000, is the principal commercial and industrial center as well as the busiest port. Wellington, the national capital, has about half as many people as Auckland. Christchurch, almost as large as Wellington, and Dunedin are the main cities of the South Island. The ethnic background of the New Zealand population is overwhelmingly British. Unlike Australia, New Zealand did not take the opportunity to attract settlers from the troubled countries of Europe after World War II, and the only non-British group of any size to be welcomed were the Dutch.

The Maori, descended from the original Polynesian settlers of New Zealand, form a significant minority in the New Zealand population—a proportion approximately the same as that of the black population in the United States. Originally perhaps totaling 200,000,

Fig. 16–8 Rotorua area, North Island. Sheep grazing on the New Zealand hillsides is perhaps the most characteristic sight in that country.

they declined after contact with the European settlers until about 1900, when only some 40,000 remained; since then, their population growth has accelerated, until now the rate is about 3.5 percent annually, and the Maori again number over 200,000 persons. Although most Maori are rural dwellers, they are moving into urban areas in ever-increasing numbers. The majority, however, live in the northern part of North Island.

The immigration of Polynesians from other Pacific islands, especially the Cook Islands, into New Zealand is now a significant stream. Auckland contains perhaps the largest Polynesian settlement in the world. Although Polynesians have no significant civil rights problems in New Zealand, and although the development of separate facilities or urban ghettos is officially and firmly discouraged, they do not match the European population economically or educationally; but progress toward removing the remaining disparities is discernible.

Oceania

The term "Oceania" may be taken to include New Guinea and other mid-Pacific islands north and east of Australia and New Zealand. In the present text, the islands of the Republic of Indonesia, the Philippine Republic, and Taiwan are discussed elsewhere in connection with Asiatic areas. It is customary to divide the islands of Oceania into three areas: Melanesia, Micronesia, and Polynesia. The Bismarck Archipelago; the Solomon, New Hebrides, and Fiji groups; and New Caledonia and associated smaller islands are called Melanesia ("black-inhabited islands") from the complexion of the natives. Micronesia ("small islands") includes the numerous but usually small islets of the Caroline, Mariana, Marshall, and Gilbert groups east of the Philippines and generally north of New Guinea and Melanesia. Polynesia ("many islands") covers a huge triangular area on both sides of the equator in the middle Pacific, from Hawaii on the north to New Zealand on the southwest and Easter Island on the east.

The native inhabitants of Oceania are as varied as the islands, in part because the peopling of the islands was accomplished over an exceedingly long period of time, with sporadic migrations of various groups ebbing, flowing, and overlapping. The general pattern was one of movement from Southeast Asia into the Pacific basin, perhaps beginning as much as 25,000 years ago. It is believed that Negritos were the first to penetrate the Pacific; they were a short, primitive people with a subsistence economy and a low level of material culture. Although they still exist in a few places, primarily in New Guinea, their imprint on the Pacific region was small, because they were usually inundated or absorbed by later migrant groups. These later migrants were quite diverse; some were primarily Negroid, others were predominantly Caucasoid, and still others had distinctively Mongoloid features. Racial mixing has taken place on a grand scale throughout the islands, resulting in a patchwork of racial types. Hence the simplistic generalizations sometimes made about Melanesians, Polynesians, and Micronesians are of questionable validity.

The varied native groups were resident in the islands for many centuries before Europeans began to penetrate the Pacific basin. For the most part, they developed a communal village way of life, with a subsistence economy based on small-scale agriculture, gathering, and fishing. Tree crops such as coconut, breadfruit, and bananas and root crops such as yams and taro were their dietary staples, and these are still the basic foods on many of the islands.

European "discovery" of the Pacific was

late, and penetration and settlement were slow. During the two and a half centuries following the crossing of the Pacific by Magellan in 1521, Dutch, English, Spanish, and French navigators made voyages of discovery. The most-renowned Pacific explorer of all was Captain James Cook, an Englishman who made three voyages between 1768 and 1779; his maps of the Pacific showed with commendable accuracy the essential features of that ocean and its islands. Other Europeans of various sorts and occupations followed the explorers. The impact of whalers, traders, missionaries, merchants, and other settlers was often disastrous to the natives. For example, in less than two centuries the number of Polynesians is believed to have been reduced by three-fourths, and Micronesians and Melanesians have suffered in a similar way.

Present-day control over the Pacific islands is politically diverse. There are four independent countries: Western Samoa (since 1962), Nauru (since 1968), the Kingdom of Tonga (formerly self-governing under British protection but completely independent since 1970), and Fiji (changed from colony to independent dominion status in 1970). The principal contemporary "colonial" powers in Oceania are the United States, the United Kingdom, France, Australia, and New Zealand. Governed by the United States are Hawaii, newest of the fifty states; American Samoa, legally an American territory since 1899; Guam, taken from Spain in 1898; and the United States Trust Territory of the Pacific Islands, which includes most of Micronesia. British possessions include the British Solomon Islands Protectorate, the Gilbert and Ellice Islands Colony, and the New Hebrides, governed as a condominium with France. France administers the far-flung islands of the Overseas Territory of French Polynesia and the Overseas Territory of New Caledonia, as well as shares the New Hebrides condominium with Britain. Australia's principal Oceanic possession is the Territory of Papua and New Guinea, which has a larger population than any other political unit in Oceania. New Zealand administers the Tokelau (or Union) Islands, Niue (or Savage Island), and a few other minor islands, as well as retaining some responsibilities in the self-governing Cook Islands. Chile governs Easter Island, at the extreme eastern end of the Polynesian "triangle." Germany and Japan were significant Pacific colonial powers in the past, but their colonies were forcibly taken from them at the time of World War I and World War II, respectively.

Polynesia

Of all the native peoples in Oceania, the Polynesians were the most admired by the European navigators. They were a large, fine-looking race of mixed ancestry and well-developed cultures. Scientists believe that the Polynesians were derived from the intermarriage of many peoples, including various strains of Asiatic, Melanesian, and Caucasian origin. Although they declined in numbers after the arrival of Europeans in islands such as Hawaii and New Zealand, the Polynesians have freely married with the English, Americans, and other nationalities, and their mixed offspring now constitute an important element in the population.

In the high volcanic islands of Polynesia the scenery is exceedingly attractive to tourists and other visitors, for it is often spectacular, with startlingly steep cliffs covered with tropical verdure and golden beaches shaded by waving palms. Paintings, photographs, motion pictures, television programs, plays, books, and stories about Polynesia, while popularizing the romantic and scenic aspects of the islands, have sometimes resulted in odd and exaggerated notions about the ease and pleasures of tropical living and the customs and mores of their inhabitants.

Polynesia lies in the tropics on both sides

of the equator, and its span is broadest from Samoa and Tonga on the west to lonely Easter Island on the east. Of the dozens of groups and hundreds of single islands scattered around this part of the Pacific, some are quite large and populous, but the greater number are small, with many being uninhabited. Overall, the extent of ocean vastly exceeds the area of land.

Most of the islands of tropical Polynesia are divisible into two types: high volcanic islands and low-lying islands built up of coral. A few of the coral islands, such as Niue, have been elevated; others, such as some of the southern Cook group, are made of a combination of volcanic and coralline material; and there are yet other examples of high basaltic islands surrounded by atoll-like islets, such as Bora Bora in the Leeward Islands. The high islands generally are larger and have more rainfall, better soils, and a greater variety of vegetation than the low islands. Having more resources, the volcanic islands can support more people and provide a greater variety of goods than the flat strips of coral thinly covered by sand and rising only a few feet above sea level. Examples of the high islands are the Hawaiian, Samoan, Society (Tahiti), and Marquesas groups and certain isolated isles such as Easter Island. In contrast, the Tuamotu Archipelago and the Ellice, Phoenix, Tokelau, and Line islands are all atolls—low, narrow strips of coral, usually of an approximately oval shape, surrounding a lagoon. The coral ring may be continuous, but more often it is broken into many "motu" (islets); sometimes an atoll has a score or more of these islets. Atolls vary in size from lagoons with a diameter of a few miles to some of 100 miles or more. The oceanic atolls are built on basalt platforms that rise from great ocean depths; these were once fringing reefs around volcanoes that have subsided beneath the sea, leaving only the coral strands. In the continental areas, raised reefs located on the continental shelf may form groups of coral islands. Portions of the Tongan group (or the Friendly Islands) have this structure.

Seabirds that resort to the low islands for nesting are everywhere. With the exception of bats and those introduced by man, land mammals are lacking. On many islands, introduced vertebrate animals such as rats and goats have caused great damage; introduced invertebrates have also created serious problems, such as the damaging of coconut palms by the rhinoceros beetle.

Coconut palms, pandanus, and a variety of bushes constitute the principal native plants on the coral islands. The coconut supplies a great variety of products: the meat is eaten fresh or pressed to make a cream; the milk of immature nuts is a refreshing drink; a fermented drink may be made from the sap; and from the dried meat, called copra, is extracted coconut oil, a major industrial raw material used in making margarine, soap, cosmetics, and many other products. The trunk of the coconut palm serves as posts for houses; the leaves are used for roofing material and fuel; the fiber, for fuel, sennit rope, and mats and as the industrial raw material coir; the shell, for charcoal and receptacles. In fact, every part of the coconut palm is put to some use. Pandanus leaves are woven into mats and clothing, and the fruit may be used as an occasional food by those islanders with strong teeth.

Coral sand is poor soil; hence if the inhabitants of low coralline islands want gardens, they first dig pits in which organic wastes are allowed to rot to enrich the soils. They can then raise root crops, vegetables, and fruits. Although the tiny coral atolls support relatively small numbers of people, they may become very densely inhabited. Settlement is usually in agglomerated villages; but, using excellent sailing canoes often equipped with outriggers for stability, the islanders range over the lagoon for fishing and travel to the other motu to gather coconuts and other products. A limiting factor of life on most low islands is the shortage of water.

Before the arrival of Europeans, the inhabitants of the high islands raised root crops, mainly taro, and breadfruit, coconuts, and such crops as bananas and arrowroot. They kept pigs and chickens, caught fish and turtles, and gathered reef creatures, seaweeds, and bird eggs.

After contact with Europeans, the number of islanders in Polynesia declined greatly. Diseases, both epidemic (such as smallpox and influenza) and environmental (including tuberculosis and typhoid), together with wars, made more ferocious by the introduction of firearms, wrought havoc with the Polynesians. Today populations in this segment of Oceania are growing very rapidly, as fast as any in the world, but only in a few island groups is it likely that numbers have again attained their precontact levels.

The islands of Samoa are generally high, and their rocks, except for the limestone reefs, are volcanic. Savaii is one of the very few Oceanic islands with active basaltic volcanoes. The sixteen islands of Samoa are divided into two separate political units. Western Samoa, an independent country, contains about 80 percent of the total Samoan population of 170,000. Its capital, chief port, and trade center is the town of Apia, on the island of Upolu. Tutuila and a few small islands to the east comprise the colony of American Samoa. The landlocked bay of Pago Pago (pronounced "Pango Pango") on Tutuila is one of the finest harbors in the Pacific, and a large jet airport has been built nearby. Most Samoans are subsistence farmers, with communal landholdings, and there is some export of copra, cacao, and bananas. Tuna fishing and tourism are the significant activities in the economy of American Samoa.

French Polynesia includes the Society Islands, the largest of which is Tahiti, and the

Fig. 16-9 Low coral islands in the South Pacific Ocean.

Tuamotu Archipelago, the Marquesas Islands, and the Tubuaï (or Austral) Islands. The principal export is copra, grown on most of the islands, along with some vanilla. Pearls are secured by divers in the lagoons of the Tuamotus. Pearl shell, tortoiseshell, and dried trepang (*bêche-de-mer*), a sort of sea slug, are other exports. The export of phosphate rock from the island of Makatéa, formerly a very important source of income, has ceased because the reserves are exhausted. The town of Papeete on Tahiti is the chief center for trade in eastern Polynesia. Artists and writers have given the world an enticing view of the attractive people, the lush tropical vegetation, and the steep-walled mountains of Tahiti and neighboring Mooréa and Bora Bora. Tahiti contains about half of the 90,000 population of the territory. Tourism now dominates the economy of French Polynesia, and French government expenditures, particularly for nuclear testing and experimentation, have also contributed a great deal to the local economy in recent years.

The populous islands of the Kingdom of Tonga are raised coral reef islands; only slightly raised in the case of Tongatabu, the main island, and considerably elevated in a series of steps in the case of Vavau, the island second in importance. The population of Tonga, nearly 80,000 persons and with a rapid rate of increase, is pressing on the limited resources of the island group. Copra and bananas are the only important exports. Tongan society has a royalty-nobility-peasantry stratification, with all land being the property of the Crown. The nobility hold large estates by grant from the king, and all Tongan males are entitled, at age sixteen, to an agricultural allotment of 8.25 acres of rural land and a house lot in a village settlement. The landholding system works less satisfactorily in practice than in theory, for the last census showed that the number of eligible males greatly exceeded the number of available lots.

The fifteen islands of the Cook group, located southeast of Samoa, were until 1965 a New Zealand dependency, but they are now self-governing internally. New Zealand retains control of their foreign affairs and defense and helps their economy with an annual subsidy. The 20,000 Cook Islanders, nearly half of whom live on the main island of Rarotonga, produce canned juices and fruits for the New Zealand market, clothing also made for sale in New Zealand (using imported cloth and low-wage local labor), some copra, and such perishables as citrus fruits and tomatoes. Pearl shell and handicrafts augment the income of the more remote atoll dwellers. There is a considerable migration of Cook Islanders into New Zealand for employment and education.

Melanesia

The land area included in Melanesia is large; but, except for Fiji and New Caledonia, the islands are little developed economically and have been among the most neglected on earth. More recently, however, their development has been substantial, and its rate is increasing.

New Guinea The second largest island in the world is New Guinea, which has an area of 306,600 square miles; it is nearly 1,500 miles long by 500 miles wide at its greatest width. The western half is controlled by Indonesia; the eastern section is governed by Australia. The Australian portion is administered as a unit, the Territory of Papua and New Guinea, but actually consists of two parts: the northeastern part, together with the Bismarck Archipelago and Bougainville Island of the Solomon group, was once a German possession but is now administered by Australia under a United Nations trusteeship; and the southeastern section, called Papua, which was taken over by Australia in 1906, although it had been a British colony, largely under the control of Queensland, since 1888.

Very high mountain ranges, several over 12,000 feet in elevation, extend the length of New Guinea; peaks reach above 16,000 feet, high enough to be snowcapped for long periods. The mountains comprise a great barrier to ground travel, but the intermontane valleys of the highlands provide the inhabitants with cooler locales for living. From the mountains descend large rivers that have the potential to produce millions of horsepower in hydroelectric energy if and when developed.

New Guinea has a variety of environments, including the dark humid jungles of the windward coasts, the cool highland valleys and plateaus, the open eucalypt savannas of the wet-and-dry zones of the south coast, and the tall tropical rain forest of the foothills. The river floodplains and many of the coastal areas are covered with mangrove swamps. Groves of sago palms that flourish in the swamps are an important source of starchy food for some groups.

The native peoples are extremely diverse, ranging from primitive Negrito hunters in the remote mountain regions to sophisticated Melanesian town dwellers along the coast. The majority of the people, however, are relatively primitive farmers, who live in small villages at the middle elevations in the highlands and practice a shifting garden type of subsistence farming. There is an increasing amount of commercial agriculture, both on indigenous small holdings and on large European-owned plantations, with coconut, coffee, cocoa, and rubber as the chief crops. Lumber is also a major export. Gold, formerly a very important revenue source, is declining. Recent discoveries of other minerals, notably copper on Bougainville, appear to be extremely valuable.

The total population of the Territory of Papua and New Guinea is about 2,300,000. Australia is attempting to develop the territory rapidly toward self-government. A legislative assembly of elected representatives and a new university have both been established in Port Moresby, the administrative center, within the last decade.

Eastern Melanesia Elsewhere in Melanesia there are a number of high islands, usually with volcanic ores, clothed with a luxuriant verdure of jungle and forest and peopled by dark-skinned natives who live in small villages and subsist on gardening and fishing. There are also a great many small coralline islands, but with a much sparser population.

The British Solomon Islands Protectorate extends for nearly 1,000 miles across the South Pacific in six island subgroups. Most of the 150,000 Solomon Islanders carry on a subsistence way of life; however, work in town, usually in the capital Honiara (on the island of Guadalcanal), or on copra plantations or in lumber mills provides occasional employment for increasing numbers. The area is thinly populated and economically underdeveloped, but recently progress has accelerated.

The condominium of the New Hebrides contains some eighty islands, although only a dozen are of considerable size. Copra, manganese, frozen fish, and cacao are the exports of importance, but most of the population still depends upon subsistence gardening and fishing. Until recently the size of the population was unknown, but in 1967 the first census was taken successfully, showing a population of about 78,000 persons, with a still modest growth rate of 2.5 percent annually. There are two small urban centers in the New Hebrides, Vila and Santo.

New Caledonia is a large island with a long mountainous backbone. It is cooler and, on the leeward side, drier than elsewhere in Melanesia. The mineral riches of the island, particularly nickel, have attracted so many immigrants, primarily from France, that Melanesians comprise less than half the total population; indeed, New Caledonia is the only South Pacific island on which the population of European origin is larger than the population of Oceanic origin. There are a number of cattle ranches and coffee plantations, but agriculture as a whole contributes much less to the economy than does mineral production. Only copra and coffee are produced on a scale sufficient to supply an export surplus. The capital city Nouméa, with a population of some 50,000, has the largest concentration of Europeans in the South Pacific.

Fiji, an independent member of the British Commonwealth of Nations, possesses aspects that are both Melanesian and Polynesian; however, it has a unique character found nowhere else in Oceania, owing to the large population of Indians inhabiting the country. More than 50 percent of the half million inhabitants of Fiji are descendants of indentured workers brought from India to labor on the sugarcane plantations. The cultural gulf between the sober, striving, landless Indians and the socially oriented, land-owning Fijians is likely to generate increasing problems, especially since both groups are increasing rapidly in numbers. Although there are more than 300 islands in the Fiji group, many are uninhabited, and 85 percent of the population inhabit the two large islands of Viti Levu and Vanua Levu. The economy of Fiji is precarious because of the erratic market prices for its principal exports: sugar, copra, and bananas. The rapid growth of tourism has been an economic advantage and a partial compensation. Fiji is the principal way station between Hawaii and Australia–New Zealand. Both its chief port, as well as capital, Suva, and its international jetport at Nadi (pronounced "Nandi") are the busiest in the South Pacific.

Micronesia

Most of Micronesia is now encompassed in the United States Trust Territory of the Pacific Islands, consisting of the Marshall, Caroline, and Mariana groups, except for the island of Guam. Spain established ownership over the islands in the sixteenth century and continued as governing power until the Spanish-American War. In 1898 Spain ceded Guam to the United States and sold the other islands to Germany. During World War I the islands were taken by Japan. In World War II they were captured by the United States, which now governs them under United Nations Trusteeship.

The Caroline Islands, situated just north of the equator, are scattered over an ocean area half the size of the United States. The Carolines include 963 islands, mostly atolls but several volcanic, having a total land area of about 830 square miles and a population of approximately 75,000 persons. The principal volcanic islands are the Palau and Yap groups toward the west, the Truk Islands near the center, and the islands of Ponape (Ascension Island) and Kusaie on the east. These five island groups account for half the area and 70 percent of the inhabitants of the Carolines. The volcanic soil is fertile, and the variety of resources on volcanic islands is greater than on the atolls, many of which are unpopulated. Subsistence gardening and fishing dominate the economy; the only significant export is copra, although frozen fish is showing promise of becoming an important economic commodity. The recent formation of an international airline, Air Micronesia, and the start of a hotel-building program point toward increasing development in the tourist industry.

The Marshall Islands are a double chain of coral atolls, having a population of about 20,000 inhabitants, who form a fairly homogeneous group both culturally and physically. The economy of the Marshalls is much like that of the Carolines, and the islands are being developed as part of the same trusteeship program. Bikini and Eniwetok atolls, where atomic experiments were conducted in 1946 and 1948, are part of the Marshall group. After having been evacuated for two decades, these two atolls are being populated again.

The Mariana Islands, including Saipan and Tinian, well known from the Pacific war, differ from other Micronesian groups in that they are all high islands. Until recently, there had been little redevelopment in these islands since World War II.

Although physically a part of the Mariana Islands, Guam has been a separate political entity since the United States took it from Spain in 1898. Its 200 square miles make it the largest island in Micronesia. About one-third of its 75,000 inhabitants consist of Americans, principally military personnel and their dependents. Most natives of Guam now work for either the military or the civilian government apparatus.

The Gilbert and Ellice Islands as well as Ocean Island (Banaba) are controlled by the United Kingdom. The Gilbert and Ellice Islands Colony contains some forty small islands that extend from Micronesia (Gilberts) into Polynesia (Ellices) and also includes phosphate-rich Ocean Island and the Line and Phoenix groups. With a population approaching 60,000 and a land area of less than 400 square miles, they include, particularly in the Gilberts, some of the most crowded and poverty-stricken islands in all of Oceania.

Nauru, the only independent nation in Micronesia, is a single island of less than 9 square miles and with fewer than 7,000 inhabitants. Its people reside along a relatively fertile coastal fringe. The export of phosphate rock from the central plateau of the island forms the bulwark of the economy.

Antarctica

The only permanently unpopulated continent is Antarctica, situated in frigid isolation almost entirely south of the Antarctic Circle. Its total area of more than 5 million square miles is nearly as large as Australia and Europe combined. The continent is so deeply covered with ice, in many localities more than 2 miles thick, that its actual configuration is not well known. In some places the ice extends well below sea level, which means that, if the ice melted, the Antarctic continent might actually consist of two or more large landmasses and a number of smaller islands. The coastline of the continent consists largely of a rugged ice cliff that rises abruptly from the sea or from the fringing shelf ice which lies on the sea surface in many places. Barren mountains rise above the ice cliffs along some parts of the coast. The interior of the continent is a high ice-covered plateau, much of it more than 8,000 feet above sea level, which is interrupted by several significant mountain ranges, the highest peaks of which reach to nearly 17,000 feet. Antarctica is a climatic desert, with the least precipitation of any continent, although there are frequent and severe blizzards, composed mostly of turbulent snow. Temperatures rarely rise above freezing anywhere on the continent and for months do not exceed 0°F; the record surface temperature of 127°F below zero is 37° lower than any temperature ever recorded on any other continent. There are virtually no terrestrial life forms, except for a few species of ticks, mites, and such simple plants as algae, moss, and lichens. The surrounding Antarctic Ocean, on the other hand, abounds with plant and animal life, ranging from microscopic diatoms to the largest of whales. Many species of seabirds range widely along the Antarctic coast, and some of these, including several kinds of penguins, nest on the continent in rocky coastal areas. Although most of Antarctica is claimed by various nations, there are no permanent settlements. Scientific research stations are maintained, however, sometimes for a duration of years, at more than a dozen localities.

In Perspective

From time immemorial, the land areas of the Pacific basin have been remote from the rest of the world, but this isolation is now disappearing. Australia and New Zealand have for many decades maintained significant, if slow, contacts with Europe in general and Great Britain in particular, but only recently have strong links with other parts of the world been forged. Now the lines of communication from these two Antipodean lands to North America and Asia are at least as important as those to Europe. Both Australia and New Zealand are active participants in international organizations such as the United Nations, the Southeast Asia Treaty Organization, and the Colombo Plan. Their foreign trade relations have been broadened, so that Australia is a major supplier of minerals and wool to Japan, sugar and meat to the United States, and wheat to China and India; and New Zealand is diversifying its traditional trade pattern so that there is less dependence upon the British market. Immigrants arrive "down under" not only from Great Britain, Holland, Germany, and Italy, but also from such countries as Spain, Malta, Greece, the United States, and Turkey. A dozen international airlines to and from Australia–New Zealand serve the growing tourist interest. The Australasian corner of the world has lost its isolation.

In the far-flung islands of Oceania, change comes slower, but even there it is coming. New mineral discoveries, more efficient fisheries, and more diversified commercial agriculture are broadening their economic base. Modern influences are felt in other spheres as well: social stratification and ethnic discrimination are diminishing; increasing measures of self-government are being granted. Perhaps most far-reaching of all changes in Oceania are the improvements in transportation and communications. Fiji has become the principal stopping point for north-south travel across the South Pacific; Tahiti and American Samoa have heavily used jetports; and with varying frequency, air service now reaches virtually every Pacific island group. The world has begun to discover Oceania; contacts will continue to expand, and, for better or for worse, the islands can expect increasing inundation by tourists.

REFERENCES

Cumberland, Kenneth B.: *Southwest Pacific*, Whitcombe and Tombs, Ltd., Christchurch, New Zealand, 1968.

Dury, G. H., and M. I. Logan (eds.): *Studies in Australian Geography*, Heinemann Publishing Company, Melbourne, 1968.

Harding, Thomas G., and Ben J. Wallace (eds.): *Cultures of the Pacific*, The Free Press, New York, 1970.

Huxley, Elspeth: *Their Shining Eldorado: A Journey through Australia*, Chatto & Windus, Ltd., London, 1967.

Kennedy, T. F.: *A Descriptive Atlas of the Pacific Islands*, Frederick A. Praeger, Inc., New York, 1968.

McKnight, Tom L.: *Australia's Corner of the World: A Geographical Summation*, Prentice-Hall, Inc., Englewood Cliffs, N.J., 1970.

Oliver, Douglas L.: *The Pacific Islands*, Doubleday & Company, Inc., Garden City, N.Y., 1961.

Robinson, K. W.: *Australia, New Zealand, and the Southwest Pacific,* University of London Press, Ltd., London, 1962.

Rose, A. J.: *Dilemmas Down Under,* D. Van Nostrand Co., Princeton, N.J., 1966.

Spate, O. H. K.: *Australia,* Frederick A. Praeger, Inc., New York, 1969.

Tudor, Judy: *Many a Green Isle,* Pacific Publications, Sydney, 1966.

Tudor, Judy (ed.): *Pacific Islands Yearbook and Who's Who,* (10th ed.), Pacific Publications, Sydney, 1968.

Watters, R. F. (ed.): *Land and Society in New Zealand,* A. H. and A. W. Reed Publishers, Wellington, New Zealand, 1965.

Wiens, Herold J.: *Atoll Environment and Ecology,* Yale University Press, New Haven, Conn., 1962.

The Far East and Australasia: A Survey and Directory of Asia and the Pacific, Europa Publications Ltd., London, 1969.

Conclusion: This Changing World

In the preceding chapters, the authors have sketched the geographical environments of the various regions of the world and pointed up many geographic relationships within and among them. Thus, this study may be considered a background against which to view current happenings and possible future developments, a study on which to base your concepts about peoples and the space they occupy.

In many ways the modern world is dynamic, a place of rapid change; yet, in other ways, changes are so slow that conditions seem almost static. It is man, his ideas and activities, not nature, who has caused most of the numerous changes in the world of the twentieth century. At the beginning of this century the nations of Europe, for all practical purposes, were recognized as the rulers or leaders of most of the world; also, the world was largely a white man's world. Since 1910, however, two major World Wars, as well as several more limited engagements, have affected every country to such an extent that it can now be said that no nation is so independent that it can remain unaffected by events outside its national boundaries. Within this century these wars, plus the virtual end of colonization, have brought numerous political changes. Every nation, even the United States because of its additional world responsibilities, has been affected by the continually changing political geography of the world.

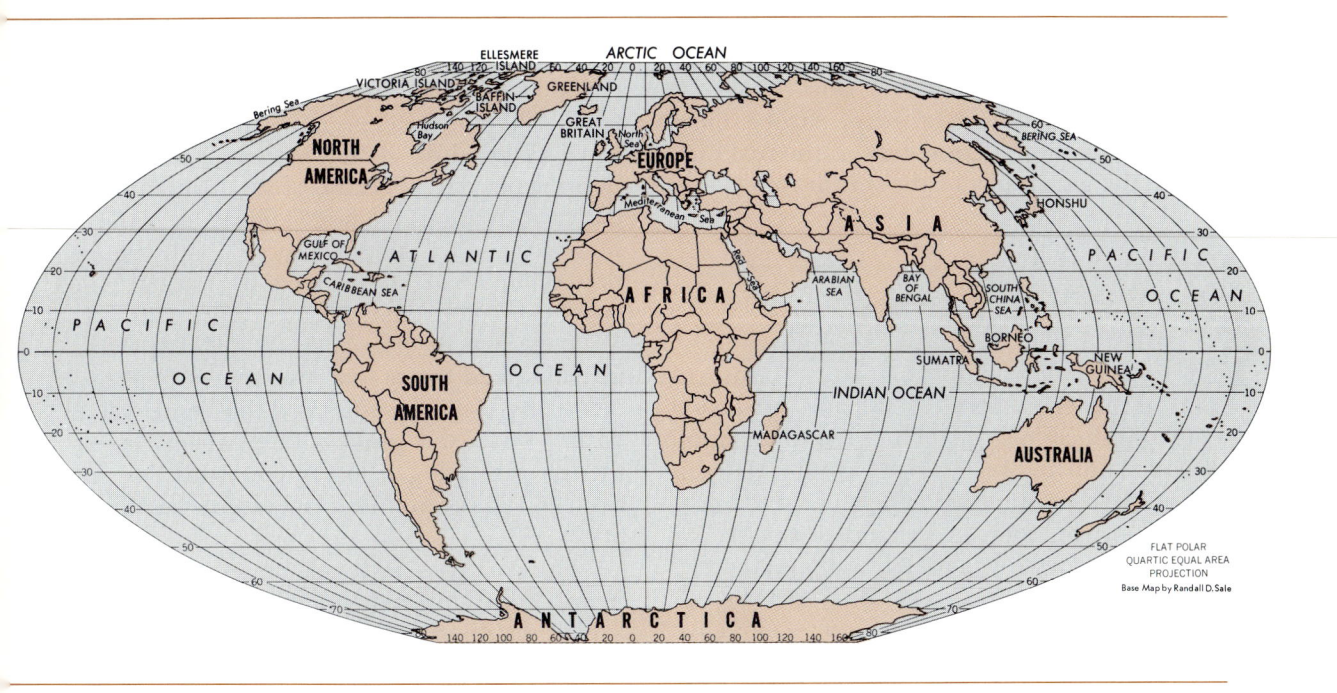

Despite changes in the boundaries of nations, large or small, much of the geography of the world remains the same. Regardless of who controls them, the mountains still have their heights and barrier effects; the major rivers and streams continue to flow in the same directions; the winds continue to follow the same general pattern and bring their life-sustaining rains; and the soils of the plains and valleys remain the principal producers of food. The earth continues to revolve and rotate, so that man may depend on a succession of seasons and know that day will follow night. Man, however, has modified his environment in many ways. Some of these effects, such as the reclamation of land, the harnessing of waterpower, and the extraction of minerals, may be beneficial. Other effects are harmful, as can be seen, for example, when man's activities cause soil erosion, forest fires, and the spread of pests and diseases or pollute the water and air. Natural changes are usually very slow, but those resulting from human activity may be quite rapid. Man, more than nature, is the real variable in the geography of the world.

International Problems

The uneven distribution of natural resources such as minerals, fertile soils, and life-giving water produces an equally uneven distribution of population and resource utilization. Man can utilize these resources to better his standard of living through work, peace, and prosperity, or eventually to destroy himself by war and famine. In the previous chapters, so-called "developed" and "underdeveloped" nations and regions have been discussed. What, then, is the world point of view?

Population

The total area of the earth is 197 million square miles. Of this area, 140 million square miles are covered by oceans, seas, and lakes; another 10 million square miles are either too cold, too dry, or too rugged for permanent settlement by large numbers of people. This leaves an area of 45 million square miles in which man does live; but, further, approximately one-half of this is subhumid or semiarid. More than 3.6 billion persons are thus confined largely to 23 million square miles of the earth's surface.

Not only is the world's population increasing more rapidly than ever before, but the normal span of life is also longer. Between 1900 and 1970, the increase in world population was approximately 2 billion persons. Since 1960 the average increase has been about 164,000 persons per day. The population of China alone is now being increased by about 45,000 live births per day. In the United States, on an average day, 10,000 babies are born, 5,000 persons die, and 1,000 more persons enter the country than leave, giving this nation a net increase of 6,000 persons per day. Because of lower infant mortality, increasing medical knowledge, and a somewhat better distribution of food, the expected life span in the United States has increased by more than 21 years since 1900. In 1970 there were more than 20 million persons in the United States sixty-five years of age and older. By 1980 it is estimated that there will be 25 million in this group. Life expectancy, although increasing in other countries, has not risen so rapidly in most places as in the United States.

The countries with the largest total populations are China and India; but people are not evenly distributed throughout their territories. Because of various physical factors, the people of China are concentrated on the narrow southeastern coastal plains, along the valleys of such rivers as the Hsi, Yangtze, and Hwang, as well as in the valleys of the many smaller streams, and in some instances on the sides of terraced highlands. Large parts of China's interior are too dry and far too elevated to support many people. In India the situation is similar; the Ganges Valley and the narrow coastal plains are densely populated, but the rougher parts of the Deccan Plateau and drier interior areas are less densely settled. When viewed in relation to the country as a whole, population density means little; when considered only for the areas of arable land, it becomes significant. For example, the overall density for China is 195 persons per square mile. Since the different parts of the country vary greatly in their ability to produce, density ranges from zero in the deserts to more than 2,500 per square mile in the arable areas. This is a density of three persons per acre—probably twice the number who could be fed by agriculture alone. In the United States, at least 2.3 acres of arable land per person are needed to maintain the present standard of living. In many ways India, China, and Italy are typical, since they emphasize the problem of too many people for too small an amount of arable land.

Part of the Northeastern United States, much of West Central Europe, and sections of South Asia, Southeastern Asia, and Central Eastern Asia have large areas in which the population density exceeds 250 persons per square mile. Each of these areas contains not only a considerable amount of productive land but other favorable features such as minerals, suitable climates for particular crops, and sufficient water supply. People in some areas do not take as much advantage of these positive factors as they might; nevertheless, the potentiality for development is theirs.

One of the prime problems of population distribution today is that of urbanization, for as fast as the world's population is growing it is urbanizing even faster. Where people are concentrated in a small area, the pressures and forces for technological, social, and cultural change increase. From this point of view, then, cities are locations for accelerating change. For the last several centuries the people of the world have been moving toward the more densely— that is, urban—areas. During this century the movement has greatly accelerated. Two centuries ago, there were no cities of a million population; today there are about 100 such places. Some twelve cities have populations in excess of 4 million persons each. Various authorities have states that (1) within the next century city populations will increase forty times at the present rate of increase; (2) if present trends continue, about 75 percent of the world's people will live in urban areas within the next twenty years; and (3) half of all Americans will live in three metropolitan areas by 1985. Such concentrations, of course, create an environment that brings together each day people of greatly varying backgrounds. The product of these contacts, good or bad, is accelerated social and cultural change.

Advancing technical knowledge will, perhaps at a date not too far in the future, enable man to use large areas of land not now in productive use. If, by using atomic power, it becomes feasible to process sea-water and pipe it to desert areas, vast amounts of now unproductive land could be put to work. New methods of cultivation, new and improved plants, better methods of processing and preserving— all these will aid, to a certain point, in caring for the increasing world population. Population pressure upon the land is, then, "world problem number one."

Natural Resources

Minerals Minerals are one of the principal bases of modern civilization, or of the modern standard of living. Iron, copper, uranium, coal, and petroleum are presently the most important of the major mineral resources. Many of the minor minerals—minor only in the sense that they are not produced in such large quantities— help to make the basic minerals far more usable and more versatile than they would be otherwise. Thus, vanadium, limestone, and other lesser minerals are as essential to producing certain kinds of steel as are the basic iron ore and coke.

Some nation, because of their large areas (the Soviet Union, the United States, China, Canada), have a variety of minerals; others, because of their smallness, location, or natural features (Italy, Togo, Bulgaria, Laos), have only a few, if any, minerals in significant amounts. Several nations produce one or two minerals (Malaysia, tin; Sweden, iron ore; South Africa, gold; Kuwait, petroleum), but because these nations lack other substances necessary for their processing, they export vast quantities of ores and other crude resources as raw materials. Minerals are of little, if any, value until put to work. Since much of the modern standard of living is based on minerals, nations have often attempted to gain control of the properties of their neighbors. Obviously, the unequal geographic distribution of mineral reserves is a problem to be solved if the world is to remain at peace.

Water Most essential of all natural resources is water. Without water, plants cannot grow even in fertile soil; unless he has water to drink, man dies within a few days. Yet overabundance can be just as disastrous to an area as a scarcity of water; the Amazon Valley is no more densely populated than large areas of the Sahara. Not all plants need the same amount of water to thrive. Some, such as the great forests of the selvas, need at least 70 to 80 inches of rainfall fairly evenly distributed throughout the year; others, like numerous xerophytes, can thrive on 5 inches or less of moisture per year. Corn does best in a region that has long, warm summer days, where the moisture is supplied by afternoon thundershowers. Wheat produces well if it has about 20 inches of rainfall, provided that it occurs at the right time. Other plants also have their peculiar water needs, which in many respects are the result of adjustment to their environment. Where a surplus of fresh water is available and irrigation can be practiced, numerous crops can be grown, and the problem of food production can be more effectively solved.

Water is as essential to life in cities as it is in rural areas. In the United States, where urban population is rapidly increasing, the limitations of the function of several major cities may be determined by the amount of water available. Cities throughout the world attempt to solve their water problems in diverse ways. Gibraltar has cemented sections of the Rock to create catchment basins; Los Angeles sometimes finds it necessary to ration its supply. New York is constantly searching for new places to build dams and thus ensure a permanent and adequate supply of water. Until the water problem is solved, approximately 9 million square miles of the earth's surface will be of little value to mankind.

In spite of man's knowledge of the importance of clean, fresh water to his daily personal life, as well as its necessity for his industries, he has ruined many sources of water supply. During the past twenty-five years, numerous streams and lakes have become so polluted by industrial waste and city sewage that they are now of little if any value as sources of usable water. Lake Erie is in such deplorable condition that much of its marine life has been killed; the Rhine River has been referred to as " the sewer of Europe." The waters near the industrial areas of Japan are just as bad, and the Soviet Union has recently pointed out certain water sources in that country which must be cleaned and improved. Many developed nations now recognize the seriousness of the problem and have enacted conservation laws that require industries and cities to dispose of waste materials in other ways.

Soils Soils capable of producing the most abundant crop yields are, like minerals and water, also unevenly distributed. Poor management and constant utilization have put hundreds of thousands of acres out of production, some permanently. Large areas in northern China, parts of North Africa, especially near Carthage, and section of Italy were in ages past good producers of food. Many farms in the southern United States are so badly gullied that the land is no longer usable. Dust storms in the subhumid parts of the Soviet Union, China, and the United States are but warning signals that grasslands are being put to improper use.

Man has gone far in solving some of his current soil problems. Rotation of crops, contour plowing, strip cropping, gully control, and fertilization are but a few of the soil conservation measures that have been developed. To carry out such programs is costly, in most instances requiring government aid in addition to what the owner contributes. Frequently, land must be taken out of production for at least a short time, and in areas where food production is at a subsistence level, the diminished crops may mean starvation. To help solve its soils problem, the world needs a better distribution of foods and a closer unity among nations.

Food Production

Each part of the world specializes in some crop or crops. The principal agricultural products of an area are usually those which do best in that particular environment. Farming in most parts of the world is thus an adjustment to the natural environment of soil, water, and climate. Irrigation and other specialized types of agriculture are, of course, exceptions.

The two great food crops of the world, wheat and rice, are basic foods for more than 90 percent of the world's people. Rice is intensively cultivated in the monsoon lands of Asia, most of which are densely populated. Wheat is an extensively cultivated plant of the semiarid and subhumid lands of central Eurasia, central North America, southeastern South America, northern China, and southwestern Australia. Most rice acreage produces four or five times as many bushels of grain as does wheat. Per capita consumption of wheat for the world as a whole is about 180 pounds; for rice, approximately 150 pounds. In general, the standard of living in wheat-growing and wheat-consuming lands is considerably higher than it is in the rice areas. Moreover, the areas growing rice depend much more upon the current crop than do the wheat-growing areas. If the rice crop in an area fails, famine and starvation commonly follow, since there is no surplus from previous years and little if any money to buy the surplus from other regions. Most wheat areas have a surplus from previous years or have the means of securing grain from another region. In some parts of the United States and Canada, beef has replaced wheat as the chief staple food.

Recent research in plant ecology and plant breeding has procreated new hybrids of both rice and wheat. Experiments in the Philippines have developed new rice strains that are believed capable of almost doubling rice production. In Mexico, similar experiments have developed wheat hybrids that will grow in drier climates and at the same time increase the number of bushels per acre by 25 to 50 percent.

Rye, corn, barley, and oats are also important grain food crops. Potatoes form the principal part of the diet in certain sections of Europe and the Americas. Supplementary foods such as vegetables and fruits vary with the locality.

Animals suitable for meat supply live in most parts of the world. In areas having the highest standard of living in the Western world, beef, mutton, pork, and poultry are essential items of diet. The consumption of meat products varies greatly from region to region, depending upon the amount of land available for pasture, the religion of the people, and the general prosperity of the area. The average annual consumption of meat per capita for the world is almost 50 pounds—ranging from 15 pounds per person in Asia to 230 pounds in Australia and New Zealand. The annual per capita consumption in Anglo-America is about 170 pounds.

Usually enough food is produced on a worldwide scale each year to feed all the people of the world if it were evenly distributed. Should production be low in some area, there is generally enough surplus in another to make up the deficiency. But conflicts and lack of transportation or money to buy food may prevent needed shipments. Food production in every part of the world can be increased by more efficient land utilization, improved methods of harvesting, and plant betterment, but improvement in transportation and in the economies of poverty-stricken peoples is also needed.

Manufacturing

The processing of raw materials and goods, in some form or other, is carried on in every part of the world. Complex manufacturing, however, is confined primarily to the northeastern United States, the most densely populated section of Western and Central Europe, Japan, and certain European areas of the Soviet Union. Most areas of America and Europe have, or did have at one time, large quantities of coal and iron ore, waterpower, access to rivers, lakes, or oceans for cheap water transportation, fairly level topography or gaps through inland barriers so that land transportation was not handicapped, access to sufficient quantities of food and necessary raw materials, and a reputation for skilled workmanship. Outside these four chief industrialized areas, there are numerous smaller developments, such as the Texas Gulf Coast and the Los Angeles area in the United States, the Ural area of the Soviet Union, the Jamshedpur and Calcutta areas of India, the Sydney area of Australia, and a few centers in China. Many smaller manufacturing centers are areas of specialization.

In most instances, the nations having the highest standards of living are those which have the easiest access to manufactured goods. Machines used in producing a specific item decrease the amount of manual labor required for making that item; ultimately, however, mechanization creates more jobs. The worker also gains more time for educational and recreational activities. Each nation attempts to increase its output of manufactured goods for sale in foreign countries, since the sale of these goods adds purchasing power; yet, simultaneously, each nation develops protective tariffs to encourage its own industries and to discourage the sale of foreign goods within its own boundaries. Many nations could use their resources to better advantage by developing agriculture, mining, or other activities rather than by trying to compete in the world market for manufactured goods. By such reasonable adjustments, manufactured goods could flow more freely to areas where they are needed and thus aid both producer and consumer.

Spheres of Influence

The three most important influential nations in the world today are the United States, the Soviet Union, and Communist China. Each is a giant in area. The Soviet Union, with 8.6 million square miles, ranks first in size; China, with 3.7 million square miles, ranks third; and the United States, with 3.6 million square miles, ranks fourth. Each has a large population: China, with about 800 million persons, ranks first; the Soviet Union, with 241 million persons, ranks third; and the United States, with over 200 million persons, is fourth. Neither the United States nor the U.S.S.R. is densely populated, and both have large areas for possible expansion of agricultural production. All three have great mineral wealth and vast industrial activities, and each of the three countries is a leader in its sphere of influence. There, however, most of their likenesses end.

The United States has attempted to build its sphere of influence by working with other nations through mutual-aid programs. The United Kingdom, France, West Germany, Greece, South Korea, South Vietnam, Thailand, and other countries have worked with the United States in the development of these programs. In addition to building joint armed forces, economic aid has often been given where needed in an attempt to better agricultural and industrial output. The U.S. Peace Corps is now assisting with education and technology in many African, Asian, and South American countries. The United States has borne most of the cost, and most of the immediate benefits have gone to the countries assisted, in an effort to improve their standards of living and increase their educational opportunities.

The Soviet Union, in building its sphere of influence, has brought many of the nations that border it partly or completely into the Communist bloc. Beyond the adjacent areas, the Communist doctrine as expounded by the Soviet Union, has been successfully spread to parts of Africa, Asia, North America, and South America. This particular bloc of nations actually controls more land area than either China or the United States. The Soviet Union has been largely responsible for the buildup of arms in the United Arab Republic, Jordan, Syria, and other Arab countries. At the same time the Soviet Union has also helped some of these countries economically, especially the United Arab Republic, by loaning money to build hydroelectric projects such as Aswan High Dam and by giving technical assistance and furnishing agricultural aid.

China has attempted to spread the Maoist doctrine of communism into neighboring Asiatic countries and to some of the underdeveloped nations of Africa. Although not so successful as either the Soviet Union or the United States in extending its influence, China has caused both the United States and the Soviet Union much concern by the development of atomic and hydrogen arms, by activities along the China-Soviet border, and by assistance to North Vietnam, as well as by its propaganda inroads elsewhere. By supplying arms to Iraq, the Palestinian resistance movement, and Cuba, China also confronts and challenges the other two leading powers. So far China has given economic and technical assistance on a much more limited scale than the other two great powers, mainly confined to certain Asian and African nations, as well as Albania and Cuba.

Another bloc of nations of special importance, although somewhat minor when compared with the three big powers, is known as the Islamic World. With the exception of the Republic of Indonesia, Malaysia, and East Pakistan, these nations form a compact group of Arab nations in North Africa and the Middle East. Their combined population of over 350 million is held together largely by a common religious belief, since more than 90 percent of the people are Moslems. Approximately 60 percent of this sphere's inhabitants live in Turkey, Pakistan, and Indonesia, three countries in which there is a variety of agricultural activities as well as some complex industrial development. In the remaining Islamic countries, the standard of living is low; most of their people eke out a substandard subsistence living by farming near an oasis or in an area where irrigation is possible. In some of these countries, many lead a nomadic life by following their flocks. Since the Middle East is the location of large oil developments and reserves, each of the three great powers would prefer that these nations remain neutral if not within their own sphere of influence. The Islamic World as a whole is not aligned with either the "Communist World" or the "Free World." Since the Six Days' War of 1967, the Arab nations bordering Israel—especially the United Arab Republic, Jordan, and Syria—have become somewhat united in trying to regain the territory they lost to Israel. Since the Soviet Union has been willing to furnish them military materials and technical assistance, the Arab nations involved in the war with Israel have tended to favor the Soviet Union. The United States, on the other hand, has not encouraged mutual relations with the Arab states by supplying arms to Israel.

Many nations, especially the relatively new

ones, are largely underdeveloped and politically nonaligned. Most are small in area and deficient in natural resources and technical knowledge, have an exceedingly low GNP, and are increasing in population faster than they can supply the necessities of life to their people. Although most are theoretically democratic, many are actually distatorships or totalitarian regimes in which the government may change within a short period of time. Too often, under present world conditions, these nations are pawns in the political games of the various major aligned groups.

Place of Geography

What, then, is the place of geography in this world of unequally distributed natural resources and unsettled peoples? Study of the many facets of human activity must not be regarded as a study of uncorrelated facts. To understand the activities of mankind, one must have knowledge of the various factors, physical and cultural, that make up man's environment. These factors may or may not determine what man will do in or about a specific area. They will of necessity, however, influence both his thinking and his actions about the problems that face him. The study of geography will also help individuals in our region of the world to understand better the people and problems of other regions. Only through sincere and sympathetic mutual understanding of one another can the people of the world hope for lasting peace and progress. An understanding of geography will definitely contribute toward this goal of good international relationships throughout the world.

Glossary

Alluvial fan A cone-shaped area of water-deposited material at the base of an abrupt upland front, created by the sharp decrease in velocity of a stream when it leaves the uplands and breaks out onto the level plains.

Alluvium Material deposited by running water, such as a floodplain or a delta.

American megalopolis The urbanized northeastern seaboard of the United States that extends from New Hampshire to Virginia and is bounded by lower Appalachia and the Atlantic Ocean. Here almost 50 million people live in less than 2 percent of the nation's land area.

Anticline An upfold in the earth's crust.

Apartheid The official, legally instituted policy of racial separation that is promulgated in the Republic of South Africa.

Arable Suitable for cultivation by plowing or tillage.

Archipelago A group of islands, or an area of ocean or sea interspersed with islands.

Arroyo A stream-cut valley in dry lands. Usually it contains water only during and immediately after rainfall.

Artesian Referring to an underground water supply, usually under enough pressure to cause the water to rise to the surface in a well.

Atoll Coral reefs of elliptical or circular shape which enclose a lagoon.

Barrens Areas of poor soil that are covered by scant or scrubby vegetation.

Basalt A dark-colored, heavy rock of volcanic origin.

Benches Elevated areas of flat land, a topographic terrace or shelf.

Bilingualism A nation or a region in which two or more languages are spoken.

Bora A violent, cold, northerly wind of the Alps and Adriatic area.

Boreal forest The extensive northern (boreal) forests, predominantly of conifers, located in subarctic North America and Eurasia. The single largest continuous forest area on earth.

Braided stream A stream or river in which there are many joining and rejoining channels of water and in which sediments are usually being deposited.

Campos A wet-and-dry tropical region in central Brazil covered with scrub and grass vegetation.

Chernozem A class of soils found along the dry margins of black prairie lands, originally covered with a thick mat of grass roots at the surface; rich soils.

Chinook A warm, dry wind that moves with high speed down the leeward side of mountains, especially the east slope of the Rocky and Cascade Mountains.

Cirque An amphitheater-shaped, steep-walled head of a glaciated valley in mountains.

Combine A machine which harvests, threshes, and cleans grain while moving over the field.

Communism A political ideology and a system

of government that is theoretically characterized by the absence of social classes and by common ownership of the means of production.

Condominium A country or region governed by two or more powers; joint dominion or sovereignty.

Confluence The place at which the tributary of a river flows into the main stream.

Coniferous Cone-bearing, like the pine tree.

Conterminous states The forty-eight states of the United States that form a compact group. The conterminous states are bordered by the Atlantic Ocean on the east, the Gulf of Mexico and Mexico on the south, the Pacific Ocean on the West, and Canada on the north.

Conurbation Cities and towns so close together as to form a large, almost continuous urban area.

Convection A process of heating the atmosphere; rapid uplift of masses of moist air in a vertical or nearly vertical stream, which may result in convectional heating and thunderstorms.

Cordillera A combination of mountain ranges or a system that forms a large unit such as the Andes or Himalayas.

Cottage industry Certain types of light industry or manufacturing activities that are carried on in the home.

Currents, ocean Movement in a definite path of large quantities of ocean water such as the Gulf Stream or the Japanese Current.

Cyclone A region of low atmospheric pressure, about which the winds blow counterclockwise in the Northern Hemisphere or clockwise in the Southern.

Deciduous Shedding leaves during the winter or dry season, as the oak or hickory does.

Delta The deposit of sediment at the mouth of a river caused by decreasing velocity of the stream.

Diurnal Daily, recurring every day.

Diversified farming Farming in which two or more crops are produced each year; a combination of stock and crop farming.

Doldrums A transition zone between the trade wind belts, characterized by calms and weak winds.

Double cropping The growing of a summer and a winter crop in the same field in one year. A common farming method in South and Eastern Asia.

Drift, ocean A movement of oceanic circulation slower than a current, such as the North Atlantic Drift.

Dual economy The existence of a commercial and a subsistence economy in close proximity to each other. Many underdeveloped countries may be so characterized.

Elevation Height above sea level.

Entrepôt Refers to such places as commercial centers and ports where goods are received for distribution, transshipment, or repackaging.

Escarpment A long, high, steep face of rock; steep cliffs such as the "Break of the Plains."

Estancia A large stock ranch in Latin America.

Estuary The drowned mouth of a river; a river mouth where the tide meets the river current.

Extensive agriculture Use of land with a minimum of labor and outlay, such as wheat farming on the Great Plains.

Fall line The break in the bedrock between the hardrock uplands, or piedmont, and the sandy coastal plains in the eastern part of the United States.

Fallow Land tilled but not planted for a season or two. Weeds and insects are destroyed and water is conserved so that one crop may be produced every two or three years.

Farm fragmentation The practice of subdividing parcels of land into increasingly smaller units that ultimately become too small to be economic.

Fathom A depth measurement of water; one fathom equals 6 feet.

Faulting Slipping or breaking of rock structure under pressure. Many faults form scarps or steep cliffs.

Fazenda   A plantation in Brazil; for example, a coffee fazenda.

Fertile triangle   That part of the Soviet Union extending along the western border from the Gulf of Finland on the north to the Black Sea on the south and from these two places eastward to the city of Irkutsk.

Finca   A farm in Latin America; for example, a coffee farm in Colombia.

Fjord   A narrow inlet of the sea between high banks or mountains which has been gouged out by glaciers, as along the coast of Norway or southern Alaska.

Floodplain   An area along the sides of a river where the river overflows and deposits its sediments.

Foehn wind   A relatively warm, dry wind which descends a mountain front when a cyclonic storm causes air to cross the range from the opposite side of the divide.

Galleria forest   A forest along the banks of rivers that flow through grasslands. Tree crowns meeting over the stream give the impression of going through a green tunnel.

Geography   Refers to the organized body of knowledge dealing with spatial relations and processes on the earth's surface, studied at scales ranging from local to global. Geography is holistic in that it touches on almost every phase of human activity. The key criterion as to whether or not something is geographic is, can it be mapped?

Gross national product (GNP)   The total monetary value of all final goods and services produced in a country during any given year.

Growing season   The period of plant growth between the last killing frost in spring and the first killing frost in fall.

Hamada   Desert areas consisting mostly of bare, wind-scoured rock.

Hogan   An earth-covered lodge of the Navaho Indians.

Humus   Partly decayed plant and animal matter in the soil; the organic portion of the soil.

Hurricane   A tropical cyclone.

Hydrophyte   A plant that grows in wet situations.

Hydrosphere   The liquid sphere of the earth (chiefly water), such as the oceans, seas, bays, and lakes.

Igneous   Formed by solidification of molten material into rocks, such as granite and lava.

Intensive agriculture   The use of the land to produce as much as possible in a given area and period of time by the expenditure of much labor and capital.

Intercropping   The growing of more than one crop in the same field at the same time. The growing of a second crop between the rows of another crop.

Irredentist   A person who advocates the recovery of territory culturally or historically related to his nation but now subject to a foreign government.

Isobar   A line on a map connecting places of equal atmospheric pressure.

Isohyet   A line on a map connecting places of equal amounts of rainfall.

Isotherm   A line on a map connecting places of equal temperature.

Jungle   Dense undergrowth or second growth in the rainy tropical forests.

Kampongs   Native villages of Indonesia.

Karroo   A dry tableland of South Africa.

Karst   A land surface formed by the solution or underground erosion of limestone rocks, as in the Highland Rim or the Karst area of Yugoslavia.

Kibbutzim   Collective farms or settlements in Israel.

Lacustrine   Formed by or in a lake; is applied to a plain formed by deposition in an old lake bed.

Land reform   The legal or illegal replacement of large landholding systems by individual farmsteads or cooperatives. Usually results in subdividing large estates into smaller holdings and placing the land in the hands of persons who can expect to receive the fruits of their own labor.

Landlocked nations Those nations that do not have a direct outlet to an ocean or sea, or to a large lake that is directly connected to an ocean or sea.

Laterite soil Reddish clay soil of the subtropics and tropics where the process of laterization is dominant.

Leaching The removal of calcium and other elements from the soil by water seeping through it.

Lithosphere The solid part of the earth.

Llanos Tropical plains covered with tall grass, located in the interior Orinoco Basin of Colombia and Venezuela.

Loess Deposits of windblown soils, usually found in the zone between dry and humid areas or in front of the former limits of glaciation.

Mediterranean climate A type of climate, occurring on the western sides of continents between latitudes 30° to 40° and characterized by dry, hot, sunny summers and warm, moist winters.

Meltwater Water from the great ice age glaciers.

Metamorphic rocks Rocks changed and formed by heat and pressure.

Metropolitan area The densely populated area around a large city, such as the greater New York and greater London areas.

Migratory agriculture Primitive agriculture, usually in the tropics, in which the larger trees are killed and the brush burned, with only a few crops grown on one field before the field is abandoned.

Milpa A term applied to plots under migratory agriculture, especially in Africa and the Americas.

Monsoons Seasonal winds that reverse their direction. For example, the summer monsoon blows toward Asia, but the winter monsoon blows away from Asia.

Moraine An accumulation of unassorted clay, earth, stones, and other materials deposited by a glacier.

Muskeg A swampy area in the subarctic or arctic regions; usually has spruce and sphagnum moss.

Nagana The African name for trypanosomiasis, a cattle disease transmitted by the tsetse fly.

Naval stores Pitch, tar, and turpentine which are extracted from the pine forests of the middle latitudes.

Oblast A province within one of the states of the Soviet Union.

Okrug A district or circuit within a state of the Soviet Union.

Orographic Referring to precipitation caused when moist air is forced to move over mountains.

Outcrop A series of rocks exposed at the surface of the earth.

Outwash Material carried from a glacier by meltwater. Laid down in stratified deposits.

Paddy A field in which flooded rice is grown; unmilled or rough rice, whether growing or cut.

Pampa Grassland of South America, especially in Argentina.

Paramo A cold, treeless zone above the tree line in the highlands of the tropical Americas.

Paramos High, bleak plateaus or similar areas in mountains.

Pedalfer A class of leached soils in which most chemical elements except aluminum and iron have been removed.

Pedocal Soils of dry areas in which little leaching has taken place; much calcium still present in a zone of accumulation in the subsoil.

Peneplain An old land surface worn down by erosion to almost a plain.

Permafrost The permanently frozen layer of the earth beneath the surface as in the polar or subpolar regions.

Placer mining The removals of minerals from an alluvial, wind, or glacier deposit by the use of water.

Podsol Leached soils developed in humid and usually cool regions, especially under cover of conifers.

Polder A tract of lowlands reclaimed from the sea by dikes and dams, as in the Netherlands.

Polyes Basin meadows or large sinkholes.

Population resource ratio The interrelationships of population, resources, and technology and how they are manifested in living standards.

Primate city A city whose functional relationships with the contiguous hinterlands makes it the dominant city in a region.

Pulses The edible seeds of legumes, usually beans and peas.

Rain shadow The leeward sides of mountain ranges, characterized by low rainfall and sparse vegetation.

Region An area on the earth's surface that is delimited or defined on the basis of sameness when contrasted with other areas. Examples are political regions, agricultural regions, urban areas, and industrial areas.

Residual soil Soil that covers the bedrock from which it was formed.

Retting The process of soaking or exposing to moisture of certain fibers, such as flax or jute.

Rift valley A valley formed by faulting, in contrast to one formed by erosion.

Sawah A term used in place of *paddy* in Indonesia.

Scablands An extremely desolate region north of the Palouse country in the state of Washington, which is made up of wide, steep-sided, interlacing, dry channels.

Scarp A steep slope caused by faulting or erosion.

Scrub Vegetation chiefly of dwarf or stunted trees and shrubs, as in the "bush" area of Australia.

Sedimentary Formed from the accumulation of sediments deposited in water; a class of rock.

Selva A rain forest in Brazil.

Sensible temperature The combination of temperature and humidity as perceived by the body.

Sericulture The production of raw silk by the raising of silkworms.

Shatter belt Refers to the nations of Central and Eastern Europe that are characterized by physical, cultural, and political fragmentation.

Sirocco A hot, dry wind blowing from the Sahara.

Site A fixed plot of land on which a building or city rests. Refers to permanent physical characteristics of a small area or place.

Skerry A rocky isle, a reef.

Steppe Usually a plains area in a semiarid or subhumid region that is covered with short grass.

Subsistence agriculture Cultivation of the soil for the immediate needs of the family; very little, if any, of the produce is left over to be sold.

Syncline A downfold in the earth's crust.

Synclinorium Folds of strata dipping toward a common line; a series of folds that create a trough.

Taiga Extensive northern forests, predominantly of conifers, in North America, Europe, and Asia.

Tierra caliente Hot lands in tropical highlands of the Americas, up to elevations of 2,500 feet.

Tierra fria Cool lands in tropical highlands of the Americas, with elevations of about 6,500 to 12,000 feet.

Tierra templada Temperate lands in tropical highlands of the Americas, with elevations of about 2,500 to 6,500 feet.

Till Unsorted and unstratified glacial drift deposited directly by the melting of ice, with sand, gravel, clay, and boulders mixed.

Timber line The elevation on a mountain above which trees do not grow; it varies with latitude and exposure. A low timber line may result from deficiency in rainfall.

Trade winds Winds blowing toward the equatorial area from the subtropical high-pressure belts.

Transhumance The seasonal movement of herds between upland and lowland pastures.

Troposphere The atmospheric layer next to the earth.

Trust territory An area whose government is administered by a country that has been appointed by, and is responsible to, the United Nations.

Tundra An area poleward of the taiga, whose vegetation is composed of lichens, mosses, and low bushes.

Typhoon A hurricane near Asia.

Underemployment A situation in which tasks are insufficient to keep a number of persons occupied all year.

Velds Semiarid grasslands which may contain scattered trees, as in South Africa.

Wadi A channel or bed of a watercourse which is dry except during or immediately after a rain, as in desert or semidesert areas.

Xerophytes Plants that are adpated to growth in areas of drought.

Index

Index

Aalesund, Norway, 230
Aberdeen, Scotland, 242
Abidjan, Ivory Coast, 426, 458
Acajutla, El Salvador, 167
Acapulco, Mexico, 165
Addis Ababa, Ethiopia, 435
Adelaide, Australia, 604-606
Adirondack Mountains, 20, 22, 25
Adriatic Sea, 269, 272, 293, 302, 308, 312, 316, 317, 325, 336, 337, 341
Aegean Sea, 272, 298, 337, 339, 346
Afghanistan, 345, 347, 388, 513, 520-522, 524, 526, 533, 536, 542, 547, 550, 551

Ahaggar Mountains, 400
Ahmadabad, India, 540, 542
Åland Islands, 237
Alaska, 20, 36, 96-99, 104
Alaska Highway, 99, 133
Alaska Range, 96
Albania, 269, 299, 308, 312, 314, 317, 318, 322, 340-341
Albany, U.S., 22
Alberta, Plateau, 124, 125
Albuquerque, U.S., 65, 70, 73
Aleppo, Syria, 407, 408
Aleutian Islands, 96
Alexandria, A. R. E., 405
Alföld, 315, 316, 318, 329, 331-334, 337

Algeria, 388, 392, 394-400
Alice Springs, Australia, 604
All-American Canal, 75
Allegheny Plateau, 22, 29
Allegheny River, 29, 35
Alpena, U.S., 41
Alps, 221, 222, 250, 252, 254, 255, 258, 264, 265, 288, 289, 295, 312
Altai Mountains, 354, 359, 360, 372, 379, 477, 491
Altiplano, 199, 200
Amazon River, 11, 178, 179, 186, 190, 198, 199
Amazonia, 177, 181, 193, 200, 215

637

American Megalopolis, 15, 20, 27, 29-34, 43
Amman, Jordan, 407
Amsterdam, Netherlands, 247
Amu Darya, 354, 371
Amur River, 354, 359, 474, 475
Anatolia, 9, 269, 298, 303, 391, 419-421
Anchorage, U.S., 99
Andaman Islands, 514, 535
Andean Cordillera, 178, 179, 181, 182, 189, 190, 195, 198-200, 203
Andorra, 304-305
Angola, 282, 285, 433, 440, 452, 454, 455, 464
Ankara, Turkey, 303, 421
Annam Plain, 572
Annamese Cordillera, 558, 561, 566, 573
Annapolis-Cornwallis Valley, 114
Antarctica, 7-9, 12, 14, 595, 617
Anti-Atlas, 395
Antigua, 156
Antofagasta, Chile, 199
Antwerp, Belgium, 247, 248
Apennines, 288, 289
Apia, Western Samoa, 613
Appalachian coalfield, 26, 40
Appalachian Highlands, 20, 22, 28, 104
Aquitaine Basin, 253, 254
Ar Riyadh, Saudi Arabia, 414
Arab Republic of Egypt, 270, 301, 388, 391, 392, 394, 400, 402-406, 411, 412, 421, 445
Arabian Peninsula, 9, 387, 388, 413-418
Arabian Sea, 521, 522, 524
Arakan Yoma, 521, 558
Aral Sea, 354, 356, 361
Aravilla Range, 522
Arawak Indians, 147

Ardenne Plateau, 221, 248, 250, 255
Argentina, 175, 178, 181, 186, 200, 202, 204-207, 245
Arica, Chile, 199
Arkansas River, 48, 53, 70
Armenian S.S.R., 364, 367
Aroostook Valley, 24
As Sukhayrah, Tunisia, 398
Astrakhan, U.S.S.R., 356
Asuncion, Paraguay, 208
Aswan High Dam, 404, 406, 455
Atacama Desert, 181, 182, 202, 204
Atbara River, 404
Athabasca River, 125, 127
Athens, Greece, 297, 299-303
Atlantic City, U.S., 27
Atlantic Coastal Plain, 20, 22, 27, 29
Atlantic Provinces, 112-116
Atlas Mountains, 9, 395-396
Auckland, New Zealand, 607, 609
Australia, 4, 7, 9, 10, 224, 491, 505, 556, 561, 595-606, 611
  economic development, 601-605
  physical setting, 597-601
  population, 605-606
Avalon Peninsula, 113
Aymara Indians, 193, 197
Azerbaidzhan S.S.R., 367, 376
Azores Islands, 279, 283, 285
Azov-Podolian Shield, 352
Aztec Indians, 148, 150, 158

Baffin Island, 106, 133
Baghdad, Iraq, 304, 413
Bahia Blanca, Argentina, 184, 205
Bahrein, 414, 415, 417
Baikal Mountains, 354
Baja California, 142, 144, 146

Bakony Forest, 315
Baku, U.S.S.R., 356, 364, 376, 379, 383
Balboa, Panama, 170
Balcones Escarpment, 59
Balkan Highlands, 311, 315, 320, 339
Balkan Peninsula, 297, 303
Balsas Valley of Chiapas, 142, 164
Baltic Lowlands, 221
Baltic Moraine, 314, 320
Baltic Sea, 220, 231, 237, 308, 314, 325, 326, 345, 347, 348, 352
Baltimore, U.S., 20, 22, 24, 28, 30, 31, 34, 189
Baluchistan, 527, 533, 547
Bamako, Mali, 433
Banda Sea, 556
Banff National Park, 128
Bangalore, India, 542
Bangkok, Thailand, 563, 565, 566
Bangla Desh, 529
Bantustans, 467
Barbados, 15, 143, 156
Barcelona, Spain, 281, 283
Barents Sea, 346, 347, 381, 382
Barossa Valley, 603
Barranquilla, Colombia, 190, 192
Basel, Switzerland, 264
Basra, Iraq, 413
Baton Rouge, U.S., 56
Bauxite, 193, 227, 250, 254, 299, 324, 338, 379, 458, 459, 505, 573, 604
Bavarian Alps, 258, 262
Bay of Bengal, 522, 524
Bay of Fundy, 115
Bay of Naples, 289
Beirut, Lebanon, 407, 409
Belem, Brazil, 215
Belfast, Ulster, 243, 244
Belfort Gap, 250
Belgium, 4, 15, 220-222, 224,

Index **639**

Belgium:
227, 245, 248-250, 253, 263, 286, 446, 460
Belgrade, Yugoslavia, 338
Beliz, British Honduras, 167
Belo Horizonte, Brazil, 210-212
Belorussian S.S.R., 352, 369
Benelux, 15, 219, 245-249
Benelux Union, 224, 249
Benghazi, Libya, 402
Benguela Current, 435
Bergen, Norway, 230
Bergslagen District, 232, 233
Bering Strait, 345
Berlin, Germany, 15, 256, 257, 260, 304
Bern, Switzerland, 264
Bhutan, 513, 516, 548, 550
Biafra, 429
Bicol Plain, 584
Big Bend National Park, 71
Big Horn Mountains, 67
Bihar Mountains, 315, 334
Bilbao, Spain, 281, 284
Bio-Bio River, 204
Biosphere, 16
Birmingham, U.K., 241
Birmingham, U.S., 56
Bismarck Archipelago, 595, 597, 610, 614
Bitterroot Valley, 67
Black Hills, 46, 48, 50
Black (Song Bo) River, 558
Black Sea, 269, 303, 308, 315, 325, 326, 335, 339, 340, 346, 348, 352, 353, 356, 361, 370, 373, 383, 419, 420, 422
Blue Grass Region, 35, 55, 59
Blue Mountains of Jamaica, 143, 154
Blue Nile River, 403, 404
Blue Ridge Mountains, 60
Bluefields, Nicaragua, 155, 168

Bogota, Colombia, 190-192
Bohemian Basin, 311, 312, 315, 329, 330
Bohemian Forest, 312, 329
Bohemian Rim System, 311, 312
Bolivia, 175, 178, 181, 193, 196, 197, 199-200
Bombay, India, 540, 542
Bonin Islands, 500, 502
Bonn, Germany, 262
Bordeaux, France, 252, 253
Boreal Forest, 108, 122
Borneo, 554, 556, 558, 560, 581
Bosporus, 302-304
Boston, U. S., 22, 24-26, 30-32, 51
Boston Mountains, 62
Botswana, 435, 440, 455, 464, 465
Brahmaputra River, 477, 522, 539
Brasilia, Brazil, 177, 210, 214, 215
Bratislava, Czechoslovakia, 329, 330
Brazil, 11, 15, 175, 177, 178, 181, 182, 184-186, 200, 207-215, 345, 452
  Amazonia, 215
  East Central Brazil, 210-213
  Interior Plateaus and Plains, 214-215
  Northeast Brazil, 213
  South Brazil, 213-214
Brazilian Highlands, 179, 181, 210
Brazzaville, People's Republic of the Congo, 460
Break of the Plains, 52
Bremen, Germany, 257, 260
Bremerhaven, Germany, 260
Brenner Pass, 265, 289
Brisbane, Australia, 604-606
British Honduras, 143, 165, 172
British Isles, 222, 238-246
  economic development, 240
  Great Britain, 240-243

British Isles:
  Ireland, 243-244
  natural resources, 239
  physical setting, 238-239
  problems and trends in, 244-245
Brittany, 220, 221, 250, 254
Brno, Czechoslovakia, 330
Broken Hill, Australia, 604
Brooks Range, 97, 106
Brunei, 575, 576, 586
Brussels, Belgium, 15, 248
Bucharest, Romania, 335
Budapest, Hungary, 309, 325, 329, 332
Buenaventura, Colombia, 191
Buenos Aires, Argentina, 178, 201, 204-207
Buffalo, U.S., 22, 28, 29, 35, 41, 43
Bug River, 326
Bulgaria, 303, 308, 315-317, 322, 339-340
Burgas, Bulgaria, 340
Burma, 12, 476, 521, 554, 558, 560, 561, 569, 573-575
Burundi, 15, 463

Cabora Bassa Dam, 455, 465
Cagayan Valley, 584
Cairo, A.R.E., 387, 405
Cajon Pass, 85
Calabozo, Venezuela, 189
Calcutta, India, 542
Calgary, Canada, 111, 128
Cali, Colombia, 191
California Current, 10
California Valleys and Coast Ranges, 82-90
  Central Valley, 83-85
  Coast Ranges, 88-90
  Southern California, 85-88
Callao, Peru, 197

Cambodia, 554, 558, 561, 566-570
Cameroun, 452, 453, 460
Canada, 11, 15, 18, 20, 102-137, 155, 250, 491, 505, 544, 596
   Atlantic Provinces, 112-116
   Canadian Shield, 119-124
   Cordillera of British Columbia and Yukon, 128-133
   Great Lakes and St. Lawrence Lowlands, 116-119
   Interior Plains, 124-128
   Northwest Territories, 133-135
   physical setting, 104-109
   population distribution, 109-112
Canadian Shield, 20, 40, 119-124, 133
Canberra, Australia, 597, 605
Cantabrian Mountains, 275
Canton, China, 475, 476, 480, 487, 488, 490, 491
Cape Agulhas, 429
Cape Breton Island, 114
Cape Chelyuskin, 345
Cape Cod, 20, 24
Cape of Good Hope, 9, 445
Cape Town, South Africa, 446, 466
Cape York, 599
Caracas, Venezuela, 188, 190
Cardiff, U.K., 243
Cardomon Mountains, 568
Caribbean countries, 186-193
   Colombia, 190-192
   Guyana, Surinam, and French Guiana, 192-193
   Venezuela, 186-190
Caribbean Free Trade Association, 172
Caribbean Lowlands, 145, 148, 169, 190
Caroline Islands, 502 610, 616

Caroni River, 187, 189
Carpathian Mountains, 220, 311, 312, 314, 318, 320, 327, 329, 332, 334, 347, 353
Carquines Strait, 83, 89
Cartagena, Colombia, 190
Casablanca, Morocco, 397
Cascade Mountains, 71, 76, 78-81, 90, 91, 106
Caspian Sea, 9, 10, 346, 352, 353, 354, 381, 389, 420
Catskill Mountains, 22, 31
Cau River, 572
Cauca Valley, 191
Caucasus Mountains, 9, 348, 353, 361, 365, 378
Cayenne, French Guiana, 192
Celebes, 581
Celebes Sea, 556
Central African Republic, 455, 460, 461
Central America, 139-143, 145, 148-150, 165-172
   British Honduras, 167
   Costa Rica, 169-170
   El Salvador, 167-168
   Guatemala, 166-167
   Honduras, 168
   Nicaragua, 168-169
   Panama and the Canal Zone, 170-171
Central American Common Market, 165, 172
Central European Lowlands, 259
Central Industrial Region, U.S.S.R., 375, 377, 378, 380, 381
Central Lowlands, Australia, 597, 598
Central Lowlands, U.S., 20, 35-45, 52, 100
   agriculture, 36-40
   forests, 41

Central Lowlands, U. S.:
   industries and cities, 42-45
   minerals, 40
   recreation, 45
   transportation, 41-42
Central Mountain Block, China, 476, 478
Central Plain, Philippines, 584
Central Plain, Thailand, 563-565
Central Plateau, Mexico, 146, 160
Central Plateau, Switzerland, 263, 264
Central Valley, California, 83-85, 88
Cerro de Pasco, Peru, 198
Ceylon, 513, 520, 521, 524, 529-531, 539, 544
Chad, 8, 442, 460, 461
Chao Phraya, 558, 564
Charlottetown, Canada, 115
Chelyabinsk, U.S.S.R., 350, 353, 378, 380
Cherrapunji, India, 526
Chesapeake Bay, 22, 25, 30
Chicago, U.S., 15, 39, 42-44, 51, 220
Chile, 177-180, 196, 200, 202-204, 505, 611
Chiloe Island, 204
Chimbote, Peru, 197
Chin Ling, 478, 480, 490
China, 4, 9, 12, 14, 15, 18, 341, 344, 345, 347, 362, 387, 463, 471-473, 491, 496-499, 504, 513, 521, 547, 549, 572
   climate and regional differentiation, 477-480
   communications and trade, 490-491
   historical development, 480-485
   manufacturing, 487-489
   relief features, 473-477

China:
    resources, 486-487
China Sea, 563, 569
Chindwin River, 558
Chittagong, Pakistan, 542
Cholon, South Vietnam, 571
Christchurch, New Zealand, 609
Chungking, China, 483, 488
Chuquicamata, Chile, 202, 203
Cincinnati, U.S., 43
Ciudad Bolivar, Venezuela, 189
Ciudad Guyana, Venezuela, 189
Clay Belt, 123, 133
Cleveland, U.S., 41-44
Cluj, Romania, 335
Clyde River, 242
Coachella Valley, 75
Coal, 29, 68, 95, 100, 114, 127, 128, 164, 192, 198, 206, 213, 219, 227, 229, 232, 233, 237, 239, 240, 242, 245, 246, 252, 257, 258, 260, 263, 265, 280, 285, 292, 299, 324, 329, 330, 332, 335, 340, 348, 350, 375, 376, 379, 422, 458, 486, 488, 494, 498, 505, 508, 539, 573, 603, 608
Coatzacoalcos, Mexico, 163
Cobán, Guatemala, 166, 167
Cochabamba, Bolivia, 199, 200
Coffee, 153-155, 167-169, 184, 187, 188, 192, 200, 210-211, 214, 452, 458-460, 462, 463, 539, 560, 571, 573, 583, 615
Colchis Lowland, 361, 371
Cologne, Germany, 262
Colombia, 8, 138, 170, 175, 178, 179, 181, 185, 186, 189-193
Colorado Plateau, 71-73
Colorado River, 70, 75, 87

Columbia Intermontane Province, 76-79
Columbia Plateau, 76, 80
Columbia River, 67, 77-81, 90-92, 131
Common Market (see European Economic Community Market)
Comodoro Rivadavia, Argentina, 206
Conarky, Guinea, 433
Concepcion, Chile, 202, 204
Concepts, geographic (see Geographic concepts)
Congo (Brazzaville), 460, 461
Congo (Kinshasa), 433, 445, 447, 448, 452, 454, 460, 461
Congo Basin, 437, 438, 442, 451, 460
Congo River, 429, 455, 460
Connecticut River, 24
Constanta, Romania, 335
Continental Drift Theory, 7-8, 430
Continental Islands, 9
Continental Shelf, 25
Cook Islands, 610-612
Copenhagen, Denmark, 232, 234, 236
Copper, 35, 67, 80, 113, 121, 131, 152, 164, 198, 200, 202-204, 243, 249, 281, 324, 330, 332, 338, 341, 379, 401, 419, 422, 454-456, 460, 461, 505
Coquimbo, Chile, 202
Cordillera of British Columbia and Yukon, 104, 106, 128-133
Cordillera Mérida, Venezuela, 181, 188, 189
Cordoba, Argentina, 206
Cordoba, Spain, 281

Corfu Island, 297, 298
Corinth Canal, 297
Corn, 24, 35, 36, 38, 52, 54, 117, 142, 168, 184, 188, 204, 205, 282, 298, 335, 337, 340, 341, 372, 404, 451, 452
Corn Belt, 36, 38-39, 44, 47, 55
Cornwall Peninsula, 238, 241
Corpus Christi, U.S., 55
Corsica, 269
Costa Rica, 139, 143, 148, 165
Cotabato Lowland, 584
Cottage Industries, 540
Cotton, 50-53, 83, 85, 142, 158, 164, 168, 188, 197, 200, 206, 211, 213, 248, 255, 263, 282, 298, 317, 322, 329, 331, 333, 335, 337, 340, 371, 372, 380, 381, 393, 404, 409, 422, 423, 452, 459, 460, 463, 487-489, 505, 508, 515, 538, 540, 542, 543, 561, 565, 573, 574, 603
Cotton Belt, 53, 55
Council for Mutual Economic Assistance, 322
Crater Lake National Park, 80
Crete, 270, 298
Crimean Peninsula, 353, 356, 361, 378
Cuba, 138, 142, 146, 150-152
Cuenca, Ecuador, 195
Cultural landscape, 5
Cultural tradition, 17
Cumberland Escarpment, 61
Cumberland Plateau, 55, 61
Cuxhaven, Germany, 260
Cuyuna Range, 40
Cuzco, Peru, 196, 198
Cyprus, 300, 388, 391, 392, 418-419

Czechoslovakia, 4, 257, 308, 311, 316, 317, 320-322, 329-331, 347

Dahomey, 459
Dairy farming, 24, 37, 39, 83, 84, 91, 92, 117, 132, 204, 211, 229, 234, 240, 242, 244, 247, 248, 254, 259, 262, 264, 299, 330, 373, 504, 599, 602, 608, 609
Dakar, Senegal, 429
Dallas, U. S., 58, 59
Damascus, Syria, 407, 408, 410
Dang Raek Mountains, 563
Danube River, 11, 226, 256, 258, 262, 265, 266, 314, 315, 320, 323, 331, 332, 334, 338, 339, 347
Danzig Bay, 314
Dar es Salaam, Tanzania, 456
Dardanelles, 269, 299, 303
Darling River, 595, 598
Darwin, Australia, 604
Davis Strait, 136
Dead Sea, 10, 407, 409
Death Valley, 64, 76
Deccan, 9, 15, 524, 537
Delaware Bay, 22, 25, 30
Delhi, India, 522
Delmarva Peninsula, 24
Democratic People's Republic of Korea (see Korea)
Denmark, 135, 153, 221, 226, 227, 229, 231, 234-236, 263
Denver, U.S., 49, 70
Detroit, U.S., 39, 43, 44, 51
Detroit River, 41, 43
Diamonds, 455, 458, 460, 463
Dinaric Alps, 311, 312, 320, 337
Dnepr River, 326, 352, 378, 379, 383

Dobrogea Platform, 316, 318, 334
Dominican Republic, 138, 139, 143, 150, 153, 154
Don River, 352
Donbas (see Donets Basin)
Donets Basin, 348, 375, 378, 380, 383
Dong Nai, 569
Donner Pass, 81, 85
Door Peninsula, 40, 43
Dortmund, Germany, 261
Douro River, 274, 275, 278, 283, 285
Drava River, 315
Dravidian, 514, 517, 529
Dresden, Germany, 257, 261
Dublin, Ireland, 240, 244
Duluth, U.S., 41, 43, 118
Dunaujváros, Hungary, 325
Dundee, U.K., 242
Dunedin, New Zealand, 609
Durango, Mexico, 164
Durban, South Africa, 466
Durëss, Albania, 341
Dusseldorf, Germany, 261
Dzungarian Basin, 476-478, 491

East Chicago, U.S., 40
East China Sea, 475
East Germany (see Germany)
Easter Island, 610-612
Eastern Europe, 308-343
  countries of, 325-341
  economic development, 322-325
  human relationships, 318-322
  physical setting, 311-318
Eastern European Communist Bloc, 257
Eastern Ghats, 524
Eastern Highlands, Australia, 9, 597-600, 602

Eastern interior coalfields, U.S., 40
Ebro River, 273, 275, 276
Ecosystems, 16
Ecuador, 178, 191, 193, 195-196, 198
Edinburgh, U.K., 242
Edjeleh, Algeria, 398
Edmonton, Canada, 111, 127, 128
Edwards Plateau, 52, 55
Egypt (see Arab Republic of Egypt)
Eire, 240, 243
El Paso, U.S., 71
El Salvador, 148, 165
El Tofo, Chile, 202
Elbe River, 222, 258-260, 315, 331
Elbruz Mountains, 419-421
Ellesmere Island, 103, 106, 133, 135
Ellice Islands, 611, 612, 617
Encarnación, Paraguay, 208
England, 219, 221, 233, 238, 240, 264
Equatorial Guinea, 460
Erie Canal, 22, 24, 28
Erzgebirge, 261
Esbjerg, Denmark, 234
Esmeraldas, Ecuador, 196
Essen, Germany, 261
Estonia, 347, 366, 367
Ethiopia, 13, 387, 388, 430, 435, 439, 442, 445, 452, 454, 462
Euphrates River, 408, 409, 412
European Coal and Steel Community, 224, 250
European Economic Community Market, 220, 224, 233, 245, 248, 249, 255-257, 259, 260, 262, 264, 266, 286, 300

European Free Trade Association, 220, 224, 245, 264, 266
European Plain, 221, 222, 311, 314, 315, 318
Europoort, Netherlands, 248
Everglades, 52, 53

Fairbanks, U.S., 98, 99
Fall Line, 22, 29, 59
Fergana Valley, 354
Fernando Po, 460
Fés, Morocco, 398
Fiji Islands, 13, 595, 610, 611, 614, 616
Finger Lakes, 22
Finland, 222, 227, 236-238, 347
Florence, Italy, 289, 297
Flores Sea, 556
Fort Peck Dam, 48
France, 4, 219, 220, 222, 227, 245, 249-256, 286, 288, 294, 446, 463, 491, 611
   economic development, 252-253
   natural resources, 250-252
   physical setting, 250
   population, 249
   regions, 253-255
   trade and problems, 255-256
Frankfurt, Germany, 262
Fraser River, 111, 130-133
Fredericton, Canada, 115
Frederikshaven, Denmark, 234
French Guiana, 186, 192-193
Frisian Islands, 246
Front Range, 70
Ft. Gouraud, Mauritania, 401

Gaberone, Botswana, 434
Gabon, 453, 454, 460, 461
Galapagos Islands, 195
Gällivare, Sweden, 233
Galveston, U.S., 55
Gambia, 442, 459
Ganges Delta, 531, 533
Ganges Plain, 520-522
Ganges River, 11, 15, 522
Garonne River, 222, 252, 253
Gary, U.S., 43
Gaspé Peninsula, 104, 112
Gdańska, Poland, 327, 328
Geneva, Switzerland, 264
Genoa, Italy, 286, 292, 293, 295
Geographic concepts, 5-7
   areal distinction, 6
   constant change, 6
   interrelationships, 6
   life layer, 6
   regionalism, 6
   resource limitation, 6
Geography, 4-5
Georgetown, Guyana, 192
Georgian Bay, 121
Georgian S.S.R., 367, 373
Germany, 15, 16, 199, 220-222, 227, 233, 245, 249, 250, 252, 255-263, 266, 326, 446
   divided, 257
   economic development, 259
   natural resources, 258-259
   physical setting, 257-258
   post World War II changes and problems, 256-257
   regions, 259-262
   trade and problems, 262-263
Ghana, 446, 447, 452, 453, 455, 459
Gibraltar, 269, 270, 280, 285-286
Gila River, 73, 75
Gilbert Islands, 610, 611, 617
Glacier National Park, 68
Glasgow, U.K., 242
Glen Canyon Dam, 71
Gloucester, U.S., 25
Gobi, 9, 476, 478, 492
Godthaab, Greenland, 136
Gold, 453-454, 458, 460, 465, 498, 539, 603, 608, 615
Gorki, U.S.S.R., 348, 358, 364, 380, 381
Göteborg, Sweden, 232
Gotland, 221, 231, 232
Gran Chaco, 179, 181, 199, 206, 208
Granada, Spain, 281
Grand Banks, 113, 279
Grand Canyon, 71
Grand Coulee, 77, 81
Grand Coulee Dam, 31, 78, 118
Grand Tetons, 68
Grapes (see Vineyards)
Great Barrier Reef, 599
Great Basin, 75-76, 81
Great Bear Lake, 120, 133
Great Britain, 9, 220, 226, 227, 234, 238-240, 244
Great Lakes, 10, 28, 29, 35, 36, 41, 45, 107, 119
   and St. Lawrence Lowland, 116-119
Great Plains, 19, 35, 36, 45-50, 52, 54, 68, 70, 100, 109
   agriculture, 47-48
   minerals, 49
   physical setting, 45-46
   recreation, 49-50
   settlement, 49
Great Rif Valley, 430
Great Slave Lake, 120, 133, 134
Great Smoky Mountains, 60, 61
Great Valley, California, 81
Great Valley, Virginia, 61
Greater Antilles, 139, 143, 147-155
   Cuba, 150-152
   Hispaniola, 153-154
   Jamaica, 154-155
   Puerto Rico and Virgin Islands, 152-153

Greece, 268, 270, 271, 297-303, 315, 387, 388, 418
  agriculture, 298-299
  cities, 301-303
  fisheries, 301
  problems of land and population, 297-298
  resources and industry, 299-300
  trade and transportation, 300-301
Greenland, 9, 106, 135-136, 279
Growing season, 37
Guadalquivir River, 275
Guadiana River, 275
Guam, 594, 611, 616, 617
Guarani Indians, 202, 208
Guatemala, 143, 148, 165, 168
Guatemala City, Guatemala, 167
Guayaquil, Ecuador, 195, 196
Guayas-Daule River, 195-196
Guiana Highlands, 179, 181, 189, 192
Guinea, 442, 459
Gulf of California, 142
Gulf Coast Oil Field, 56
Gulf Coastal Plain, 35, 51, 52, 58
Gulf of St. Lawrence, 115, 120, 121
Gulf of Tonkin, 561
Gulf Stream, 10, 112
Guri Dam, 189
Guyana, 186, 192-193

Hadhramaut, 414
Hague, Netherlands, 246
Hainan Island, 477, 480
Haiphong, North Vietnam, 571, 572
Haiti, 138, 143, 150, 153, 154
Hamburg, Germany, 257, 260
Hamilton, Canada, 117-119
Han River, 475
Hanoi, North Vietnam, 571, 572
Harz Mountains, 258, 261
Havana, Cuba, 150-152
Hawaii, 9, 13, 20, 36, 93-95
Helsinki, Finland, 237
Hercynian Mountains, 258
High Plains, 46, 52-54
Himalaya Mountains, 477, 514, 521, 522, 531, 536, 547-549
Hindu Kush, 521
Hindustan, 12
Hispaniola, 150, 153-154
Hobart, Australia, 604-606
Honduras, 139, 148, 165, 166
Hong Kong, 475, 477, 480, 488, 491, 508-509, 605
Honolulu, U.S., 95
Hoover Dam, 72, 118
Houston, U.S., 53, 58
Houston Ship Canal, 58
Hsi Kiang, 14
Hsiang River, 475
Huachipato, Chile, 202
Hudson Bay, 104, 106, 120, 133, 134
Hudson Bay Lowland, 105
Hudson-Mohawk Lowland, 35
Hudson River, 21, 22, 28, 30
Hugoton Gas Field, 49
Humboldt (Peruvian) Current, 10, 196, 197
Humboldt River, 76
Hunedoára, Romania, 325
Hungarian Plain, 311, 312, 314, 315
Hungary, 308, 316-318, 320-322, 331-333, 347
Hwang Ho, 11, 14, 475, 480, 485, 487
Hydrosphere, 10-11

Iberian Peninsula (Spain and Portugal). 9, 268, 269, 273-285
  agricultural land use, 276-279
  fisheries, 279
  industry, 280-283
  minerals, 279-280
  overseas territories, 285
  relief features, 275-276
  towns and cities, 283-285
  trade and transportation, 283
Iceland, 9, 227, 236
Ifni, 285, 400
Iguassu Falls, 206
Ili Valley, 477
Imperial Valley, 71, 75
Inchon, Korea, 499
India, 14, 15, 18, 362, 505, 513-544, 550, 551
  early peoples and cultures, 514-517
  economic development, 535-544
  Hinduism and Buddhism, 517-519
  human mosaic, 526-535
  Indo-Islamic synthesis, 519-521
  physical stage, 521-526
Indianapolis, U.S., 43
Indonesia, 13, 558, 570, 580-584, 610
Indonesian Archipelago, 554
Indus Plain, 514, 521, 522
Indus River, 11, 517, 522, 546
Indus Valley, 514, 515, 517
Industrial Revolution, 2, 4
Inland Sea of Japan, 15, 500, 501, 505
Inner Mongolia, 478, 485, 487
Interior highlands, 62-63
  Arkansas River Valley, 62
  Ouachita Mountains, 62
  Ozark Plateau, 62-63

Index **645**

Interior Plains, 35, 52, 105, 106, 124-128
Intermontane plateaus, basins, and ranges, 71-79
   Colorado Plateau, 71-73
   Columbia Intermontane Province, 76-79
   Great Basin, 75-76
   Southwestern Basin and Range Province, 73-75
Ionian Sea, 269, 298
Iquique, Chile, 182
Iran, 9, 347, 387, 388, 392, 394, 417, 419-423
Iraq, 388, 391, 392, 394, 412-413, 419, 422
Ireland, 221, 238, 243-244
Irian Barat, 556, 558, 560, 580
Irkutsk, U.S.S.R., 376
Iron Curtain, 259, 260
Iron ore, 25, 29, 35, 40, 41, 56, 100, 113, 118, 121, 131, 149, 152, 164, 165, 187, 189, 192, 198, 202, 206, 212, 227, 229, 231-233, 239-241, 243, 245, 248, 250, 252, 254, 255, 258, 263, 265, 279, 284, 292, 299, 324, 328, 329-331, 335, 348, 350, 378, 379, 401, 422, 458, 459, 465, 486, 498, 505, 508, 540, 542, 573, 604, 608
Irrawaddy River, 558, 574
Irtysh River, 354
Isfahan, Iran, 421
Ishikari Plain, 500
Isle Royal National Park, 45
Israel, 388, 391, 392, 394, 406-412
Istanbul, Turkey, 303, 304, 421
Isthmus of Panama, 8, 170
Itabira, Brazil, 212

Italy, 4, 249, 253, 268, 269, 271, 272, 275, 286-297, 312, 446
   agriculture, 289-291
   cities, 294-297
   industry, 291-294
   relief features, 288-289
Ivory Coast, 446, 452, 453, 455, 459

Jablonica Pass, 312
Jalapa, Mexico, 163
Jamaica, 141, 150, 152, 154-155, 157
James Bay Lowland, 105
Jamshedpur, India, 540, 543
Jämtland, 233
Japan, 12, 15, 16, 131, 199, 204, 347, 354, 387, 448, 471, 472, 482, 483, 487, 499-508, 566, 581, 604, 605, 616
   agriculture, 504-507
   climate, 500-501
   communications and commerce, 507-508
   historical development, 501-502
   industry, 504-507
   population and livelihood, 502-503
   relief features, 499-500
Japanese Current, 10, 496, 501
Jasper National Park, 128
Java, 556, 560, 581-583
Jelep La Pass, 550
Jerusalem, Israel, 411, 412
Jhelum River, 546
Jiddah, Saudi Arabia, 416
Joban coalfield, 505
Johannesburg, South Africa, 426, 442, 443, 454, 466
Joplin District, 63

Jordan, 388, 406-412
Jordan Valley, 407, 408, 412
Juan de Fuca Strait, 90
Juneau, U.S., 98, 99
Juras, 249, 250, 254, 264
Jute, 537, 538, 542, 543, 565, 571
Jutland Peninsula, 220, 234

Kabul Basin, 522
Kagoshima, Japan, 501
Kaima Plateau, 496, 499
Kalahari Desert, 433, 435, 440, 442, 464
Kalimantan, 578, 581
Kaliningrad, U.S.S.R., 345
Kamchatka Peninsula, 354, 500
Kanpur, India, 542
Kansas City, U.S., 43
Kara Kum, 354
Karaganda Basin, 375, 376, 379
Karakoram Range, 521
Karbala, Iraq, 392
Karelian Isthmus, 347, 352
Karst Highlands, 312
Kashmir, 521, 524, 533, 536, 545-548
Katanga, 454, 460
Katmandu, Nepal, 548, 549
Katmandu Basin, 522
Kazakh S. S. R., 363, 364, 367
Kazakh Upland, 354, 371, 373, 375
Kenya, 433, 435, 446, 448, 452, 462
Khangai Mountains, 491-492
Kharkov, U.S.S.R., 348, 364, 380
Khartoum, Sudan, 403, 404
Khobdo Basin, 491, 492
Khyber Pass, 547
Kiev, U.S.S.R., 348, 359, 364, 381

Kii Peninsula, 501
King William Islands, 106
Kingdom of Tonga, 611, 612, 614
Kinshasa, Congo, 426, 455
Kingston, Jamaica, 155
Kirgiz S.S.R., 367
Kirkenes, Norway, 231
Kirkland Lake, Canada, 121, 123
Kiruna, Sweden, 233
Kitimat, Canada, 131
Klamath Mountains, 85, 90
Kobe, Japan, 508
Kola Peninsula, 352, 379
Kompong Som, Cambodia, 568
Königsberg, U.S.S.R., 347
Korea, 4, 347, 471, 495-499
   agriculture, 496-497
   communications, 499
   fisheries, 497
   historical development, 497-498
   relief features, 495-496
   resources and industry, 498-499
Kota Kinabalu, Malaysia, 578
Kraków, Poland, 326, 328
Kristiansund, Norway, 230
Krivoi Rog, 378
Kuibyshev, U.S.S.R., 348, 364, 378
Kunlun Range, 478, 521
Kurile Islands, 354, 500, 502
Kuwait, 391, 413, 415, 417, 418
Kuznetsk Basin, 350, 354, 375, 379, 380, 383
Kwanto Plain, 500, 503-505
Kweichow Plateau, 480
Kyoto, Japan, 504
Kyzyl Kum, 354

La Guaria, Venezuela, 188
La Paz, Bolivia, 199, 200
La Plata, Argentina, 206
La Serena, Chile, 182, 202, 203
Labrador, 106, 108, 113, 120
Labrador Current, 10
Lagōa dos Patos, 214
Lake Albert, 403
Lake Baikal, 354
Lake Balaton, 315, 332
Lake Biwa, 504
Lake Bonneville, 76
Lake Champlain, 20, 22
Lake Champlain Lowland, 21, 24
Lake Erie, 22, 28, 35, 40, 41, 43, 104, 116-118
Lake Eyre, 597
Lake Huron, 104, 116
Lake Malawi, 463
Lake Maracaibo, 186, 187
Lake Mead, 72
Lake Michigan, 40-43, 118
Lake Nasser, 404
Lake Nicaragua, 143, 169
Lake Ontario, 20, 22, 104, 116-118
Lake Poopo, 199
Lake Powell, 71
Lake St. Clair, 43
Lake St. John, 122-124
Lake Superior, 41, 120, 121
Lake Tanganyika, 430
Lake Titicaca, 10, 199
Lake Victoria, 403, 439, 442, 463
Lake Winnipeg, 124, 127
Lancashire, U.K., 239, 241
Landes District, 252, 253
Laos, 476, 554, 558, 560, 561, 566-570
Lassen Peak, 79-81
Latin American Free Trade Association, 172
Latrobe Valley, 603
Latvia, 347, 366, 367
Lebanon, 388, 392, 394, 406-412
Leeds, U.K., 241
Leichou Peninsula, 477, 490
Leipzig, Germany, 261
Lena River, 354, 376
Leningrad, U.S.S.R., 347, 348, 356, 358, 359, 364, 376-378, 381
Lesotho, 464, 465
Lesser Antilles, 139, 143, 148, 150, 152, 155-158
Liao Ho Plain, 475
Liaotung Peninsula, 476
Liberia, 438, 453, 455
Libya, 8, 301, 387, 388, 391, 392, 400-402
Liechtenstein, 219, 222, 264-265
Life layer, 16-17
Lille, France, 253
Lim Fjord, Denmark, 234
Lima, Peru, 197
Limpopo River, 464
Linyu, China, 475
Lisbon, Portugal, 270, 276, 279, 281, 283, 285
Lithosphere, 7-10
   Continental Drift Theory, 7-8
   continental topography, 9
   islands, 9-10
Lithuania, 347, 366, 367
Liverpool, U.K., 241
Ljubljana, Yugoslavia, 338
Llanos, 177, 192
Łódz, Poland, 328
Loess Plateau, 476, 478, 479, 483
Lofoten Islands, 230
Loire River, 222, 252, 253
London, Canada, 118
London, U.K., 15, 239-241, 244, 283, 309
Longs Peak, 70
Lorraine Iron Ore Field, 246, 250, 252, 258, 261
Los Angeles, U.S., 76, 83, 85-88, 162, 220
Luang Prabang, Laos, 566
Lubbock, U.S., 53

Luleå, Sweden, 231-233
Lusaka, Zambia, 433
Luta, China, 483, 491
Luxembourg, 15, 222, 224, 227, 245, 249, 250
Luzon, 584-586
Lyon, France, 252, 254

Ma River, 571, 572
Macao, 285
Mackenzie Mountains, 106, 128
Mackenzie River, 106, 125, 133-135
Madagascar, 13, 440, 461
Madras, India, 527, 542
Madrid, Spain, 276, 280, 281, 283
Magdalena River, 179, 186, 190, 192
Maghrib, 388, 389, 394-400
Magnitogorsk, U.S.S.R., 350, 378-380
Malabar Coast, 526, 527, 531, 533, 539
Malagasy Republic, 13, 447, 452, 463
Malawi, 433, 463
Malay Peninsula, 12, 554, 557, 558
Malaysia, 554, 561, 575-579
Mali, 442, 458, 459
Malmo, Sweden, 232
Malta, 305-306
Man-land ratio, 14
Managua, Nicaragua, 169
Manaus, Brazil, 215
Manchester Ship Canal, 241
Manchuria, 475, 480, 485, 486, 490
Manchurian Plain, 474, 479, 483
Mandalay, Burma, 574
Manila, Philippines, 586
Manitoba Lowland, 106, 124

Maracaibo Lowlands, 179, 187, 188
Mariana Deep, 7
Mariana Islands, 502, 610, 616
Maritime Alps, 289
Maritsa Valley, 311, 315, 317, 339, 340
Marquesas Islands, 612, 614
Marseille, France, 254
Marshall Islands, 502, 610, 616
Martinique, 9, 156
Massif Central, France, 250, 252, 254
Matto Grosso, 215
Mauritania, 388, 391, 400-402
Mauritius, 461, 463
Maya Indians, 148, 150, 158, 163, 166
Mecca, Saudi Arabia, 392, 414, 416, 458
Medellin, Colombia, 191
Medina, Saudi Arabia, 416
Megalopolis (see American Megalopolis)
Mekong River, 11, 558, 563, 564, 566, 568, 569, 573
Melanesia, 610, 614-616
Melbourne, Australia, 604-606
Mérida, Mexico, 163
Mesa Verde National Park, 72
Mesabi Range, 40, 44
Meseta Central of Costa Rica, 169
Metropolitan Aqueduct, 87
Mexico, 8, 138-149, 158-165, 171-173
  Central Mexico, 160-162
  Gulf Coast, 162-163
  North, 163-164
  North Pacific, 164
  South Pacific, 164-165
Mexico City, Mexico, 158, 160-162, 165
Mezzogiorno, 287, 288, 295
Miami Beach, U.S., 59, 153

Micronesia, 610, 616-617
Mid-Continent petroleum field, 40, 49, 56
Middle America, 138-173, 177
  Central America, 165-171
  common characteristics, 140-142
  cultural differences, 147-150
  Greater Antilles, 150-155
  Lesser Antilles and Trinidad, 155-158
  Mexico, 158-165
  physical setting, 142-146
Milan, Italy, 286, 292, 295, 296
Milk River Plain, 48
Milwaukee, U.S., 43, 51
Mindanao, 584
Minneapolis, U.S., 39, 43
Minusinsk Basin, 354
Miskolc, Hungary, 332
Mississippi Embayment, 52
Mississippi River, 11, 35, 42, 51-53
Missouri Plateau, 46
Missouri River, 11, 35, 43, 48, 67
Moffat Tunnel, 70
Mohawk River, 22
Mohawk Valley and Lowlands, 20, 22, 27, 29, 35
Mo-i-Rana, Norway, 229
Mojave Desert, 75, 76
Moldau River, 315
Moldavian S.S.R., 347, 352, 360, 367
Monaco, 15, 305
Mongolia, 347, 471, 472, 477, 478, 491-495
  historical development, 494-495
  pastoral economy, 492-494
  relief features, 491-492
Mongolian People's Republic (see Mongolia)
Monongahela River, 29, 35

Mont Blanc Tunnel, 288
Monte Carlo, Monaco, 305
Montego Bay, Jamaica, 155
Monterrey, Mexico, 164, 165
Montevideo, Uruguay, 202, 207-208
Montreal, Canada, 109, 116, 118, 122
Moravian Corridor, 315, 329, 338
Morocco, 8, 285, 388, 391, 392, 394, 396-400
Moscow, U. S. S. R., 309, 325, 348, 356, 364, 367, 376, 380, 381, 383
Moulmein, Burma, 575
Mt. Aconcagua, 178, 202
Mt. Blanc, 221
Mt. Chimborazo, 195
Mt. Cotopaxi, 195
Mt. Everest, 7, 521
Mt. Herman, 407
Mt. Kilimanjaro, 430
Mt. Lebanon, 408-410
Mt. McKinley, 64, 96
Mt. Matterhorn, 264
Mt. Mitchell, 60
Mt. Orizaba, 142, 162
Mt. Rainier, 79, 80
Mt. Snowdon, 242
Mt. Whitney, 81
Mozambique, 282, 285, 430, 433, 440, 455, 464
Mukden, China, 483
Munich, Germany, 262
Murmansk, U.S.S.R., 346
Murray River, 597, 598, 603
Murrumbidgee River, 597, 598, 603
Muscat, 391, 414

Nadi, Fiji, 616
Naga Hills, 558
Nagoya, Japan, 507, 508
Nairobi, Kenya, 426, 463
Namib Desert, 435, 464
Nan Ling Range, 476, 480, 484
Nancy, France, 255
Nanking, China, 478, 488
Naples, Italy, 287, 288, 291, 292, 295, 296
Narvik, Norway, 231, 233
Nashville Basin, 35, 52, 55
Natal, 442
National Institute of Industry, 280
Natu La Pass, 550
Nauru, 611, 617
Nebraska Sand Hills, 46, 48
Nepal, 513, 520, 530, 544, 548-550
Netherlands, 4, 15, 220-222, 224, 226, 245, 248
New Caledonia, 595, 610, 614, 615
New Guinea, 13, 556, 558, 561, 580, 581, 595, 597, 610, 611, 614, 615
New Hebrides, 595, 610, 611, 615
New Orleans, U.S., 53, 58, 153
New York, U.S., 15, 20, 22, 24, 26, 28-33, 109, 153, 241
New York State Barge Canal, 22, 28, 29
New Zealand, 13, 207, 226, 595, 602, 607-611, 614
Newcastle, U.K., 239-240
Niagara escarpment, 115, 117
Niagara Falls, 29, 116, 118
Niagara Peninsula, 117, 132
Niagara River, 22, 116
Nicaragua, 139, 142, 143, 165
Nicosia, Cyprus, 418
Niger, 442, 458, 459
Niger River, 429, 454, 456, 460
Nigeria, 4, 429, 430, 442, 447, 448, 451, 453, 454, 456, 458-460
Nile River, 11, 388, 402-404, 429, 445, 461
Nile Valley, 402, 404
Nilgiri Range, 524, 539
Niue Island, 611, 612
Nizhni Tagil, U.S.S.R., 353, 378, 380
Noranda, Canada, 121
Nordic Union, 233
North Atlantic Drift, 10, 136, 236
North Atlantic Free Trade Area, 224
North Atlantic Treaty Organization, 224, 256, 257, 286, 306
North Island, New Zealand, 607, 608
North Korea (*see* Korea)
North Vietnam, 554, 557, 558, 561, 567, 571-573
Northeastern United States, 15, 20-34
    cities and industries, 27-34
    economic development, 23-26
    physical setting, 20-23
    recreation, 26-27
Northern Ireland (*see* Ulster)
Northern Rocky Mountains, 66-68
Northumberland-Durham coalfields, 239
Northwest Frontier, India, 527, 547
Northwest Territories, 104, 111, 120, 133-135
Northwestern and Central Europe, 12, 218-267
    Benelux countries, 249-256
    British Isles, 238-245
    France, 249-256
    Germany, 256-263
    land-locked countries, 263-267
    physical setting, 220-222

Northwestern and Central Europe:
   population distribution, 224-227
   Scandinavian countries and Finland, 227-238
   significance of diversity of cultures, 222-224
Norway, 220, 227, 229-231, 234, 347
Nouakchott, Mauritania, 401
Novosibirsk, U.S.S.R., 364, 381
Nullarbor Plain, 600

Ob River, 354, 376, 378
Ocean Island, 617
Oceania, 610-617
   Melanesia, 614-616
   Micronesia, 616-617
   Polynesia, 611-614
Oder River, 222, 258, 314, 315, 326, 327, 331
Oder-Neisse River, 256, 314, 326, 347
Odessa, U.S.S.R., 348
Ohio River, 11, 28, 29, 35, 42, 50, 59
Oil (see Petroleum)
Okefenokee, 52
Öland, 221, 231, 232
Olives, 253, 270, 272, 273, 278, 283, 290, 291, 298, 299, 307, 341, 393, 397, 402, 409
Olympic Mountains, 90, 91
Olympic National Forest, 64
Omaha, U.S., 43
Oman, 414, 415, 417
Omsk, U.S.S.R., 356, 381
Oporto, Portugal, 278, 281, 283, 285
Orange Free State, 465

Orange River, 429, 455, 466
Ore Mountains, 312, 315, 329, 330
Oregon Trail, 68, 79, 91
Orinoco Lowlands, 177, 181 188
Orinoco River, 179, 186, 189, 190
Orurú, Bolivia, 200
Osage Plains, 35, 52
Osaka, Japan, 504, 506, 508
Oslo, Norway, 230
Ostende, Belgium, 246
Ottawa River, 122
Ouachita Mountains, 62
Owens Valley, 76, 87
Ozark Plateau, 35, 50, 62-63

Pacific Northwest Coastal Province, 90-93
   Coast ranges, 90-91
   Puget-Willamette Lowland, 91-93
Pago Pago, American Samoa, 613
Pakhtunistan, 546-547
Pakistan, 12, 513-544, 550, 551
Palestine, 391, 407, 410-412
Palouse Hills, 77, 78
Pampas, 179, 200, 202, 205-207, 213
Pan-American Highway, 170, 192, 196, 198
Panama, 8, 138, 143, 146, 150, 155, 165, 172
Panama Canal, 31, 143, 170, 196, 202, 204
Panama City, Panama, 171
Paotow, China, 487
Papeete, Tahiti, 614
Papua, 597, 611, 614
Paraguay, 200, 202, 208, 214

Paraguay-Paraná River, 179, 186, 199, 206-208, 210, 214, 215
Paraiba Valley, 211, 212
Paramaribo, Surinam, 192
Paris, France, 252, 253, 255, 283, 288
Paris Basin, 250, 253
Patagonia, 179, 182, 206
Peace River, 125, 127, 132
Pechanga District (see Petsamo)
Pechora Basin, 375, 376
Pécs, Hungary, 332
Pegu Yoma, 574
Peking, China, 483, 487
Peninsular Massif, India, 521, 522, 524, 535, 536
Pennines, 239-242
People's Republic of Southern Yemen, 414
Pernik, Bulgaria, 325, 340
Persian Gulf, 387, 412, 417, 418, 423
Perth, Australia, 604-606
Peru, 11, 175, 178, 181, 185, 186, 193, 196-199, 504, 505
Peruvian Coastal Desert, 182
Peruvian Montana, 177, 197, 198
Petén, Guatemala, 166
Petroleum, 26, 35, 40, 49, 55-57, 62, 68, 73, 82, 85-87, 89, 95, 97, 100, 114, 118, 127, 128, 142, 157, 163, 164, 178, 186, 189-192, 197, 198, 200, 204, 206, 227, 229, 233, 237, 239, 240, 244, 246, 248, 250, 254, 259, 262, 263, 280, 291 293, 324, 328-330, 332, 333, 335, 340, 376, 380, 398, 401, 402, 406, 412, 413, 417, 422, 423, 454, 458, 486, 494, 505, 508,

Index **649**

Petroleum:
    540, 574, 580, 583, 604, 608
Petsamo, 236, 347
Philadelphia, U.S., 20, 24, 28, 31, 34, 189
Philippines, 13, 505, 554, 557, 560, 561, 563, 569, 576, 584-586, 605, 610
Phnom Penh, Cambodia, 566, 568, 569
Phoenix, U.S., 73
Piedmont, 22, 53
Pikes Peak, 70
Pilcomayo River, 199
Pindus Mountains, 297
Pineapples, 94, 95
Piraeus, Greece, 297, 300-302
Pittsburgh, U.S., 29, 35, 44
Plains of the Southland, 51-59
    agriculture, 52-55
    cities and towns, 58-59
    fisheries, 55
    forests, 55
    manufacturing and transportation, 57-58
    minerals, 55-56
    physical setting, 51-52
    recreation, 59
Platte River, 48
Ploesti, Romania, 335
Plzen, Czechoslovakia, 330
Po River, 271, 273, 286, 289, 290, 293
Pocono Mountains, 22
Poland, 221, 256, 257, 308, 314, 316, 318, 320-322, 326-329, 347
Poleyse, 352, 353, 359, 371
Polynesia, 610-614
Polynesians, 13, 95, 608, 609
Pomeranian Bay, 314
Popocatepetl, 142
Population, 12-16
    density, 14-16

Population:
    distribution, 14-16
    increase, 13-14
Population explosion, 13-14
Port Augusta, Australia, 603
Port Étienne, Mauritania, 401
Port-of-Spain, Trinidad, 157
Portland, Maine, U.S., 27
Portland, Oregon, U.S., 65, 92
Porto Alegre, Brazil, 214
Portugal, 270, 271, 273-285, 429, 446
Portuguese Guinea, 282, 459
Portuguese Timor, 575, 586
Potomac River, 20, 22, 30, 50
Potosi, Bolivia, 200
Prague, Czechoslovakia, 320, 325, 329, 330
Prai, Malaysia, 578
Prairie Provinces, 104, 106, 108, 111, 118
Prince of Wales Island, 106
Pripyat Marshes, 314, 359
Pueblo, U.S., 70
Puerto Barrios, Guatemala, 166, 167
Puerto Cabezas, Nicaragua, 168
Puerto Limón, Costa Rica, 169
Puerto Montt, Chile, 204
Puerto Rico, 141, 143, 150, 152-153
Puget Sound, 90-92
Puget-Willamette Lowland, 91-93
Punta Arenas, Chile, 204
Punta del Este, Uruguay, 208
Pusan, Korea, 499
Pyonyang, Korea, 496, 499
Pyrenees Mountains, 221, 249, 250, 254, 274

Qatar, 391, 414, 415, 417
Quebec, Canada, 109, 111, 122
Quebec Lowland, 116-117, 120, 122

Quechua Indians, 193, 197
Quezaltenango, Guatemala, 166
Quito, Ecuador, 195, 196

Ramadan, 392
Rangoon, Burma, 574
Red (Song Koi) River, 558, 572
Red River Plain, 125, 126
Red Sea, 9, 400, 404, 412, 413, 433
Regina, Canada, 128
Regional geography, 6-7
Reno, U.S., 75
Republic of Maldive Islands, 513
Republic of South Africa (see South Africa)
Réunion, 461, 463
Reykjavik, Iceland, 236
Rhine River, 11, 222, 224, 248, 250, 255, 256, 258, 261
Rhodesia, 440, 445, 464, 465
Rhodope Mountains, 311, 312, 315, 317, 320, 329, 340,
Rhone River, 11, 222, 250, 252, 254, 272
Rice, 164, 188, 193, 197, 211, 273, 290, 298, 307, 317, 322, 332, 335, 337, 404, 413, 423, 456, 480, 483, 485, 487, 496, 503, 504, 515, 526, 536, 537, 560, 561, 563, 564, 566, 568-571, 573-577, 579, 583, 603
Richelieu River, 22
Ridge and Valley Area, 22
Rif Atlas, 395
Rimac Valley, 197
Rio de Janeiro, Brazil, 178, 209-212, 214
Rio de la Plata, 179, 186, 201, 202, 205-207
Rio Grande, 35, 52, 70, 71, 163, 164

Rio Grande Valley, 53
Rio Muni, 460
Rio Negro, 215
Rio Papaloapan, 162
Rio San Juan, 143
Rocky Mountain National Park, 70
Rocky Mountain Trench, 67, 128
Rocky Mountains, 19, 35, 65-71, 76, 106, 124, 128
   Northern, 66-68
   Southern, 70-71
   Wyoming Basin, 68-70
Rogers Pass, 64
Romania, 308, 315, 316, 322, 333-336, 347
Rome, Italy, 270, 283, 287-289, 291, 295, 296, 305
Rosario, Argentina, 205, 206
Rostov, U.S.S.R., 348, 380, 383
Rotorua, New Zealand, 608
Rotterdam, Netherlands, 247, 248
Rubber, 215, 565, 566, 569, 571, 576, 579, 580, 583, 615
Ruhr, 255, 257, 258, 260, 261
Russian Plain, 352, 353
Russian S.F.S.R., 367, 373, 381
Rwanda, 441, 463
Ryukyu Islands, 500, 502

Saar, 250
Sabah, 575, 578
Sable Island Banks, 114
Sacramento River, 82-85, 88
Saguenay Valley, 122, 124
Sahara Desert, 9, 14, 272, 397, 404, 426, 429
Saida, Lebanon, 409, 417
Saigon, South Vietnam, 558, 566, 568, 571
St. Croix Island, 153
St. Elias Mountains, 130
St-Étienne, France, 252, 254

St. Gotthard Pass, 289
Saint John, Canada, 112, 115
St. John River, 109, 112, 114, 115
St. John's, Canada, 114
St. Kitts, 156
St. Lawrence Lowland, 20, 108, 109, 122
St. Lawrence River, 11, 22, 116, 118-120
St. Lawrence Seaway, 29, 42, 118
St. Louis, U.S., 35, 43
St-Nazaire, France, 255
St. Paul, U.S., 43
St. Thomas, 153
Sakhalin Island, 354, 376
Salinas Valley, 88
Salonika, Greece, 297, 302
Salpauselka, 237
Salt Lake City, U.S., 70, 76
Salton Sea, 75
Salween River, 11, 558, 574
Salzburg, Austria, 265
Sambre-Meuse coalfields, 246, 248, 252, 253
Samoa, 594, 612, 613
San Antonio, U.S., 59
San Diego, U.S., 85, 86, 88
San Felix, Venezuela, 189
San Francisco, U.S., 10, 64, 65, 82, 83, 88, 89
San Joaquin River, 83-85, 88
San José, Costa Rica, 143, 169
San Juan, Puerto Rico, 153
San Juan Mountains, 70
San Luis Park, 70
San Marino, 305
San Pedro Sula, Honduras, 168
San Salvador, El Salvador, 168
Sanmen Gorge, 480, 487
Santa Clara Valley, 88
Santa Cruz, Bolivia, 200
Santa Elena Peninsula, 196
Santa Fe, Argentina, 205
Santa Fe, U.S., 70

Santa Marta Uplands, 190
Santiago, Chile, 202, 204
Santo Domingo, Dominican Republic, 154
Santos, Brazil, 210
São Paulo, Brazil, 209-211, 214
Saône River, 250, 252-254
Sarawak, 578, 579, 586
Sardinia, 269, 287, 291, 292
Saskatchewan Plain, 124, 125
Saudi Arabia, 391, 413-418
Sava River, 315, 337, 338
Sayan Mountains, 354, 491
Schefferville, Canada, 121
Schuman Plan (*see* European Coal and Steel Community)
Scotland, 221, 238, 240
Scottish Lowlands, 226, 239, 242
Sea of Japan, 247, 496
Sea of Marmara, 303, 339
Seattle, U.S., 93, 97
Seine River, 11, 222, 252
Seistan Desert, 526
Senegal, 442, 456, 459
Senegal River, 401
Sennar Dam, 404
Seoul, Korea, 496, 499
Serra do Mar, 179, 210
Seville, Spain, 281, 283, 284
Shan Plateau, 476
Shanghai, China, 475-477, 483, 487, 488, 489, 491
Shannon River, 222, 238, 243, 244
Shantung Peninsula, 484
Shasta Dam, 82
Shatt-al-Arab, 421
Sheffield, U.K., 241
Shenandoah Valley, 61
Shickshock Mountains, 104
Shkodër, Albania, 341
Sicily, 268, 269, 287-289, 292
Sieg Valley iron ore area, 261
Sierra de Guadarrama, 275
Sierra Leone, 438, 455, 459

Sierra Madre, 142, 162, 163
Sierra Madre del Chiapas, 142
Sierra Maestra of Cuba, 143, 151, 152
Sierra Nevada, Spain, 275
Sierra Nevada, U.S., 71, 75, 76, 81-83
Si-kiang, 475, 476, 487
Sikkim, 513, 548, 549
Silesia, 327
Silesian coalfields, 256, 327, 328, 330
Sinai Peninsula, 9
Sind, 526, 527
Singapore, 15, 502, 554-557, 561, 566, 570, 576, 579-580, 587, 605
Sinkiang, 478, 485, 494, 521
Sinú Valley, Colombia, 190
Skåne, 221, 231, 232
Skoumtsa, Greece, 299
Småland, 232
Smolensk-Moscow Ridge, 352
Snake River, 67, 76-78
Snowy River, 597
Society Islands, 612, 613
Solomon Islands, 595, 597, 610, 611, 615
Somalia, 429, 430, 435, 439, 442, 446, 449, 462
Song Be, 569
Soo Canal, 41, 118
South Africa, 429, 435, 441, 447, 453-455, 464-467
South Dakota Badlands, 46
South Island, New Zealand, 607, 608
South Korea (see Korea)
South Vietnam, 554, 558, 560, 561, 566, 567, 569-571
South Wales coalfields, 239, 243
Southern Appalachian Highlands, 59-62
Southern California, 85-88

Southern Peninsular Europe, 268-307
 cultural landscape, 272-273
 European Turkey, 303-304
 Gibraltar, 285-286
 Greece, 297-303
 Iberian Peninsula, 273-285
 Italy, 386-397
 minor Mediterranean states, 304-306
 physical setting, 271-272
Southern Rocky Mountains, 49, 70-71
Southern Scottish Uplands, 242
Southern South America, 200-208
 Argentina, 204-207
 Chile, 202-204
 Paraguay, 208
 Uruguay, 207-208
South-West Africa, 14, 435, 440, 454, 455, 464, 467
Southwestern Basin and Range Province, 73-75
Soviet Union (see Union of Soviet Socialist Republics)
Spain, 4, 147-149, 270, 271, 273-285, 446
Spanish-Americans, 51, 58, 86-89
Spanish Meseta, 275, 276, 278, 281, 283
Spanish Sahara, 285, 388, 401-402
Spitsbergen, 229
Spokane, U.S., 78
Stanovi Mountains, 354
Stavanger, Norway, 230
Steep Rock Lake, 121
Stockholm, Sweden, 232, 236
Strait of Gibraltar, 285
Strait of Mackinac, 41
Strait of Magellan, 204-206
Strait of Messina, 288
Strait of Otranto, 269
Sucre, Bolivia, 199

Sudan, 13, 388, 390-392, 400, 402-406, 439
Sudbury, Canada, 121, 123
Sudd, 403
Sudeten Mountains, 312, 314, 318, 320, 327, 329
Suez Canal, 269, 270, 405, 406, 412, 456, 463, 464
Sugarcane, 94, 142, 150, 152, 153, 156, 158, 169, 184, 188, 191, 193, 197, 200, 210, 213, 452, 463, 484, 537, 538, 561, 565, 573, 574, 583, 586, 603
Suitcase farmer, 47
Sulu Sea, 556, 561, 578
Sumatra, 554, 556, 558, 581
Sundarbans, 522
Sundra Platform, 556
Sungari River, 474, 475, 479, 487
Superior Highlands, 40
Surinam, 186, 192-193
Sutleg River, 517, 546
Suva, Fiji, 616
Sverdlovsk, U.S.S.R., 348, 350, 353, 364, 380, 381
Swansea, U.K., 243
Swaziland, 464, 465
Sweden, 227, 231-234, 237, 258
Swiss Plateau, 226
Switzerland, 221, 222, 224, 263-264, 275, 286, 295
Sydney, Australia, 603-606
Sydney, Canada, 114
Syr Darya, 354, 371
Syracuse, U.S., 25, 29
Syria, 388, 391, 394, 406-412, 419, 422
Systematic geography, 6
Szczecin, Poland, 327, 328
Szechwan Basin, 476, 478, 480, 483-487
Szeged, Hungary, 332, 333

Ta Yu Range, 476
Tadzhik S.S.R., 367
Taebaek Range, 496
Tagus River, 273-275, 281, 285
Tahiti, 9, 612, 613
Taiwan, 480-482, 484, 488, 490, 500, 502, 610
Takla Makan Desert, 14
Talara, Peru, 197
Talysh Lowland, 361
Tampere, Finland, 237
Tampico, Mexico, 163
Tanezrouft, 400
Tannu Tuva, 494
TanZam Railway, 456, 463
Tanzania, 433, 442, 448, 452, 455, 463
Taranto, Italy, 292
Tarbagatai Mountains, 491
Tarim Basin, 476-478
Tashkent, U.S.S.R., 361, 364
Tea, 542, 583
Tees River, 240
Tegucigalpa, Honduras, 168
Tehachapi Pass, 81, 85
Tehran, Iran, 421
Tehuantepec, 142, 143, 162, 164
Tel Aviv, Israel, 410
Tell Atlas, 396
Tenasserim Mountains, 574
Tennessee Valley Authority, 61
Thailand, 554, 558, 560-567, 576
Thames River, 222, 241
Thar Desert, 522, 526, 533, 536
Thessalonika, Greece, 300, 302
Thunder Bay, Canada, 118, 123
Tiber River, 289, 295, 305
Tibesti Mountains, 400
Tibetan High Plateau, 9, 14, 476-478, 548
Tien Shan, 354, 476, 478
Tientsin, China, 483, 487, 491
Tierra del Fuego, 178, 202
Tigris River, 412

Tijuana, Mexico, 164
Timor, 285
Tin, 200, 460, 505, 565, 566, 580
Tiranë, Albania, 320, 341
Tisza River, 315, 338, 347
Tobago, 157
Togo, 459
Tokelau Islands, 611, 612
Tokyo, Japan, 15, 241, 500, 501, 506, 508
Toledo, U.S., 43, 44
Tonkin Hills, 572
Tonle Sap. 568, 569
Toronto, Canada, 111, 118, 119
Toulouse, France, 254
Transcaucasia, 350, 363, 364, 371, 379-381
Trans-Siberian Railway, 348, 363, 383, 494
Transvaal, 465
Transylvania, 333-335
Trieste, Italy, 297
Trinidad, 139, 142, 150, 155-158
Tripoli, Libya, 402
Tromsö, Norway, 231
Trondheim Fjord, 229, 230
Truk Islands, 616
Tsaidam Basin, 476
Tsingtao, China, 477, 487, 491
Tuamotu Archipelago, 612, 614
Tucson, U.S., 73
Tucumán, Argentina, 206
Tularosa Basin, 70
Tundzha Valley, 340
Tungting Hu, 476, 488
Tunisia, 301, 388, 392, 394, 396-400
Turfan Depression, 476
Turgay Gap, 354
Turin, Italy, 286, 292, 295, 296
Turkey, 268, 269, 298, 303-304, 315, 346, 347, 388, 394, 418-423
Turkmen, S.S.R., 367

Turku, Finland, 237
Tuy River, 188
Tyrrhenian Sea, 269

Uganda, 433, 445, 448, 452, 454, 463
Uinta Mountains, 70
Ukrainian S.S.R., 312, 347, 348, 359, 360, 363, 366, 367, 370-373, 376, 378, 381
Ulan Bator, Mongolia, 491, 494
Ulster, 240, 243-244
Unaka Mountains, 60
Ungava-Labrador Region, 113
Union of Soviet Socialist Republics, 4, 9, 11, 15, 18, 163, 207, 218, 220, 231, 237, 250, 257, 259, 263, 266, 294, 304, 309, 315, 324-327, 331, 334, 336, 339, 344-385, 429, 454, 462, 471, 472, 491, 495, 498, 504, 505, 544, 547, 549
  economic development, 368-383
  goals, 350-352
  minority groups, 365-368
  natural regions, 357-361
  physical setting, 352-357
  population, 362-364
United Kingdom, 4, 103, 155, 167, 224, 227, 237, 238, 240, 242, 252, 253, 255, 260, 327, 336, 339, 446, 465, 497, 581, 611
United States, 4, 8, 10, 11, 14-16, 18-102, 104, 107, 116, 141, 151, 152, 155, 162, 170, 199, 210, 213, 218, 220, 233, 241, 250, 255, 264, 278, 282, 283, 286, 288, 300, 327, 336, 339, 345, 368, 373, 387, 429,

United States:
448, 452, 454, 459, 472, 497, 499, 503, 505, 531, 544, 549, 556, 566, 594, 603, 607, 611
  east of Rocky Mountains, 18-63
  west of Great Plains, 64-101
United States Trust Territory, 502, 611, 616
Upper Volta, 459
Ural Mountains, 9, 314, 348-350, 353, 354, 359, 360, 363, 364, 370, 375-377, 379, 380
Uranium, 73, 261
Uruguay, 175, 179, 200, 207-208
Uruguay River, 206
Urumchi, China, 487, 490
Usküdar, Turkey, 304
Uspallata Pass, 204, 206
Ussuri River, 354
Uzbek S.S.R., 367
Uzhok Pass, 312

Vaduz, Liechenstein, 264
Valdai Hills, 352
Valdiva, Chile, 203
Vale of Kashmir, 522, 545
Vale of Manipur, 522
Valencia, Ireland, 238
Valencia, Spain, 273, 281, 283, 284
Valencia, Venezuela, 188
Valletta, Malta, 306
Valparaiso, Chile, 204
Vancouver, Canada, 111, 131-133
Vancouver Island, 90, 91, 111, 130, 132
Vardar River, 302, 337
Varna, Bulgaria, 340
Vasyugan Swamp, 354, 363

Vatican City, 17, 295, 305
Venezuela, 10, 163, 178, 179, 185-190, 192, 193
Venice, Italy, 296
Veracruz, Mexico, 162
Veretski Pass, 312
Verkhoyansk, U.S.S.R., 356
Verkhoyansk Range, 354
Vermilion Range, 40
Victoria, Canada, 133
Victoria Island, 106
Vienna, Austria, 265, 326
Vientiane, Laos, 566, 568
Villarrica, Paraguay, 208
Viña del Mar, Chile, 204
Vineyards, 270, 273, 278, 281, 290, 291, 298, 307, 393, 397, 402, 452, 598, 603
Virgin Islands, 153, 156, 157
Vistula River, 220, 314, 326-328
Vitoria, Brazil, 212
Vladivostok, U.S.S.R., 347, 363
Vltava, Czechoslovakia, 329
Volcanic National Park, 95
Volga River, 356, 360, 361, 372, 373, 376, 378, 380-383
Volgograd, U.S.S.R., 348, 356, 378, 380, 381, 383
Volta Redonda, Brazil, 212
Volta River, 459
Vosges Mountains, 221, 250, 252, 255

Waikato, New Zealand, 608
Walachian Plain, 311, 315, 318, 334
Warsaw, Poland, 316, 320, 328
Wasatch Range, 70, 75, 76
Washington, U.S., 20, 22, 26, 28, 30, 31, 34, 51
Wei River, 483
Welland Canal, 118
Wellington, New Zealand, 609

Weser River, 258-260
West Central South America, 193-200
  Bolivia, 199-200
  Ecuador, 195-196
  Peru, 196-199
West Germany (see Germany)
Western Ghats, 524, 535, 539
Western Plateau, Australia, 597
Western Samoa, 611, 613
Westphalian coalfield, 258
Wheat, 36, 38, 39, 47, 126, 149, 158, 184, 188, 200, 204, 205, 232, 240, 253, 254, 260, 262, 270, 272, 273, 278, 290, 298, 306, 322, 327, 330, 335-337, 340, 341, 360, 371, 372, 393, 397, 404, 409, 413, 422, 483-485, 487, 496, 504, 515, 536, 597, 598, 601-603, 609
White Nile, 403
Whiting, U.S., 40
Willamette River, 90, 91
Willamette Valley, 79, 91, 92
Williston Basin, 49
Wind River Mountains, 67
Winnipeg, Canada, 111, 128
Winter Wheat Belt, 47
Witwatersrand, 454
Wonsan, Korea, 496, 499
Wuhan, China, 483, 487
Wyoming Basin, 68-70

Yablonovio Mountains, 354
Yaila Mountains, 353
Yalu River, 487, 496, 499
Yaounde, Cameroun, 458
Yangtze Gorge, 487
Yangtze Kiang, 11, 14, 475, 476, 478, 480, 485-487, 481

Yangtze Plain, 475, 476, 483, 488
Yaqui Valley, 164
Yazoo River, 53
Yellow Knife, Canada, 134
Yellow Plain, 475, 480, 483, 491
Yellow River (*see* Hwang Ho)
Yellow Sea, 496
Yellowstone National Park, 68, 77
Yellowstone River, 48, 67
Yemen Arab Republic, 390, 414, 415

Yenisey River, 352, 363, 378, 381, 383
Yokohama, Japan, 508
Yosemite National Park, 81, 82
Yucatan Peninsula, 146, 148, 162, 163
Yugoslavia, 269, 303, 308, 312, 315-318, 322, 336-339
Yukon Plateau, 130
Yukon River, 96, 97, 131
Yuma, U.S., 64, 74
Yunnan Plateau, 476, 480

Zagreb, Yugoslavia, 338
Zagros Mountains, 419-421
Zambezi River, 429, 455, 465
Zambia, 433, 453, 454, 456, 461, 463, 465
Zanzibar, 449, 463
Zaragoza, Spain, 280
Zeeland Islands, 246
Zenica, Yugoslavia, 325, 338
Zinder, Niger, 433
Zuider Zee, 247
Zurich, Switzerland, 264

50.80
260773